Cisco Nexus 交换机与 NX-OS 排错指南

Troubleshooting
Cisco Nexus Switches
and NX-OS

[印] 维尼特·贾恩（Vinit Jain）
[美] 布拉德·埃奇沃斯（Brad Edgeworth） 著
[美] 理查德·弗尔（Richard Furr）

夏俊杰 译

人民邮电出版社
北京

图书在版编目（CIP）数据

Cisco Nexus交换机与NX-OS排错指南 /（印）维尼特·贾恩（Vinit Jain），（美）布拉德·埃奇沃斯（Brad Edgeworth），（美）理查德·弗尔（Richard Furr）著；夏俊杰译. -- 北京：人民邮电出版社，2021.5
 ISBN 978-7-115-55449-9

Ⅰ. ①C… Ⅱ. ①维… ②布… ③理… ④夏… Ⅲ. ①计算机网络－信息交换机－指南 Ⅳ. ①TN915.05-62

中国版本图书馆CIP数据核字（2020）第236248号

版权声明

Authorized translation from the English language edition, entitled Troubleshooting Cisco Nexus Switches and NX-OS by Vinit Jain , Brad Edgeworth , Richard Furr, published by Pearson Education, Inc, publishing as Cisco Press, Copyright © 2020 Pearson Education, Inc.

All rights reserved. No part of this book may be reproduced or transmitted in any form or by any means, electronic or mechanical, including photocopying, recording or by any information storage retrieval system, without permission from Pearson Education, Inc.

CHINESE SIMPLIFIED language edition published by POSTS AND TELECOM PRESS CO., LTD., Copyright ©2020.

本书中文简体字版由美国 Pearson Education 集团授权人民邮电出版社有限公司出版。未经出版者书面许可，对本书任何部分不得以任何方式复制或抄袭。

版权所有，侵权必究。

◆ 著　　[印] 维尼特·贾恩（Vinit Jain）
　　　　[美] 布拉德·埃奇沃斯（Brad Edgeworth）
　　　　[美] 理查德·弗尔（Richard Furr）
　译　　夏俊杰
　责任编辑　陈聪聪
　责任印制　王 郁　彭志环

◆ 人民邮电出版社出版发行　北京市丰台区成寿寺路11号
　邮编　100164　电子邮件　315@ptpress.com.cn
　网址　https://www.ptpress.com.cn
　北京市艺辉印刷有限公司印刷

◆ 开本：787×1092　1/16
　印张：44.25
　字数：1257 千字　　　　　　2021年5月第1版
　印数：1－2 000 册　　　　　　2021年5月北京第1次印刷

著作权合同登记号　图字：01-2019-0416 号

定价：189.90 元
读者服务热线：（010）81055410　印装质量热线：（010）81055316
反盗版热线：（010）81055315
广告经营许可证：京东市监广登字 20170147 号

内容提要

NX-OS 是一款功能强大、特性丰富的网络操作系统，其充分融入了二层交换、三层路由和 SAN 交换功能，广泛应用于各种复杂网络环境当中。本书详细讨论了思科 Nexus 交换机和 NX-OS 的功能特性及故障排查技术，全面分析了二层功能特性（包括 VLAN、PVLAN、SPT、vPC、vPC+、FabricPath 等）和三层功能特性（包括 EIGRP、OSPF、IS-IS、BGP 和路由映射等）的故障排查技术，同时还讨论了 IP 多播、高可用性、OTV 等关键技术和关键应用。为便于理解，本书提供了大量配置示例，对于读者理解并掌握 NX-OS 和 Nexus 交换机的配置及故障排查操作极为有益。

本书适合所有希望学习并掌握思科 Nexus 平台和 NX-OS 操作系统的网络工程师、架构师或技术顾问学习与参考，通过阅读本书读者能够掌握通过 NX-OS 排查复杂网络故障的知识和技能。此外，本书也可以为准备 CCIE 数据中心认证的人员提供有益参考。

关于作者

维尼特·贾恩（CCIE No.22854，R&S、SP、Security&DC）是思科 TAC（Technical Assistance Center，技术支持中心）技术负责人，负责提供路由和数据中心领域的技术支持。维尼特是多个网络论坛的演讲嘉宾，包括在全球各地举办的各类 Cisco Live 活动。在加入思科之前，维尼特曾担任 CCIE 培训师和网络顾问。除 CCIE 之外，维尼特还拥有编程和数据库方面的多项认证。维尼特毕业于德里大学数学系，获得了印度库韦姆普大学的信息技术硕士学位。

布拉德·埃奇沃斯（CCIE No. 31574，R&S、SP）是思科系统工程师，也是 Cisco Live 的杰出演讲嘉宾，曾在多个主题发表演讲。加入思科之前，布拉德曾在多家财富 500 强企业担任网络架构师和顾问，擅长企业和服务提供商的网络架构设计与运维优化工作。布拉德拥有得克萨斯州奥斯汀圣爱德华大学计算机系统管理专业的学士学位。

理查德·弗尔（CCIE No. 9173，R&S、SP）是思科 TAC（Technical Assistance Center，技术支持中心）技术负责人，负责为全球客户及 TAC 团队提供技术支持。在过去的 17 年当中，理查德曾在思科 TAC 和 HTTS（High Touch Technical Support，深入接触技术支持）部门工作，为服务提供商、企业和数据中心环境提供技术支持。理查德擅长路由协议、MPLS、多播和网络叠加技术等领域的故障排查。

关于审校者

布赖恩·萨克（Brian Sak）（CCIE No.14441）是思科解决方案架构师，主要为思科安全产品和安全服务提供方案支持与开发。布赖恩在信息安全领域拥有 20 多年的工作经验，包括咨询服务、评估服务和渗透测试、实施、架构设计及开发等。布赖恩拥有信息安全与保护专业的硕士学位，拥有多项安全和行业认证，曾在互联网安全中心、Packt 出版社和思科出版社出版或撰写过相关图书。

拉米罗·加尔扎·里奥斯（Ramiro Garza Rios）（CCIE No. 15469，R&S、SP、Security）是思科高级服务解决方案集成架构师，负责规划、设计、部署和优化 IP NGN 服务提供商网络。在 2005 年加入思科之前，拉米罗是墨西哥思科金牌合作伙伴的网络咨询与售前工程师，负责规划、设计和部署企业及服务提供商网络。

马特·埃索（Matt Esau）（CCIE No. 18586，R&S）毕业于北卡罗来纳大学教堂山分校，目前与妻子和两个孩子（分别是 3 岁和 1 岁）住在俄亥俄州。马特是 Cisco Live 的杰出演讲者，于 2002 年加入思科，在过去 15 年的客户服务生涯里，他一直致力于帮助客户解决网络故障和产品可用性问题。在过去的 8 年当中，马特始终致力于数据中心领域的 Nexus 平台的技术支持工作。

献辞

谨将本书献给我生命中重要的 3 个女人：我的母亲、我的妻子胡什布（Khushboo）以及索纳尔（Sonal）。妈妈，感谢您在我人生的不同阶段始终作为我的朋友和导师，给予我巨大的勇气，让我能够直面人生当中的各种挑战。胡什布，感谢你在我写作期间给予的无与伦比的耐心，没有你的支持，我将无法完成本书或任何其他项目，我无法用言语表达我对你的爱和感激，谨以本书略表对你无尽的爱与感激。索纳尔，感谢你始终激励我，每次见面，你都会给我设定新的目标，激发我不断挑战新的高度。谨以本书表达我对你（为我）所做的一切的感激之情。

此外，还要将本书献给我的父亲和我的兄弟，在我面对生活中的挑战时，你们一如既往地站在我的身后，没有你们的支持，就不会有如今的我。

——维尼特·贾恩

谨将本书献给大卫·凯尔（David Kyle）。谢谢您给予我的种种机会，您不仅仅是我的前任老板，在我的职业生涯初期，您还教会了我正确的工作态度和基本技能。

除使用 Quake 对网络进行压力测试之外，您还教会我坚持正确的人生理念，激励我不断前行。看一看我现在的状态吧！

——布拉德·埃奇沃斯

谨将本书献给我挚爱的妻子桑德拉（Sandra）和可爱的女儿卡连纳（Calianna）。你们是我的灵感源泉，你们的爱和支持是我走向成功的源动力，让我不断超越自我。卡连纳，现在你才两岁，等你长到能够读懂这段话时，相信你早已忘记爸爸在这个项目上度过的漫漫长夜，但希望你看到本书时能够始终记住：奉献和勤劳能够让一切皆有可能。

此外，还要将这本书献给我的父母。妈妈，感谢您一直鼓励我，教会我一切事情。爸爸，谢谢您一直以来的支持和理解，教会我专注和努力。你们给我的一切都是最棒的！

——理查德·弗尔

致谢

维尼特·贾恩

布拉德和理查德，感谢你们全力参与持续一年的写作旅程，没有你们的支持，本书将无法完成。这是一个伟大的团队，与你们一起工作非常愉快。

感谢技术编辑拉米罗（Ramiro）和马特（Matt），感谢你们对本书所做的全面检查，是你们提供的大量的见解，使这个项目能够最终成功。

感谢我的经理奇普·利特尔（Chip Little）和迈克·斯托林斯（Mike Stallings），没有你们的大力支持，我将无法完成这样一个里程碑事件。感谢你们提供的大量资源，这为我们创造了一个充满机遇的环境。

感谢大卫·詹森（David Jansen）、卢卡斯·克拉蒂格（Lukas Krattiger）、维纳亚克·苏达梅（Vinayak Sudame）、什里达·多达普卡（Shridhar Dhodapkar）和瑞安·麦肯纳（Ryan McKenna），感谢你们对本书提出的宝贵意见。

尤其要感谢布雷特·巴托（Brett Bartow）和玛丽安·巴托（Marianne Bartow），感谢你们对本书写作过程的大力支持，没有你们的支持，我们将无法完成这个项目。

布拉德·埃奇沃斯

维尼特，再次感谢您让我有机会与您合写本书。理查德，再次感谢您无与伦比的洞察力。我一直都很喜欢我们的深夜电话会议。

拉米罗（Ramiro）和马特（Matt），谢谢你们检查出的所有错误，特别是在本书付诸印刷之前指出这些错误！

还有很多人为本书的写作和出版提供了有益的反馈、建议和支持，使本书的内容更加完善。感谢所有在此过程中提供过帮助的人，特别是布雷特·巴托（Brett Bartow）、玛丽安·巴托（Marianne Bartow）、杰伊·富兰克林（Jay Franklin）、凯瑟琳·麦克纳马拉（Katherine McNamara）、达斯汀·舒曼（Dustin Schuemann）、克雷格·史密斯（Craig Smith）以及我的经理。

附言：蒂甘（Teagan），虽然本书没有飞龙或公主，不过下一本书很可能会有哦！

理查德·弗尔

首先感谢我的合著者维尼特·贾恩和布拉德·埃奇沃斯，是你们让我有机会参与本项目，虽然写作图书极具挑战，但我也同样收获满满。

布拉德，感谢您为我撰写的第1章内容提供了大量有益的指导和严格的指正，教会我如何把词语和句子变成一本书。维尼特，您的干劲和信心极具感染力！期待未来有机会能与你们再次合作。

感谢技术编辑马特·埃索（Matt Esau）和拉米罗·加尔扎·里奥斯（Ramiro Garza Rios），感谢你们丰富的专业知识和细心指导，没有你们的贡献，本书将无法完成。

感谢我的经理迈克·斯托林斯（Mike Stallings），没有你的支持和鼓励，我将不可能完成这个项目。迈克，感谢您允许我发挥创造力并从事类似的项目，并为我们创造了最佳工作环境。

序

　　数据中心是数字时代所有企业的核心资源，负责处理提供给客户的产品和服务的比特和数据。现代企业的数据存储和处理能力已成为创造营收的代名词。对于各类垂直行业来说（如建筑、医疗、娱乐等），所有业务领域的企业每年都要以数字化方式存储和处理越来越多的信息，这就意味着必须从速度、容量和灵活性等角度设计出更具匹配性的通信网络。Nexus 平台在设计之初就全面考虑了速度和带宽容量问题，Nexus 7000 于 2008 年推出，它以低廉的端口成本提供了高密度的 10Gbit/s 接口。Nexus 交换机操作系统 NX-OS 提供了包括 vPC（virtual Port Channel，虚拟端口通道）在内的多种面向未来的先进技术，不但增加了可用带宽和冗余能力，而且还克服了 STP（Spanning-Tree Protocol，生成树协议）存在的低效问题。此外，NX-OS 还引入了 OTV（Overlay Transport Virtualization，叠加传输虚拟化）技术，实现了站点之间的主机移动性以及全数据中心的冗余能力，从根本上改变了数据中心网络的设计模式。目前，Nexus 平台以高密度紧凑型外形支持了大量 25/40/100 千兆接口，推动了平台的持续发展，并将 VXLAN 和 ACI（Application Centric Infrastructure，以应用为中心的基础设施）等创新技术推向市场。

　　NX-OS 在设计之初就以简化操作为基础，提供了大量便捷的实用工具和功能特性，大大提高了网络的整体运行效率。当前的应用环境迫切要求网站和应用程序实现每天 24 小时、每周 7 天和每年 365 天的高可用性。数据中心的宕机时间会直接转化为财务损失，企业的数字化转型发展以及网络对业务的潜在影响，都要求网络工程师必须掌握有效的数据中心网络故障排查技能，这一点比以往任何时候都更为重要。

　　作为思科 25 年来的技术服务领导者，有机会与业内优秀的网络专家携手合作，使我受益匪浅。本书由维尼特、布拉德和理查德共同撰写，他们都是我所在部门的"网络摇滚明星"，工作经验丰富，为多个思科客户提供了卓有成效的技术支持工作。本书为 Nexus 交换机和 NX-OS 操作系统的故障排查提供了全面的参考指南，书中提到的方法也与思科技术服务部门长期解决复杂网络故障问题采用的方法相同。

——约瑟夫·平托（Joseph Pinto）
思科技术服务高级副总裁，圣何塞

前言

NX-OS（Nexus Operating System，Nexus 操作系统）采用了模块化软件架构，主要面向高速/高密度网络应用环境（如数据中心）。NX-OS 为 Nexus 交换机提供了虚拟化、高可用性、可扩展性和可升级性等诸多功能特性。

特别是 NX-OS 可以在软件升级或硬件升级（故障切换、OIR）期间提供高效弹性机制，能够在完全不影响不间断转发的情况下完成软硬件升级操作。市场要求 NX-OS 必须扩展以支持大规模多机架系统，而且在面对中断故障时能够提供强大的系统恢复能力。NX-OS 提供了包含大量功能和协议的丰富功能特性集，通过 vPC（virtual Port Channel，虚拟端口通道）、OTV（Overlay Transport Virtualization，叠加传输虚拟化）和当前的 VxLAN（Virtual extensible LAN，虚拟可扩展 LAN）等创新技术彻底改变了数据中心设计模式。

Nexus 7000 交换机于 2008 年首次亮相，提供了超过 512 个 10Gbit/s 端口。多年以来，思科持续发布了一系列 Nexus 交换机，包括 Nexus 5000、Nexus 2000、Nexus 9000 和 VirtualNexus 1000。NX-OS 的功能特性也在持续丰富当中，这使 Nexus 交换机可以越来越多地部署在企业路由和交换领域。

本书帮助读者学习和理解基于 NX-OS 操作系统的 Nexus 平台故障排查技术，将此前分布在多个相关信息来源和思科技术社区里的大量信息进行了高效整合，涵盖了各种功能特性在 Nexus 平台的工作方式，以及利用 NX-OS 各种功能特性进行故障排查所需的知识和技能，包括功能特性和体系架构层面的各种信息。

本书对象

本书适合所有希望进一步理解底层 Nexus 平台和 NX-OS 操作系统的网络工程师、架构师或技术顾问学习与参考，通过阅读本书，读者能够掌握通过 NX-OS 排查复杂网络故障的相关知识和技能。此外，本书也可以为准备 CCIE 数据中心认证的人员提供有益参考。

本书组织方式

本书在章节设计上充分考虑了读者灵活阅读的需要，允许读者在不同的章节之间轻松切换，从而快速发现所需的有用信息。当然，读者也可以按部就班地逐章阅读本书。

第一部分"Nexus 交换机故障排查概述"：简要介绍了 Nexus 平台以及排查网络事件故障所需的各类 NX-OS 组件。

- 第 1 章"NX-OS 概述"：本章介绍了 Nexus 平台和 Nexus 操作系统（NX-OS）的主要功能组件，包括弹性、虚拟化、效率和可扩展性共四大 NX-OS 基本支柱。
- 第 2 章"NX-OS 故障排查工具"：本章讨论了常见故障排查工具，包括抓包、NetFlow、EEM、日志以及事件历史记录等。
- 第 3 章"Nexus 平台故障排查"：本章解释了 Nexus 平台组件以及排查管理引擎卡、线卡、硬件丢包和交换矩阵故障所需的相关命令，讨论了线卡接口和 PLIM 层面的故障排查问题。此外，本章还讨论了与 CoPP 策略相关的知识以及 CoPP 故障排查技术。

第二部分"二层转发故障排查"：讨论了如何在网络数据包交换期间开展 Nexus 交换机特定组件的故障排查技术。

- **第 4 章"Nexus 交换技术"**：本章讨论了 Nexus 转发数据包的方式以及交换端口类型、私有 VLAN 和 STP 等技术。
- **第 5 章"端口通道、vPC 和 FabricPath"**：本章详细介绍了 vPC、FabricPath 和 vPC+ 的工作原理以及对下一代 DC 进行设计的价值。此外，本章还重点讨论了与 vPC 和 vPC+ 相关的设计、部署和故障排查问题。

第三部分"三层路由故障排查"：讨论了 NX-OS 的底层 IP 组件，包括 EIGRP、OSPF、IS-IS、BGP 等路由协议以及实现路由过滤或路径控制的路由选择机制。

- **第 6 章"IP 和 IPv6 服务故障排查"**：本章解释了各种 IPv4 和 IPv6 服务的工作方式以及在 Nexus 平台上排查这些服务故障的方式。此外，本章还讨论了 FHRP 协议，如 HSRP、VRRP 和 GLBP。
- **第 7 章"EIGRP 故障排查"**：本章讨论了与 EIGRP 故障排查相关的各种问题，包括建立 EIGRP 邻接关系、次优路由选择以及其他常见 EIGRP 问题。
- **第 8 章"OSPF 故障排查"**：本章讨论了与 OSPF 故障排查相关的各种问题，包括建立 OSPF 邻接关系、次优路由选择以及其他常见 OSPF 问题。
- **第 9 章"IS-IS 故障排查"**：本章讨论了与 IS-IS 故障排查相关的各种问题，包括建立 IS-IS 邻接关系、次优路由选择以及其他常见 IS-IS 问题。
- **第 10 章"Nexus 路由映射故障排查"**：本章讨论了与路由过滤或度量控制相关的多种网络选择技术，解释了基于 ACL、前缀列表和路由映射实现路由的条件匹配机制。
- **第 11 章"BGP 故障排查"**：本章讨论了与 BGP 故障排查相关的各种问题，包括建立 BGP 邻接关系、路径选择以及其他常见 BGP 问题。

第四部分"高可用性故障排查"：讨论并解释了 NX-OS 的高可用性组件。

- **第 12 章"高可用性"**：本章讨论了高可用性组件的故障排查技术，包括 BFD（Bidirectional Forward Detection，双向转发检测）、SSO（Stateful Switchover，状态化切换）、ISSU（In-Service Software Upgrade，不中断软件升级）和 GIR（Graceful Insertion and Removal，平滑插拔）等。

第五部分"多播网络流量"：解释了 Nexus 交换机的多播网络流量操作组件。

- **第 13 章"多播故障排查"**：本章讨论了与多播相关的各种组件以及识别和排查多播网络故障的相关技术。

第六部分"Nexus 隧道故障排查"：讨论了 NX-OS 提供的多种隧道技术。

- **第 14 章"OTV 故障排查"**：本章讨论了革命性的 OTV（Overlay Transport Virtualization，叠加传输虚拟化）技术及其运行方式，以及故障排查过程。

第七部分"网络可编程性"：详细介绍了通过 API 和自动化机制配置 NX-OS 的方法。

- **第 15 章"可编程性与自动化"**：本章讨论了 NX-OS 提供的各种 API 以及通过这些 API 实现网络操作自动化的相关机制。

资源与支持

本书由异步社区出品，社区（https://www.epubit.com/）为您提供相关资源和后续服务。

提交勘误

作者和编辑尽最大努力来确保书中内容的准确性，但难免会存在疏漏。欢迎您将发现的问题反馈给我们，帮助我们提升图书的质量。

当您发现错误时，请登录异步社区，按书名搜索，进入本书页面，单击"提交勘误"，输入勘误信息，单击"提交"按钮即可（见下图）。本书的作者和编辑会对您提交的勘误进行审核，确认并接受后，您将获赠异步社区的 100 积分。积分可用于在异步社区兑换优惠券、样书或奖品。

扫码关注本书

扫描下方二维码，您将会在异步社区微信服务号中看到本书信息及相关的服务提示。

与我们联系

我们的联系邮箱是 chencongcong@ptpress.com.cn。

如果您对本书有任何疑问或建议,请您发邮件给我们,并请在邮件标题中注明本书书名,以便我们更高效地做出反馈。

如果您有兴趣出版图书、录制教学视频,或者参与图书翻译、技术审校等工作,可以发邮件给我们。

如果您所在的学校、培训机构或企业,想批量购买本书或异步社区出版的其他图书,也可以发邮件给我们。

如果您在网上发现有针对异步社区出品图书的各种形式的盗版行为,包括对图书全部或部分内容的非授权传播,请您将怀疑有侵权行为的链接发邮件给我们。您的这一举动是对作者权益的保护,也是我们持续为您提供有价值的内容的动力之源。

关于异步社区和异步图书

"异步社区"是人民邮电出版社旗下 IT 专业图书社区,致力于出版精品 IT 技术图书和相关学习产品,为作译者提供优质出版服务。异步社区创办于 2015 年 8 月,提供大量精品 IT 技术图书和电子书,以及高品质技术文章和视频课程。更多详情请访问异步社区官网 https://www.epubit.com。

"异步图书"是由异步社区编辑团队策划出版的精品 IT 专业图书的品牌,依托于人民邮电出版社近 30 年的计算机图书出版积累和专业编辑团队,相关图书在封面上印有异步图书的 LOGO。异步图书的出版领域包括软件开发、大数据、AI、测试、前端、网络技术等。

异步社区

微信服务号

目录

第一部分 Nexus 交换机故障排查概述

第 1 章 NX-OS 概述 ……………………… 2
1.1 Nexus 平台概述 ………………………… 2
1.2 NX-OS 架构 ……………………………… 6
1.3 理解 NX-OS 软件的发布与打包 …… 19
1.4 NX-OS 高可用性基础设施 …………… 21
1.5 NX-OS 虚拟化功能 …………………… 26
1.6 管理和操作功能 ……………………… 29
1.7 本章小结 ……………………………… 37

第 2 章 NX-OS 故障排查工具 …………… 38
2.1 抓包：网络嗅探 ……………………… 38
2.2 Nexus 平台工具 ……………………… 44
2.3 NetFlow ………………………………… 51
2.4 网络时间协议 ………………………… 57
2.5 嵌入式事件管理器 …………………… 59
2.6 日志记录 ……………………………… 62
2.7 本章小结 ……………………………… 66

第 3 章 Nexus 平台故障排查 …………… 67
3.1 硬件故障排查 ………………………… 67
3.2 VDC …………………………………… 92
3.3 NX-OS 系统组件故障排查 ………… 101
3.4 HWRL、CoPP 和系统 QoS ………… 127
3.5 本章小结 ……………………………… 139

第二部分 二层转发故障排查

第 4 章 Nexus 交换技术 ………………… 142
4.1 二层网络通信概述 …………………… 142
4.2 VLAN ………………………………… 144
4.3 生成树协议基础 ……………………… 157
4.4 检测和修复转发环路 ………………… 173
4.5 本章小结 ……………………………… 181

第 5 章 端口通道、vPC 和 FabricPath ………………………… 183
5.1 端口通道 ……………………………… 183
5.2 vPC …………………………………… 197
5.3 FabricPath …………………………… 211
5.4 仿真交换机和 vPC+ ………………… 223
5.5 本章小结 ……………………………… 230

第三部分 三层路由故障排查

第 6 章 IP 和 IPv6 服务故障排查 ……… 232
6.1 IP SLA ………………………………… 232
6.2 对象跟踪 ……………………………… 237
6.3 IPv4 服务 ……………………………… 242
6.4 IPv6 服务 ……………………………… 255
6.5 FHRP …………………………………… 268
6.6 本章小结 ……………………………… 283

第 7 章 EIGRP 故障排查 ………………… 284
7.1 EIGRP 基础知识 ……………………… 284
7.2 EIGRP 邻居邻接关系故障排查 …… 289
7.3 EIGRP 路径选择和路由丢失故障排查 ……………………………… 303
7.4 收敛问题 ……………………………… 317
7.5 本章小结 ……………………………… 322

第 8 章 OSPF 故障排查 ………………… 323
8.1 OSPF 基础知识 ……………………… 323
8.2 OSPF 邻居邻接关系故障排查 …… 327
8.3 OSPF 路由丢失故障排查 …………… 345
8.4 OSPF 路径选择故障排查 …………… 354
8.5 本章小结 ……………………………… 361

第 9 章 IS-IS 故障排查 ………………… 362
9.1 IS-IS 基础知识 ……………………… 362
9.2 IS-IS 邻居邻接关系故障排查 …… 370
9.3 IS-IS 路由丢失故障排查 …………… 390
9.4 本章小结 ……………………………… 405

第 10 章 Nexus 路由映射故障排查 …… 406
10.1 条件匹配 …………………………… 406
10.2 路由映射 …………………………… 414

10.3 策略路由 420
10.4 本章小结 423

第 11 章 BGP 故障排查 424
11.1 BGP 基础知识 424
11.2 BGP 会话 426
11.3 BGP 对等关系故障排查 432
11.4 BGP 路由处理与路由传播 447
11.5 BGP 的扩展性 460
11.6 BGP 路由过滤和路由策略 469
11.7 BGP 路由映射 477
11.8 Looking Glass 和路由服务器 487
11.9 日志采集 487
11.10 本章小结 487

第四部分 高可用性故障排查

第 12 章 高可用性 490
12.1 BFD 490
12.2 Nexus 高可用性 503
12.3 GIR 511
12.4 本章小结 520

第五部分 多播网络流量

第 13 章 多播故障排查 522
13.1 多播基础知识 522
13.2 NX-OS 多播体系架构 527
13.3 IGMP 533
13.4 PIM 多播 548
13.5 多播和 vPC 604
13.6 多播 Ethanalyzer 案例 621
13.7 本章小结 621

第六部分 Nexus 隧道故障排查

第 14 章 OTV 故障排查 624
14.1 OTV 基础知识 624
14.2 理解和验证 OTV 控制平面 630
14.3 理解和验证 OTV 数据平面 650
14.4 OTV 高级功能特性 668
14.5 本章小结 674

第七部分 网络可编程性

第 15 章 可编程性与自动化 676
15.1 可编程性与自动化概述 676
15.2 Open NX-OS 简介 677
15.3 NX-SDK 687
15.4 NX-API 689
15.5 本章小结 694

第一部分

Nexus 交换机故障排查概述

第 1 章　NX-OS 概述
第 2 章　NX-OS 故障排查工具
第 3 章　Nexus 平台故障排查

第 1 章

NX-OS 概述

本章主要讨论如下主题。
- Nexus 平台。
- NX-OS 架构。
- NX-OS 虚拟化功能。
- 管理和操作功能。

2008 年发布 NX-OS（Nexus Operating System，Nexus 操作系统）和 Nexus 7000 平台的时候，与当时的其他交换产品相比，NX-OS 在弹性、可扩展性、虚拟化和系统架构等方面都取得了长足进步。原先裸机服务器资源存在的过剩容量浪费问题，目前已经用高效的虚拟机所取代，当前虚拟化浪潮已席卷网络领域。网络正在从传统的三层设计模式（接入层、分发层、核心层）演变为需要更大容量、更大规模和更高可用性的设计模式。业界已经无法忍受因 STP（Spanning Tree Protocol，生成树协议）阻塞机制导致的链路空闲，希望充分利用这些空闲容量来提升网络可用性。

随着网络拓扑结构的不断发展，业界对连接主机和网段的网络基础设施的期望也不断提高。网络运营商们正在寻找能够更加灵活应对网络故障的平台，希望获取更大的交换容量，并在网络设计过程中引入网络虚拟化，从而更有效地利用宝贵的物理硬件资源。此外，随着数据中心规模的不断扩大，运营商们还希望业界能够在降低能耗和制冷需求方面取得更大的进步，实现更高的效率。

Nexus 7000 系列交换机是思科 Nexus 交换机产品线的第一个平台，该系列交换机的设计目的是满足不断变化的数据中心市场需求。NX-OS 将二层交换、三层路由和 SAN 交换功能融入一个统一的操作系统中。

从最初的发行版本开始，NX-OS 操作系统一直都在不断发展当中，Nexus 交换机产品也逐步扩展为多个系列交换机，以更好地满足现代网络需求。在持续发展过程中，NX-OS 始终坚持以下 4 个基本特性。
- 弹性。
- 虚拟化。
- 效率。
- 可扩展性。

本章将详细介绍各种类型的 Nexus 平台及其在现代网络架构中的位置，以及 NX-OS 的主要功能组件。此外，还将介绍一些高级的可服务性和可用性功能，为后续故障排查章节做好准备。掌握了 NX-OS 和 Nexus 交换机的基础知识之后，就可以更好地学习后面的故障排查章节。

1.1 Nexus 平台概述

思科 Nexus 交换产品线包含以下平台。
- Nexus 2000 系列。

- Nexus 3000 系列。
- Nexus 5000 系列。
- Nexus 6000 系列。
- Nexus 7000 系列。
- Nexus 9000 系列。

下面将详细介绍这些 Nexus 平台，并从高层视角就常见部署场景对这些平台的功能特性的要求及部署位置进行讨论。

1.1.1 Nexus 2000 系列

Nexus 2000 系列产品是一组被称为 FEX（Fabric EXtender，交换矩阵扩展器）的设备，FEX 在本质上是母交换机的远程线卡，可以将母交换机的交换矩阵扩展到服务器接入层。

FEX 架构的主要优点如下。

- 可以将交换矩阵延伸至主机，无须生成树。
- FEX 架构具有高度可扩展性（与主机类型无关）。
- 可以从母交换机实施单点管理。
- 能够升级母交换机并保留 FEX 硬件。

Nexus 2000 FEX 产品不能作为独立设备运行，需要母级交换机才能运行为模块化系统。Nexus 2000 提供了多种模块以满足不同的主机端口物理连接需求，包括 GE、10GE 以及 FCoE（Fiber Channel over Ethernet，以太网光纤通道）等多种连接选项。在回连母交换机的 FEX 交换矩阵侧，Nexus 2000 FEX 提供了 GE、10GE 和 40GE 等多种上行接口。当前提供的 FEX 型号包括以下几种。

- GE 交换矩阵扩展器：2224TP、2248TP、2248TP-E。
- 10GBase-T 交换矩阵扩展器：2332TQ、2348TQ、2348TQ-E、2232TM-E、2232TM。
- 10G SFP+交换矩阵扩展器：2348UPQ、2248PQ、2232PP。

在决定使用 FEX 平台之前，必须全面考虑主机的连接需求、母交换机的连接需求以及母交换机的兼容性。同时，还要考虑主机的预期吞吐量和性能需求，因为增加了 FEX 之后，可以基于主机的可用前面板带宽，为交换矩阵侧接口分配超额带宽。

1.1.2 Nexus 3000 系列

Nexus 3000 系列由多种型号的高性能、低时延且配置固定的交换机组成，属于紧凑型产品，大小为 1RU 或 2RU（Rack Unit，机架单元），前面板端口密度高，可以提供 GE、10GE、40GE 到 100GE 等多种速率。Nexus 3000 系列交换机不仅具有高性能，还具有非常好的通用性，支持多种二层功能及三层路由协议和 IP 多播。Nexus 3000 系列交换机的型号由平台系列、端口数量或端口总带宽以及接口类型组成。

目前提供的 Nexus 3000 设备型号包括以下几种。

- Nexus 3000：3064X、3064-32T、3064T、3048。
- Nexus 3100：3132Q/3132Q-X、3164Q、3172PQ、3172TQ、31128PQ。
- Nexus 3100-V：31108PC-V、31108TC-V、3132Q-V。
- Nexus 3200：3232C、3264Q。
- Nexus 3500：3524/3524-X、3548/3548-X。
- Nexus 3600：36180YC-R。

根据不同的预期应用场景，这些型号的交换机各有特色。例如，Nexus 3500 系列交换机支持超低时延交换功能（低于 250ns），在高性能计算和高频股票交易环境中极受欢迎。Nexus 3100-V

交换机支持 VXLAN（Virtual Extensible Local Area Network，虚拟可扩展局域网）路由，Nexus 3200 交换机可以提供低时延和大缓冲区，而 Nexus 3000 和 Nexus 3100 系列交换机则是优秀的全能型的线速 ToR（Top of Rack，架顶式）交换机。

> **注**：除 Nexus 3500 系列之外，所有的 Nexus 3000 系列交换机运行的 NX-OS 软件版本均与 Nexus 9000 系列交换机相同。

1.1.3 Nexus 5000 系列

Nexus 5000 系列支持广泛的二层和三层功能，能够根据网络设计需求实现多样化的能力。Nexus 5500 系列需要安装额外的硬件和软件许可，以获得完整的三层能力支持，而 Nexus 5600 系列则提供了一个处理性能为 160Gbit/s 的原生三层路由引擎。与 Nexus 5500 系列相比，Nexus 5600 还支持 VXLAN 和更大的表容量。

目前提供的 Nexus 5000 型号包括以下几种。

- Nexus 5500：5548UP、5596UP、5596T。
- Nexus 5600：5672 UP、5672UP-16G、56128P、5624Q、5648Q、5696Q。

Nexus 5000 系列非常适合作为 ToR 或 EoR（End of Row，行尾式）交换机，用于高密度和大规模网络环境，支持 GE、10GE 和 40GE 以太网和 FCoE 连接。用作 FEX 聚合的母交换机时，可以获得更高的端口密度。5696Q 支持 100GE 上行链路（带有扩展模块）。Nexus 5000 平台的命名规则首先是型号系列，然后是支持的 10GE 或 40GE 端口数（取决于具体型号）。Nexus 5672 是一种 5600 平台交换机，支持 72 个 10GE 以太网端口，字母 UP 表示存在 40GE 上行链路端口。

Nexus 5000 系列由于支持丰富的三层功能，可以提供大量端口、FEX 聚合，并在单个平台中灵活支持以太网、FCoE 和光纤通道，因而成为很多网络环境的首选 ToR 或 EoR 设备。

1.1.4 Nexus 6000 系列

Nexus 6001 和 Nexus 6004 交换机适用于高密度数据中心网络 ToR 和 EoR 场景。Nexus 6001 是一个 1RU 机箱，支持 GE、10GE 服务器连接；Nexus 6004 是一个 4RU 机箱，适用于 10GE、40GE 服务器连接或 FCoE。FEX 聚合也是 Nexus 6000 系列交换机的常见应用。Nexus 6000 系列交换机提供了大缓冲区和低时延交换能力，可以满足高性能计算环境的需求。此外，在安装了适当功能特性许可的情况下，Nexus 6000 还支持健壮的二层、三层以及存储功能特性集。不过，Nexus 6000 系列产品已在 2017 年 4 月 30 日停止生产，不再销售。思科将 Nexus 5600 平台作为替代平台，因为 Nexus 5600 在数据中心能够提供类似的优势、端口密度及适用性。

1.1.5 Nexus 7000 系列

Nexus 7000 系列约在 10 年前首发，目前仍是全球企业网络、数据中心及服务提供商网络的主流设备选择。Nexus 7000 系列的成功因素众多，是一个真正的模块化平台，基于完全分布式的 Crossbar（交叉开关矩阵）架构，提供了丰富的功能特性。Nexus 7000 系列分为两个机箱系列：Nexus 7000 和 Nexus 7700。Nexus 7000 系列机箱支持以下配置（其中，平台名称的最后两位数字表示机箱的插槽数）。

- Nexus 7000：7004、7009、7010、7018。
- Nexus 7700：7702、7706、7710、7718。

不同的机箱配置主要考虑了不同网络环境下的尺寸优化问题。Nexus 7000 系列有 5 个交换矩阵模块插槽，Nexus 7700 有 6 个交换矩阵模块插槽。Nexus 7004 和 Nexus 7702 不使用单独的交换矩阵模块，因为 I/O（Input/Output，输入/输出）模块上的 Crossbar 架构能够完全满足平台的处理需求。对交换矩阵的访问由管理引擎（Supervisor）上的集中式仲裁单元控制，允许访问入站模块的交换矩阵，从而向出站模块发送数据包。VOQ（Virtual Output Queues，虚拟输出队列）在入站 I/O 模块（代表出站 I/O 模块的交换矩阵容量）上实现，可以实现队首阻塞（Head-of-line blocking）的最小化，其中，设备在拥塞期间等待出站板卡接收数据包的时候可能会出现队首阻塞。

Nexus 7000 和 Nexus 7700 使用一块管理引擎模块，负责运行平台的管理和控制平面，并监督平台的运行状况。从 Supervisor 1 到 Supervisor 2 再到 Supervisor 2E，管理引擎模块的 CPU 功率、内存容量和交换性能都得到了持续提升。

由于 Nexus 7000 是一个分布式系统，因此 I/O 模块都运行自己的软件并处理所有数据平面流量。Nexus 7000 I/O 模块支持两类转发引擎：M 系列或 F 系列。这两类线卡的端口速率范围均从 GE、10GE、40GE 到 100GE，通常以它们的转发引擎版本（M1、M2、M3 和 F1、F2、F3）加以区分，每一代新版本均比前一代版本拥有更强大的转发能力和更丰富的功能特性。M 系列通常拥有较大的转发表容量和较大的包缓冲区。虽然以前的 M 系列比 F 系列支持更丰富的三层功能，但随着 F3 板卡的发布，这方面的功能差距已经大大缩小，F3 板卡已经全面支持 LISP（Locator-ID Separation Protocol，位置与身份分离协议）和 MPLS 等功能特性。图 1-1 解释了 Nexus 7000 系列的 I/O 模块命名约定。

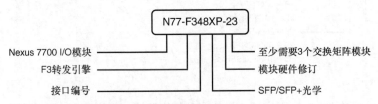

图 1-1 Nexus 7000 系列 I/O 模块命名约定

Nexus 7000 通常部署在汇聚层或核心层。不过，将 FEX 与 Nexus 7000 联合使用，也可以为主机提供高密度的接入连接。此外，由于 Nexus 7000 功能特性丰富且性能优异，因此它也是 MPLS、LISP、OTV（Overlay Transport Virtualization，叠加传输虚拟化）和 VxLAN 等叠加技术的流行选择。

1.1.6 Nexus 9000 系列

Nexus 9000 系列于 2013 年底发布。Nexus 9500 是一款模块化交换机，也是第一款搭载多种创新功能的机型。采用模块化机箱设计的目的是减少组件数量，因而没有中间背板（mid-plane），线卡模块直接与机箱后部的交换矩阵模块相连。整个机箱的交换容量最多可以达到 6 个设计为全线速、所有端口无阻塞的交换矩阵。最近，思科刚刚发布了 R 系列线卡和交换矩阵模块，该系列具有更深的缓冲能力以及更大的转发表，适用于更加严苛的应用环境。Nexus 9500 是一个模块化交换平台，拥有管理引擎模块、交换矩阵模块和多种线卡选项。Nexus 9500 支持两种管理引擎模块。

- Supervisor A：拥有 4 核 1.8 GHz CPU、16 GB RAM 和 64 GB SSD 存储。
- Supervisor B：拥有 6 核 2.2 GHz CPU、24 GB RAM 和 256 GB SSD 存储。

Nexus 9000 系列综合采用了商用交换 ASIC（Application-Specific Integrated Circuit，专用集成电路）和思科自己开发的 ASIC，有效降低了设备成本。Nexus 9500 之后是 Nexus 9300 和 Nexus 9200 系列，支持 GE、10GE、25GE、40GE 和 100GE 等接口速率（取决于具体型号），而且某些型号还支持 FCoE 和 FEX 聚合功能。Nexus 9500 是一款灵活的模块化设备，可以充当叶/汇聚层或脊/核心层交换机，具体取决于应用环境的规模。

Nexus 9300 和 Nexus 9200 主要充当高性能 ToR/EoR/叶交换机。Nexus 9000 系列的大小从 1RU 到 21RU 不等，支持多种模块和连接选项，符合绝大多数连接和性能要求。当前提供的设备型号包括以下几种。

- Nexus 9500：9504、9508、9516。
- Nexus 9300 100M/1GBase-T：9348GC-FXP。
- Nexus 9300 10GBase-T：9372TX、9396TX、93108TC-FX、93120TX、93128TX、93108TC-EX。
- Nexus 9300 10/25GE Fiber：9372px、9396px、93180yc-fx、93180yc-ex。
- Nexus 9300 40GE：9332PQ、9336PQ、9364C、93180LC-EX。
- Nexus 9200：92160YC-X、9272Q、92304QC、9236C、92300YC。

Nexus 9000 平台命名约定如图 1-2 所示。

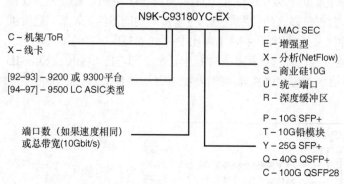

图 1-2　Nexus 9000 平台命名约定

Nexus 9000 系列因速率高、功能特性丰富而广泛应用于多种网络环境，主要用于高频交易、高性能计算、大规模叶/脊架构，是目前实现 VxLAN 主流的思科 Nexus 平台。

注：Nexus 9000 系列可以运行为独立的 NX-OS 模式或 ACI（Application-Centric Infrastructure，以应用为中心的基础设施）模式，具体取决于安装的软件和许可情况。本书仅讨论独立模式下的 Nexus 配置及故障排查。

由于 Nexus 系列交换产品一直都在持续发展当中，因此有关各类产品的最新信息，请查阅思科官网的产品手册和相关文档。

1.2　NX-OS 架构

自面世以来，弹性、虚拟化、效率和可扩展性一直都是 NX-OS 的四大基本支柱。此外，设计人员还希望提供一个类似于 IOS 风格的用户界面，以便从传统产品迁移到 NX-OS 的客户能够很容易地部署和操作这些新产品。相对于 IOS，NX-OS 对核心操作系统的较大改进表现在以下领域。

- 进程调度。
- 内存管理。
- 进程隔离。
- 功能特性进程的管理。

对于 NX-OS 中的功能特性进程来说，其仅在用户配置了该功能特性之后才会启动，这样做的好处是可以节省系统资源，实现更强的扩展性和更高的效率。这些功能特性使用自己的内存和系统资源，有助于增强操作系统的稳定性。虽然在风格上很相似，但是与思科 IOS 操作系统相比，NX-OS 操作系统在很多领域都有了全面改进。

NX-OS 的模块化架构如图 1-3 所示。

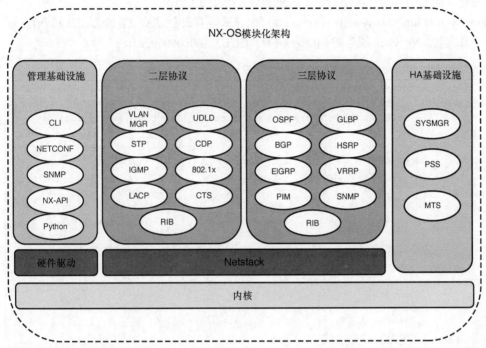

图 1-3 NX-OS 模块化架构

> **注**：下面将介绍一些重要的 NX-OS 组件，其余 NX-OS 服务和组件将在本书的其余部分，结合特定案例进行解释。

1.2.1 内核

内核的主要职责是管理系统资源以及系统硬件组件之间的接口。NX-OS 操作系统继承了 Linux 内核提供的大量重要优势，如支持 SMP（Symmetric Multiprocessor，对称多处理器）和抢占式多任务处理。可以在多个处理器之间调度和分发多线程进程，以提高可扩展性。操作系统的每个组件进程都被设计成模块化、自包含进程，而且能够保护内存不受其他组件进程的影响。这种架构模式可以构建高度弹性化的系统，实现进程故障的有效隔离，发生故障后也更容易恢复，这种自包含、自修复的架构模式能够以中断最小甚至毫无中断的方式实现故障恢复，因为只有个别进程重启，因而系统可以在无须重新加载的情况下实现自我修复。

> **注**：以前访问 NX-OS 的 Linux 模块时，还需要思科支持人员安装"调试插件"，不过目前的 NX-OS 已经在某些平台上提供了 **feature bash-shell**，允许用户访问 NX-OS 的底层 Linux 模块。

1.2.2 系统管理器（sysmgr）

系统管理器是 NX-OS 的组件之一，负责管理系统运行的进程。也就是说，系统管理器负责启动进程并监控进程的运行状况，以确保这些进程始终保持正常工作。如果进程出现故障，那么系统管理器就会采取相应的措施进行恢复，根据进程的性质，系统管理器可以按照有状态或无状态方式重启进程，也可以根据需要启动系统切换操作（切换到冗余管理引擎）以恢复系统。

NX-OS 以 UUID（Universally Unique Identifier，通用唯一标识符）来标识每个进程，由该标

识符来标识进程所代表的 NX-OS 服务。NX-OS 使用 UUID 的原因是 PID（Process ID，进程 ID）可能会发生变化，而 UUID 始终保持不变（即便 PID 发生了变化）。

命令 **show system internal sysmgr service all** 可以显示所有服务以及服务的 UUID 和 PID，如例 1-1 所示。请注意，NetStack 服务的 PID 为 6427，UUID 为 0x00000221。

例 1-1　show system internal sysmgr service all 命令

```
NX-1# show system internal sysmgr service all
! Output omitted for brevity
Name                UUID          PID     SAP     state       Start   Tag     Plugin ID
count
----------------    ----------    ------  -----   ---------   ------  ---     -----   ---
aaa                 0x000000B5    6227    111     s0009       1       N/A     0
ospf                0x41000119    13198   320     s0009       2       32      1
psshelper_gsvc      0x0000021A    6147    398     s0009       1       N/A     0
platform            0x00000018    5817    39      s0009       1       N/A     0
radius              0x000000B7    6455    113     s0009       1       N/A     0
securityd           0x0000002A    6225    55      s0009       1       N/A     0
tacacs              0x000000B6    6509    112     s0009       1       N/A     0
eigrp               0x41000130    [NA]    [NA]    s0075       1       N/A     1
mpls                0x00000115    6936    274     s0009       1       N/A     1
mpls_oam            0x000002EF    6935    226     s0009       1       N/A     1
mpls_te             0x00000120    6934    289     s0009       1       N/A     1
mrib                0x00000113    6825    255     s0009       1       N/A     1
netstack            0x00000221    6427    262     s0009       1       N/A     0
nfm                 0x00000195    6824    306     s0009       1       N/A     1
ntp                 0x00000047    6462    72      s0009       1       N/A     0
obfl                0x0000012A    6228    1018    s0009       1       N/A     0
```

如果在这条命令中指定了具体的 UUID，那么就可以查看该服务的详细信息，如当前状态、重启次数以及崩溃次数等信息，例 1-2 给出了命令 **show system internal sysmgr service uuid** *uuid* 的语法示例。

例 1-2　show system internal sysmgr service uuid 命令

```
NX-1# show system internal sysmgr service uuid 0x00000221
UUID = 0x221.
Service "netstack" ("netstack", 182):
        UUID = 0x221, PID = 6427, SAP = 262
        State: SRV_STATE_HANDSHAKED (entered at time Fri Feb 17 23:56:39 2017).
        Restart count: 1
        Time of last restart: Fri Feb 17 23:56:39 2017.
        The service never crashed since the last reboot.
        Tag = N/A
        Plugin ID: 0
```

> **注：** 如果服务崩溃，那么就可以在 **show cores** 命令的输出结果中看到相应的进程名、PID 以及事件发生的日期/时间。

对于拥有冗余管理引擎模块的 NX-OS 平台来说，系统管理器的另一个重要角色就是负责协调主用和备用管理引擎上的服务之间的状态，确保主用设备发生故障并切换到备用设备之后的状态同步。

1.2.3　消息和事务服务

NX-OS 服务的模块化和自包含性要求进程之间能够交换消息和数据，同时维护前面提到的自包含式体系架构。操作系统组件负责提供服务间通信能力的组件是 MTS（Message and Transactional

Service，消息和事务服务）。

顾名思义，MTS 用于 NX-OS 中的进程间通信，通过 SAP（Service Access Point，服务访问点）实现服务之间的消息交换。打个比方，如果将 MTS 视为邮政服务，那么就可以将 SAP 视为进程的邮政信箱，消息的收发操作则由进程使用 SAP 通过 MTS 实现。

这里仍然采用前面的系统管理器表输出结果来查看服务名及其 UUID、PID 和 SAP 等信息，可以通过 SAP 号从 MTS 获取交换的消息数以及 MTS 缓冲区状态等详细信息。例如，假设某 OSPF 进程配置的进程标记为 32，那么就可以通过命令 **show system internal sysmgr service all** 的输出结果显示该 OSPF 进程信息，如例 1-3 所示。可以看出，UUID 为 0x41000119、PID 13198、SAP 为 320。

例 1-3　查找服务名对应的 UUID

```
NX-1# show system internal sysmgr service all
! Output omitted for brevity
Name                UUID         PID    SAP    state      Start Tag Plugin ID
count
----------------    ----------   ------ -----  ---------  ----- --- ---------
aaa                 0x000000B5   6227   111    s0009      1     N/A 0
ospf                0x41000119   13198  320    s0009      2     32  1
psshelper_gsvc      0x0000021A   6147   398    s0009      1     N/A 0
platform            0x00000018   5817   39     s0009      1     N/A 0
```

从例 1-4 可以看出，**show system internal mts sup sap** *sap-id* **[description | uuid | stats]** 命令的作用是获取特定 SAP 的详细信息。如果要分析特定 SAP，那么需要先确认服务名和 UUID 与 **show system internal sysmgr services all** 命令输出结果中的数值完全匹配，这是一项健康性检查，目的是确保正在检查正确的 SAP。**show system internal mts sup sap** *sap-id* **[description]** 命令的输出结果应与服务名相匹配，**show system internal mts sup sap** *sap-id* **[UUID]** 命令的输出结果应与 sysmgr 输出结果中的 UUID 相匹配。接下来检查 SAP 的 MTS 统计信息，该输出结果有助于确定 MTS 队列的最大值（高水平线）以及该服务交换的消息数，如果 *max_q_size ever reached* 等于 *hard q limit*，那么就表明 MTS 可能丢弃了该服务的消息。

例 1-4　检查 SAP 的 MTS 队列

```
NX-1# show system internal mts sup sap 320 description
Below shows sap on default-VDC, to show saps on non-default VDC, run
       show system internal mts node sup-<vnode-id> sap ...
ospf-32
NX-1# show system internal mts sup sap 320 uuid
Below shows sap on default-VDC, to show saps on non-default VDC, run
       show system internal mts node sup-<vnode-id> sap ...
1090519321
NX-1# show system internal mts sup sap 320 stats
Below shows sap on default-VDC, to show saps on non-default VDC, run
       show system internal mts node sup-<vnode-id> sap ...
msg tx: 40
byte tx: 6829
msg rx: 20
byte rx: 2910

opc sent to myself: 32768
max_q_size q_len limit (soft q limit): 1024
max_q_size q_bytes limit (soft q limit): 50%
max_q_size ever reached: 13
max_fast_q_size (hard q limit): 4096
```

```
rebind count: 0
Waiting for response: none
buf in transit: 40
bytes in transit: 6829

NX-1# hex 1090519321
0x41000119
NX-1# dec 0x41000119
1090519321
```

> **注：** 例 1-4 的输出结果将 UUID 显示为十进制数值，而系统管理器的输出结果将 UUID 显示为十六进制数值。NX-OS 提供了一个内置实用程序，可以使用 **hex** *value* 或 **dec** *value* 命令进行不同格式之间的转换。

有关 NX-OS MTS 服务的详细信息将在第 3 章进行讨论，同时还将提供详细的故障排查案例。

1.2.4 持久性存储服务

为了实现 NX-OS 所需的弹性能力水平，设计人员需要采取某种方法，以中断影响最小化的方式恢复系统服务。该方法不但要监控和重启故障服务，而且还要恢复服务的运行时状态，以便重启后的服务能够在重启或故障发生后恢复正常工作。系统管理器、MTS 和 PSS（Persistent Storage Service，持久性存储服务）提供了实现这种高可用性所需的 NX-OS 基础设施。系统管理器负责启动、终止和监控服务的心跳信息，以确保服务的正常工作。PSS 服务提供了一种存储运行时数据的方法，确保在故障恢复或重启进程时可用。

PSS 以轻量级键/值对数据库方式为 NX-OS 服务提供了可靠和持久存储。PSS 提供了两种存储方式：易失性存储和非易失性存储。易失性存储将数据存储在 RAM 中，用于存储需要在进程重启或崩溃后存活的服务状态。第二种存储方式是非易失性存储，将数据存储在闪存中，用于存储需要在系统重新加载后存活的服务状态。例 1-5 利用 **show system internal flash** 命令显示了闪存文件系统，并给出了验证非易失性 PSS 当前可用空间的方式。

例 1-5 核查闪存文件系统中的 PSS 大小和位置

```
NX-1# show system internal flash
Mount-on                  1K-blocks       Used   Available   Use%  Filesystem
/                            409600      65624      343976     17  /dev/root
/proc                             0          0           0      0  proc
/sys                              0          0           0      0  none
/isan                       1572864     679068      893796     44  none
/var                          51200        488       50712      1  none
/etc                           5120       1856        3264     37  none
/nxos/tmp                    102400       2496       99904      3  none
/var/log                      51200       1032       50168      3  none
/var/home                      5120         36        5084      1  none
/var/tmp                     307200        744      306456      1  none
/var/sysmgr                 3670016        664     3669352      1  none
/var/sysmgr/ftp              819200     219536      599664     27  none
/var/sysmgr/srv_logs         102400          0      102400      0  srv
/var/sysmgr/ftp/debug_logs    10240          0       10240      0  none
/dev/shm                    3145728     964468     2181260     31  none
/volatile                    512000          0      512000      0  none
/debug                         5120         32        5088      1  none
/dev/mqueue                       0          0           0      0  none
/debugfs                          0          0           0      0  nodev
/mnt/plog                    242342       5908      223921      3  /dev/sdc1
```

/mnt/fwimg	121171	4127	110788	4	/dev/sdc3
/mnt/cfg/0	75917	5580	66417	8	/dev/md5
/mnt/cfg/1	75415	5580	65941	8	/dev/md6
/bootflash	1773912	1046944	636856	63	/dev/md3
/cgroup	0	0	0	0	vdccontrol
/var/sysmgr/startup-cfg	409600	15276	394324	4	none
/dev/pts	0	0	0	0	devpts
/mnt/pss	38172	9391	26810	26	/dev/md4
/usbslot1	7817248	5750464	2066784	74	/dev/sdb1
/fwimg_tmp	131072	508	130564	1	tmpfs

利用易失性和非易失性 PSS, NX-OS 服务可以根据需要检查其运行时数据。与 NX-OS 的模块化特性一致, PSS 并不规定应该将哪些内容存储到何种类型的 PSS 中, 而是将决策交给服务。PSS 仅提供基础设施, 允许服务按需存储和检索其数据。

1.2.5 功能特性管理器

可以按需启用 NX-OS 中的功能特性, 且仅在启用后消耗系统资源, 如内存、CPU 时间、MTS 队列和 PSS 等。如果正在使用某种功能特性, 后来又被操作人员关闭了, 那么与该功能特性相关联的系统资源将由系统进行释放并回收。启用或禁用功能特性的任务由被称为功能特性管理器 (Feature Manager) 的 NX-OS 基础设施组件进行处理, 功能特性管理器还负责维护和跟踪系统中的所有功能特性的运行状态。

为了更好地理解功能特性管理器的作用及其与其他服务之间的交互关系, 下面将以一个特定案例加以说明。某操作员希望在特定 Nexus 交换机上启用 BGP, 由于 NX-OS 中的服务仅在启用后才能启动, 因而用户必须首先在配置模式下输入 **feature bgp** 命令, 由功能特性管理器执行该请求 (需要确保已经安装了该功能特性的许可), 功能特性管理器会向系统管理器发送消息以启动该服务。BGP 服务启动之后, 会绑定到某个 MTS SAP, 创建其 PSS 表项以存储运行时状态, 然后再通知系统管理器。接下来, BGP 服务会将自己注册到功能特性管理器中, 此时的操作状态将变为启用状态 (enabled)。

如果用户禁用了某功能特性, 那么就会以相反顺序发生上述事件。功能特性管理器要求服务禁用自己, 禁用之后, 功能特性会清空其 MTS 缓冲区并销毁其 PSS 数据, 然后与系统管理器及功能特性管理器进行通信, 后者将操作状态设置为禁用状态 (disabled)。

需要注意的是, 有些服务可能依赖于其他服务。如果启动了某个服务, 但并没有满足其依赖性, 那么就会启动其他相关服务, 以保证该功能特性的正常运行, 如依赖 RPM (Route Policy Manager, 路由策略管理器) 的 BGP 功能特性。从这里可以看出, 服务实现了一种或多种功能特性, 且不同的功能特性之间存在一定的依赖性。除用户必须启用功能特性之外, 其余操作对于用户来说都是透明的, NX-OS 可以自动处理这方面的依赖关系。

某些复杂功能特性要求用户在启用关联功能特性之前安装专门的功能特性集, 如 MPLS、FEX 和 FabricPath 等功能特性。如果要启用这些功能特性, 则用户必须首先通过命令 **install feature-set**[*feature*] 安装这些功能特性集, 然后再通过 **feature-set**[*feature*] 命令启用这些功能特性集。

> **注:** 许可管理器可以跟踪系统中的所有功能特性许可。一旦许可过期, 许可管理器就会通知功能特性管理器关闭该功能特性。

可以通过 **show system internal feature-mgr feature state** 命令验证功能特性的当前状态, 如例 1-6 所示。输出结果以表格形式显示, 列出了功能特性名称及其 UUID、状态及当前状态的原因。从例 1-6 可以看出, 功能特性管理器成功启用了多个功能特性, 包括两个 EIGRP 实例。此外,该输出结果还显示了尚未启用的功能特性实例,如 EIGRP 实例 3 ~ 实例 16。

例 1-6 检查功能特性管理器的功能特性状态

```
NX-1# show system internal feature-mgr feature state
! Output omitted for brevity
Feature              UUID          State     Reason
-------------------- ----------    --------  --------------------
bfd                  0x000002c2    enabled   SUCCESS
bfd_app              0x000002c9    enabled   SUCCESS
bgp                  0x0000011b    disabled  feature never enabled
cts                  0x0000021e    disabled  feature never enabled
dhcp                 0x000001ba    enabled   SUCCESS
dot1x                0x0000017d    disabled  feature never enabled
__inst_1__eigrp      0x41000130    enabled   SUCCESS
__inst_2__eigrp      0x42000130    enabled   SUCCESS
__inst_3__eigrp      0x43000130    disabled  feature never enabled
__inst_4__eigrp      0x44000130    disabled  feature never enabled
__inst_5__eigrp      0x45000130    disabled  feature never enabled
__inst_6__eigrp      0x46000130    disabled  feature never enabled
__inst_7__eigrp      0x47000130    disabled  feature never enabled
__inst_8__eigrp      0x48000130    disabled  feature never enabled
__inst_9__eigrp      0x49000130    disabled  feature never enabled
__inst_10__eigrp     0x4a000130    disabled  feature never enabled
__inst_11__eigrp     0x4b000130    disabled  feature never enabled
__inst_12__eigrp     0x4c000130    disabled  feature never enabled
__inst_13__eigrp     0x4d000130    disabled  feature never enabled
__inst_14__eigrp     0x4e000130    disabled  feature never enabled
__inst_15__eigrp     0x4f000130    disabled  feature never enabled
__inst_16__eigrp     0x50000130    disabled  feature never enabled
..
```

虽然与功能特性管理器相关的故障并不常见,但 NX-OS 仍然提供了一种使用 CLI(Command-Line Interface,命令行接口)验证差错的方式。例 1-7 显示了使用 **show system internal feature-mgr feature action** 命令获取特定功能特性错误代码(如果存在)的方法(虽然该输出结果并没有错误代码)。

例 1-7 检查功能特性管理器的差错情况

```
NX-1# show system internal feature-mgr feature action
Feature              Action     Status    Error-code
-------------------- --------   --------  --------------------
tacacs               none       none      SUCCESS
scheduler            none       none      SUCCESS
bgp                  none       none      SUCCESS
pim                  enable     none      SUCCESS
msdp                 none       none      SUCCESS
pim6                 none       none      SUCCESS
__inst_1__eigrp      enable     none      SUCCESS
__inst_2__eigrp      enable     none      SUCCESS
__inst_3__eigrp      none       none      SUCCESS
__inst_4__eigrp      none       none      SUCCESS
__inst_5__eigrp      none       none      SUCCESS
__inst_6__eigrp      none       none      SUCCESS
__inst_7__eigrp      none       none      SUCCESS
__inst_8__eigrp      none       none      SUCCESS
__inst_9__eigrp      none       none      SUCCESS
```

```
__inst_10__eigrp        none    none    SUCCESS
__inst_11__eigrp        none    none    SUCCESS
__inst_12__eigrp        none    none    SUCCESS
__inst_13__eigrp        none    none    SUCCESS
__inst_14__eigrp        none    none    SUCCESS
__inst_15__eigrp        none    none    SUCCESS
__inst_16__eigrp        none    none    SUCCESS
lacp                    none    none    SUCCESS
dot1x                   none    none    SUCCESS
glbp                    none    none    SUCCESS
```

注：NX-OS 为众多功能特性和服务维护了一个事件运行日志，称为事件历史记录日志（event-history log），本章将在后面讨论这些日志，本书也将一直使用这些日志。功能特性管理器提供了两种事件历史记录日志（差错日志和消息日志），可以为故障排查操作提供大量详细信息。可以使用 **show system internal feature-mgr event-history**[*msgs | errors*]命令获得相应的输出结果。

1.2.6　NX-OS 线卡微码

分布式线卡运行了微码版本的 NX-OS 操作系统，如图 1-4 所示。NX-OS 的模块化架构可以将软件的基本概念和组件一致化地应用到各个线卡和整个系统。

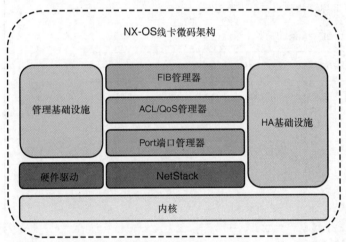

图 1-4　NX-OS 模块化线卡微码体系架构

在系统引导或者将线卡插入机箱的过程当中，由管理引擎来确定是否要为线卡加电，此时需要检查线卡类型并验证所需的电源、软件和硬件资源是否到位以保证线卡正确运行。如果一切正常，那么就会给线卡加电。线卡加电后将启动并执行 BIOS（Basic Input/Output System，基本输入/输出系统）、通电自检等操作，同时启动系统管理器。接下来，将启动正常运行所需的所有线卡服务，与管理引擎建立通信和消息传递通道，允许管理引擎推送配置数据并升级线卡软件（根据需要）。此外，还会启动一些其他服务，用于实现异常日志的本地化处理、环境传感器的管理、线卡 LED、运行状况监测等。启动了关键系统服务之后，将启动各个 ASIC，以允许线卡转发流量。

线卡处于操作状态之后，就可以转发数据包，并根据需要与管理引擎进行通信，以更新计数器、统计数据和环境数据。线卡拥有 PSS 和 OBFL（On-Board Failure Logging，板载故障日志）的本地化存储。OBFL 数据存储在非易失性存储器中，可以在重新加载之后存活下来，是排查线卡特定故障的极好数据来源。异常历史、线卡引导历史、环境历史等信息均存储在 OBFL 存储器中。

对于日常操作来说，通常不需要输入线卡 CLI。NX-OS 操作系统和分布式平台被设计为从管理引擎模块进行配置和管理。当然，在某些特殊情况下，也可能需要直接访问线卡的 CLI，此时可能还要与思科 TAC 相结合，以收集数据并对各种线卡子系统进行故障排查。从例 1-8 可以看出，从管理引擎模块使用 **attach module** 命令之后就可以进入线卡 CLI。请注意，此时的命令提示符发生了变化，指示了用户当前连接的模块信息。用户进入线卡 CLI 之后，执行 **show hardware internal dev-port-map** 命令，可以显示 Nexus 7000 M2 系列线卡上的前面板端口与线卡上的各个 ASIC 之间的映射关系。

例 1-8　从管理引擎模块使用 attach module CLI

```
NX-1# attach module 10
Attaching to module 10 ...
To exit type 'exit', to abort type '$.'
module-10# show hardware internal dev-port-map
--------------------------------------------------------------
CARD_TYPE:         24 port 10G
>Front Panel ports:24
--------------------------------------------------------------
Device name              Dev role                 Abbr num_inst:
--------------------------------------------------------------
> Skytrain               DEV_QUEUEING             QUEUE   4
> Valkyrie               DEV_REWRITE              RWR_0   4
> Eureka                 DEV_LAYER_2_LOOKUP       L2LKP   2
> Lamira                 DEV_LAYER_3_LOOKUP       L3LKP   2
> Garuda                 DEV_ETHERNET_MAC         MAC_0   2
> EDC                    DEV_PHY                  PHYS    6
> Sacramento Xbar ASIC   DEV_SWITCH_FABRIC        SWICHF  1
+-------------------------------------------------------------+
+-----------------+++FRONT PANEL PORT TO ASIC INSTANCE MAP+++-------------+
+-------------------------------------------------------------+
FP port | PHYS | SECUR | MAC_0 | RWR_0 | L2LKP | L3LKP | QUEUE |SWICHF
   1       0      0       0      0,1      0       0      0,1      0
   2       0      0       0      0,1      0       0      0,1      0
   3       0      0       0      0,1      0       0      0,1      0
   4       0      0       0      0,1      0       0      0,1      0
   5       1      0       0      0,1      0       0      0,1      0
   6       1      0       0      0,1      0       0      0,1      0
   7       1      0       0      0,1      0       0      0,1      0
   8       1      0       0      0,1      0       0      0,1      0
   9       2      0       0      0,1      0       0      0,1      0
  10       2      0       0      0,1      0       0      0,1      0
  11       2      0       0      0,1      0       0      0,1      0
  12       2      0       0      0,1      0       0      0,1      0
  13       3      1       1      2,3      1       1      2,3      0
  14       3      1       1      2,3      1       1      2,3      0
  15       3      1       1      2,3      1       1      2,3      0
  16       3      1       1      2,3      1       1      2,3      0
  17       4      1       1      2,3      1       1      2,3      0
  18       4      1       1      2,3      1       1      2,3      0
  19       4      1       1      2,3      1       1      2,3      0
  20       4      1       1      2,3      1       1      2,3      0
  21       5      1       1      2,3      1       1      2,3      0
  22       5      1       1      2,3      1       1      2,3      0
  23       5      1       1      2,3      1       1      2,3      0
  24       5      1       1      2,3      1       1      2,3      0
+-------------------------------------------------------------+
+-------------------------------------------------------------+
```

> **注：** 访问线卡 CLI 的一个常见原因就是在本地转发引擎上运行了 ELAM（Embedded Logic Analyzer Module，嵌入式逻辑分析仪模块）抓包操作，其中，ELAM 是一种解决数据平面转发和硬件转发表编程故障的常用工具。有关 ELAM 抓包操作的详细内容不在本书写作范围之内。

1.2.7 文件系统

文件系统对于所有操作系统来说都是关键组件，NX-OS 也不例外。文件系统包含了启动操作系统、记录事件和存储用户生成的数据所需的目录和文件，如支持文件、调试输出和脚本。此外，文件系统还负责存储配置数据以及各种服务存储在非易失性 PSS 中的所有数据，用于故障后的系统恢复。

NX-OS 文件系统的使用方式与思科 IOS 中的文件相似，同时也进行了一些改进。NX-OS 的文件和目录是从 *bootflash*:或被称为 *slot0*:的外部 USB 存储器进行创建和删除的，系统会创建归档文件并压缩大型文件（如 **show tech** 文件）以节省空间。表 1-1 列出了管理和排查 NX-OS 交换机故障所需的文件系统命令列表。

表 1-1　　　　　　　　　　　　　　文件系统命令

命令	作用
pwd	显示当前目录名称
cd {*directory* | *filesystem*:[*//module/*][*directory*]}	更改为新的当前目录
dir [*directory* | *filesystem*:[*//module/*][*directory*]]	显示目录内容
mkdir [*filesystem*:[*//module/*]]*directory*	创建新目录
rmdir [*filesystem* :[*//module/*]]*directory*	删除目录
move [*filesystem*:[*//module/*][*directory* /] | *directory*/]*source-filename* {{*filesystem*:[*//module/*][*directory* /] |*directory*/}[*target-filename*] | *target-filename*}	移动文件
copy [*filesystem*:[*//module/*][*directory*/] | *directory*/]*source-filename* | {*filesystem*:[*//module/*][*directory*/]] |*directory*/}[*target-filename*]	复制文件
delete {*filesystem*:[*//module/*][*directory*/] | *directory*/}*filename*	删除文件
show file [*filesystem*:[*//module/*]][*directory*/]*filename*	显示文件内容
gzip [*filesystem*:[*//module/*][*directory*/] | *directory*/]*filename*	压缩文件
gunzip [*filesystem*:[*//module/*][*directory*/] | *directory*/]*filename*.gz	解压缩文件
tar create {**bootflash:** | **volatile:**}*archive-filename* [**absolute**] [**bz2-compress**] [**gz-compress**] [**remove**] [**uncompressed**] [**verbose**] *filename-list*	创建归档文件并向其增加文件
tar append {**bootflash:** | **volatile:**}*archive-filename* [**absolute**] [**remove**] [**verbose**] *filename-list*	向现有归档文件增加文件
tar extract {**bootflash:** | **volatile:**}*archive-filename* [**keep-old**] [**screen**] [**to** {**bootflash:** | **volatile:**} [*/directory-name*]] [**verbose**]	解压现有归档文件

> **注：** 在故障排查过程中，如果需要采集数据，那么就可以使用 gzip 和 tar 选项，该选项可以将多个文件组合成一个归档文件并进行压缩，从而很方便地被导出到集中式服务器中进行分析。

1.2.8 闪存文件系统

闪存文件系统用于存储系统映像和用户生成的文件。如果要查看文件系统中的目录内容，则可以使用 **dir** [*directory* \| *filesystem*:[*//module/*] [*directory*]]命令。从例 1-9 可以看出，NX-OS 映像文件位于 *bootflash*:目录中。

例 1-9　dir bootflash:命令输出结果

```
NX-1# dir bootflash:
       4096    May 02 18:57:24 2017  .patch/
       7334    Jan 26 00:57:28 2017  LDP.txt
       1135    Mar 02 02:00:38 2016  MDS201309060745595990.lic
        580    Mar 02 02:00:12 2016  MDS201309060748159200.lic
        584    Mar 02 01:59:01 2016  MDS201309060749036210.lic
        552    Mar 02 01:56:02 2016  MDS201309071119059040.lic
       1558    Apr 21 05:21:39 2017  eigrp_route_clear.txt
       4096    Apr 29 09:37:44 2017  lost+found/
  425228450    Jun 30 01:27:40 2017  n7000-s2-dk9.6.2.12.bin
  580426199    Apr 06 20:08:12 2017  n7000-s2-dk9.7.3.1.D1.1.bin
   67492267    Dec 06 02:00:13 2016  n7000-s2-epld.6.2.14.img
   36633088    Jun 30 01:29:42 2017  n7000-s2-kickstart.6.2.12.bin
   36708352    May 24 01:43:48 2017  n7000-s2-kickstart.6.2.18.bin
   37997056    Apr 18 22:37:46 2017  n7000-s2-kickstart.7.2.2.D1.2.bin
   46800896    Apr 06 20:07:20 2017  n7000-s2-kickstart.7.3.1.D1.1.bin
       3028    Jun 13 00:06:22 2017  netflow_cap.pcap
          0    Apr 21 02:11:19 2017  script_out.log
         13    Apr 21 03:15:32 2017  script_output.txt
       4096    Apr 18 19:35:28 2016  scripts/
      17755    Mar 20 05:36:52 2016  startup-config-defaultconfig-DONOTDELETE
       4096    Nov 11 00:30:10 2016  vdc_2/
       4096    Apr 21 02:25:04 2017  vdc_3/
       4096    Dec 05 19:07:18 2016  vdc_4/
       4096    Apr 03 04:31:36 2016  vdc_5/
       4096    Apr 12 22:26:42 2013  vdc_6/
       4096    Apr 12 22:26:42 2013  vdc_7/
       4096    Apr 12 22:26:42 2013  vdc_8/
       4096    Apr 12 22:26:42 2013  vdc_9/
       4096    Apr 18 19:33:57 2016  virtual-instance/
       4096    Apr 18 20:36:58 2016  virtual-instance-stby-sync/
        664    Jun 30 02:17:45 2017  vlan.dat
      45137    Jun 30 01:33:45 2017  vtp_debug.log

Usage for bootflash://sup-local
 1370902528 bytes used
  445583360 bytes free
 1816485888 bytes total
```

该输出结果显示了当前主用管理引擎上的文件和子目录列表。对于拥有冗余管理引擎的平台来说，可以在目录路径上附加**//sup-standby**/来访问备用管理引擎目录，如例 1-10 所示。

例 1-10　列出备用管理引擎上的文件

```
NX-1# dir bootflash://sup-standby/
       4096    Jun 12 20:31:56 2017  .patch/
       1135    Mar 03 00:07:58 2016  MDS201309060745595990.lic
        580    Mar 03 00:08:09 2016  MDS201309060748159200.lic
        584    Mar 03 00:08:20 2016  MDS201309060749036210.lic
        552    Mar 03 00:08:32 2016  MDS201309071119059040.lic
       4096    May 24 01:27:09 2017  lost+found/
  580426199    Apr 14 20:53:14 2017  n7000-s2-dk9.7.3.1.D1.1.bin
  579340490    Jun 13 19:40:12 2017  n7000-s2-dk9.8.0.1.bin
   36633088    Jun 30 01:49:14 2017  n7000-s2-kickstart.6.2.12.bin
   36708352    May 24 01:17:23 2017  n7000-s2-kickstart.6.2.18.bin
   37997056    Apr 18 22:37:46 2017  n7000-s2-kickstart.7.2.2.D1.2.bin
```

```
    46800896        Apr 14 20:50:38 2017  n7000-s2-kickstart.7.3.1.D1.1.bin
        8556        May 05 10:57:35 2017  pim-2nd.pcap
           0        May 05 10:05:45 2017  pim-first
        3184        May 05 10:12:24 2017  pim-first.pcap
        4096        Apr 18 20:28:05 2016  scripts/
        4096        May 19 22:42:12 2017  vdc_2/
        4096        Jul 18 21:22:15 2016  vdc_3/
        4096        Jul 18 21:22:49 2016  vdc_4/
        4096        Mar 02 08:23:14 2016  vdc_5/
        4096        Nov 28 05:52:06 2014  vdc_6/
        4096        Nov 28 05:52:06 2014  vdc_7/
        4096        Nov 28 05:52:06 2014  vdc_8/
        4096        Nov 28 05:52:06 2014  vdc_9/
        4096        Apr 18 19:46:52 2016  virtual-instance/
        4096        Apr 18 20:40:29 2016  virtual-instance-stby-sync/
         664        Jun 30 02:17:45 2017  vlan.dat
       12888        Jun 12 20:34:40 2017  vtp_debug.log

Usage for bootflash://sup-standby
 1458462720 bytes used
  315793408 bytes free
 1774256128 bytes total
```

1.2.9 OBFL

OBFL（On-Board Failure Logging，板载故障日志）是 Nexus 平台提供的持久存储能力，用于存储线卡的本地操作信息。例 1-11 显示了 Nexus 7000 平台为 M2 I/O 模块默认启用的选项信息，OBFL 存储的持久性历史信息对于排查模块故障来说非常有用。

例 1-11 确认模块已启用 OBFL

```
NX-1# show logging onboard module 10 status
----------------------------
OBFL Status
----------------------------
    Switch OBFL Log:                                        Enabled

    Module: 10 OBFL Log:                                    Enabled
    counter-stats                                           Enabled
    cpu-hog                                                 Enabled
    credit-loss                                             Enabled
    environmental-history                                   Enabled
    error-stats                                             Enabled
    exception-log                                           Enabled
    interrupt-stats                                         Enabled
    mem-leak                                                Enabled
    miscellaneous-error                                     Enabled
    obfl-log (boot-uptime/device-version/obfl-history)      Enabled
    register-log                                            Enabled
    request-timeout                                         Enabled
    system-health                                           Enabled
    stack-trace                                             Enabled
```

注：虽然例 1-11 的输出结果来自分布式平台，但非分布式平台也有 OBFL 数据，具体启用选项取决于特定平台。可以使用带有各种子命令选项的配置命令 **hw-module logging onboard** 配置 OBFL 选项。请注意，通常没有任何理由禁用 OBFL。

1.2.10 Logflash

Logflash 是一种持久存储设施,用于存储系统日志、系统日志消息、调试输出以及核心文件。对于某些 Nexus 平台来说,Logflash 是一个外部紧凑型闪存或 USB,有可能并未安装或者在某些时候被拔除了,此时系统将会生成一条周期性消息,指示 Logflash 缺失,以提醒操作人员注意这种情况并加以纠正。建议正确安装 Logflash 并供系统使用,从而能够将操作数据存储在 Logflash 中。系统出现问题后,Logflash 的持久存储能力意味着可以利用这些数据进行故障分析。例 1-12 使用 **show system internal flash** 命令来验证是否已挂载 *logflash:* 以及当前可用空间。

例 1-12 验证 logflash:的状态及可用空间

```
NX-1# show system internal flash | egrep Filesystem|logflash
Filesystem       1K-blocks     Used Available Use% Mounted on
/dev/sda7         8256952   164288   7673236   3% /logflash
```

可以利用 **dir logflash:** 命令来检查 Logflash 目录的内容,如例 1-13 所示。

例 1-13 验证 logflash:目录的内容

```
NX-1# dir logflash:
      4096     Jun 05 17:43:10 2017  ISSU_debug_logs/
      4096     May 19 13:00:36 2017  controller/
      4096     Mar 30 14:03:38 2017  core/
      4096     Mar 30 14:03:38 2017  debug/
      4096     Jul 10 16:43:33 2017  debug_logs/
    413807     Mar 30 14:02:21 2017  dme.log.2017.03.30.21.02.21.tar.gz
    148751     Mar 31 12:21:01 2017  dme.log.2017.03.31.19.21.01.tar.gz
    144588     May 19 12:58:31 2017  dme.log.2017.05.19.19.58.31.tar.gz
      4096     Mar 30 14:03:38 2017  generic/
      4096     Mar 30 13:58:28 2017  log/
     16384     Mar 30 13:57:52 2017  lost+found/
      4096     Jun 13 21:29:33 2017  vdc_1/

Usage for logflash://sup-local
  597725184 bytes used
 7857393664 bytes free
 8455118848 bytes total
```

例 1-14 利用命令显示了 *logflash:* 中的文件内容。

例 1-14 显示 logflash:中的特定文件内容

```
NX-1# show file logflash://sup-local/log/messages
2017 Mar 30 20:58:30  %VDC_MGR-5-VDC_STATE_CHANGE:
vdc 1 state changed to create in progress
2017 Mar 30 20:58:30  %VDC_MGR-5-VDC_STATE_CHANGE:
vdc 1 state changed to create pending
2017 Mar 30 20:58:31  Mar 30 20:58:30 %KERN-3-SYSTEM_MSG: [ 2726.358042]
biosinfo checksum failed expected ff Got 8   - kernel
2017 Mar 30 20:58:31  Mar 30 20:58:30 %KERN-3-SYSTEM_MSG: [ 2726.358044]
read_from_biosinfo: No Valid biosinfo - kernel
2017 Mar 30 20:58:31  %VMAN-2-INSTALL_STATE: Installing virtual service
'guestshell+'
2017 Mar 30 20:58:33  netstack: Registration with cli server complete
2017 Mar 30 20:58:48  %USER-2-SYSTEM_MSG:
ssnmgr_app_init called on ssnmgr up - aclmgr
2017 Mar 30 20:58:54  %USER-0-SYSTEM_MSG: end of default policer - copp
2017 Mar 30 20:58:54  %COPP-2-COPP_NO_POLICY: Control-plane is unprotected.
```

```
2017 Mar 30 20:58:56  %CARDCLIENT-2-FPGA_BOOT_PRIMARY: IOFPGA booted from Primary
2017 Mar 30 20:58:56  %CARDCLIENT-2-FPGA_BOOT_PRIMARY: MIFPGA booted from Primary
```

1.3 理解 NX-OS 软件的发布与打包

新的 NX-OS 软件版本在发布时通常可以分为 3 种类型（主版本/次版本/维护版本）。
- 主版本会引入重要的新特性、功能和平台。
- 次版本会增强现有主版本的特性和功能。
- 维护版本则解决次版本中的某些产品缺陷。

根据 Nexus 平台和版本的不同，软件版本的命名约定也有所不同。对于早期的 NX-OS 版本来说，每个平台都建立在自己的 NX-OS 操作系统代码库之上。目前大多数平台使用的是 NX-OS 公共基础操作系统，在此基础上根据需要修改或扩充公共基础代码，以满足特定平台的特性需求或硬件支持。这种方式的优点是，修复与平台无关的基础代码中的软件缺陷之后，可以合并回公共基础代码当中，所有平台都能从这些缺陷修复中获益。

图 1-5 解释了 NX-OS 软件命名约定方式，以 Nexus 7000 平台的 NX-OS 6.2 版本映像名为例说明了主/次/维护版本号。

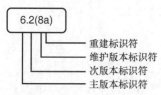

图 1-5 NX-OS 软件命名约定

图 1-6 解释了如何以 Nexus 7000 平台的通用平台无关基本代码和平台相关版本信息来理解 NX-OS 的软件命名约定。

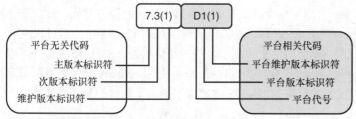

图 1-6 带有平台标识符的 NX-OS 软件命名约定

目前的 NX-OS 平台标识符信息如下。
- I：Nexus 9000 和 3000 通用代码库。
- D：Nexus 7000/7700。
- N：Nexus 5000/6000。
- U：Nexus 3000（用于公共代码库之前的版本）。
- A：Nexus 3548。

注： 目前 Nexus 3000 和 Nexus 9000 系列平台共享同一个与平台相关的通用软件库，映像名称以 *nxos* 开头，如 *nxos.7.0.3.i6.1.bin*。

除主/次/维护版本之外，还可以将版本分为长期（long-lived）版本和短期（short-lived）版本，用于描述版本的开发生命周期。对于绝大多数部署场景来说，在支持所需的功能特性和硬件的情况下，通常建议使用长期版本而非短期版本。长期版本通常经历了更多的维护和重新编译，每次新的维护和重新编译都会对软件进行加固并修复相应的软件缺陷。短期版本通常是引入新的硬件或软件功能特性，并根据需要接受维护版本。与长期版本相比，短期版本的支持时间通常较短，维护重建也较少。只要长期版本支持了所需的功能特性和硬件，那么就建议从短期版本迁移到长期版本。

思科官网提供的 NX-OS 操作系统，由 kickstart（快速启动）映像和系统映像组成，不过，Nexus 9000 和 Nexus 3000 是例外，它们使用的是单一系统映像文件。对于使用 kickstart 映像的平台来说，首先由 BIOS 引导 kickstart 映像，然后再由 kickstart 映像加载系统映像。对于 Nexus 3000 和 Nexus 9000 平台来说，则是直接启动单个二进制文件。

某些平台还可以升级 EPLD（Erasable Programmable Logic Device，可擦除编程逻辑器件）映像。EPLD 映像与 NX-OS 操作系统分开打包，EPLD 映像可以升级 I/O 模块或线卡硬件组件上的固件，在无须更换硬件的情况下提供新的硬件功能或解决已知缺陷。

> 注：并非所有的 NX-OS 系统升级都需要升级 EPLD。思科官网提供了每种 Nexus 平台的 NX-OS 软件和 EPLD 映像安装示例，有关详细信息请参阅软件升级和安装指南。

1.3.1 软件维护升级

NX-OS 的新增功能是将特定的错误修正作为补丁应用到当前安装的软件当中。这个概念对于有思科 IOS-XR 平台操作经验的用户来说应该非常熟悉，NX-OS 也使用了相同的术语。

SMU（Software Maintenance Upgrade，软件维护升级）功能允许网络运营商修正 Nexus 交换机的特定错误，而不需要重新加载系统或进行 ISSU（In-Service Software Upgrade，在线软件升级）。对于关键性的网络环境来说，在进行软件升级之前必须对系统架构及所要配置的功能特性进行全面的合格性测试。

以前修正软件错误的时候，必须要对新的 NX-OS 维护版本进行合格性测试，然后再发布到网络中。很明显，这样做不但会增加在 NX-OS 维护版本中提供错误修正的等待时间，而且还会增加网络打补丁以最终消除软件错误之前所必须等待的合格性测试时间。这种延迟可以利用 SMU 的概念进行解决，因为此时仅将 SMU 的变更应用于合格的基础映像。SMU 安装过程可以尽可能地利用进程重启或 ISSU 来最小化安装过程对网络的影响。此后，Nexus 交换机就可以运行已应用的 SMU，直到 Cisco 官网发布的合格的 NX-OS 维护版本提供了相应的错误修正为止。

> 注：SMU 仅对专门创建的映像有效。如果 NX-OS 软件升级到另一个版本，那么就会停用该 SMU。在执行升级操作之前，必须确保新软件版本已经修复了所有适用的软件缺陷。

SMU 文件被打包为二进制文件和 README.txt（详细说明了该 SMU 解决的所有错误）。SMU 文件的命名方式为 *platform-package_type.release_version.bug_id.file_type*，如 *n7700-s2¬dk9.7.3.1.d1.1.csvc44582.bin*。SMU 的安装步骤如下。

- 第 1 步：将打包文件复制到本地存储设备或文件服务器中。
- 第 2 步：使用 **install add** 命令在设备中添加一个或多个软件包。
- 第 3 步：使用 **install activate** 命令激活设备上的软件包。
- 第 4 步：使用 **install commit** 命令提交当前的软件包集。不过，在重新加载或 ISSU SMU 的情况下，需要在重新加载或 ISSU 之后提交软件包。
- 第 5 步：（可选）根据需要停用并删除软件包。

> **注：** 准备安装 SMU 之前，请查阅思科官网上有关该平台的详细案例信息。

1.3.2 许可

NX-OS 要求操作人员必须在启用特定功能特性之前获取并安装适当的许可文件。Nexus 平台通常支持基本功能集，无额外的许可要求，基本功能集通常包括大多数二层功能和部分三层路由功能。如果要启用某种高级功能特性，如 MPLS、OTV、FabricPath、FCoE、高级路由或 VxLAN，那么可能需要安装特定许可（取决于具体平台）。除功能特性许可之外，有些 Nexus 平台还提供了其他许可，以提供额外的硬件功能。如 Nexus 7000 系列的 SCALEABLE_SERVICES_PKT 允许支持 XL 能力的 I/O 模块运行在 XL 模式下，而且还可以使用更大的表空间。另一个案例就是，某些 Nexus 3000 平台可以使用端口升级许可。

NX-OS 操作系统内置许可的检查机制，由功能特性管理器执行，如果没有安装相应的许可，那么就会禁用该服务。如果无法配置某些功能特性，那么最可能的原因就是缺少相应的许可。思科允许在没有许可的情况下进行功能特性测试（在全局配置模式下配置 license grace-period 命令），这样就可以在未安装许可的情况下最长运行 120 天。不过，该功能并未涵盖所有平台的所有功能特性许可。需要注意的是，Nexus 9000 和 Nexus 3000 不支持 license grace-period。

可以从思科官网下载许可文件。如果获取了许可文件，那么就可以通过 show license host-id 命令查找相应的序列号，然后再使用软件许可声明中的 PAK（Product Authorization Key，产品授权密钥）检索许可文件并复制到交换机中。安装许可是一项非中断式任务，可以通过 install license 命令安装。支持 VDC（Virtual Device Context，虚拟设备上下文）的平台将许可安装在默认 VDC 上并加以管理，这样许可就能够同时应用于机箱上的所有 VDC。可以通过命令 show license 验证许可的安装情况。

1.4 NX-OS 高可用性基础设施

本章前面介绍的系统管理器、MTS 和 PSS 基础设施组件为 NX-OS 提供了高可用性基础设施的核心能力，这种高可用性基础设施使 NX-OS 能够从大多数故障场景（如管理引擎切换或进程重启）中实现无缝恢复。

NX-OS 可以在对数据平面转发流量影响最小化的情况下，重启服务并恢复正常操作。该进程重启事件可以是状态化事件，也可以是无状态事件，可以由用户手工发起，也可以由系统管理器在发现进程故障后自动发起。

如果是无状态重启，那么与故障进程相关联的所有运行时数据结构都将丢失，系统管理器会快速生成一个新进程来替换故障进程。状态化重启意味着可以利用一部分运行时数据来恢复进程，并在前一个进程出现故障或重启之后的中断位置无缝地重新提供正确功能。状态化重启是可以实现的，因为服务可以在正常运行时更新 PSS 中的服务状态，在故障发生后从 PSS 恢复重要的运行时数据结构。重启后的服务将提取进程队列中的持久性 MTS 消息，以实现无缝恢复。由于能够重新处理 MTS 队列中的持久性消息，因而服务重启操作对于与故障进程进行通信的其他服务来说完全透明。

NX-OS 为各种进程都提供了基础设施，允许进程选择所要实现的恢复机制类型。但是在某些情况下，状态化恢复并没有什么意义，因为这些进程的恢复机制内置在协议的较高层级。例如，路由协议进程（如 OSPF 或 BGP）拥有协议级的平滑重启或不间断转发机制。对于这些协议来说，将路由更新存储到 PSS 基础设施中并没有任何实际意义，因为协议可以恢复这些更新信息。

> 注：可以通过命令 **show system reset-reason** 查看重置原因，通过 **show processes log pid** 和 **show cores** 命令查看进程崩溃或重启的详细信息。

1.4.1 管理引擎冗余

拥有冗余管理引擎模块的 Nexus 平台以主用/备用（Active/Standby）冗余模式运行，意味着每次只有一块管理引擎处于活动状态，备用管理引擎处于就绪状态，等到主用管理引擎出现致命故障后就能接管主用角色。主用/备用冗余管理引擎为设备提供了完全冗余的控制平面，能够实现 SSO（Stateful Switchover，状态化切换）和 ISSU（In-Service Software Upgrade，在线软件升级）。可以在 **show module** 和 **show system redundancy status** 命令的输出结果中查看当前冗余状态以及处于主用状态的管理引擎，如例 1-15 所示。

例 1-15 确定当前管理引擎的冗余状态

```
NX-1# show system redundancy status
Redundancy mode
---------------
       administrative:   HA
         operational:   HA
This supervisor (sup-1)
-----------------------
   Redundancy state:    Active
   Supervisor state:    Active
     Internal state:    Active with HA standby

Other supervisor (sup-2)
-----------------------
   Redundancy state:    Standby
   Supervisor state:    HA standby
     Internal state:    HA standby
NX-1# show module
Mod   Ports   Module-Type                        Model              Status
---   -----   -------------------------------    ----------------   ----------
3     32      10 Gbps Ethernet Module            N7K-M132XP-12      ok
5     0       Supervisor Module-2                N7K-SUP2E          active *
6     0       Supervisor Module-2                N7K-SUP2E          ha-standby
8     48      1000 Mbps Optical Ethernet Module  N7K-M148GS-11      ok
9     48      1/10 Gbps Ethernet Module          N7K-F248XP-25E     ok
10    24      10 Gbps Ethernet Module            N7K-M224XP-23L     ok
```

冗余管理引擎启动后，将发生如下事件。

1. 管理引擎的主用/备用选举过程已经完成。
2. 备用管理引擎上的系统管理器进程将自己宣告给主用管理引擎上的系统管理器进程。
3. 备用管理引擎的系统管理器从主用管理引擎同步启动配置（Startup Configuration），并启动备用管理引擎上的所有服务以镜像主用管理引擎。
4. 备用管理引擎上的服务与主用管理引擎上的服务状态快照进行同步。
5. 将主用管理引擎上的服务的 MTS 消息复制到备用管理引擎。
6. 此时备用管理引擎上的服务状态已经与主用管理引擎实现同步。
7. 此时已将进程事件复制到备用管理引擎，因而两块管理引擎上的服务在正常运行情况下保持同步（基于事件的同步）。

如果管理引擎出现切换，那么系统管理器将通知备用管理引擎上的服务恢复状态并准备接管主用角色。由于正常运行期间，MTS 已经将进程事件同步给备用管理引擎，因而可以实现快速恢复。切换操作完成后，将重启此前处于主用状态的管理引擎，并执行正常的启动诊断测试。如果诊断测试通过且启动成功，那么将按照前述步骤与当前主用管理引擎进行同步。图 1-7 给出了管理引擎冗余模型中的各种 NX-OS 服务之间的关系。

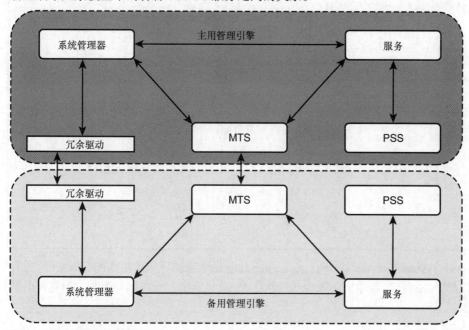

图 1-7　管理引擎冗余模型

在极少数情况下，备用管理引擎可能无法达到 HA 备用状态。一个可能的原因就是备用管理引擎上的服务无法与主用管理引擎实现状态同步。如果要检查该问题，需要验证主用和备用管理引擎的 sysmgr 状态，以确认哪些服务无法实现状态同步。如果配置了多个 VDC，那么就需要对每个 VDC 都执行该验证操作。如果要验证管理引擎的同步状态，那么就可以使用 **show system internal sysmgr state** 命令，如例 1-16 所示。

例 1-16　确认冗余和同步状态

```
NX-1# show system internal sysmgr state
The master System Manager has PID 4862 and UUID 0x1.
Last time System Manager was gracefully shutdown.
The state is SRV_STATE_MASTER_ACTIVE_HOTSTDBY entered at time Fri Jun 30
01:48:40 2017.
The '-b' option (disable heartbeat) is currently disabled.
The '-n' (don't use rlimit) option is currently disabled.
Hap-reset is currently enabled.
Watchdog checking is currently enabled.
Watchdog kgdb setting is currently disabled.
        Debugging info:
The trace mask is 0x00000000, the syslog priority enabled is 3.
The '-d' option is currently disabled.
The statistics generation is currently enabled.

        HA info:

slotid = 5     supid = 0
cardstate = SYSMGR_CARDSTATE_ACTIVE .
```

```
cardstate = SYSMGR_CARDSTATE_ACTIVE (hot switchover is configured enabled).
Configured to use the real platform manager.
Configured to use the real redundancy driver.
Redundancy register: this_sup = RDN_ST_AC, other_sup = RDN_ST_SB.
EOBC device name: veobc.
Remote addresses:   MTS - 0x00000601/3        IP - 127.1.1.6
MSYNC done.
Remote MSYNC not done.
Module online notification received.
Local super-state is: SYSMGR_SUPERSTATE_STABLE
Standby super-state is: SYSMGR_SUPERSTATE_STABLE
Swover Reason : SYSMGR_UNKNOWN_SWOVER
Total number of Switchovers: 0
Swover threshold settings: 5 switchovers within 4800 seconds
Switchovers within threshold interval: 0
Last switchover time: 0 seconds after system start time
Cumulative time between last 0 switchovers: 0
Start done received for 1 plugins, Total number of plugins = 1
        Statistics:
Message count:           0
Total latency:           0              Max latency:            0
Total exec:              0              Max exec:               0
```

可以利用 **show system internal sysmgr gsync-pending** 命令来验证同步操作的完成情况，所有未完成同步操作的服务都将列在输出结果中。从例 1-17 可以看出，所有服务都与主用管理引擎实现了状态同步。

例 1-17 验证是否有服务等待同步

```
NX-1# show system internal sysmgr gsync-pending
Gsync is not pending for any service
```

sysmgr 的输出结果表明两块管理引擎的状态都已经稳定，无任何问题。如果存在问题，那么管理引擎的状态（superstate）将显示为 *unstable*（不稳定）。只要连接到备用管理引擎模块上，即可验证备用管理引擎的状态，如例 1-18 所示。

例 1-18 验证备用管理引擎的 sysmgr 状态

```
NX-1# attach module 6
Attaching to module 6 ...
<output removed for brevity>
NX-1(standby)# show system internal sysmgr state
The master System Manager has PID 4708 and UUID 0x1.
Last time System Manager was gracefully shutdown.
The state is SRV_STATE_MASTER_HOTSTDBY entered at time Fri Jun 30 01:49:50 2017.
The '-b' option (disable heartbeat) is currently disabled.
The '-n' (don't use rlimit) option is currently disabled.
Hap-reset is currently enabled.
Watchdog checking is currently enabled.
Watchdog kgdb setting is currently disabled.


        Debugging info:

The trace mask is 0x00000000, the syslog priority enabled is 3.
The '-d' option is currently disabled.
The statistics generation is currently enabled.
```

```
        HA info:

slotid = 6      supid = 0
cardstate = SYSMGR_CARDSTATE_STANDBY .
cardstate = SYSMGR_CARDSTATE_STANDBY (hot switchover is configured enabled).
Configured to use the real platform manager.
Configured to use the real redundancy driver.
Redundancy register: this_sup = RDN_ST_SB, other_sup = RDN_ST_AC.
EOBC device name: veobc.
Remote addresses:   MTS - 0x00000501/3        IP - 127.1.1.5
MSYNC done.
Remote MSYNC done.
Module online notification received.
Local super-state is: SYSMGR_SUPERSTATE_STABLE
Standby super-state is: SYSMGR_SUPERSTATE_STABLE
Swover Reason : SYSMGR_UNKNOWN_SWOVER
Total number of Switchovers: 0
Swover threshold settings: 5 switchovers within 4800 seconds
Switchovers within threshold interval: 0
Last switchover time: 0 seconds after system start time
Cumulative time between last 0 switchovers: 0
Start done received for 1 plugins, Total number of plugins = 1

        Statistics:

Message count:              0
Total latency:              0            Max latency:              0
Total exec:                 0            Max exec:                 0
```

可以看出该管理引擎的状态稳定，冗余寄存器表明该管理引擎的冗余状态是备用状态（RDN_ST_SB）。从例1-19可以看出，备用管理引擎没有待同步的服务。

例1-19 验证没有服务等待同步

```
NX-1(standby)# show system internal sysmgr gsync-pending
Gsync is not pending for any service
```

如果输出结果显示存在待同步的服务，那么接下来就需要验证该特定服务的MTS队列。验证特定服务的MTS队列的方式已在本章前面做了介绍，第3章还会做进一步说明。如果MTS队列有待同步服务的消息，那么接下来就要分析这些消息未处理的原因。一般来说，网络或设备不稳定可能会造成服务的频繁MTS更新，这是妨碍同步完成的主要原因。

1.4.2 ISSU

NX-OS将ISSU作为一种高可用功能特性。ISSU利用冗余管理引擎的NX-OS SSO功能，可以在不影响数据流量的情况下实现系统软件的更新操作。在ISSU期间，机箱的所有组件都能进行升级。

可以利用**install all**命令启动ISSU，该命令执行以下步骤来升级系统。

- 第1步：确定升级操作是否是中断性操作，并询问是否继续。
- 第2步：确保备用管理引擎的bootflash拥有足够的可用空间。
- 第3步：将kickstart映像和系统映像复制到备用管理引擎模块。
- 第4步：设置KICKSTART和SYSTEM引导变量。
- 第5步：用新的思科NX-OS软件重新加载备用管理引擎模块。

- **第 6 步**：使用新的思科 NX-OS 软件重新加载主用管理引擎模块，从而切换到新升级的备用管理引擎模块。
- **第 7 步**：升级线卡。
- **第 8 步**：升级两个管理引擎上的 CMP（Connectivity Management Processor，连接管理处理器）。请注意，Nexus 7000 仅升级 Sup1。

没有冗余管理引擎的平台（如 Nexus 5000 系列）采用其他方法来实现 ISSU。由于数据平面在持续转发数据包时，控制平面处于非活动状态，因而可以重启管理引擎 CPU，而不会导致流量中断并加载新的 NX-OS 软件版本。在新软件版本上启动 CPU 之后，控制平面将恢复到先前的配置和运行时状态。此后，交换机将控制平面状态同步到数据平面。

Nexus 9000 和 Nexus 3000 平台在 Release 7.0(3)I5(1) 引入了增强型 ISSU 功能。NX-OS 软件通常直接运行在硬件上，不过，有了增强型 ISSU 功能之后，NX-OS 软件就可以分别运行在管理引擎和线卡的 LXC（Linux Container，Linux 容器）中。增强型 ISSU 在操作期间，会创建第三个容器以充当备用管理引擎，这样就可以在不中断数据流量的情况下升级主用管理引擎和线卡。可以利用命令 **boot mode lxc** 在支持该功能特性的平台上启用该功能特性。

> **注：** ISSU 在某些平台上有一定的限制，而且某些 NX-OS 版本可能不支持 ISSU。使用 ISSU 升级之前，请参考思科官网上的文档以确保支持 ISSU。

1.5 NX-OS 虚拟化功能

作为现代数据中心级操作系统，NX-OS 和 Nexus 交换机平台必须支持硬件和软件资源的虚拟化，以满足当前网络架构的普遍需求。本节将简要介绍这些功能，后续还会在其他章节进行详细讨论。

1.5.1 VDC

Nexus 7000 平台允许运营商将物理交换机划分为多个虚拟交换机，称为 VDC（Virtual Device Context，虚拟设备上下文）。VDC 是一种非常重要的虚拟化功能特性，可以将物理交换机划分为多个逻辑交换机，而且每个逻辑交换机在拓扑结构中的作用各不相同。

常见的 VDC 用例就是 OTV 或 LISP，此时需要为叠加封装协议配置专用 VDC，由另一个 VDC 充当执行传统二层和三层功能的分发层交换机。另一种常见的 VDC 应用就是配置生产性 VDC 和测试/开发 VDC，从而在单个机箱中分离这些不同的应用环境。合理规划并创建了 VDC 之后，操作人员就可以将端口分配给每个 VDC，然后互连这些端口，从而能够在 VDC 之间交换控制平面协议和数据平面流量。

从本质上来说，VDC 架构意味着某些资源对交换机来说是全局性的，其他资源则是在 VDC 之间共享或专用于特定 VDC。例如，VDC-1 中的 OSPF 进程与 VDC-2 中的 OSPF 进程虽然共享交换机的公共 CPU 资源，但却相互独立。管理引擎上的管理以太网由所有 VDC 共享，I/O 模块上的端口专用于特定 VDC，而 NX-OS 内核则属于交换机全局资源。

虽然 VDC 之间的逻辑分离一直延伸到协议栈，但交换机上的所有 VDC 共享相同的内核资源和基础设施。在设计上，系统基础设施作为共享资源，可以实现公平的资源分配，从内核到每个 VDC 协议栈的控制平面队列也同样能够做到公平分配。其他资源则专用于特定 VDC，如 VLAN 和路由表空间。图 1-8 给出了 Nexus 7000 系列 VDC 架构的直观示意图。

有了适当的许可之后，Supervisor 1 和 Supervisor 2 可以支持 4 个 VDC 和 1 个管理 VDC（admin

VDC），Supervisor 2E 支持 8 个 VDC 和 1 个管理 VDC。管理 VDC 不处理任何数据平面流量，仅提供交换机管理功能。需要注意的是，在 VDC 环境下执行操作或故障排查任务时，某些任务只能从默认 VDC 发起。

1. ISSU/ISSD（In-Service Software Upgrade/Downgrade，在线软件升级/降级）。
2. EPLD（Erasable Programmable Logic Device，可擦除可编程逻辑器件）升级。
3. CoPP（Control-Plane Policing，控制平面策略）配置。
4. 许可操作。
5. VDC 配置，包括创建、挂起、删除和资源分配。
6. 系统范围内的 QoS 策略和端口通道负载均衡配置。
7. GOLD（Generic Online Diagnostic，通用在线诊断）配置。
8. Ethanalyzer 抓包操作。

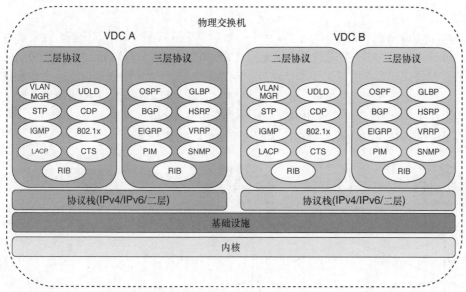

图 1-8　Nexus 7000 VDC 架构

虽然 VDC 支持很多额外的功能特性，但是也存在一些限制。例如，所有 VDC 都必须运行在相同的 NX-OS 版本和内核上。此外还存在其他一些限制，如哪些 I/O 模块能够位于同一个 VDC 中，哪些线卡的端口可以分配给 VDC，这些都与 I/O 模块的硬件 ASCI（Application-Specific Integrated Circuit，专用集成电路）架构和转发引擎相关。创建 VDC 之前，需要仔细查阅交换机安装的管理引擎和 I/O 模块的技术文档，确保在设计和规划阶段就处理好相关限制因素。

注：截至本书写作之时，仅 Nexus 7000 系列交换机支持多 VDC。

1.5.2　VRF

事实证明，VRF（Virtual Routing and Forwarding，虚拟路由转发）实例对于网络资源的逻辑分离来说非常有用。VRF 的设计目的就是允许多个控制平面和数据平面转发表实例同时运行在单台设备上，同时还能保持逻辑分离。

VRF-lite 可以在无 MPLS（Multiprotocol Label Switching，多协议标签交换）传输网络的情况下，在单台设备上定义多个逻辑分离的路由和转发表。MPLS VPN 在 PE（Provider Edge，提供商边缘）节点上使用 VRF，从逻辑上将 MPLS 传输网络中的 PE 之间的多个路由和转发表进行分离。

NX-OS 支持 VRF-lite 和 MPLS VPN，实现路由表和转发状态的虚拟化和逻辑分离；支持在 VRF 上下文之间导入和导出路由，同时还可以从全局路由表导入和导出到 VRF 表。除用户定义的 VRF 之外，NX-OS 默认将交换机的管理以太网接口放到自己的管理 VRF 中，从而实现了数据平面与管理平面服务的完美分离。

在虚拟化层次架构中，VDC 在本地拥有 VRF，而且同一台物理交换机可以存在多个 VDC。如果不同 VDC 中配置的 VRF 之间需要进行通信，那么就需要利用控制平面路由协议，在不同 VDC 中的 VRF 之间交换信息。这一点与使用默认 VRF 在 VDC 之间进行路由的操作方式相同，利用控制平面协议在 VRF 之间交换路由信息，如果需要通信的 VRF 位于同一个 VDC 中，那么就可以利用路由泄露机制在这些 VRF 之间交换路由信息。

注：对 MPLS VPN 功能特性的支持取决于具体平台的能力以及安装的功能特性许可。

1.5.3 vPC

vPC（virtual Port Channel，虚拟端口通道）允许一对对等交换机连接第三方设备，并显示为单台交换机。对第三方设备的唯一要求就是必须支持 IEEE 802.3ad 端口通道功能。由于使用 vPC 交换机对不需要任何特殊配置，因而该技术已成为业界解决接入层 STP 阻塞端口的一个非常有吸引力的选择。将两台交换机配置为 vPC 对之后，其中一台交换机将被选为主用交换机（优先级较低的交换机胜出），主用角色将在 STP 及某些故障场景中发挥作用。

图 1-9 给出了一对 vPC 交换机通过 vPC 连接两台交换机的配置示例。

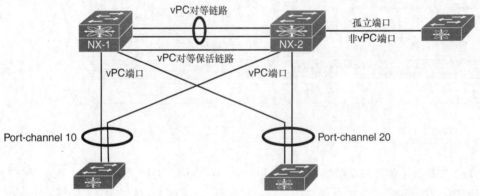

图 1-9 vPC 体系架构

从图 1-9 可以看出，vPC 对使用 vPC Port-channel 10 和 vPC Port-channel 20 与两台接入交换机相连。NX-2 右侧的第三台交换机未采用 vPC 模式进行连接，vPC 术语将该非 vPC 接口称为孤立端口（Orphan Port）。图 1-9 中的 vPC 术语如下。

- **vPC 对等链路（vPC Peer-link）**：负责承载用户定义的 VLAN，并在 vPC 对等交换机之间转发 BPDU、HSRP Hello 消息以及 CFS（Cisco Fabric Service，思科交换矩阵服务）协议包。为了实现冗余性，该链路应该是不同模块上的成员链路的端口通道。
- **vPC 对等保活链路（vPC Peer-Keepalive Link）**：该链路与对等链路应该是不同的路径，不要求是点对点链路，可以穿越路由基础设施。对等保活链路用于确保 vPC 对等交换机的活跃性。
- **孤立端口**：连接 vPC 拓扑结构设备的非 vPC 端口。
- **vPC 端口（vPC 成员端口）**：分配给 vPC 端口通道组（Port-channel group）的端口，vPC 端口在 vPC 对等体之间进行拆分。

由于所有链路均处于 Up 状态且以 vPC 方式进行转发，因此决定使用哪个 vPC 成员链路接口转发数据包是由发送数据包的设备的端口通道负载均衡散列算法决定的。发送端交换机查看流量流的数据帧源地址和目的地址，并将这些信息提供给算法，由算法执行散列函数并返回选定的端口通道成员端口，数据帧将通过该端口进行转发。因此，端口通道的所有成员链路接口都可以共享流量负载。

以前，路由协议邻接关系在 vPC 拓扑结构中存在一定的局限性，不过最新的 NX-OS 版本对此进行了修正，允许动态单播路由协议运行在 vPC 之上，不再受此前的限制影响。在具体配置之前，请检查交换机平台的 vPC 配置指南，以确保平台支持该特性。

由于每个 vPC 对等体都可以独立做出帧转发决定，因而需要对等体之间实现状态同步。CFS 是实现二层（L2）状态同步的协议，运行在对等链路之上且自动启用。CFS 的作用是确保 vPC 对等体之间的 vPC 成员端口的兼容性，同时还可以在 vPC 对等体之间同步 MAC 地址表和 IGMP 监听状态，从而确保两个 vPC 对等体拥有完全相同的表项。需要注意的是，每个 vPC 对等体上的三层（L3）转发表和协议状态都是独立的。

注：这里仅将 vPC 作为虚拟化概念进行简单介绍，后续还将做进一步讨论。

1.6 管理和操作功能

接下来将介绍 Nexus 交换平台的操作、可维护性和可用性功能，这些功能旨在改善用户体验并简化 Nexus 交换机的配置、故障排查及运行维护等任务。

1.6.1 NX-OS 高级 CLI

NX-OS CLI 的界面在设计上最初与 IOS 相似（两种 CLI 接口在高层视角上非常相似）。例如，IOS 和 NX-OS 都拥有执行模式、不同的用户权限、运行配置和启动配置等概念。在很多情况下，NX-OS 和 IOS 的 **show** 命令和调试操作都完全相同。不过，NX-OS 提供了额外的命令行增强功能和实用程序，使 NX-OS 的操作和故障排查更加容易，也更加高效。

在故障调查分析阶段，通常需要收集大量数据进行离线分析。这就意味着用户需要执行各种命令、抓取输出结果，然后再将输出结果从设备传输到其他位置以供查看。NX-OS 提供了操作符>和>>（见例 1-20），可以将输出结果分别重定向到新文件或者追加到现有文件中，这对于收集冗长的 **show tech-support** 文件来说尤为有用。

例 1-20　show 命令输出结果重定向

```
NX-1# show tech netstack > bootflash:tech_netstack.txt
NX-1# dir bootflash: | inc netstack
   23144888    Jul 13 16:09:09 2017  tech_netstack.txt
```

如果使用了解析工具（"| **count**" 和 "| **wc**"），那么还可以统计正在执行的 **show** 命令中的行数或单词数。例 1-21 给出了这类工具的应用示例，该场景可以通过简单计数来验证当前状态，如对比配置变更前后的 **show ip ospf neighbor** 命令的输出行数。

例 1-21　使用 count 或 wc 工具

```
NX-1# show ip ospf neighbor | count
6
```

很多故障排查场景需要多次提取命令输出结果并进行对比，以确定自上次执行以来发生变化的内容或增加的计数器。例 1-22 给出了 **diff** 工具的应用示例。需要注意的是，千万不要在大量输出结果中（如 **show tech**）使用该工具，因为这样做会消耗大量系统资源以保留输出结果进行对比。如果要对比大量输出结果，最好先将这些输出结果离线传输出去。

例 1-22　使用 diff 工具

```
NX-1# show int e3/19 | diff
22,24c22,24
<     30 seconds input rate 360 bits/sec, 0 packets/sec
<     30 seconds output rate 112 bits/sec, 0 packets/sec
<     input rate 360 bps, 0 pps; output rate 112 bps, 0 pps
---
>     30 seconds input rate 336 bits/sec, 0 packets/sec
>     30 seconds output rate 72 bits/sec, 0 packets/sec
>     input rate 336 bps, 0 pps; output rate 72 bps, 0 pps
34,35c34,35
<     468 unicast packets  183033 multicast packets 1 broadcast packets
<     183506 input packets  18852532 bytes
---
>     468 unicast packets  183034 multicast packets 1 broadcast packets
>     183507 input packets  18852610 bytes
```

有时仅希望显示输出结果的最后几行，而不是成页地输出结果或通过 include/exclude 工具过滤输出结果。**last** *count* 工具可以仅显示最后的少量信息，通常用于解析记账日志、系统日志缓存或事件历史记录日志。例 1-23 仅打印了日志缓存中的最后一行。

例 1-23　使用 last 工具

```
NX-1# show logging logfile | last 1
2017 Jul 13 21:57:15 F340-35-02-N7K-7009-A %VSHD-5-VSHD_SYSLOG_CONFIG_I:
Configured from vty by admin on console0
```

另一个非常有用的功能特性就是 "," 工具，该工具可以执行具有多个参数的命令。例 1-24 给出了同时检查两个不同端口上的接口速率的配置示例（配合使用 **egrep** 工具）。

例 1-24　执行具有多个参数的命令

```
NX-1# show interface e3/18 , ethernet 3/19 | egrep "30 seconds"
  Load-Interval #1: 30 seconds
    30 seconds input rate 64680 bits/sec, 11 packets/sec
    30 seconds output rate 67304 bits/sec, 11 packets/sec
  Load-Interval #1: 30 seconds
    30 seconds input rate 66288 bits/sec, 11 packets/sec
    30 seconds output rate 64931 bits/sec, 11 packets/sec
```

egrep 和 **grep** 工具对于解决输出结果的混乱现象来说非常有用，它们可以仅显示感兴趣的字符模式。例 1-25 给出了 **egrep** 的常见用法，即查看事件历史记录日志并查找特定事件。该例中的 **egrep** 用于查看 OSPF 每次运行其 SPF（Shortest Path First，最短路径优先）算法的情况，选项 **prev 1** 的作用是打印模式匹配的前一行（表示运行全部或部分 SPF），选项 **next** 的作用是获取 **egrep** 模式匹配之后的所有输出行。

例 1-25　使用 egrep 解析事件历史记录日志

```
NX-1# show ip ospf event-history spf | egrep prev 1 STARTED
2017 Jul  9 23:46:00.646652 ospf 12 [16161]: : Examining summaries
2017 Jul  9 23:46:00.646644 ospf 12 [16161]: : SPF run 12 STARTED with flags
0x4, vpn superbackbone changed flag is FALSE
--
2017 Jul  9 23:44:05.194089 ospf 12 [16161]: : This is a full SPF
2017 Jul  9 23:44:05.194087 ospf 12 [16161]: : SPF run 11 STARTED with flags
0x1, vpn superbackbone changed flag is FALSE
--
```

```
2017 Jul  9 23:44:00.094088 ospf 12 [16161]: : This is a full SPF
2017 Jul  9 23:44:00.094085 ospf 12 [16161]: : SPF run 10 STARTED with flags
0x1, vpn superbackbone changed flag is FALSE
--
2017 Jul  9 23:43:56.074094 ospf 12 [16161]: : This is a full SPF
2017 Jul  9 23:43:56.074091 ospf 12 [16161]: : SPF run 9 STARTED with flags
0x1, vpn superbackbone changed flag is FALSE
--
```

　　egrep 还有一些值得关注的有用选项，包括 **count**（返回匹配次数的计数值）、**invert-match**（仅打印与模式不匹配的行）和 **line-number**（为每一行附加匹配的行号）。

　　表 1-2 列出了 NX-OS 提供的一些常用命令工具。

表 1-2　　　　　　　　　　　　　　常用命令工具

工具	作用
no-more	达到终端长度时，无须输入回车键或空格键即可对命令的输出结果进行分页
json	将命令输出结果打印为 JSON 格式。由于此时的输出结果是结构化数据，因而收集数据后如果需要供脚本或软件应用程序使用，那么该工具将非常有用
xml	以 XML 格式打印输出结果
email	通过电子邮件发送命令的输出结果
include	仅打印与模式相匹配的行
exclude	仅打印与模式不匹配的行
section	仅打印输出结果中包含模式的部分

　　例 1-26 给出了 **show cli list** [*string*]输出结果示例，该命令返回了与给定字符串输入相匹配的所有 CLI 命令。该命令可以节省使用 "**?**" 来确定存在哪些命令的时间。

例 1-26　使用 show cli list 命令

```
NX-1# show cli list ospf
<! output omitted for brevity>
MODE exec
show logging level ospf
show tech-support ospf brief
show tech-support ospf
show ip ospf <str> vrf <str>
show ip ospf <str> vrf <str>
show ip ospf <str> vrf all
show ip ospf <str>
show ip ospf <str> neighbors <if> <ip>
show ip ospf <str> neighbors <if> <str>
show ip ospf <str> neighbors <if>
show ip ospf <str> neighbors <ip> vrf <str>
show ip ospf <str> neighbors <ip> vrf <str>
show ip ospf <str> neighbors <ip> vrf all
```

　　作为前一个示例的补充，可以利用 **show cli syntax** [*string*]命令显示已识别 CLI 命令的语法信息，如例 1-27 所示。

例 1-27　使用 show cli syntax 命令

```
NX-1# show cli syntax ospf
<! output omitted for brevity>
MODE exec
```

```
(0) show logging level ospf
(1) show tech-support ospf [ brief ]
(2) show ip ospf [ <tag> ] [ vrf { <vrf-name> | <vrf-known-name> | all } ] [ ]
(3) show ip ospf [ <tag> ] ha [ vrf { <vrf-name> | <vrf-known-name> | all } ] [ ]
(4) show ip ospf [ <tag> ] neighbors [ { { <interface> [ <neighbor> | <neighborname>
    ] } | { [ <neighbor> | <neighbor-name> ] [ vrf { <vrf-name> | <vrf-knownname>
    | all } ] } } ] [ ]
```

show running-config diff 命令可快速对比交换机的运行配置（running-configuration）和启动配置（startup-configuration）。例 1-28 配置了一个日志记录文件，并在输出结果中以 "!" 突出显示两个文件之间的差异。

例 1-28 使用 show running-config diff 命令

```
NX-1# show running-config diff
*** Startup-config
--- Running-config
***************
*** 318,328 ****
  boot kickstart bootflash:/n7000-s2-kickstart.7.2.2.D1.2.gbin sup-2
  boot system bootflash:/n7000-s2-dk9.7.2.2.D1.2.gbin sup-2
  mpls ldp configuration
    router-id Lo0 force
  no system auto-upgrade epld
! no logging logfile
  logging level user 5
  router ospf 12
    router-id 172.16.0.1
--- 317,327 ----
  boot kickstart bootflash:/n7000-s2-kickstart.7.2.2.D1.2.gbin sup-2
  boot system bootflash:/n7000-s2-dk9.7.2.2.D1.2.gbin sup-2
  mpls ldp configuration
    router-id Lo0 force
  no system auto-upgrade epld
! logging logfile log 6 size 4194304
  logging level user 5
  router ospf 12
    router-id 172.16.0.1
```

思科 IOS 要求在配置模式下使用 **do** 命令时必须附加用户级命令，而 NX-OS 的 **do** 命令无此要求，因为 NX-OS 允许在配置模式下执行用户级命令。

1.6.2 技术支持文件

经常操作思科路由器或交换机平台的人员应该对 CLI 命令 **show tech-support** 非常熟悉，该命令的基本思想是获取与故障问题相关的通用输出结果，以供离线分析。NX-OS 提供了非常有用的层次化的 **show tech-support** 命令，最高层级的命令是 **show tech details**，可以通过单个 CLI 命令获取并聚合很多常用功能特性的 **show tech** 文件、事件历史以及内部数据结构。另一个非常有用的命令是 **tac-pac**，该命令可以收集 **show tech details** 的输出结果并将输出结果自动存储为 *bootflash:* 中的压缩文件。

NX-OS 启用的每个功能特性都能够提供技术支持文件，从而获取与特定功能特性相关的有用信息。**show tech-support** [*feature*]命令可以获得与指定功能特性相关的功能配置、**show** 命令、数据结构和事件历史，以供离线分析。执行数据收集操作时，必须注意功能特性的依赖性，从而收集与故障问题有关的所有相关信息。

例如，处理路由协议 OSPF 的单播路由故障时，不但需要收集 **show tech-support ospf** 命令的输出结果，而且还需要从全面分析出发，收集 **show tech-support routing ip unicast** 命令的输出结果，获得 URIB（Unicast Routing Information Base，单播路由信息库）事件信息。需要注意的是，必须根据实际案例来确定功能特性的依赖性以及所要收集的信息。很多功能特性的 **show tech-support** 输出结果中并不包括完整的 **show running-config**。

收集完整的 **show running-config** 信息以及与特定功能特性相关的 **show tech** 信息，是一种非常好的做法。例 1-29 给出了收集 **show tech- support** 命令输出结果以及运行配置以排查 OSPF 单播路由故障的示例。

例 1-29　收集 show tech- support 命令输出结果以排查 OSPF 故障

```
NX-1# show tech-support forwarding l3 unicast detail > tech_l3_unicast.txt
NX-1# show tech-support routing ip unicast > tech_routing.txt
NX-1# show tech-support ospf > tech_ospf.txt
NX-1# show running-config > bootflash:running-config
NX-1# tar create bootflash:routing-issue-data bootflash:tech_l3_unicast.txt
  bootflash:tech_ospf.txt bootflash:tech_routing.txt bootflash:running-config
NX-1# dir | inc routing|tech|running
   578819     Jul 13 00:15:24 2017  routing-issue-data.tar.gz
     2743     Jul 13 00:12:38 2017  running-config
  8357805     Jul 12 23:57:46 2017  tech_l3_unicast.txt
   151081     Jul 12 23:58:04 2017  tech_ospf.txt
   396743     Jul 12 23:58:39 2017  tech_routing.txt
```

需要注意的是，例 1-29 将每个 **show tech** 输出结果都收集为单个文件（利用 ">" 操作符将输出结果重定向到 *bootflash:*）。收集了所有相关功能特性的 **show tech** 输出结果之后，利用 **tar** 命令将它们组合成一个存档文件，从而便于从交换机复制文件，供后续分析。

> **注**：除 **show tech-support** [*feature*]命令之外，NX-OS 还提供了 **show running-config** [*feature*]命令，该命令仅打印指定功能特性的运行配置。

1.6.3　记账日志

NX-OS 在记账日志（Accounting Log）中记录设备的所有配置变更历史，这些信息对于确定交换机的配置变更内容以及变更人员来说非常有用。一般来说，故障排查都要从故障发生的时间开始，记账日志可以回答"交换机发生了哪些变更？"。例 1-30 给出了查看记账日志的示例，由于仅关注最后几行，因此使用了 **start-seqnum** 选项，直接跳转到列表末尾。

例 1-30　查看记账日志

```
NX-1# show accounting log ?
  <CR>
  <0-250000>    Log Size(in bytes)
  >             Redirect it to a file
  >>            Redirect it to a file in append mode
  all           Display accounting log including show commands (Use <terminal
                log-all> to enable show command accounting)
  last-index    Show accounting log last index information
  nvram         Present in nvram
  start-seqnum  Show messages starting from a given sequence number
  start-time    Show messages from a given start-time
  |             Pipe command output to filter
NX-01# show accounting log last-index
accounting-log last-index : 25712
```

```
NX-01# show accounting log start-seqnum 25709
Last Log cleared/wrapped time is : Fri Sep 9 10:14:48 2016
25709:Thu Jul 13 00:04:41 2017:update:console0:admin:switchto ; dir | inc routing
|tech (SUCCESS)
25710:Thu Jul 13 00:13:30 2017:update:console0:admin:switchto ; dir | inc routing
|tech|running (SUCCESS)
25711:Thu Jul 13 00:15:24 2017:update:console0:admin:switchto ; tar create bootf
lash:/routing-issue-data bootflash:/tech_l3_unicast.txt bootflash:/tech_ospf.txt
bootflash:/tech_routing.txt bootflash:/running-config (SUCCESS)
25712:Thu Jul 13 00:15:30 2017:update:console0:admin:switchto ; dir | inc routing
|tech|running (SUCCESS)
```

记账日志永久存储在 *logflash:* 中,始终可用(即使交换机重启)。

注:**terminal log-all configuration** 命令的作用是在记账日志中记录 **show** 命令。

1.6.4 功能特性事件历史记录

NX-OS 的一个非常有用的可维护性功能就是,为每个已配置的功能特性保留环形事件历史(Event-History)记录缓冲区。事件历史记录在很多方面都与始终在线(always-on)调试功能相似,但是对交换机的 CPU 没有任何负面影响。虽然存储在事件历史记录中的事件颗粒度取决于具体的功能特性,但大多数功能特性的事件历史记录与调试输出结果相同。在很多情况下,事件历史记录能提供足够多的数据,可以确定特定功能特性发生过的事件序列(无须其他调试日志),因而事件历史记录是一个非常好的故障排查资源。

由于事件历史记录缓冲区是一种环形缓冲区,因而在识别故障状况时,历史事件可能会被覆盖,导致故障排查时找不到相应的事件历史证据。某些功能特性支持将事件历史记录大小配置为 [**small** | **medium** | **large**]。如果特定功能特性总是周期性地出现故障,那么就可以增大事件历史记录,以提高在缓冲区中抓取故障事件的机会。可以利用 **show** *{feature}* **internal event-history** 命令查看大多数功能特性的事件历史记录,如例 1-31 所示。

例 1-31 查看 OSPF 邻接性的事件历史记录

```
NX-1# show ip ospf internal event-history adjacency
Adjacency events for OSPF Process "ospf-1"
2017 Jul 12 23:30:55.816540 ospf 1 [5817]: : Nbr 192.168.1.2: EXCHANGE --> FULL,
 event EXCHDONE
2017 Jul 12 23:30:55.816386 ospf 1 [5817]: :   seqnr 0x385a2ae7, dbdbits 0, mtu
1500, options 0x42
2017 Jul 12 23:30:55.816381 ospf 1 [5817]: : Got DBD from 192.168.1.2 with 0
 entries
2017 Jul 12 23:30:55.816372 ospf 1 [5817]: :   seqnr 0x385a2ae7, dbdbits 0, mtu
1500, options 0x42
2017 Jul 12 23:30:55.816366 ospf 1 [5817]: : Got DBD from 192.168.1.2 with 0 ent ries
2017 Jul 12 23:30:55.814575 ospf 1 [5817]: :    mtu 1500, opts: 0x42, ddbits: 0x1,
 seq: 0x385a2ae7
2017 Jul 12 23:30:55.814572 ospf 1 [5817]: : Sent DBD with 0 entries to
192.168.1.2 on Vlan2
2017 Jul 12 23:30:55.814567 ospf 1 [5817]: : Sending DBD to 192.168.1.2 on Vlan2
```

注:某些故障排查场景可能需要定期转储特定功能特性的技术支持文件,该操作可以通过 EEM(Embedded Event Manager,嵌入式事件管理器)或编写数据采集脚本来完成。**bloggerd** 是一款解决这类故障排查场景的有效工具,但是,建议最好在思科 TAC(Technical Assistance Center,技术支持中心)的指导下使用。

1.6.5 调试选项：日志文件和过滤器

如果某功能特性的事件历史记录不够精细，无法解决故障问题，或者在因缓冲区被覆盖而导致事件丢失之前无法抓取所需的事件历史记录，那么就可以启用针对该功能特性的调试操作。启用调试操作时，需要配置过滤器并将调试结果记录到调试日志文件中。例 1-32 给出了一个调试过滤器和调试日志文件示例，经过滤后仅显示特定 OSPF 邻居的输出结果。

例 1-32 调试日志文件和调试过滤器

```
NX-1# debug-filter ip ospf neighbor 192.168.1.2
NX-1# debug logfile ospf-debug-log
NX-1# debug ip ospf adjacency detail
NX-1# dir log: | inc ospf
      1792     Jul 13 00:49:40 2017  ospf-debug-log
NX-1# show debug logfile ospf-debug-log
2017 Jul 13 00:49:14.896199 ospf: 1 [5817] (default)   Nbr 192.168.1.2
FSM start: old state FULL, event HELLORCVD
2017 Jul 13 00:49:14.896376 ospf: 1 [5817] (default)   Nbr 192.168.1.2:
FULL --> FULL, event HELLORCVD
2017 Jul 13 00:49:14.896404 ospf: 1 [5817] (default)   Nbr 192.168.1.2
FSM start: old state FULL, event TWOWAYRCVD
2017 Jul 13 00:49:14.896431 ospf: 1 [5817] (default)   Nbr 192.168.1.2:
FULL --> FULL, event TWOWAYRCVD
```

1.6.6 配置检查点和回滚

NX-OS 支持创建检查点配置功能。创建检查点是存储系统已知正常配置的一种方式，如果系统配置变更后产生了非期望行为，那么就可以利用检查点快速回滚到已知正常配置，而无须重新加载系统。该方法解决了容易出错的人为因素问题，不必将之前的正常配置复制并粘贴回设备。NX-OS 支持以下 4 类配置回滚模式。

- 原子模式（Atomic）：仅在无错误时执行回滚操作，该模式是默认选项。
- 尽力而为模式（Best-effort）：执行回滚操作并跳过任何错误。
- 首次故障停止模式（Stop-at-first-failure）：执行回滚操作，但在错误发生时停止。
- 详细模式（Verbose mode）：在回滚操作期间显示详细的执行日志。

执行配置回滚操作时，可以利用 **show diff rollback-patch checkpoint** 命令查看将要应用的配置变更，从而将检查点文件与其他检查点文件或运行配置进行对比。回滚期间，如果遇到错误，那么就需要决定是否取消或继续回滚操作。如果取消回滚，那么就会提供已应用的变更列表，需要手动回退这些变更以返回到回滚前配置。例 1-33 给出了配置检查点并回滚的操作示例。

例 1-33 配置检查点并回滚

```
NX-1# checkpoint known_good
...............Done
NX-1# show diff rollback-patch checkpoint known_good running-config
Collecting Running-Config
#Generating Rollback Patch
Rollback Patch is Empty
NX-1# conf t
Enter configuration commands, one per line. End with CNTL/Z.
NX-1(config)# no router ospfv3 1
NX-1(config)# end
NX-1# show diff rollback-patch checkpoint known_good running-config
Collecting Running-Config
```

```
#Generating Rollback Patch
!!
no router ospfv3 1
NX-1# rollback running-config checkpoint known_good
Note: Applying config parallelly may fail Rollback verification
Collecting Running-Config
#Generating Rollback Patch
Executing Rollback Patch
Generating Running-config for verification
Generating Patch for verification
Verification is Successful.

Rollback completed successfully.
NX-1# show diff rollback-patch checkpoint known_good running-config
Collecting Running-Config
#Generating Rollback Patch
Rollback Patch is Empty
```

从例 1-33 可以看出，配置检查点之后删除了 OSPFv3 进程，通过 **show diff rollback- patch checkpoint** 命令可以显示检查点与运行配置之间的差异。此后，配置变更回滚成功，恢复了 OSPFv3 进程。

1.6.7 一致性检查器

一致性检查器是 NX-OS 平台在所有版本中提高服务可用性的一种工具。某些软件错误或竞态条件可能会导致控制平面、数据平面或转发 ASIC 之间出现状态不匹配问题。这些问题的发现对于平台来说至关重要，但通常要求对平台有深入了解。因此，NX-OS 平台引入了一致性检查器来解决这些问题，这些不匹配问题都是 TAC 和客户反馈的结果。例 1-34 给出了在 Nexus 3172 平台上执行一致性检查器的应用示例。

例 1-34 执行一致性检查器

```
NX-1# test consistency-checker forwarding ipv4
NX-1# show consistency-checker forwarding ipv4
IPV4 Consistency check : table_id(0x1)
Execution time : 73 ms ()
No inconsistent adjacencies.
No inconsistent routes.
Consistency-Checker: PASS for ALL
```

不同的平台提供的一致性检查器类型也有所不同，而且一致性检查器的功能在不断发展当中，每个版本都会增加一些新协议和新平台支持能力。如果已经通过故障定位到特定设备，那么在排查具体故障时，就可以根据需要运行一致性检查器，以快速确认故障问题是否由平台中的状态不匹配问题引起。

1.6.8 调度器、Python 和 EEM

NX-OS 支持调度器功能，它可以按照预定义的启动时间和频率以非交互方式执行命令或脚本。调度器功能不需要任何特殊许可即可运行。

调度器是一种非常有用的功能特性，可以按照指定时间间隔备份配置、复制文件或收集数据。调度器功能与 NX-OS Python 或 EEM（Embedded Event Manager，嵌入式事件管理器）结合使用，可以提供一种强大的自动执行任务方法。例 1-35 定义了一个调度器作业，要求在每天午夜执行 Python 脚本。调度器的配置要求首先启用 **feature scheduler**，然后再配置该作业的任务和计划，以确定作业执行时间。

例 1-35　配置调度器任务

```
NX-1# show run | sec scheduler
feature scheduler
scheduler job name run_script
source /bootflash/scripts/snapshot_compare.py
end-job
scheduler schedule name at_midnight
  job name run_script
  time daily 00:00
```

NX-OS 允许通过 **python** 命令从 exec 模式以 CLI 方式访问 Python 解释器，如例 1-36 所示。

例 1-36　NX-OS Python 解释器

```
NX-1# python
Python 2.7.5 (default, Jun  3 2016, 03:57:06)
[GCC 4.6.3] on linux2
Type "help", "copyright", "credits" or "license" for more information.
>>> print "hello world!"
hello world!
>>> quit()
```

除 Python 之外，NX-OS 还支持 EEM（另一种自动执行数据收集任务的方式）或动态修改配置（如果发生了已定义事件）。

注：有关 NX-OS 编程能力和自动化能力的详细信息，请参阅本书第 15 章。

1.6.9　Bash Shell

Bash Shell（外壳）是 NX-OS 提供的一项较新的功能特性，启用了 **feature bash-shell** 之后，用户就可以进入 NX-OS 操作系统的 Linux Bash Shell。例 1-37 给出了从 Exec 提示符访问 Bash Shell 的方式。

例 1-37　进入 Bash Shell

```
NX-1# run bash
bash-4.2$ pwd
/bootflash/home/admin
bash-4.2$ uname -srvo
Linux 3.4.43-WR5.0.1.13_standard #1 SMP Thu Apr 7 08:39:09 PDT 2016 GNU/Linux
bash-4.2$
```

访问 Bash Shell 时，用户账户必须与 *dev-ops* 或 *network-admin* 角色相关联。也可以使用 **run bash** [*command*] 选项，从交换机的 Exec 模式以 CLI 方式运行 Bash 命令，该命令采用指定参数，在 Bash Shell 中运行并返回运行结果。此外，也可以在 Bash Shell 中运行 Python 脚本。需要注意的是，通过 Bash Shell 管理设备时务必小心操作。

1.7　本章小结

NX-OS 是一种功能强大、特性丰富的网络操作系统，广泛应用于世界各地的网络当中。在过去的十多年里，NX-OS 一直都在持续发展，以满足大量网络因时间推移而产生的各种多样化的需求。NX-OS 模块化架构能够实现功能特性的快速迭代开发，而且还可以在不同的 Nexus 交换平台之间共享源代码。NX-OS 和 Nexus 交换机在设计之初就支持强大的硬件和软件弹性能力，即使组件出现故障，也能实现不间断服务。

本章详细介绍了 NX-OS 的主要管理和操作功能，这些知识对于后续章节的故障排查及网络应用来说，都极为重要。

第 2 章

NX-OS 故障排查工具

本章主要讨论如下主题。
- 抓包：网络嗅探。
- Nexus 平台工具。
- NetFlow。
- NTP（Network Time Protocol，网络时间协议）。
- EEM（Embedded Event Manager，嵌入式事件管理器）。

故障排查是一门艺术，需要深入了解故障涉及的相关主题以及验证操作和隔离故障行为的能力。如果工程师排查网络故障时，面临的拓扑结构包含了数百或数千台设备，那么故障排查操作就会显得比较困难。但是，如果仅需要面对出现故障的部分拓扑结构，那么网络故障排查操作就会变得较为容易。适当的故障排查工具和正确的故障拓扑结构视图，有助于快速隔离故障问题，能够有效减少大规模网络造成的复杂性影响。本章将重点介绍 Nexus 平台可用的各种故障排查工具，这些工具能够有效提高故障排查和日常操作效率。

2.1 抓包：网络嗅探

NX-OS 提供了 CLI（Command-Line Interface，命令行接口），可以帮助解决各种复杂的网络问题。但是在某些情况下，**show** 和 **debug** 命令可能无法提供足够的信息来隔离数据包流的故障方向，此时执行抓包操作就显得非常有必要。对于转发故障来说，需要隔离出现故障的方向并了解数据包是否确实到达了远端设备。如果要了解两台直连设备之间的数据包流情况，就必须实施以下 3 个方面的操作。
- 确定发送端路由器是否正在通过网络介质传输数据包。
- 确定目的端路由器是否正在接收数据包。
- 检查流经网络介质的数据包。

这就是网络嗅探的主要作用，网络嗅探是一种拦截穿越传输介质的流量以分析流量协议并执行深度包分析的技术。数据包嗅探技术不但有助于解决数据包转发问题，而且还能帮助网络安全专家进行深度网络分析并找出安全漏洞。

网络嗅探抓包操作要求计算机（PC）必须安装抓包工具（如 Wireshark）并连接到交换机上，复制相关流量的镜像副本并发送给目的端接口，抓包工具将抓取这些数据并执行相关分析。图 2-1 中的 Nexus 交换机连接在两台路由器与抓包 PC 之间，PC 安装了 Wireshark 以抓取路由器 R1 与 R2 之间的流量。

思科设备将嗅探功能称为 SPAN（Switched Port Analyzer，交换端口分析器）功能，将源端口称为受监控端口，将目的端口称为监控端口。NX-OS 的 SPAN 功能与思科 IOS 类似，但不同的 Nexus 交换机拥有不同的 SPAN 功能（具体取决于硬件支持能力）。下列源接口都可以用作 SPAN 源接口。
- 以太网。

- FEX（Fabric Expander，交换矩阵扩展器）端口/交换矩阵端口通道。
- 端口通道。
- VLAN 或 VSPAN（VLAN-based SPAN，基于 VLAN 的 SPAN）。
- RSPAN（Remote SPAN，远程 SPAN）VLAN。
- 去往控制平面 CPU 的带内接口（对于 Nexus 7000 来说，仅默认 VDC[Virtual Device Context，虚拟设备上下文]支持该功能特性）。
- FCoE 端口。

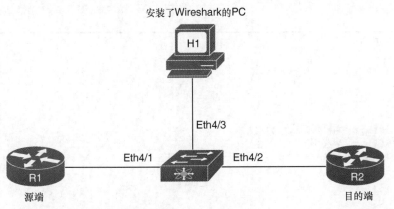

图 2-1 Nexus 交换机的嗅探设置

注：这些功能特性可能因不同的 Nexus 平台而异（具体取决于硬件支持能力），活动会话数以及每个会话的源接口和目的接口数取决于不同的 Nexus 平台。在 Nexus 交换机上配置 SPAN 会话之前，请务必查阅相关的思科文档。

为了(让端口能够)将穿越端口的流量转发给抓包PC,需要通过接口参数命令 **switchport monitor** 为接口启用监控能力，目的接口可以是配置为接入或中继模式的以太网接口或端口通道接口。可以通过 **monitor session** *session-number* 命令配置 SPAN 会话，通过 **source interface** *interface-id* [**rx** | **tx** | **both**]命令指定源接口，选项 **rx** 的作用是抓取入站（输入）流量，选项 **tx** 的作用是抓取出站（输出）流量，该命令的默认选项是 **both**，即同时抓取源接口上的入站和出站流量。目的接口由命令 **destination interface** *interface-id* 指定。默认情况下，监控会话处于关闭状态，必须手工取消关闭状态才能使 SPAN 会话正常运行。

注：SPAN 功能可能因不同的 Nexus 平台而异。如 Nexus 5000 和 Nexus 6000 系列支持 SPAN-on-Drop 和 SPAN-on-Latency 等功能特性，而 Nexus 7000 系列不支持这些功能特性。有关功能特性的具体支持信息，请参阅相关平台的技术文档。

例 2-1 给出了 Nexus 交换机的 SPAN 会话配置示例。请注意，本例通过 **source interface** 命令定义了一系列接口及抓包方向。

例 2-1 NX-OS 的 SPAN 配置

```
NX-1(config)# interface Ethernet4/3
NX-1(config-if)# switchport
NX-1(config-if)# switchport monitor
NX-1(config-if)# no shut
NX-1(config)# monitor session 1
NX-1(config-monitor)# source interface Ethernet4/1-2 both
NX-1(config-monitor)# source interface Ethernet5/1 rx
```

```
NX-1(config-monitor)# destination interface Ethernet4/3
NX-1(config-monitor)# no shut
NX-1(config-monitor)# exit
```

注：对于 FCoE 端口来说，需要通过 **switchport mode SD** 命令来配置 SPAN 目的接口，该命令与 **switchport monitor** 类似。

例 2-2 显示了监控会话的状态，可以看出，接口 Eth4/1 和 Eth4/2 配置了 **rx**、**tx** 和 **both** 字段，而接口 Eth5/1 仅配置了 **rx** 字段。此外，还可以在监控会话下通过 **filter vlan** *vlan-id* 命令过滤 VLAN。

例 2-2 验证 SPAN 会话

```
NX-1# show monitor session 1
   session 1
---------------
type                  : local
state                 : up
source intf           :
   rx                 : Eth4/1    Eth4/2    Eth5/1
   tx                 : Eth4/1    Eth4/2
   both               : Eth4/1    Eth4/2
source VLANs          :
   rx                 :
   tx                 :
   both               :
filter VLANs          : filter not specified
destination ports     : Eth4/3

Legend: f = forwarding enabled, l = learning enabled
```

SPAN 会话的默认行为是将所有流量都镜像到目的端口，不过，NX-OS 提供了流量过滤功能，可以对被镜像到目的端口的流量进行过滤。如果要过滤相关流量，则需要创建一个 ACL（Access Control List，访问控制列表），由 **filter access-group** *acl* 命令在 SPAN 会话配置中引用该 ACL。例 2-3 解释了 SPAN 会话的过滤配置方式，并通过 **show monitor session** 命令加以验证。

注：ACL 过滤功能因不同的 Nexus 平台而异。有关 Nexus 平台的 ACL 过滤功能的具体支持情况，请查阅相关 CCO 文档。

例 2-3 过滤 SPAN 流量：配置和验证

```
NX-1(config)# ip access-list TEST-ACL
NX-1(config-acl)# permit ip 100.1.1.0/24 200.1.1.0/24
NX-1(config-)# exit
NX-1(config)# monitor session 1
NX-1(config-monitor)# filter access-group TEST-ACL
NX-1(config-monitor)# exit

NX-1# show monitor session 1
   session 1
---------------
type                  : local
state                 : up
acl-name              : TEST-ACL
source intf           :
   rx                 : Eth4/1    Eth4/2    Eth5/1
   tx                 : Eth4/1    Eth4/2
   both               : Eth4/1    Eth4/2
```

```
source VLANs              :
    rx                    :
    tx                    :
    both                  :
filter VLANs              : filter not specified
destination ports         : Eth4/3

Legend: f = forwarding enabled, l = learning enabled
```

注：Nexus 平台不支持 RSPAN（Remote SPAN，远程 SPAN）。

2.1.1 封装远程 SPAN

ERSPAN（Encapsulated Remote SPAN，封装远程 SPAN）是一种 SPAN 功能特性，将 SPAN 流量封装为 IP-GRE 帧格式，以支持 IP 网络的远程监控流量。ERSPAN 可以监控网络中的多台远程交换机，也就是说，ERSPAN 可以跨越多台交换机将流量从源端口镜像到网络分析器所连接的目的交换机。ERSPAN 会话包括如下组件。

- ERSPAN ID。
- ERSPAN 源会话。
- GRE 封装的流量。
- ERSPAN 目的会话。

其中，ERSPAN ID 负责区分多个源设备，将跨区流量发送到单个集中式服务器。

图 2-2 的网络拓扑配置了 ERSPAN 功能，两台 Nexus 交换机通过路由网络进行连接。N6k-1 交换机被配置为 ERSPAN 源端，拥有本地源 SPAN 端口，目的端口位于 IP 网络中的 N7k-1 交换机。GRE 封装的数据包通过 IP 网传输到目的交换机，由目的交换机进行解封装并发送给流量分析器。

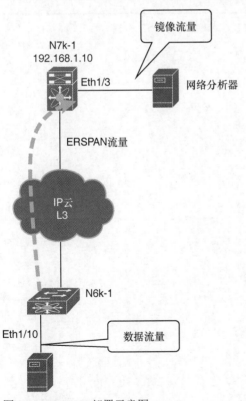

图 2-2　ERSPAN 部署示意图

可以在不同的交换机上分别配置源会话和目的会话，包括抓取入站、出站或双向源流量。ERSPAN 被配置为跨以太网端口、VLAN、VSAN 和 FEX 端口镜像流量。目的端口保持监控状态，不参与生成树或任何三层协议。例 2-4 给出了两台 Nexus 交换机的源端口和目的端口配置示例，需要注意的是，这两台交换机的 ERSPAN-ID 应该相同。

例 2-4　ERSPAN 配置

```
! ERSPAN Source Configuration
N6k-1(config)# monitor session 10 type erspan-source
N6k-1(config-erspan-src)# erspan-id 20
N6k-1(config-erspan-src)# vrf default
N6k-1(config-erspan-src)# destination ip 192.168.1.10
N6k-1(config-erspan-src)# source interface ethernet 1/10
N6k-1(config-erspan-src)# no shut
N6k-1(config-erspan-src)# exit
N6k-1(config)# monitor erspan origin ip-address 192.168.1.1 global
! ERSPAN Destination Configuration
N7k-1(config)# monitor session 10 type erspan-destination
N7k-1(config-erspan-dst)# erspan-id 10
N7k-1(config-erspan-dst)# source ip 192.168.1.10
N7k-1(config-erspan-dst)# destination interface e1/3
N7k-1(config-erspan-dst)# no shut
```

如果要启用 ERSPAN 源会话，那么就必须保证目的 IP 位于路由表中。使用 **show monitor session** *session-id* 命令可以验证 ERSPAN 会话的状态。例 2-5 给出了源 ERSPAN 会话和目的 ERSPAN 会话的验证示例。

例 2-5　验证 ERSPAN 会话

```
N6k-1# show monitor session 10
   session 10
---------------
type              : erspan-source
state             : up
erspan-id         : 20
vrf-name          : default
destination-ip    : 192.168.1.10
ip-ttl            : 255
ip-dscp           : 0
acl-name          : acl-name not specified
origin-ip         : 192.168.1.1 (global)
source intf       :
   rx             : Eth1/10
   tx             : Eth1/10
   both           : Eth1/10
source VLANs      :
   rx             :
source VSANs      :
   rx             :
N7k-1# show monitor session 10
   session 10
---------------
type              : erspan-destination
state             : up
erspan-id         : 10
source-ip         : 192.168.1.10
```

```
destination ports : Eth1/3

Legend: f = forwarding enabled, l = learning enabled
```

注： 在 Nexus 交换机上配置 ERSPAN 之前，请参阅相关的思科文档，以了解特定平台的限制条件。

2.1.2 基于时延和丢包的 SPAN

SPAN 和 ERSPAN 可以根据协议和 IP 地址对 SPAN 流量进行过滤。用户或应用程序会报告源端与目的端之间的高时延或流量丢包问题，但很难确定丢包发生的位置。在这种情况下，分析对用户造成影响的流量状况对于故障排查操作来说就显得非常重要，这不但能够尽量减少故障造成的影响，而且还能提高故障排查速度。

NX-OS 可以根据指定的时延阈值或路径中发现的丢包信息来镜像流量，SPAN 和 ERSPAN 都支持该功能。

1. SPAN-on-Latency

SOL（SPAN-on-Latency，基于时延的 SPAN）功能特性与常规的 SPAN 会话的工作方式略有不同。SOL 的源端口是监控时延的出站端口，目的端口仍然是交换机连接网络分析器的端口。可以通过 **packet latency threshold** *threshold-value* 命令来定义被监控接口的时延阈值，如果数据包的时延超出了指定阈值，那么就会触发 SPAN 会话并抓取数据包。如果未在接口下指定阈值，那么就会将阈值设定为最接近的 8 的倍数。

例 2-6 给出了 SOL 配置示例，仅在出站接口 Eth1/1 和 Eth1/2 对时延超过 1μs 的流量执行数据包嗅探操作。**packet latency threshold** 的配置用于 40G 接口的所有端口，如果有 4x10G 接口，那么这些接口将共享相同配置。从例 2-6 的日志消息可以看出，接口 Eth1/1 ~ Eth1/4 配置的时延阈值均为 1000 ns。

例 2-6 SPAN-on-Latency 配置

```
N6k-1(config)# monitor session 20 type span-on-latency
N6k-1(config-span-on-latency)# source interface ethernet 1/1-2
N6k-1(config-span-on-latency)# destination interface ethernet 1/3
N6k-1(config-span-on-latency)# no shut
N6k-1(config-span-on-latency)# exit
N6k-1(config)# interface eth1/1-2
N6k-1(config-if-range)# packet latency threshold 1000

Interfaces Eth1/1, Eth1/2, Eth1/3 and Eth1/4 are configured with latency
  threshold 1000
```

只要在 **monitor session** 命令中将监控类型指定为 **span-on-latency-erspan**，即可配置 SOL-ERSPAN。

SOL 或 SOL-ERSPAN 的限制如下。
- 仅支持以太网源端口，不能将端口通道作为源端口。
- 源端口不能是其他 SPAN 会话的一部分。
- SOL 不支持 SPAN 方向。
- SOL 不支持 ACL 过滤。

2. SPAN-on-Drop

SPAN-on-Drop（基于丢包的 SPAN）是一项新功能特性，可以镜像因缓冲区或队列空间不足而丢弃的入站数据包。该功能特性可以镜像这些被丢弃的数据包，并将镜像流量的副本发送到指定目的端口。只要在 **monitor session** 配置中将监控类型指定为 **span-on-drop**，即可配置 SPAN-on-Drop 会话。例 2-7 给出了 SPAN-on-Drop 监控会话的配置示例，配置中指定的源接口 Eth1/1 是出现拥塞的接口。

例 2-7　SPAN-on-Drop 配置

```
N6k-1(config)# monitor session 30 type span-on-drop
N6k-1(config-span-on-latency)# source interface ethernet 1/1
N6k-1(config-span-on-latency)# destination interface ethernet 1/3
N6k-1(config-span-on-latency)# no shut
N6k-1(config-span-on-latency)# exit
```

注：SPAN-on-Drop 功能仅抓取因缓冲区拥塞而丢弃的单播流中的数据包。

　　与其他 SPAN 功能特性不同，SPAN-on-Drop 不涉及 TCAM（Ternary Content Addressable Memory，三重内容可寻址存储器）编程，源端的编程操作都集中在缓冲区或队列空间。除此以外，交换机只能启用一个 SPAN-on-Drop 实例，如果启用第二个实例，那么就会关闭会话并显示系统日志消息"No hardware resource error"（无硬件资源错误）。如果 SPAN-on-Drop 会话已启动但没有镜像数据包，那么就需要验证单播流中是否出现了丢包现象，此时可以使用命令 **show plat software qd info interface** *interface-id*，检查计数器 IG_RX_SPAN_ON_DROP 是否正在递增且非零。例 2-8 显示了计数器 IG_RX_SPAN_ON_DROP 的输出结果，表明该单播流并未出现丢包问题。

例 2-8　验证入站 L3 单播流丢包情况

```
N6k-1# show plat software qd info interface ethernet 1/1 | begin BM-INGRESS
BM-INGRESS                                BM-EGRESS
------------------------------------------------------------------------
IG_RX                         364763|TX                          390032
SP_RX                           1491|TX_MCAST                         0
LB_RX                          15689|CRC_BAD                          0
IG_RX_SPAN_ON_DROP                 0|CRC_STOMP                        0
IG_RX_MCAST                    14657|DQ_ABORT_MM_XOFF_DROP            0
LB_RX_SPAN                     15689|MTU_VIO                          0
IG_FRAME_DROP                      0|
SP_FRAME_DROP                      0|
LB_FRAME_DROP                      0|
IG_FRAME_QS_EARLY_DROP             0|
ERR_IG_MTU_VIO                     0|
ERR_SP_MTU_VIO                     0|
ERR_LB_MTU_VIO                     0|
```

　　SPAN-on-Drop ERSPAN 是 SPAN-on-Drop 功能特性的扩展，可以将丢弃的数据帧镜像到连接了网络分析器的远程 IP 地址。

注：截至本书写作之时，仅 Nexus 5600 系列和 Nexus 6000 系列交换机支持 SOL 和 SPAN-on-Drop。

2.2　Nexus 平台工具

　　Nexus 交换机是目前业界功能强大的数据中心交换机之一，一部分原因是交换机提供的可用 CPU 和内存；另一部分原因是 NX-OS 提供的大量集成式工具，可以通过这些工具抓取交换机上到不同 ASIC 层级的数据包，有助于验证硬件编程以及硬件或软件对被调查数据包采取的操作。常见的主要工具有以下几种。

- Ethanalyzer。
- ELAM（Embedded Logic Analyzer Module，嵌入式逻辑分析器模块）。
- 数据包跟踪器（Packet Tracer）。

这些工具能够抓取去往 CPU 的流量中的数据包或传输硬件交换流量，有助于用户理解数据包在交换机中经历的各个处理过程以及快速缩小故障范围。这些功能特性的主要优势在于不需要花费时间部署外部嗅探设备。

> 注：虽然 Nexus 交换机都支持 ELAM 抓包功能，但由于使用 ELAM 的时候，必须对 ASIC 有深入理解，而且不同的 Nexus 平台的配置也有所差别，因此相关内容已超出了本书写作范围。此外，建议最好在思科 TAC（Technical Assistance Center，技术支持中心）工程师的指导下使用 ELAM。Nexus 5000 或 Nexus 5500 交换机不支持 ELAM。

2.2.1 Ethanalyzer

Ethanalyzer 是 TShark 的 NX-OS 实现，是 Wireshark 的终端版本。TShark 使用 libpcap 库，因而 Ethanalyzer 能够抓取和解码数据包，可以抓取所有 Nexus 平台上的带内和管理流量。Ethanalyzer 为用户提供了以下功能。

- 抓取交换机管理引擎 CPU 收发的数据包。
- 定义要抓取的数据包数量。
- 定义要抓取的数据包长度。
- 显示数据包的详细协议信息或仅显示单行摘要信息。
- 打开并保存抓取的数据包数据。
- 根据多种条件过滤抓取的数据包（抓包过滤器）。
- 根据多种条件过滤所要显示的数据包（显示过滤器）。
- 解码控制包的内部报头。
- 无须使用外部嗅探设备抓取流量。

Ethanalyzer 无法在交换机的数据端口之间抓取硬件交换流量。如果要抓取这类数据包，就需要使用 SPAN 或 ELAM。如果接口配置了 ACL，且 ACE 配置了日志选项，那么就可以将硬件交换流量转出（Punt）到 CPU，进而通过 Ethanalyzer 抓取这些数据包。但是，不建议在生产性网络中这么做，因为数据包可能因 CoPP 策略而被丢弃，或者转出到 CPU 的流量过多而影响设备上的其他服务。

Ethanalyzer 的配置步骤如下。

- **第 1 步**：定义抓包接口。
- **第 2 步**：定义过滤器——设置抓包过滤器或显示过滤器。
- **第 3 步**：定义停止抓包的规则。

首先定义抓包接口，抓包接口通常包括以下 3 类。

- **Mgmt**：抓取交换机 Mgmt0 接口上的流量。
- **Inbound-hi**：抓取带内高优先级控制包，如 STP（Spanning Tree Protocol，生成树协议）、LACP（Link Aggregation Control Protocol，链路聚合控制协议）、CDP（Cisco Discovery Protocol，思科发现协议）、DCBX（Data Center Bridging Exchange，数据中心桥接交换）、光纤通道（Fiber Channel）和 FCOE（Fiber Channel Over Ethernet，以太网光纤通道）流量。
- **Inbound-low**：抓取带内低优先级控制包，如 IGMP（Internet Group Management Protocol，Internet 组管理协议）、TCP（Transmission Control Protocol，传输控制协议）、UDP（User

Datagram Protocol，用户数据报协议）、IP（Internet Protocol，Internet 协议）和 ARP（Address Resolution Protocol，地址解析协议）流量。

接下来需要设置过滤器。如果读者熟悉 Wireshark，那么设置 Ethanalyzer 过滤器就非常简单。Ethanalyzer 支持两类过滤器：抓包过滤器和显示过滤器。顾名思义，如果设置了抓包过滤器，那么将仅抓取与过滤器相匹配的帧。显示过滤器用于显示抓取的数据包中与过滤器相匹配的数据包，也就是说，Ethanalyzer 虽然会抓取与显示过滤器不匹配的数据包，但是不会在输出结果中显示这些数据包。Ethanalyzer 默认最多抓取 10 帧，此后将自动停止抓包，可以通过选项 **limit-captured-frames** 修改该默认值，0 表示无限制。

> **注：** 设置了选项 **inbound-hi** 或 **inbound-low** 之后，将抓取与交换机管理引擎之间收发数据的所有带内以太网端口。当然，也可以应用显示过滤器或抓包过滤器。

由于 Ethanalyzer 是运行在管理引擎上的一种软件，因此必须了解 Ethanalyzer 对管理引擎 CPU 的影响。一般来说，Ethanalyzer 对管理引擎的 CPU 没有太大影响，但有时也可能会增加 5%的 CPU 利用率。不过，也可以使用 Ethanalyzer 的 **write** 选项，将抓取的数据包保存到文件中，这样就可以将 CPU 的利用率降低 1%~2%。

如果要启用 Ethanalyzer 执行抓包操作，那么就要使用 **ethanalyzer local interface [inbound-hi | inbound-low | mgmt]** *options* 命令，该命令提供了如下选项。

- **autostop**：自动停止抓包。
- **capture-filter**：对 Ethanalyzer 应用抓包过滤器。
- **capture-ring-buffer**：抓取环行缓冲区。
- **decode-internal**：解码时包含内部系统报头。
- **detail**：显示详细的协议信息。
- **display-filter**：对抓取的帧应用显示过滤器。
- **limit-captured-frames**：定义最大抓取帧数。
- **limit-frame-size**：仅抓取帧的部分子集。
- **write**：定义保存抓包数据的文件名。

使用 Ethanalyzer 时，如果熟悉 Wireshark 过滤器，那么在设置 Ethanalyzer 的时候就比较简单。抓包过滤器和显示过滤器的语法不同，表 2-1 列出了常见抓包过滤器和显示过滤器及其语法。

表 2-1　　　　　　　　　Ethanalyzer 抓包过滤器和显示过滤器

	抓包过滤器	显示过滤器
运算符		与 - && 或 - || 等于 - == 不等于 - !=
VLAN	**vlan** *vlan-id*	**vlan.id**==*vlan-id*
二层	**ether host** 00:AA:BB:CC:DD:EE **ether dst** 00:AA:BB:CC:DD:EE **ether src** 00:AA:BB:CC:DD:EE **ether broadcast** **ether multicast** **ether proto** *protocol*	**eth.addr**==*00:AA:BB:CC:DD:EE* **eth.src**==*00:AA:BB:CC:DD:EE* **eth.dst**==*00:AA:BB:CC:DD:EE* 匹配前 2 字节： **eth.src[0:1]**==*00:AA* 对制造商进行过滤： **eth.src[0:2]**==*vendor-mac-addr* 如，Cisco：**eth.src[0:2]**==**00.00.0c** **eth.addr** 包含 aa:bb:cc

续表

	抓包过滤器	显示过滤器
三层	ip（过滤低层协议，如 ARP 和 STP） host *192.168.1.1* dst host *192.168.1.1* src host *192.168.1.1* net *192.168.1.0/24* net *192.168.1.0* netmask *24* src net *192.168.1.0/24* dst net *192.168.1.0/24* ip broadcast ip multicast not broadcast not multicast icmp udp tcp ip proto 6 (udp) ip proto 17 (tcp) ip proto 1 (icmp) 数据包长度： less *length* greater *length*	IP 地址： ip.addr==*192.168.1.1* 源 IP 地址： ip.src==*192.168.1.1* 目的 IP 地址： ip.dst==*192.168.10.1* 子网： ip.addr==*192.168.1.0/24* 分段： 基于 DF 比特集的过滤器（0=可以分段） ip.flags.df==*1* TCP 序列号： tcp.seq==*TCP-Seq-Num*
四层	udp port *53* udp dst port *53* udp src port *53* tcp port *179* tcp portrange *2000-2100*	tcp.port==*53* udp.port==*53*
FabricPath	proto 0x8903	目的 HMAC/MC 目的端： cfp.d_hmac==*mac* cfp.d_hmac_mc==*mac* EID/FTAG/IG 比特： cfp.eid== cfp.ftag== cfp.ig== 源 LID/OOO/DL 比特/源 HMAC： cfp.lid== cfp.ooodl== cfp.s_hmac== Subswitch ID/Switch ID/TTL： cfp.sswid== cfp.swid== cfp.ttl==
ICMP	icmp	icmp==*icmp-type* ICMP 类型： *icmp-echoreply* *icmp-unreach* *icmp-sourcequench* *icmp-redirect* *icmp-echo* *icmp-routeradvert* *icmp-routersolicit* *icmp-timxceed* *icmp-paramprob* *icmp-tstamp* *icmp-tstampreply* *icmp-ireq* *icmp-ireqreply* *icmp-maskreq* *icmp-maskreply*

例 2-9 给出了 Ethanalyzer 抓包示例，本例将抓取 Nexus 6000 上所有符合 **inbound-low** 和 **inbound-hi** 队列的数据包。从输出结果可以看出，虽然用于 BGP 对等连接的 TCP SYN/SYN ACK 包属于 **inbound-low** 队列，但常规的 BGP 更新和保持激活包（如 BGP 对等连接建立之后的 TCP 包）以及确认包仍然属于 **inbound-hi** 队列。

例 2-9　Ethanalyzer 抓包

```
N6k-1# ethanalyzer local interface inbound-low limit-captured-frames 20
Capturing on inband
2017-05-21 21:26:22.972623 10.162.223.33 -> 10.162.223.34 TCP bgp > 45912 [SYN,
ACK] Seq=0 Ack=0 Win=16616 Len=0 MSS=1460
2017-05-21 21:26:33.214254 10.162.223.33 -> 10.162.223.34 TCP bgp > 14779 [SYN,
ACK] Seq=0 Ack=0 Win=16616 Len=0 MSS=1460
2017-05-21 21:26:44.892236 8c:60:4f:a7:9a:6b -> 01:00:0c:cc:cc:cc CDP Device ID:
 N6k-1(FOC1934R1BF)  Port ID: Ethernet1/4
2017-05-21 21:26:44.892337 8c:60:4f:a7:9a:68 -> 01:00:0c:cc:cc:cc CDP Device ID:
 N6k-1(FOC1934R1BF)  Port ID: Ethernet1/1
2017-05-21 21:27:42.965431 00:25:45:e7:d0:00 -> 8c:60:4f:a7:9a:bc ARP 10.162.223
.34 is at 00:25:45:e7:d0:00
! Output omitted for brevity
N6k-1# ethanalyzer local interface inbound-hi limit-captured-frames 10

Capturing on inband
2017-05-21 21:34:42.821141 10.162.223.34 -> 10.162.223.33 BGP KEEPALIVE Message
2017-05-21 21:34:42.932217 10.162.223.33 -> 10.162.223.34 TCP bgp > 14779 [ACK]
Seq=1 Ack=20 Win=17520 Len=0
2017-05-21 21:34:43.613048 10.162.223.33 -> 10.162.223.34 BGP KEEPALIVE Message
2017-05-21 21:34:43.814804 10.162.223.34 -> 10.162.223.33 TCP 14779 > bgp [ACK]
Seq=20 Ack=20 Win=15339 Len=0
2017-05-21 21:34:46.005039    10.1.12.2 -> 224.0.0.5    OSPF Hello Packet
2017-05-21 21:34:46.919884 10.162.223.34 -> 10.162.223.33 BGP KEEPALIVE Message
2017-05-21 21:34:47.032215 10.162.223.33 -> 10.162.223.34 TCP bgp > 14779 [ACK]
Seq=20 Ack=39 Win=17520 Len=0
! Output omitted for brevity
```

前面曾经说过，最佳抓包实践是将抓取的帧写到文件中，然后再读取抓到的数据帧。使用命令 **ethanalyzer local read** *location* **[detail]** 可以读取存储在本地 bootflash 中的文件。

Nexus 7000 不支持选项 **inbound-hi** 或 **inbound-low**，可以通过 CLI 抓取 mgmt 接口或带内接口，带内接口抓包操作将同时抓取高优先级和低优先级数据包。例 2-10 给出了读写已保存的抓包数据的方式，本例中的 Ethanalyzer 应用了抓包过滤器，仅抓取 STP 数据包。

例 2-10　Ethanalyzer 读写操作

```
N7k-Admin# ethanalyzer local interface inband capture-filter "stp" write
  bootflash:stp.pcap
Capturing on inband
10
N7k-Admin# ethanalyzer local read bootflash:stp.pcap
2017-05-21 23:48:30.216952 5c:fc:66:6c:f3:f6 -> Spanning-tree-(for-bridges)_00
 STP 60 RST. Root = 4096/1/50:87:89:4b:bb:42  Cost = 0  Port = 0x9000
2017-05-21 23:48:30.426556 38:ed:18:a2:27:b0 -> Spanning-tree-(for-bridges)_00
 STP 60 RST. Root = 4096/1/50:87:89:4b:bb:42  Cost = 1  Port = 0x8201
2017-05-21 23:48:30.426690 38:ed:18:a2:27:b0 -> Spanning-tree-(for-bridges)_00
 STP 60 RST. Root = 4096/1/50:87:89:4b:bb:42  Cost = 1  Port = 0x8201
2017-05-21 23:48:30.426714 38:ed:18:a2:17:a6 -> Spanning-tree-(for-bridges)_00
! Output omitted for brevity
```

```
! Detailed output of ethanalyzer
N7k-Admin# ethanalyzer local read bootflash:stp.pcap detail
Frame 1: 60 bytes on wire (480 bits), 60 bytes captured (480 bits)
    Encapsulation type: Ethernet (1)
    Arrival Time: May 21, 2017 23:48:30.216952000 UTC
    [Time shift for this packet: 0.000000000 seconds]
    Epoch Time: 1495410510.216952000 seconds
    [Time delta from previous captured frame: 0.000000000 seconds]
    [Time delta from previous displayed frame: 0.000000000 seconds]
    [Time since reference or first frame: 0.000000000 seconds]
    Frame Number: 1
    Frame Length: 60 bytes (480 bits)
    Capture Length: 60 bytes (480 bits)
    [Frame is marked: False]
    [Frame is ignored: False]
    [Protocols in frame: eth:llc:stp]
IEEE 802.3 Ethernet
    Destination: Spanning-tree-(for-bridges)_00 (01:80:c2:00:00:00)
        Address: Spanning-tree-(for-bridges)_00 (01:80:c2:00:00:00)
        .... ..0. .... .... .... .... = LG bit: Globally unique address (factory
  default)
        .... ...1 .... .... .... .... = IG bit: Group address (multicast/broadcast)
    Source: 5c:fc:66:6c:f3:f6 (5c:fc:66:6c:f3:f6)
        Address: 5c:fc:66:6c:f3:f6 (5c:fc:66:6c:f3:f6)
        .... ..0. .... .... .... .... = LG bit: Globally unique address (factory
  default)
        .... ...0 .... .... .... .... = IG bit: Individual address (unicast)
    Length: 39
    Padding: 00000000000000
Logical-Link Control
    DSAP: Spanning Tree BPDU (0x42)
    IG Bit: Individual
    SSAP: Spanning Tree BPDU (0x42)
    CR Bit: Command
    Control field: U, func=UI (0x03)
        000. 00.. = Command: Unnumbered Information (0x00)
        .... ..11 = Frame type: Unnumbered frame (0x03)
Spanning Tree Protocol
    Protocol Identifier: Spanning Tree Protocol (0x0000)
    Protocol Version Identifier: Rapid Spanning Tree (2)
    BPDU Type: Rapid/Multiple Spanning Tree (0x02)
    BPDU flags: 0x3c (Forwarding, Learning, Port Role: Designated)
        0... .... = Topology Change Acknowledgment: No
        .0.. .... = Agreement: No
        ..1. .... = Forwarding: Yes
        ...1 .... = Learning: Yes
        .... 11.. = Port Role: Designated (3)
        .... ..0. = Proposal: No
        .... ...0 = Topology Change: No
    Root Identifier: 4096 / 1 / 50:87:89:4b:bb:42
        Root Bridge Priority: 4096
        Root Bridge System ID Extension: 1
        Root Bridge System ID: 50:87:89:4b:bb:42 (50:87:89:4b:bb:42)
    Root Path Cost: 0
    Bridge Identifier: 4096 / 1 / 50:87:89:4b:bb:42
        Bridge Priority: 4096
```

```
                Bridge System ID Extension: 1
                Bridge System ID: 50:87:89:4b:bb:42 (50:87:89:4b:bb:42)
        Port identifier: 0x9000
        Message Age: 0
        Max Age: 20
        Hello Time: 2
        Forward Delay: 15
        Version 1 Length: 0
! Output omitted for brevity
```

可以将保存的.pcap 文件通过 FTP（File Transfer Protocol，文件传输协议）、TFTP（Trivial File Transfer Protocol，简单文件传输协议）、SCP（Secure Copy Protocol，安全复制协议）、SFTP（Secure FTP，安全 FTP）和 USB（Universal Serial Bus，通用串行总线）等方式传输到远程服务器，此后可以通过 Wireshark 等数据包分析工具进行详细分析。

> **注**：如果 Nexus 7000 部署了多个 VDC，那么 Ethanalyzer 仅运行在管理 VDC 或默认 VDC 上。此外，从版本 7.2 开始，可以在 Nexus 7000 上通过选项以 VDC 为基础进行过滤。

2.2.2 数据包跟踪器

排查网络故障时，通常很难理解系统对特定数据包或数据流采取的操作，此时可以考虑使用数据包跟踪器（Packet Tracer）功能。NX-OS 从版本 7.0(3)I2(2a)开始，在 Nexus 9000 交换机上引入了数据包跟踪器，可以在网络出现间歇性丢包或全面丢包时使用该工具。

> **注**：截至本书写作之时，只有基于 Broadcom Trident II ASIC 的线卡或交换矩阵模块才支持数据包跟踪器。有关思科 Nexus 9000 ASIC 的详细信息，请参阅思科官网。

数据包跟踪器的配置包括以下两个简单步骤。
- **第 1 步**：定义过滤器。
- **第 2 步**：启动数据包跟踪器。

设置数据包跟踪器时，需要使用命令 **test packet-tracer** [**src-ip** *src-ip* | **dst-ip** *dst-ip*] [**protocol** *protocol-num* | **l4-src-port** *src-port* | **l4-dst-port** *dst-port*]，然后再通过命令 **test packet-tracer start** 启动数据包跟踪器。如果要查看指定流量的统计信息及其操作情况，可以使用命令 **test packet-tracer show**。最后，可以通过命令 **test packet-tracer stop** 停止数据包跟踪器。例 2-11 给出了通过数据包跟踪器分析两台主机之间的 ICMP 统计信息的配置方式。

例 2-11 配置和验证数据包跟踪器

```
! Defining the Filter in Packet-Tracer
N9000-1# test packet-tracer src-ip 192.168.2.2 dst-ip 192.168.1.1 protocol 1

! Starting the Packet-Tracer
N9000-1# test packet-tracer start

! Verifying the statistics
N9000-1# test packet-tracer show

 Packet-tracer stats
 --------------------

Module 1:
Filter 1 installed: src-ip 192.168.2.2 dst-ip 192.168.1.1 protocol 1
ASIC instance 0:
```

```
Entry 0: id = 9473, count = 120, active, fp,
Entry 1: id = 9474, count = 0, active, hg,
Filter 2 uninstalled:
Filter 3 uninstalled:
Filter 4 uninstalled:
Filter 5 uninstalled:
! Second iteration of the Output
N9000-1# test packet-tracer show

 Packet-tracer stats
 --------------------

Module 1:
Filter 1 installed: src-ip 192.168.2.2 dst-ip 192.168.1.1 protocol 1
ASIC instance 0:
Entry 0: id = 9473, count = 181, active, fp,
Entry 1: id = 9474, count = 0, active, hg,
Filter 2 uninstalled:
Filter 3 uninstalled:
Filter 4 uninstalled:
Filter 5 uninstalled:
! Stopping the Packet-Tracer
N9000-1# test packet-tracer stop
```

即使入站流量被 ACL 丢弃了，也能通过数据包跟踪器来确定这些数据包是否到达了路由器的入站接口。如果要删除数据包跟踪器中的所有过滤器，那么可以使用命令 **test packet-tracer remove-all**。

2.3 NetFlow

NetFlow 是思科的一项重要功能特性，可以在 IP 流量进入或离开接口时采集流量的统计信息和详细信息。NetFlow 为运营商提供了网络和安全监控、网络规划、流量分析和 IP 记账等功能。网络流量通常是不对称的（即使是小型网络），而探针通常要求部署为对称方式。NetFlow 不需要借助仪表就能优化网络，可以通过流量的自然路径来跟踪流量在网络中的传输情况。除流量速率之外，NetFlow 还能在网络中的每个位置为特定应用、服务和流量流提供 QoS 标记、TCP 标志等信息。NetFlow 能够帮助验证拓扑结构中任意位置的流量工程或策略实施情况。

虽然思科 NX-OS 支持传统 NetFlow（版本 5）和 Flexible NetFlow（版本 9）导出格式，但建议在 Nexus 平台上使用 Flexible NetFlow。因为传统 NetFlow 导出的所有关键字和字段都是固定的，而且仅支持 IPv4 流。系统默认采用下列 7 个关键字来定义流。

- 源 IP 地址。
- 目的 IP 地址。
- 源端口。
- 目的端口。
- 三层协议类型。
- TOS 字节（DSCP 标记）。
- 输入逻辑接口（ifindex）。

虽然用户也可以选择其他字段，但 NetFlow 版本 5 在所能提供的细节信息上有所限制。FNF（Flexible NetFlow，灵活 NetFlow）在 NetFlow 版本 9 的基础上实现了标准化，允许用户更加灵活地定义流，而且还能为每个流灵活地定义导出字段。Flexible NetFlow 支持 IPv6 和二层 NetFlow 记录。由于 NetFlow 版本 9 基于模板，因而用户可以指定所要导出的数据。

FNF 的优点如下。
- 灵活选择流的定义（关键字和非关键字字段）。
- 灵活选择为不同的接口应用不同的流定义。
- 灵活选择导出器接口。
- 未来扩展具有灵活性，如 IPFIX。

网络运营商和架构师常常希望知道应该在何处连接 NetFlow 监控器。对于这类需求来说，可以考虑以下问题。
- 用户希望了解什么类型的信息？MAC 字段或 IPv4/v6 字段？
- 设备有哪些接口？三层或二层接口？
- 设备在 VLAN 内交换数据包还是通过 SVI（Switched Virtual Interface，交换式虚接口）在 VLAN 间路由数据包？

2.3.1 NetFlow 配置

上述问题可以帮助用户正确选择三层或二层 NetFlow 配置。Nexus 交换机的 NetFlow 配置步骤如下。
- 第 1 步：启用 NetFlow 功能特性。
- 第 2 步：通过指定的关键字和非关键字字段来定义流记录。
- 第 3 步：通过指定导出格式、协议、目的端及其他参数来定义一个或多个流导出器（Flow Exporter）。
- 第 4 步：根据前面的流记录和流导出器定义流监控器（Flow Monitor）。
- 第 5 步：通过指定的采样方法将流监控器应用到接口上。

1. 启用 NetFlow 功能特性

可以在 NX-OS 中通过 **feature netflow** 命令启用 NetFlow 功能特性。启用 NetFlow 功能特性之后，就可以使用与 NetFlow 相关的所有 CLI。

> 注：NetFlow 需要使用 TCAM 和 CPU 等硬件资源，因而在启用 NetFlow 之前，应了解设备的资源利用情况。

2. 定义流记录

可以通过指定 NetFlow 标识数据包所用的各种匹配关键字和参数来定义流记录，创建流记录的命令是 **flow record** *name*。NX-OS 在创建流记录时默认启用以下匹配字段。
- **match interface input**
- **match interface output**
- **match flow direction**

此外，还可以在流记录中指定必须采集的感兴趣的流字段。NetFlow 支持以下匹配关键字来标识 NetFlow 中的流。
- IPv4 源/目的地址。
- IPv6 源/目的地址。
- IPv6 流标签。
- IPv6 选项。
- ToS 字段。
- L4 协议。
- L4 源/目的端口。

二层 NetFlow 提供了以下匹配关键字。
- 源/目的 MAC 地址。
- Ethertype。
- VLAN。

用户可以在 NetFlow 版本 5 和版本 9 中灵活选择可用的采集参数（但 IPv6 参数除外，因为仅版本 9 支持 IPv6 参数）。

- L3 字节数（32 比特或 64 比特）。
- 数据包数（32 比特或 64 比特）。
- 流的方向。
- 流的采样器 ID（Sampler ID）。
- 接口信息（输入和/或输出）。
- 本地设备或对等体的源/目的 AS 号。
- 下一跳 IPv4/IPv6 地址。
- 第一个或最后一个数据包的系统正常运行时间。
- TCP 标志。

例 2-12 给出了三层和二层流量的流记录配置示例，本例中的流记录创建了多个匹配项及采集参数。

例 2-12　NetFlow 流记录

```
! Flow Record for Layer 3 Traffic
flow record FR_V4
  match ipv4 source address
  match ipv4 destination address
  match ip protocol
  match ip tos
  collect timestamp sys-uptime last
  collect flow sampler id
  collect ip version
! Flow record for Layer 2 Traffic
flow record FR_L2
  match datalink mac source-address
  match datalink mac destination-address
  match datalink vlan
  match datalink ethertype
  collect counter packets
  collect flow sampler id
```

3. 定义流导出器

现在定义流导出器。NetFlow 通过 UDP 帧将 NetFlow 数据导出给远程采集器。流超时时间（用户可配置）到期后，就会定期导出 NetFlow 数据。默认的流超时值为 30min。定义流导出器时应定义以下字段。

- 采集器 IPv4/IPv6 地址。
- 源接口。
- VRF（Virtual Routing and Forwarding，虚拟路由和转发）。
- 版本。
- UDP 端口号。

例 2-13 给出了流导出器的配置示例。

例 2-13 NetFlow 流导出器

```
flow exporter FL_Exp
  destination 100.1.1.1 use-vrf management
  transport udp 3000
  source mgmt0
  version 9
```

4. 定义和应用流监控器

定义了流导出器之后，需要将流记录和流导出器绑定到流监控器上。定义了流监控器之后，就可以将流监控器连接到接口上，采集 NetFlow 统计信息。例 2-14 给出了流监控器配置示例，并说明了如何在接口下启用 NetFlow。

例 2-14 NetFlow 流监控器和接口配置

```
flow monitor FL_MON
  record FR_V4
  exporter FL_Exp
!
interface Eth3/31-32
ip flow monitor FL_MON input
ip flow monitor FL_MON output
```

如果要为 IPv4/IPv6/二层流量应用 NetFlow，那么就可以使用命令 **[ip | ipv6 | layer2- switched] flow monitor** *name* **[input | output]**。

可以通过 **show run netflow** 命令查看 NetFlow 配置，通过 **show flow [record** *record-name* | **exporter** *exporter-name* | **monitor** *monitor-name***]** 命令验证 NetFlow 配置。

如果要查看前面示例中配置的流入流出接口 E1/4 的流统计信息，那么就可以使用命令 **show hardware flow [ip | ipv6] [detail]**。例 2-15 显示了穿越接口 Eth3/31-32 的流入流出流量的统计信息，本例显示了入站（I）和出站（O）流量。NetFlow 显示了 OSPF 及其他 ICMP 流量统计信息，以及相应的协议号和数据包数量。

例 2-15 NetFlow 统计信息

```
N7k-1# show hardware flow ip
slot  3
=======
D - Direction; L4 Info - Protocol:Source Port:Destination Port
IF - Interface: ()ethernet, (S)vi, (V)lan, (P)ortchannel, (T)unnel
TCP Flags: Ack, Flush, Push, Reset, Syn, Urgent


D IF       SrcAddr           DstAddr           L4 Info           PktCnt       TCPFlags
-+---------+-----------------+-----------------+-----------------+------------+--------
I 3/31     010.012.001.002   224.000.000.005   089:00000:00000   0000000159   ......
I 3/32     010.013.001.003   224.000.000.005   089:00000:00000   0000000128   ......
I 3/32     003.003.003.003   002.002.002.002   001:00000:00000   0000000100   ......
I 3/31     002.002.002.002   003.003.003.003   001:00000:00000   0000000100   ......
O 3/31     003.003.003.003   002.002.002.002   001:00000:00000   0000000100   ......
O 3/32     002.002.002.002   003.003.003.003   001:00000:00000   0000000100   ......
```

例 2-15 的 NetFlow 统计信息来自 N7k 平台，该平台支持基于硬件的流匹配功能。但是，并非所有 Nexus 平台均支持基于硬件的流匹配功能，如 Nexus 6000 等 Nexus 交换机就不支持基于硬件的流匹配功能，而是必须执行基于软件的流匹配操作，这样可能会消耗系统资源并影响平台性能。因此，这类平台仅支持采样式 NetFlow（Sampled NetFlow）。

注：Nexus 5600 和 Nexus 6000 仅支持接口的入站 NetFlow，Nexus 7000 同时支持入站和出站 NetFlow 统计信息采集。

2.3.2 NetFlow 采样

NetFlow 支持数据点采样，从而有效减少数据采集量。通常将这种 NetFlow 称为 SNF（Sampled NetFlow，采样 NetFlow）。SNF 支持 *M:N* 数据包采样，即仅采样 *N* 个数据包中的 *M* 个数据包。

采集器的配置命令是 **sampler** *name*。在采样器配置下，可以通过命令 **mode** *sample-number* **out-of** *packet-number* 定义采样器模式，其中，*sample-number* 的范围是 1~64，*packet-number* 的范围是 1~65536。定义了采样器之后，就可以在接口配置下与流监控器联合使用，如例 2-16 所示。

例 2-16　NetFlow 采样器和接口配置

```
sampler NF-SAMPLER1
 mode 1 out-of 1000
!
interface Eth3/31-32
ip flow monitor FL_MON input sampler NF-SAMPLER1
```

此外，用户还可以通过 **flow timeout [active | inactive]** *time-in-seconds* 命令为流定义活动定时器和非活动定时器。

NX-OS 从版本 7.3(0)D1(1) 开始，CoPP（Control Plane Policing，控制平面策略）接口也支持 NetFlow。CoPP 接口上的 NetFlow 功能允许用户监控和采集去往交换机管理引擎模块的各种数据包的统计信息。NX-OS 允许 IPv4 流监控器和采样器在输出方向上连接控制平面接口。例 2-17 给出了 CoPP 接口的 NetFlow 配置以及 Nexus 7000 平台的 NetFlow 统计信息。

例 2-17　配置和验证 CoPP NetFlow

```
Control-plane
ip flow monitor FL_MON output sampler NF-SAMPLER1
```

注：如果 NetFlow 出现了问题，那么就可以在故障状态下采集 **show tech-support netflow** 命令的输出结果。

2.3.3 sFlow

sFlow 定义在 RFC 3176 中，是一种通过采样机制监控流量的技术，该采样机制在包含交换机和路由器的数据网络中是 sFlow 代理的一部分。sFlow 代理是 Nexus 9000 和 Nexus 3000 平台提供的一种新的软件功能特性，这些平台上的 sFlow 代理负责采集入站和出站端口上的采样数据包，并转发给集中式采集器（称为 sFlow 分析器）。sFlow 代理可以周期性地采样或轮询与被采样数据包的数据源相关联的计数器。

在接口上启用 sFlow 之后，将同时启用入站和出站方向采集功能。只能为以太网和端口通道接口配置 sFlow。可以通过配置命令 **feature sflow** 启用 sFlow，同时还可以在配置命令中定义各种参数，如表 2-2 所示。

表 2-2　　　　　　　　　　　　　　　　sFlow 参数

sFlow 参数配置	描述
sflow sampling rate *rate*	数据包采样速率，默认值为 4096，值为 0 时表示禁用采样功能
sflow max-sampled-size *sampling-size*	数据包的最大采样尺寸，默认值为 128 字节，取值范围是 64~256 字节
sflow counter-poll-interval *poll-interval*	接口的轮询间隔，默认值为 20s

sFlow 参数配置	描述
sflow max-datagram-size *size*	数据报的最大尺寸，默认值为 1400 字节
sflow collector-ip *ip-address* vrf *vrf-context*	sFlow 采集器/分析器的 IP 地址
sflow collector-port *port-number*	sFlow 分析器的 UDP 端口号
sflow agent-ip *ip-address*	sFlow 代理的地址，是交换机上的本地有效 IP 地址
sflow data-source interface *interfacetype interface-num*	sFlow 采样数据源

例 2-18 给出了 Nexus 3000 交换机的 sFlow 配置示例，通过 **show run sflow** 命令可以查看 sFlow 的运行配置。

例 2-18 sFlow 运行配置

```
feature sflow
sflow sampling-rate 1000
sflow max-sampled-size 200
sflow counter-poll-interval 100
sflow max-datagram-size 2000
sflow collector-ip 172.16.1.100 vrf management
sflow collector-port 2020
sflow agent-ip 170.16.1.130
sflow data-source interface ethernet 1/1-2
```

如果要验证 sFlow 配置，则可以使用 **show sflow** 命令，该命令可以显示 sFlow 的所有配置信息，如例 2-19 所示。

例 2-19 show sflow 命令输出结果

```
N3K-1# show sflow
sflow sampling-rate : 1000
sflow max-sampled-size : 200
sflow counter-poll-interval : 100
sflow max-datagram-size : 2000
sflow collector-ip : 172.16.1.100 , vrf : management
sflow collector-port : 2020
sflow agent-ip : 172.16.1.130
sflow data-source interface Ethernet1/1
sflow data-source interface Ethernet1/2
```

配置了 sFlow 后，sFlow 代理就可以开始采集统计信息。虽然需要通过 sFlow 采集器工具查看实际的流量，但是仍然可以使用 **show sflow statistics** 命令查看交换机的 sFlow 统计信息，而且还可以使用命令 **show system internal sflow info** 查看与 sFlow 相关的内部信息和统计信息。例 2-20 显示了 sFlow 的统计信息。请注意，虽然数据包的总数很多，但采样数据包的数量却非常少，这是因为 sFlow 在配置中定义了采样比例，即每 1000 个数据包采样 1 次。此外，sFlow 的系统内部命令还显示了资源利用率以及当前状态等信息。

例 2-20 sFlow 统计信息和内部信息

```
N3K-1# show sflow statistics
Total Packets       : 1053973
Total Samples       : 11
Processed Samples   : 11
Dropped Samples     : 0
Sent Datagrams      : 56
```

```
Dropped Datagrams         : 13
N3K-1# show system internal sflow info
sflow probe state -> RUN
sflow inband sflow is valid
sflow inband driver -> UP
sflow IPv4 socket fd 47
number of sflow sampled packets : 11
number of sflow dropped packets : 0
number of sflow datagram sent : 56
number of sflow datagram dropped : 13
sflow process cpu usage 0.86
sflow process cpu limit 50
```

注：如果 sFlow 出现了问题，那么就可以在故障状态下采集 **show tech-support sflow** 命令的输出结果。

2.4 网络时间协议

排查网络故障时，网络操作人员通常会利用日志来了解故障症状和故障时间信息，从而寻找触发故障的原因。如果设备上的时间不同步，那么就很难在多台设备之间关联事件。为了解决这个问题，建议将 Nexus 设备的时钟与整个网络的 NTP（Network Time Protocol，网络时间协议）进行同步。NTP 的作用是同步网络中的设备时钟，使用预定义的 UDP 端口号 123。

在设备上配置 NTP 时，可以使用 **feature ntp** 命令启用 NTP。Nexus 设备根据其在网络中的角色（NTP 客户端或 NTP 服务器）来配置 NTP，NTP 服务器负责从时间源（时间源连接在时间服务器上）接收时间并通过网络分配时间。与距离矢量协议一样，NTP 也使用层级（Stratum）值来描述网络设备与权威时间源之间的距离。

可以将 NX-OS 设备配置为 NTP 服务器，也可以配置为 NTP 对等体。网络工程师可以通过 NTP 对等体指定其他主机，在 NTP 服务器出现故障时进行时间同步。可以通过 **ntp [server | peer]** [*ip-address* | *ipv6-address*] [**prefer**] [**use-vrf** *vrf-name*]命令定义 NTP 服务器或 NTP 对等体，通过 **show ntp peers** 命令查看已配置的 NTP 服务器和 NTP 对等体。

除此以外，用户还可以通过 **ntp master** 命令将 Nexus 交换机配置为权威时间服务器，而且 NX-OS 还允许用户在交换机上指定 NTP 数据包的源接口或 IP 地址，可以通过 **ntp source** *ip-address* 和 **ntp source-interface** *interface-id* 命令定义 NTP 数据包的源端，如果指定了源接口或 IP 地址，那么必须确保能够通过指定的 IP 地址访问 NTP 服务器。这些配置对于配置了多个 VDC 的 Nexus 7000 交换机来说非常有效，如果 Nexus 7000 交换机配置了多个 VDC，那么将在默认 VDC 中同步其硬件时钟，非默认 VDC 可以从默认 VDC 获取时间更新。在 VDC 中配置了 **ntp master** 命令之后，非默认 VDC 也可以充当网络中其他客户端的时间服务器。例 2-21 给出了 Nexus 设备作为 NTP 服务器和客户端的 NTP 配置示例。

例 2-21 NTP 配置

```
! NTP Server Configuration
ntp peer 172.16.1.11 use-vrf management
ntp source-interface mgmt0
ntp master 8
! NTP Client Configuration
ntp server 172.16.1.10 prefer use-vrf management
ntp server 172.16.1.11 use-vrf management
ntp source-interface  mgmt0
```

配置了 NTP 之后，客户端将自动通过 NTP 与服务器进行时间同步。如果要检查 NTP 服务器或对等体的状态，可以使用 **show ntp peer-status** 命令。对等体地址旁边的"*"表示 NTP 已与服务器同步。例 2-22 显示了 NTP 服务器和客户端的输出结果，从 NTP 服务器的输出结果可以看出，对等体地址为 127.127.1.0，意味着该设备就是 NTP 服务器。从 NTP 客户端的输出结果可以看出，"*"位于 172.16.1.10 的旁边，该地址被配置为首选 NTP 服务器。请注意，例中的所有设备都属于同一管理子网。

例 2-22　NTP 配置验证

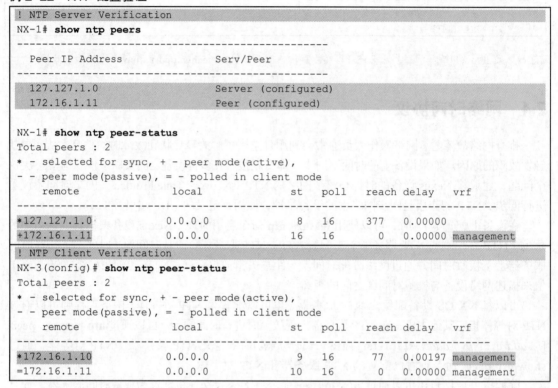

完成了 NTP 同步之后，可以通过 **show clock** 命令验证设备的时间信息。

此外，NX-OS 还内置了一项名为 CFS（Cisco Fabric Service，思科交换矩阵服务）的专有功能特性，可以向所有 Nexus 设备分发数据和配置变更信息。CFS 可以在网络中向所有 Nexus 设备分发全部本地 NTP 配置。启动 NTP 配置之后，CFS 可以在整个网络范围内锁定 NTP，出现配置变更之后，用户可以丢弃或提交变更，而且提交的配置信息可以复制到所有的 Nexus 设备上。使用 **ntp distribute** 命令可以为 NTP 启用 CFS 功能。使用 **ntp commit** 命令可以将配置提交给所有 Nexus 设备，使用 **ntp abort** 命令则可以终止该操作。无论执行哪一条命令，CFS 都会释放对网络设备的 NTP 锁定。如果要检查是否为 NTP 启用了交换矩阵分发服务，则可以使用 **show ntp status** 命令。

NX-OS 支持通过 CLI 来验证 NTP 数据包的统计信息。用户可以查看 NTP 数据包的输入—输出统计信息、NTP 维护的本地计数器、与内存相关的 NTP 计数器（在排查 NTP 进程引发的内存泄露故障时尤为有用）以及每个对等体的 NTP 统计信息。如果由某种原因导致 NTP 数据包丢失，那么就可以通过 CLI 查看这些统计信息。如果要查看这些统计信息，可以使用 **show ntp statistics [io | local | memory | peer ipaddr** *ip-address*] 命令。例 2-23 给出了 NTP 数据包的 I/O 和本地统计信息，如果收到了错误的 NTP 数据包或错误的认证请求，那么就可以在本地统计信息中查看这些计数器。

例 2-23 NTP 统计信息

```
NX-1# show ntp statistics io
  time since reset:       91281
  receive buffers:        10
  free receive buffers:   9
  used receive buffers:   0
  low water refills:      1
  dropped packets:        0
  ignored packets:        0
  received packets:       9342
  packets sent:           9369
  packets not sent:       0
  interrupts handled:     9342
  received by int:        9342
NX-1# show ntp statistics local
  system uptime:          91294
  time since reset:       91294
  old version packets:    9195
  new version packets:    0
  unknown version number: 0
  bad packet format:      0
  packets processed:      4
  bad authentication:     0
```

2.5 嵌入式事件管理器

EEM（Embedded Event Manager，嵌入式事件管理器）是集成在 NX-OS 中的一种强大的设备和系统管理技术。EEM 可以帮助客户更好地利用思科软件提供的网络智能，从而能够根据网络事件的发生情况自定义操作行为。EEM 是一种事件驱动型工具，支持多种类型的触发输入，允许用户自定义响应操作，包括抓取各种 **show** 命令输出结果或者在事件触发后执行 TCL（Tool Command Language，工具命令语言）或 Python 脚本。

EEM 包括以下两个主要组件。

- 事件（Event）：定义其他 NX-OS 组件监控到的事件。
- 操作（Action）：定义事件触发后要采取的操作。

其他的 EEM 组件还有 EEM 策略，EEM 策略与一个或多个操作相匹配，以帮助排查网络故障或者从事件中恢复正常。系统定义的某些策略会查找特定的系统级事件（如线卡重新加载或管理引擎切换），并根据这些事件执行预定义操作。**show event manager system-policy** 命令可以查看这些系统级策略，而且这些策略都可以覆盖，可以通过该命令验证策略覆盖情况。系统策略可以避免对设备或网络造成更大的影响，例如，如果某个模块出现了故障且崩溃，那么就可能会给服务造成严重影响并导致中断，此时，设置 N 次崩溃后直接关闭故障模块的系统策略就能有效降低故障影响。

例 2-24 列出了一些常见的系统策略事件以及相应的事件响应操作，**show event manager policy-state** *system-policy-name* 命令可以检查事件发生的次数。

例 2-24 EEM 系统策略

```
NX-1# show event manager system-policy
         Name : __lcm_module_failure
  Description : Power-cycle 2 times then power-down
  Overridable : Yes
```

```
              Name : __pfm_fanabsent_any_singlefan
       Description : Shutdown if any fanabsent for 5 minute(s)
       Overridable : Yes

              Name : __pfm_fanbad_any_singlefan
       Description : Syslog when fan goes bad
       Overridable : Yes

              Name : __pfm_power_over_budget
       Description : Syslog warning for insufficient power overbudget
       Overridable : Yes

              Name : __pfm_tempev_major
       Description : TempSensor Major Threshold. Action: Shutdown
       Overridable : Yes

              Name : __pfm_tempev_minor
       Description : TempSensor Minor Threshold. Action: Syslog.
       Overridable : Yes
NX-1# show event manager policy-state __lcm_module_failure
Policy __lcm_module_failure
  Cfg count :    3
    Hash         Count       Policy will trigger if
-----------------------------------------------------------------
    default        0         3 more event(s) occur
```

EEM 中的事件可以是系统事件，也可以是用户触发的事件（如配置变更）。操作定义的则是事件发生后应触发的解决措施或通知，EEM 支持以下操作（在 **action** 语句中定义）。

- 执行 CLI 命令（配置命令或 **show** 命令）。
- 更新计数器。
- 记录异常事件。
- 重新加载设备。
- 打印系统日志消息。
- 发送 SNMP 通告。
- 设置系统策略的默认操作策略。
- 执行 TCL 或 Python 脚本。

例如，路由器出现高 CPU 利用率时执行指定操作，或者 BGP 会话出现振荡时记录日志。例 2-25 给出了 Nexus 平台的 EEM 配置示例，该 EEM 将事件触发条件设置为高 CPU 利用率（如 70%或更高的 CPU 利用率），采取的操作则是在发生高 CPU 利用率时抓取一系列 BGP **show** 命令的输出结果。使用 **show event manager policy internal** *policy-name* 命令可以查看策略信息。

例 2-25　EEM 配置和验证

```
event manager applet HIGH-CPU
event snmp oid 1.3.6.1.4.1.9.9.109.1.1.1.1.6.1 get-type exact entry-op ge
    entry-val 70 exit-val 30 poll-interval 1
action 1.0 syslog msg High CPU hit $_event_pub_time
action 2.0 cli command enable
action 3.0 cli command "show clock >> bootflash:high-cpu.txt"
action 4.0 cli command "show processes cpu sort >> bootflash:high-cpu.txt"
action 5.0 cli command "show bgp vrf all all summary >> bootflash:high-cpu.txt"
action 6.0 cli command "show clock >> bootflash:high-cpu.txt"
```

2.5 嵌入式事件管理器

```
action 7.0 cli command "show bgp vrf all all summary >> bootflash:high-cpu.txt"
NX-1# show event  manager policy internal HIGH-CPU
                       Name : HIGH-CPU
                Policy Type : applet
  action 1.0 syslog msg "High CPU hit $_event_pub_time"
  action 1.1 cli command "enable"
  action 3.0 cli command "show clock >> bootflash:high-cpu.txt"
  action 4.0 cli command "show processes cpu sort >> bootflash:high-cpu.txt"
  action 5.0 cli command "show bgp vrf all all summary >> bootflash:high-cpu.txt"
  action 6.0 cli command "show clock >> bootflash:high-cpu.txt"
  action 7.0 cli command "show bgp vrf all all summary >> bootflash:high-cpu.txt"
```

在某些情况下，必须在事件触发后执行重复的配置或 **show** 命令。此外，使用外部脚本可能难以持续监控设备事件并在事件发生后触发脚本。对于这种情况来说，一种更好的解决方案是使用 NX-OS 提供的自动化脚本和工具。NX-OS 为 EEM 提供了内置的 TCL 和 Python 脚本功能，可以仅在事件触发后执行这些脚本。

下面考虑一个软件故障示例，假设交换机出现任意链路中断之后，都会禁用交换机上的所有 VLAN。例 2-26 给出了因链路中断而触发的 TCL 脚本示例，该 TCL 以扩展名.tcl 保存在 bootflash 上，该 TCL 文件将遍历所有 VLAN 数据库，并在 VLAN 配置模式下执行 **no shutdown** 命令。

例 2-26　EEM 使用 TCL 脚本

```
! Save the file in bootflash with the .tcl extension
set i 1
while {$i<10} {
cli configure terminal
cli vlan $i
cli no shutdown
cli exit
incr i
}

! EEM Configuration referencing TCL Script
event manager applet TCL
event cli match "shutdown"
 action 1.0 syslog msg "Triggering TCL Script on Module Failure Event"
 action 2.0 cli local tclsh EEM.tcl
```

与此相似，也可以在 EEM 脚本中引用 Python 脚本。Python 脚本也保存在 bootflash 中，扩展名为.py。例 2-27 给出了 Python 脚本以及在 EEM 脚本中引用 Python 脚本的配置示例，只要接口上的流量超过已配置的 storm-control 阈值，那么就会触发 EEM 脚本。对于本例来说，触发的 Python 脚本将收集多条命令的输出结果。

例 2-27　EEM 使用 Python 脚本

```
! Save the Python script in bootflash:
import re
import cisco
cisco.cli ("show module >> bootflash:EEM.txt")
cisco.cli ("show redundancy >> bootflash:EEM.txt")
cisco.cli ("show interface >> bootflash:EEM.txt")

! EEM Configuration referencing Python Script
event manager applet Py_EEM
event storm-control
```

```
action 1.0 syslog msg "Triggering TCL Script on Module Failure Event"
action 2.0 cli local python EEM.py
```

> **注：**有关在各种思科操作系统上配置 EEM 的详细信息，请查阅思科官网上的 CCO 文档。如果 EEM 出现了操作行为故障，那么就可以采集设备的 **show tech-support eem** 命令输出结果。

2.6 日志记录

如果设备提供不了任何信息，那么就很难分析和排查网络故障。例如，如果 OSPF 邻接关系中断且不存在任何相关告警，那么就很难确定故障的发生时间和发生原因。因此，日志记录非常重要，所有的思科路由器和交换机都支持日志记录功能，也可以为特定的功能特性和协议应用日志记录功能，如可以为 BGP 会话状态或 OSPF 邻接关系状态变更启用日志记录功能。

表 2-3 列出了可配置的各种日志记录级别。

表 2-3　　　　　　　　　　　日志记录级别

级别号	级别名称
0	紧急（Emergency）
1	警报（Alert）
2	关键（Critical）
3	错误（Error）
4	告警（Warning）
5	通告（Notification）
6	信息性（Informational）
7	调试（Debugging）

如果设置了较高的日志记录级别值，那么默认将启用所有比该设置值低的日志记录级别。例如，如果将日志记录级别设置为 5（通告），那么就会记录 0～5（紧急到通告）所有级别的事件。从故障排查角度来说，将日志记录级别设置为 7（调试）是一种非常好的做法。

思科设备提供了多种日志记录选项。

- 控制台日志记录。
- 缓冲区日志记录。
- 记录到 syslog 服务器。

如果设备出现崩溃或高 CPU 利用率状况，且无法通过 Telnet 或 SSH（Secure Shell，安全外壳）访问终端会话，那么控制台日志记录就显得极为重要。不过，在执行调试操作期间启用控制台日志记录并不是一种好的做法，因为某些调试输出非常繁杂，可能会给设备控制台带来巨大压力。根据最佳实践，建议在调试操作期间始终禁用控制台日志记录。例 2-28 给出了在 Nexus 平台上启用控制台日志记录的配置示例。

例 2-28　配置控制台日志记录

```
NX-1(config)# logging console ?
  <CR>
  <0-7>  0-emerg;1-alert;2-crit;3-err;4-warn;5-notif;6-inform;7-debug
NX-1(config)# logging console 6
```

NX-OS 不但提供了强大的日志记录功能，而且能在重新加载后保持日志记录。所有缓存的日志记录都永久保存在/var/log/external/目录中。如果要查看内部目录，可以使用 **show system internal flash** 命令，该命令可以列出闪存中的所有内部目录及其使用情况。**show logging log** 命令可以查看缓存的日志消息。

例 2-29 显示了闪存中的目录信息以及/var/log/external/目录中的内容。如果 **show logging log** 命令未显示输出结果或日志记录已停止，那么就应该检查/var/log/目录以确保该目录有足够的可用空间。

例 2-29　内部闪存目录

```
NX-1# show system internal flash
Mount-on                      1K-blocks         Used    Available     Use%   Filesystem
/                                409600        69476       340124       17   /dev/root
/proc                                 0            0            0        0   proc
/sys                                  0            0            0        0   none
/debugfs                              0            0            0        0   nodev
/cgroup                               0            0            0        0   vdccontrol
/isan                            716800       519548       197252       73   none
/etc                               5120         1632         3488       32   none
/nxos/tmp                         20480         1536        18944        8   none
/var/log                          51200          108        51092        1   none
/var/home                          5120            0         5120        0   none
/var/tmp                         307200          460       306740        1   none
/var/sysmgr                     1048576          144      1048432        1   none
/var/sysmgr/ftp                  409600           80       409520        1   none
/dev/shm                        1048576       353832       694744       34   none
/volatile                        204800            0       204800        0   none
/debug                             2048           28         2020        2   none
/dev/mqueue                           0            0            0        0   none
/mnt/cfg/0                       325029        12351       295897        5   /dev/sda5
/mnt/cfg/1                       325029        12349       295899        5   /dev/sda6
/mnt/cdrom                          350          350            0      100   /dev/scd0
/var/sysmgr/startup-cfg           40960         4192        36768       11   none
/dev/pts                              0            0            0        0   devpts
/mnt/pss                         325061         8898       299380        3   /dev/sda3
/bootflash                      3134728       202048      2773444        7   /dev/sda4
/smack                                0            0            0        0   smackfs
NX-1# show system internal dir /var/log/external/
                                                        ./         240
                                                       ../         300
                                              libfipf.5834           0
                                              l2fm_ut.txt         774
                                              plcmgr.dbg           21
                                                 snmp_log         180
                                              libfipf.3884           0
                                              libfipf.3855           0
                                         syslogd_ha_debug       11221
                                                 messages       25153
                                             startupdebug        3710
                                                   dmesg@          31
```

此外，还可以为不同的 NX-OS 组件定义不同的日志记录级别，从而允许用户更好地控制复杂组件的日志记录，或者为较为简单或不太重要的组件禁用某些日志消息。具体配置方式是使用 **logging level** *component-name level* 命令来设置组件的日志记录级别。例 2-30 将 ARP 和 Ethpm 组件的日志记录级别设置为 3，目的是减少不必要的日志消息。

例 2-30 NX-OS 组件的日志记录级别

```
NX-1(config)# logging level arp 3
NX-1(config)# logging level ethpm 3
```

最持久的日志记录形式是使用 syslog 服务器来记录所有设备日志。syslog 服务器可以是文本文件，也可以是自定义的应用程序，能够主动将设备日志信息存储到数据库中。

例 2-31 给出了 syslog 日志记录配置示例。在 NX-OS 上配置基于 syslog 的日志记录之前，必须先为日志消息启用 **logging timestamp [microseconds | milliseconds | seconds]** 命令，以确保所有日志消息都带有时间戳，从而能够更好地分析日志消息。一般情况下，会为管理接口配置一个管理 VRF，此时需要在 NX-OS 中通过 **logging server** *ip-address* **use-vrf** *vrf-name* 命令指定 syslog 主机，以便路由器知道能够通过哪个 VRF 路由表到达 syslog 服务器。如果未指定 VRF 选项，那么系统将在默认 VRF（全局路由表）中执行查找操作。

例 2-31 syslog 日志记录配置

```
NX-1(config)# logging timestamp milliseconds
NX-1(config)# logging server 10.1.1.100 7 use-vrf management
```

2.6.1 调试日志文件

NX-OS 允许用户将调试输出重定向到文件中，该选项在运行调试操作并将调试输出与常规日志消息相隔离的时候非常有用，配置命令是 **debug logfile** *file-name size*。例 2-32 给出了通过 **debug logfile** 命令将调试输出抓取到日志文件中的配置示例，本例创建了一个名为 bgp_dbg 的调试日志文件，大小为 10000 字节。日志文件的大小范围是 4096～4194304 字节，所有的调试结果都记录在日志文件中。如果要进一步过滤调试输出以抓取更精确的调试输出结果，则可以使用 **debug-filter** 选项。下面的示例启用了 BGP 更新调试，并在 VRF 上下文 VPN_A 中过滤邻居 10.12.1.2 的更新调试日志。

例 2-32 在日志文件中抓取调试输出结果

```
NX-1# debug logfile bgp_dbg size 100000
NX-1# debug ip bgp updates
NX-1# debug-filter bgp neighbor 10.12.1.2
NX-1# debug-filter bgp vrf VPN_A
```

NX-OS 软件在 log:文件系统的根目录下创建日志文件，因而可以通过 **dir log:** 命令查看所有已创建的日志文件。创建了调试日志文件之后，就可以启用相应的调试，并将所有的调试输出结果重定向到调试日志文件。如果要查看日志文件的内容，则可以使用 **show debug logfile** *file-name* 命令。

2.6.2 记账日志

在故障排查期间，确定故障问题的触发事件是一件非常重要的工作（可能是常规的 **show** 命令或配置变更）。此时，检查故障期间的所有配置和 **show** 命令能够提供非常重要的信息。

NX-OS 将这些信息都记录到记账日志文件中，用户可以随时查阅该文件。使用命令 **show accounting log**，用户可以抓取系统已经执行和配置的所有命令，包括时间戳和用户信息。记账日志能够在系统重新加载后保持持久性。在默认情况下，记账日志仅抓取配置命令。如果希望同时抓取 **show** 命令和配置命令，那么就需要使用配置命令 **terminal log-all**。例 2-33 显示了记账日志的输出结果，例中突出显示了设备的配置变更情况。

> 注：记账日志和 show 命令的日志文件都存储在 logflash 中，并且可以在重新加载后访问。

例 2-33　记账日志

```
NX-1# show accounting log
Sun Apr  2 01:09:02 2017:type=update:id=vsh.12412:user=admin:cmd=configure terminal ;
  version 6.0(2)U6(9) (SUCCESS)
Sun Apr  2 01:09:03 2017:type=update:id=vsh.12412:user=admin:cmd=interface-vlan
  enable
Sun Apr  2 01:09:03 2017:type=update:id=vsh.12412:user=admin:cmd=configure terminal ;
  feature interface-vlan (SUCCESS)
Sun Apr  2 01:09:38 2017:type=update:id=vsh.12963:user=admin:cmd=configure terminal ;
  control-plane (SUCCESS)
Sun Apr  2 01:09:38 2017:type=update:id=vsh.12963:user=admin:cmd=configure terminal ;
  control-plane ; service-policy input copp-system-policy (SUCCESS
)
Sun Apr  2 01:09:38 2017:type=update:id=vsh.12963:user=admin:cmd=configure terminal ;
  hardware profile tcam region arpacl 128 (SUCCESS)
Sun Apr  2 01:09:38 2017:type=update:id=vsh.12963:user=admin:cmd=configure terminal ;
  hardware profile tcam region ifacl 256 (SUCCESS)
Sun Apr  2 01:09:38 2017:type=update:id=vsh.12963:user=admin:cmd=configure terminal ;
  ip ftp source-interface mgmt0 (SUCCESS)
! Output omitted for brevity
```

2.6.3　事件历史记录日志

NX-OS 可以持续记录系统中的硬件和软件组件发生的所有事件，并形成事件历史记录（event-history）日志。事件历史记录日志都是 VDC 本地化的，且基于每个组件进行维护。事件历史记录日志可以减少在实时生产环境中运行调试操作的需要，有助于排查服务中断故障（即使服务恢复了正常）。由于系统是在后台抓取每个组件的事件历史记录日志，因而执行该任务不会对 CPU 利用率造成任何影响。

可以将事件历史记录日志配置为以下 3 种大小。

- 大（Large）。
- 中（Medium）。
- 小（Small）。

可以从每个组件的 CLI 查看事件历史记录日志。例如，可以使用 **show ip arp internal event-history event** 命令查看所有 ARP 事件的事件历史记录。例 2-34 显示了 ARP 的事件历史日志，并给出了修改事件历史记录大小的配置示例。定义事件历史记录大小的时候，需要利用 **disabled** 关键字禁用事件历史记录日志。不过，通常并不建议禁用事件历史记录，因为这样做可能会降低发现故障根源的概率，而且难以理解事件的发生顺序。

例 2-34　ARP 事件历史记录日志和缓冲区大小

```
NX-1# show ip arp internal event-history event
1) Event:E_DEBUG, length:143, at 449547 usecs after Mon May 29 11:11:38 2017
    [116] [4201]: Adj info: iod: 2, phy-iod: 2, ip: 172.16.1.11, mac: fa16.3ee2.
b6d3, type: 0, sync: FALSE, suppress-mode: ARP Suppression Disabled

2) Event:E_DEBUG, length:193, at 449514 usecs after Mon May 29 11:11:38 2017
    [116] [4201]: Entry added to ARP pt, added to AM for 172.16.1.11, fa16.3ee2.
b6d3, state 2 on interface mgmt0, physical interface mgmt0, ismct 0. R
earp (interval: 0, count: 0), TTL: 1500 seconds
```

```
3) Event:E_DEBUG, length:79, at 449432 usecs after Mon May 29 11:11:38 2017
   [116] [4201]: arp_add_adj: Updating MAC on interface mgmt0, phy-interface mgmt0
! Output omitted for brevity
NX-1(config)# ip arp event-history event size ?
  disabled  Disabled
            *Default value is small
  large     Large buffer
  medium    Medium buffer
  small     Small buffer
NX-1(config)# ip arp event-history event size large
```

2.7 本章小结

本章重点介绍了 NX-OS 提供的一些有效的故障排查工具,分析了 Nexus 平台的各种抓包功能,包括 SPAN 和 ERSPAN。NX-OS 提供的以下功能特性能够有效排查因缓冲区拥塞造成的时延和丢包问题。

- SPAN-on-Latency。
- SPAN-on-Drop。

本章介绍了包括 Ethanalyzer 和数据包跟踪器在内的内部平台工具的使用方式,描述了用于采集统计信息和实施网络规划的 NetFlow 和 sFlow 用例、部署及配置方式。NTP 可以确保多台设备之间的时钟同步,从而正确关联跨设备的事件时序。EEM 脚本对于日常故障排查或事件发生后的信息采集来说都非常方便。最后,本章还介绍了 NX-OS 提供的多种日志记录方法,包括记账日志和事件历史记录日志。

第 3 章

Nexus 平台故障排查

本章主要讨论如下主题。
- 线卡故障排查。
- Nexus 交换矩阵故障排查。
- 硬件丢包故障排查。
- VDC（Virtual Device Context，虚拟设备上下文）。
- 系统 QoS 和 CoPP。
- NX-OS。

第 1 章讨论了各种 Nexus 平台及其支持的线卡情况。除平台和架构之外，了解所有的系统组件对于排查 Nexus 平台各种硬件级组件故障来说也非常重要。本章将重点介绍 Nexus 平台级别的故障排查技术。

3.1 硬件故障排查

Nexus 是一种模块化平台，可以是单插槽或多插槽机箱。单插槽机箱中的 Nexus 交换机配置了一块集成了物理接口的管理引擎卡。多插槽机箱支持管理引擎卡（SUP 卡）、线卡和交换矩阵卡，每种类型的板卡在 Nexus 转发架构中都发挥着至关重要的作用，使其成为高度可用的分布式架构平台。这些板卡的任何故障都会导致部分网络甚至整个数据中心的服务质量下降或服务中断。因此，理解平台的架构并隔离 Nexus 设备本身的故障问题极为重要，这可以最大限度地降低故障对服务造成的影响。

在深入研究 Nexus 平台的硬件故障排查之前，必须首先了解正在排查的 Nexus 设备的型号以及机箱中配置的板卡类型。第一步就是查看机箱中配置的板卡信息，可以通过命令 **show module** [*module-number*]查看 Nexus 设备上的所有板卡，其中，可选项 *module-number* 可以查看指定线卡的详细信息。例 3-1 给出了 Nexus 7009 和 Nexus 3548P 的 **show module** 命令输出结果，输出结果的第一部分来自 Nexus 7000，可以看出，该交换机配置了主用和备用 SUP 卡以及 3 块其他板卡：一块正常运行，另外两块断电。该命令的输出结果还显示了每块板卡的软件和硬件版本信息以及这些板卡的在线诊断状态，展示了设备处于断电状态的原因。在输出结果的最后，显示该机箱配置了交换矩阵模块，同时还显示了相应的软件和硬件版本及状态信息。

输出结果的第二部分来自 Nexus 3500 交换机，可以看出，该交换机只有一块 SUP 卡，因为 Nexus 3548P 是一款单 RU（Rack Unit，机架单元）交换机。请注意，机箱中的模块数量取决于设备型号及其支持的板卡类型。

例 3-1 show module 命令输出结果

```
Nexus 7000
N7K1# show module
Mod  Ports  Module-Type                                    Model              Status
```

```
---   -----   ---------------------------------   -------------------   ----------
1     0       Supervisor Module-2                 N7K-SUP2E             active *
2     0       Supervisor Module-2                 N7K-SUP2E             ha-standby
5     48      10/100/1000 Mbps Ethernet XL Module                       powered-dn
6     48      1/10 Gbps Ethernet Module           N7K-F248XP-25E        ok
7     32      10 Gbps Ethernet XL Module                                powered-dn

Mod   Power-Status    Reason
---   ------------    --------------------------
5     powered-dn      Unsupported/Unknown Module
7     powered-dn      Unsupported/Unknown Module
Mod   Sw                    Hw
---   ----------------      ------
1     8.0(1)                0.403
2     8.0(1)                1.0
6     8.0(1)                1.2

Mod   MAC-Address(es)                         Serial-Num
---   --------------------------------------  ----------
1     6c-9c-ed-48-0d-9f to 6c-9c-ed-48-0d-b1  JAF1608AAPL
2     84-78-ac-10-99-cf to 84-78-ac-10-99-e1  JAF1710ACHA
5     00-00-00-00-00-00 to 00-00-00-00-00-00  JAF1803AMGR
6     b0-7d-47-da-fb-04 to b0-7d-47-da-fb-37  JAE191908QG
7     00-00-00-00-00-00 to 00-00-00-00-00-00  JAF1553ASRE

Mod   Online Diag Status
---   ------------------
1     Pass
2     Pass
6     Pass

Xbar  Ports  Module-Type                        Model                 Status
---   -----  ---------------------------------  -------------------   ----------
1     0      Fabric Module 2                    N7K-C7009-FAB-2       ok
2     0      Fabric Module 2                    N7K-C7009-FAB-2       ok
3     0      Fabric Module 2                    N7K-C7009-FAB-2       ok
4     0      Fabric Module 2                    N7K-C7009-FAB-2       ok
5     0      Fabric Module 2                    N7K-C7009-FAB-2       ok

Xbar  Sw                    Hw
---   ----------------      ------
1     NA                    2.0
2     NA                    3.0
3     NA                    2.0
4     NA                    2.0
5     NA                    2.0

Xbar  MAC-Address(es)                         Serial-Num
---   --------------------------------------  ----------
1     NA                                      JAF1621BCDA
2     NA                                      JAF1631APEH
3     NA                                      JAF1621BBTF
4     NA                                      JAF1621BCEM
5     NA                                      JAF1621BCFJ

Nexus 3500
N3K1# show module
```

```
Mod  Ports  Module-Type                              Model                    Status
---  -----  ---------------------                    ---------------------    ----------
1    48     48x10GE Supervisor                       N3K-C3548P-10G-SUP       active *

Mod  Sw             Hw       World-Wide-Name(s) (WWN)
---  -------------  ------   ------------------------
1    6.0(2)A6(8)    1.1      --

Mod  MAC-Address(es)                          Serial-Num
---  ---------------------------------------  ----------
1    f872.ea99.6468 to f872.ea99.64a7         FOC17263D71
```

注：虽然所有的 Nexus 7000 机箱交换机都不需要交换矩阵模块（如 Nexus 7004 机箱就没有交换矩阵模块），但是较高的插槽机箱类型仍然需要使用交换矩阵模块，才能确保 Nexus 7000 交换机的正常运行。

Nexus 7000/7700 在设备安装或硬件升级过程中的常见故障问题是互操作性问题。如网络操作人员可能会尝试在 VDC 中安装线卡，但是该线卡与现有线卡的配合使用效果不好。例如，在同一个 VDC 中，M3 卡只能与 M2 或 F3 卡配合使用。同样，Nexus FEX（Fabric Extender，交换矩阵扩展器）卡与某些线卡配合使用可能是不支持的。在实际应用中，请务必查阅相关的兼容性矩阵列表，以免出现互操作性问题。例 3-1 给出了 Nexus 7000 交换机的 **show module** 命令输出结果以高亮方式显示了这类问题，例中的两块线卡因不兼容而导致断电。

注：如果希望查阅 Nexus I/O 模块兼容性矩阵的 CCO 文档，那么可以访问思科官网。

可以通过命令 **show hardware** 查看 Nexus 设备的详细软件和硬件信息，该命令可以显示 Nexus 交换机的状态、正常运行时间、板卡运行状况（包括线卡和交换矩阵卡）以及设备机箱中的电源和风扇信息。

3.1.1 通用在线诊断测试

与思科 6500 系列交换机相似，Nexus 设备也支持 GOLD（Generic Online Diagnostic Test，通用在线诊断测试）工具，GOLD 是一种独立于平台的故障检测工具，可以在启动和运行时隔离系统中的硬件和资源故障。诊断测试可以是中断性的，也可以是非中断性的，中断性测试可能会影响部分或全部系统功能，非中断性测试则不会影响系统运行的功能。

1. 启动诊断

启动诊断（Bootup Diagnostics）可以检测焊接错误、连接松动和模块故障等硬件故障，这些测试操作是在系统启动之后、硬件启用之前运行的。表 3-1 列出了一些常见的启动诊断测试。

表 3-1　　　　　　　　　　　　　　Nexus 启动诊断测试

测试名称	描述	属性	硬件
ASIC 寄存器测试	该测试将访问 ASIC 中的所有寄存器	中断性测试	SUP 和线卡
ASIC 内存测试	该测试将访问 ASIC 中的所有内存	中断性测试	SUP 和线卡
EOBC（Ethernet Out-of-Band Channel，以太网带外通道）端口环回测试	测试 EOBC 的环回接口	中断性测试	SUP 和线卡
端口环回测试	进行内部环回测试并检查转发路径（通过在同一个端口上发送和接收数据进行测试）	中断性测试	线卡
引导 ROM（Read-Only Memory，只读内存）测试	测试 SUP 板卡上的主用和备用引导设备的完整性	非中断性测试	SUP

测试名称	描述	属性	硬件
USB（Universal Serial Bus，通用串行总线）测试	验证 SUP 板卡上的 USB 控制器的初始化	非中断性测试	SUP
管理端口环回测试	对 SUP 板卡上的管理接口进行环回测试	中断性测试	SUP
OBFL（Onboard Failure Logging，板载故障记录）测试	测试 OBFL 闪存的完整性	非中断性测试	SUP 和线卡
FIPS（Federal Information Processing Standard，联邦信息处理标准）测试	验证模块上的安全设备	中断性测试	线卡

注：Nexus 7000 的 F1 系列模块不支持 FIPS 测试。

启动诊断支持不同级别的配置和执行方式。

- **无（旁路）（None[Bypass]）诊断方式**：模块在不运行任何启动诊断测试的情况下实现启动运行，目的是快速启动板卡。
- **全面（Complete）诊断方式**：对模块执行全面启动诊断测试，该选项是默认和推荐的启动诊断等级。

可以在全局配置模式下使用 **diagnostic bootup level [bypass | complete]** 命令配置诊断级别。对于 VDC 场景来说，必须为每个 VDC 配置诊断级别，此时，可以通过 **show diagnostic bootup level** 命令验证启动诊断级别。

2．运行时诊断

运行时诊断（Runtime Diagnostics）指的是系统处于运行状态时（也就是在活动节点上）执行的诊断操作。运行时诊断测试有助于检测运行时硬件差错，如内存错误、资源耗尽和硬件故障/降级。运行时诊断可以分为两类。

- 运行状况监测诊断（Health-monitoring diagnostics）。
- 按需诊断（On-demand diagnostics）。

HM（Health-monitoring，运行状况监测）测试是非中断性测试，在模块的后台运行。HM 测试的主要目的是确保交换机的硬件和软件组件在处理网络流量时的健康性。某些特定的 HM 测试（被标记为 HM-always）在模块上线时默认自动启用，用户可以通过 CLI 在所有模块上轻松启用和禁用所有的 HM 测试（HM-always 测试除外）。除此以外，用户还可以变更所有 HM 测试的时间间隔（标记为 HM-fixed 的测试除外，这些测试的时间间隔固定）。表 3-2 列出了 SUP 和线卡模块可用的 HM 测试情况。

表 3-2   Nexus HM 诊断测试

测试名称	描述	属性	硬件
ASIC 暂存寄存器测试	该测试将访问 ASIC 的暂存寄存器	非中断性测试	SUP 和线卡（支持暂存寄存器的所有 ASIC）
RTC（Realtime Clock，实时时钟）测试	验证管理引擎上的 RTC 运行正常	非中断性测试	SUP
NVRAM（Nonvolatile Random Access Memory，非易失性随机存取存储器）可用性测试	测试 SUP 模块上的 NVRAM 区块的可用性	非中断性测试	SUP
端口环回测试	通过环回数据包的方式，在不中断端口流量的情况下周期性地检查转发路径	非中断性测试	线卡
主用引导 ROM 测试	测试板卡上的主用引导设备的完整性	非中断性测试	SUP 和线卡

测试名称	描述	属性	硬件
备用引导 ROM 测试	测试板卡上的备用引导设备的完整性	非中断性测试	SUP 和线卡
CompactFlash 测试	验证 SUP 卡上的内部 CompactFlash 的可访问性	非中断性测试	SUP
外部 CompactFlash 测试	验证 SUP 卡上的外部 CompactFlash 的可访问性	非中断性测试	SUP
电源管理总线测试	测试 SUP 卡上的备用电源管理控制总线	非中断性测试	SUP
Spine（脊）控制总线测试	测试和验证备用 Spine 模块控制总线的可用性	非中断性测试	SUP
备用交换矩阵环回测试	测试备用 SUP 与交换矩阵之间的数据包路径	非中断性测试	SUP
状态总线（两线）测试	检查将各种模块（包括交换矩阵卡）连接到 SUP 模块上的两线接口（Two Wire Interface）	非中断性测试	SUP

可以通过全局配置命令 **diagnostic monitor interval module** *slot* **test** [*name* | *test-id* | **all**] **hour** **min** *minutes* **second** *sec* 设置 HM 测试的时间间隔。需要注意的是，命令中的测试名称区分大小写。如果要启用或禁用 HM 测试，那么就可以使用全局配置命令[**no**] **diagnostic monitor module** *slot* **test** [*name* | *test-id* | **all**]。命令 **show diagnostic content module** [*slot* | **all**]可以显示指定线卡的诊断信息和属性信息。例 3-2 给出了查看 Nexus 7000 交换机线卡诊断信息以及禁用 HM 测试的配置示例。例 3-2 中的线卡是 SUP 卡，因而列出的测试名称仅与 SUP 卡有关（与线卡无关）。例如，对于外部 CompactFlash 测试来说，第一块输出结果中的属性被设置为 *A*，表示该测试处于活动状态。如果在配置模式下禁用了该测试，那么输出结果显示的属性就是 *I*，表示该测试处于非活动状态。

例 3-2 show diagnostic content module 命令输出结果

```
Nexus 7000
N7K1# show diagnostic content module 1
Diagnostics test suite attributes:
B/C/*  - Bypass bootup level test / Complete bootup level test / NA
P/*    - Per port test / NA
M/S/*  - Only applicable to active / standby unit / NA
D/N/*  - Disruptive test / Non-disruptive test / NA
H/O/*  - Always enabled monitoring test / Conditionally enabled test / NA
F/*    - Fixed monitoring interval test / NA
X/*    - Not a health monitoring test / NA
E/*    - Sup to line card test / NA
L/*    - Exclusively run this test / NA
T/*    - Not an ondemand test / NA
A/I/*  - Monitoring is active / Monitoring is inactive / NA
Z/D/*  - Corrective Action is enabled / Corrective Action is disabled / NA


Module 1: Supervisor Module-2 (Active)
                                                         Testing Interval
ID     Name                                 Attributes   (hh:mm:ss)
___    _____          _____   _____

 1)    ASICRegisterCheck-------------->     ***N******A*    00:00:20
 2)    USB---------------------------->     C**N**X**T**    -NA-
 3)    NVRAM-------------------------->     ***N******A*    00:05:00
 4)    RealTimeClock------------------>     ***N******A*    00:05:00
 5)    PrimaryBootROM----------------->     ***N******A*    00:30:00
 6)    SecondaryBootROM--------------->     ***N******A*    00:30:00
 7)    CompactFlash------------------->     ***N******A*    00:30:00
 8)    ExternalCompactFlash----------->     ***N******A*    00:30:00
 9)    PwrMgmtBus--------------------->     **MN******A*    00:00:30
10)    SpineControlBus--------------->      ***N******A*    00:00:30
```

```
 11)     SystemMgmtBus----------------->       **MN******A*        00:00:30
 12)     StatusBus--------------------->       **MN******A*        00:00:30
 13)     PCIeBus------------------------>      ***N******A*        00:00:30
 14)     StandbyFabricLoopback--------->       **SN******A*        00:00:30
 15)     ManagementPortLoopback-------->       C**D**X**T**        -NA-
 16)     EOBCPortLoopback-------------->       C**D**X**T**        -NA-
 17)     OBFL-------------------------->       C**N**X**T**        -NA-
N7K1# config t
N7K1(config)# no diagnostic monitor module 1 test ExternalCompactFlash
N7K1# show diagnostic content module 1
! Output omitted for brevity
Module 1: Supervisor Module-2 (Active)

                                                            Testing Interval
 ID      Name                                 Attributes    (hh:mm:ss)
 ____    ____                                 _____    _____

 1)      ASICRegisterCheck------------->      ***N******A*        00:00:20
 2)      USB--------------------------->      C**N**X**T**        -NA-
 3)      NVRAM------------------------->      ***N******A*        00:05:00
 4)      RealTimeClock----------------->      ***N******A*        00:05:00
 5)      PrimaryBootROM---------------->      ***N******A*        00:30:00
 6)      SecondaryBootROM-------------->      ***N******A*        00:30:00
 7)      CompactFlash------------------>      ***N******A*        00:30:00
 8)      ExternalCompactFlash---------->      ***N******I*        00:30:00
 9)      PwrMgmtBus-------------------->      **MN******A*        00:00:30
 10)     SpineControlBus--------------->       ***N******A*        00:00:30
 11)     SystemMgmtBus----------------->       **MN******A*        00:00:30
 12)     StatusBus--------------------->       **MN******A*        00:00:30
 13)     PCIeBus------------------------>      ***N******A*        00:00:30
 14)     StandbyFabricLoopback--------->       **SN******A*        00:00:30
 15)     ManagementPortLoopback-------->       C**D**X**T**        -NA-
 16)     EOBCPortLoopback-------------->       C**D**X**T**        -NA-
 17)     OBFL-------------------------->       C**N**X**T**        -NA
```

show diagnostic content module [*slot* | **all**]命令不但能显示 HM 测试，而且还能显示启动诊断测试。从例 3-2 可以看出，输出结果中的测试属性以 *C* 开头，表示这些测试是完整的启动级别测试。如果要查看所有测试结果和统计信息，那么就可以使用 **show diagnostic result module** [*slot* | **all**] [**detail**]命令。验证诊断结果时，请务必确保没有任何测试出现失败（F）或错误（E）结果。例 3-3 以摘要和详细方式显示了 SUP 卡的诊断测试结果，从输出结果可以看出，启动诊断测试的级别已被设置成 "*complete*"（全面）。第一部分输出结果列出了 SUP 模块经历的所有测试及测试结果，"."表示测试已通过。输出结果的详细版本列出了更具体的细节信息，如错误代码、上一次执行时间、下一次执行时间以及失败原因。如果模块出现了故障且需要排查以隔离瞬态故障或硬件故障，那么这些信息将非常有用。

例 3-3 诊断测试结果

```
N7K1# show diagnostic result module 1
Current bootup diagnostic level: complete
Module 1: Supervisor Module-2 (Active)

        Test results: (. = Pass, F = Fail, I = Incomplete,
        U = Untested, A = Abort, E = Error disabled)

          1) ASICRegisterCheck-------------> .
          2) USB-------------------------->  .
          3) NVRAM------------------------> .
          4) RealTimeClock----------------> .
```

```
             5) PrimaryBootROM----------------> .
             6) SecondaryBootROM--------------> .
             7) CompactFlash------------------> .
             8) ExternalCompactFlash----------> U
             9) PwrMgmtBus--------------------> .
            10) SpineControlBus---------------> .
            11) SystemMgmtBus-----------------> .
            12) StatusBus--------------------->  .
            13) PCIeBus----------------------->  .
            14) StandbyFabricLoopback--------->  U
            15) ManagementPortLoopback-------->  .
            16) EOBCPortLoopback-------------->  .
            17) OBFL-------------------------->  .
N7K1# show diagnostic result module 1 detail
Current bootup diagnostic level: complete
Module 1: Supervisor Module-2 (Active)

  Diagnostic level at card bootup: complete

        Test results: (. = Pass, F = Fail, I = Incomplete,
        U = Untested, A = Abort, E = Error disabled)


        _____


        1) ASICRegisterCheck .

                Error code ------------------> DIAG TEST SUCCESS
                Total run count -------------> 38807
                Last test execution time ----> Thu May 7 18:24:16 2015
                First test failure time -----> n/a
                Last test failure time ------> n/a
                Last test pass time ---------> Thu May 7 18:24:16 2015
                Total failure count ---------> 0
                Consecutive failure count ---> 0
                Last failure reason ---------> No failures yet
                Next Execution time ---------> Thu May 7 18:24:36 2015


        2) USB .

                Error code ------------------> DIAG TEST SUCCESS
                Total run count -------------> 1
                Last test execution time ----> Tue Apr 28 18:44:36 2015
                First test failure time -----> n/a
                Last test failure time ------> n/a
                Last test pass time ---------> Tue Apr 28 18:44:36 2015
                Total failure count ---------> 0
                Consecutive failure count ---> 0
                Last failure reason ---------> No failures yet
                Next Execution time ---------> n/a
! Output omitted for brevity
```

按需诊断的侧重点有所不同。这些测试虽然不必定期运行，但是可能需要在出现某些事件（如故障）或预期出现某些事件（如资源超量）时运行。按需测试对于定位故障和应用故障抑制解决方案来说非常有用。

中断性和非中断性按需诊断测试都是通过 CLI 方式执行的，通过 **diagnostic start module** *slot*

test [*test-id* | *name* | **all** | **non-disruptive**] [**port** *port-number* | **all**]命令即可执行按需诊断测试，其中的变量 *test-id* 是给定模块支持的测试数量。按需诊断也能以端口为基础执行测试（取决于测试类型），可以通过可选关键字 **port** 来指定测试端口。可以通过 **diagnostic stop module slot test** [*test-id* | *name* | **all**]命令终止按需测试。按需测试默认仅执行单次测试操作，也可以通过命令 **diagnostic ondemand iteration** *number* 增加迭代次数，其中的 *number* 表示迭代次数。需要注意的是，在生产性网络中执行中断性按需诊断测试时务必小心谨慎。

例 3-4 给出了在 Nexus 7000 交换机模块上执行按需端口环回测试的配置示例。

例 3-4 按需诊断测试

```
N7K1# diagnostic ondemand iteration 3
N7K1# diagnostic start module 6 test PortLoopback
N7K1# show diagnostic status module 6
            <BU>-Bootup Diagnostics, <HM>-Health Monitoring Diagnostics
            <OD>-OnDemand Diagnostics, <SCH>-Scheduled Diagnostics

==============================================
Card:(6) 1/10 Gbps Ethernet Module
==============================================
Current running test                Run by
PortLoopback                        OD

Currently Enqueued Test             Run by
PortLoopback                        OD (Remaining Iteration: 2)
N7K1# show diagnostic result module 6 test PortLoopback detail
Current bootup diagnostic level: complete
Module 6: 1/10 Gbps Ethernet Module

  Diagnostic level at card bootup: complete

        Test results: (. = Pass, F = Fail, I = Incomplete,
        U = Untested, A = Abort, E = Error disabled)


        _____

   6) PortLoopback:
        Port  1  2  3  4  5  6  7  8  9 10 11 12 13 14 15 16
        -----------------------------------------------------
              U  U  U  U  U  U  U  U  U  U  U  U  U  .  .  .

        Port 17 18 19 20 21 22 23 24 25 26 27 28 29 30 31 32
        -----------------------------------------------------
              U  U  .  .  U  U  U  U  U  U  U  U  U  U  U  U

        Port 33 34 35 36 37 38 39 40 41 42 43 44 45 46 47 48
        -----------------------------------------------------
              U  U  U  U  U  U  U  U  U  U  U  U  U  U  U  U


        Error code ------------------> DIAG TEST SUCCESS
        Total run count -------------> 879
        Last test execution time ----> Thu May  7 21:25:48 2015
        First test failure time -----> n/a
        Last test failure time ------> n/a
        Last test pass time ---------> Thu May  7 21:26:00 2015
```

```
              Total failure count ---------> 0
              Consecutive failure count ---> 0
              Last failure reason ---------> No failures yet
              Next Execution time ---------> Thu May  7 21:40:48 2015
```

在故障排查期间，如果迭代次数设置得比较高，且测试失败后需要执行某些操作，那么就可以使用 **diagnostic ondemand action-on-failure [continue failure-count** *num-fails* | **stop]** 命令。如果使用了关键字 **continue**，那么就可以通过 **failure-count** 参数设置终止测试前允许的失败次数，默认值为 0，表示即使出现失败也不终止测试。**show diagnostic ondemand setting** 命令可以验证按需诊断的配置情况。例 3-5 定义了按需诊断测试失败后应执行的操作，将 **action-on-failure**（测试失败后的操作）设置为继续，直至故障计数器达到数值 2。

例 3-5 按需诊断测试的 action-on-failure

```
! Setting the action-on-failure to continue till 2 failure counts.
N7K1# diagnostic ondemand action-on-failure continue failure-count 2
N7K1# show diagnostic ondemand setting
       Test iterations = 3
       Action on test failure = continue until test failure limit reaches 2
```

> 注：也可以在离线模式下执行诊断测试。使用 **hardware module** *slot* **offline** 命令可以将模块置于离线模式，然后使用 **diagnostic start module** *slot* **test** [*test-id* | *name* | **all**] **offline** 命令以 *offline*（离线）属性执行诊断测试。

3. GOLD 测试和 EEM 支持

虽然诊断测试有助于识别 SUP 和线卡的硬件故障，但操作人员希望在发现问题之后能够采取必需的纠正措施。因此，NX-OS 将 GOLD 测试与 EEM（Embedded Event Manager，嵌入式事件管理器）结合在一起，可以在诊断测试失败后执行纠正操作。常见的 GOLD 测试用例之一就是将设备纳入生产环境之前进行老化测试或分阶段引入新设备。老化测试类似于负荷测试：设备通常处于一定的负荷状况下，测试设备随时间推移的资源利用率情况，包括内存、CPU 和缓冲区。该测试能够有效防止设备开始处理生产流量之前因硬件故障而导致的重大中断。

NX-OS 可以针对以下 HM 测试执行纠正操作。

- RewriteEngineLoopback（重写引擎环回）测试。
- StandbyFabricLoopback（备用交换矩阵环回）测试。
- 内部 PortLoopback（端口环回）测试。
- SnakeLoopback（蛇形环回）测试。

如果在管理引擎模块上的 StandbyFabricLoopback 测试失败，那么系统将重新加载备用管理引擎卡。如果备用管理引擎卡在 3 次重试的情况下都无法恢复，那么就会关闭备用管理引擎卡。重新加载备用管理引擎卡之后，默认将启动 HM 诊断。系统默认禁用纠正操作，可以通过 **diagnostic eem action conservative** 命令启用纠正操作。

> 注：不能以每项测试为基础单独配置 **diagnostic eem action conservative** 命令，该命令适用于前面提到的所有 4 项 GOLD 测试。

3.1.2 Nexus 设备运行状况检查

对于所有网络环境来说，网络管理员和操作员都必须执行常规的设备运行状况检查，以确保网络的稳定性，并在产生重大网络影响之前发现问题。可以采取手工方式或自动化工具来执行设备的运行状况检查。虽然不同的 Nexus 平台可能拥有不同的命令行，但是都需要定期验证以下信息。

- 模块状态和诊断。
- 硬件和进程崩溃与重置。
- 丢包。
- 接口差错和丢包。

前面已经讨论了模块状态和诊断问题，本节将接着讨论不同 Nexus 平台执行其他运行状况检查时的配置方式。

1. 硬件和进程崩溃

线卡和管理引擎卡的重新加载或崩溃可能会给网络造成严重中断，硬件或软件问题可能导致板卡崩溃或重新加载。由于 NX-OS 采取的是分布式架构，因而崩溃问题也可能发生在进程上。大多数硬件或进程崩溃后会生成一个核心文件（core file），思科 TAC 通常利用该核心文件来确定崩溃的发生原因，使用 **show cores vdc-all** 命令可以找到该核心文件。对于 Nexus 7000 交换机来说，需要从默认 VDC 运行 **show cores vdc-all** 命令。例 3-6 显示了 Nexus 7000 交换机生成的核心文件示例，该例为 VDC 1 模块 6 和进程 RPM 生成了核心文件。

例 3-6 Nexus 核心文件

```
N7k-1# show cores vdc-all
VDC  Module  Instance  Process-name  PID   Date(Year-Month-Day Time)
---  ------  --------  ------------  ----  -------------------------
1    6       1         rpm           4298  2017-02-08 15:08:48
```

找到核心文件之后，就可以将其复制到 Bootflash 或其他外部位置，如 FTP（File Transfer Protocol，文件传输协议）或 TFTP（Trivial FTP，简单 FTP）服务器。Nexus 7000 的核心文件位于 *core:* 文件系统。可以通过以下 URL 定位核心文件：

core://<module-number>/<process-id>/<instance-number>

例如，例 3-6 中的核心文件的位置是 *core://6/4298/1*。如果 Nexus 7000 交换机出现重启或切换操作，那么核心文件将位于 **logflash://[sup-1 | sup-2]/core**。其他 Nexus 平台（如 Nexus 5000、4000 或 3000）的核心文件位于 **volatile:** 文件系统（而非 *logflash:* 文件系统）中，因而设备重新加载后，核心文件可能会丢失。不过，这些将核心文件存储在 *volatile:* 文件系统中的平台在较新的软件版本中增加了一项新功能，可以将核心文件写入 *bootflash:* 或远程文件位置。

如果出现了进程崩溃，但是没有为该崩溃事件生成核心文件，那么有可能为该进程生成了协议栈轨迹（stack trace）。但是，如果既没有为崩溃服务生成核心文件，也没有生成协议栈轨迹，那么就需要使用 **show processes log vdc-all** 命令来定位受影响的进程，这类崩溃进程通常带有标志 *N*。使用上一条命令中的 PID（Process ID，进程 ID）值和 **show processes log pid** *pid* 命令，可以确定服务中断的原因，该命令的输出结果会在 *Death reason* 字段显示进程失败的原因。例 3-7 给出了通过 **show processes log** 和 **show processes log pid** 命令识别 Nexus 平台进程崩溃情况的配置示例。

例 3-7 Nexus 进程崩溃

```
N7k-1# show processes log
VDC Process           PID    Normal-exit  Stack  Core   Log-create-time
--- ----------------  -----  -----------  -----  -----  ---------------
 1  ascii-cfg         5656        N         Y      N    Thu Feb 23 17:10:43 2017
 1  ascii_cfg_serve   7811        N         N      N    Thu Feb 23 17:10:43 2017
 1  installer         23457       N         N      N    Tue May 23 02:00:00 2017
 1  installer         25885       N         N      N    Tue May 23 02:28:23 2017
 1  installer         26212       N         N      N    Tue May 23 15:51:19 2017
! Output omitted for brevity
N7k-1# show processes log pid 5656
```

```
===================================================
Service: ascii-cfg
Description: Ascii Cfg Server
Executable: /isan/bin/ascii_cfg_server

Started at Thu Feb 23 17:06:20 2017 (155074 us)
Stopped at Thu Feb 23 17:10:43 2017 (738171 us)
Uptime: 4 minutes 23 seconds

Start type: SRV_OPTION_RESTART_STATELESS (23)
Death reason: SYSMGR_DEATH_REASON_FAILURE_HEARTBEAT (9)
Last heartbeat 40.01 secs ago
RLIMIT_AS: 1936268083
System image name: n7000-s2-dk9.7.3.1.D1.1.bin
System image version: 7.3(1)D1(1) S19

PID: 5656
Exit code: signal 6 (core dumped)

cgroup: 1:devices,memory,cpuacct,cpu:/1


CWD: /var/sysmgr/work

RLIMIT_AS:         1936268083
! Output omitted for brevity
```

如果希望快速验证重置原因，那么就可以使用 **show system reset-reason** 命令。如果未生成核心文件，那么可以通过以下命令来识别重置原因。

- **show system exception-info**。
- **show module internal exceptionlog module** *slot*。
- **show logging onboard** [**module** *slot*]。
- **show process log details**。

2. 丢包

对于任何网络环境来说，丢包都是一项非常复杂的故障排查难题。丢包发生的可能原因很多。

- 硬件损坏。
- 平台导致的丢包。
- 路由或交换问题。

路由和交换问题导致的丢包可以通过修正配置来解决，而硬件损坏则可能会影响局部端口或整个线卡的所有流量。Nexus 平台提供了多种计数器，查看这些计数器有助于确定设备的丢包原因（详见以下内容）。

3. 接口差错和丢包

除平台或硬件丢包之外，接口故障也可能会给数据中心带来丢包和服务质量下降问题。链路振荡、链路未启动、接口错误以及输入或输出丢包等问题只是部分可能给服务造成严重影响的差错情景。一般来说，排查交换机链路故障通常都比较困难，不过 NX-OS 提供了一些有用的 CLI 和内部平台命令来帮助解决该问题。

show interface *interface-number* 命令可以显示指定接口的详细信息，如接口流量速率、输入和输出统计信息、统计输入/输出错误的差错计数器、CRC 差错以及溢出计数器等。此外，NX-OS CLI 还为验证接口能力、收发器信息、计数器、流控制、MAC 地址信息以及交换机端口和中继信息提供了多种命令选项（包括 **show interface** 命令）。例 3-8 显示了 **show interface** 命令的输出结果，以高亮方式突出显示了需要在接口上验证的信息，输出结果的第二部分则列出了接口的各种能力信息。

例 3-8　Nexus 接口的详细信息和能力信息

```
N9k-1# show interface Eth2/1
Ethernet2/1 is up
admin state is up, Dedicated Interface
  Hardware: 40000 Ethernet, address: 1005.ca57.287f (bia 88f0.31f9.5710)
  Internet Address is 192.168.10.1/24
  MTU 1500 bytes, BW 40000000 Kbit, DLY 10 usec
  reliability 255/255, txload 1/255, rxload 1/255
  Encapsulation ARPA, medium is broadcast
  full-duplex, 40 Gb/s, media type is 40G
  Beacon is turned off
  Auto-Negotiation is turned on
  Input flow-control is off, output flow-control is off
  Auto-mdix is turned off
  Rate mode is dedicated
  Switchport monitor is off
  EtherType is 0x8100
  EEE (efficient-ethernet) : n/a
  Last link flapped 2d01h
  Last clearing of "show interface" counters never
  2 interface resets
  30 seconds input rate 64 bits/sec, 0 packets/sec
  30 seconds output rate 0 bits/sec, 0 packets/sec
  Load-Interval #2: 5 minute (300 seconds)
    input rate 32 bps, 0 pps; output rate 32 bps, 0 pps
  RX
    950396 unicast packets  345788 multicast packets  15 broadcast packets
    1296199 input packets  121222244 bytes
    0 jumbo packets  0 storm suppression packets
    0 runts  0 giants  0 CRC  0 no buffer
    0 input error  0 short frame  0 overrun  0 underrun  0 ignored
    0 watchdog  0 bad etype drop  0 bad proto drop  0 if down drop
    0 input with dribble  0 input discard
    0 Rx pause
  TX
    950398 unicast packets  2951181 multicast packets  19 broadcast packets
    3901598 output packets  396283422 bytes
    0 jumbo packets
    0 output error  0 collision  0 deferred  0 late collision
    0 lost carrier  0 no carrier  0 babble  0 output discard
    0 Tx pause
N9k-1# show interface Eth2/1 capabilities
Ethernet2/1
  Model:                  N9K-X9636PQ
  Type (SFP capable):     QSFP-40G-CR4
  Speed:                  40000
  Duplex:                 full
  Trunk encap. type:      802.1Q
  FabricPath capable:     no
  Channel:                yes
  Broadcast suppression:  percentage(0-100)
  Flowcontrol:            rx-(off/on/desired),tx-(off/on/desired)
  Rate mode:              dedicated
```

```
Port mode:              Routed,Switched
QOS scheduling:         rx-(none),tx-(4q)
CoS rewrite:            yes
ToS rewrite:            yes
SPAN:                   yes
UDLD:                   yes
MDIX:                   no
TDR capable:            no
Link Debounce:          yes
Link Debounce Time:     yes
FEX Fabric:             yes
dot1Q-tunnel mode:      yes
Pvlan Trunk capable:    yes
Port Group Members:     1
EEE (efficient-eth):    no
PFC capable:            yes
Buffer Boost capable:   no
Speed group capable:    yes
```

如果希望仅查看接口的各种计数器信息，那么就可以使用 **show interface counters errors** 命令，选项 **counters errors** 也可以与特定的 **show interface** *interface-number* 命令结合使用。例 3-9 显示了接口的差错计数器，如果发现计数器在增加，那么就需要根据接收到的差错类型进一步排查故障。差错可以指向第一层故障、端口损坏甚至是缓冲区问题。输出结果中的某些计数器表示的并不是差错，而是其他问题，如巨包计数器（Giants Counter）表示正在接收的数据包的 MTU 大于接口的配置值。

例 3-9　接口差错计数器

```
N9k-1# show interface Eth 2/1 counters errors

--------------------------------------------------------------------------------
Port         Align-Err    FCS-Err    Xmit-Err    Rcv-Err    UnderSize  OutDiscards
--------------------------------------------------------------------------------
Eth2/1       0            0          0           0          0          0

--------------------------------------------------------------------------------
Port         Single-Col   Multi-Col  Late-Col    Exces-Col  Carri-Sen  Runts
--------------------------------------------------------------------------------
Eth2/1       0            0          0           0          0          0

--------------------------------------------------------------------------------
Port         Giants   SQETest-Err  Deferred-Tx  IntMacTx-Er  IntMacRx-Er  Symbol-Err
--------------------------------------------------------------------------------
Eth2/1       0        --           0            0            0            0
```

如果要查看硬件接口资源和利用率信息，那么就可以使用 **show hardware capacity interface** 命令，该命令不但能够显示缓冲区信息，而且还能显示每块线卡上多个端口的入站和出站方向的丢包情况。虽然不同的 Nexus 平台（如 Nexus 7000 和 Nexus 9000）的输出结果略有不同，但是该命令对于识别交换机上丢包数最多的接口来说确实非常有用。例 3-10 显示了 Nexus 7000 交换机的硬件接口资源。

例 3-10　硬件接口资源和丢包信息

```
N7k-1# show hardware capacity interface
Interface Resources

  Interface drops:
    Module   Total drops                 Highest drop ports
```

```
      3    Tx: 0                           -
      3    Rx: 101850                      Ethernet3/37
      4    Tx: 0                           -
      4    Rx: 64928                       Ethernet4/4

 Interface buffer sizes:
    Module     Bytes:  Tx buffer          Rx buffer
       3               705024             1572864
       4               705024             1572864
```

常见的接口故障就是输入和输出丢包问题，出现这些差错的原因主要是端口拥塞。前面的接口命令和 **show hardware internal errors** [**module** *slot*]命令能够有效识别输入或输出丢包问题，如果识别出输入丢包，那么就必须定位出站端口的拥塞问题。如果出站端口出现了超量分配带宽，那么即便在设备上配置了 SPAN，也会出现输入丢包问题，因而需要确保未在设备上配置 SPAN，除非要求执行 SPAN 抓包操作，如果是这种情况，那么就应该删除 SPAN。如果拥塞的出站端口是千兆端口，那么故障原因可能是多对一的单播流量流导致的拥塞，此时可以考虑将端口升级为万兆端口或者将多个千兆端口捆绑成端口通道接口来解决该问题。

输出丢包通常是由接口排队策略引起的，可以通过 **show system internal qos queuing stats int** *interface-id* 命令加以验证，命令 **show queuing interface** *interface-id* 或 **show policy-map interface** *interface-id* [**input** | **output**]可以查看排队策略的配置信息，调整 QoS 策略能够有效解决输出丢包问题。例 3-11 显示了接口 Ethernet1/5 的排队统计信息，表明接口上的多个队列都出现了丢包问题。

例 3-11 接口排队统计信息

```
N7k-1# show system internal qos queuing stats int eth1/5
Interface Ethernet1/5 statistics

Transmit queues
----------------------------------------
    Queue 1p7q4t-out-q-default
        Total bytes                    0
        Total packets                  0
        Current depth in bytes         0
        Min pg drops                   0
        No desc drops                  0
        WRED drops                     0
        Taildrop drops                 0
    Queue 1p7q4t-out-q2
        Total bytes                    0
        Total packets                  0
        Current depth in bytes         0
        Min pg drops                   0
        No desc drops                  0
        WRED drops                     0
        Taildrop drops                 0
    Queue 1p7q4t-out-q3
        Total bytes                    0
        Total packets                  0
        Current depth in bytes         0
        Min pg drops                   0
        No desc drops                  0
        WRED drops                     0
        Taildrop drops                 0
    Queue 1p7q4t-out-q4
```

```
            Total bytes                  0
            Total packets                0
            Current depth in bytes       0
            Min pg drops                 0
            No desc drops                0
            WRED drops                   0
            Taildrop drops               81653
        Queue 1p7q4t-out-q5
            Total bytes                  0
            Total packets                0
            Current depth in bytes       0
            Min pg drops                 0
            No desc drops                0
            WRED drops                   0
            Taildrop drops               35096
        Queue 1p7q4t-out-q6
            Total bytes                  0
            Total packets                0
            Current depth in bytes       0
            Min pg drops                 0
            No desc drops                0
            WRED drops                   0
            Taildrop drops               245191
        Queue 1p7q4t-out-q7
            Total bytes                  0
            Total packets                0
            Current depth in bytes       0
            Min pg drops                 0
            No desc drops                0
            WRED drops                   0
            Taildrop drops               657759
        Queue 1p7q4t-out-pq1
            Total bytes                  0
            Total packets                0
            Current depth in bytes       0
            Min pg drops                 0
            No desc drops                0
            WRED drops                   0
            Taildrop drops               0
```

4. 与平台相关的丢包

Nexus 平台可以提供各种平台级计数器的详细信息，以帮助识别硬件和软件组件故障。如果在特定接口或线卡上发现了丢包问题，那么平台级命令可以为排查丢包原因提供大量有用信息。例如，可以在 Nexus 7000 交换机上执行 **show hardware internal statistics [module** *slot* | **module-all] pktflow dropped** 命令来识别丢包原因，该命令详细列出了每个线卡模块的信息以及线卡上所有接口的丢包信息。例 3-12 显示了插槽 3 中的线卡上的所有端口的丢包信息，输出结果表明丢包原因主要有包长错误、数据包 MAC（Media Access Control，媒体访问控制）错误、CRC（Cyclic Redundancy Check，循环冗余校验）错误等。将关键字 **diff** 与该命令一起使用，有助于识别特定接口的丢包增加情况及其原因，从而可以展开进一步的故障排查操作。

例 3-12 Nexus 7000 分组流丢包计数器

```
N7k-1# show hardware internal statistics module 3 pktflow dropped
|----------------------------------------|
```

```
|Executed at : 2017-06-02 10:09:16.914   |
|---------------------------------------|
Hardware statistics on module 03:
|----------------------------------------------------------------------|
| Device:Flanker Eth Mac Driver     Role:MAC                Mod: 3   |
| Last cleared @ Fri Jun  2 00:28:46 2017                            |
|----------------------------------------------+---------------------|
Instance:0
Cntr  Name                                               Value              Ports
----- -----                                              -----              -----
    0 igr in upm: pkts rcvd, len(>= 64B, <= mtu) with bad crc 0000000000000001
3 -
    1 igr rx pl:  received error pkts from mac           0000000000000001   3 -
    2 igr rx pl: EM-IPL i/f dropped pkts cnt             0000000000000004   3 -
    3 igr rx pl: cbl drops                               0000000000002818   3 -
    4 igr rx pl: EM-IPL i/f dropped pkts cnt             0000000000000002   4 -


Instance:1
Cntr  Name                                               Value              Ports
----- -----                                              -----              -----
    5 igr in upm: pkts rcvd, len > MTU with bad CRC      0000000000000001   10 -
    6 igr in upm: pkts rcvd, len > MTU with bad CRC      0000000000000001   11 -
    7 igr rx pl: EM-IPL i/f dropped pkts cnt             0000000000000002   9 -
    8 igr rx pl: EM-IPL i/f dropped pkts cnt             0000000000000011   10 -
    9 igr rx pl: cbl drops                               0000000000000004   10 -
   10 igr rx pl:  received error pkts from mac           0000000000000001   11 -
   11 igr rx pl: EM-IPL i/f dropped pkts cnt             0000000000000017   11 -
   12 igr rx pl: cbl drops                               0000000000002812   11 -


Instance:3
Cntr  Name                                               Value              Ports
----- -----                                              -----              -----
   13 igr rx pl: EM-IPL i/f dropped pkts cnt             0000000000000003   26 -
   14 igr rx pl: cbl drops                               0000000000000008   26 -
   15 igr rx pl: EM-IPL i/f dropped pkts cnt             0000000000000001   31 -


Instance:4
Cntr  Name                                               Value              Ports
----- -----                                              -----              -----
   16 igr in upm: pkts rcvd, len > MTU with bad CRC      0000000000000027   35 -
   17 igr in upm: pkts rcvd, len > MTU with bad CRC      0000000000000044   36 -
   18 igr in upm: pkts rcvd, len(>= 64B, <= mtu) with bad crc 0000000000000001
36 -
   19 igr in upm: pkts rcvd, len > MTU with bad CRC      0000000000005795   37 -
   20 igr in upm: pkts rcvd, len > MTU with bad CRC      0000000000000034   38 -
   21 igr rx pl: EM-IPL i/f dropped pkts cnt             0000000000000008   33 -
   22 igr rx pl: cbl drops                               0000000000002801   33 -
   23 igr rx pl: EM-IPL i/f dropped pkts cnt             0000000000000004   34 -
   24 egr out pl: total pkts dropped due to cbl          0000000000001769   34 -
   25 igr rx pl: received error pkts from mac            0000000000000003   35 -
   26 igr rx pl: EM-IPL i/f dropped pkts cnt             0000000000000200   35 -
   27 igr rx pl: cbl drops                               0000000000002813   35 -
   28 igr rx pl: dropped pkts cnt                        0000000000000017   35 -
   29 igr rx pl: received error pkts from mac            0000000000000093   36 -
```

```
   30 igr rx pl: EM-IPL i/f dropped pkts cnt          0000000000002515        36 -
   31 igr rx pl: cbl drops                            0000000000002894        36 -
   32 igr rx pl: dropped pkts cnt                     0000000000000166        36 -
   33 igr rx pl: EM-IPL i/f dropped pkts cnt          0000000000047337        37 -
   34 igr rx pl: dropped pkts cnt                     0000000000001371        37 -
   35 igr rx pl: EM-IPL i/f dropped pkts cnt          0000000000000212        38 -
   36 igr rx pl: dropped pkts cnt                     0000000000000012        38 -

! Output omitted for brevity
|---------------------------------------------------------------------------|
| Device:Flanker Xbar Driver         Role:XBR-INTF              Mod: 3      |
| Last cleared @ Fri Jun  2 00:28:46 2017                                   |
|---------------------------------------------------------------------------|
|---------------------------------------------------------------------------|
| Device:Flanker Queue Driver        Role:QUE                   Mod: 3      |
| Last cleared @ Fri Jun  2 00:28:46 2017                                   |
|---------------------------------------------------------------------------|
Instance:4
Cntr  Name                                           Value                Ports
----- -----                                          -----                -----
    0 igr ib_500: pkt drops                          0000000000000003     35 -
    1 igr ib_500: pkt drops                          0000000000000010     36 -
    2 igr ib_500: vq ib pkt drops                    0000000000000013     33-40 -
    3 igr vq: l2 pkt drop count                      0000000000000013     33-40 -
    4 igr vq: total pkts dropped                     0000000000000013     33-40 -


Instance:5
Cntr  Name                                           Value                Ports
----- -----                                          -----                -----
    5 igr ib_500: de drops, shared by parser and de  0000000000000004     41-48 -
    6 igr ib_500: vq ib pkt drops                    0000000000000004     41-48 -
    7 igr vq: l2 pkt drop count                      0000000000000004     41-48 -
    8 igr vq: total pkts dropped                     0000000000000004     41-48 -



|---------------------------------------------------------------------------|
| Device:Lightning                   Role:ARB-MUX               Mod: 3      |
| Last cleared @ Fri Jun  2 00:28:46 2017                                   |
|---------------------------------------------------------------------------|
```

管理引擎卡、线卡和交换矩阵卡可以通过 EOBC（Ethernet Out-of-Band Channel，以太网带外通道）进行通信。如果 EOBC 通道出现了差错，那么 Nexus 交换机就可能出现丢包和重大服务中断。使用 **show hardware internal cpu-mac eobc stats** 命令可以验证 EOBC 的差错情况，在输出结果的"Error Counters"（差错计数器）部分显示了 EOBC 接口的差错列表。在大多数情况下，只要在物理上重新安装线卡就可以修复 EOBC 差错。例 3-13 显示了 Nexus 7000 交换机的 EOBC 差错计数器（Error Counter）统计信息，该例通过关键字 **grep** 来过滤输出结果，从而仅检查差错计数器，如例 3-13 所示。

例 3-13 EOBC 统计信息和差错计数器

```
N7k-1# show hardware internal cpu-mac eobc stats | grep -a 26 Error.counters
Error counters
--------------------------------+--
CRC errors ................... 0
Alignment errors ............. 0
```

```
Symbol errors .................. 0
Sequence errors ................ 0
RX errors ...................... 0
Missed packets (FIFO overflow)   0
Single collisions .............. 0
Excessive collisions ........... 0
Multiple collisions ............ 0
Late collisions ................ 0
Collisions ..................... 0
Defers ......................... 0
Tx no CRS ...................... 0
Carrier extension errors ....... 0
Rx length errors ............... 0
FC Rx unsupported .............. 0
Rx no buffers .................. 0
Rx undersize ................... 0
Rx fragments ................... 0
Rx oversize .................... 0
Rx jabbers ..................... 0
Rx management packets dropped .. 0
Tx TCP segmentation context .... 0
Tx TCP segmentation context fail 0
```

此外，Nexus 平台还为 CPU 处理的数据包提供了带内统计信息。如果差错计数器显示带内统计信息频繁增加，那么就可能是管理引擎卡出现了问题，导致出现了丢包。使用 **show hardware internal cpu-mac inband stats** 命令可以查看 CPU 带内统计信息，该命令可以显示 CPU 接收或发送的数据包以及包长等统计信息、中断计数器、差错计数器以及当前和最大转存统计数据。例 3-14 的输出结果显示了 Nexus 7000 交换机的带内统计信息，该命令也可以用在 Nexus 9000 交换机上（如第二部分输出结果所示）。

例 3-14　Nexus 7000/Nexus 9000 带内统计信息

```
N7k-1# show hardware internal cpu-mac inband stats

RMON counters                         Rx                   Tx
----------------------+--------------------+--------------------
total packets                    1154193               995903
good packets                     1154193               995903
64 bytes packets                       0                    0
65-127 bytes packets              432847               656132
128-255 bytes packets             429319                 8775
256-511 bytes packets             236194               328244
512-1023 bytes packets               619                   18
1024-max bytes packets             55214                 2734
broadcast packets                      0                    0
multicast packets                      0                    0
good octets                    262167681            201434260
total octets                           0                    0
XON packets                            0                    0
XOFF packets                           0                    0
management packets                     0                    0

! Output omitted for brevity

Interrupt counters
-------------------+--
```

```
Assertions              1176322
Rx packet timer         1154193
Rx absolute timer       0
Rx overrun              0
Rx descr min thresh     0
Tx packet timer         0
Tx absolute timer       1154193
Tx queue empty          995903
Tx descr thresh low     0
```

Error counters
```
--------------------------------+--
CRC errors .................... 0
Alignment errors .............. 0
Symbol errors ................. 0
Sequence errors ............... 0
RX errors ..................... 0
Missed packets (FIFO overflow)  0
Single collisions ............. 0
Excessive collisions .......... 0
Multiple collisions ........... 0
Late collisions ............... 0
Collisions .................... 0
Defers ........................ 0
Tx no CRS ..................... 0
Carrier extension errors ...... 0
Rx length errors .............. 0
FC Rx unsupported ............. 0
Rx no buffers ................. 0
Rx undersize .................. 0
Rx fragments .................. 0
Rx oversize ................... 0
Rx jabbers .................... 0
Rx management packets dropped . 0
Tx TCP segmentation context ... 0
Tx TCP segmentation context fail 0
```

Throttle statistics
```
------------------------------+---------
Throttle interval ........... 2 * 100ms
Packet rate limit ........... 64000 pps
Rate limit reached counter .. 0
Tick counter ................ 193078
Active ...................... 0
Rx packet rate (current/max) 3 / 182 pps
Tx packet rate (current/max) 2 / 396 pps
```

NAPI statistics
```
----------------+---------
Weight Queue 0 ......... 512
Weight Queue 1 ......... 256
Weight Queue 2 ......... 128
Weight Queue 3 ......... 16
Weight Queue 4 ......... 64
Weight Queue 5 ......... 64
Weight Queue 6 ......... 64
Weight Queue 7 ......... 64
```

```
Poll scheduled . 1176329
Poll rescheduled 0
Poll invoked ... 1176329
Weight reached . 0
Tx packets ..... 995903
Rx packets ..... 1154193
Rx congested ... 0
Rx redelivered . 0

qdisc stats:
----------------+---------
Tx queue depth . 10000
qlen ........... 0
packets ........ 995903
bytes .......... 197450648
drops .......... 0

Inband stats
----------------+---------
Tx src_p stamp . 0
N9396PX-5# show hardware internal cpu-mac inband stats
================ Packet Statistics ======================
Packets received:                58021524
Bytes received:                  412371530221
Packets sent:                    57160641
Bytes sent:                      409590752550
Rx packet rate (current/peak):   0 / 281 pps
Peak rx rate time:               2017-03-08 19:03:21
Tx packet rate (current/peak):   0 / 289 pps
Peak tx rate time:               2017-04-24 14:26:36
```

注：不同的 Nexus 平台的输出结果有所不同。例如，前面的输出结果很简单，来自 Nexus 9396 PX 交换机。相同的命令在 Nexus 9508 交换机上的输出结果与 Nexus 7000 交换机相似，该命令适用于所有 Nexus 平台。

从上面的输出结果可以看出，Nexus 9396 的带内统计命令显示了流量达到峰值速率的时间（虽然很简洁）。不过，Nexus 7000 交换机的带内统计命令不显示此类信息，Nexus 7000 提供了 **show hardware internal cpu-mac inband events** 命令，该命令可以显示 CPU 入站（Rx）或出站（Tx）方向流量速率的事件历史记录，包括峰值速率。例 3-15 显示了 CPU 入站或出站方向流量速率的带内事件历史记录，峰值流量速率的时间戳对于排查 Nexus 7000 交换机的高 CPU 利用率或丢包问题来说非常有用。

例 3-15　Nexus 7000 带内事件

```
N7k-1# show hardware internal cpu-mac inband events

1) Event:TX_PPS_MAX, length:4, at 546891 usecs after Fri Jun  2 01:34:38 2017
    new maximum = 396


2) Event:TX_PPS_MAX, length:4, at 526888 usecs after Fri Jun  2 01:31:57 2017
    new maximum = 219


3) Event:TX_PPS_MAX, length:4, at 866931 usecs after Fri Jun  2 00:31:30 2017
    new maximum = 180
```

```
 4) Event:RX_PPS_MAX, length:4, at 866930 usecs after Fri Jun  2 00:31:30 2017
    new maximum = 182

 5) Event:TX_PPS_MAX, length:4, at 826891 usecs after Fri Jun  2 00:30:47 2017
    new maximum = 151

 6) Event:RX_PPS_MAX, length:4, at 826890 usecs after Fri Jun  2 00:30:47 2017
    new maximum = 152
! Output omitted for brevity
```

NX-OS 提供了简短的带内计数器 CLI，可以显示入站（Rx）和出站（Tx）方向的带内数据包数量、差错、丢包计数器、溢出计数器等信息。这些信息可以快速确定带内流量是否出现丢包情况。例 3-16 显示了 **show hardware internal cpu-mac inband counters** 命令的输出结果，如果发现差错、丢包或溢出计数器存在非零情况，那么就需要使用关键字 **diff** 来进一步确定这些计数器是否存在频繁增加现象，该命令适用于所有 Nexus 平台。

例 3-16　Nexus 带内计数器

```
N7k-1# show hardware internal cpu-mac inband counters
eth0      Link encap:Ethernet   HWaddr 00:0E:0C:FF:FF:FF
          inet addr:127.5.1.5  Bcast:127.5.1.255  Mask:255.255.255.0
          inet6 addr: fe80::20e:cff:feff:ffff/64 Scope:Link
          UP BROADCAST RUNNING PROMISC MULTICAST  MTU:9338  Metric:1
          RX packets:2475891 errors:0 dropped:0 overruns:0 frame:0
          TX packets:5678434 errors:0 dropped:0 overruns:0 carrier:0
          collisions:0 txqueuelen:10000
          RX bytes:799218439 (762.1 MiB)  TX bytes:1099385202 (1.0 GiB)
```

Nexus 交换机出现丢包问题的主要原因是各种硬件故障，线卡或管理引擎模块本身都可能出现丢包问题。如果要查看 Nexus 交换机所有模块的差错及计数器信息，则可以使用 **show hardware internal errors [all | module** *slot***]** 命令。例 3-17 显示了 Nexus 7000 交换机的硬件内部差错信息，请注意，该命令适用于所有 Nexus 平台。

例 3-17　硬件内部差错

```
N7k-1# show hardware internal errors
|-----------------------------------------------------------------------|
| Device:Clipper MAC              Role:MAC                Mod: 1        |
| Last cleared @ Wed May 31 12:59:42 2017                               |
| Device Statistics Category :: ERROR                                   |
|-----------------------------------------------------------------------|
Instance:0
Cntr Name                                         Value           Ports
---- ----                                         -----           -----
 148 GD GMAC rx_config_word change interrupt      0000000000000001  - I1
2196 GD GMAC rx_config_word change interrupt      0000000000000003  - I2
2202 GD GMAC symbol error interrupt               0000000000000002  - I2
2203 GD GMAC sequence error interrupt             0000000000000002  - I2
2207 GD GMAC transition from sync to nosync int   0000000000000002  - I2



|-----------------------------------------------------------------------|
```

```
| Device:Clipper XBAR                    Role:QUE                  Mod: 1       |
| Last cleared @ Wed May 31 12:59:42 2017                                       |
| Device Statistics Category :: ERROR                                           |
|-------------------------------------------------------------------------------|
|-------------------------------------------------------------------------------|
| Device:Clipper FWD                     Role:L2                   Mod: 1       |
| Last cleared @ Wed May 31 12:59:42 2017                                       |
| Device Statistics Category :: ERROR                                           |
|-------------------------------------------------------------------------------|
! Output omitted for brevity
```

注：每种 Nexus 平台都有各自不同的可以观察差错或丢包情况的 ASIC，不过，这部分内容不在本书写作范围内。分析故障统计信息时，建议收集 **show tech-support detail** 和 **tac-pac** 命令的输出结果，这些信息有助于识别导致丢包问题的平台级故障。

3.1.3 Nexus FEX

FEX（Fabric Extender，交换矩阵扩展器）是一种固定配置的 1RU 机箱，旨在为服务器提供架顶式连接。顾名思义，FEX 本身并不能单独工作，主要用来扩展 Nexus 交换机的架构和功能。FEX 可以连接 Nexus 9000、7000、6000 和 5000 系列父交换机，将 FEX 连接到父交换机的上行链路端口称为交换矩阵端口或 NIF（Network-facing Interface，面向网络的接口）端口，将 FEX 模块连接服务器的端口（前面板端口）称为卫星端口或 HFI（Host-facing Interface，面向主机的接口）端口。思科按功能和容量发布了 3 种 FEX 模块。

- 1GE FEX。
  - N2224TP，24 端口。
  - N2248TP，48 端口。
  - N2248TP-E，48 端口。
- 10GBASE-T FEX。
  - N2332TQ，32 端口。
  - N2348TQ，48 端口。
  - N2348TQ-E，48 端口。
  - N2232TM，32 端口。
  - N2232TM-E，32 端口。
- 10G SFP + FEX。
  - N2348UPQ，48 端口。
  - N2248PQ，48 端口。
  - N2232PP，48 端口。

注：FEX 与父交换机之间的兼容性基于 Nexus 交换机使用的软件版本的软件发行说明。

父交换机与 FEX 之间支持以下 3 种连接模式。

- **固定模式（Pinning）**：在固定模式下，HIF 端口与上行链路端口是一对一的映射关系，因而来自特定 HIF 端口的流量只能穿越特定上行链路。上行链路端口出现故障后，会关闭所映射的 HIF 端口。
- **端口通道模式（Port-channeling）**：端口通道模式将上行链路视为一个逻辑接口，父交换机与 FEX 之间的所有流量都以散列方式分布到端口通道的不同链路上。

- **混合模式（Hybrid）**：该模式是固定模式和端口通道模式的组合。该模式将上行链路端口分成两个端口通道，HIF 端口固定连接特定的上行链路端口通道。

> **注**：有关 FEX 支持和不支持的功能特性信息，请参阅第 4 章。

NX-OS 在启用 FEX 的时候，首先需要使用 **install feature-set fex** 命令启用 FEX 功能特性集。如果在 Nexus 7000 上启用 FEX，那么 FEX 功能特性集将安装在默认 VDC 中，并应用 **no hardware ip verify address reserved** 命令，然后在相关 VDC 下配置 **feature-set fex** 命令。仅在启用 IDS（Intrusion Detection System，入侵检测系统）保留地址检查特性时，才需要应用 **no hardware ip verify address reserved** 命令，可以通过 **show hardware ip verify** 命令加以验证，如果已禁用该检查，那么就不需要配置 **no hardware ip verify address reserved** 命令。

执行了 **feature-set fex** 命令之后，可以通过 **switchport mode fex-fabric** 命令将接口启用为 FEX 交换矩阵端口。接下来需要为 FEX 分配 ID（可以利用该 ID 区分交换机上的 FEX）。例 3-18 给出了 Nexus 交换机连接 FEX 的配置示例。

例 3-18 FEX 配置

```
N9k-1(config)# install feature-set fex
N9k-1(config)# feature-set fex
N9k-1(config)# interface Eth3/41-44
N9k-1(config-if)# channel-group 1
N9k-1(config-if)# no shutdown
N9k-1(config-if)# exit
N9k-1(config)# interface port-channel1
N9k-1(config-if)# switchport
N9k-1(config-if)# switchport mode fex-fabric
N9k-1(config-if)# fex associate 101
N9k-1(config-if)# no shutdown
```

完成了 FEX 配置之后，就可以在父交换机上访问 FEX，并对其接口进行进一步配置。如果要验证 FEX 的状态，可以使用 **show fex** 命令，该命令可以显示 FEX 的状态、FEX 模块号以及与父交换机相关联的 ID。如果要确定父交换机能够访问的 FEX 接口，可以使用 **show interface** *interface-id* **fex-intf** 命令。请注意，该命令中的 *interface-id* 是 NIF 端口通道接口。例 3-19 给出了 **show fex** 和 **show interface fex-intf** 命令的输出结果示例，验证了 FEX 的状态及其接口信息。

例 3-19 FEX 验证

```
Leaf1# show fex
  FEX         FEX              FEX                        FEX
Number      Description      State            Model              Serial
---------------------------------------------------------------------------
101         FEX0101          Online           N2K-C2248TP-1GE    JAF1424AARL
Leaf1# show interface port-channel 1 fex-intf
Fabric          FEX
Interface       Interfaces
---------------------------------------------------------------------
Po1             Eth101/1/48    Eth101/1/47    Eth101/1/46    Eth101/1/45
                Eth101/1/44    Eth101/1/43    Eth101/1/42    Eth101/1/41
                Eth101/1/40    Eth101/1/39    Eth101/1/38    Eth101/1/37
                Eth101/1/36    Eth101/1/35    Eth101/1/34    Eth101/1/33
                Eth101/1/32    Eth101/1/31    Eth101/1/30    Eth101/1/29
                Eth101/1/28    Eth101/1/27    Eth101/1/26    Eth101/1/25
                Eth101/1/24    Eth101/1/23    Eth101/1/22    Eth101/1/21
```

```
                        Eth101/1/20    Eth101/1/19    Eth101/1/18    Eth101/1/17
                        Eth101/1/16    Eth101/1/15    Eth101/1/14    Eth101/1/13
                        Eth101/1/12    Eth101/1/11    Eth101/1/10    Eth101/1/9
                        Eth101/1/8     Eth101/1/7     Eth101/1/6     Eth101/1/5
                        Eth101/1/4     Eth101/1/3     Eth101/1/2     Eth101/1/1
```

show fex *fex-number* **detail** 命令可以查看有关 FEX 的更多详细信息，该命令可以显示 FEX 的状态和所有 FEX 接口信息，同时还能显示固定模式的详细信息以及与 FEX 交换矩阵端口相关的信息。例 3-20 显示了 FEX 101 的详细信息。

例 3-20　FEX 详细信息

```
Leaf1# show fex 101 detail
FEX: 101 Description: FEX0101    state: Online
  FEX version: 6.2(12) [Switch version: 6.2(12)]
  FEX Interim version: (12)FH_0_171
  Switch Interim version: 6.2(12)
  Extender Serial: FOC1710R0JF
  Extender Model: N2K-C2248PQ-10GE,   Part No: 73-14775-03
  Card Id: 207, Mac Addr: f0:29:29:ff:8e:c2, Num Macs: 64
  Module Sw Gen: 21   [Switch Sw Gen: 21]
  Pinning-mode: static     Max-links: 1
  Fabric port for control traffic: Eth3/41
  FCoE Admin: false
  FCoE Oper: false
  FCoE FEX AA Configured: false
  Fabric interface state:
    Po1 - Interface Up. State: Active
    Eth3/41 - Interface Up. State: Active
    Eth3/42 - Interface Up. State: Active
    Eth3/43 - Interface Up. State: Active
    Eth3/44 - Interface Up. State: Active
  Fex Port         State   Fabric Port
       Eth101/1/1    Down        Po1
       Eth101/1/2    Up          Po1
       Eth101/1/3    Down        Po1
       Eth101/1/4    Down        Po1
       Eth101/1/5    Down        Po1
       Eth101/1/6    Down        Po1
       Eth101/1/7    Down        Po1
       Eth101/1/8    Down        Po1
       Eth101/1/9    Down        Po1
       Eth101/1/10   Down        Po1
! Output omitted for brevity
```

FEX 卫星端口可用之后，就可以将这些端口配置为二层或三层端口。此外，如果将这些端口作为 vPC 配置的一部分，那么就可以让它们充当双活（active-active）端口。

如果交换矩阵端口或卫星端口出现了问题，那么就可以使用 **show system internal fex info fport** [**all** | *interface-number*]或 **show system internal fex info satport** [**all** | *interface-number*]命令查看状态变更信息。例 3-21 显示了 Nexus 7000 交换机的卫星端口和交换矩阵端口的内部信息，输出结果的第一部分显示了系统启用 FEX 所经历的事件列表，列出了所有的有限状态机事件，这些信息对于排查 FEX 未正常启动而陷入某种状态的故障来说非常有用。输出结果的第二部分显示了卫星端口及其状态信息等内容。

例 3-21 FEX 内部信息

```
Leaf1# show system internal fex info fport all
  intf      ifindex    Oper chass module-id      Sdp Rx Sdp Tx State  AA mode
       Po1 0x16000000   Up  101  0x000000000000      0      0 Active      0
Interface :      Po1 - 0x16000000 Up Remote chassis: 101
    satellite: 0x0,  SDP state Init, Rx:0, Tx:0
    Not Fabric mode. satellite Not Bound. Fport state: Active
    fabric slot:33, SDP module id:0x0, rlink: 0x0
    parent:0x0 num mem: 4 num mem up: 4
    Active members(4): Eth3/41, Eth3/42, Eth3/43, Eth3/44,
    Flags: , , ,
    Fcot: Not checked, Not valid, Not present
 Fex AA Mode: 0
 Err disable Mode: 0
 Oper fabric mode: 0
Logs:
06/04/2017 15:30:19.553797: Remote-chassis configured
    Eth3/41 0x1a128000   Up  101  0xc08eff2929f0    169    175 Active      0
Interface : Eth3/41 - 0x1a128000 Up Remote chassis: 101
    satellite: 0xc08eff2929f0,  SDP state Active, Rx:169, Tx:175
    Fabric mode. satellite Bound. Fport state: Active
    fabric slot:33, SDP module id:0xc08eff2929f0, rlink: 0x20000000
    parent:0x16000000 num mem: 0 num mem up: 0
    Active members(0):
    Flags: , Bundle membup rcvd, , Switchport fabric,
    Fcot: Checked, Valid, Present
 Fex AA Mode: 0
 Err disable Mode: 0
 Oper fabric mode: 2
Logs:
06/04/2017 15:29:32.706998: pre config: is not a port-channel member
06/04/2017 15:29:32.777929: Interface Up
06/04/2017 15:29:32.908528: Fcot message sent to Ethpm
06/04/2017 15:29:32.908649: Satellite discovered msg sent
06/04/2017 15:29:32.908744: State changed to: Discovered
06/04/2017 15:29:32.909163: Fcot response received. SFP valid
06/04/2017 15:29:38.931664: Interface Down
06/04/2017 15:29:38.931787: State changed to: Created
06/04/2017 15:29:40.852076: Interface Up
06/04/2017 15:29:42.967594: Fcot message sent to Ethpm
06/04/2017 15:29:42.967661: Satellite discovered msg sent
06/04/2017 15:29:42.967930: State changed to: Discovered
06/04/2017 15:29:42.968363: Fcot response received. SFP valid
06/04/2017 15:29:45.306713: Interface Down
06/04/2017 15:29:45.306852: State changed to: Created
06/04/2017 15:29:45.462260: pre config: is not a port-channel member
06/04/2017 15:30:15.798370: Interface Up
06/04/2017 15:30:15.801215: Port Bringup rcvd
06/04/2017 15:30:15.802072: Suspending Fabric port. reason: Fex not configured
06/04/2017 15:30:15.802106: fport bringup retry end: sending out resp
06/04/2017 15:30:17.413620: Fcot message sent to Ethpm
06/04/2017 15:30:17.413687: Satellite discovered msg sent
06/04/2017 15:30:17.413938: State changed to: Discovered
06/04/2017 15:30:17.414382: Fcot response received. SFP valid
06/04/2017 15:30:19.554112: Port added to port-channel
06/04/2017 15:30:19.554266: State changed to: Configured
```

```
06/04/2017 15:30:19.554874: Remote-chassis configured
06/04/2017 15:30:19.568677: Interface Down
06/04/2017 15:30:19.685945: Port removed from port-channel
06/04/2017 15:30:19.686854: fport phy cleanup retry end: sending out resp
06/04/2017 15:30:19.689911: pre config: is a port-channel member
06/04/2017 15:30:19.689944: Port added to port-channel
06/04/2017 15:30:19.690170: Remote-chassis configured
06/04/2017 15:30:19.690383: Port changed to fabric mode
06/04/2017 15:30:19.817093: Interface Up
06/04/2017 15:30:19.817438: Started SDP
06/04/2017 15:30:19.817495: State changed to: Fabric Up
06/04/2017 15:30:19.817991: Port Bringup rcvd
06/04/2017 15:30:19.923327: Fcot message sent to Ethpm
06/04/2017 15:30:19.923502: Fcot response received. SFP valid
06/04/2017 15:30:19.923793: Advertizing Vntag
06/04/2017 15:30:19.924329: State changed to: Connecting
06/04/2017 15:30:21.531270: Satellite connected. Bind msg sent
06/04/2017 15:30:21.532110: fport bringup retry end: sending out resp
06/04/2017 15:30:21.534074: State changed to: Active
06/04/2017 15:30:21.640543: Bundle member bringup rcvd
! Output omitted for brevity
N7kA-1-N7KA-LEAF1# show system internal fex info satport ethernet 101/1/1
  Interface-Name   ifindex    State Fabric-if  Pri-fabric Expl-Pinned
      Eth101/1/1 0x1f640000 Down          Po1        Po1    NoConf
  Port Phy Not Up. Port dn req: Not pending
```

注：如果 FEX 出现了故障，那么建议在故障状态下收集 **show tech-support fex** *fex-number* 命令的输出结果。故障可能来自于 Nexus 的 EthPM（Ethernet Port Manager，以太网端口管理器）组件，因为 FEX 会将状态变更消息发送给 EthPM，所以在故障状态下收集 **show tech-support ethpm** 命令的输出结果也很有必要。有关 EthPM 的详细内容将在本章后面讨论。

3.2 VDC

VDC（Virtual Device Context，虚拟设备上下文）是物理设备的逻辑分区，可以提供软件故障隔离和独立管理每个分区的能力。由于每个 VDC 实例都运行自己的路由协议服务实例，因而能够更好地利用系统资源。创建 VDC 之前，必须记住以下几点。

- 只有拥有 network-admin 角色的用户才能创建 VDC 并为其分配资源。
- VDC1（默认 VDC）始终处于活动状态且无法删除。
- VDC 的名称不区分大小写。
- 仅 Nexus 7000 或 7700 系列交换机支持 VDC。
- Supervisor 1（管理引擎卡）和 Supervisor 2 最多支持 4 个 VDC，Supervisor 2E 最多支持 8 个 VDC。
- 运行 Supervisor 2 或 Supervisor 2E 卡、NX-OS 6.1(1) 及以后版本的 Nexus 交换机支持管理 VDC（Admin VDC）。

Nexus 7000 平台支持以下 3 类 VDC。

- **以太网 VDC**：支持传统的 L2/L3 协议。
- **存储 VDC**：支持与 FCoE（Fiber Channel over Ethernet，以太网光纤通道）相关的特定协议，如 FIP（FCoE Initialization Protocol，FCoE 初始化协议）。
- **管理 VDC**：为整个系统提供管理控制功能，帮助管理系统上配置的其他 VDC。

3.2.1 VDC 资源模板

VDC 资源模板允许用户为具有资源需求相同的 VDC 分配资源。除非将资源模板分配给 VDC，否则资源模板不会生效。使用资源模板可以最大限度地减少配置工作量，同时还能简化 Nexus 平台的管理操作。可以通过以下参数来限制每个 VDC 资源模板中的资源。

- **Monitor-session**：SPAN 会话数。
- **Port-channel**：端口通道数。
- **U4route-mem**：IPv4 路由内存限制。
- **U6route-mem**：IPv6 路由内存限制。
- **M4route-mem**：IPv4 多播内存限制。
- **M6route-mem**：IPv6 多播内存限制。
- **Vlan**：VLAN 数量。
- **Vrf**：VRF（Virtual Routing and Forwarding，虚拟路由和转发）实例数量。

可以通过 **vdc resource template** *name* 命令配置 VDC 资源模板，该命令将进入资源模板配置模式，此时可以通过 **limit-resource** *resource* **minimum** *value* **maximum** *value* 命令限制前面提到的资源，其中，*resource* 可以是上面列出的资源中的任何一个。如果要查看资源模板中配置的资源，可以使用 **show vdc resource template** [**vdc-default** | *name*] 命令，其中，**vdc-default** 用于默认 VDC 模板。例 3-22 给出了 VDC 模板的配置示例，**show vdc resource template** 命令的输出结果显示了资源模板中配置的资源信息。

例 3-22　VDC 资源模板

```
! Default VDC Template
N7K-1# show vdc resource template vdc-default

  vdc-default
  --------------
    Resource                              Min          Max
    ---------                             -----        -----
    monitor-rbs-product                   0            12
    monitor-rbs-filter                    0            12
    monitor-session-extended              0            12
    monitor-session-mx-exception-src      0            1
    monitor-session-inband-src            0            1
    port-channel                          0            768
    monitor-session-erspan-dst            0            23
    monitor-session                       0            2
    vlan                                  16           4094
    anycast_bundleid                      0            16
    m6route-mem                           5            20
    m4route-mem                           8            90
    u6route-mem                           4            4
    u4route-mem                           8            8
    vrf                                   2            4096
N7K-1(config)# vdc resource template DEMO-TEMPLATE
N7K-1(config-vdc-template)# limit-resource port-channel minimum 1 maximum 4
N7K-1(config-vdc-template)# limit-resource vrf minimum 5 maximum 100
N7K-1(config-vdc-template)# limit-resource vlan minimum 20 maximum 200
N7K-1# show vdc resource template DEMO-TEMPLATE

  DEMO-TEMPLATE
  --------------
```

```
   Resource                  Min         Max
   ----------                -----       -----
     vlan                     20         200
     vrf                       5         100
     port-channel              1           4
```

如果网络要求 Nexus 的所有 VDC 执行不同的任务并分配不同类型的资源,那么最好不要配置 VDC 模板,可以在 **vdc** 配置模式下使用 **limit-resource** 命令限制 VDC 资源。

3.2.2 配置 VDC

可以将 VDC 的创建过程分为以下 4 个步骤。

- **第 1 步:定义 VDC**。使用命令 **vdc** *name* [**id** *id*] [**type Ethernet | storage**]定义 VDC。默认情况下,创建的 VDC 是以太网 VDC。
- **第 2 步:分配接口**。将单个或多个接口分配给 VDC,可以通过配置命令 **allocate interface** *interface-id* 分配接口。需要注意的是,**allocate interface** 配置是强制项,不能省略接口分配过程。接口只能从一个 VDC 分配到另一个 VDC,而且无法释放回默认 VDC。如果用户删除了 VDC,那么分配给该 VDC 的接口就会变成未分配接口,成为 VDC ID 0 的一部分。对于 10G 接口来说,某些模块要求同时移动与端口 ASIC 相关联的所有端口,这样做的目的是保持完整性,使每个端口组都能在专用模式与共享模式之间进行切换。如果不是同时分配同一端口组的所有成员,那么就会显示出错消息。从 NX-OS 版本 5.2(1)开始,如果仅将端口组的一个成员添加到 VDC 中,那么该端口组的所有成员都将自动分配给 VDC。
- **第 3 步:定义 HA(High Availability,高可用性)策略**。配置 HA 策略的依据是 Nexus 运行的是单管理引擎卡还是双管理引擎卡。在 VDC 配置下使用 **ha-policy [single-sup | dual-sup]** *policy* 命令可以配置 HA 策略。表 3-3 基于单管理引擎卡或双管理引擎卡列出了不同的 HA 策略。

表 3-3   HA 策略

单 SUP	双 SUP
关闭(Bringdown)	关闭(Bringdown)
重启(Restart)(默认策略)	重启(Restart)
重置(Reset)	切换(Switchover)(默认策略)

- **第 4 步:限制资源**。可以通过应用 VDC 资源模板或通过 **limit-resource** 命令手工分配资源来限制 VDC 资源。由于某些资源无法作为资源模板的一部分进行分配,因而 **limit-resource** 命令是必需的。此外,**limit-resource** 命令还允许用户定义 VDC 支持的模块类型。VDC 初始化之后,只能通过 **limit-resource** 命令修改该 VDC 的资源,此时的资源模板选项无效。

例 3-23 给出了以太网 VDC 的创建示例。请注意,如果将某接口添加到 VDC 中,而端口组的其他成员不是列表的一部分,那么 NX-OS 就会自动尝试将其余端口也添加到 VDC 中。例 3-23 定义的 VDC 仅限于 F3 系列模块,例如,如果添加 F2 或 M2 系列模块的端口,那么就会出现差错。

例 3-23   VDC 配置

```
N7K-1(config)# vdc N7K-2
Note: Creating VDC, one moment please ...
2017 Apr 21 03:51:55 %$ VDC-5 %$ %SYSLOG-2-SYSTEM_MSG : Syslogs wont be logged into
  logflash until logflash is online
```

```
N7K-1(config-vdc)#
N7K-1(config-vdc)# limit-resource module-type f3
This will cause all ports of unallowed types to be removed from this vdc. Continue
   (y/n)? [yes] yes
N7K-1(config-vdc)# allocate interface ethernet 3/1
Entire port-group is not present in the command. Missing ports will be included
   automatically
Additional Interfaces Included are :
    Ethernet3/2
    Ethernet3/3
    Ethernet3/4
    Ethernet3/5
    Ethernet3/6
    Ethernet3/7
    Ethernet3/8
Moving ports will cause all config associated to them in source vdc to be removed.
   Are you sure you want to move the ports (y/n)?  [yes] yes
N7K-1(config-vdc)# ha-policy dual-sup ?
  bringdown    Bring down the vdc
  restart      Bring down the vdc, then bring the vdc back up
  switchover   Switchover the supervisor
N7K-1(config-vdc)# ha-policy dual-sup restart
N7K-1(config-vdc)# ha-policy single-sup bringdown
N7K-1(config-vdc)# limit-resource port-channel minimum 3 maximum 5
N7K-1(config-vdc)# limit-resource vlan minimum 20 maximum 100
N7K-1(config-vdc)# limit-resource vrf minimum 5 maximum 10
```

3.2.3 VDC 初始化

在应用与 VDC 相关的特定配置之前，需要先初始化 VDC。在初始化 VDC 之前，需要先创建 VDC 并执行 **copy run start** 命令，从而使新创建的 VDC 成为启动配置的一部分。可以从默认 VDC 或管理 VDC 执行 **switchto vdc** *name* 命令来初始化 VDC，如例 3-24 所示。VDC 的初始化过程与新 Nexus 交换机的启动过程相似，首先提示输入管理员密码，然后根据基本配置对话框进行配置。对话框方式可以对 VDC 进行基本的配置设定，当然，也可以在基本配置对话框中选择 "*no*" 来执行手动配置。命令 **switchback** 的作用是切换回默认或管理 VDC。

例 3-24   VDC 初始化

```
N7k-1# switchto vdc N7k-2
         ---- System Admin Account Setup ----

Do you want to enforce secure password standard (yes/no) [y]:

  Enter the password for "admin":
  Confirm the password for "admin":

         ---- Basic System Configuration Dialog VDC: 2 ----

This setup utility will guide you through the basic configuration of
the system. Setup configures only enough connectivity for management
of the system.

Please register Cisco Nexus7000 Family devices promptly with your
supplier. Failure to register may affect response times for initial
```

```
service calls. Nexus7000 devices must be registered to receive
entitled support services.

Press Enter at anytime to skip a dialog. Use ctrl-c at anytime
to skip the remaining dialogs.

Would you like to enter the basic configuration dialog (yes/no): yes


  Create another login account (yes/no) [n]:

  Configure read-only SNMP community string (yes/no) [n]:

  Configure read-write SNMP community string (yes/no) [n]:

  Enter the switch name : N7k-2

  Continue with Out-of-band (mgmt0) management configuration? (yes/no) [y]:

    Mgmt0 IPv4 address : 192.168.1.10

    Mgmt0 IPv4 netmask : 255.255.255.0

  Configure the default gateway? (yes/no) [y]:

    IPv4 address of the default gateway : 192.168.1.1

  Configure advanced IP options? (yes/no) [n]:

  Enable the telnet service? (yes/no) [n]: yes

  Enable the ssh service? (yes/no) [y]: yes

    Type of ssh key you would like to generate (dsa/rsa) [rsa]:

    Number of rsa key bits <1024-2048> [1024]:

  Configure default interface layer (L3/L2) [L3]:

  Configure default switchport interface state (shut/noshut) [shut]:

The following configuration will be applied:
  password strength-check
  switchname N7k-2
vrf context management
ip route 0.0.0.0/0 192.168.1.100
exit
  feature telnet
  ssh key rsa 1024 force
  feature ssh
  no system default switchport
  system default switchport shutdown
interface mgmt0
ip address 192.168.1.1 255.255.255.0
no shutdown

Would you like to edit the configuration? (yes/no) [n]:
```

```
Use this configuration and save it? (yes/no) [y]:
! Output omitted for brevity
N7k-1-N7k-2#
N7k-1-N7k-2# switchback
N7k-1#
```

从例 3-24 可以看出，VDC 初始化之后，VDC 的主机名变成了 N7k-1-N7k-2，也就是将默认 VDC 和新 VDC 的主机名串在了一起。如果要避免该行为，可以在默认或管理 VDC 中配置 **no vdc combined-hostname** 命令。

3.2.4 带外和带内管理

思科 NX-OS 软件提供了一个虚拟管理接口，用于每个 VDC 的带外管理。每个虚拟管理接口都配置了一个独立的 IP 地址，可以通过物理 mgmt0 接口进行访问。使用虚拟管理接口的作用是允许用户仅使用一个管理网络，该管理网络在多个 VDC 之间共享 AAA 服务器和 syslog 服务器。

此外，VDC 还支持带内管理。可以通过分配给 VDC 的以太网接口访问 VDC。采样带内管理机制需要使用唯一的独立管理网络，以确保不同 VDC 的 AAA 服务器和 syslog 服务器的分离。

3.2.5 VDC 管理

NX-OS 软件提供了强大的 CLI，允许用户在故障排查期间轻松管理 VDC。可以从默认或管理 VDC 查看所有 VDC 的配置信息，使用 **show run vdc** 命令可以查看所有与 VDC 相关的配置。此外，保存配置时，可以通过命令 **copy run start vdc-all** 复制所有 VDC 已经完成的配置。

NX-OS 提供的 CLI 可以在不查看配置信息的情况下查看 VDC 的详细信息。**show vdc [detail]** 命令可以查看每个 VDC 的详细信息，**show vdc detail** 命令可以显示每个 VDC 的各种信息列表，如 ID、名称、状态、HA 策略、CPU 份额、VDC 的创建时间和正常运行时间、VDC 类型以及每个 VDC 支持的线卡信息，如例 3-25 所示。对于 Nexus 7000 交换机来说，某些 VDC 可能运行了关键服务。NX-OS 默认为所有 VDC 都分配相同的 CPU 份额（CPU 资源）。对于管理引擎卡 SUP2 和 SUP2E 来说，NX-OS 允许用户为 VDC 分配特定数量的交换机 CPU，以便为更重要的 VDC 提供更高的优先级。

例 3-25 show vdc detail 命令输出结果

```
N7k-1# show vdc detail
Switchwide mode is m1 f1 m1xl f2 m2xl f2e f3 m3

vdc id: 1
vdc name: N7k-1
vdc state: active
vdc mac address: 50:87:89:4b:c0:c1
vdc ha policy: RELOAD
vdc dual-sup ha policy: SWITCHOVER
vdc boot Order: 1
CPU Share: 5
CPU Share Percentage: 50%
vdc create time: Fri Apr 21 05:57:30 2017
vdc reload count: 0
vdc uptime: 1 day(s), 0 hour(s), 35 minute(s), 41 second(s)
vdc restart count: 1
vdc restart time: Fri Apr 21 05:57:30 2017
vdc type: Ethernet
vdc supported linecards: f3
```

```
vdc id: 2
vdc name: N7k-2
vdc state: active
vdc mac address: 50:87:89:4b:c0:c2
vdc ha policy: RESTART
vdc dual-sup ha policy: SWITCHOVER
vdc boot Order: 1
CPU Share: 5
CPU Share Percentage: 50%
vdc create time: Sat Apr 22 05:05:59 2017
vdc reload count: 0
vdc uptime: 0 day(s), 1 hour(s), 28 minute(s), 12 second(s)
vdc restart count: 1
vdc restart time: Sat Apr 22 05:05:59 2017
vdc type: Ethernet
vdc supported linecards: f3
```

如果要进一步查看分配给每个 VDC 的详细资源信息，可以使用 **show vdc resource [detail]** 命令，该命令可以显示已配置的每种资源的最小值和最大值，以及每种资源的已使用值、未使用值和可用值。使用命令 **show vdc** *name* **resource [detail]** 可以显示各个 VDC 的详细资源信息。例 3-26 显示了 Nexus 7000 上的所有 VDC 的资源配置情况和利用率信息，该交换机运行了两个 VDC（如 N7k-1 和 N7k-2）。

例 3-26 show vdc resource detail 命令输出结果

```
N7k-1# show vdc resource detail

  vlan             34 used     8 unused   16349 free   16341 avail   16383 total
  ------
          Vdc              Min         Max          Used        Unused         Avail
          ---              ---         ---          ----        ------         -----
          N7k-1            16          4094         26          0              4068
          N7k-2            16          4094         8           8              4086

  monitor-session  0 used     0 unused      2 free       2 avail       2 total
  ---------------
          Vdc              Min         Max          Used        Unused         Avail
          ---              ---         ---          ----        ------         -----
          N7k-1            0           2            0           0              2
          N7k-2            0           2            0           0              2
  vrf              5 used     0 unused   4091 free    4091 avail    4096 total
  -----
          Vdc              Min         Max          Used        Unused         Avail
          ---              ---         ---          ----        ------         -----
          N7k-1            2           4096         3           0              4091
          N7k-2            2           4096         2           0              4091

  port-channel     5 used     0 unused    763 free     763 avail     768 total
  ------------
          Vdc              Min         Max          Used        Unused         Avail
          ---              ---         ---          ----        ------         -----
          N7k-1            0           768          5           0              763
          N7k-2            0           768          0           0              763
```

```
u4route-mem              2 used      102 unused    514 free    412 avail    516 total
-------------
          Vdc            Min          Max          Used        Unused       Avail
          ---            ---          ---          ----        ------       -----
          N7k-1          96           96           1           95           95
          N7k-2          8            8            1           7            7
! Output omitted for brevity
```

NX-OS 根据 VDC 支持的线卡类型，为每个 VDC 分配相应的接口。如果要查看每个 VDC 的成员接口，那么可以使用命令 **show vdc membership**。例 3-27 给出了 **show vdc membership** 命令的输出结果示例，请注意例 3-27 中属于 VDC 1（N7k-1）和 VDC 2（N7k-2）的不同接口。如果删除了特定 VDC，那么该 VDC 的接口将变成未分配接口，显示在 VDC ID 0 下面。

例 3-27　show vdc membership 命令输出结果

```
N7k-1# show vdc membership
Flags : b - breakout port
-------------------------------

vdc_id: 0 vdc_name: Unallocated interfaces:

vdc_id: 1 vdc_name: N7k-1 interfaces:
        Ethernet3/9            Ethernet3/10           Ethernet3/11
        Ethernet3/12           Ethernet3/13           Ethernet3/14
        Ethernet3/15           Ethernet3/16           Ethernet3/17
        Ethernet3/18           Ethernet3/19           Ethernet3/20
        Ethernet3/21           Ethernet3/22           Ethernet3/23
        Ethernet3/24           Ethernet3/25           Ethernet3/26
        Ethernet3/27           Ethernet3/28           Ethernet3/29
        Ethernet3/30           Ethernet3/31           Ethernet3/32
        Ethernet3/33           Ethernet3/34           Ethernet3/35
        Ethernet3/36           Ethernet3/37           Ethernet3/38
        Ethernet3/39           Ethernet3/40           Ethernet3/41
        Ethernet3/42           Ethernet3/43           Ethernet3/44
        Ethernet3/45           Ethernet3/46           Ethernet3/47
        Ethernet3/48

vdc_id: 2 vdc_name: N7k-2 interfaces:
        Ethernet3/1            Ethernet3/2            Ethernet3/3
        Ethernet3/4            Ethernet3/5            Ethernet3/6
        Ethernet3/7            Ethernet3/8
```

此外，NX-OS 还提供了内部事件历史记录日志，可以查看与 VDC 相关的差错或其他消息。使用 **show vdc internal event-history [errors |msgs | vdc_id** *id*]命令可以查看与 VDC 相关的调试信息。例 3-28 给出了创建新 VDC（N7k-3）的配置示例，同时还显示了相关的事件历史记录日志，这些事件历史记录日志显示了 VDC 创建过程所经历的事件（在创建并激活 VDC 以供使用之前）。例 3-28 中的事件表明正在创建 VDC，后面显示该 VDC 已变成激活状态。

例 3-28　VDC 内部事件历史记录日志

```
N7k-1(config)# vdc N7k-3
Note: Creating VDC, one moment please ...
2017 Apr 25 04:19:03  %$ VDC-3 %$ %SYSLOG-2-SYSTEM_MSG : Syslogs wont be logged into
  logflash until logflash is online
N7k-1(config-vdc)#
N7k-1# show vdc internal event-history vdc_id 3
```

```
1) Event:VDC_SEQ_CONFIG, length:170, at 74647 usecs after Tue Apr 25 04:20:31 2017
   vdc_id = 3    vdc_name = N7k-3    vdc_state = VDC_ACTIVE
   desc = VDC_CR_EV_SEQ_DONE

2) Event:VDC_SEQ_CONFIG, length:170, at 74200 usecs after Tue Apr 25 04:20:31 2017
   vdc_id = 3    vdc_name = N7k-3    vdc_state = VDC_CREATE_IN_PROGRESS
   desc = VDC_SHARE_SEQ_CHECK

3) Event:VDC_SEQ_PORT_CONFIG, length:216, at 74130 usecs after Tue Apr 25 04:20:31 2017
   vdc_id = 3    vdc_name = N7k-3    vdc_state = VDC_CREATE_IN_PROGRESS
   Dest_vdc_id = 3   Source_vdcs =    Num of Ports = 0

4) Event:E_MTS_RX, length:48, at 73920 usecs after Tue Apr 25 04:20:31 2017
   [RSP] Opc:MTS_OPC_VDC_PRE_CREATE(20491), Id:0X0047D41A, Ret:SUCCESS
   Src:0x00000101/179, Dst:0x00000101/357, Flags:None
   HA_SEQNO:0X00000000, RRtoken:0x0047D40D, Sync:UNKNOWN, Payloadsize:4
   Payload:
   0x0000:  00 00 00 00

5) Event:E_MTS_TX, length:50, at 36406 usecs after Tue Apr 25 04:20:31 2017
   [REQ] Opc:MTS_OPC_VDC_PRE_CREATE(20491), Id:0X0047D40D, Ret:SUCCESS
   Src:0x00000101/357, Dst:0x00000101/179, Flags:None
   HA_SEQNO:0X00000000, RRtoken:0x0047D40D, Sync:UNKNOWN, Payloadsize:6
   Payload:
   0x0000:  00 03 00 00 00 05
```

注：如果 VDC 出现了问题，那么就可以在故障状态下收集 **show tech-support vdc** 和 **show tech-support detail** 命令的输出结果，以开启 TAC 案例。

线卡互操作限制

虽然 VDC 的创建过程很简单，但是将机箱中安装的不同类型的模块接口分配给 VDC 时，可能就存在一定的挑战性。由于机箱中安装的线卡的组合方式不同，线卡的操作模式也有所不同，因而在限制 VDC 的模块类型资源时，务必注意 M 系列线卡和 F 系列线卡之间的兼容性。如果机箱中同时存在 F 和 M 系列线卡，那么请记住以下准则。

- F2E 和 M3 系列线卡的接口不能共存。
- 如果 M2 模块接口与 M3 模块接口一起使用，那么就不能将 M2 模块的接口分配给其他 VDC。
- 如果 VDC 中存在 M2 和 M3 系列线卡的接口，那么 M2 模块必须运行在 M2-M3 互操作模式下。
- 如果 VDC 中存在 F2E 和 M2 系列线卡的接口，那么 M2 模块必须运行在 M2-F2E 模式下。
- M2 模块必须处于 M2-F2E 模式下，才能运行在其他 VDC 中。

M2 系列线卡支持 M2-F2E 和 M2-M3 互操作模式，默认运行 M2-F2E 模式。另外，M3 系列线卡仅支持 M2-M3 互操作模式。如果要从属于同一 VDC 的 M2 和 M3 模块分配接口，那么就需要使用 **system interop-mode m2-m3 module** *slot* 命令将 M2 线卡的操作模式变更为 M2-M3。如果使用了选项 **no**，那么将禁用 M2-M3 模式，并返回 M2 线卡的默认 M2-F2E 模式。

如果要在同一个 VDC 支持 M 和 F2E 系列模块，那么就必须让 F2E 系列模块运行在代理模式下，该模式下的所有三层流量都将发送给同一 VDC 中的 M 系列线卡。

表 3-4 列出了以太网 VDC 支持的模块类型组合。

表 3-4　　　　　　　　　　以太网 VDC 支持的模块类型组合

模块	M1	F1	M1XL	M2	M3	F2	F2e	F3
M1	是	是	是	是	否	否	是	否
F1	是	是	是	是	否	否	否	否
M1XL	是	是	是	是	否	否	是	否
M2	是	是	是	是	是	否	是	否
M3	否	否	否	是	是	否	否	否
F2	否	否	否	否	否	是	是	否
F2e	是	否	是	是	否	是	是	是
F3	否	否	否	否	否	是	是	是

3.3　NX-OS 系统组件故障排查

由于 Nexus 是一种分布式架构平台，因而同时运行了 PI（Platform Independent，平台无关）和 PD（Platform Dependent，平台相关）功能特性。排查 PI 功能特性故障时（如路由协议控制平面），了解这些功能特性对于轻松隔离故障问题来说非常重要。不过，如果需要排查 PD 功能特性故障，那么就必须了解 NX-OS 的系统组件。

排查 PD 故障时，不但要了解各种系统组件，而且还要了解相关服务或组件。例如，RPM（Route Policy Manager，路由策略管理器）是一种依赖于 ARP（Address Resolution Protocol，地址解析协议）和 Netstack 进程的进程（见例 3-29），而这些进程又进一步依赖于其他进程。可以通过命令 **show system internal sysmgr service dependency srvname** *name* 查看功能特性的依赖关系层次结构。

例 3-29　功能特性依赖关系层次结构

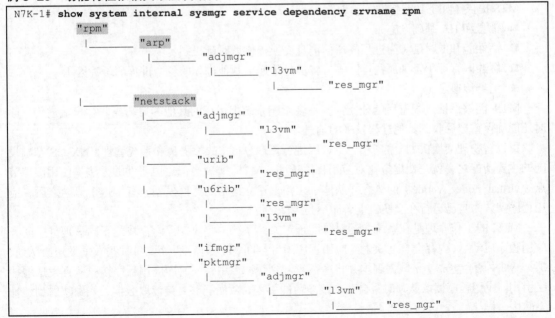

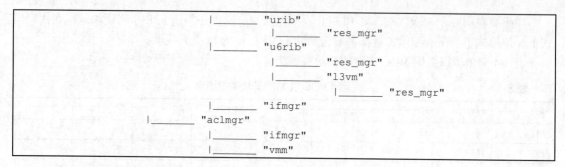

虽然了解所有组件并不现实，但是了解 NX-OS 平台执行重要任务的一些主要系统组件，能够有效帮助用户隔离故障问题。本节将重点介绍以下组件。

- MTS（Message and Transaction Service，消息和事务服务）。
- Netstack 和数据包管理器（Packet Manager）。
- ARP 和 AdjMgr。
- 转发组件。
  - URIB（Unicast Routing Information Base，单播路由信息库）、UFIB（Unicast Forwarding Information Base，单播转发信息库）和 UFDM（Unicast Forwarding Distribution Manager，单播转发分发管理器）。
- EthPM（Ethernet Port Manager，以太网端口管理器）和 Port-Client（端口客户端）。

3.3.1 MTS

MTS（Message and Transaction Service，消息和事务服务）是管理引擎和线卡进行进程间通信的基本通信方式，也就是说，MTS 是一种 IPC（Inter-Process Communication，进程间通信）代理，负责处理系统内的服务与硬件之间的消息路由和排队。另外，节点间通信（如管理引擎上的进程 A 与线卡上的进程 B 之间的通信）由 AIPC（Asynchronous Inter-Process Communication，异步进程间通信）进行处理，AIPC 提供了跨 EOBC（Ethernet Out of Band Channel，以太网带外通道）的可靠传输以及数据包的分段重组等功能特性。

MTS 提供了如下功能特性。

- 消息和 HA 基础设施。
- 高性能和低时延（为进程间通信消息交换提供低时延机制）。
- 缓冲区管理（管理那些排队等待将消息传递给其他进程的各个进程的缓冲区）。
- 消息传递。

MTS 能够保证进程重启的独立性，不会影响系统上运行的其他客户端或非客户端进程，同时还能确保重启后收到其他进程发来的消息。

可以将物理交换机划分为多个 VDC 以进行资源分区、故障隔离和系统管理。NX-OS 基础设施的主要功能之一就是让虚拟化对应用程序透明。MTS 实现虚拟化透明性的方法是使用虚拟节点（virtual node，vnode）的概念并从体系架构上建立清晰的通信模型，有了这个概念之后，应用程序就认为自己运行在交换机上（而不是 VDC 上）。

MTS 的工作原理是在系统启动时分配预定义的系统内存块，该内存块位于内核地址空间。应用程序启动后，内存会自动映射到应用程序地址空间，如果应用程序要将某些数据发送给队列，那么 MTS 就会生成一份数据副本并将净荷复制到缓冲区中，同时向应用程序接收队列发送缓冲区引用。应用程序读取队列时，会获得该净荷的引用，从而能够直接读取数据，就像映射到地址空间中一样。

接下来考虑一个简单案例。假设 OSPF 从邻接邻居的 LSA 更新中学到了一条新路由，OSPF 进程要求将该路由安装到路由表中，OSPF 进程将所需信息（包括前缀、下一跳等）都放到 MTS 消息中，由 MTS 发送给 URIB。例中的 MTS 负责在 OSPF 与 URIB 组件之间交换信息。

MTS 利用 SAP（Service Access Point，服务访问点）实现进程间通信，允许服务相互交换消息。交换机中的每块板卡都至少运行一个 MTS 实例，也称为 MTS 域。节点地址用于标识消息处理所涉及的 MTS 域。MTS 域是一种逻辑节点，仅为域内进程提供服务。在 MTS 域内，SAP 表示访问服务的地址。如果进程需要与其他 SAP 进行通信，那么就必须先绑定一个 SAP。SAP 分为以下 3 类。

- **静态 SAP**：范围是 1～1023。
- **动态 SAP**：范围是 1024～65535。
- **注册表 SAP**：0（保留）。

注：客户端与服务器进行通信之前，必须首先知道服务器的 SAP（通常是静态 SAP）。

MTS 地址包括两部分：一个 4 字节节点地址和一个 2 字节 SAP 编号。由于 MTS 域负责为与该域相关联的进程提供服务，因而 MTS 地址中的节点地址用于确定目的 MTS 域。因此，SAP 编号位于 MTS 域中（MTS 域由节点地址标识）。如果 Nexus 交换机配置了多个 VDC，那么每个 VDC 都有自己的 MTS 域，即 VDC1 的 SUP、VDC2 的 SUP-1 和 VDC3 的 SUP-2 等。

此外，MTS 拥有不同的操作代码，用于识别 MTS 消息中不同类型的净荷。

- **sync**：用于同步信息以做备用。
- **notification**：该操作代码用于单向通告。
- **request_response**：消息携带了一个令牌以匹配请求和响应。
- **switchover_send**：可以在切换过程中发送操作代码。
- **switchover_recv**：可以在切换过程中接收操作代码。
- **seqno**：该操作代码携带了序列号。

很多现象可能表明 MTS 出现了故障问题，不同的现象意味着不同的故障。如果某项功能特性或进程未按预期运行，那么就可能会在 Nexus 交换机上出现高 CPU 利用率，或者交换机端口出现无故振荡，导致 MTS 消息卡在队列中。最简单的检查方法就是使用 **show system internal mts buffers summary** 命令检查 MTS 缓冲区的利用率，需要多次检查该命令的输出结果，以查看哪些队列未被清除。例 3-30 给出了队列未清除时的 MTS 缓冲区摘要示例，从输出结果可以看出，SAP 号为 2938 的进程似乎被卡住了（因为 MTS 消息被卡在接收队列中），而另一个 SAP 号为 2592 的进程似乎已经清除了接收队列中的消息。

例 3-30　卡在队列中的 MTS 消息

```
N7k-1# show system internal mts buffers summary
node   sapno   recv_q   pers_q   npers_q   log_q
sup    2938    367      0        0         0
sup    2592    89       0        0         0
sup    284     0        10       0         0
N7k-1# show system internal mts buffers summary
node   sapno   recv_q   pers_q   npers_q   log_q
sup    2938    367      0        0         0
sup    2592    27       0        0         0
sup    284     0        10       0         0
```

表 3-5 列出了 MTS 队列名称及其功能。

表 3-5 MTS 队列名称和功能

缩写	队列名称	功能
recv_q	接收队列（Receive Queue）	
pers_q	持久性队列（Persistent Queue）	该队列中的消息在崩溃后仍然存活，MTS 可以在崩溃后重现该队列中的消息
npers_q	非持久性队列（Nonpersistent Queue）	该队列中的消息在崩溃后无法存活
log_q	日志队列（Log Queue）	MTS 在应用程序发送或接收消息时记录消息，应用程序利用日志记录在重启时恢复事务，应用程序可以在重启后显式检索记录的消息

消息卡在队列中可能会给设备带来不同的影响。例如，假设某设备运行了 BGP，即便 BGP 对等体拥有可达性信息和正确配置，也可能会偶然出现 BGP 翻动或 BGP 对等会话未启动问题，或者用户无法执行配置变更操作，如增加新的邻居配置。

如果确定消息卡在了某个队列中，那么就要确定与该 SAP 号相关联的进程。此时，可以通过 **show system internal mts sup sap** *sapno* **description** 命令获取该信息，也可以通过 **show system internal sysmgr service all** 命令从 sysmgr 的输出结果中查看该信息。如果希望了解所有排队消息的详细信息，可以使用 **show system internal mts buffers detail** 命令。例 3-31 显示了 SAP 2938 的描述信息，该描述信息显示了 statsclient 进程的相关信息，statsclient 进程负责收集管理引擎或线卡模块的统计信息。输出结果的第二部分显示了队列中的所有消息。

例 3-31 SAP 描述和排队的 MTS 消息

```
N7k-1# show system internal mts sup sap 2938 description
Below shows sap on default-VDC, to show saps on non-default VDC, run
       show system internal mts node sup-<vnode-id> sap ...
statscl_lib4320
N7k-1# show system internal mts buffers detail
Node/Sap/queue   Age(ms)         SrcNode    SrcSAP   DstNode   OPC    MsgSize
sup/3570/nper    5               0x601      3570     0x601     7679   30205
sup/2938/recv    50917934468     0x802      980      0x601     26     840
sup/2938/recv    50899918777     0x802      980      0x601     26     840
sup/2938/recv    50880095050     0x902      980      0x601     26     840
sup/2938/recv    46604123941     0x802      980      0x601     26     840
sup/2938/recv    46586081502     0x902      980      0x601     26     840
sup/2938/recv    46569929011     0x802      980      0x601     26     840
! Output omitted for brevity
N7k-1# show system internal mts sup sap 980 description
Below shows sap on default-VDC, to show saps on non-default VDC, run
       show system internal mts node sup-<vnode-id> sap ...
statsclient
```

注： 例 3-31 中的 SAP 描述信息来自默认 VDC。如果希望了解非默认 DVC 的信息，则可以使用 **show system internal mts node sup-** [*vnode-id*] **sap** *sapno* **description** 命令。

检查上述输出结果时，首先需要检查的重要字段就是 SAP 号及其时间（Age）。如果消息卡在队列中的持续时间非常长，那么就需要调查这些消息，因为它们可能会导致服务在 Nexus 平台上的行为异常。接下来要查看的一个字段就是 OPC（OPerational Code，操作代码），从缓冲区的输出结果中验证了队列消息之后，就可以通过 **show system internal sup opcodes** 命令来确定与该消息相关联的操作代码，以了解进程的状态。

同时还要查看 SAP 统计信息，以验证各种 SAP 的不同队列限制，并检查进程达到的最大队列限制，实现方式是使用 **show system internal mts sup sap** *sapno* **stats** 命令，如例 3-32 所示。

例 3-32　MTS SAP 统计信息

```
N7k-1# show system internal mts sup sap 980 stats
Below shows sap on default-VDC, to show saps on non-default VDC, run
        show system internal mts node sup-<vnode-id> sap ...
msg  tx: 14
byte tx: 1286
msg  rx: 30
byte rx: 6883

opc sent to myself: 0
max_q_size q_len limit (soft q limit): 4096
max_q_size q_bytes limit (soft q limit): 15%
max_q_size ever reached: 3
max_fast_q_size (hard q limit): 4096
rebind count: 0
Waiting for response: none
buf in transit: 14
bytes in transit: 1286
```

除上述验证检查之外，还可以在 OBFL 日志或系统日志中查看 MTS 差错消息。如果 MTS 队列满了，那么就会出现例 3-33 所示的差错日志，使用 **show logging onboard internal kernel** 命令可以确定没有因 MTS 而报告差错日志。

例 3-33　MTS OBFL 日志

```
2017 Apr 30 18:23:05.413 n7k 30 18:23:05 %KERN-2-SYSTEM_MSG: mts_is_q_space_
  available_old():1641: regular+fast mesg total = 48079,
 soft limit = 1024  - kernel
2017 Apr 30 18:23:05.415 n7k 30 18:23:05 %KERN-2-SYSTEM_MSG: mts_is_q_space_
  available_old(): NO SPACE - node=0, sap=27, uuid=26, pid=26549,
  sap_opt = 0x1, hdr_opt = 0x10, rq=48080(11530072), lq=0(0), pq=0(0), nq=0(0),
  sq=0(0), fast:rq=0, lq=0, pq=0, nq=0, sq=0 - kernel
```

MTS 事件历史记录日志也可以显示 MTS 差错信息，可以通过 **show system internal mts event-history errors** 命令查看详细信息。

如果消息卡在 MTS 队列中，或者发现 MTS 缓冲区泄露，那么就可以执行管理引擎切换操作以清除 MTS 队列，并恢复因 MTS 队列卡顿而中断的服务。

> **注：** 如果 SAP 284 出现在 MTS 缓冲区队列中，那么就直接忽略，因为 SAP 284 属于 TCP/UDP 进程客户端，是预期结果。

3.3.2　Netstack 和数据包管理器

Netstack 是用户模式 TCP/IP 协议栈的 NX-OS 实现，仅运行在管理引擎模块上。Netstack 组件运行在用户空间进程，每个 Netstack 组件都以多线程方式运行为独立进程。带内数据包以及与 NX-OS 相关的功能特性（如 vPC 和 VDC 感知能力）都必须以软件方式进行处理，而 Netstack 则是执行软件交换数据包操作的 NX-OS 组件。如前所述，Netstack 进程的主要作用包括以下几种。

- 将带内数据包传递给正确的控制平面进程应用程序。
- 通过软件以期望方式转发带内转存的数据包。
- 维护带内网络协议栈的配置数据。

Netstack 由 KLM（Kernel Loadable Module，内核可加载模块）和用户空间组件组成。用户空间组件是 VDC 本地进程，包含数据包管理器（是二层处理组件）、IP 输入（是三层处理组件）和 TCP/UDP

功能（负责处理四层数据包）。数据包管理器（PktMgr）组件通常与 IP 输入和 TCP/UDP 组件相隔离，即使这些组件共享相同的进程空间。图 3-1 显示了 Netstack 架构以及 KLM 和用户空间组件。

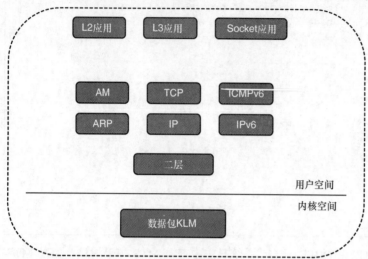

图 3-1　Netstack 架构

排查 Netstack 故障的一种简单方式就是了解 Netstack 处理数据包的方式。数据包被硬件交换到管理引擎带内接口，由数据包 KLM 处理数据帧，数据包 KLM 执行数据总线（DBUS）报头的最小化处理，并执行源接口索引查找操作以识别数据包属于哪个 VDC。由于 KLM 仅对数据包执行最小化处理，因而崩溃仅限于内核级别，而不会出现升级。大多数数据包处理操作发生在用户空间，支持多个 Netstack 进程实例（每个 VDC 一个）以及进程崩溃后的重启能力。

Netstack 使用多个软件队列来支持关键功能的优先级排序。在这些队列当中，BPDU（Bridge Protocol Data Unit，桥接协议数据单元）由专用队列进行处理，其他所有带内流量则在内核驱动程序中区分为 Hi（高优先级）或 Low（低优先级）队列。如果要查看 KLM 统计信息以及不同队列处理的数据包数量，可以使用 **show system inband queuing statistics** 命令，如例 3-34 所示。请注意，KLM 分别映射 ARP 和 BPDU 数据包，如果 BPDU 队列或其他队列出现丢包行为，那么就会在输出结果的"Inband Queues"（带内队列）字段标识这些丢包计数器。

例 3-34　带内 Netstack KLM 统计信息

```
N7k-1# show system inband queuing statistics
 Inband packets unmapped to a queue: 0
 Inband packets mapped to bpdu queue: 259025
 Inband packets mapped to q0: 448
 Inband packets mapped to q1: 0
 In KLM packets mapped to bpdu: 0
 In KLM packets mapped to arp : 0
 In KLM packets mapped to q0   : 0
 In KLM packets mapped to q1   : 0
 In KLM packets mapped to veobc : 0
 Inband Queues:
 bpdu: recv 259025, drop 0, congested 0 rcvbuf 33554432, sndbuf 33554432 no drop 1
 (q0): recv 448, drop 0, congested 0 rcvbuf 33554432, sndbuf 33554432 no drop 0
 (q1): recv 0, drop 0, congested 0 rcvbuf 2097152, sndbuf 4194304 no drop 0
```

PktMgr 是 Netstack 架构中的低层组件，负责处理 KLM 收发的所有带内或管理帧。PktMgr 基于二层（L2）数据包和平台报头信息对数据包进行解复用，并发送给 L2 客户端。同时，PktMgr 还负责将 L2 客户端的数据包出列，并发送给相应的驱动程序。所有的 L2 或非 IP 协议都直接注册到 PktMgr 进程，如 STP（Spanning Tree Protocol，生成树协议）、CDP（Cisco Discovery Protocol，思科发现协议）、UDLD（Unidirectional Link Detection，单向链路检测）、CFS（Cisco Fabric Service，思科交换矩阵服务）、LACP（Link Aggregation Control Protocol，链路聚合控制协议）和 ARP，所有的 IP 协议则直接注册到 IP 输入（IP Input）进程。

由于 Netstack 进程运行在管理引擎上，因而下列数据包必须发送给管理引擎进行处理。

- L2 客户端——BPDU 地址：STP、CDP 等。
- EIGRP、OSPF、ICMP、PIM、HSRP 和 GLBP 协议包。
- 网关 MAC 地址。
- 异常数据包。
    - 探查邻接关系（Glean Adjacency）数据包。
    - 管理引擎终结的数据包。
    - 带有 IP 选项的 IPv4/IPv6 数据包。
    - 相同接口（IF）检查。
    - RPF（Reverse Path Forwarding，反向路径转发）检查失败。
    - TTL（Time To Live，生存时间）到期数据包。

Netstack 进程的重启和切换操作都是状态化操作，Netstack 进程的引导过程依赖于 URIB（Unicast Routing Information Base，单播路由信息库）、U6RIB（IPv6 Unicast Routing Information Base，IPv6 单播路由信息库）和 ADJMGR（Adjacency Manager，邻接管理器）进程。Netstack 利用 CLI 服务器进程来还原配置并通过 PSS（Persistent Storage Service，持久存储服务）来还原被重启进程的状态。此外，Netstack 进程使用 RIB 共享内存来执行 L3 查找操作，使用 AM SDB（Shared DataBase，共享数据库）执行 L3 到 L2 的查找操作。为了实现故障排查操作，Netstack 提供了多种内部 show 命令和调试工具，可以帮助确定与 Netstack 相关的各种进程故障。

- 数据包管理器（Packet Manager）。
- IP/IPv6。
- TCP/UDP。
- ARP。
- AM（Adjacency Manager，邻接管理器）。

为了更好地理解数据包管理器组件的工作原理，下面将以 ICMPv6 为例加以分析。ICMPv6 是 PktMgr 客户端，ICMPv6 进程首次初始化时，会向 PktMgr 注册并分配一个客户端 ID、控制（Ctrl）SAP ID 和数据 SAP ID。MTS 负责处理 PktMgr 与 ICMPv6 之间的通信，从 PktMgr 到 ICMPv6 的 Rx 流量以数据 SAP ID 为目的端传递给 MTS，从 ICMPv6 到 PktMgr 的 Tx 流量则发送给 Ctrl SAP ID。PktMgr 从 ICMPv6 接收数据帧、构建正确的报头并发送给 KLM，从而传输给硬件设备。

排查 PktMgr 客户端故障时，需要找出作为 PktMgr 组件客户端的进程。可以通过 show system internal pktmgr client 命令来完成，该命令将显示 PktMgr 客户端的 UUID 和 Ctrl SAP ID。接下来需要查看服务管理器（Service Manager）下面的进程，以获得 UUID（Universally Unique Identifier，通用唯一标识符）和 SAP ID 信息。例 3-35 给出了上述排查步骤示例，本例在确定了正确的进程之后，通过 show system internal pktmgr client uuid 命令验证了 PktMgr 客户端的统计信息（包括丢包情况）。

例 3-35 带内 Netstack KLM 统计信息

```
N7k-1# show system internal pktmgr client | in Client|SAP
Client uuid: 263, 2 filters, pid 4000
  Ctrl SAP: 246
 Total Data SAPs : 1 Data SAP 1: 247
```

```
  Client uuid: 268, 4 filters, pid 3998
    Ctrl SAP: 278
   Total Data SAPs : 2 Data SAP 1: 2270    Data SAP 2: 2271
  Client uuid: 270, 1 filters, pid 3999
    Ctrl SAP: 281
   Total Data SAPs : 1 Data SAP 1: 283
  Client uuid: 545, 3 filters, pid 4054
    Ctrl SAP: 262
   Total Data SAPs : 1 Data SAP 1: 265
  Client uuid: 303, 2 filters, pid 4186
    Ctrl SAP: 171
   Total Data SAPs : 1 Data SAP 1: 177
  Client uuid: 572, 1 filters, pid 4098
    Ctrl SAP: 425
   Total Data SAPs : 1 Data SAP 1: 426
! Output omitted fore brevity
N7k-1# show system internal sysmgr service all | ex NA | in icmpv6|Name|--
Name          UUID         PID    SAP     state      Start count   Tag      Plugin ID
-------       ----------   ------ -----   ---------  -----------   ------   ----------
icmpv6        0x0000010E   3999   281     s0009          1         N/A          0
! Using the UUID value of 0x10E from above output
N7k-1# show system internal pktmgr client 0x10E
Client uuid: 270, 1 filters, pid 3999
  Filter 1: EthType 0x86dd, DstIf 0x150b0000, Excl. Any
  Rx: 0, Drop: 0
  Options: TO 0, Flags 0x18040, AppId 0, Epid 0
  Ctrl SAP: 281
   Total Data SAPs : 1 Data SAP 1: 283
  Total Rx: 0, Drop: 0, Tx: 0, Drop: 0
  Recirc Rx: 0, Drop: 0
  Input Rx: 0, Drop: 0
  Rx pps Inst/Max: 0/0
  Tx pps Inst/Max: 0/0
  COS=0 Rx: 0, Tx: 0     COS=1 Rx: 0, Tx: 0
  COS=2 Rx: 0, Tx: 0     COS=3 Rx: 0, Tx: 0
  COS=4 Rx: 0, Tx: 0     COS=5 Rx: 0, Tx: 0
  COS=6 Rx: 0, Tx: 0     COS=7 Rx: 0, Tx: 0
```

如果发送到管理引擎的数据包来自特定接口，那么就可以通过 **show system internal pktmgr interface** *interface-id* 命令验证该接口的 PktMgr 统计信息，例 3-36 显示了已经发送和接收的单播、多播和广播数据包的数量。

例 3-36 接口 PktMgr 的统计信息

```
N7k-1# show system internal pktmgr interface ethernet 1/1
Ethernet1/1, ordinal: 10  Hash_type: 0
  SUP-traffic statistics: (sent/received)
    Packets: 355174 / 331146
    Bytes: 32179675 / 27355507
    Instant packet rate: 0 pps / 0 pps
    Packet rate limiter (Out/In): 0 pps / 0 pps
    Average packet rates(1min/5min/15min/EWMA):
    Packet statistics:
      Tx: Unicast 322117, Multicast 33054
          Broadcast 3
      Rx: Unicast 318902, Multicast 12240
          Broadcast 4
```

PktMgr 记账信息（统计信息）对于确定是否出现了低层丢包故障来说非常有用，其故障原因是封装差错或其他内核交互问题。可以通过 **show system internal pktmgr stats [brief]** 命令加以验证（见例 3-37），该命令可以显示去往 KLM 的 PktMgr 驱动程序接口。省略的输出结果显示了其他差错和管理驱动程序的详细信息。

例 3-37　PktMgr 记账信息

```
N7k-1# show system internal pktmgr stats
Route Processor Layer-2 frame statistics

  Inband driver: valid 1, state 0, rd-thr 1, wr-thr 0, Q-count 0
  Inband sent: 1454421, copy_drop: 0, ioctl_drop: 0, unavailable_buffer_hdr_drop: 0
  Inband standby_sent: 0
  Inband encap_drop: 0, linecard_down_drop: 0
  Inband sent by priority [0=1041723,6=412698]
  Inband max output queue depth 0
  Inband recv: 345442, copy_drop: 0, ioctl_drop: 0, unavailable_buffer_hdr_drop: 0
  Inband decap_drop: 0, crc_drop: 0, recv by priority: [0=345442]
  Inband bad_si 0, bad_if 0, if_down 0
  Inband last_bad_si 0, last_bad_if 0, bad_di 0
  Inband kernel recv 85821, drop 0, rcvbuf 33554432, sndbuf 33554432


-----------------------------------------
  Driver:
-----------------------------------------
  State:           Up
  Filter:          0x0


! Output omitted for brevity
```

如果是 IP 处理进程，那么 Netstack 将查询 URIB（路由表和 RPM[Route Policy Manager，路由策略管理器]等其他所有必要组件）以做出数据包转发决策。Netstack 会在 **show ip traffic** 命令的输出结果中执行所有记账操作，利用 IP 流量统计信息来跟踪分段、IGMP（Internet Control Message Protocol，Internet 控制报文协议）、TTL 和其他异常数据包。此外，该命令还可以显示 RFC 4293 流量统计信息。确定 IP 数据包是否正在访问 NX-OS Netstack 组件的一种简便方法就是观察异常转存流量（如分段数据包）的统计信息。例 3-38 显示了 **show ip traffic** 命令的输出结果示例。

例 3-38　PktMgr 记账信息

```
N7k-1# show ip traffic

IP Software Processed Traffic Statistics
----------------------------------------
Transmission and reception:
  Packets received: 0, sent: 0, consumed: 0,
  Forwarded, unicast: 0, multicast: 0, Label: 0
  Ingress mcec forward: 0
Opts:
  end: 0, nop: 0, basic security: 0, loose source route: 0
  timestamp: 0, record route: 0
  strict source route: 0, alert: 0,
  other: 0
Errors:
  Bad checksum: 0, packet too small: 0, bad version: 0,
  Bad header length: 0, bad packet length: 0, bad destination: 0,
```

```
      Bad ttl: 0, could not forward: 0, no buffer dropped: 0,
      Bad encapsulation: 0, no route: 0, non-existent protocol: 0
      Bad options: 0
      Vinci Migration Packets : 0
       Total packet snooped : 0
       Total packet on down svi : 0
       Stateful Restart Recovery: 0,  MBUF pull up fail: 0
      Bad context id: 0, rpf drops: 0 Bad GW MAC 0
      Ingress option processing failed: 0
      NAT inside drop: 0, NAT outside drop: 0
      Ingress option processing failed: 0  Ingress mforward failed: 0
      Ingress lisp drop: 0
      Ingress lisp decap drop: 0
      Ingress lisp encap drop: 0
      Ingress lisp encap: 0
      Ingress Mfwd copy drop: 0
      Ingress RA/Reass drop: 0
      Ingress ICMP Redirect processing drop: 0
      Ingress Drop (ifmgr init): 0,
      Ingress Drop (invalid filter): 0
      Ingress Drop (Invalid L2 msg): 0
      ACL Filter Drops :
           Ingress - 0
           Egree -   0
           Directed Broadcast - 0
    Fragmentation/reassembly:
      Fragments received: 0, fragments sent: 0, fragments created: 0,
      Fragments dropped: 0, packets with DF: 0, packets reassembled: 0,
      Fragments timed out: 0
    Fragments created per protocol

    ICMP Software Processed Traffic Statistics
    ------------------------------------------
    Transmission:
      Redirect: 0, unreachable: 0, echo request: 0, echo reply: 0,
      Mask request: 0, mask reply: 0, info request: 0, info reply: 0,
      Parameter problem: 0, source quench: 0, timestamp: 0,
      Timestamp response: 0, time exceeded: 0,
      Irdp solicitation: 0, irdp advertisement: 0
      Output Drops - badlen: 0, encap fail: 0, xmit fail: 0
      ICMP originate Req: 0, Redirects Originate Req: 0
      Originate deny - Resource fail: 0, short ip: 0, icmp: 0, others: 0
    Reception:
      Redirect: 0, unreachable: 0, echo request: 0, echo reply: 0,
      Mask request: 0, mask reply: 0, info request: 0, info reply: 0,
      Parameter problem: 0, source quench: 0, timestamp: 0,
      Timestamp response: 0, time exceeded: 0,
      Irdp solicitation: 0, irdp advertisement: 0,
      Format error: 0, checksum error: 0
      Lisp processed: 0, No clients: 0: Consumed: 0
      Replies: 0, Reply drops - bad addr: 0, inactive addr: 0

    Statistics last reset: never

    RFC 4293: IP Software Processed Traffic Statistics
    --------------------------------------------------
    Reception
```

```
  Pkts recv: 0, Bytes recv: 0,
  inhdrerrors: 0, innoroutes: 0, inaddrerrors: 0,
  inunknownprotos: 0, intruncatedpkts: 0, inforwdgrams: 0,
  reasmreqds: 0, reasmoks: 0, reasmfails: 0,
  indiscards: 0, indelivers: 0,
  inmcastpkts: 0, inmcastbytes: 0,
  inbcastpkts: 0,
Transmission
  outrequests: 0, outnoroutes: 0, outforwdgrams: 0,
  outdiscards: 0, outfragreqds: 0, outfragoks: 0,
  outfragfails: 0, outfragcreates: 0, outtransmits: 0,
  bytes sent: 0, outmcastpkts: 0, outmcastbytes: 0,
  outbcastpkts: 0, outbcastbytes: 0
```

Netstack TCPUDP 组件

TCPUDP 进程包含以下功能。

- TCP。
- UDP。
- 原始数据包处理。
- 套接字层和套接字库。

TCP/UDP 协议栈基于 BSD，支持符合标准的 TCP 和 UDP 实现，支持窗口缩放、慢启动和时延确认等功能特性，不支持 TCP 选择性 ACK 和报头压缩。套接字库兼容 POSIX（Portable Operating System Interface，可移植操作系统接口），支持所有的标准套接字系统调用以及基于文件系统的系统调用。INPCB（Internet Protocol Control Block，Internet 协议控制块）散列表存储套接字连接数据，Netstack 重启能够保留套接字数据，而管理引擎切换则无法保留套接字数据。该进程拥有 16 个 TCP/UDP 工作线程，负责提供所有功能。

接下来考虑 NX-OS 的 TCP 套接字创建过程。Netstack 收到 TCP SYN 数据包之后，会在散列表中构建一个存根 INPCB 表项，然后将部分信息填充到 PCB（Protocol Control Block，协议控制块）中。TCP 三次握手完成之后，就可以填充所有的 TCP 套接字信息以创建完整的套接字。如果要验证该进程，可以查看调试命令 **debug sockets tcp pcb** 的输出结果。例 3-39 说明了在 **debug** 命令的帮助下创建套接字的过程以及 Netstack 的交互情况。从调试输出可以看出，收到 SYN 数据包之后，会将其添加到缓存中；三次握手完成之后，就可以创建一个完整的套接字。

例 3-39　创建 TCP 套接字和 Netstack 交互过程

```
N7k-1# debug sockets tcp pcb
2017 May  4 00:52:03.432086 netstack: syncache_insert: SYN added for
   L:10.162.223.34.20608 F:10.162.223.33.179, tp:0x701ff01c inp:0x701fef54
2017 May  4 00:52:03.434633 netstack: in_pcballoc: PCB: Allocated pcb, ipi_count:6
2017 May  4 00:52:03.434704 netstack: syncache_socket: Created full blown socket
   with F:10.162.223.34.20608 L:10.162.223.33.179 peer_mss 1460
2017 May  4 00:52:03.434930 netstack: in_setpeeraddr: PCB: in_setpeeraddr
   L 10.162.223.33.179 F 10.162.223.34.20608 C: 3
2017 May  4 00:52:03.435200 netstack: in_setsockaddr: PCB: in_setsockaddr
   L 10.162.223.33.179 F 10.162.223.34.20608 C: 3
```

使用命令 **show sockets connection tcp [detail]** 可以验证 TCP 套接字连接的详细信息。如果使用了 **detail** 选项，那么输出结果将显示 TCP 窗口信息、会话的 MSS 值和套接字状态等详细信息。此外，该命令的输出结果还提供了 MTS SAP ID。如果 TCP 套接字出现了问题，那么就可以在缓冲区中查找 MTS SAP ID 以查看其是否卡在队列中。例 3-40 显示了两台路由器之间的 BGP 对等连接的套接字信息。

例 3-40　TCP 套接字连接信息

```
N7k-1# show sockets connection tcp detail
Total number of tcp sockets: 6
Local host: 10.162.223.33 (179), Foreign host: 10.162.223.34 (20608)
  Protocol: tcp, type: stream, ttl: 1, tos: 0xc0, Id: 15
  Options:  REUSEADR, pcb flags none, state: | NBIO
  MTS: sap 14545
  Receive buffer:
    cc: 0, hiwat: 17520, lowat: 1, flags: none
  Send buffer:
    cc: 19, hiwat: 17520, lowat: 2048, flags: none
  Sequence number state:
    iss: 1129891008, snduna: 1129891468, sndnxt: 1129891487, sndwnd: 15925
    irs: 3132858499, rcvnxt: 3132858925, rcvwnd: 17520, sndcwnd: 65535
  Timing parameters:
    srtt: 3500 ms, rtt: 0 ms, rttv: 1000 ms, krtt: 1000 ms
    rttmin: 1000 ms, mss: 1460, duration: 49500 ms
  State: ESTABLISHED
  Flags:   NODELAY
No MD5 peers  Context: devl-user-1
! Output omitted for brevity
```

使用 **show sockets client detail** 命令可以监控 Netstack 套接字客户端，该命令可以显示套接字客户端的行为以及客户端执行的套接字库调用次数。该命令对于识别特定套接字客户端故障来说非常有用，因为输出结果包含了"Errors"字段，可以显示故障客户端的差错信息，如例 3-41 所示。输出结果显示了两个客户端 syslog 和 bgp，列出了与客户端相关联的 SAP ID 以及进程执行的套接字调用次数等统计信息。输出结果中的"Errors"字段为空，原因是所显示的套接字没有出错。

例 3-41　Netstack 套接字客户端详细信息

```
N7k-1# show sockets client detail
Total number of clients: 7
client: syslogd, pid: 3765, sockets: 2
  cancel requests:    0
  cancel unblocks:    0
  cancel misses:      0
  select drops:       0
  select wakes:       0
  sockets: 27:1(mts sap: 2336), 28:2(mts sap: 2339)
  Statistics:
    socket calls: 2    fcntl calls: 6    setsockopt calls: 6
    socket_ha_update calls: 6
  Errors:

! Output omitted for brevity

client: bgp, pid: 4639, sockets: 3
  fast_tcp_mts_ctrl_q: sap 2734
  cancel requests:    0
  cancel unblocks:    0
  cancel misses:      0
  select drops:       0
  select wakes:       0
  sockets: 49:13(mts sap: 2894), 51:14(mts sap: 2896), 54:15(mts sap: 14545)
  Statistics:
    socket calls: 5    bind calls: 5    listen calls: 2
```

```
       accept calls: 14        accept_dispatch errors: 14    connect_dispatch: 3
       close calls: 16         fcntl calls: 9      setsockopt calls: 31
       getsockname calls: 11   socket_ha_update calls: 38    Fast tcp send requests:
   207802
       Fast tcp send success: 207802      Fast tcp ACK rcvd: 203546
    Errors:
       connect errors: 3
       pconnect_einprogress errors: 3    pclose_sock_null errors: 14

   Statistics: Cancels 100811, Cancel-unblocks 100808, Cancel-misses 1
               Select-drops 2, Select-wakes 100808.
```

Netstack 还具有记账功能，可以提供有关 UDP、TCP、原始套接字以及各种内部表格的统计信息。使用 **show sockets statistics all** 命令可以查看 Netstack 套接字的统计信息，该命令可以查看 TCP 的丢包、无序包或重复包信息，Netstack 基于每个实例来维护这些统计信息。在输出结果的末尾，还可以看到 INPCB 和 IN6PCB 表的统计信息和差错计数器，可以通过这些表统计信息深入了解 Netstack 创建和删除的套接字连接数。INPCB 或 IN6PCB 表的"Errors"字段表明分配套接字信息时出现了问题。例 3-42 显示了 Netstack 的套接字记账信息。

例 3-42　Netstack 套接字记账信息

```
N7k-1# show sockets statistics all

TCP v4 Received:
     402528 total packets received,     203911 packets received in sequence,
     3875047 bytes received in sequence,   8 out-of-order packets received,
     10 rcvd duplicate acks,      208189 rcvd ack packets,
     3957631 bytes acked by rcvd acks,    287 Dropped no inpcb,
     203911 Fast recv packets enqueued,   16 Fast TCP can not recv more,
     208156 Fast TCP data ACK to app,
TCP v4 Sent:
     406332 total packets sent,    20 control (SYN|FIN|RST) packets sent,
     208162 data packets sent,     3957601 data bytes sent,
     198150 ack-only packets sent,

! Output omitted for brevity

INPCB Statistics:
in_pcballoc: 38  in_pcbbind: 9
in_pcbladdr: 18  in_pcbconnect: 14
in_pcbdetach: 19         in_pcbdetach_no_rt: 19
in_setsockaddr: 13       in_setpeeraddr: 14
in_pcbnotify: 1  in_pcbinshash_ipv4: 23
in_pcbinshash_ipv6: 5    in_pcbrehash_ipv4: 18
in_pcbremhash: 23
INPCB Errors:

IN6PCB Statistics:
in6_pcbbind: 5
in6_pcbdetach: 4         in6_setsockaddr: 1
in6_pcblookup_local: 2
IN6PCB Errors:
```

由于很多客户端（ARP、STP、BGP、EIGRP 和 OSPF 等）都要与 Netstack 组件进行交互，因而在排查控制平面故障时，如果能在 Ethanalyzer 中看到数据包，而客户端组件未收到数据包，那么就表明故障可能与 Netstack 或数据包管理器（Pktmgr）有关。图 3-2 显示了控制平面数据包流的路径情况以及 Netstack 和 Pktmgr 组件在系统中的位置。

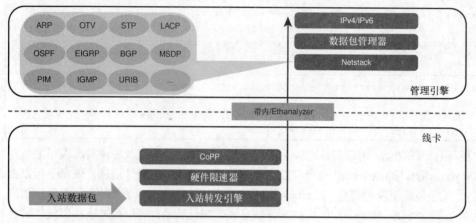

图 3-2 控制平面故障排查：流量路径

> **注**：如果 Netstack 组件或 Netstack 组件客户端出现了问题（如 OSPF 或 TCP 故障），那么就可以收集 show tech-support net-stack 和 show tech-support pktmgr 命令的输出结果，以及相关客户端的 show tech-support 命令输出结果，以协助思科 TAC 执行进一步排查。

3.3.3 ARP 和邻接管理器

ARP 组件负责处理 Nexus 交换机接口的 ARP 功能。ARP 组件在 PktMgr 中被注册为二层组件并提供如下功能。

- 管理三层到二层邻接关系的学习和定时器。
- 管理静态 ARP 表项。
- 将探查邻接关系数据包发转存到 CPU，从而触发 ARP 解析操作。
- 将 ARP 表项添加到 AM（Adjacency Manager，邻接管理器）数据库中。
- 管理 FHRP（First-Hop Redundancy Protocol，第一跳冗余协议）注册的虚拟地址。常见的 FHRP 主要有 VRRP（Virtual Router Redundancy Protocol，虚拟路由器冗余协议）、HSRP（Hot Standby Router Protocol，热备份路由器协议）和 GLBP（Gateway Load-Balancing Protocol，网关负载平衡协议）等。
- 让客户端侦听 ARP 数据包（如 ARP 监听、HSRP、VRRP 和 GLBP 等）。

所有与 ARP 组件交互的消息传递和通信过程都要借助 MTS，ARP 数据包通过 MTS 发送给 PktMgr。虽然 ARP 组件不支持 RARP（Reverse ARP，反向 ARP）功能，但支持代理 ARP（Proxy ARP）、本地代理 ARP（Local Proxy ARP）和黏性 ARP（Sticky ARP）等功能。

> **注**：如果路由器收到去往同一子网的其他主机的数据包，且接口启用了本地代理 ARP 功能，那么路由器就不会发送 ICMP 重定向消息。系统默认禁用本地代理 ARP。
> 如果接口设置了黏性 ARP 选项，那么就会标记所有新学到的 ARP 表项，以确保新的邻接关系（如无故 ARP[Gratuitous ARP]）不会覆盖这些表项，而且这些表项也不会老化。该功能有助于防止恶意用户欺骗 ARP 表项。

探查邻接关系可能会导致丢包，并导致转存到 CPU 的数据包过多。如果发现探查邻接关系数据包，那么理解数据包的处理方式将至关重要。假设某交换机收到了 IP 数据包，下一跳是直连网络，如果存在 ARP 表项，但 FIB 或 AM 共享数据库未安装主机路由（/32 路由），那么 FIB 查找将指向探查邻接关系，探查邻接关系数据包会被限速。如果未在 FIB 中找到匹配网络，那么硬件将以静默方式丢弃数据包（称为 FIB 未命中）。

为了保护 CPU 在没有 ARP 表项或硬件编程的邻接关系的情况下，免受高带宽流量流的影响，NX-OS 为 Nexus 7000 和 9000 平台的探查邻接关系流量提供了限速器，可以通过 **show run all | in glean** 命令查看为探查邻接关系流量预设的硬件限速器。例 3-43 显示了探查邻接关系流量的硬件限速器配置信息。

例 3-43 探查邻接关系流量的硬件限速器

```
N7k-1# show run all | in glean
hardware rate-limiter layer-3 glean 100
hardware rate-limiter layer-3 glean-fast 100
hardware rate-limiter layer-3 glean 100 module 3
hardware rate-limiter layer-3 glean-fast 100 module 3
hardware rate-limiter layer-3 glean 100 module 4
hardware rate-limiter layer-3 glean-fast 100 module 4
```

控制平面在解析 ARP 的时候，会在硬件中安装临时邻接关系丢弃表项，后续所有数据包在完成 ARP 解析之前都将被丢弃。临时邻接关系在探查定时器到期之前始终存在，定时器到期之后，将再次开始常规的转存/丢弃进程。

可以通过 **show ip arp** [*interface-type interface-num*]命令查看 NX-OS 的 ARP 表项，该命令的输出结果不但显示已学到的 ARP 表项，而且显示探查表项（标记为 INCOMPLETE[不完整]表项）。例 3-44 给出了 VLAN 10 SVI 接口的 ARP 表示例，包括已学到的 ARP 表项和 INCOMPLETE 表项。

例 3-44 ARP 表

```
N7k-1# show ip arp vlan 10
Flags: * - Adjacencies learnt on non-active FHRP router
       + - Adjacencies synced via CFSoE
       # - Adjacencies Throttled for Glean
       D - Static Adjacencies attached to down interface

IP ARP Table
Total number of entries: 2
Address         Age         MAC Address      Interface
10.1.12.10      00:10:20    5087.894b.bb41   Vlan10
10.1.12.2       00:00:09    INCOMPLETE       Vlan10
```

如果发现不完整 ARP，那么就可以利用内部跟踪历史记录来确定问题是否出在了 ARP 组件或其他组件上。填充 ARP 表项时，会通过两项操作（Create[创建]和 Update[更新]）在 FIB 中填充信息。如果 ARP 组件出现了问题，那么就可能仅出现 Create 操作，而没有 Update 操作。可以通过 **show forwarding internal trace v4-adj-history** [**module** *slot*] 命令来查看实际的操作序列（见例 3-45），可以看出，对于 10.1.12.2 的下一跳来说，Destroy（销毁）操作之后仅出现了 Create 操作，此后没有出现 Update 操作，从而导致 ARP 表项被标记为探查邻接关系表项。

例 3-45 邻接关系内部转发跟踪情况

```
N7k-1# show forwarding internal trace v4-adj-history module 4
HH 0x80000018
       Time                     if         NH             operation
 Sun May  7 06:43:10 2017      Vlan10     10.1.12.10      Create
 Sun May  7 06:43:10 2017      Vlan10     10.1.12.10      Update
! History for Non-Working host i.e. 10.1.12.2
 Sun May  7 06:43:10 2017      Vlan10     10.1.12.2       Create
```

```
Sun May  7 06:43:10 2017    Vlan10    10.1.12.2    Update
Sun May  7 06:53:54 2017    Vlan10    10.1.12.2    Destroy
Sun May  7 06:56:03 2017    Vlan10    10.1.12.2    Create
```

如果要查看转发邻接关系（Forwarding Adjacency），那么就可以使用 **show forwarding ipv4 adjacency** *interface-type interface-num* [**module** *slot*] 命令。如果特定下一跳的邻接关系显示为未解析，那么就表明无邻接关系；此后，FIB 将匹配网络探查邻接关系并执行转存操作。例 3-46 给出的 **show forwarding ipv4 adjacency** 命令输出结果中包含了未解析的邻接关系表项。

例 3-46　验证转发邻接关系

```
N7k-1# show forwarding ipv4 adjacency vlan 10 module 4
IPv4 adjacency information

next-hop          rewrite info       interface
--------------    ---------------    -------------
10.1.12.10        5087.894b.bb41     Vlan10
10.1.12.2         unresolved         Vlan10
```

ARP 组件提供了事件历史记录，可以帮助排查是否存在可能导致 ARP 和邻接关系故障的差错状况。如果要查看 ARP 事件历史记录，可以使用 **show ip arp internal event-history** [**events** | **errors**] 命令。例 3-47 显示了 **show ip arp internal event-history events** 命令的输出结果，列出了主机 10.1.12.2/24 的 ARP 解析情况。从事件历史记录可以看出，交换机发出了 ARP 请求，此后根据应答信息，建立邻接关系并进一步更新到 AM 数据库中。

例 3-47　ARP 事件历史记录

```
N7k-1# show ip arp internal event-history events
1) Event:E_DEBUG, length:144, at 720940 usecs after Sun May  7 17:31:30 2017
    [116] [4196]: Adj info: iod: 181, phy-iod: 36, ip: 10.1.12.2, mac: fa16.3e29
.5f82, type: 0, sync: FALSE, suppress-mode: ARP Suppression Disabled

2) Event:E_DEBUG, length:198, at 720916 usecs after Sun May  7 17:31:30 2017
    [116] [4196]: Entry added to ARP pt, added to AM for 10.1.12.2, fa16.3e29.5f
82, state 2 on interface Vlan10, physical interface Ethernet2/1, ismct 0. Rearp
(interval: 0, count: 0), TTL: 1500 seconds

3) Event:E_DEBUG, length:86, at 718187 usecs after Sun May  7 17:31:30 2017
    [116] [4196]: arp_add_adj: Updating MAC on interface Vlan10, phy-interface
Ethernet2/1

4) Event:E_DEBUG, length:145, at 713312 usecs after Sun May  7 17:31:30 2017
    [116] [4200]: Adj info: iod: 181, phy-iod: 181, ip: 10.1.12.2, mac: 0000.000
0.0000, type: 0, sync: FALSE, suppress-mode: ARP Suppression Disabled

5) Event:E_DEBUG, length:181, at 713280 usecs after Sun May  7 17:31:30 2017
    [116] [4200]: Entry added to ARP pt, added to AM for 10.1.12.2, NULL, state
1 on interface Vlan10, physical interface Vlan10, ismct 0. Rearp (interval: 2,
count: 4), TTL: 30 seconds

6) Event:E_DEBUG, length:40, at 713195 usecs after Sun May  7 17:31:30 2017
    [116] [4200]: Parameters l2_addr is null

7) Event:E_DEBUG, length:40, at 713154 usecs after Sun May  7 17:31:30 2017
    [116] [4200]: Parameters l2_addr is null
```

```
 8) Event:E_DEBUG, length:59, at 713141 usecs after Sun May  7 17:31:30 2017
    [116] [4200]: Create adjacency, interface Vlan10, 10.1.12.2

 9) Event:E_DEBUG, length:81, at 713074 usecs after Sun May  7 17:31:30 2017
    [116] [4200]: arp_add_adj: Updating MAC on interface Vlan10, phy-interface
Vlan10

10) Event:E_DEBUG, length:49, at 713054 usecs after Sun May  7 17:31:30 2017
    [116] [4200]: ARP request for 10.1.12.2 on Vlan10
```

注: 可以在入站和出站方向通过 Ethanalyzer 抓取 ARP 数据包。

ARP 组件和 AM（Adjacency Manager，邻接管理器）组件紧密耦合，AM 负责在硬件中对/32 主机路由进行编程。AM 主要提供如下功能。

- 通过共享内存导出三层到二层邻接关系信息。
- 生成邻接关系变更通告（包括接口删除通告），并通过 MTS 发送更新。
- 学到邻接关系信息之后，将主机路由（/32 路由）添加到 URIB/U6RIB 中。
- 从接口向外转发数据包时执行 AM 数据库的 IP/IPv6 查找操作。
- 处理邻接关系重启（通过维护邻接关系 SDB 来恢复 AM 状态）。
- 为 URIB/UFDM 提供单一接口，从多个源端学习路由。

学到新 ARP 表项之后，AM 会将该 ARP 表项添加到 AM SDB 中。接下来 AM 将与 URIB 和 UFDM 进行直接通信，在硬件中安装/32 邻接关系表项。AM 数据库会查询活动 ARP 表项的状态，由于进程重启后无法保持 ARP 表，因而必须重新查询 AM SDB。AM 负责注册各种能够安装邻接关系表项的客户端，如果要查看已注册的客户端，可以使用 **show system internal adjmgr client** 命令，如例 3-48 所示。常见的 AM 客户端就是 ARP。

例 3-48 AM 客户端

```
N7k-1# show system internal adjmgr client
Protocol Name    Alias    UUID
netstack         Static   545
rpm              rpm      305
IPv4             Static   268
arp              arp      268
IP               IP       545
icmpv6           icmpv6   270
```

可以通过 **show ip adjacency** *ip-address* **detail** 命令验证所有未解析的邻接关系表项。如果已经解析，那么输出结果就会为指定 IP 填充正确的 MAC 地址，否则，MAC 地址字段就会显示 0000.0000.0000。例 3-49 给出了已解析和未解析邻接关系表项之间的对比情况。

例 3-49 已解析和未解析邻接关系表项

```
! Resolved Adjacency
N7k-1# show ip adjacency 10.1.12.10 detail
No. of Adjacency hit with type INVALID: Packet count 0, Byte count 0
No. of Adjacency hit with type GLOBAL DROP: Packet count 0, Byte count 0
No. of Adjacency hit with type GLOBAL PUNT: Packet count 0, Byte count 0
No. of Adjacency hit with type GLOBAL GLEAN: Packet count 0, Byte count 0
No. of Adjacency hit with type GLEAN: Packet count 0, Byte count 0
No. of Adjacency hit with type NORMAL: Packet count 0, Byte count 0

Adjacency statistics last updated before: never
```

```
IP Adjacency Table for VRF default
Total number of entries: 1

Address :            10.1.12.10
MacAddr :            5087.894b.bb41
Preference :         50
Source :             arp
Interface :          Vlan10
Physical Interface : Ethernet2/1
Packet Count :       0
Byte Count :         0
Best :               Yes
Throttled :          No
! Unresolved Adjacency
N7k-1# show ip adjacency 10.1.12.2 detail
! Output omitted for brevity

Adjacency statistics last updated before: never

IP Adjacency Table for VRF default
Total number of entries: 1

Address :            10.1.12.10
MacAddr :            5087.894b.bb41
Preference :         50
Source :             arp
Interface :          Vlan10
Physical Interface : Ethernet2/1
Packet Count :       0
Byte Count :         0
Best :               Yes
Throttled :          No
! Unresolved Adjacency
N7k-1# show ip adjacency 10.1.12.2 detail
! Output omitted for brevity

Adjacency statistics last updated before: never

IP Adjacency Table for VRF default
Total number of entries: 1

Address :            10.1.12.2
MacAddr :            0000.0000.0000
Preference :         255
Source :             arp
Interface :          Vlan10
Physical Interface : Vlan10
Packet Count :       0
Byte Count :         0
Best :               Yes
Throttled :          No
```

在 URIB 中安装 AM 邻接关系表项的步骤如下。

- **第 1 步**：AM 对添加邻接关系表项的请求进行排队。
- **第 2 步**：AM 调用 URIB 来安装路由。
- **第 3 步**：AM 将新邻接关系表项附加到 Add（添加）列表中。

- **第 4 步**：URIB 添加路由。
- **第 5 步**：AM 独立调用 UFDM API，在硬件中安装邻接关系表项。

可以通过 **show system internal adjmgr internal event-history events** 命令查看 AM 组件发生的一系列事件。例 3-50 给出了该命令的输出结果，可以看到安装主机 10.1.12.2 的邻接关系表项时发生的一系列事件。请注意，前缀 10.1.12.2 已经被添加到 IPv4 地址族的 RIB 缓冲区中了。

例 3-50　邻接关系表项安装事件

```
N7k-1# show system internal adjmgr internal event-history events
1) Event:E_DEBUG, length:101, at 865034 usecs after Tue May  9 05:21:19 2017
    [117] [4017]: Appending ADD 10.1.12.2 on Vlan10 (TBL:1) AD 250 to rib buffer
 for Address Family :IPv4

2) Event:E_DEBUG, length:84, at 845226 usecs after Tue May  9 05:21:19 2017
    [117] [4043]: Add 10.1.12.2 on Vlan10 to rib work queue for afi: IPv4with wo
rk bit: 1

3) Event:E_DEBUG, length:61, at 845128 usecs after Tue May  9 05:21:19 2017
    [117] [4043]: is_mct 0, entry_exists 1, iod 0x85 phy_iod 0x85

4) Event:E_DEBUG, length:61, at 840347 usecs after Tue May  9 05:21:19 2017
    [117] [4043]: is_mct 0, entry_exists 0, iod 0x85 phy_iod 0x85
Adjacency related errors could be verified using the event-history logs as well by
using the command show system internal adjmgr internal event-history errors.
```

> **注**：如果 ARP 或 AM 组件出现了问题，那么就可以在故障状态下收集 **show tech arp** 和 **show tech adjmgr** 命令的输出结果。

1. 单播转发组件

设备的 IP/IPv6 包转发决策是由 RIB（Routing Information Base，路由信息库）和 FIB（Forwarding Information Base，转发信息库）决定的。NX-OS 中的 RIB 由 URIB（Unicast Routing Information Base，单播路由信息库）组件进行管理，FIB 由 IPFIB（IP Forwarding Information Base，IP 转发信息库）组件进行管理。URIB 是管理引擎上的路由信息的软件视图，而 IPFIB 则是线卡上的路由信息的软件视图。本节将详细讨论这些负责管理 NX-OS 平台转发决策的组件信息。

2. 单播路由信息库

NX-OS 的 URIB 组件负责维护所有路由协议安装的三层单播路由的 SDB。URIB 是 VDC 本地进程，也就是说，不能跨多个 VDC 共享路由，除非这些 VDC 之间存在路由邻接关系。URIB 进程使用多个客户端，可以通过 **show routing clients** 命令查看这些客户端，如例 3-51 所示。

- 路由协议：EIGRP（Enhanced Interior Gateway Routing Protocol，增强型内部网关路由协议）、OSPF（Open Shortest Path First，开放最短路径优先）、BGP（Border Gateway Protocol，边界网关协议）等。
- Netstack（更新静态路由的 URIB）。
- AM。
- RPM。

例 3-51　URIB 客户端

```
N7k-1# show routing clients
CLIENT: static
 index mask: 0x0000000000000080
 epid: 4059      MTS SAP: 266       MRU cache hits/misses:      1/1
 Stale Time: 30
```

```
Routing Instances:
 VRF: "default" routes: 0, rnhs: 0, labels: 0
Messages received:
 Register            : 1    Convergence-all-nfy: 1
Messages sent:

CLIENT: ospf-100
 index mask: 0x0000000000008000
 epid: 23091 MTS SAP: 320       MRU cache hits/misses:          2/1
 Stale Time: 2100
 Routing Instances:
  VRF: "default" routes: 1, rnhs: 0, labels: 0
 Messages received:
  Register           : 1    Convergence-notify: 1      Modify-route     : 1

 Messages sent:
  Modify-route-ack   : 1

! Output omitted for brevity
```

每种路由协议都有自己的共享 URIB 内存空间区域。路由协议从邻居学习路由时，会将这些学到的路由安装到自己的共享 URIB 内存空间区域中。接下来，URIB 会将更新后的路由复制到自己的共享内存保护区域，该区域是只读内存，仅供 Netstack 和其他组件读取。路由决策由 URIB 共享内存中的表项做出，需要注意的是，URIB 本身并不会在路由表中执行任何添加、修改或删除操作，这一点至关重要。由 URIB 客户端（路由协议和 Netstack）处理所有更新操作，除非 URIB 客户端进程崩溃，此时 URIB 可能会删除废弃路由。

OSPF CLI 为用户提供了 **show ip ospf internal txlist urib** 命令，可以查看发送给 URIB 的 OSPF 路由。如果是其他路由协议，就需要使用事件历史记录命令来查看该信息。例 3-52 给出了该命令的输出结果，显示 OSPF 进程的源 SAP ID 和 MTS 消息的目标 SAP ID。

例 3-52　分发给 URIB 的 OSPF 路由

```
N7k-1# show ip ospf internal txlist urib

ospf 100 VRF default
ospf process tag 100
ospf process instance number 1
ospf process uuid 1090519321
ospf process linux pid 23091
ospf process state running
System uptime 4d04h
SUP uptime 2 4d04h

Server up         : L3VM|IFMGR|RPM|AM|CLIS|URIB|U6RIB|IP|IPv6|SNMP
Server required   : L3VM|IFMGR|RPM|AM|CLIS|URIB|IP|SNMP
Server registered : L3VM|IFMGR|RPM|AM|CLIS|URIB|IP|SNMP
Server optional   : none
Early hello : OFF
Force write PSS: FALSE
OSPF mts pkt sap 324
OSPF mts base sap 320

 OSPFv2->URIB transmit list: version 0xb

      9: 10.1.12.0/24
```

```
            10: 1.1.1.1/32
            11: 2.2.2.2/32
            11: RIB marker
N7k-1# show system internal mts sup sap 320 description
ospf-100
N7k-1# show system internal mts sup sap 324 description
OSPF pkt MTS queue
```

所有从 OSPF 进程或其他路由进程发送给 URIB 的路由更新，都被记录在事件历史记录日志中。如果要查看 OSPF 从 OSPF 进程内存复制到 URIB 共享内存的路由更新，可以使用命令 **show ip ospf internal event-history rib**。此外，还可以通过 **show routing internal event-history msgs** 命令检查 URIB 全局可读共享内存的路由更新情况。例 3-53 显示了将学到的 OSPF 路由更新到 URIB 的处理过程，同时，路由事件历史记录还显示了更新到共享内存的路由信息。

例 3-53 路由协议和 URIB 更新

```
N7k-1# show ip ospf internal event-history rib
OSPF RIB events for Process "ospf-100"
2017 May 14 03:12:14.711449 ospf 100 [23091]: : Done sending routes to URIB
2017 May 14 03:12:14.711447 ospf 100 [23091]: : Examined 3 OSPF routes
2017 May 14 03:12:14.710532 ospf 100 [23091]: : Route (mbest) does not have any
  next-hop
2017 May 14 03:12:14.710531 ospf 100 [23091]: : Path type changed from nopath to
  intra
2017 May 14 03:12:14.710530 ospf 100 [23091]: : Admin distance changed from 255
  to 110
2017 May 14 03:12:14.710529 ospf 100 [23091]: : Mbest metric changed from 429496
  7295 to 41
2017 May 14 03:12:14.710527 ospf 100 [23091]: : Processing route 2.2.2.2/32
  (mbest)
2017 May 14 03:12:14.710525 ospf 100 [23091]: : Done processing next-hops for
  2.2.2.2/32
2017 May 14 03:12:14.710522 ospf 100 [23091]: : Route 2.2.2.2/32 next-hop
  10.1.12.2 added to RIB.
2017 May 14 03:12:14.710515 ospf 100 [23091]: : Path type changed from nopath to
  intra
2017 May 14 03:12:14.710513 ospf 100 [23091]: : Admin distance changed from 255
  to 110
2017 May 14 03:12:14.710511 ospf 100 [23091]: : Ubest metric changed from 429496
  7295 to 41
2017 May 14 03:12:14.710509 ospf 100 [23091]: : Processing route 2.2.2.2/32 (ubest)
! Output omitted for brevity
2017 May 14 03:12:14.710430 ospf 100 [23091]: : Start sending routes to URIB and
summarize
N7k-1# show routing internal event-history msgs
! Output omitted for brevity
6) Event:E_MTS_TX, length:60, at 710812 usecs after Sun May 14 03:12:14 2017
    [NOT] Opc:MTS_OPC_URIB(52225), Id:0X0036283B, Ret:SUCCESS
    Src:0x00000101/253, Dst:0x00000101/320, Flags:None
    HA_SEQNO:0X00000000, RRtoken:0x00000000, Sync:NONE, Payloadsize:312
    Payload:
    0x0000:  04 00 1a 00 53 0f 00 00 53 0f 00 00 ba 49 07 00
7) Event:E_MTS_RX, length:60, at 710608 usecs after Sun May 14 03:12:14 2017
    [NOT] Opc:MTS_OPC_URIB(52225), Id:0X00362839, Ret:SUCCESS
    Src:0x00000101/320, Dst:0x00000101/253, Flags:None
    HA_SEQNO:0X00000000, RRtoken:0x00000000, Sync:NONE, Payloadsize:276
```

```
        Payload:
        0x0000:  04 00 19 00 33 5a 00 00 33 5a 00 00 ba 49 07 00
N7k-1# show system internal mts sup sap 253 description
URIB queue
```

在 URIB 中安装路由之后,可以使用 **show ip route** *routing-process* **detail** 命令查看这些路由,其中,*routing-process* 是各种路由协议的 NX-OS 进程,如例 3-53 所示的 ospf-100。

> **注:** URIB 将所有路由信息都存储在共享内存中。由于内存空间是共享的,因而可能会被大规模路由或内存泄露故障所耗尽。使用 **show routing memory statistics** 命令可以查看 URIB 的共享内存空间。

3. UFDM 和 IPFIB

更新了 URIB 的路由之后,需要接着更新 FIB,此时需要用到 UFDM。UFDM 是一种 VDC 本地进程,主要功能是将路由、邻接关系信息和 uRPF(unicast Reverse Path Forwarding,单播反向路径转发)信息可靠地分发给 Nexus 机箱中的所有线卡。UFDM 负责维护前缀、邻接关系和 ECMP(Equal Cost Multipath,等价多路径)数据库,交换机硬件利用这些信息做出转发决策。UFDM 运行在管理引擎模块上,并与每块线卡上的 IPFIB 进行通信。IPFIB 进程负责对每块线卡上的 FE(Forwarding Engine,转发引擎)和硬件邻接关系进行编程。

UFDM 通过 4 组 API 在系统中执行不同的任务。

- **FIB API**:URIB 和 U6RIB 模块使用该 API 添加、更新和删除 FIB 中的路由。
- **AdjMgr 通告**:AM 与 UFDM AM API 进行直接交互以安装/32 主机路由。
- **uRPF 通告**:IP 模块发送通告以启用或禁用每个接口的不同 RPF 检查模式。
- **统计信息采集 API**:用于从平台采集邻接关系统计信息。

上述任务中的前 3 组以自上而下的方式执行(从管理引擎到线卡),第 4 组则是以自下而上的方式执行(从线卡到管理引擎)。

> **注:** NX-OS 不再支持 CEF(Cisco Express Forwarding,思科快速转发),转由硬件 FIB 提供相关功能,硬件 FIB 基于 AVL 树,其中,AVL 树是一种自平衡二叉查找树。

UFDM 组件负责将 AM、FIB 和 RPF 更新分发给 VDC 中的每块线卡上的 IPFIB,然后向 URIB 发送一条确认消息 *route-ack*,可以通过 **show system internal ufdm event-history debugs** 命令加以验证,如例 3-54 所示。

例 3-54 UFDM 将路由分发给 IPFIB 并进行确认

```
N7k-1# show system internal ufdm event-history debugs
! Output omitted for brevity
807) Event:E_DEBUG, length:94, at 711536 usecs after Sun May 14 03:12:14 2017
    [104] ufdm_route_send_ack(185):TRACE: sent route nack, xid: 0x58f059ec,
v4_ack: 0, v4_nack: 24


808) Event:E_DEBUG, length:129, at 711230 usecs after Sun May 14 03:12:14 2017
    [104] ufdm_route_distribute(615):TRACE: v4_rt_upd # 24 rt_count: 1, urib_xid
: 0x58f059ec, fib_xid: 0x58f059ec recp_cnt: 0 rmask: 0


809) Event:E_DEBUG, length:94, at 652231 usecs after Sun May 14 03:12:09 2017
    [104] ufdm_route_send_ack(185):TRACE: sent route nack, xid: 0x58f059ec,
v4_ack: 0, v4_nack: 23
```

```
810) Event:E_DEBUG, length:129, at 651602 usecs after Sun May 14 03:12:09 2017
    [104] ufdm_route_distribute(615):TRACE: v4_rt_upd # 23 rt_count: 1, urib_xid
: 0x58f059ec, fib_xid: 0x58f059ec recp_cnt: 0 rmask: 0
```

与平台相关的 FIB 负责管理硬件专用结构，如硬件表索引和设备实例。NX-OS 命令 **show forwarding internal trace v4-pfx-history** 可以显示 FIB 路由数据的创建和销毁历史。例3-55 显示了前缀 2.2.2.2/32（通过 OSPF 学到）转发 IPv4 前缀的历史记录，历史记录显示了该前缀的 Create（创建）、Destroy（销毁）以及另一个 Create（创建）操作，同时还显示了相应的时间戳信息，这些信息对于解决因路由未安装在硬件 FIB 中而引起的转发故障来说非常有用。

例 3-55　FIB 路由的历史信息

```
N7k-1# show forwarding internal trace v4-pfx-history
PREFIX 1.1.1.1/32 TABLE_ID 0x1
    Time                    ha_handle   next_obj   next_obj_HH   NH_cnt   operation
  Sun May 14 16:42:47 2017  0x23d6b     V4 adj     0xb           1        Create
  Sun May 14 16:42:47 2017  0x23d6b     V4 adj     0xb           1        Update

PREFIX 10.1.12.1/32 TABLE_ID 0x1
    Time                    ha_handle   next_obj   next_obj_HH   NH_cnt   operation
  Sun May 14 16:42:39 2017  0x21d24     V4 adj     0xb           1        Create
PREFIX 2.2.2.2/32 TABLE_ID 0x1
    Time                    ha_handle   next_obj   next_obj_HH   NH_cnt   operation
  Sun May 14 16:44:08 2017  0x23f55     V4 adj     0x10000       1        Create
  Sun May 14 16:44:17 2017  0x23f55     V4 adj     0x10000       1        Destroy
  Sun May 14 16:45:02 2017  0x23f55     V4 adj     0x10000       1        Create

PREFIX 10.1.12.2/32 TABLE_ID 0x1
    Time                    ha_handle   next_obj   next_obj_HH   NH_cnt   operation
  Sun May 14 16:43:58 2017  0x21601     V4 adj     0x10000       1        Create
```

完成了硬件 FIB 编程之后，可以通过 **show forwarding route** *ip-address/len* [**detail**] 命令验证转发信息，该命令的输出结果可以显示到达目的前缀和出站接口的下一跳信息，以及目的 MAC 信息。此外，还可以通过 **show forwarding ipv4 route** *ip-address/len* **platform** [**module** *slot*] 命令在平台层面验证该信息，从硬件/平台层面获得更多的详细信息。

接下来需要将这些转发信息传播给相关联的线卡，可以通过 **show system internal forwarding route** *ip-address/len* [**detail**] 命令加以验证。该命令的输出结果还提供了接口硬件的邻接关系信息，可以通过 **show system internal forwarding adjacency entry** *adj* 命令进行验证，其中的 *adj* 是从上一条命令收到的邻接关系值。

注：也可以在管理引擎卡和线卡层面收集前面的输出结果，操作方式是通过 **attach module** *slot* 命令登录线卡控制台，然后再执行前面描述的转发命令。

例 3-56 给出了在线卡层面对编程在 FIB 中的路由进行验证的详细步骤。

例 3-56　验证平台 FIB

```
N7k-1# show forwarding route 2.2.2.2/32 detail
slot  3
=======

Prefix 2.2.2.2/32, No of paths: 1, Update time: Sun May 14 21:29:43 2017
   10.1.12.2          Vlan10               DMAC: 5087.894b.c0c2
      packets: 0             bytes: 0
```

```
N7k-1# show forwarding ipv4 route 2.2.2.2/32 platform module 3

Prefix 2.2.2.2/32, No of paths: 1, Update time: Sun May 14 21:16:20 2017
   10.1.12.2          Vlan10             DMAC: 5087.894b.c0c2
     packets: 0              bytes: 0
HH:0x80000026  Flags:0x0  Holder:0x1  Next_obj_type:5
Inst :     0    1    2    3    4    5    6    7    8    9    10   11
Hw_idx:  6320  N/A  N/A  N/A  N/A  N/A
N7k-1# show system internal forwarding route 2.2.2.2/32
slot 3
=======


Routes for table default/base

----+----------------------+----------+----------+-----------
Dev | Prefix               | PfxIndex | AdjIndex | LIF
----+----------------------+----------+----------+-----------
  0   2.2.2.2/32             0x6320     0x5f       0x3

N7k-1# show system internal forwarding route 2.2.2.2/32 detail
slot 3
=======
 RPF Flags legend:
        S - Directly attached route (S_Star)
        V - RPF valid
        M - SMAC IP check enabled
        G - SGT valid
        E - RPF External table valid
     2.2.2.2/32         ,  Vlan10
    Dev: 0 , Idx: 0x6320 , Prio: 0x8507  , RPF Flags: V   , DGT: 0 , VPN: 9
        RPF_Intf_5:   Vlan10       (0x3     )
        AdjIdx: 0x5f   , LIFB: 0   , LIF: Vlan10      (0x3     ), DI: 0x0
        DMAC: 5087.894b.c0c2 SMAC: 5087.894b.c0c5
N7k-1# show system internal forwarding adjacency entry 0x5f
slot 3
=======

Device: 0    Index: 0x5f    dmac: 5087.894b.c0c2 smac: 5087.894b.c0c5
                  e-lif: 0x3   packets: 0           bytes: 0
```

注：如果出现了转发问题，那么就可以在故障状态下收集以下 **show tech** 命令的输出结果。

- **show tech routing ip unicast**。
- **show tech-support forwarding l3 unicast [module** *slot*]。
- **show tech-support detail**。

3.3.4 EthPM 和 Port-Client

NX-OS 提供了一个名为 EthPM（Ethernet Port Manager，以太网端口管理器）的 VDC 本地进程来管理 Nexus 平台上的所有以太网接口，包括物理和逻辑接口（仅服务器接口，不包括 SVI）、带内接口和管理接口。EthPM 组件负责执行以下两项主要功能。

- **抽象**：为所有与 EthPM 管理的接口进行交互的其他组件提供一个抽象层。
- **FSM（Port Finite State Machine，端口有限状态机）**：为其管理的接口提供 FSM，并处理接口的创建和删除操作。

EthPM 组件与其他组件（如端口通道管理器[Port-Channel Manager]、VxLAN 管理器和 STP）进行交互以编程接口状态。同时，EthPM 进程还负责管理接口的配置（如双工、速率、MTU、允许的 VLAN 等）。

Port-Client（端口客户端）是一种与 EthPM 进程进行密切交互的线卡全局进程（仅 Nexus 7000 和 Nexus 9000 交换机），负责维护从 EthPM 收到的跨不同 VDC 的全局信息。Port-Client 从本地硬件端口 ASIC 接收更新信息并更新 EthPM。Port-Client 拥有 PI（Platform Independent，平台无关）和 PD（Platform Dependent，平台相关）组件，Port-Client 进程的 PI 组件与 EthPM 进行交互（EthPM 也是 PI 组件），PD 组件专门用于线卡硬件编程。

可以通过 EthPM 组件 CLI 查看从 IM（Interface Manager，接口管理器）组件收到的平台级信息（如 EthPM 接口索引）、接口管理状态和操作状态、接口能力以及接口 VLAN 状态等信息，相应的查询命令是 **show system internal ethpm info interface** *interface-type interface-num*。例 3-57 显示了接口 Ethernet 3/1 的 EthPM 信息，该接口被配置为 VLAN 10 的接入端口。

例 3-57　验证接口 EthPM 信息

```
N7k-1# show system internal ethpm info interface ethernet 3/1

Ethernet3/1 - if_index: 0x1A100000
Backplane MAC address: 38:ed:18:a2:17:84
Router MAC address:    50:87:89:4b:c0:c5

Admin Config Information:
  state(up), mode(access), speed(auto), duplex(Auto), medium_db(0)
  layer(L2), dce-mode(edge), description(),
  auto neg(on), auto mdix(on), beacon(off), num_of_si(0)
  medium(broadcast), snmp trap(on), MTU(1500),
  flowcontrol rx(off) tx(off), link debounce(100),
  storm-control bcast:100.00% mcast:100.00% ucast:100.00%
  span mode(0 - not a span-destination)
  delay(1), bw(10000000), rate-mode(dedicated)
  eee(n/a), eee_lpi(Normal), eee_latency(Constant)
  fabricpath enforce (DCE Core)(0)
  load interval [1-3]: 30, 300, 0 (sec).
  lacp mode(on)
  graceful convergence state(enabled)
  Ethertype 0x8100
  Slowdrain Congestion : mode core timeout[500], mode edge [500]
  Slowdrain Pause : mode core enabled [y] timeout[500]
  Slowdrain Pause : mode edge enabled [y] timeout[500]
  Slowdrain Slow-speed : mode core enabled [n] percent[10]
  Slowdrain Slow-speed : mode edge enabled [n] percent[10]
  Monitor fp header(included)
  shut lan (disabled)
  Tag Native Mode (disabled)

Operational (Runtime) Information:
  state(up), mode(access), speed(10 Gbps), duplex(Full)
  state reason(None), error(no error)
  dce-mode(edge), intf_type(0), parent_info(0-1-5)
  port-flags-bitmask(0x0) reset_cntr(4)
  last intf reset time is 0 usecs after Thu Jan 1 00:00:00 1970
  secs flowcontrol rx(off) tx(off), vrf(disabled)
  mdix mode(mdix), primary vlan(10), cfg_acc_vlan(10)
```

```
    access vlan(10), cfg_native vlan(1), native vlan(1)
    eee(n/a), eee_wake_time_tx(0), eee_wake_time_rx(0)

    bundle_bringup_id(5)
    service_xconnect(0)
    current state [ETH_PORT_FSM_ST_L2_UP]
    xfp(inserted), status(ok) Extended info (present and valid)

Operational (Runtime) ETHPM_LIM Cache Information:
    Num of EFP(0), EFP port mode (0x100000), EFP rewrite(0),
    PORT_CMD_ENCAP(9), PORT_CMD_PORT_MODE(0),
    PORT_CMD_SET_BPDU_MATCH(2)
    port_mem_of_es_and_lacp_suspend_disable(0)

MTS Node Identifier: 0x302

Platform Information:
    Local IOD(0xd7), Global IOD(0) Runtime IOD(0xd7)

Capabilities:
    Speed(0xc), Duplex(0x1), Flowctrl(r:0x3,t:0x3), LinkDebounce(0x1)
    udld(0x1), SFPCapable(0x1), TrunkEncap(0x1), AutoNeg(0x1)
    channel(0x1), suppression(0x1), cos_rewrite(0x1), tos_rewrite(0x1)
    dce capable(0x4), l2 capable(0x1), l3 capable(0x2) qinq capable(0x10)
     ethertype capable(0x1000000), Fabric capable (y), EFP capable (n)
     slowdrain congestion capable(y), slowdrain pause capable (y)
     slowdrain slow-speed capable(y)
    Num rewrites allowed(104)
    eee capable speeds () and eee flap flags (0)
    eee max wk_time rx(0) tx(0) fb(0)

Information from GLDB Query:
    Platform Information:
      Slot(0x2), Port(0), Phy(0x2)
        LTL(0), VQI(0xc), LDI(0), IOD(0xd7)
      Backplane MAC address in GLDB: 38:ed:18:a2:17:84
      Router MAC address in GLDB:    50:87:89:4b:c0:c5

Operational Vlans: 10

Operational Bits:  3-4,13,53
    is_link_up(1), pre_cfg_done(1), l3_to_l2(1), pre_cfg_ph1_done(1),
Keep-Port-Down Type:0 Opc:0 RRToken:0X00000000, gwrap:(nil)
    Multiple  Reinit: 0 Reinit when shut: 0
    Last   SetTs: 487184 usecs after Sun May 14 18:54:29 2017
    Last ResetTs: 717229 usecs after Sun May 14 18:54:30 2017

DCX LAN LLS enabled: FALSE
MCEC LLS down: FALSE
Breakout mapid 0

    User config flags:  0x3
    admin_state(1), admin_layer(1), admin_router_mac(0) admin_monitor_fp_header(0)

Lock Info: resource [Ethernet3/1]
    type[0] p_gwrap[(nil)]
        FREE @ 528277 usecs after Sun May 14 21:29:05 2017
```

```
      type[1] p_gwrap[(nil)]
          FREE @ 528406 usecs after Sun May 14 21:29:05 2017
      type[2] p_gwrap[(nil)]
          FREE @ 381980 usecs after Sun May 14 18:54:28 2017
0x1a100000

Pacer Information:
  Pacer State: released credits
  ISSU Pacer State: initialized

Data structure info:
  Context: 0xa2f1108
  Pacer credit granted after:    4294967295 sec 49227 usecs
  Pacer credit held for:   1 sec 4294935903 usecs
```

Port-client 命令 **show system internal port-client link-event** 可以从软件角度跟踪线卡的接口链路事件，该命令是一个线卡级命令，需要进入线卡控制台。例 3-58 显示了 Module 3 上的端口的 Port-client 链路事件，可以看到因链路启用和停用而产生的不同时间戳的链路事件。

例 3-58　Port-client 链路事件

```
N7k-1# attach module 3
Attaching to module 3 ...
To exit type 'exit', to abort type '$.'
module-3# show system internal port-client link-event
*************** Port Client Link Events Log ***************
----                       ------         -----  -----  -------
Time                       PortNo         Speed  Event  Stsinfo
----                       ------         -----  -----  -------
May 15 05:53:01 2017 00879553  Ethernet3/1    10G    UP     Autonegotiation
  completed(0x40e50008)

May 15 05:52:58 2017 00871071  Ethernet3/1    ----   DOWN   SUCCESS(0x0)
May 15 05:47:35 2017 00553866  Ethernet3/11   ----   DOWN   Link down debounce
  timer stopped and link is down

May 15 05:47:35 2017 00550650  Ethernet3/11   ----   DOWN   SUCCESS(0x0)

May 15 05:47:35 2017 00454119  Ethernet3/11   ----   DOWN   Link down debounce
  timer started(0x40e50006)
```

对于这些链路事件来说，可以通过链路级命令 **show system internal port-client event-history port** *port-num* 在 Port-client 事件历史记录日志中查看指定端口的相关消息。

注：如果出现了因 Nexus 机箱端口停用而导致的故障问题，那么就可以在故障状态下收集命令 **show tech ethpm** 的输出结果。

3.4　HWRL、CoPP 和系统 QoS

DoS（Denial of Service，拒绝服务）的攻击形式有很多，会给网络环境中的服务器和基础设施（特别是数据中心）造成严重影响。针对基础设施设备的攻击会以非常高的数据速率生成 IP 流量流，这些 IP 数据流包含了大量需要由 RP（Route Processor，路由处理器）控制平面处理的数据包。由于发送给 RP 的非法数据包速率非常高，致使控制平面不得不花费大量时间来处理这些 DoS 流量，从而产生如下故障问题。

- 线路协议保活消息丢失，致使线路中断并产生路由翻动和严重的网络切换。
- 需要处理过量数据包，因为这些数据包正源源不断地转存到 CPU。
- 路由协议更新丢失，导致路由翻动和严重的网络切换。
- 二层网络不稳定。
- 将近 100%的 CPU 利用率会锁定路由器，致使路由器无法完成高优先级处理操作（导致其他负面影响）。
- RP 利用率接近 100%，导致 CLI 响应迟缓或 CLI 锁定，致使用户无法采取纠正措施来响应 DoS 攻击。
- 消耗大量内存、缓冲区和数据结构等资源，从而造成严重负面影响。
- 备份数据包队列，导致重要数据包被不加区别地丢弃。
- 路由器崩溃。

为了解决 DoS/DDoS 攻击和过量数据包处理问题，NX-OS 为用户提供了两阶段策略机制。

- 在将数据包发送给 CPU 之前，以每个模块为基础在硬件中对数据包进行限速。
- 利用 CoPP（Control Plane Policing，控制平面策略），对通过限速器的流量进行基于策略的流量监管。

硬件限速器和 CoPP 策略相结合，可以保护 CPU（路由处理器）免受不必要的流量或 DoS 攻击，并优先处理发送给 CPU 的相关流量，从而有效提高了设备的安全性。需要注意的是，硬件限速器仅限于 Nexus 7000 和 Nexus 9000 系列交换机，其他 Nexus 平台不可用。

可以将到达 CPU 或控制平面的数据包分为以下几类。

- **收到的数据包**：这些数据包去往路由器（如保持激活消息）。
- **多播数据包**：这些数据包又可以进一步分为以下两种。
  - 直连多播源的数据包。
  - 多播控制包。
- **副本数据包**：为了支持 ACL-log（ACL 日志）等功能特性，需要生成一份原始数据包的副本并发送给管理引擎，因而将这些数据包称为副本数据包。
  - ACL-log 副本。
  - FIB 单播副本。
  - 多播副本。
  - NetFlow 副本。
- **异常数据包**：这些数据包需要进行特殊处理，硬件无法处理这些数据包或检测异常，因而需要将这些数据包发送给管理引擎执行进一步处理。这类数据包归属异常类别，常见的异常数据包有以下几种。
  - 相同接口检查。
  - TTL 到期。
  - MTU 故障。
  - DHCP（Dynamic Host Control Protocol，动态主机控制协议）ACL 重定向。
  - ARP ACL 重定向。
  - 源 MAC IP 检查失败。
  - 不支持的重写。
  - 邻接关系过期错误。
- **探查数据包**：如果 FIB 中没有目的 IP 或下一跳的二层 MAC 地址，那么就会将数据包发送给管理引擎，由管理引擎会为目的主机或下一跳生成 ARP 请求。

- **广播及非 IP 数据包**：以下数据包属于该类别。
    - 广播 MAC + 非 IP 数据包。
    - 广播 MAC + IP 单播数据包。
    - 组播 MAC + IP 单播数据包。

需要记住的是，CoPP 策略和限速器均以每个模块、每个 FE（Forwarding Engine，转发引擎）为基础加以应用。

> **注**：对于 Nexus 7000 平台来说，除 F1 线卡之外的所有线卡均支持 CoPP 策略，F1 系列线卡完全使用限速器来保护 CPU。Nexus 7000/7700 和 Nexus 9000 系列平台均支持 HWRL。

例 3-59 显示了 **show hardware rate-limiter [module** *slot***]** 命令的输出结果，可以查看机箱中每个线卡模块的限速器配置和统计信息。

例 3-59 验证 N7k 和 N9k 交换机的硬件限速器

```
n7k-1# show hardware rate-limiter module 3

Units for Config: packets per second
Allowed, Dropped & Total: aggregated since last clear counters
rl-1: STP and Fabricpath-ISIS
rl-2: L3-ISIS and OTV-ISIS
rl-3: UDLD, LACP, CDP and LLDP
rl-4: Q-in-Q and ARP request
rl-5: IGMP, NTP, DHCP-Snoop, Port-Security, Mgmt and Copy traffic

Module: 3

Rate-limiter PG Multiplier: 1.00

   R-L Class           Config       Allowed         Dropped          Total
 +-------------------+--------+----------------+---------------+----------------+
   L3 mtu              500          0               0                0
   L3 ttl              500          0               0                0
   L3 control          10000        0               0                0
   L3 glean            100          0               0                0
   L3 mcast dirconn    3000         1               0                1
   L3 mcast loc-grp    3000         0               0                0
   L3 mcast rpf-leak   500          0               0                0
   L2 storm-ctrl       Disable
   access-list-log     100          0               0                0
   copy                30000        54649           0                54649
   receive             30000        292600          0                292600
   L2 port-sec         500          0               0                0
   L2 mcast-snoop      10000        2242            0                2242
   L2 vpc-low          4000         0               0                0
   L2 l2pt             500          0               0                0
   L2 vpc-peer-gw      5000         0               0                0
   L2 lisp-map-cache   5000         0               0                0
   L2 dpss             100          0               0                0
   L3 glean-fast       100          0               0                0
   L2 otv              100          0               0                0
   L2 netflow          500          0               0                0


Port group with configuration same as default configuration
    Eth3/1-32
```

```
N9K-1# show hardware rate-limiter module 2

Units for Config: packets per second
Allowed, Dropped & Total: aggregated since last clear counters

Module: 2
  R-L Class            Config        Allowed         Dropped           Total
  +------------------+---------+----------------+----------------+----------------+
  L3 glean             100           0               0                 0
  L3 mcast loc-grp     3000          0               0                 0
  access-list-log      100           0               0                 0
  bfd                  10000         0               0                 0
  exception            50            0               0                 0
  fex                  3000          0               0                 0
  span                 50            0               0                 0
  dpss                 6400          0               0                 0
  sflow                40000         0               0                 0
For verifying the rate-limiter statistics on F1 module on Nexus 7000 switches, use
the command show hardware rate-limiter [f1 rl-1 | rl-2 | rl-3 | rl-4 | rl-5].
```

此外，还可以在 Nexus 7000 系列交换机上查看为 SUP 流量应用的限速器情况，可以通过不同的模块来确定与每个限速器相匹配的异常情况，可以通过 **show hardware internal forwarding rate-limiter usage** [**module** *slot*]命令来查看这些差异信息。例 3-60 给出了该命令的输出结果，不但可以看到各种限速器信息，而且还能看出哪些数据包流或限速器是由 CoPP、L2 限速器或 L3 限速器处理的。

例 3-60 限速器使用情况

```
N7K-1# show hardware internal forwarding rate-limiter usage module 3

Note: The rate-limiter names have been abbreviated to fit the display.

-------------------------+------+------+--------+------+--------+--------
Packet streams           | CAP1 | CAP2 | DI     | CoPP | L3 RL  | L2 RL
-------------------------+------+------+--------+------+--------+--------
L3 control (224.0.0.0/24)| Yes  | x    | sup-hi | x    | control| copy
L2 broadcast             | x    | x    | flood  | x    | x      | strm-ctl
ARP request              | Yes  | x    | sup-lo | Yes  | x      | copy
Mcast direct-con         | Yes  | x    | x      | Yes  | m-dircon| copy
ISIS                     | Yes  | x    | sup-lo | x    | x      | x
L2 non-IP multicast      | x    | x    | x      | x    | x      | x
Access-list log          | x    | Yes  | acl-log| x    | x      | acl-log
L3 unicast control       | x    | x    | sup-hi | Yes  | x      | receive
L2 control               | x    | x    | x      | x    | x      | x
Glean                    | x    | x    | sup-lo | x    | x      | glean
Port-security            | x    | x    | port-sec| x   | x      | port-sec
IGMP-Snoop               | x    | x    | m-snoop| x    | x      | m-snoop
-------------------------+------+------+--------+------+--------+--------
Exceptions               | CAP1 | CAP2 | DI     | CoPP | L3 RL  | L2 RL
-------------------------+------+------+--------+------+--------+--------
IPv4 header options      | 0    | 0    | x      | Yes  |        | x
FIB TCAM no route        | 0    | 0    | x      | Yes  |        | x
Same interface check     | 0    | 0    | x      | x    | ttl    | x
IPv6 scope check fail    | 0    | 0    | drop   | x    |        | x
Unicast RPF more fail    | 0    | 0    | drop   | x    |        | x
Unicast RPF fail         | 0    | 0    | drop   | Yes  |        | x
```

```
Multicast RPF fail       0    0    drop    x            x
Multicast DF fail        0    0    drop    x            x
TTL expiry               0    0    x       x     ttl    x
Drop                     0    0    drop    x            x
L3 ACL deny              0    0    drop    x            x
L2 ACL deny              0    0    drop    x            x
IPv6 header options      0    0    drop    Yes          x
MTU fail                 0    0    x       x     mtu    x
DHCP ACL redirect        0    0    x       Yes   mtu    x
ARP ACL redirect         0    0    x       Yes   mtu    x
Smac IP check fail       0    0    x       x     mtu    x
Hardware drop            0    0    drop    x            x
Software drop            0    0    drop    x            x
Unsupported RW           0    0    x       x     ttl    x
Invalid packet           0    0    drop    x            x
L3 proto filter fail     0    0    drop    x            x
Netflow error            0    0    drop    x            x
Stale adjacency error    0    0    x       x     ttl    x
Result-bus drop          0    0    drop    x            x
Policer drop             0    0    x       x            x
```

可以通过 **show hardware internal forwarding l3 asic exceptions** *exception* **detail [module** *slot***]** 命令查看指定异常行为的相关信息。

可以通过 **show hardware internal forwarding [l2 | l3] asic rate-limiter** *rl-name* **detail [module** *slot***]** 命令可以查看 L2 和 L3 ASIC 限速器的配置信息，其中的变量 *rl-name* 是限速器的名称。例 3-61 显示了 L3 ASIC 异常以及 L2 和 L3 限速器信息。第一部分输出结果显示了未通过 RPF 检查的数据包的配置和统计信息，第二部分和第三部分输出结果显示了未通过 MTU 检查的数据包的限速器和异常配置信息。

例 3-61　L2 和 L3 限速器及异常配置

```
! L2 Rate-Limiter
N7K-1# show hardware internal forwarding l2 asic rate-limiter layer-3-glean detail
Device: 1
Device: 1
       Enabled:  0
   Packets/sec:  0

Match fields:
     Cap1 bit: 0
     Cap2 bit: 0
    DI select: 0
           DI: 0
    Flood bit: 0

Replaced result fields:
     Cap1 bit: 0
     Cap2 bit: 0
           DI: 0
! L3 Rate-Limiter
N7K-1# show hardware internal forwarding l3 asic rate-limiter layer-3-mtu detail
   slot   3
   ========
Dev-id: 0
Rate-limiter configuration: layer-3 mtu
```

```
         Enabled:   1
     Packets/sec:   500
    Packet burst:   325 [burst period of 1 msec]
L3 Exceptions
N7K-1# show hardware internal forwarding l3 asic exceptions mtu-fail detail
slot  3
=======
Egress exception priority table programming:
              Reserved: 0
     Disable LIF stats: 0
               Trigger: 0
               Mask RP: 0x1
         Dest info sel: 0
  Clear exception flag: 0x1
             Egress L3 : 0
   Same IF copy disable: 0x1
      Mcast copy disable: 0x1
      Ucast copy disable: 0
    Exception dest sel: 0x6
       Enable copy mask: 0
      Disable copy mask: 0x1

Unicast destination table programming:
      Reserved: 0
       L2 fwd: 0x1
      Redirect: 0x1
 Rate-limiter: 0x6
         Flood: 0
    Dest index: 0x10c7
           CCC: 0

Multicast destination table programming:
      Reserved: 0
        L2 fwd: 0
      Redirect: 0
 Rate-limiter: 0
         Flood: 0
    Dest index: 0x285f
           CCC: 0
```

Nexus 平台也在硬件中实现了 CoPP，这样做的好处是帮助保护管理引擎免受 DoS 攻击，可以控制数据包到达管理引擎 CPU 的允许速率。需要记住的是，到达管理引擎模块 CPU 的流量主要来自以下 4 种途径。

- 线卡发送流量的带内接口。
- 管理接口。
- CMP（Control and Monitoring Processor，控制和监控处理器）接口（用于控制台）。
- EOBC（Ethernet Out of Band Channel，以太网带外通道）。

系统仅将带内接口发送的流量发送给 CoPP，因为这是通过线卡上的不同转发引擎（FE）到达管理引擎模块的唯一流量。CoPP 策略是对逐个 FE 单独实施的。

Nexus 平台启动时，NX-OS 会安装一个名为 *copp-system-policy* 的默认 CoPP 策略。此外，NX-OS 还为 CoPP 提供了不同的配置文件设置，从而为系统提供不同的保护等级。目前支持的 CoPP 配置文件包括以下几种。

- **严格模式（Strict）**：定义的常规类别 BC 值为 250ms，重要类别 BC 值为 1000ms。

- **中度模式（Moderate）**：定义的常规类别 BC 值为 375ms，重要类别 BC 值为 1250ms。
- **宽松模式（Lenient）**：定义的常规类别 BC 值为 375ms，重要类别 BC 值为 1500ms。
- **密集模式（Dense）**：如果机箱中配置的 F2 线卡比其他 I/O 模块多，那么就建议采用该配置文件。NX-OS 从版本 6.0(1)开始引入。

如果未在初始设置期间选择上述 CoPP 策略，那么 NX-OS 就会为控制平面应用严格模式的配置文件。当然，也可以不选择上述配置文件，而是为 CoPP 创建一个自定义策略。NX-OS 的默认 CoPP 策略将策略分为以下预定义类别。

- **关键（Critical）**：IP 优先级为 6 的路由协议数据包。
- **重要（Important）**：冗余协议（如 GLBP、VRRP 和 HSRP）。
- **管理（Management）**：所有管理流量（如 Telnet、SSH、FTP、NTP 和 Radius）。
- **监控（Monitoring）**：Ping 和 traceroute 流量。
- **异常（Exception）**：ICMP 不可达和 IP 选项。
- **非期望（Undesirable）**：所有不需要的流量。

例 3-62 给出了系统初次启动时的严格 CoPP 策略示例，可以通过 **show run copp all** 命令查看 CoPP 配置信息。

例 3-62　Nexus 的 CoPP 严格策略

```
class-map type control-plane match-any copp-system-p-class-critical
  match access-group name copp-system-p-acl-bgp
  match access-group name copp-system-p-acl-rip
  match access-group name copp-system-p-acl-vpc
  match access-group name copp-system-p-acl-bgp6
  match access-group name copp-system-p-acl-lisp
  match access-group name copp-system-p-acl-ospf
  ! Output omitted for brevity
class-map type control-plane match-any copp-system-p-class-exception
  match exception ip option
  match exception ip icmp unreachable
  match exception ipv6 option
  match exception ipv6 icmp unreachable
class-map type control-plane match-any copp-system-p-class-important
  match access-group name copp-system-p-acl-cts
  match access-group name copp-system-p-acl-glbp
  match access-group name copp-system-p-acl-hsrp
  match access-group name copp-system-p-acl-vrrp
  match access-group name copp-system-p-acl-wccp
  ! Output omitted for brevity
class-map type control-plane match-any copp-system-p-class-management
  match access-group name copp-system-p-acl-ftp
  match access-group name copp-system-p-acl-ntp
  match access-group name copp-system-p-acl-ssh
  match access-group name copp-system-p-acl-ntp6
  match access-group name copp-system-p-acl-sftp
  match access-group name copp-system-p-acl-snmp
  match access-group name copp-system-p-acl-ssh6
  ! Output omitted for brevity
class-map type control-plane match-any copp-system-p-class-monitoring
  match access-group name copp-system-p-acl-icmp
  match access-group name copp-system-p-acl-icmp6
  match access-group name copp-system-p-acl-mpls-oam
  match access-group name copp-system-p-acl-traceroute
```

```
    match access-group name copp-system-p-acl-http-response
! Output omitted for brevity
class-map type control-plane match-any copp-system-p-class-normal
    match access-group name copp-system-p-acl-mac-dot1x
    match exception ip multicast directly-connected-sources
    match exception ipv6 multicast directly-connected-sources
    match protocol arp
class-map type control-plane match-any copp-system-p-class-undesirable
    match access-group name copp-system-p-acl-undesirable
    match exception fcoe-fib-miss

policy-map type control-plane copp-system-p-policy-strict
  class copp-system-p-class-critical
    set cos 7
    police cir 36000 kbps bc 250 ms conform transmit violate drop
  class copp-system-p-class-important
    set cos 6
    police cir 1400 kbps bc 1500 ms conform transmit violate drop
  class copp-system-p-class-management
    set cos 2
    police cir 10000 kbps bc 250 ms conform transmit violate drop
  class copp-system-p-class-normal
    set cos 1
    police cir 680 kbps bc 250 ms conform transmit violate drop
  class copp-system-p-class-exception
    set cos 1
    police cir 360 kbps bc 250 ms conform transmit violate drop
  class copp-system-p-class-monitoring
    set cos 1
    police cir 130 kbps bc 1000 ms conform transmit violate drop
  class class-default
    set cos 0
    police cir 100 kbps bc 250 ms conform transmit violate drop
```

如果要查看不同 CoPP 配置文件之间的差异,可以使用命令 **show copp diff profile** *profile-type* **profile** *profile-type*,该命令可以显示两个指定配置文件的策略映射(policy-map)配置差异。

> **注:** NX-OS 从版本 6.2(2)开始,为多播流量增加了 *copp-system-p-class-multicast-router*、*copp-system-p-class-multicast-host* 和 *copp-system-p-class-normal* 策略类别。在版本 6.2(2)之前,这些类别是通过自定义用户配置实现的。

HWRL 和 CoPP 都是在转发引擎(FE)层面完成的功能特性。由于来自多个 FE 的聚合流量仍然有可能导致 CPU 超负荷,因而 HWRL 和 CoPP 都是尽力而为策略。另外,需要记住的是,CoPP 策略不应过于激进,应根据实际的网络规划和配置情况设计适当的 CoPP 策略。例如,如果命中 CoPP 策略的路由协议包的速率高于监管速率,那么就可能会丢弃合法会话并出现协议振荡。如果必须修改预定义 CoPP 策略,那么就可以复制预定义 CoPP 策略,在此基础上通过编辑操作创建自定义 CoPP 策略。请注意,所有预定义 CoPP 配置文件都是不可编辑的。此外,**show running-config** 命令无法显示 CoPP 策略,可以通过 **show running-config all** 或 **show running-config copp all** 命令查看 CoPP 策略。例 3-63 给出了使用 CoPP 策略配置的方式以及创建自定义严格策略的配置方式。

例 3-63 查看 CoPP 策略并创建自定义 CoPP 策略

```
R1# show running-config copp
copp profile strict
```

```
R1# show running-config copp all
class-map type control-plane match-any copp-system-p-class-critical
  match access-group name copp-system-p-acl-bgp
  match access-group name copp-system-p-acl-rip
  match access-group name copp-system-p-acl-vpc
  match access-group name copp-system-p-acl-bgp6
! Output omitted for brevity

R1# copp copy profile strict ?
  prefix Prefix for the copied policy
  suffix Suffix for the copied policy
R1# copp copy profile strict prefix custom

R1# configure terminal
R1(config)# control-plane
R1(config-cp)# service-policy input custom-copp-policy-strict
```

命令 **show policy-map interface control-plane** 可以显示 CoPP 策略的计数器。对于汇总视图来说，可以在该命令中使用过滤器 **include"class | conform | violated"**，以查看符合策略的数据包数量以及违反策略和丢弃的数据包数量，如例 3-64 所示。

例 3-64 show policy-map interface control-plane 命令输出结果

```
R1# show policy-map interface control-plane | include "class|conform|violated"
    class-map custom-copp-class-critical (match-any)
        conformed 123126534 bytes; action: transmit
        violated 0 bytes; action: drop
        conformed 0 bytes; action: transmit

        violated 0 bytes; action: drop
        conformed 107272597 bytes; action: transmit
        violated 0 bytes; action: drop
        conformed 0 bytes; action: transmit
        violated 0 bytes; action: drop
    class-map custom-copp-class-important (match-any)
        conformed 0 bytes; action: transmit
        violated 0 bytes; action: drop
        conformed 0 bytes; action: transmit
        violated 0 bytes; action: drop
        conformed 0 bytes; action: transmit
        violated 0 bytes; action: drop
        conformed 0 bytes; action: transmit
        violated 0 bytes; action: drop
! Output omitted for brevity
```

对于 CoPP 策略的访问列表来说，面临的一个主要问题就是 IP 和 MAC ACL（Access Control List，访问控制列表）不支持 **statistics per-entry** 命令，因而在 ACL 下应用该命令将毫无作用。如果要在输入/输出（I/O）模块上查看 CoPP 策略引用的 IP 和 MAC ACL 计数器，那么就可以使用 **show system internal access-list input entries detail** 命令。例 3-65 给出了 **show system internal access-list input entries detail** 命令的输出结果，显示了 FabricPath MAC 地址 0180.c200.0041 命中 MAC ACL 的情况。

例 3-65 TCAM 中的 IP 和 MAC ACL 计数器

```
n7k-1# show system internal access-list input entries detail | grep 0180.c200.0041
[020c:4344:020a] qos 0000.0000.0000 0000.0000.0000 0180.c200.0041 ffff.ffff.ffff
  [0]
```

```
[020c:4344:020a] qos 0000.0000.0000 0000.0000.0000 0180.c200.0041 ffff.ffff.ffff
  [20034]
[020c:4344:020a] qos 0000.0000.0000 0000.0000.0000 0180.c200.0041 ffff.ffff.ffff
  [19923]
[020c:4344:020a] qos 0000.0000.0000 0000.0000.0000 0180.c200.0041 ffff.ffff.ffff
  [0]
```

NX-OS 从版本 5.1 开始可以配置日志记录阈值,对满足特定类别的 CoPP 策略而产生的丢包行为生成 syslog 消息。如果某流量类别丢弃的字节数超出了用户配置的阈值,那么就会生成 syslog 消息。可以使用 **logging drop threshold** *dropped-bytes-count* [**level** *logging-level*]命令配置该阈值。例 3-66 将日志记录阈值设置为丢弃 100 个字节且日志记录级别为 7,同时还显示了超出丢弃字节数阈值之后生成 syslog 消息的配置方式。

例 3-66　记录 syslog 的丢弃阈值

```
R1(config)# policy-map type control-plane custom-copp-policy-strict
R1(config-pmap)# class custom-copp-class-critical
R1(config-pmap-c)# logging drop threshold ?
  <1-80000000000>  Dropped byte count
R1(config-pmap-c)# logging drop threshold 100 ?
  <CR>
  level  Syslog level

R1(config-pmap-c)# logging drop threshold 100 level ?
  <1-7>  Specify the logging level between 1-7

R1(config-pmap-c)# logging drop threshold 100 level 7
%COPP-5-COPP_DROPS5: CoPP drops exceed threshold in class:
custom-copp-class-critical,
check show policy-map interface control-plane for more info.
```

NX-OS 从版本 6.0 开始引入了缩放因子(Scale factor)配置机制。缩放因子的作用是在不改变实际 CoPP 策略配置的情况下,以每块线卡为基础缩放所应用的 CoPP 策略的策略器速率。缩放因子配置范围是 0.1~2.0。配置缩放因子时,需要在控制平面配置模式下使用命令 **scale-factor** *value* [**module** *slot*]。例 3-67 说明了为 Nexus 机箱中的各种线卡配置缩放因子的方式,命令 **show system internal copp info** 可以查看缩放因子的设置情况,同时还能显示其他信息,包括上次执行的操作及其状态、CoPP 数据库信息和 CoPP 运行时状态,这些信息对于 CoPP 策略的故障排查来说非常有用。

例 3-67　配置缩放因子

```
n7k-1(config)# control-plane
n7k-1(config-cp)# scale-factor 0.5 module 3
n7k-1(config-cp)# scale-factor 1.0 module 4
n7k-1# show system internal copp info

Active Session Details:
----------------------
There isn't any active session

Last operation status:
---------------------
    Last operation: Show Command
    Last operation details: show policy-map interface
    Last operation Time stamp: 16:58:14 UTC May 14 2015
```

```
    Operation Status: Success

! Output omitted for brevity
Runtime Info:
--------------
    Config FSM current state: IDLE
    Modules online: 3 4 5 7

Linecard Configuration:
----------------------
Scale Factors
Module 1: 1.00
Module 2: 1.00
Module 3: 0.50
Module 4: 1.00
Module 5: 1.00
Module 6: 1.00
Module 7: 1.00
Module 8: 1.00
Module 9: 1.00
```

注： 有关 Nexus 7000 机箱缩放因子的配置建议，可参阅 CCO 文档。

配置 NX-OS CoPP 策略时，应记住以下最佳实践。

- 使用严格模式的 CoPP 配置文件。
- 每次 NX-OS 升级之后或至少每次主要 NX-OS 版本升级之后，都要执行 **copp profile strict** 命令。如果此前修改了 CoPP 策略，那么就必须在升级后重新应用该策略。
- 如果机箱加载的都是 F2 系列模块或加载的 F2 系列模块多于所有其他 I/O 模块，那么就建议使用密集模式的 CoPP 配置文件。
- 不建议禁用 CoPP，应根据需要调整默认 CoPP。
- 监控意外丢包行为，根据期望流量增加或修改默认 CoPP 策略。

由于数据中心的流量模式一直都处于变化当中，因而定制 CoPP 策略也是一个持续性过程。

MTU 设置

Nexus 平台的 MTU 设置与其他思科平台的作用不同。Nexus 支持两种 MTU 设置：二层（L2）MTU 和三层（L3）MTU。可以在接口配置下使用 **mtu** *value* 命令手工配置 L3 MTU，L2 MTU 可以通过网络 QoS 策略进行配置，也可以在支持逐端口 MTU 设置的 Nexus 交换机的接口上直接配置 MTU。需要在 **network-qos** 策略类型下定义 L2 MTU，然后在 **system qos** 策略配置下应用 L2 MTU。例 3-68 给出了在 Nexus 平台上启用巨帧（Jumbo）L2 MTU 的配置示例。

例 3-68 巨帧 MTU 系统配置

```
N7K-1(config)# policy-map type network-qos policy-MTU
N7K-1(config-pmap-nqos)# class type network-qos class-default
N7K-1(config-pmap-nqos-c)# mtu 9216
N7K-1(config-pmap-nqos-c)# exit
N7K-1(config-pmap-nqos)# exit
N7K-1(config)# system qos
N7K-1(config-sys-qos)# service-policy type network-qos policy-MTU
```

建议在接口上应用巨帧 L3 MTU 之前，首先启用巨帧 L2 MTU。

> 注：并非所有平台都支持端口级别的巨帧 L2 MTU，仅 Nexus 7000、7700、9300 和 9500 平台支持端口级别的 L2 MTU 配置，其他平台（如 Nexus 3048、3064、3100、3500、5000、5500 和 6000）仅支持基于网络 QoS 策略的巨帧 L2 MTU 配置。

可以通过 **show interface** *interface-type x/y* 命令查看 Nexus 3000、7000、7700 和 9000（支持逐端口 MTU 设置的平台）的 MTU 设置，使用 **show queuing interface** *interface-type x/y* 命令查看 Nexus 3100、3500、5000、5500 和 6000（支持基于网络 QoS 策略的 MTU 设置的平台）的 MTU 设置。

1. FEX 巨帧 MTU 设置

可以在父交换机上配置 Nexus 2000 FEX 的巨帧 MTU。如果父交换机支持逐端口设置 MTU，那么就可以在 FEX 交换矩阵端口通道接口上配置 MTU。如果父交换机不支持逐端口 MTU 设置，那么就可以在网络 QoS 策略下完成该配置。例 3-69 给出了支持逐端口 MTU 设置的 Nexus 交换机和支持网络 QoS 策略的 Nexus 交换机的 FEX MTU 配置示例。

例 3-69　FEX 巨帧 MTU 设置

```
! Per-Port Basis Configuration
NX-1(config)# interface port-channel101
NX-1(config-if)# switchport mode fex-fabric
NX-1(config-if)# fex associate 101
NX-1(config-if)# vpc 101
NX-1(config-if)# mtu 9216
! Network QoS based MTU Configuration
NX-1(conf)# class-map type network-qos match-any c-MTU-custom
(config-cmap-nqos)# match cos 0-7

NX-1(config)# policy-map type network-qos MTU-custom template 8e
NX-1(config-pmap-nqos)# class type network-qos c-MTU-custom
! Below command configures the congestion mechanism as tail-drop
NX-1(config-pmap-nqos-c)# congestion-control tail-drop
NX-1(config-pmap-nqos-c)# mtu 9216

NX-1(config)# system qos
NX-1(config-sys-qos)# service-policy type network-qos MTU-custom
```

> 注：从 NX-OS 版本 6.2 开始，Nexus 7000 交换机不再支持 FEX 端口的逐端口 MTU 配置，需要通过自定义网络 QoS 策略来配置这些 MTU，如例 3-69 所示。

2. MTU 故障排查

MTU 故障的常见原因是配置差错或网络设计不当、未在接口或系统层面正确设置 MTU，此时只要更新配置并查看网络设计方案即可轻松解决这类配置差错。如果接口或系统层面的 MTU 配置没问题，但软件或硬件未正确编程，那么故障排查就显得较为麻烦。此时，需要执行一些检查操作以确认 MTU 是否正确编程。

MTU 故障排查的第一步就是使用 **show interface** 或 **show queuing interface** *interface-type x/y* 命令验证接口的 MTU 设置。对于支持基于网络 QoS 策略设置 MTU 的设备来说，可以使用 **show policy-map system type network-qos** 命令来验证 MTU 的设置情况，如例 3-70 所示。

例 3-70　网络 QoS 策略验证

```
N7K-1# show policy-map system type network-qos
  Type network-qos policy-maps
  ===============================
  policy-map type network-qos policy-MTU template 8e
```

```
    class type network-qos class-default
      mtu 9216
      congestion-control tail-drop threshold burst-optimized
```

NX-OS 中的 EthPM 进程可以管理端口级别的 MTU 配置，通过 show system internal ethpm info interface *interface-type x/y* 命令可以验证 EthPM 进程的 MTU 信息，如例 3-71 所示。

例 3-71　验证 EthPM 进程的 MTU 配置

```
NX-1# show system internal ethpm info interface ethernet 2/1 | egrep MTU
    medium(broadcast), snmp trap(on), MTU(9216),
```

此外，还可以在 ELIM（EARL LIF Table Manager，EARL LIF 表管理器）进程上验证 MTU 设置信息，该进程负责维护以太网状态信息。此外，ELTM 进程还负责管理逻辑接口，如 SVI（Switch Virtual Interface，交换机虚拟接口）。如果要在特定接口上验证 ELTM 进程的 MTU 设置，那么就可以使用 show system internal eltm info interface *interface-type x/y* 命令，如例 3-72 所示。

例 3-72　验证 ELTM 进程的 MTU 配置

```
NX-1# show system internal eltm info interface e2/1 | in mtu
    mtu = 9216 (0x2400), f_index = 0 (0x0)
```

注：如果多台设备都出现 MTU 问题，或者 EthPM 进程或 MTU 设置出现了软件故障，那么就可以将 **show tech-support ethpm** 和 **show tech-support eltm [detail]** 的输出结果保存到文件中，并开启 TAC 案例进行深入排查。

3.5　本章小结

本章重点讨论了 Nexus 平台的硬件和软件故障排查问题。从硬件故障排查角度来看，本章涵盖了以下内容。

- GOLD 测试。
- 线卡和进程崩溃。
- 丢包和平台差错。
- 接口差错和丢包。
- FEX 故障排查。

本章详细介绍了 VDC 的工作原理，并探讨了 VDC 的故障排查问题，VDC 的不同模块组合可能会产生各种故障问题。此外，本章还介绍了限制 VDC 资源的方式以及各种 NX-OS 组件，如 Netstack、UFDM 和 IPFIB、EthPM 和 Port-Client。最后，本章还讨论了 CoPP 以及 CoPP 策略下的丢包故障排查问题，包括解决以太网和 FEX 端口的 MTU 问题。

第二部分

二层转发故障排查

第 4 章 Nexus 交换技术
第 5 章 端口通道、vPC 和 FabricPath

第 4 章

Nexus 交换技术

本章主要讨论如下主题。
- 二层网络通信概述。
- VLAN。
- PVLAN（Private VLAN，私有 VLAN）。
- STP（Spanning Tree Protocol，生成树协议）。
- 检测和修复转发环路。
- 端口安全。

思科在发布 Nexus 产品线时，推出了一类新型网络设备，称为数据中心交换产品。数据中心交换产品提供了高密度、高速交换能力，以满足数据中心服务器（物理和虚拟服务器）的特性化需求。本章将重点介绍网络交换产品的核心组件并验证这些组件是否处于正常工作状态的方式，从而更好地隔离和排查二层转发故障。

4.1 二层网络通信概述

以太网协议最初使用的是 Thinnet（10Base2）或 Thicknet（10Base5）技术，这些技术通过同一条电缆连接所有的网络设备。如果两台设备试图同时通信，那么就会产生问题，因为以太网设备采用 CSMA/CD（Carrier Sense Multiple Access/Collision Detect，载波侦听多路访问/冲突检测）机制来确保冲突域中每个时刻只有一台设备在通信。如果设备检测到另一台设备正在传输数据，那么就会延迟发送数据包，直至电缆处于空闲状态。

随着加入电缆的设备越来越多，网络效率也变得越来越低。所有设备都位于相同的 CD（Collision Domain，冲突域）中。网络集线器进一步加剧了该问题，因为集线器的端口密度非常高，而且直接将流量中继到所有端口上，网络集线器没有任何智能机制来引导网络流量。

网络交换机通过创建虚拟通道来增强网络的可扩展性和稳定性。交换机会维护一张表格，将主机的以太网 MAC 地址与发送网络流量的端口相关联。交换机使用 MAC 地址表将网络流量仅转发给与数据包目的 MAC 地址相关联的目的端口，而不是通过所有端口向外泛洪所有流量。仅当交换机无法识别目的 MAC 地址时，才会从 LAN 的所有网络端口向外转发数据包（称为单播泛洪）。

网络广播（MAC 地址 ff:ff:ff:ff:ff:ff）会导致交换机从所有的 LAN 交换端口向外广播数据包，该通信方式具有相当大的破坏性，因为这样做的后果是导致网络交换机的效率与集线器相当，网络设备之间的通信会因为 CSMA/CD 而停止。网络广播无法跨越三层边界（从一个子网到另一个子网），位于同一个二层（L2）网段中的所有设备都被视为位于同一个广播域中。

图 4-1 中的 PC-A 正在向网络上的所有设备（包括 PC-B、PC-C 和 R1）发布广播流量，R1 不会将广播流量从一个广播域（192.168.1.0/24）转发到另一个广播域（192.168.2.0/24）。

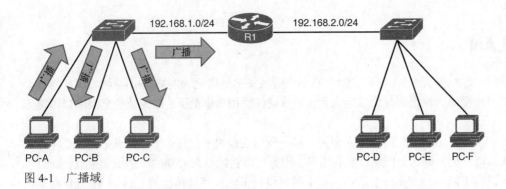

图 4-1 广播域

本地 MAC 地址表包含了 MAC 地址列表以及学到这些 MAC 地址的端口,可以通过 **show mac address-table** [**address** *mac-address*]命令显示 MAC 地址表。为了正确编程交换机的硬件 ASIC,可以通过 **show hardware mac address-table** *module* [**dynamic**] [**address** *mac-address*] 命令显示硬件 MAC 地址表。

例 4-1 显示了 Nexus 交换机的 MAC 地址表。排查 L2 转发故障的第一步就是要找到网络设备所连接的交换端口。如果同一端口上出现了多个 MAC 地址,那么就表示该端口连接了一台交换机,作为故障排查过程的一部分,此时可能需要连接该交换机,以识别网络设备所连接的端口。

例 4-1 查看 Nexus 交换机的 MAC 地址

```
NX-1# show mac address-table
Legend:
        * - primary entry, G - Gateway MAC, (R) - Routed MAC, O - Overlay MAC
        age - seconds since last seen,+ - primary entry using vPC Peer-Link,
        (T) - True, (F) - False, C - ControlPlane MAC
   VLAN     MAC Address      Type       age     Secure NTFY Ports
---------+-----------------+--------+---------+------+----+------------------
*  1        0007.b35b.c420   dynamic   0           F    F    Eth1/7
*  1        0011.2122.2370   dynamic   0           F    F    Eth1/6
*  1        0027.e398.5481   dynamic   0           F    F    Eth1/1
*  1        0027.e398.54c0   dynamic   0           F    F    Eth1/1
*  1        0035.1a93.e4c2   dynamic   0           F    F    Eth1/2
*  1        286f.7fa3.e401   dynamic   0           F    F    Eth1/3
*  1        9caf.ca2e.76c1   dynamic   0           F    F    Eth1/5
*  1        9caf.ca2e.9041   dynamic   0           F    F    Eth1/4
G  -        885a.92de.617c   static    -           F    F    sup-eth1(R)
NX-1# show hardware mac address-table 1 dynamic
FE |PI| VLAN |      MAC       |Trunk|  TGID  |Mod|Port|Virt|Static|Hit|Hit|CPU|Pend
   |  |      |                |     |        |   |    |Port|      | SA| DA|   |
---+--+------+----------------+-----+--------+---+----+----+------+---+---+---+----+
0   1  1      286f.7fa3.e401   0     0        1   3    0    1      0   0   0
0   1  1      9caf.ca2e.76c1   0     0        1   5    0    1      1   0   0
0   1  1      0027.e398.5481   0     0        1   1    0    1      0   0   0
0   1  1      0035.1a93.e4c2   0     0        1   2    0    1      0   0   0
0   1  1      0027.e398.54c0   0     0        1   1    0    0      0   0   0
0   1  1      0011.2122.2370   0     0        1   6    0    1      0   0   0
0   1  1      0007.b35b.c420   0     0        1   7    0    1      0   0   0
0   1  1      9caf.ca2e.9041   0     0        1   4    0    1      1   0   0
```

注:术语网络设备(network device)和主机(host)在本书可以互换。

4.2 VLAN

在 LAN 网段之间增加路由器，有助于缩小广播域并提供优化的网络通信能力。LAN 网段上的主机位置可能会因为网络编址而发生变化，导致硬件使用效率低下（因为某些交换端口可能未被使用）。

VLAN（Virtual LAN，虚拟 LAN）通过在同一网络交换机上创建多个广播域来提供逻辑分段功能。VLAN 可以大大提高交换端口的利用率，因为可以将端口与必要的广播域相关联，而且多个广播域可以位于同一台交换机上。VLAN 中的网络设备无法通过传统的 L2 或广播流量与其他 VLAN 中的设备进行通信。

VLAN 标准由 IEEE（Institute of Electrical and Electronics Engineers，电气和电子工程师协会）802.1Q 定义，该标准在数据包的报头增加了 32 比特字段信息。

- TPID（Tag Protocol Identifier，标记协议标识符）：16 比特字段，被设置为 0x8100，表示该数据包是 802.1Q 数据包。
- PCP（Priority Code Point，优先级代码点）：3 比特字段，表示 CoS（Class of Service，服务类别），是交换机之间的二层 QoS（Quality of Service，服务质量）的一部分。
- DEI（Drop Eligible Indicator，丢弃合格指示符）：1 比特字段，指示存在带宽争用时是否可以丢弃数据包。
- VID（VLAN Identifier，VLAN 标识符）：12 比特字段，指定与网络数据包相关联的 VLAN。

图 4-2 给出了 VLAN 报文结构。

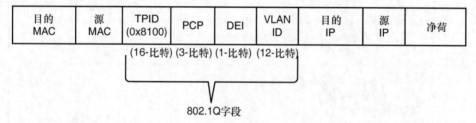

图 4-2 VLAN 报文结构

VLAN 标识符只有 12 比特，可以提供 4094 个唯一 VLAN。NX-OS 在 VLAN 标识符的使用上遵循以下逻辑。

- VLAN 0 保留用于 802.1P 流量，无法修改或删除。
- VLAN 1 是默认 VLAN，无法修改或删除。
- VLAN 2～1005 属于常规 VLAN 范围，可以根据需要添加、删除或修改。
- VLAN 1006～3967 和 4048～4093 属于扩展 VLAN 范围，可以根据需要添加、删除或修改。
- VLAN 3968～4047 和 4094 是内部 VLAN，归 NX-OS 内部使用，无法添加、删除或修改。
- VLAN 4095 由 802.1Q 标准保留，无法使用。

4.2.1 VLAN 创建

可以通过全局配置命令 **vlan** *vlan-id* 创建 VLAN，通过 VLAN 子模式配置命令 **name** *name* 为 VLAN 配置好记的名称（32 个字符）。创建 VLAN 的时候，必须保证 CLI 已经回到全局配置环境或其他 VLAN 标识符。例 4-2 给出了在 NX-1 上创建 VLAN 10（Accounting）、VLAN 20（HR）和 VLAN 30（Security）的配置示例。

例 4-2 VLAN 创建

```
NX-1(config)# vlan 10
NX-1(config-vlan)# name Accounting
NX-1(config-vlan)# vlan 20
NX-1(config-vlan)# name HR
NX-1(config-vlan)# vlan 30
NX-1(config-vlan)# name Security
```

可以通过 **show vlan** [**id** *vlan-id*]命令验证 VLAN 及其端口分配情况（见例 4-3），使用可选关键字 **id** 可以将输出结果过滤为指定 VLAN。请注意，输出结果分为 3 块：传统 VLAN、RSPAN（Remote Switched Port Analyzer，远程交换端口分析器）VLAN 和 PVLAN。

例 4-3 show vlan 命令输出结果

```
NX-1# show vlan
! Traditional and common VLANs will be listed in this section. The ports
! associated to these VLANs are displayed to the right.
VLAN Name                             Status    Ports
---- -------------------------------- --------- -------------------------------
1    default                          active    Eth1/1, Eth1/21, Eth1/22
                                                Eth1/23, Eth1/24, Eth1/25
                                                Eth1/26, Eth1/27, Eth1/28
                                                Eth1/29, Eth1/30, Eth1/31
                                                Eth1/32, Eth1/33, Eth1/34
                                                Eth1/35, Eth1/36, Eth1/37
                                                Eth1/38, Eth1/39, Eth1/40
                                                Eth1/41, Eth1/42, Eth1/43
                                                Eth1/44, Eth1/45, Eth1/46
                                                Eth1/47, Eth1/48, Eth1/49
                                                Eth1/50, Eth1/51, Eth1/52
10   Accounting                       active    Eth1/2, Eth1/3, Eth1/4, Eth1/5
                                                Eth1/6, Eth1/7, Eth1/8, Eth1/9
                                                Eth1/10, Eth1/11, Eth1/12
20   HR                               active    Eth1/13, Eth1/14, Eth1/15
                                                Eth1/16, Eth1/17, Eth1/18
                                                Eth1/19, Eth1/20
30   Security                         active

VLAN Type    Vlan-mode
---- ------  ----------
1    enet    CE
10   enet    CE
20   enet    CE
30   enet    CE

! If a Remote SPAN VLAN is configured, it will be displayed in this section.
! Remote SPAN VLANs were explained in Chapter 2
Remote SPAN VLANs
-------------------------------------------------------------------------------

! If Private VLANs are configured, they will be displayed in this section.
! Private VLANs are covered later in this chapter.
Primary  Secondary  Type             Ports
-------  ---------  ---------------  -----------------------------------------

NX-1# show vlan id 10
```

```
VLAN Name                             Status    Ports
----  ----                            ------    -----
10    Accounting                      active    Eth1/2, Eth1/3, Eth1/4, Eth1/5
                                                Eth1/6, Eth1/7, Eth1/8, Eth1/9
                                                Eth1/10, Eth1/11, Eth1/12

VLAN Type      Vlan-mode
---- ----      ---------
10   enet      CE

Remote SPAN VLAN
----------------
Disabled


Primary  Secondary  Type            Ports
-------  ---------  ---------       -----
```

> **注**：大多数工程师假定子网与 VLAN 之间的比率关系是 1:1。不过，多个子网也可以位于同一个 VLAN 中，实现方法是为路由器接口分配辅助 IP 地址（Secondary IP address）或者将多台路由器连接到同一个 VLAN 上，此时，这些子网都属于同一个广播域。

4.2.2 接入端口

接入端口（Access port）是交换机的基本模块，接入端口仅分配给一个 VLAN，负责将流量从 VLAN 传送到与其相连的设备，或者将流量从一台设备传送到该交换机上同一 VLAN 中的其他设备。

NX-OS 默认将 L2 交换端口作为接入端口，可以通过命令 **switchport mode access** 将交换端口配置为接入端口，通过命令 **switchport access vlan** *vlan-id* 将接入端口与指定 VLAN 相关联。如果未指定 VLAN 号，那么就默认为 VLAN 1。接入端口发送或接收的数据包都不包含 802.1Q 标记。

查看传统运行配置时，无法看到 **switchport mode access** 命令，此时需要使用可选关键字 **all**，如例 4-4 所示。

例 4-4 查看接入端口配置命令

```
NX-1# show run interface e1/2
! Output omitted for brevity
interface Ethernet1/2
  switchport access vlan 10

NX-1# show run interface eth1/2 all | include access
  switchport mode access
  switchport access vlan 10
```

命令 **show interface** *interface-id* 可以显示端口正在使用的模式，命令 **show vlan** 可以查看端口分配的 VLAN（见例 4-2），也可以使用命令 **show interface status**。例 4-5 给出了接入端口和相关 VLAN 的验证方式，为了确保 L2 转发的正常操作，必须确认两台主机都位于同一个 VLAN 中。

例 4-5 验证接入端口模式

```
NX-1# show interface eth1/2 | include Port
  Port mode is access
NX-1# show interface status

--------------------------------------------------------------------------------
Port          Name              Status    Vlan      Duplex  Speed   Type
--------------------------------------------------------------------------------
```

```
mgmt0              --              connected routed   full    1000    --
Eth1/1             --              connected trunk    full    1000    10g
Eth1/2             --              connected 10       full    1000    10g
```

4.2.3 中继端口

中继端口可以承载多个 VLAN。如果有多个 VLAN 需要在交换机与其他交换机、路由器或防火墙之间建立连接，那么通常需要使用中继端口。数据包通过链路进行传输时，可以在数据包中增加 802.1Q 报头来标识 VLAN。交换机收到数据包之后，将检查报头并与正确的 VLAN 相关联，然后再删除报头信息。

必须在 Nexus 交换机上通过接口命令 **switchport mode trunk** 静态定义中继端口。例 4-6 给出了将 Eth1/1 设置为中继端口的配置方式。

例 4-6 中继端口的配置和验证

```
NX-1# config t
Enter configuration commands, one per line. End with CNTL/Z.
NX-1(config)# int eth1/1
NX-1(config-if)# switchport mode trunk
NX-1# show interface eth1/1 | include Port
  Port mode is trunk
```

排查网络设备之间的连接性故障时，可以通过命令 **show interface trunk** 提供大量有价值的重要信息。

- 输出结果的第一部分列出了所有中继端口及其状态、与端口通道的关联关系以及本征 VLAN（Native VLAN）。
- 输出结果的第二部分显示了中继端口允许的 VLAN 列表。可以在中继端口上最小化流量，将 VLAN 范围限定为指定交换机，进而限制广播流量。其他用例还包括在多条网络链路之间实现某种形式的负载均衡，让选定的 VLAN 流量通过其中的某条中继链路，其他的 VLAN 流量则通过其他中继链路。
- 输出结果的第三部分显示了处于差错禁用（Err-disabled）状态的端口或 VLAN 信息。一般来说，这些差错都与 vPC（virtual Port Channel，虚拟端口信道）配置不完整有关。有关 vPC 的详细内容请参阅第 5 章。
- 输出结果的第四部分显示了处于转发状态的 VLAN 信息，处于阻塞状态的端口则不在这部分显示。

例 4-7 给出了 **show interface trunk** 命令的使用示例。

例 4-7 show interface trunk 命令输出结果

```
NX-1# show interface trunk
! Section 1 displays the native VLAN associated on this port, the status and
! if the port is associated to a port-channel
--------------------------------------------------------------------------------
Port            Native  Status          Port
                Vlan                    Channel
--------------------------------------------------------------------------------
Eth1/1          1       trunking        --


--------------------------------------------------------------------------------
! Section 2 displays all of the VLANs that are allowed to be transmitted across
! the trunk port
```

```
Port              Vlans Allowed on Trunk
---------------------------------------------------------------------------
Eth1/1            1-4094

---------------------------------------------------------------------------
! Section 3 displays ports that are disabled due to an error.
Port              Vlans Err-disabled on Trunk
---------------------------------------------------------------------------
Eth1/1            none

---------------------------------------------------------------------------
! Section 4 displays all of the VLANs that are allowed across the trunk and are
! in a spanning tree forwarding state
Port              STP Forwarding
---------------------------------------------------------------------------
Eth1/1            1,10,20,30,99

---------------------------------------------------------------------------
Port              Vlans in spanning tree forwarding state and not pruned
---------------------------------------------------------------------------
Feature VTP is not enabled
Eth1/1            1,10,20,30,99
```

1. 本征 VLAN

中继端口上的本征 VLAN 中的流量没有 802.1Q 标记。本征 VLAN 是与端口相关联的配置，可以通过接口命令 **switchport trunk native vlan** *vlan-id* 加以更改。

两个端口上的本征 VLAN 必须匹配，否则流量可能会更改 VLAN。虽然可以在主机之间建立连接（假设它们位于不同的 VLAN），但是这样做会给大多数网络工程师造成混淆，不是最佳实践。

> 注：所有的交换机控制平面流量都通过 VLAN 1 进行宣告。思科的安全加固指南建议将本征 VLAN 改为 VLAN 1 之外的其他 VLAN，具体来说，就是将本征 VLAN 设置为未使用的 VLAN，以防止 VLAN 跳跃（VLAN hopping）攻击。

2. 允许的 VLAN

如前所述，作为流量工程的一种方式，可以将 VLAN 限制在某些中继端口上。如果希望两台主机之间的流量穿越中继链路，而中继端口不允许该 VLAN 流量穿越，那么可能会产生故障。此时，可以通过接口命令 **switchport trunk allowed** *vlan-ids* 指定允许穿越中继链路的 VLAN。例 4-8 将允许穿越 Eth1/1 中继链路的 VLAN 限制为 1、10、30 和 99。

例 4-8 查看中继链路允许的 VLAN

```
NX-1# show run interface eth1/1
! Output omitted for brevity
interface Ethernet1/1
  switchport mode trunk
  switchport trunk allowed vlan 1,10,30,99
```

> 注：完整的命令语法是 **switchport trunk allowed** {*vlan-ids* | **all** | **none** |**add** *vlan-ids* | **remove** *vlan-ids* | **except** *vlan-ids*}，可以在单一命令中提供大量有用功能。
>
> 如果配置脚本出现了变化，那么最好使用关键字 **add** 或 **remove**，因为这样做的规范性更强。常见差错是使用 **switchport trunk allowed** *vlan-ids* 命令，仅在命令中列出所要增加的 VLAN，这样做的后果是会覆盖当前 VLAN 列表，从而导致未在命令中列出的 VLAN 的流量被丢弃。

4.2.4 PVLAN

某些网络设计方案需要在网络设备之间进行分段，此时可以通过以下两种方式实现。

- 为每个安全域创建一个唯一的子网，并通过 ACL 限制网络流量。该技术可能会在主机数量超出子网范围时浪费 IP 地址（如拥有 65 台主机的安全区域需要 /25 地址，从而浪费了 63 个 IP 地址，此处未考虑广播地址和网络地址）。
- 使用 PVLAN（Private VLAN，私有 VLAN）。

为了从 L2 的角度限制端口之间的流量，PVLAN 采用了两层架构（主 VLAN 和辅助 VLAN）。为了实现 PVLAN 外部通信，必须在主 VLAN 与辅助 VLAN 之间建立显式映射。PVLAN 的端口包括 3 类。

- **混杂端口（Promiscuous Port）**：与该 VLAN 相关联的端口是主 PVLAN（第一层），能够与所有主机进行通信。这类端口通常分配给提供集中式服务（如 DHCP、DNS 等）的路由器、防火墙或服务器。
- **隔离端口（Isolated Port）**：这类端口位于辅助 PVLAN（分层结构的第二层），只能同与混杂 PVLAN 相关联的端口进行通信，同一隔离 VLAN 中的端口之间不相互传输流量。
- **团体端口（Community Port）**：这类端口位于辅助 PVLAN，可以与该 VLAN 中的其他端口以及与混杂 VLAN 相关联的端口进行通信。

图 4-3 给出了 PVLAN 在服务提供商网络中的应用示例。R1 是 10.0.0.0/24 网段中所有主机的路由器，与混杂 PVLAN 相连。主机 2 和主机 3 属于不同的公司，不应该与任何主机进行通信，而只能与 R1 进行通信。

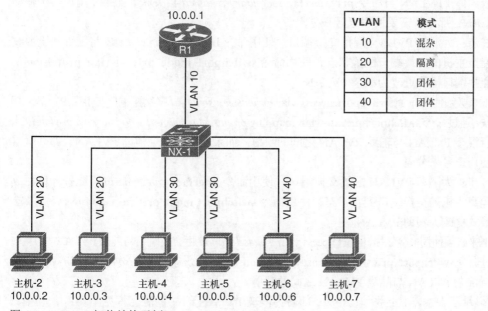

图 4-3 PVLAN 拓扑结构示例

主机 4 和主机 5 均来自第三家公司，需要与 R1 进行通信。主机 6 和主机 7 均来自第四家公司，需要与 R1 进行通信。其余通信均不允许。

表 4-1 列出了主机间的通信能力。请注意，主机 4 与主机 5 可以进行相互通信，但无法与主机 2、主机 3、主机 6 和主机 7 进行通信。

表 4-1　　　　　　　　　　　　　　　　PVLAN 通信能力

	R1	主机 2	主机 3	主机 4	主机 5	主机 6	主机 7
R1	-	√	√	√	√	√	√
主机 2	√	-	×	×	×	×	×
主机 3	√	×	-	×	×	×	×
主机 4	√	×	×	-	√	×	×
主机 5	√	×	×	√	-	×	×
主机 6	√	×	×	×	×	-	√
主机 7	√	×	×	×	×	√	-

1. 隔离 PVLAN

隔离 PVLAN 只能与混杂端口进行通信，因而每个 L3 域只需要一个隔离 PVLAN。在 Nexus 交换机上部署隔离 PVLAN 的步骤如下。

- **第 1 步**：启用 PVLAN 功能。在全局配置模式下使用 **feature private-vlan** 命令启用 PVLAN 功能。
- **第 2 步**：定义隔离 PVLAN。使用 **vlan** *vlan-id* 命令创建隔离 PVLAN。在 VLAN 配置环境下，使用命令 **private-vlan isolated** 将 VLAN 标识为隔离 PVLAN。
- **第 3 步**：定义混杂 PVLAN。使用命令 **vlan** *vlan-id* 创建混杂 PVLAN。在 VLAN 配置环境下，使用命令 **private-vlan primary** 将 VLAN 标识为混杂 PVLAN。
- **第 4 步**：将隔离 PVLAN 与混杂 PVLAN 相关联。在混杂 PVLAN 配置环境下，将辅助（隔离或团体）PVLAN 与命令 **private-vlan** *secondary-pvlan-id* 相关联。如果使用了多个辅助 PVLAN，那么就需要使用逗号来分隔。
- **第 5 步**：为混杂 PVLAN 配置交换端口。使用命令 **interface** *interface-id* 将配置环境更改为混杂主机的交换端口配置状态，使用命令 **switchport mode private-vlan promiscuous** 将交换端口模式更改为混杂 PVLAN。
  接下来必须使用命令 **switchport access vlan** *promiscuous-vlan-id* 将交换端口与混杂 PVLAN 相关联，通过命令 **switchport private-vlan mapping** *promiscuous-vlan-id secondary-pvlan-vlan-id* 来执行混杂 PVLAN 与辅助 PVLAN 之间的映射。如果使用了多个辅助 PVLAN，那么就需要使用逗号来分隔。
- **第 6 步**：为隔离 PVLAN 配置交换端口。使用命令 **interface** *interface-id* 将配置环境更改为隔离主机的交换端口配置状态，使用命令 **switchport mode private-vlan host** 将交换端口模式更改为辅助 PVLAN。
  接下来必须使用命令 **switchport access vlan** *isolated-vlan-id* 将交换端口与混杂 PVLAN 相关联，通过命令 **switchport private-vlan host-association** *promiscuous-vlan-id isolated-pvlan-vlan-id* 来执行混杂 PVLAN 与隔离 PVLAN 之间的映射。

例 4-9 显示了在 NX-1 上将 VLAN 20 部署为隔离 PVLAN 的配置情况（见图 4-3），其中，VLAN 10 是混杂 PVLAN。

例 4-9 在 NX-1 上部署隔离 PVLAN

```
NX-1(config)# feature private-vlan
Warning: Private-VLAN CLI entered...
Please disable multicast on this Private-VLAN by removing Multicast related
config(IGMP, PIM, etc.)   Please remove any VACL related config on Private-VLANs.
```

```
VLAN QOS needs to have atleast one port per ASIC instance.
NX-1(config)# vlan 20
NX-1(config-vlan)#   name PVLAN-ISOLATED
NX-1(config-vlan)#   private-vlan isolated
NX-1(config-vlan)# vlan 10
NX-1(config-vlan)#   name PVLAN-PROMISCOUS
NX-1(config-vlan)#   private-vlan primary
NX-1(config-vlan)#   private-vlan association 20
NX-1(config-vlan)# exit
NX-1(config)# interface Ethernet1/1
NX-1(config-if)#   switchport mode private-vlan promiscuous
NX-1(config-if)#   switchport access vlan 10
NX-1(config-if)#   switchport private-vlan mapping 10 20
NX-1(config-if)# interface Ethernet1/2
NX-1(config-if)#   switchport mode private-vlan host
NX-1(config-if)#   switchport access vlan 20
NX-1(config-if)#   switchport private-vlan host-association 10 20
NX-1(config-if)# interface Ethernet1/3
NX-1(config-if)#   switchport mode private-vlan host
NX-1(config-if)#   switchport access vlan 20
NX-1(config-if)#   switchport private-vlan host-association 10 20
```

配置了交换机的 PVLAN 之后，建议使用命令 **show vlan [private-vlan]** 验证配置，可选关键字 **private-vlan** 可以仅显示 show vlan 命令输出结果中的 PVLAN 部分。

从例 4-10 可以看出，主 VLAN 与混杂 PVLAN 相关联，辅助 VLAN 与隔离 PVLAN 相关联，且 PVLAN 类型已得到确认，所有活动端口均列在一旁。输出结果始终包含混杂端口，如果发现缺少某些端口，那么就需要重新检查接口配置，因为 PVLAN 的映射配置可能存在差错。

例 4-10 验证隔离 PVLAN 配置

```
NX-1# show vlan private-vlan
Primary   Secondary   Type              Ports
-------   ---------   ---------------   ---------------------------------------
10        20          isolated          Eth1/1, Eth1/2, Eth1/3

NX-1# show vlan
! Output omitted for brevity

! Notice how there are not any ports listed in the regular VLAN section because
! they are all in the PVLAN section.
VLAN Name                               Status    Ports
---- -------------------------------    --------- -------------------------------
1    default                            active    Eth1/4, Eth1/5, Eth1/6, Eth1/7
10   PVLAN-PROMISCOUS                   active
20   PVLAN-ISOLATED                     active
..
Primary   Secondary   Type              Ports
-------   ---------   ---------------   ---------------------------------------
10        20          isolated          Eth1/1, Eth1/2, Eth1/3
```

注： 隔离或团体 VLAN 只能与一个主 VLAN 相关联。

PVLAN 端口支持多种端口类型，可以通过 **switchport mode private-vlan {promiscuous | host}** 命令进行设置。如果要验证配置信息，可以使用 **show interface** 命令检查接口配置。例 4-11 给出了验证 PVLAN 交换端口类型设置的配置示例。

例 4-11 验证 PVLAN 交换端口类型

```
NX-1# show interface Eth1/1 | i Port
  Port mode is Private-vlan promiscuous
```
```
NX-1# show interface Eth1/2 | i Port
  Port mode is Private-vlan host
```
```
NX-1# show interface Eth1/3 | i Port
  Port mode is Private-vlan host
```

另一种方式是验证隔离 PVLAN 的主机设备是否可以到达混杂主机设备，只要通过简单的 ping 测试即可实现，如例 4-12 所示。

例 4-12 验证隔离 PVLAN 通信

```
! Verification that both hosts can ping R1
Host-2# ping 10.0.0.1
Sending 5, 100-byte ICMP Echos to 10.0.0.1, timeout is 2 seconds:
!!!!!
Success rate is 100 percent (5/5), round-trip min/avg/max = 1/5/9 ms

Host-3# ping 10.0.0.1
Sending 5, 100-byte ICMP Echos to 10.0.0.1, timeout is 2 seconds:
!!!!!
Success rate is 100 percent (5/5), round-trip min/avg/max = 1/5/9 ms

! Verification that both hosts cannot ping each other
Host-2# ping 10.0.0.3
Sending 5, 100-byte ICMP Echos to 10.0.0.3, timeout is 2 seconds:
.....
Success rate is 0 percent (0/5)

Host-3# ping 10.0.0.2
Sending 5, 100-byte ICMP Echos to 10.0.0.2, timeout is 2 seconds:
.....
Success rate is 0 percent (0/5)
```

2. 团体 PVLAN

团体 PVLAN 仅允许与同一团体 PVLAN 中的混杂端口及其他端口进行通信。在 Nexus 交换机上部署团体 PVLAN 的过程如下。

- 第 1 步：启用 PVLAN 功能。在全局配置模式下使用命令 **feature private-vlan** 启用 PVLAN 功能。
- 第 2 步：定义团体 PVLAN。使用命令 **vlan** *vlan-id* 创建团体 PVLAN。在 VLAN 配置环境下，使用命令 **private-vlan community** 将 VLAN 标识为团体 PVLAN。
- 第 3 步：定义混杂 PVLAN。使用命令 **vlan** *vlan-id* 创建混杂 PVLAN。在 VLAN 配置环境下，使用命令 **private-vlan primary** 将 VLAN 标识为混杂 PVLAN。
- 第 4 步：将团体 PVLAN 与混杂 PVLAN 相关联。在混杂 PVLAN 配置环境下，通过命令 **private-vlan** *secondary-pvlan-id* 关联辅助（隔离或团体）PVLAN。如果使用了多个辅助 PVLAN，那么就需要使用逗号来分隔。
- 第 5 步：为混杂 PVLAN 配置交换端口。使用命令 **interface** *interface-id* 将配置环境更改为混杂主机的交换端口。通过命令 **switch port mode private vlan promiscuous** 将交换端口模式更改为混杂 PVLAN。
接下来必须使用命令 **switchport access vlan** *promiscuous-vlan-id* 将交换端口与混杂 PVLAN 进行关联，通过命令 **switchport private vlan mapping** *promiscuous-vlan-id*

secondary-pvlan-vlan-id 执行混杂 PVLAN 与辅助 PVLAN 之间的映射。如果使用了多个辅助 PVLAN，那么就需要使用逗号来分隔。

- **第 6 步**：为团体 PVLAN 配置交换端口。使用命令 **interface** *interface-id* 将配置环境更改为隔离主机的交换端口配置状态，使用命令 **switchport mode private-vlan host** 将交换端口模式更改为辅助 PVLAN。

    接下来必须使用命令 **switchport access vlan** *isolated-vlan-id* 将交换端口与混杂 PVLAN 相关联，通过命令 **switchport private-vlan host-association** *promiscuous-vlan-id isolated-pvlan-vlan-id* 来执行混杂 PVLAN 与团体 PVLAN 之间的映射。

例 4-13 显示了将 VLAN 30 配置为主机 4 和主机 5 的团体 PVLAN 以及将 VLAN 40 配置为主机 6 和主机 7 的团体 PVLAN 的情况（见图 4-3），其中，VLAN 10 是混杂 PVLAN。

例 4-13　在 NX-1 上部署团体 PVLAN

```
NX-1(config)# vlan 30
NX-1(config-vlan)#    name PVLAN-COMMUNITY1 10 40
NX-1(config-vlan)#    private-vlan community
NX-1(config-vlan)# vlan 40
NX-1(config-vlan)#    name PVLAN-COMMUNITY2
NX-1(config-vlan)#    private-vlan community
NX-1(config-vlan)# vlan 10
NX-1(config-vlan)#    name PVLAN-PROMISCOUS
NX-1(config-vlan)#    private-vlan primary
NX-1(config-vlan)#    private-vlan association 20,30,40
NX-1(config-vlan)# exit
NX-1(config)# interface Ethernet1/1
NX-1(config-if)#    switchport mode private-vlan promiscuous
NX-1(config-if)#    switchport access vlan 10
NX-1(config-if)#    switchport private-vlan mapping 10 20,30,40
NX-1(config-if)# interface Ethernet1/4
NX-1(config-if)#    switchport mode private-vlan host
NX-1(config-if)#    switchport access vlan 30
NX-1(config-if)#    switchport private-vlan host-association 10 30
NX-1(config-if)# interface Ethernet1/5
NX-1(config-if)#    switchport mode private-vlan host
NX-1(config-if)#    switchport access vlan 30
NX-1(config-if)#    switchport private-vlan host-association 10 30
NX-1(config-if)# interface Ethernet1/6
NX-1(config-if)#    switchport mode private-vlan host
NX-1(config-if)#    switchport access vlan 40
NX-1(config-if)#    switchport private-vlan host-association 10 40
NX-1(config-if)# interface Ethernet1/7
NX-1(config-if)#    switchport mode private-vlan host
NX-1(config-if)#    switchport access vlan 40
NX-1(config-if)#    switchport private-vlan host-association 10 40
```

注：例中的 VLAN 20 是混杂端口配置的一部分，目的是说明隔离 PVLAN 与团体 PVLAN 的共存方式。本例是前述配置示例的延续，共同实现图 4-3 的解决方案。

可以通过命令 **show vlan** [*private-vlan*] 验证团体 PVLAN 的配置情况。需要记住的是，如果输出结果的 PVLAN 部分缺少端口，那么就需要重新检查接口配置，因为映射配置可能存在差错。

例 4-14 显示了所有的 PVLAN 及相关端口信息。请注意，VLAN 10 是 VLAN 20、VLAN 30 和 VLAN 40 的主 VLAN。

例 4-14 验证团体 PVLAN 配置

```
NX-1# show vlan
! Output omitted for brevity

VLAN Name                             Status      Ports
---- -------------------------------- --------- -------------------------------
1    default                          active
10   PVLAN-PROMISCOUS                 active
20   PVLAN-ISOLATED                   active
30   PVLAN-COMMUNITY1                 active
40   PVLAN-COMMUNITY2                 active
..
Primary   Secondary   Type              Ports
-------   ---------   ---------------   -----------------------------------------
10        20          isolated          Eth1/1, Eth1/2, Eth1/3
10        30          community         Eth1/1, Eth1/4, Eth1/5
10        40          community         Eth1/1, Eth1/6, Eth1/7
```

从例 4-15 的验证情况可以看出，隔离和团体 PVLAN 中的所有主机都能到达 R1。不允许任何主机到达隔离 PVLAN 中的主机，团体 PVLAN 中的主机只能到达相同团体 PVLAN 中的主机。

例 4-15 验证 PVLAN 之间的连接情况

```
! Verification that hosts in both communities can ping R1
Host-4# ping 10.0.0.1
Sending 5, 100-byte ICMP Echos to 10.0.0.1, timeout is 2 seconds:
!!!!!
Success rate is 100 percent (5/5), round-trip min/avg/max = 1/2/9 ms

Host-6# ping 10.0.0.1
Sending 5, 100-byte ICMP Echos to 10.0.0.1, timeout is 2 seconds:
!!!!!
Success rate is 100 percent (5/5), round-trip min/avg/max = 1/2/4 ms

! Verification that both hosts can ping other hosts in the same community PVLAN
Host-4# ping 10.0.0.5
Sending 5, 100-byte ICMP Echos to 10.0.0.5, timeout is 2 seconds:
!!!!!
Success rate is 100 percent (5/5), round-trip min/avg/max = 1/5/9 ms

Host-6# ping 10.0.0.7
Type escape sequence to abort.
Sending 5, 100-byte ICMP Echos to 10.0.0.7, timeout is 2 seconds:
!!!!!
Success rate is 100 percent (5/5), round-trip min/avg/max = 1/1/1 ms

! Verification that both hosts cannot ping hosts in the other community PVLAN
Host-4# ping 10.0.0.6
Sending 5, 100-byte ICMP Echos to 10.0.0.6, timeout is 2 seconds:
.....
Success rate is 0 percent (0/5)

Host-6# ping 10.0.0.4
Sending 5, 100-byte ICMP Echos to 10.0.0.4, timeout is 2 seconds:
.....
Success rate is 0 percent (0/5)

! Verification that both hosts cannot ping hosts in the isolated PVLAN
Host-4# ping 10.0.0.2
```

```
Sending 5, 100-byte ICMP Echos to 10.0.0.2, timeout is 2 seconds:
.....
Success rate is 0 percent (0/5)

Host-6# ping 10.0.0.2
Sending 5, 100-byte ICMP Echos to 10.0.0.2, timeout is 2 seconds:
.....
Success rate is 0 percent (0/5)
```

3. 在 SVI 上使用混杂 PVLAN 端口

可以将 SVI（ Switched Virtual Interface，交换式虚接口 ）配置为 PVLAN 的混杂端口。主 PVLAN 与辅助 PVLAN 之间的映射如前所述，但 VLAN 接口需要使用命令 **private-vlan mapping** *secondary-vlan-id*。例 4-16 给出了 NX-1 为 VLAN 10 添加 SVI（IP 地址为 10.0.0.10/24）的配置示例。

例 4-16 配置混杂 PVLAN SVI

```
NX-1# conf t
NX-1(config)# interface vlan 10
NX-1(config-if)# ip address 10.0.0.10/24
NX-1(config-if)# private-vlan mapping 20,30,40
NX-1(config-if)# no shut
NX-1(config-if)# do show run vlan
! Output omitted for brevity

vlan 10
  name PVLAN-PROMISCOUS
  private-vlan primary
  private-vlan association 20,30,40
vlan 20
  name PVLAN-ISOLATED
  private-vlan isolated
vlan 30
  name PVLAN-COMMUNITY1
  private-vlan community
vlan 40
  name PVLAN-COMMUNITY2
  private-vlan community
```

可以通过 **show interface vlan** *promiscuous-vlan-id* **private-vlan mapping** 命令验证混杂 PVLAN SVI 端口映射情况，如例 4-17 所示。

例 4-17 验证混杂 PVLAN SVI 映射

```
NX-1# show interface vlan 10 private-vlan mapping
Interface Secondary VLAN
--------- ---------------------------------------------------------------
vlan10    20   30   40
```

例 4-18 给出了通过混杂 PVLAN SVI 连接主机的验证情况，两台混杂设备（NX-1 和 R1）可以相互 ping 通。此外，所有主机（由主机 2 表示）均能 ping 通 NX-1 和 R1，而且不影响分配给隔离或团体 PVLAN 端口的 PVLAN 功能。

例 4-18 验证基于混杂 PVLAN SVI 的连接

```
! Verification that both the promiscuous SVI can ping the other promiscuous

! host (R1)
```

```
NX-1# ping 10.0.0.1
PING 10.0.0.1 (10.0.0.1): 56 data bytes
64 bytes from 10.0.0.1: icmp_seq=0 ttl=254 time=2.608 ms
64 bytes from 10.0.0.1: icmp_seq=1 ttl=254 time=2.069 ms
64 bytes from 10.0.0.1: icmp_seq=2 ttl=254 time=2.241 ms
64 bytes from 10.0.0.1: icmp_seq=3 ttl=254 time=2.157 ms
64 bytes from 10.0.0.1: icmp_seq=4 ttl=254 time=2.283 ms

--- 10.0.0.1 ping statistics ---
5 packets transmitted, 5 packets received, 0.00% packet loss
round-trip min/avg/max = 2.069/2.271/2.608 ms

! Verification that a isolated PVLAN host can ping the physical and SVI
! promiscuous ports
Host-2# ping 10.0.0.1
Sending 5, 100-byte ICMP Echos to 10.0.0.1, timeout is 2 seconds:
!!!!!
Success rate is 100 percent (5/5), round-trip min/avg/max = 1/8/25 ms
Host-2# ping 10.0.0.10
Sending 5, 100-byte ICMP Echos to 10.0.0.10, timeout is 2 seconds:
!!!!!
Success rate is 100 percent (5/5), round-trip min/avg/max = 1/1/1 ms

! Verification that an isolated host cannot ping another host in the isolated PVLAN

Host-2# ping 10.0.0.3
Type escape sequence to abort.
Sending 5, 100-byte ICMP Echos to 10.0.0.3, timeout is 2 seconds:
.....
Success rate is 0 percent (0/5)
```

4．中继交换机之间的 PVLAN

某些拓扑结构要求 PVLAN 跨越多台交换机，此时可能存在 3 种场景（取决于上游或下游交换机的能力）。

- **所有交换机都支持 PVLAN**：该场景必须同时在上游和下游交换机上配置全部 PVLAN 以及相应的主/辅助映射，要求在设备之间建立常规的 802.1Q 中继链路。拥有混杂端口的交换机负责将流量引导进/引导出混杂端口。该场景中的生成树需要为每个 PVLAN 维护一个独立实例。
- **上游交换机不支持 PVLAN**：该场景必须在下游交换机上配置 PVLAN 以及相应的主/辅助映射。由于上游交换机不支持 PVLAN，因而下游交换机必须将辅助 PVLAN 合并/分离到主 PVLAN 中，以便上游交换机上的设备只需使用主 PVLAN-ID。此外，还要在上游中继交换端口上配置 **switchport mode private-vlan trunk promiscuous** 命令，通常将这些中继端口称为混杂 PVLAN 中继端口。
- **下游交换机不支持 PVLAN**：该场景必须在上游交换机上配置 PVLAN 以及相应的主/辅助映射。由于下游交换机不支持 PVLAN，因而上游交换机必须将辅助 PVLAN 合并/分离到主 PVLAN 中，以便下游交换机上的设备只需要使用辅助 PVLAN-ID。需要在下游中继交换端口上配置 **switchport mode private-vlan trunk secondary** 命令，通常将这些中继端口称为隔离 PVLAN 中继端口。

注：上述场景中的常规 VLAN 均通过中继链路进行传输。

> 注：并非所有的 Nexus 平台都支持混杂或隔离 PVLAN 中继端口，详情可访问思科官网。

4.3 生成树协议基础

好的网络设计方案可以提供设备和网络链路（路径）冗余能力，一种简单的解决方案是在交换机之间增加第二链路以解决潜在的网络链路故障问题，或者确保交换机至少连接拓扑结构中的两台交换机。

不过，如果交换机必须转发广播包或者出现未知的单播泛洪，那么这类拓扑结构就会出现问题，网络将以持续环路的方式进行广播，直至链路拥塞、交换机强制丢包。此外，如果数据包出现环路，那么 MAC 地址表将不断改变端口，从而增大了 CPU 和内存消耗，导致交换机崩溃。

生成树协议可以临时阻塞特定端口上的流量，从而在网络环境中构建二层无环拓扑结构。目前存在多种形式的生成树协议。

- 802.1D 是原始的生成树协议规范。
- PVST（Per-VLAN Spanning Tree，每 VLAN 生成树）。
- PVST+（Per-VLAN Spanning Tree Plus，增强型每 VLAN 生成树）。
- 802.1W RSTP（Rapid Spanning Tree Protocol，快速生成树协议）。
- 802.1S MST（Multiple Spanning Tree Protocol，多生成树协议）。

Nexus 交换机仅支持 RSTP 或 MST 模式，这两种模式均与 802.1D 标准后向兼容。

4.3.1 IEEE 802.1D 生成树协议

原始的 STP（Spanning Tree Protocol，生成树协议）版本来自 IEEE 802.1D 标准，可以确保 VLAN 实现无环拓扑。STP 中的每个端口都会经历以下状态。

- **禁用状态**：端口处于管理性关闭状态（关闭）。
- **阻塞状态**：交换端口已启用，但端口尚未转发任何流量以确保创建无环拓扑。此时的交换机不修改 MAC 地址表，只能从其他交换机接收 BPDU（Bridge Protocol Data Unit，桥接协议数据单元）。
- **侦听状态**：此时交换端口已从阻塞状态切换为侦听状态，可以发送或接收 BPDU，但无法转发其他网络流量。
- **学习状态**：此时交换端口可以利用接收到的网络流量修改 MAC 地址表。除 BPDU 之外，交换机仍然不转发其他网络流量。转发时延到期后，交换端口将进入学习状态。
- **转发状态**：此时交换端口可以转发所有网络流量，且能够按预期更新 MAC 地址表。该状态是交换端口转发网络流量的最终状态。
- **损坏状态**：交换机检测到端口可能存在导致重大影响的配置或操作问题，在问题修复之前，端口将一直丢弃数据包。

最初的生成树协议定义了以下 3 种端口类型。

- **指派端口（Designated port）**：该网络端口负责接收并将帧转发给其他交换机，指派端口提供与下游设备及交换机的连接。
- **根端口（Root port）**：该网络端口负责连接生成树拓扑结构中的根交换机或上游交换机。
- **阻塞端口（Blocking port）**：该网络端口因为生成树协议而不转发流量。

对于生成树协议来说，必须理解以下关键术语。

- **根网桥（Root bridge）**：根网桥是 L2 拓扑结构中的一类重要交换机，所有端口均处于转发状态。其他执行路径计算的交换机都将该交换机视为生成树的顶端，根网桥上的所有端口都被归类为指派端口。
- **BPDU（Bridge Protocol Data Unit，桥接协议数据单元）**：该网络帧负责检测 STP 拓扑，允许交换机识别根网桥、根端口、指派端口和阻塞端口。BPDU 包含字段 STP 类型、根路径开销、根网桥标识符、本地网桥标识符、最大老化时间、Hello 时间、转发时延。BPDU 使用目的 MAC 地址 01:80:c2:00:00:00。
- **根路径开销（Root Path Cost）**：指的是去往根交换机的特定路径的开销之和。
- **根网桥标识符（Root Bridge Identifier）**：是根网桥系统 MAC、系统 ID 扩展以及根网桥系统优先级的组合。
- **本地网桥标识符（Local Bridge Identifier）**：是宣告交换机的网桥系统 MAC、系统 ID 扩展以及根网桥系统优先级的组合。
- **最大老化时间（Max Age）**：该定时器是控制桥接端口保存 BPDU 信息之前所要经过的最长时间。对于 Nexus 交换机来说，最大老化时间与使用传统 802.1D STP 的交换机的向后兼容性相关。
- **Hello 时间（Hello Time）**：指的是从端口向外宣告 BPDU 的时间。可以使用命令 **spanning-tree vlan** *vlan-id* **hello-time** *hello-time* 将默认值设置为 2s，可配置值为 1～10s。
- **转发时延（Forward Delay）**：指的是端口保持在侦听和学习状态的时间。默认值为 15s，可以使用命令 **spanning-tree vlan** *vlan-id* **forward-time** *forward-time* 将转发时延更改为 15～30s。

> **注：** 很多 STP 术语都用到了术语网桥（即使 STP 运行在交换机上），此处的术语网桥与交换机可以互换。

4.3.2 快速生成树协议

802.1D（生成树协议的第一个版本）仅创建一棵拓扑结构树。对于拥有多个 VLAN 的大型网络环境来说，创建不同的 STP 拓扑可以让不同的 VLAN 使用不同的链路。思科开发了 PVST（Per-VLAN Spanning Tree，每 VLAN 生成树）和 PVST+（Per-VLAN Spanning Tree Plus，增强型每 VLAN 生成树），以实现这种灵活性。

PVST 和 PVST+ 是思科专有生成树协议，这些协议中的概念与其他增强型功能相结合，就形成了 IEEE 802.1W 规范。802.1W 规范融入了一些增强型功能以实现快速融合，被称为 RSTP（Rapid Spanning Tree Protocol，快速生成树协议）。

RSTP 定义了以下端口角色。
- **指派端口（Designated port）**：该网络端口负责接收并将帧转发给其他交换机，指派端口提供与下游设备及交换机的连接。
- **根端口（Root port）**：该网络端口负责连接生成树拓扑结构中的根交换机或上游交换机。
- **可选端口（Alternate port）**：该网络端口提供经其他交换机去往根交换机的可选连接。
- **备份端口（Backup port）**：该网络端口提供去往当前根交换机的冗余链路。如果上游交换机出现了故障，那么备用端口将无法保证与根网桥的连接。仅当相同的交换机之间存在多条链路时，才存在备用端口。

RSTP 协议下的交换机与其他交换机交换握手信息时，可以更快地经历如下 STP 状态。
- **丢弃（Discarding）状态**：此时交换端口已启用，但端口未转发任何流量以确保创建无环拓扑。该状态结合了传统 STP 的"禁用""阻塞"和"侦听"状态。

- **学习（Learning）状态**：此时交换端口开始通过接收到的网络流量修改 MAC 地址表，不过，除 BPDU 之外，交换机不转发任何网络流量。
- **转发（Forwarding）状态**：此时交换端口将转发所有网络流量并按预期更新 MAC 地址表，该状态是交换端口转发网络流量的最终状态。

> **注**：交换机会尝试与连接在端口上的设备建立 RSTP 握手过程，如果没有发生握手，那么就认为该设备不兼容 RSTP，并默认将端口设置为常规 802.1D。这就意味着在建立了网络链路之后，计算机和打印机等主机设备仍然会遇到非常明显的传输时延（约 50s）。

> **注**：拥有基本配置的 L2 交换端口默认启用 RSTP，如果希望进一步调整 RSTP 配置，还可以应用更多额外配置。

1．生成树路径开销

接口的 STP 开销是根路径计算的基本组件，因为确定根路径的依据是到达根网桥的累计接口 STP 开销。接口 STP 开销以参考带宽 20Gbit/s 为基础进行计算，表 4-2 列出了常见接口速率以及相对应的接口 STP 开销。

表 4-2　　　　　　　　　　　　　默认的接口 STP 开销

链路速率	接口 STP 开销
10Mbit/s	100
100Mbit/s	19
1Gbit/s	4
10Gbit/s	2
20Gbit/s 或更高	1

2．选举根网桥

STP 操作的第一步就是识别根网桥。交换机在初始化的时候，将自己假定为根网桥并使用本地网桥标识符作为根网桥标识符，然后侦听邻居的 BPDU 并执行以下操作。

- 如果邻居的 BPDU 次于自己的 BPDU，那么交换机就会忽略该 BPDU。
- 如果邻居的 BPDU 优于自己的 BPDU，那么交换机就会更新其 BPDU 以包含新的根网桥标识符以及新的根路径开销（交换机将利用该开销计算到达新根网桥的总路径开销）。该过程将持续进行，直至拓扑结构中的所有交换机都识别出根网桥交换机。

如果网桥标识符中的优先级低于其他 BPDU，那么生成树协议就认为该交换机更优。如果优先级相同，那么交换机将优选系统 MAC 较小的 BPDU。

> **注**：一般来说，越早的交换机的 MAC 地址越小，因而通常被认为较优。当然，也可以根据需要调整网络配置，以优化根交换机在 L2 拓扑结构中的位置。

图 4-4 以一个简单的拓扑结构为例解释了生成树的一些重要概念。该拓扑结构中的 NX-1、NX-2、NX-3、NX-4 和 NX-5 相互连接，所有交换机的 STP 均没有自定义配置。虽然本例主要关注的是 VLAN 1，但拓扑结构还包含了 VLAN 10、VLAN 20 和 VLAN 30。NX-1 已被确定为根网桥，因为其系统 MAC 地址（5e00.4000.0007）在拓扑结构中最小。

可以通过 **show spanning-tree root** 命令识别根网桥。例 4-19 列出了 NX-1 的输出结果，可以看出包含了 VLAN 号、根网桥标识符、根路径开销、Hello 时间、最大老化时间和转发时延。由于 NX-1 是根网桥，所有端口都是指派端口，因而 Root Port 字段显示 "This bridge is root"（*该网桥是根网桥*）。

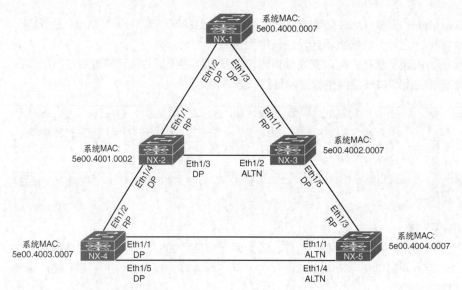

图 4-4 简单的 STP 拓扑结构

例 4-19 验证 STP 根网桥

```
NX-1# show spanning-tree root
                                    Root Hello Max Fwd
Vlan            Root ID              Cost Time Age Dly Root Port
---------------- -------------------- ---- ---- --- --- ----------------
VLAN0001         32769 5e00.4000.0007   0    2   20  15  This bridge is root
VLAN0010         32778 5e00.4000.0007   0    2   20  15  This bridge is root
VLAN0020         32788 5e00.4000.0007   0    2   20  15  This bridge is root
VLAN0030         32798 5e00.4000.0007   0    2   20  15  This bridge is root
```

例 4-20 显示了 NX-2 和 NX-3 的同一命令输出结果，可以看出，Root ID 字段与 NX-1 相同，但根路径开销已经更改为 2，因为这两台交换机都需要通过 10Gbit/s 链路到达 NX-1。此外，这两台交换机均已将 Eth 1/1 确定为根端口。

例 4-20 确定根端口

```
NX-2# show spanning-tree root
                                    Root Hello Max Fwd
Vlan            Root ID              Cost Time Age Dly Root Port
---------------- -------------------- ---- ---- --- --- ----------------
VLAN0001         32769 5e00.4000.0007   2    2   20  15  Ethernet1/1
VLAN0010         32778 5e00.4000.0007   2    2   20  15  Ethernet1/1
VLAN0020         32788 5e00.4000.0007   2    2   20  15  Ethernet1/1
VLAN0030         32798 5e00.4000.0007   2    2   20  15  Ethernet1/1

NX-3# show spanning-tree root
                                    Root Hello Max Fwd
Vlan            Root ID              Cost Time Age Dly Root Port
---------------- -------------------- ---- ---- --- --- ----------------
VLAN0001         32769 5e00.4000.0007   2    2   20  15  Ethernet1/1
VLAN0010         32778 5e00.4000.0007   2    2   20  15  Ethernet1/1
VLAN0020         32788 5e00.4000.0007   2    2   20  15  Ethernet1/1
VLAN0030         32798 5e00.4000.0007   2    2   20  15  Ethernet1/1
```

3. 定位根端口

交换机识别出根网桥之后，就必须确定它们的 RP（Root Port，根端口）。根网桥会持续向所有端口宣告 BPDU，交换机则通过比较 BPDU 信息来识别 RP。RP 的选择规则如下（如果当前规则相同，那么就使用下一条规则）。

- 优选路径开销最小的接口。
- 优选宣告交换机系统优先级最低的接口。
- 优选宣告交换机系统 MAC 地址最小的接口。
- 如果有多条链路关联到同一台交换机，那么就优选来自宣告路由器端口的优先级最低的接口。
- 如果有多条链路关联到同一台交换机，那么优选来自宣告路由器端口号较低的接口。

例 4-21 给出了 NX-4 和 NX-5 的 **show spanning-tree root** 命令输出结果，可以看出，Root ID 字段与例 4-20 的 NX-1 相同，但根路径开销已更改为 4，因为这两台交换机都必须穿越两条 10Gbit/s 链路才能到达 NX-1。此外，这两台交换机均已将 Eth1/3 确定为 RP。

例 4-21 确定 NX-4 和 NX-5 的根端口

```
NX-4# show spanning-tree root

                                        Root  Hello Max Fwd
Vlan                 Root ID             Cost  Time  Age Dly  Root Port
---------------- -------------------- ------- ----- --- ---  ----------------
VLAN0001          32769 5e00.4000.0007     4    2    20  15   Ethernet1/3

NX-5# show spanning-tree root

                                        Root  Hello Max Fwd
Vlan                 Root ID             Cost  Time  Age Dly  Root Port
---------------- -------------------- ------- ----- --- ---  ----------------
VLAN0001          32769 5e00.4000.0007     4    2    20  15   Ethernet1/3
```

4. 定位阻塞的交换端口

确定了根网桥和 RP 之后，其他端口均被认为是指派端口。但是，如果两台交换机通过指派端口进行互联，那么就必须将其中的某个交换端口设置为阻塞状态，以避免出现转发环路。为了确定应该阻塞两台交换机之间的哪些端口，相应的判断规则如下。

- 该接口不能被视为 RP。
- 将本地交换机的系统优先级与远程交换机的系统优先级进行比较，如果远程交换机的系统优先级低于本地交换机，那么就将本地端口设置为阻塞状态。
- 将本地交换机的系统 MAC 地址与远程交换机的系统 MAC 进行比较。如果远程交换机的系统 MAC 地址低于本地交换机，那么就将本地端口设置为阻塞状态。

注：第 3 步是上述判断过程的最后一步。如果交换机有多条去往根交换机的链路，那么就会始终将下游交换机标识为 RP，其余端口都将符合第 2 步或第 3 步条件而处于阻塞状态。

命令 **show spanning-tree [vlan** *vlan-id*]可以为定位端口的 STP 状态提供非常有用的信息。例 4-22 给出了 NX-1 关于 VLAN 1 的 STP 信息，输出结果的第一部分显示了与根网桥相关的信息，然后是本地网桥信息，同时还显示了关联接口的 STP 端口开销、端口优先级和端口类型。由于 NX-1 是根网桥，因而其端口都是指派端口（Desg）。

例 4-22 NX-1 的 STP 信息

```
NX-1# show spanning-tree vlan 1

VLAN0001
  Spanning tree enabled protocol rstp
! The section displays the relevant information for the STP Root Bridge
```

```
   Root ID    Priority    32769
              Address     5e00.4000.0007
              This bridge is the root
              Hello Time  2  sec  Max Age 20 sec  Forward Delay 15 sec
! The section displays the relevant information for the Local STP Bridge
   Bridge ID  Priority    32769  (priority 32768 sys-id-ext 1)
              Address     5e00.4000.0007
              Hello Time  2  sec  Max Age 20 sec  Forward Delay 15 sec

Interface        Role Sts Cost      Prio.Nbr Type
---------------- ---- --- --------- -------- --------------------------------
Eth1/1           Desg FWD 2         128.1    P2p
Eth1/2           Desg FWD 2         128.2    P2p
Eth1/3           Desg FWD 2         128.3    P2p
```

NX-OS 交换机支持以下端口类型。

- **P2P（Point-to-Point，点对点）端口**：此类端口负责连接其他网络设备（PC 或 RSTP 交换机）。
- **P2P 对等体（STP）端口**：此类端口检测到所连接的是 802.1D 交换机，以后向兼容方式运行。
- **网络 P2P 端口**：此类端口专门连接其他 RSTP 交换机，将提供网桥保障功能。
- **边缘 P2P 端口**：此类端口专门连接其他主机设备（PC 而非交换机），将启用 Portfast 功能。

注：如果 Type 字段包含了 * TYPE_Inc-，那么就表示 Nexus 交换机与其连接的交换机之间的端口配置不匹配，可能是端口类型配置错误，也可能是端口模式（接入与中继）配置错误。

例 4-23 给出了 NX-2 和 NX-3 的 STP 拓扑结构信息。可以看出，第一段根网桥输出结果显示了根路径总开销以及被确定为 RP 的交换机端口。

例 4-23　验证 VLAN 的根端口和阻塞端口

```
NX-2# show spanning-tree vlan 1

VLAN0001
  Spanning tree enabled protocol rstp
  Root ID    Priority    32769
             Address     5e00.4000.0007
             Cost        2
             Port        1 (Ethernet1/1)
             Hello Time  2  sec  Max Age 20 sec  Forward Delay 15 sec

  Bridge ID  Priority    32769  (priority 32768 sys-id-ext 1)
             Address     5e00.4001.0007
             Hello Time  2  sec  Max Age 20 sec  Forward Delay 15 sec

Interface        Role Sts Cost      Prio.Nbr Type
---------------- ---- --- --------- -------- --------------------------------
Eth1/1           Root FWD 2         128.1    P2p
Eth1/3           Desg FWD 2         128.3    P2p
Eth1/4           Desg FWD 2         128.4    P2p

NX-3# show spanning-tree vlan 1
! Output omitted for brevity

  Bridge ID  Priority    32769  (priority 32768 sys-id-ext 1)
```

```
                  Address         5e00.4002.0007
                  Hello Time   2  sec  Max Age 20 sec  Forward Delay 15 sec

Interface        Role Sts Cost      Prio.Nbr Type
---------------- ---- --- --------- -------- --------------------------------
Eth1/1           Root FWD 2         128.1    P2p
Eth1/2           Altn BLK 2         128.2    P2p
Eth1/5           Desg FWD 2         128.5    P2p
```

NX-2 的所有端口均处于转发状态，而 NX-3 的端口 Eth1/2 处于阻塞（BLK）状态，也就是说，在端口 Eth1/1 出现连接故障的情况下，端口 Eth1/2 被指派为到达根网桥的可选端口。

与 NX-2 的 Eth1/3 端口相比，NX-3 的 Eth1/2 端口处于阻塞状态的原因是 NX-2 的系统 MAC 地址（5e00.4001.0007）小于 NX-3 的系统 MAC 地址（5e00.4002.0007）。这一点可以从图 4-4 以及输出结果中的系统 MAC 地址看出来。

5．验证中继链路上的 VLAN

show spanning-tree 命令的输出结果可以显示加入该 VLAN 的所有接口，不过，对于承载了多个 VLAN 的中继端口来说，这是一项非常烦琐的任务。此时，可以通过 **show spanning-tree interface** *interface-id* 命令检查接口的 STP 状态，从而实现快速检查，如例 4-24 所示。如果发现中继端口缺失了某个 VLAN，那么就可以检查该中继端口所允许的 VLAN 列表，因为该 VLAN 可能不在列表当中。

例 4-24 查看接口上参与 STP 的 VLAN

```
NX-1# show spanning-tree interface e1/1
Vlan              Role Sts Cost      Prio.Nbr Type
----------------- ---- --- --------- -------- --------------------------------
VLAN0001          Desg FWD 2         128.1    P2p
VLAN0010          Desg FWD 2         128.1    P2p
VLAN0020          Desg FWD 2         128.1    P2p
VLAN0030          Desg FWD 2         128.1    P2p
```

6．生成树协议调优

有时，正确设计的网络可能会从策略上将指定交换机设置为根网桥，将某些特定端口修改为指派端口（转发状态），将某些特定端口设置为可选端口（丢弃/阻塞状态）。这方面的设计考虑因素主要包括硬件平台、弹性机制及网络拓扑。

7．调整根网桥的位置

在理想情况下，根网桥应该是核心交换机，同时，为了最大限度地降低生成树的变更影响，还应该指定辅助根网桥。为了将特定交换机设置为根网桥，可以将其系统优先级设置为最低可能值，将辅助根网桥的系统优先级设置得略高于根网桥，同时增大其他交换机的系统优先级，这样就能确保根网桥位置的一致性。可以通过以下命令来设置系统优先级。

- **spanning-tree vlan** *vlan-id* **priority** *priority*

  优先级的取值范围是 0～61440，增量为 4096。

- **spanning-tree vlan** *vlan-id* **root** {**primary** | **secondary**}

  选择关键字 **primary**，优先级将设置为 24576；选择关键字 **secondary**，优先级将设置为 28672。

注：为了避免非期望设备接管根网桥的角色，最好的方法就是将期望根网桥交换机的系统优先级设置为零。

例 4-25 将 NX-1 设置为主用根网桥，将 NX-2 设置为辅助根网桥。请注意，NX-2 的输出结果显示了根网桥的系统优先级，与 NX-2 的系统优先级不同。

例 4-25 修改 STP 的系统优先级

```
NX-1(config)# spanning-tree vlan 1 root primary
NX-2(config)# spanning-tree vlan 1 root secondary
NX-1(config)# do show spanning-tree vlan 1
! Output omitted for brevity
VLAN0001
  Spanning tree enabled protocol rstp
  Root ID    Priority    24577
             Address     5e00.4000.0007
             This bridge is the root

  Bridge ID  Priority    24577   (priority 24576 sys-id-ext 1)
             Address     5e00.4000.0007
NX-2# show spanning-tree vlan 1
! Output omitted for brevity
VLAN0001
  Spanning tree enabled protocol rstp
  Root ID    Priority    24577
             Address     5e00.4000.0007
             Cost        2
             Port        1 (Ethernet1/1)

  Bridge ID  Priority    28673   (priority 28672 sys-id-ext 1)
             Address     5e00.4001.0007
```

注：请注意，NX-1 的优先级需要减 1。这是因为 BPDU 数据包中的优先级是优先级加上 Sys-Id-Ext 值（VLAN 号），因而 VLAN 1 的优先级是 24577，VLAN 10 的优先级是 24586。

8. 根保护

根保护（Root Guard）是 STP 的一种功能特性。在端口上配置了该功能特性之后，如果端口收到了一个上级 BPDU，那么就会将端口置入 ErrDisabled（差错禁用）状态，从而防止该端口成为根端口。根保护特性可以防止下游交换机（通常是配置差错或非法设备）成为拓扑结构中的根网桥。

可以通过接口命令 **spanning-tree guard root** 逐个对接口启用根保护特性，应该在面向永远也不应该成为根网桥的交换机的指派端口上配置根保护特性。以前面的拓扑结构为例，应该在 NX-2 的 Eth1/4 端口和 NX3 的 Eth1/5 端口上配置根保护特性，这样就可以防止 NX-4 和 NX-5 成为根网桥，但是在 NX-1←→NX-2 链接出现问题的情况下，仍然允许 NX-2 经由 NX-3 与 NX-1 保持连接。

9. 调整 STP 根端口和阻塞端口的位置

STP 的端口开销用于计算 STP 树。交换机生成 BPDU 时，总路径开销仅包含其到根网桥计算出的度量，而不包含向外宣告 BPDU 的端口开销，收端路由器收到 BPDU 之后，会将收到该 BPDU 的接口的端口开销与 BPDU 中的总路径开销值相加。

从图 4-4 可以看出，NX-1 将 BPDU 宣告给 NX-3，总路径开销为零。NX-3 收到 BPDU 之后，将自己的 STP 端口开销 2 加到该 BPDU 的总路径开销（零）上，使总路径开销值为 2。此后，NX-3 以总路径开销值 2 将 BPDU 宣告给 NX-5，NX-5 又加上了自己的端口开销 2。此时，NX-5 宣告其通过 NX-3 到达根网桥的开销为 4。上述过程可以从例 4-26 的输出结果中得到确认。请注意，NX-1 的输出结果中没有总路径开销。

例 4-26 验证总路径开销

```
NX-1# show spanning-tree vlan 1
! Output omitted for brevity
```

```
VLAN0001
  Spanning tree enabled protocol rstp
  Root ID    Priority    32769
             Address     5e00.4000.0007
             This bridge is the root
             Hello Time  2  sec  Max Age 20 sec  Forward Delay 15 sec

  Bridge ID  Priority    32769  (priority 32768 sys-id-ext 1)
             Address     5e00.4000.0007


Interface        Role Sts Cost      Prio.Nbr Type
---------------- ---- --- --------- -------- --------------------------------
Eth1/1           Desg FWD 2         128.1    P2p
Eth1/2           Desg FWD 2         128.2    P2p
Eth1/3           Desg FWD 2         128.3    P2p

NX-3# show spanning-tree vlan 1
! Output omitted for brevity
VLAN0001
  Spanning tree enabled protocol rstp
  Root ID    Priority    32769
             Address     5e00.4000.0007
             Cost        2

Bridge ID    Priority    32769  (priority 32768 sys-id-ext 1)
             Address     5e00.4002.0007


Interface        Role Sts Cost      Prio.Nbr Type
---------------- ---- --- --------- -------- --------------------------------
Eth1/1           Root FWD 2         128.1    P2p
Eth1/2           Altn BLK 2         128.2    P2p
Eth1/5           Desg FWD 2         128.5    P2p

NX-5# show spanning-tree vlan 1
! Output omitted for brevity
VLAN0001
  Spanning tree enabled protocol rstp
  Root ID    Priority    32769
             Address     5e00.4000.0007
             Cost        4

  Bridge ID  Priority    32769  (priority 32768 sys-id-ext 1)
             Address     5e00.4004.0007
Interface        Role Sts Cost      Prio.Nbr Type
---------------- ---- --- --------- -------- --------------------------------
Eth1/1           Altn BLK 2         128.1    P2p
Eth1/3           Root FWD 2         128.3    P2p
Eth1/4           Altn BLK 2         128.4    P2p
```

可以通过接口配置命令 **spanning-tree** [**vlan** *vlan-id*] **cost** *cost* 修改接口路径，从而影响指派端口或可选端口。如果省略了可选关键字 **vlan**，那么该设置将应用于所有 VLAN，当然，也可以指定具体 VLAN。

例 4-27 的 NX-3 修改了 Eth1/1 的端口开销，由于 Eth1/2 端口不再是备用端口，而是指派端口，因而最终影响了生成树协议的拓扑结构。NX-2 的 Eth1/3 端口则从指派端口更改为可选端口。

例 4-27 修改 STP 端口开销

```
NX-3(config)# interface ethernet 1/1
NX-3(config-if)# spanning-tree cost 1

NX-3# show spanning-tree vlan 1
! Output omitted for brevity
VLAN0001
  Root ID    Priority    32769
             Address     5e00.4000.0007
             Cost        1
             Port        1 (Ethernet1/1)

  Bridge ID  Priority    32769  (priority 32768 sys-id-ext 1)
             Address     5e00.4002.0007

Interface        Role Sts Cost      Prio.Nbr  Type
---------------- ---- --- --------- --------- --------------------------------
Eth1/1           Root FWD 1         128.1     P2p
Eth1/2           Desg FWD 2         128.2     P2p
Eth1/5           Desg FWD 2         128.5     P2p

NX-2# show span vlan 1
! Output omitted for brevity
VLAN0001
  Root ID    Priority    32769
             Address     5e00.4000.0007
             Cost        2
             Port        1 (Ethernet1/1)

  Bridge ID  Priority    32769  (priority 32768 sys-id-ext 1)
             Address     5e00.4001.0007

Interface        Role Sts Cost      Prio.Nbr  Type
---------------- ---- --- --------- --------- --------------------------------
Eth1/1           Root FWD 2         128.1     P2p
Eth1/3           Altn BLK 2         128.3     P2p
Eth1/4           Desg FWD 2         128.4     P2p
```

10. 调整 STP 端口优先级

如果交换机之间存在多条链路，那么调整 STP 端口优先级就可以影响可选端口。图 4-4 中的 NX-4 与 NX-5 之间拥有两条链路，Eth1/3（连接 NX-3）的 STP 端口开销被修改为 1234，因而连接 NX-4 的某条链路将成为 RP。

从例 4-28 可以看出，调整端口优先级之后，NX-5 的 Eth1/1 成为 NX-4 的 RP。需要注意的是，system-Id 与端口开销相同，因而接下来需要检查端口优先级，然后是端口号。端口优先级和端口号均由上游交换机控制。

例 4-28 查看 STP 端口优先级

```
NX-5# show spanning-tree vlan 1
! Output omitted for brevity
VLAN0001

Interface        Role Sts Cost      Prio.Nbr  Type
---------------- ---- --- --------- --------- --------------------------------
Eth1/1           Root FWD 2         128.1     P2p
Eth1/3           Altn BLK 1234      128.3     P2p
Eth1/4           Altn BLK 2         128.4     P2p
```

可以通过 **spanning-tree [vlan** *vlan-id***] port-priority** *priority* 命令修改 NX-4 的端口优先级，可选关键字 **vlan** 允许对逐个 VLAN 修改优先级。例 4-29 将 NX-4 的 Eth1/5 端口的优先级修改为 64，可以看出修改操作对 NX-5 造成的影响，此时 NX-5 的端口 Eth1/4 已成为 RP。

例 4-29 验证端口优先级对 STP 拓扑结构的影响

```
NX-4(config)# int eth1/5
NX-4(config-if)# spanning-tree port-priority 64

NX-4# show spanning-tree vlan 1
! Output omitted for brevity
VLAN0001
Interface       Role Sts Cost      Prio.Nbr  Type
--------------- ---- --- --------- --------- --------------------------------
Eth1/1          Desg FWD 2         128.1     P2p
Eth1/2          Root FWD 2         128.2     P2p
Eth1/5          Desg FWD 2          64.5     P2p

NX-5# show spanning-tree vlan 1
! Output omitted for brevity
VLAN0001
Interface       Role Sts Cost      Prio.Nbr  Type
--------------- ---- --- --------- --------- --------------------------------
Eth1/1          Altn BLK 2         128.1     P2p
Eth1/3          Altn BLK 1234      128.3     P2p
Eth1/4          Root FWD 2          64.4     P2p
```

11. 拓扑变更与 STP Portfast

从传统意义上来说，STP 拓扑结构中的 BPDU 都是从根网桥流向边缘交换机。在正常操作条件下，BPDU 不会发送给根网桥。但是，在链路被激活或去活之后，拓扑结构的变化会对 L2 拓扑结构中的所有交换机产生影响。

检测到链路状态变更的交换机会向根网桥发送 TCN（Topology Change Notification，拓扑变更通告），根网桥则创建一个新的 TCN，并泛洪给 L2 转发域中的所有交换机。收到根网桥的 TCN 之后，所有交换机都将清除自己的 MAC 地址表，因而在重建 MAC 地址表的过程中，流量会泛洪到所有端口。需要记住的是，主机使用 CSMA/CD 进行通信，因而在交换机重建 MAC 地址表时，该行为会导致一定的通信时延。

由于 TCN 的生成以 VLAN 为基础，因而 TCN 的影响与 VLAN 中的主机数量直接相关。主机数量越多，生成 TCN 的概率就越大，受广播影响的主机也越多。因此，在故障排查过程中应注意检查拓扑结构的变化情况。

在根网桥上运行 **show spanning-tree [vlan** *vlan-id***] detail** 命令可以查看拓扑结构的变化情况，该命令可以显示自上次变化以来拓扑结构再次发生变化的次数和时间。如果发现 TCN 突然增加或持续增加，那么就表明可能存在一定的故障，应该做进一步检查。

例 4-30 显示了 **show spanning-tree vlan 10 detail** 命令的输出结果，包含了检测到上次 TCN 以来的时间以及发起 TCN 的接口等信息。接下来需要定位与发起 TCN 的端口相连接的交换机，此时可以查看 CDP 表或者网络文档，然后还要再次执行 **show spanning-tree [vlan** *vlan-id***] detail** 命令，以找到拓扑结构中的最后一台交换机，从而识别故障端口。

例 4-30 查看生成树状态的详细信息

```
NX-1# show spanning-tree vlan 10 detail
 VLAN0010 is executing the rstp compatible Spanning Tree protocol
  Bridge Identifier has priority 32768, sysid 10, address 5e00.4000.0007
```

```
    Configured hello time 2, max age 20, forward delay 15
    We are the root of the spanning tree
    Topology change flag set, detected flag not set
    Number of topology changes 11 last change occurred 0:00:04 ago
            from Ethernet1/2
    Times:  hold 1, topology change 35, notification 2
            hello 2, max age 20, forward delay 15
    Timers: hello 0, topology change 30, notification 0
..
```

查看 NX-OS 事件历史记录可以从另一个侧面来了解交换机的 STP 活动信息，可以通过 **show spanning-tree internal event-history all** 命令显示 STP 事件历史记录，如例 4-31 所示。

例 4-31 查看 STP 事件历史记录

```
NX-1# show spanning-tree internal event-history all
-------------------- All the active STPs -----------
VDC01 VLAN0001
0) Transition at 917636 usecs after Tue Aug 29 02:31:43 2017
    Root: 0000.0000.0000.0000 Cost: 0 Age:  0 Root Port: none Port: none
    [STP_TREE_EV_UP]

1) Transition at 703663 usecs after Tue Aug 29 02:31:44 2017
    Root: 7001.885a.92de.617c Cost: 0 Age:  0 Root Port: none Port: Ethernet1/51
    [STP_TREE_EV_MULTI_FLUSH_RCVD]

2) Transition at 723529 usecs after Tue Aug 29 02:31:44 2017
    Root: 7001.885a.92de.617c Cost: 0 Age:  0 Root Port: none Port: Ethernet1/51
    [STP_TREE_EV_MULTI_FLUSH_RCVD]

3) Transition at 609383 usecs after Tue Aug 29 02:31:45 2017
    Root: 7001.885a.92de.617c Cost: 0 Age:  0 Root Port: none Port: Ethernet1/51
    [STP_TREE_EV_MULTI_FLUSH_RCVD]

4) Transition at 601588 usecs after Tue Aug 29 02:31:47 2017
    Root: 7001.885a.92de.617c Cost: 0 Age:  0 Root Port: none Port: none
    [STP_TREE_EV_DOWN]
```

由于主机通常只有一条去往网络的连接，因而为主机生成 TCN 并没有任何意义。因此，为了提高 L2 网络的稳定性和运行效率，可以限定仅为那些连接了其他交换机和网络设备的端口生成 TCN。STP 的 Portfast 功能可以禁止接入端口生成 TCN。

STP Portfast 功能特性的另一个好处是允许接入端口避开 802.1D 的早期 STP 状态（学习状态和侦听状态）并立即转发流量，该功能对于计算机使用 DHCP（Dynamic Host Configuration Protocol，动态主机配置协议）或 PXE（Preboot eXecution Environment，预引导执行环境）的网络环境来说非常有用。

可以通过 **spanning-tree port type edge** 命令在特定端口上启用 Portfast 功能，也可以通过 **spanning-tree port type edge default** 命令在所有接入端口上全局启用 Portfast 功能。

例 4-32 在 NX-1 的 Eth1/6 端口上启用了 Portfast 功能特性并加以验证。可以看出，Portfast 端口被显示为 *Edge P2P*。最后一条配置命令则在全局为所有接入端口都启用了 Portfast 功能特性。

例 4-32 Spanning Tree Protocol Portfast Enablement

```
NX-1(config-if)# int eth1/6
NX-1(config-if)# spanning-tree port type edge
```

```
Warning: Edge port type (portfast) should only be enabled on ports connected to
 a single host. Connecting hubs, concentrators, switches, bridges, etc... to this
 interface  when edge port type (portfast) is enabled, can cause temporary
 bridging loops.
 Use with CAUTION
NX-1# show spanning-tree vlan 1
! Output omitted for brevity
VLAN0001
  Spanning tree enabled protocol rstp
  Root ID    Priority    32769
             Address     5e00.4000.0007
             This bridge is the root
             Hello Time  2  sec  Max Age 20 sec  Forward Delay 15 sec

  Bridge ID  Priority    32769   (priority 32768 sys-id-ext 1)
             Address     5e00.4000.0007

Interface         Role Sts Cost       Prio.Nbr Type
----------------- ---- --- ---------- -------- --------------------------------
Eth1/1            Desg FWD 2          128.1    P2p
Eth1/2            Desg FWD 2          128.2    P2p
Eth1/3            Desg FWD 2          128.3    P2p
Eth1/4            Desg FWD 2          128.3    P2p
Eth1/5            Desg FWD 2          128.3    P2p
Eth1/6            Desg FWD 2          128.3    Edge P2p
NX-1(config)# spanning-tree port type edge default
```

4.3.3 多生成树协议

对于拥有成千上万个 VLAN 的网络环境来说，为所有 VLAN 维护 STP 状态会给交换机带来严重负担。MST 可以将多个 VLAN 放到单个 STP 树中，称为 MST 实例，修改 MST 实例参数（而不需要逐一修改每个 VLAN 的参数），就可以执行期望的流量工程操作。

1. MST 配置

MST 的配置过程如下。

- 第 1 步：将 STP 模式设置为 MST。可以通过命令 **spanning-tree mode mst** 将生成树协议定义为 MST。
- 第 2 步：定义 MST 实例优先级（可选）。可以通过以下两种方式为 MST 区域定义 MST 实例优先级。
  - **spanning-tree mst** *instance-number* **priority** *priority*
    优先级的取值范围为 0 ~ 61440，增量为 4096。
  - **spanning-tree mst** *instance-number* **root** {**primary** | **secondary**}
    选择关键字 **primary**，优先级将设置为 24576；选择关键字 **secondary**，优先级将设置为 28672。
- 第 3 步：将 VLAN 关联到 MST 实例。默认情况下，所有 VLAN 都与 MST 0 实例相关联。必须通过命令 **spanning-tree mst configuration** 进入 MST 配置子模式，然后通过命令 **instance** *instance-number* **vlan** *vlan-id* 将 VLAN 分配给不同的 MST 实例。
- 第 4 步：指定 MST 版本号。同一个 MST 区域内的所有交换机的 MST 版本号必须完全匹配，可以通过子配置命令 **revision** *version* 配置 MST 版本号。
- 第 5 步：定义 MST 区域名（可选）。交换机可以通过名称来识别 MST 区域。默认情况下，域名是一个空字符串，可以通过命令 **name** *mst-region-name* 设置 MST 域名。

例 4-33 给出了 NX-1 的 MST 配置示例。可以看出，MST 实例 2 包含了 VLAN 30，MST 实例 1 包含了 VLAN 10 和 VLAN 20，MST 实例 0 包含了所有其他 VLAN。

例 4-33　NX-1 的 MST 配置示例

```
NX-1(config)# spanning-tree mode mst
NX-1(config)# spanning-tree mst 0 root primary
NX-1(config)# spanning-tree mst 1 root primary
NX-1(config)# spanning-tree mst 2 root primary
NX-1(config)# spanning-tree mst configuration
NX-1(config-mst)#   name NX-OS
NX-1(config-mst)#   revision 3
NX-1(config-mst)#   instance 1 vlan 10,20
NX-1(config-mst)#   instance 2 vlan 30
```

可以通过命令 **show spanning-tree mst configuration** 快速验证交换机的 MST 配置。从例 4-34 可以看出，MST 实例 0 包含了除 VLAN 10、VLAN 20 和 VLAN 30 之外的所有 VLAN（无论交换机是否配置了这些 VLAN）。

例 4-34　验证 MST 配置

```
NX-2# show spanning-tree mst configuration
Name       [NX-OS]
Revision   3        Instances configured 3
Instance   Vlans mapped
--------   ---------------------------------------------------------------
0          1-9,11-19,21-29,31-4094
1          10,20
2          30
```

2. MST 验证

也可以通过 **show spanning-tree** 命令获取生成树信息，主要区别在于该命令不显示 VLAN 号，但是显示 MST 实例。同样，此时交换机的优先级数值是 MST 实例号加上交换机的优先级，如例 4-35 所示。

例 4-35　MST 状态信息

```
NX-1# show spanning-tree
! Output omitted for brevity
! Spanning Tree information for Instance 0 (All VLANs but 10,20, and 30)
MST0000
  Spanning tree enabled protocol mstp
  Root ID    Priority    0
             Address     5e00.0000.0007
             This bridge is the root
             Hello Time  2  sec  Max Age 20 sec  Forward Delay 15 sec

  Bridge ID  Priority    32768    (priority 32768 sys-id-ext 0)
             Address     5e00.0000.0007
             Hello Time  2  sec  Max Age 20 sec  Forward Delay 15 sec

Interface         Role Sts Cost      Prio.Nbr Type
----------------  ---- --- --------- -------- --------------------------
Eth1/1            Desg FWD 2         128.1    P2p
Eth1/2            Desg FWD 2         128.2    P2p
Eth1/3            Desg FWD 2         128.3    P2p
```

```
! Spanning Tree information for Instance 1 (VLANs 10 and 20)
MST0001
  Spanning tree enabled protocol mstp
  Root ID    Priority    24577
             Address     5e00.0000.0007
             This bridge is the root

  Bridge ID  Priority    32769  (priority 32768 sys-id-ext 1)
             Address     5e00.0000.0007

Interface        Role Sts Cost      Prio.Nbr Type
---------------- ---- --- --------- -------- --------------------------------
Eth1/1           Desg FWD 2         128.1    P2p
Eth1/2           Desg FWD 2         128.2    P2p
Eth1/3           Desg FWD 2         128.3    P2p

! Spanning Tree information for Instance 0 (VLAN 30)
MST0002
  Spanning tree enabled protocol mstp
  Root ID    Priority    24578
             Address     5e00.0000.0007
             This bridge is the root
             Hello Time  2  sec  Max Age 20 sec  Forward Delay 15 sec

  Bridge ID  Priority    32770  (priority 32768 sys-id-ext 2)
             Address     5e00.0000.0007
             Hello Time  2  sec  Max Age 20 sec  Forward Delay 15 sec

Interface        Role Sts Cost      Prio.Nbr Type
---------------- ---- --- --------- -------- --------------------------------
Eth1/1           Desg FWD 2         128.1    P2p
Eth1/2           Desg FWD 2         128.2    P2p
Eth1/3           Desg FWD 2         128.3    P2p
```

可以通过命令 **show spanning-tree mst** [*instance-number*] 显示 MST 拓扑结构表的完整信息，如果使用可选关键字 *instance-number*，那么就可以将输出结果限定为指定实例。从例 4-36 可以看出，MST 实例的旁边显示了 VLAN 信息，这一点对于故障排查操作来说非常有用。

例 4-36 MST 拓扑结构的详细信息

```
NX-1# show spanning-tree mst
! Output omitted for brevity

##### MST0    vlans mapped:   1-9,11-19,21-29,31-4094
Bridge        address 5e00.0000.0007  priority      0      (0 sysid 0)
Root          this switch for the CIST

Regional Root this switch
Operational   hello time 2 , forward delay 15, max age 20, txholdcount 6
Configured    hello time 2 , forward delay 15, max age 20, max hops    20

Interface        Role Sts Cost      Prio.Nbr Type
---------------- ---- --- --------- -------- --------------------------------
Eth1/1           Desg FWD 2         128.1    P2p
Eth1/2           Desg FWD 2         128.2    P2p
Eth1/3           Desg FWD 2         128.3    P2p
```

```
##### MST1    vlans mapped:   10,20
Bridge        address 5e00.0000.0007  priority       24577 (24576 sysid 1)
Root          this switch for MST1

Interface        Role Sts Cost       Prio.Nbr Type
---------------- ---- --- ---------  -------- --------------------------------
Eth1/1           Desg FWD 2          128.1    P2p
Eth1/2           Desg FWD 2          128.2    P2p
Eth1/3           Desg FWD 2          128.3    P2p


##### MST2    vlans mapped:   30
Bridge        address 5e00.0000.0007  priority       24578 (24576 sysid 2)
Root          this switch for MST2

Interface        Role Sts Cost       Prio.Nbr Type
---------------- ---- --- ---------  -------- --------------------------------
Eth1/1           Desg FWD 2          128.1    P2p
Eth1/2           Desg FWD 2          128.2    P2p
Eth1/3           Desg FWD 2          128.3    P2p
```

可以通过命令 **show spanning-tree mst interface** *interface-id* 查看指定接口的 MST 配置信息。从例 4-37 可以看出，输出结果还显示了一些可选的 STP 功能特性信息，如 BPDU 过滤器（BPDU Filter）和 BPDU 保护（BPDU Guard）。

例 4-37　查看接口的 MST 配置信息

```
NX-2# show spanning-tree mst interface ethernet 1/1

Eth1/1 of MST0 is root forwarding
Port Type: normal            (default)     port guard : none       (default)
Link type: point-to-point    (auto)        bpdu filter: disable    (default)
Boundary : internal                        bpdu guard : disable    (default)
Bpdus sent 9, received 119

Instance Role Sts Cost       Prio.Nbr Vlans mapped
-------- ---- --- ---------  -------- --------------------------------
0        Root FWD 2          128.1    1-9,11-19,21-29,31-4094
1        Root FWD 2          128.1    10,20
2        Root FWD 2          128.1    30
```

3. MST 调整

MST 可以调整端口开销和端口优先级，可以通过接口配置命令 **spanning-tree mst** *instance-number* **cost** *cost* 设置接口开销。例 4-38 将 NX-3 的 Eth1/1 端口开销更改为 1，同时还验证了更改前后的接口开销情况。

例 4-38　更改 MST 接口开销

```
NX-3# show spanning-tree mst 0
! Output omitted for brevity
Interface        Role Sts Cost       Prio.Nbr Type
---------------- ---- --- ---------  -------- --------------------------------
Eth1/1           Desg FWD 2          128.1    P2p
Eth1/2           Root FWD 2          128.2    P2p
Eth1/5           Desg FWD 2          128.5    P2p
```

```
NX-3(config)# interface eth1/1
NX-3(config-if)# spanning-tree mst 0 cost 1
```

```
NX-3# show spanning-tree mst 0
! Output omitted for brevity
Interface        Role Sts Cost      Prio.Nbr  Type
---------------- ---- --- --------- --------- --------------------------------
Eth1/1           Desg FWD 1         128.1     P2p
Eth1/2           Root FWD 2         128.2     P2p
Eth1/5           Desg FWD 2         128.5     P2p
```

可以通过接口配置命令 **spanning-tree mst** *instance-number* **port-priority** *priority* 设置接口优先级。例 4-39 将 NX-4 的 Eth1/5 端口优先级更改为 64，同时还验证了更改前后的接口优先级情况。

例 4-39 更改 MST 接口优先级

```
NX-4# show spanning-tree mst 0
! Output omitted for brevity
##### MST0      vlans mapped:   1-9,11-19,21-29,31-4094
Interface        Role Sts Cost      Prio.Nbr  Type
---------------- ---- --- --------- --------- --------------------------------
Eth1/1           Desg FWD 2         128.1     P2p
Eth1/2           Root FWD 2         128.2     P2p
Eth1/5           Desg FWD 2         128.5     P2p

NX-4(config)# interface eth1/5
NX-4(config-if)# spanning-tree mst 0 port-priority 64

NX-4# show spanning-tree mst 0
! Output omitted for brevity
##### MST0      vlans mapped:   1-9,11-19,21-29,31-4094
Interface        Role Sts Cost      Prio.Nbr  Type
---------------- ---- --- --------- --------- --------------------------------
Eth1/1           Desg FWD 2         128.1     P2p
Eth1/2           Root FWD 2         128.2     P2p
Eth1/5           Desg FWD 2         64.5      P2p
```

4.4 检测和修复转发环路

L2 转发环路的常见特征就是高 CPU 利用率和低可用内存空间。网络数据包在 L2 拓扑结构中转发时，其报头的 TTL（Time-To-Live，生存时间）不会递减。数据包在 L2 拓扑结构中持续转发时，不但要消耗交换机的带宽，而且极有可能导致 CPU 出现利用率峰值，最终导致拓扑结构中的交换机因 CPU 和内存资源耗尽而崩溃。

常见的 L2 转发环路场景如下。

- 负载均衡器配置出错。负载均衡器负责将流量向外传送给拥有相同 MAC 地址的多个端口。
- 桥接了两个物理端口的虚拟交换机配置出错。虚拟交换机通常并不参与 STP。
- 最终用户使用了哑网络交换机或集线器。
- 虚拟机迁移到其他物理主机（虽然该场景不是 L2 环路，但确实会显示端口之间的 MAC 地址出现变化）。

幸运的是，NX-OS 提供了一些行之有效的保护措施来解决 L2 转发环路问题。

4.4.1 MAC 地址通告

一项重要的保护措施就是能够识别出哪些数据包和接口出现了环路问题。Nexus 交换机可以检测 MAC 地址变化并通过 MAC 地址变化 syslog 进行通告。较新的 NX-OS 软件版本均默认启用 MAC 地址通告功能，也可以通过全局配置命令 **mac address-table notification mac-move** 手动启用该功能特性。

思科提供了一种保护机制，可以在 L2 转发环路期间防止 CPU 利用率不断升高。如果 MAC 地址翻动超出了阈值（10s 内在一组端口之间来回翻动 3 次），那么 Nexus 交换机就会清除 MAC 地址表并在指定时间段内停止学习 MAC 地址。虽然此时的数据包仍然以环路方式持续进行转发，但 CPU 利用率并不会无限升高，从而允许操作人员执行必要的诊断命令以修复该故障。

从例 4-40 可以看出，交换机检测出 VLAN 1 存在转发环路并清除了 MAC 地址表。

例 4-40　NX-OS 检测转发环路

```
07:39:56 NX-1 %$ VDC-1 %$ %L2FM-2-L2FM_MAC_FLAP_DISABLE_LEARN_N3K: Loops detected
  in the network for mac 9caf.ca2e.9040 among ports Eth1/49 and Eth1/51 vlan 1 -
  Disabling dynamic learning notifications for a period between 120 and 240 seconds
  on vlan 1
07:39:56 NX-1 %$ VDC-1 %$ %-SLOT1-5-BCM_L2_LEARN_DISABLE: MAC Learning Disabled
  unit=0
07:42:13 NX-1 %$ VDC-1 %$ %-SLOT1-5-BCM_L2_LEARN_ENABLE: MAC Learning Enabled
  unit=0
```

NX-OS 为环路检测操作提供了增强功能，一旦检测到 MAC 地址翻动，就会将端口置入关闭状态。可以通过命令 **mac address-table loop-detect port-down** 启用该功能特性，例 4-41 给出了该功能特性的配置以及配置了该功能特性之后验证接口是否处于关闭状态的示例。

例 4-41　出现 MAC 地址翻动通告之后将端口置入关闭状态

```
NX-1(config)# mac address-table loop-detect port-down
06:27:37 NX-1 %$ VDC-1 %$ %L2FM-2-L2FM_MAC_MOVE_PORT_DOWN: Loops detected in the
  network for mac 9caf.ca2e.9040 among ports Eth1/2 and Eth1/3 vlan 1 - Port Eth1/2
  Disabled on loop detection
06:27:37 NX-1 %$ VDC-1 %$ last message repeated 9 times
NX-1# show interface status
--------------------------------------------------------------------------------
Port          Name              Status    Vlan      Duplex  Speed   Type
--------------------------------------------------------------------------------
mgmt0         --                notconnec routed    auto    auto    --
Eth1/1        --                disabled  1         full    10G     10Gbase-SR
Eth1/2        --                errDisabl 1         full    10G     10Gbase-SR
Eth1/3        --                connected 1         full    10G     10Gbase-SR
```

注：有些平台默认并不显示 MAC 地址通告，因而需要配置以下命令：
```
logging level spanning-tree 6
logging level fwm 6
logging monitor 6
```

4.4.2 BPDU 保护

BPDU 保护是一种安全机制，可以在收到 BPDU 后关闭配置了 STP Portfast 功能的端口。这样就能确保未授权交换机加入拓扑结构之后，不会意外地产生转发环路。

可以通过命令 **spanning-tree port type edge bpduguard default** 在所有 STP Portfast 端口全局

启用 BPDU 保护机制，可以使用命令 **spanning-tree bpduguard {enable |disable}** 命令在特定接口上启用或禁用 BPDU 保护机制。例 4-42 显示了在特定端口上启用 BPDU 保护机制以及在所有接入端口上全局启用 BPDU 保护机制的配置示例，检查生成树端口详细信息时，关键字 *by default* 表示全局配置已经为该端口启用了 BPDU 保护机制。

例 4-42　配置 BPDU 保护机制

```
NX-1(config)# interface Ethernet1/6
NX-1(config-if)# spanning-tree bpduguard enable
NX-1(config)# spanning-tree port type edge bpduguard default
NX-1# show spanning-tree interface ethernet 1/1 detail
 Port 1 (Ethernet1/1) of VLAN0001 is designated forwarding
   Port path cost 4, Port priority 128, Port Identifier 128.1
   Designated root has priority 28673, address 885a.92de.617c
   Designated bridge has priority 28673, address 885a.92de.617c
   Designated port id is 128.1, designated path cost 0
   Timers: message age 0, forward delay 0, hold 0
   Number of transitions to forwarding state: 1
   The port type is edge by default
   Link type is point-to-point by default
   Bpdu guard is enabled by default
   BPDU: sent 32151, received 0
```

> **注：** 应该在所有面向主机的端口上配置 BPDU 保护机制，但是不要在 PVLAN 混杂端口上启用 BPDU 保护机制。

在默认情况下，因 BPDU 保护而进入 ErrDisabled 状态的端口无法自动恢复。此时，可以通过差错恢复服务（Error Recovery Service）重新激活因故障而关闭的端口，从而有效减少管理开销。差错恢复服务通过命令 **errdisable recovery cause bpduguard** 恢复因 BPDU 保护而关闭的端口，可以通过命令 **errdisable recovery interval** *time-seconds* 配置差错恢复服务检查端口的时间间隔。

例 4-43 为 BPDU 保护机制配置了差错恢复服务，同时还显示了差错恢复服务响应情况。

例 4-43　配置和验证差错恢复服务

```
NX-1(config)# errdisable recovery cause bpduguard
NX-1(config)# errdisable recovery interval 60
11:16:17 NX-1 %$ VDC-1 %$ %STP-2-BLOCK_BPDUGUARD: Received BPDU on port Ethernet1/6
  with BPDU Guard enabled. Disabling port.
11:16:17 NX-1 %$ VDC-1 %$ %ETHPORT-5-IF_DOWN_ERROR_DISABLED: Interface Ethernet1/6
  is down (Error disabled. Reason:BPDUGuard)
11:21:17 NX-1 %$ VDC-1 %$ %ETHPORT-5-IF_ERRDIS_RECOVERY: Interface Ethernet1/6 is
  being recovered from error disabled state (Last Reason:BPDUGuard)
```

4.4.3　BPDU 过滤器

BPDU 过滤器可以很简单地阻止端口向外发送 BPDU，可以在全局或指定接口上启用 BPDU 过滤器。请注意，BPDU 过滤器的处理行为与具体配置有关。

- 如果使用命令 **spanning-tree port type edge bpdufilter enable** 在全局启用 BPDU 过滤器，那么端口将会发送一系列至少 10 个 BPDU。如果远程端口启用了 BPDU 保护机制，那么通常会因为环路预防机制而关闭端口。
- 如果使用命令 **spanning-tree bpdufilter enable** 在特定接口上启用 BPDU 过滤器，那么端口将不会持续发送任何 BPDU。但是，在端口首次激活之后，交换机将会发送一系列至少 10 个 BPDU。如果远程端口启用了 BPDU 保护机制，那么通常会因为环路预防机制而关闭端口。

> **注：** 部署 BPDU 过滤器时请务必谨慎，因为配置出错可能会导致严重故障。大多数网络设计方案不需要 BPDU 过滤器，BPDU 过滤器可能会在引入风险的同时增加不必要的复杂性。

从例 4-44 可以看出，Eth1/1 接口在全局启用了 BPDU 过滤器，端口在首次激活时发送了 10 个 BPDU。

例 4-44　验证 BPDU 过滤器

```
NX-1# show spanning-tree interface ethernet 1/1 detail
 Port 1 (Ethernet1/1) of VLAN0001 is designated forwarding
   Port path cost 4, Port priority 128, Port Identifier 128.1
   Designated root has priority 28673, address 885a.92de.617c
   Designated bridge has priority 28673, address 885a.92de.617c
   Designated port id is 128.1, designated path cost 0
   Timers: message age 0, forward delay 0, hold 0
   Number of transitions to forwarding state: 1
   The port type is edge by default
   Link type is point-to-point by default
   Bpdu filter is enabled by default
   BPDU: sent 32151, received 0
```

4.4.4　单向链路问题

使用光纤进行连接的网络拓扑可能会出现单向流量流问题，因为其中的一条光纤负责从下游向上游交换机传输数据，而另一条光纤则从上游向下游交换机传输数据。此时无法传送 BPDU，导致下游交换机的当前根端口最终超时，从而将其他端口标识为根端口。这样一来，某个端口收到的流量将会通过其他端口向外转发，从而产生转发环路问题。

可以通过以下措施来解决上述问题。

- STP 环路保护（LoopGuard）。
- UDLD（UniDirectional Link Detection，单向链路检测）。
- BA（Bridge Assurance，网桥保障）。

1. STP 环路保护

环路保护机制可以防止因根端口上的 BPDU 丢失而导致可选端口或根端口成为指派端口（面向下游交换机的端口）。如果没有收到 BPDU，那么环路保护机制就会将发送端口置入 ErrDisabled 状态，如果端口重新开始发送 BPDU，那么端口将恢复并再次经历 STP 状态。

可以通过命令 **spanning-tree loopguard default** 在全局范围内启用环路防护机制，也可以通过接口命令 **spanning-tree guard loop** 以接口为基础启用环路防护机制。请注意，不要在启用了 Portfast 的端口上启用环路防护机制（与根端口/可选端口逻辑相冲突），也不要在 vPC（virtual Port-Channel，虚拟端口通道）端口上启用环路防护机制。

例 4-45 为 NX-2 的 Eth1/1 端口配置了环路保护机制。

例 4-45　配置环路保护机制

```
NX-2(config)# interface Eth1/1
NX-2(config-if)# spanning-tree guard loop
```

在 NX-2 连接 NX-1（根网桥）的 Eth1/1 端口上设置了 BPDU 过滤器之后，将触发环路保护机制，如例 4-46 所示。

例 4-46 触发了环路保护机制的 syslog

```
NX-2(config-if)# interface Eth1/1
NX-2(config-if)# spanning-tree bpdufilter enable
18:46:06 NX-2 %$ VDC-1 %$ %STP-2-LOOPGUARD_BLOCK: Loop guard blocking port
         Ethernet1/1 on VLAN0001.
18:46:07 NX-2 %$ VDC-1 %$ %STP-2-LOOPGUARD_BLOCK: Loop guard blocking port Ethernet1/1
         on VLAN0010.
18:46:07 NX-2 %$ VDC-1 %$ %STP-2-LOOPGUARD_BLOCK: Loop guard blocking port Ethernet1/1
         on VLAN0020.
18:46:07 NX-2 %$ VDC-1 %$ %STP-2-LOOPGUARD_BLOCK: Loop guard blocking port
  Ethernet1/1
         on VLAN0030.
```

此时，端口处于不一致（Inconsistent）状态。可以通过命令 **show spanning-tree inconsistentports** 查看处于不一致状态的端口，例 4-47 显示了端口 Eth1/1 承载的所有 VLAN 的表项信息。

例 4-47 查看处于不一致状态的 STP 端口

```
NX-2# show spanning-tree inconsistentports
Name                    Interface               Inconsistency
--------------------    --------------------    ------------------
VLAN0001                Eth1/1                  Loop Inconsistent
VLAN0010                Eth1/1                  Loop Inconsistent
VLAN0020                Eth1/1                  Loop Inconsistent
VLAN0030                Eth1/1                  Loop Inconsistent

Number of inconsistent ports (segments) in the system : 4
```

2. UDLD

UDLD（UniDirectional Link Detection，单向链路检测）可以实现光纤和以太网铜缆的双向监控，实现方式是将 UDLD 数据包（包含了发送该 UDLD 数据包的接口的 System-ID 和 Port-ID）发送给邻居设备，收端路由器会将该信息回送给发端路由器，其中包含了自己的 System-ID 和 Port-ID。该过程将持续进行，如果数据帧未得到确认，那么端口将进入 ErrDisabled 状态。

请注意，首先必须通过命令 **feature udld** 启用 UDLD 功能特性，然后通过命令 **udld enable** 在指定接口下启用 UDLD。例 4-48 在 NX-1 连接 NX-2 的链路上配置了 UDLD。

例 4-48 配置 UDLD

```
NX-1(config)# feature udld
NX-1(config)# interface e1/2
NX-1(config-if)# udld enable
```

此外，还必须在远程交换机上启用 UDLD。配置完成之后，可以通过 **show udld** *interface-id* 命令检查接口的 UDLD 状态。例 4-49 显示了接口的 UDLD 状态输出结果，包含了当前状态、Device-ID（序列号）、发端接口 ID 和回送接口 ID。

例 4-49 验证 UDLD 交换端口状态

```
NX-1# show udld ethernet 1/2

Interface Ethernet1/49
--------------------------------
Port enable administrative configuration setting: enabled
Port enable operational state: enabled
Current bidirectional state: bidirectional
Current operational state:   advertisement - Single neighbor detected
```

```
       Message interval: 15
       Timeout interval: 5

            Entry 1
            ---------------
            Expiration time: 35
            Cache Device index: 1
            Current neighbor state: bidirectional
            Device ID: FDO1348R0VM
            Port ID: Eth1/2
            Neighbor echo 1 devices: FOC1813R0C
            Neighbor echo 1 port: Ethernet1/1

            Message interval: 15
            Timeout interval: 5
            CDP Device name: NX-2

            Last pkt send on: 291908, 19:07:51 2017
                  Probe pkt send on: 291908, Sep  1 19:07:51 2017
                  Echo  pkt send on: 177683, Sep  1 19:06:21 2017
                  Flush pkt send on: None.

            Last pkt recv on: 469579, 19:07:50 2017
                  Probe pkt recv on: 469579, Sep  1 19:07:50 2017
                  Echo  pkt recv on: 470536, Sep  1 19:06:21 2017
                  Flush pkt recv on: None.

            Deep pkt inspections done: None.
            Mismatched if index found: None.
            Deep pkt inspection drops: None.
```

UDLD 出现故障后，接口状态将显示该端口因 UDLD 故障而处于关闭状态，如例 4-50 所示。

例 4-50 接口状态反映 UDLD 故障

```
NX-1# show interface brief
! Output omitted for brevity
Ethernet     VLAN    Type Mode   Status  Reason              Speed      Port
Interface                                                                Ch #
--------------------------------------------------------------------------------
Eth1/1       --      eth  routed up      none                10G(D)     --
Eth1/2       --      eth  trunk  down    UDLD empty echo     auto(D)    --
```

事件历史记录可以为排查 UDLD 故障提供有用信息，可以通过命令 **show udld internal event-history errors** 查看事件历史记录，包括时间戳以及故障根源的初步指示，如例 4-51 所示。

例 4-51 UDLD 事件历史记录

```
NX-1# show udld internal event-history errors

1) Event:E_DEBUG, length:75, at 177895 usecs after NX-011 09:42:23 2014
    [102] udld_demux(646): (646): (Warning) unexpected mts msg (opcode - 61467)

2) Event:E_DEBUG, length:70, at 983485 usecs after NX-011 09:42:22 2014
    [102] udldDisablePort(3985): Ethernet1/2: Port UDLD set error disabled

3) Event:E_DEBUG, length:70, at 983415 usecs after NX-011 09:42:22 2014
    [102] udldDisablePort(3975): calling mts_msg_send_recv_ethpm() w/ f01b
```

```
4) Event:E_DEBUG, length:77, at 983387 usecs after NX-011 09:42:22 2014
   [102] udldDisablePort(3915): current bidirdetect_flag (error): udld_empty_echo

5) Event:E_DEBUG, length:73, at 983180 usecs after NX-011 09:42:22 2014
   [102] udldDisablePort(3888): calling udld_send_flush_msg for: Ethernet1/2

6) Event:E_DEBUG, length:151, at 983036 usecs after NX-011 09:42:22 2014
   [102] udld_recv_det1(5640): Ethernet1/2: UDLD error Unidirection detected
   during extended detection, sent UDLD_MAIN_EV_DETECTION_WINDOW_CONCLUSION_OTHER
```

UDLD 的常见故障包括以下两种。
- 空回显（Empty Echo）。
- Tx-Rx 环路（Tx-Rx Loop）。

3. 空回显

下列场景可能会出现空回显 UDLD 问题。
- UDLD 会话超时。
- 远程交换机不处理 UDLD 数据包。
- 本地交换机不传输 UDLD 数据包。

例 4-52 显示了与 UDLD 空回显检测相关的系统日志消息。

例 4-52　UDLD 空回显检测

```
11:57:56.155 NX-1 ETHPORT-2-IF_DOWN_ERROR_DISABLED Interface Ethernet1/2 is down
   (Error disabled. Reason:UDLD empty echo)
11:57:56.186 NX-1 ETH_PORT_CHANNEL-5-PORT_INDIVIDUAL_DOWN individual port
   Ethernet1/2 is down
11:57:56.336 NX-1 ETHPORT-2-IF_DOWN_ERROR_DISABLED Interface Ethernet1/2 is down
   (Error disabled. Reason:UDLD empty echo)
```

4. Tx-Rx 环路

如果收到 UDLD 帧的端口与宣告该 UDLD 帧的端口相同，那么就表明出现了 Tx-Rx 环路问题，意味着收到的 UDLD 数据包中的 System-ID 和 Port-ID 与接收端交换机的 System-ID 和 Port-ID 相同。下列场景可能会出现 Tx-Rx 环路问题。
- 中间设备（光传输设备）配置出错或配线错误。
- 配线或介质有问题。

例 4-53 显示了与 UDLD Tx-Rx 环路相关的系统日志消息。

例 4-53　Tx-Rx 环路检测

```
14:52:30 NX-1 %ETHPORT-2-IF_DOWN_ERROR_DISABLED: Interface Ethernet17/5  is down
   (Error disabled. Reason:UDLD Tx-Rx Loop)
14:52:30 NX-1 %ETHPORT-2-IF_DOWN_ERROR_DISABLED: Interface Ethernet17/5  is down
   (Error disabled. Reason:UDLD Tx-Rx Loop)
```

5. BA

BA（Bridge Assurance，网桥保障）解决了环路保护机制和 UDLD 机制存在的一些限制条件，BA 工作在 STP 指派端口上（环路保护机制不行），可以解决端口在单向状态下启动所存在的问题。BA 要求 STP 采取类似于路由协议（EIGRP/OSPF 等）的操作方式，要求双向发送运行状况检查数据包。

虽然系统默认启用 BA 进程，但仍然需要通过 **spanning-tree port type network** 命令显式配置中继端口。例 4-54 在 NX-1、NX-2 和 NX-3 的互联接口上配置了 BA 功能特性。

例 4-54　配置 BA

```
NX-1(config)# interface eth1/2,eth1/3
NX-1(config-if-range)# spanning-tree port type network
```

```
NX-2(config)# interface eth1/1,eth1/3
NX-2(config-if-range)# spanning-tree port type network
```

```
NX-3(config)# interface eth1/1,eth1/2
NX-3(config-if-range)# spanning-tree port type network
```

例 4-55 显示配置了 BA 功能特性之后的 STP 端口类型，请注意，端口类型 P2P 的前面增加了关键字 *Network*。

例 4-55　查看配置了 BA 特性之后的 STP 端口类型

```
NX-1# show spanning-tree vlan 10 | b Interface
Interface        Role Sts Cost      Prio.Nbr Type
---------------- ---- --- --------- -------- --------------------------------
Eth1/2           Desg FWD 2         128.2    Network P2p
Eth1/3           Desg FWD 2         128.3    Network P2p

NX-2# show spanning-tree vlan 10 | b Interface
Interface        Role Sts Cost      Prio.Nbr Type
---------------- ---- --- --------- -------- --------------------------------
Eth1/1           Root FWD 2         128.1    Network P2p
Eth1/3           Desg FWD 2         128.3    Network P2p
Eth1/4           Desg FWD 2         128.4    P2p

NX-3# show spanning-tree vlan 10 | b Interface
Interface        Role Sts Cost      Prio.Nbr Type
---------------- ---- --- --------- -------- --------------------------------
Eth1/1           Root FWD 2         128.1    Network P2p
Eth1/2           Altn BLK 2         128.2    Network P2p
Eth1/5           Desg FWD 2         128.5    P2p
```

例 4-56 在 NX-2 连接 NX-3 的链路上应用了 BPDU 过滤器，由于无法维持 BPDU 数据包的双向握手，因而 BA 几乎瞬间就在 NX-2 和 NX-3 上发挥了作用。

例 4-56　BA 发挥作用

```
NX-2(config-if)# interface Eth1/3
NX-2(config-if)# spanning-tree bpdufilter enable
20:46:34 NX-2 %$ VDC-1 %$ %STP-2-BRIDGE_ASSURANCE_BLOCK: Bridge Assurance blocking
  port Ethernet1/3 VLAN0001.
20:46:35 NX-2 %$ VDC-1 %$ %STP-2-BRIDGE_ASSURANCE_BLOCK: Bridge Assurance blocking
  port Ethernet1/3 VLAN0002.
20:46:35 NX-2 %$ VDC-1 %$ %STP-2-BRIDGE_ASSURANCE_BLOCK: Bridge Assurance blocking
  port Ethernet1/3 VLAN0010.
20:46:35 NX-2 %$ VDC-1 %$ %STP-2-BRIDGE_ASSURANCE_BLOCK: Bridge Assurance blocking
  port Ethernet1/3 VLAN0020.
20:46:35 NX-2 %$ VDC-1 %$ %STP-2-BRIDGE_ASSURANCE_BLOCK: Bridge Assurance blocking
  port Ethernet1/3 VLAN0030.
```

此时的 STP 端口类型包含了注释*BA_Inc*，表明这些接口当前正处于 BA 不一致端口状态。例 4-57 显示了这种新的接口类型。

例 4-57　Detecting Inconsistent Port State

```
NX-2# show spanning-tree vlan 10 | b Interface
Interface        Role Sts Cost      Prio.Nbr Type
```

```
Eth1/1           Root FWD 2          128.1      Network P2p
Eth1/3           Desg BKN*2          128.3      Network P2p *BA_Inc
Eth1/4           Desg FWD 2          128.4      P2p

NX-3# show spanning-tree vlan 10 | b Interface
Interface        Role Sts Cost       Prio.Nbr Type
--------------- ---- --- --------- -------- --------------------------------
Eth1/1           Root FWD 2          128.1      Network P2p
Eth1/2           Desg BKN*2          128.2      Network P2p *BA_Inc
Eth1/5           Desg FWD 2          128.5      P2p
```

可以通过命令 **show spanning-tree inconsistentports** 查看所有处于不一致状态的接口及其原因。例 4-58 给出了 NX-2 的输出结果，与前面提到的事件历史记录交叉引用相关。

例 4-58 查看处于不一致状态的端口

```
NX-2# show spanning-tree inconsistentports
Name                    Interface              Inconsistency
--------------------- -------------------- ------------------
VLAN0001                Eth1/3                 Bridge Assurance Inconsistent
VLAN0002                Eth1/3                 Bridge Assurance Inconsistent
VLAN0010                Eth1/3                 Bridge Assurance Inconsistent
VLAN0020                Eth1/3                 Bridge Assurance Inconsistent
VLAN0030                Eth1/3                 Bridge Assurance Inconsistent


Number of inconsistent ports (segments) in the system : 5
```

删除 BPDU 过滤器之后，就会解除网桥保障，端口也将返回转发状态，如例 4-59 所示。

例 4-59 通过允许处理 BPDU 来恢复连接

```
NX-2(config)# int Eth1/3
NX-2(config-if)# no spanning-tree bpdufilter
NX-2(config-if)#
20:48:30 NX-2 %$ VDC-1 %$ %STP-2-BRIDGE_ASSURANCE_UNBLOCK: Bridge Assurance
  unblocking port Ethernet1/3 VLAN0001.
20:48:31 NX-2 %$ VDC-1 %$ %STP-2-BRIDGE_ASSURANCE_UNBLOCK: Bridge Assurance
  unblocking port Ethernet1/3 VLAN0002.
20:48:31 NX-2 %$ VDC-1 %$ %STP-2-BRIDGE_ASSURANCE_UNBLOCK: Bridge Assurance
  unblocking port Ethernet1/3 VLAN0010.
20:48:32 NX-2 %$ VDC-1 %$ %STP-2-BRIDGE_ASSURANCE_UNBLOCK: Bridge Assurance
  unblocking port Ethernet1/3 VLAN0020.
20:48:32 NX-2 %$ VDC-1 %$ %STP-2-BRIDGE_ASSURANCE_UNBLOCK: Bridge Assurance
  unblocking port Ethernet1/3 VLAN0030.
```

注： 网桥保障是检测单向链路的首选方法，如果所有平台都支持网桥保障，那么就应该使用该功能特性。

4.5 本章小结

本章简要回顾了以太网通信标准以及管理型交换机为 L2 拓扑结构带来的优势。L2 转发故障的排查操作包含了大量组件，排查 L2 转发故障的第一步就是识别源端和目的端交换端口，然后按照图 4-5 所示流程开展故障排查操作，按照本章讨论的内容即可解决不同排查阶段的操作需求。

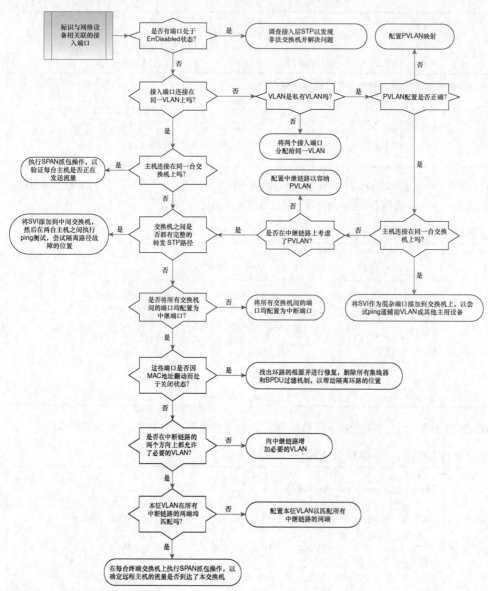

图 4-5 L2 转发故障排查流程

第 5 章

端口通道、vPC 和 FabricPath

本章主要讨论如下主题。
- 端口通道（Port-Channel）。
- vPC（virtual Port-Channel，虚拟端口通道）。
- FabricPath。
- vPC+（virtual Port-Channel Plus，增强型虚拟端口通道）。

正确的网络设计可以确保备用路径和备用设备能够在故障状态下转发流量，从而解决单点故障问题。路由协议可以通过 ECMP（Equal-Cost MultiPath，等价多路径）机制来使用冗余路径，但 STP（Spanning Tree Protocol，生成树协议）为了防止转发环路，不允许在交换机之间的冗余链路上转发流量。

虽然 STP 很有用，但同时也限制了交换机之间能够提供的带宽量。端口通道提供了一种增加可用带宽的有效机制，可以将多条物理链路整合为单一虚拟链路，所有成员接口都能转发网络流量。本章将介绍端口通道的操作方式以及端口通道出现问题后的故障排查技术。

5.1 端口通道

端口通道是由一条或多条物理成员链路组成的逻辑链路。端口通道定义在 IEEE 803.3AD 链路聚合规范中，有时也称为 EtherChannel。用于构建逻辑端口通道的物理接口被称为成员接口。端口通道可以是二层（L2）交换或三层（L3）路由链路。

图 5-1 以形象化的方式显示了端口通道的关键组件（成员接口和逻辑接口），同时也显示了端口通道与单一链路相比所具备的优势。

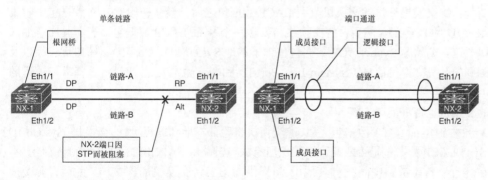

图 5-1 端口通道相对于单一链路的优势

端口通道的主要优势在于增加或删除端口通道的成员链路时，不会影响网络的拓扑结构。虽然变更可能会触发 L2 STP 树计算或 L3 SPF 计算，但端口通道的设备之间仍能继续转发流量。Nexus 交换机可以通过两种方式构建端口通道，一种是以静态方式将端口通道设置为"on"

状态,另一种是使用 LACP(Link-Aggregation Control Packet,链路聚合控制包)检测设备之间的连接性。大多数网络工程师更喜欢使用 LACP,因为 LACP 可以确保设备之间的端到端连接性。LACP 可以为成员接口检测单向链路故障或者识别路径中是否存在其他设备(如不支持链路传播的 DWDM[Dense Wavelength-Division Multiplexing,密集波分复用]设备)。

图 5-2 中的 NX-1 和 NX-2 将 Eth1/1 和 Eth1/2 接口组合成 port-channel 1。光传输设备 DWDM-1 与 DWDM-2 之间的链路-A 故障不会传播给 NX-1 或 NX-2 的 Eth1/1 接口,Nexus 交换机仍然能够通过 Eth1/1 接口向外转发流量,因为这些端口与 DWDM-1 或 DWDM-2 仍然保持物理连接状态,被静态设置为 on 状态的端口—通道端口没有运行状况检查机制。不过,如果配置了 LACP,那么 NX-1 和 NX-2 就能检测出流量无法通过上层路径进行端到端地传输,从而将该链路从逻辑端口—通道中删除。

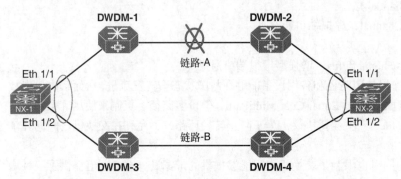

图 5-2 端口通道链路状态的传播与检测

使用下列消息建立 LACP 之后,端口通道内的成员链路将变为活动状态。

- **Sync(S,同步)消息**:第一个标志,表示本地交换机将成员接口作为端口通道的一部分。
- **Collecting(C,收集)消息**:第二个标志,表示本地交换机处理该接口收到的网络流量。
- **Distributing(D,分发)消息**:第三个标志,表示本地交换机利用该成员接口传输网络流量。

端口启动后,将按照以下步骤交换消息。

- **第 1 步**:两台交换机(源和目的交换机)将同步、收集、分发标志设置为零(off)并宣告 LACP 数据包。
- **第 2 步**:源交换机从目的交换机收到 LACP 数据包之后,会从初始 LACP 数据包中收集 System-ID 和 Port-ID。然后,源交换机将发送一个 Sync LACP 数据包,表示其愿意加入端口通道。初始的 Sync LACP 数据包包含了本地 System-ID、Port-ID、端口优先级以及检测到的远程交换机信息(System-ID、Port-ID 和端口优先级)。此时将选择端口通道的 LACP 成员。

  目的交换机也将重复该步骤。
- **第 3 步**:收到 Sync LACP 数据包之后,源交换机将验证本地和远程(目的交换机)System-ID 与 Sync LACP 数据包是否匹配,以确保所有成员链路的 Switch-ID 相同,而且链路上不存在多台设备(链路中间没有其他设备[提供与第三台交换机的连接])。此后,源交换机将发送一个 Collecting LACP 数据包,指示该源交换机已经准备好在接口上接收流量。
- **第 4 步**:目的交换机对照源交换机执行的操作验证源交换机的 Sync LACP 数据包的准确性。然后目的交换机将发送一个 Collecting LACP 数据包,指示目的交换机已经准备好在该接口上接收流量。

- **第 5 步**：源交换机从目的交换机收到 Collecting LACP 数据包之后，向目的交换机发送一个 Distributing LACP 数据包，指示其正通过该成员链路传输数据。
- **第 6 步**：目的交换机从源交换机收到 Collecting LACP 数据包之后，向源交换机发送 Distributing LACP 数据包，指示其正通过该成员链路传输数据。
- **第 7 步**：两台交换机通过成功完成上述步骤的成员链路接口传输数据。

注：在满足需求的情况下，第 7 步中的 LACP 数据包与其他交换机无关。

图 5-3 解释了 NX-1（源交换机）与 NX-2（目的交换机）之间的 LACP 消息交换过程。

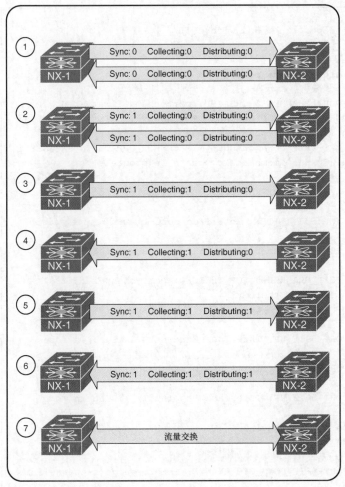

图 5-3　LACP 协商过程

注：成员链路加入端口通道接口的时候，都要经历上述协商操作。

5.1.1　基本的端口通道配置

端口通道的配置方式是进入成员接口的接口配置模式，然后将这些成员接口分配给端口通道并将其静态设置为"on"，或者通过 LACP 动态协商方式。LACP 支持两种操作模式。

- **被动模式**：接口不会发起建立端口通道，也不会向外发送 LACP 数据包。如果远程交换机收到 LACP 数据包，那么接口就会做出响应并建立 LACP 邻接关系。如果两台设备都是 LACP 被动设备，那么就不会建立 LACP 邻接关系。

- **主动模式**：接口会尝试发起建立端口通道并向外发送 LACP 数据包。仅当远程接口被配置为主动或被动模式时，主动 LACP 接口才能建立 LACP 邻接关系。

首先必须使用全局命令 **feature lacp** 启用 LACP 功能，然后利用接口参数命令 **channel-group** *port-channel-number* **mode** {**on** | **active** | **passive**}将常规接口转换为成员接口。

例 5-1 以成员接口 Eth1/1 和 Eth1/2 为例解释了 port-channel 1 的配置情况。请注意，port-channel 1 被配置为中继接口，而不是单个成员接口。

例 5-1 端口通道配置示例

```
NX-1# conf t
Enter configuration commands, one per line. End with CNTL/Z.
NX-1(config)# feature lacp
NX-1(config)# interface ethernet 1/1-2
NX-1(config-if-range)# channel-group 1 mode active
! Output omitted for brevity
03:53:14 NX-1 %$ VDC-1 %$ %ETH_PORT_CHANNEL-5-CREATED: port-channel1 created
03:53:14 NX-1 %$ VDC-1 %$ %ETHPORT-5-IF_DOWN_CHANNEL_MEMBERSHIP_UPDATE_IN_PROGRESS:
   Interface Ethernet1/2 is down (Channel membership update in progress)
03:53:14 NX-1 %$ VDC-1 %$ %ETHPORT-5-IF_DOWN_CHANNEL_MEMBERSHIP_UPDATE_IN_PROGRESS:
   Interface Ethernet1/1 is down (Channel membership update in progress)
..
03:53:16 NX-1 %$ VDC-1 %$ %ETHPORT-5-SPEED: Interface port-channel1, operational
   speed changed to 10 Gbps
03:53:16 NX-1 %$ VDC-1 %$ %ETHPORT-5-IF_DUPLEX: Interface port-channel1, operational
   duplex mode changed to Full
03:53:21 NX-1 %$ VDC-1 %$ %ETH_PORT_CHANNEL-5-PORT_UP: port-channel1: Ethernet1/1
   is up
03:53:21 NX-1 %$ VDC-1 %$ %ETH_PORT_CHANNEL-5-FOP_CHANGED: port-channel1: first
   operational port changed from none to Ethernet1/1
03:53:21 NX-1 %$ VDC-1 %$ %ETH_PORT_CHANNEL-5-PORT_UP: port-channel1: Ethernet1/2
   is up
03:53:21 NX-1 %$ VDC-1 %$ %ETHPORT-5-IF_UP: Interface Ethernet1/1 is up in mode
   access
03:53:21 NX-1 %$ VDC-1 %$ %ETHPORT-5-IF_UP: Interface port-channel1 is up in mode
   access
03:53:21 NX-1 %$ VDC-1 %$ %ETHPORT-5-IF_UP: Interface Ethernet1/2 is up in mode
   access
NX-1(config-if-range)# interface port-channel 1
NX-1(config-if)# switchport mode trunk
! Output omitted for brevity
03:53:21 NX-1 %$ VDC-1 %$ %ETHPORT-5-IF_DOWN_CFG_CHANGE: Interface port-channel1 is
   down(Config change)
03:53:21 NX-1 %$ VDC-1 %$ %ETH_PORT_CHANNEL-5-PORT_DOWN: port-channel1: Ethernet1/1
   is down
03:53:21 NX-1 %$ VDC-1 %$ %ETH_PORT_CHANNEL-5-PORT_DOWN: port-channel1: Ethernet1/2
   is down
..
03:53:29 NX-1 %$ VDC-1 %$ %ETH_PORT_CHANNEL-5-PORT_UP: port-channel1: Ethernet1/1
   is up
03:53:29 NX-1 %$ VDC-1 %$ %ETH_PORT_CHANNEL-5-FOP_CHANGED: port-channel1: first
   operational port changed from none to Ethernet1/1
03:53:29 NX-1 %$ VDC-1 %$ %ETH_PORT_CHANNEL-5-PORT_UP: port-channel1: Ethernet1/2
   is up
03:53:29 NX-1 %$ VDC-1 %$ %ETHPORT-5-IF_UP: Interface Ethernet1/1 is up in mode
```

```
  trunk
03:53:29 NX-1 %$ VDC-1 %$ %ETHPORT-5-IF_UP: Interface port-channel1 is up in mode
  trunk
03:53:29 NX-1 %$ VDC-1 %$ %ETHPORT-5-IF_UP: Interface Ethernet1/2 is up in mode
  trunk
```

5.1.2 验证端口通道状态

配置了端口通道之后，接下来就要验证端口通道的建立情况。命令 **show port-channel summary** 可以显示所有已配置的端口通道及其状态信息，如例 5-2 所示。

例 5-2 查看端口通道的摘要状态

```
NX-1# show port-channel summary
Flags:  D - Down         P - Up in port-channel (members)
        I - Individual   H - Hot-standby (LACP only)
        s - Suspended    r - Module-removed
        S - Switched     R - Routed
        U - Up (port-channel)
        p - Up in delay-lacp mode (member)
        M - Not in use. Min-links not met
--------------------------------------------------------------------------------
Group Port-        Type     Protocol  Member Ports
      Channel
--------------------------------------------------------------------------------
1     Po1(SU)      Eth      LACP      Eth1/1(P)    Eth1/2(P)
```

查看 **show port-channel summary** 命令的输出结果时，请注意检查端口通道的状态（位于端口通道接口的下方），端口通道的状态应该显示为"U"，如例 5-2 所示。

表 5-1 列出了端口通道的标志信息。

表 5-1　　　　　　　　　　　逻辑端口通道接口状态字段

字段	描述
U	端口通道接口工作正常
D	端口通道接口关闭
M	端口通道接口已成功建立了至少一个 LACP 邻接关系，但端口通道配置了最少活动接口数量，该数量超过了实际参与的活动成员接口的数量。流量不会通过该端口通道转发。该端口通道接口配置了命令 **lacp min-links** *number-member-interfaces*
S	端口通道接口被配置为二层（L2）交换
R	端口通道接口被配置为三层（L3）路由

表 5-2 简要说明了与成员接口相关的字段信息。

表 5-2　　　　　　　　　　　端口通道的成员接口状态字段

字段	描述
P	成员接口主动参与并转发该端口通道流量
H	端口通道配置了最大活动接口数，虽然该接口与远程对等体通过 LACP 进行了协商，但仅作热备份，并不转发流量。该端口通道接口配置了命令 **lacp max-bundle** *number-member-interfaces*
I	成员接口被视为单个接口且未在接口上检测到任何 LACP 活动
w	该字段表示从邻居接收数据包所剩余的时间，以确保其仍然处于活动状态
s	成员接口处于挂起状态
r	与该接口相关联的交换模块已从机箱中卸下

可以利用命令 **show interface port-channel** *port-channel-id* 查看逻辑接口信息，除成员接口之外，输出结果将显示传统以太网接口的相关数据字段，带宽反映了所有活动成员接口的合并吞吐量。这些信息发生变化之后，QoS 策略和路由协议的接口开销等因素也会做出相应调整。

例 5-3 给出了 NX-1 运行该命令后的输出结果示例。可以看出，端口通道的带宽为 20Gbit/s，且与端口通道接口中的两个 10Gbit/s 接口相关。

例 5-3 查看端口通道接口的状态

```
NX-1# show interface port-channel 1
! Output omitted for brevity
port-channel1 is up
admin state is up,
  Hardware: Port-Channel, address: 885a.92de.6158 (bia 885a.92de.6158)
  MTU 1500 bytes, BW 20000000 Kbit, DLY 10 usec
  reliability 255/255, txload 1/255, rxload 1/255
  Encapsulation ARPA, medium is broadcast
  Port mode is trunk
  full-duplex, 10 Gb/s
  Input flow-control is off, output flow-control is off
  Auto-mdix is turned off
  Switchport monitor is off
  EtherType is 0x8100
  Members in this channel: Eth1/1, Eth1/2
..
```

5.1.3 验证 LACP 数据包

NX-OS 的日志记录为识别配置不兼容问题提供了很多有用的系统日志消息。排查端口通道建立故障的重要步骤就是验证设备之间是否传输了 LACP 数据包。故障排查的第一步就是利用命令 **show lacp counters** [**interface port-channel** *port-channel-number*]验证 LACP 计数器。

该命令的输出结果包含了端口通道接口及其相关联的成员接口、发送/接收的 LACP 数据包计数器以及各种差错信息。接口的 Send 和 Recv 列应该在一定的时间间隔内不断增加。如果计数器不增加，那么就表明出现了问题。问题的原因可能与物理链路或与远程设备的配置不完整/不兼容有关。检查设备的 LACP 计数器，就可以确定该设备是否正在发送 LACP 数据包。

例 5-4 给出了该命令的输出结果示例。可以看出，port-channel 1 中的 Ethernet1/2 的 Recv 列没有递增，但 Sent 列却在递增。

例 5-4 查看 LACP 数据包计数器

```
NX-1# show lacp counters
NOTE: Clear lacp counters to get accurate statistics

-------------------------------------------------------------------------------
                        LACPDUs              Markers/Resp LACPDUs
Port            Sent        Recv             Recv Sent    Pkts Err
-------------------------------------------------------------------------------
port-channel1
Ethernet1/1     5753        5660             0    0       0
Ethernet1/2     5319        0                0    0       0

NX-1# show lacp counters
NOTE: Clear lacp counters to get accurate statistics

-------------------------------------------------------------------------------
```

```
                          LACPDUs                   Markers/Resp LACPDUs
Port            Sent            Recv                Recv Sent   Pkts Err
--------------------------------------------------------------------------------
port-channel1
Ethernet1/1     5755            5662                0    0      0
Ethernet1/2     5321            0                   0    0      0
```

另一种方法是使用命令 **show lacp internal info interface** *interface-id*。该命令可以显示接口最后一次发送或收到数据包时的时间戳，如例 5-5 所示。

例 5-5　查看接口上的 LACP 传输时间戳

```
NX-1# show lacp internal info interface ethernet 1/1
Interface Ethernet1/1(0x1a030000) info
---------------------------------------
  port_pr 0x8000
  rid type IF-Rid: ifidx 0x1a030000: ch_num 0
  cfg_pc_if_idx 0x16000000: oper_pc_if_idx 0x16000000
  is state_change_notif_pending 0
  lacp detected link down 0
  lag [(8000, 0-62-ec-9d-c5-0, 1, 8000, 11c), (8000, 88-5a-92-de-61-7c, 8000,
  8000, 131)]
  aggr_id 0x0
  LACP last pkt sent at      : Mon Oct 23 03:50:41 2017, 153348 usecs
      LACP PDU sent at       : Mon Oct 23 03:50:41 2017, 153348 usecs
      MARKER RESP sent at: None.
      ERROR PDU sent at      : None.
  LACP last pkt recv at      : Mon Oct 23 03:50:35 2017, 864017 usecs
      LACP PDU recv at       : Mon Oct 23 03:50:35 2017, 864017 usecs
      MARKER PDU recv at : None.
      ERROR PDU recv at      : None.
```

命令 **show lacp neighbor [interface port-channel** *port-channel-number*]能够以链路为基础显示端口通道接口、成员接口以及远程设备的更多额外信息。

例 5-6 给出了该命令的输出结果示例。可以看出，输出结果包含了邻居的 System-ID、系统优先级、远程端口号、远程端口优先级以及是否正在使用快速或慢速 LACP 数据包间隔等详细信息。

例 5-6　查看 LACP 邻居信息

```
NX-1# show lacp neighbor
Flags:  S - Device is sending Slow LACPDUs F - Device is sending Fast LACPDUs
        A - Device is in Active mode        P - Device is in Passive mode
port-channel1 neighbors
Partner's information
! The following section provides the remote neighbors system priority, system-id,
! port number, LACP state, and fast/slow LACP interval in that order.
            Partner                 Partner                     Partner
Port        System ID               Port Number      Age        Flags
Eth1/1      32768,18-9c-5d-11-99-80 0x138            985        SA
! The following section includes the remote device LACP port-priority
            LACP Partner            Partner                     Partner
            Port Priority           Oper Key                    Port State
            32768                   0x1                         0x3d


Partner's information
            Partner                 Partner                     Partner
Port        System ID               Port Number      Age        Flags
```

```
Eth1/2       32768,18-9c-5d-11-99-800x139           985           SA

             LACP Partner              Partner                Partner
             Port Priority             Oper Key               Port State
             32768                     0x1                    0x3d
```

> **注**：可以利用 LACP System-ID 来验证成员接口是否连接在相同设备上（而不是拆分到不同的设备上），可以使用 **show lacp system-identifier** 命令查看本地 LACP System-ID。

可以通过 NX-OS Ethanalyzer 工具查看本地 Nexus 交换机发送和接收的 LACP 数据包，实现方式是抓取携带 LACP MAC 目的地址的数据包。命令 **ethanalyzer local interface inband capture-filter "ether host 0180.c200.0002" [detail]** 可以抓取接收到的 LACP 数据包。可选关键字 **detail** 可以提供更多有用信息，如例 5-7 所示。

例 5-7　利用 Ethanalyzer 抓取 LACP 数据包

```
NX-1# ethanalyzer local interface inband capture-filter "ether host 0180.c200.0002"
Capturing on inband
2017-10-23 03:58:11.213625 88:5a:92:de:61:58 -> 01:80:c2:00:00:02 LACP Link Aggr
egation Control Protocol
2017-10-23 03:58:11.869668 88:5a:92:de:61:59 -> 01:80:c2:00:00:02 LACP Link Aggr
egation Control Protocol
2017-10-23 03:58:23.381249 00:62:ec:9d:c5:1c -> 01:80:c2:00:00:02 LACP Link Aggr
egation Control Protocol
2017-10-23 03:58:24.262746 00:62:ec:9d:c5:1b -> 01:80:c2:00:00:02 LACP Link Aggr
egation Control Protocol
2017-10-23 03:58:41.218262 88:5a:92:de:61:58 -> 01:80:c2:00:00:02 LACP Link Aggr
egation Control Protocol
```

5.1.4　LACP 高级配置选项

接下来将介绍一些 LACP 高级配置选项以及这些配置选项对端口通道成员接口选择所产生的影响。

1. 端口通道成员接口的最低数量

仅当一个及以上成员接口与远程设备成功建立了 LACP 邻接关系之后，端口通道接口才处于活动状态并正常启动。某些设计方案要求端口通道接口在变为活动状态之前，必须建立最低数量的 LACP 邻接关系，可以通过端口通道接口命令 **lacp min-links** 配置该选项。

例 5-8 将端口通道接口的最小数量设置为 2，然后关闭了 NX-1 上的某个成员接口，可以看出，此时的端口通道处于"Not in use"状态。

例 5-8　配置端口通道成员接口的最低数量

```
NX-1# conf t
NX-1(config)# interface port-channel 1
NX-1(config-if)# lacp min-links 2
NX-1(config-if)# interface Eth1/1
NX-1(config-if)# shut
04:22:45 NX-1 %$ VDC-1 %$ %ETH_PORT_CHANNEL-5-PORT_DOWN: port-channel1: Ethernet1/1
  is down
04:22:45 NX-1 %$ VDC-1 %$ %ETHPORT-5-IF_DOWN_CFG_CHANGE: Interface Ethernet1/1 is
  down(Config change)
04:22:45 NX-1 %$ VDC-1 %$ %ETHPORT-5-IF_DOWN_ADMIN_DOWN: Interface Ethernet1/1 is
```

```
  down (Administratively down)
04:22:47 NX-1 %$ VDC-1 %$ %ETHPORT-5-IF_DOWN_PORT_CHANNEL_MEMBERS_DOWN: Interface
  port-channel1 is down (No operational members)
04:22:47 NX-1 %$ VDC-1 %$ %ETH_PORT_CHANNEL-5-PORT_DOWN: port-channel1: Ethernet1/2
  is down
04:22:47 NX-1 %$ VDC-1 %$ %ETH_PORT_CHANNEL-5-FOP_CHANGED: port-channel1: first
  operational port changed from Ethernet1/2 to none
04:22:47 NX-1 %$ VDC-1 %$ %ETHPORT-5-IF_DOWN_INITIALIZING: Interface Ethernet1/2 is
  down (Initializing)
04:22:47 NX-1 %$ VDC-1 %$ %ETHPORT-5-IF_DOWN_PORT_CHANNEL_MEMBERS_DOWN: Interface
  port-channel1 is down (No operational members)
04:22:47 NX-1 %$ VDC-1 %$ %ETHPORT-5-SPEED: Interface port-channel1, operational
  speed changed to 10 Gbps
04:22:47 NX-1 %$ VDC-1 %$ %ETHPORT-5-IF_DUPLEX: Interface port-channel1, operational
  duplex mode changed to Full
04:22:47 NX-1 %$ VDC-1 %$ %ETHPORT-5-IF_RX_FLOW_CONTROL: Interface port-channel1,
  operational Receive Flow Control state changed to off
04:22:47 NX-1 %$ VDC-1 %$ %ETHPORT-5-IF_TX_FLOW_CONTROL: Interface port-channel1,
  operational Transmit Flow Control state changed to off
04:22:50 NX-1 %$ VDC-1 %$ %ETH_PORT_CHANNEL-5-PORT_SUSPENDED: Ethernet1/2:
  Ethernet1/2 is suspended

NX-1# show port-channel summary
Flags:  D - Down        P - Up in port-channel (members)
        I - Individual  H - Hot-standby (LACP only)
        s - Suspended   r - Module-removed
        S - Switched    R - Routed
        U - Up (port-channel)
        p - Up in delay-lacp mode (member)
        M - Not in use. Min-links not met
--------------------------------------------------------------------------------
Group Port-       Type     Protocol  Member Ports
      Channel
--------------------------------------------------------------------------------
1     Po1(SM)     Eth      LACP      Eth1/1(D)   Eth1/2(s)
```

注：虽然不需要在两台设备上同时配置端口通道成员接口的最低数量，也能实现正常工作。但是，为了加快故障排查速度，并为操作人员提供有效帮助，建议同时在两台交换机上配置该选项。

2．端口通道成员接口的最大数量

可以为端口通道配置其所能拥有的最多成员接口数量。一种常见的设计方案就是确保活动成员接口的数量保持为 2 的 N 次方（2、4、8、16），以适应负载均衡散列。可以通过端口通道接口配置命令 **lacp max-bundle** *max-links* 配置端口通道中的最大成员接口数。

例 5-9 给出了配置端口通道最大活动成员接口数的配置示例，这些接口目前均显示处于"Hot-Standby"（热备份）状态。

例 5-9 配置并验证最大链路数

```
NX-1# configure terminal
Enter configuration commands, one per line. End with CNTL/Z.
NX-1(config)# interface port-channel 2
NX-1(config-if)# lacp max-bundle 4
04:44:04 NX-1 %$ VDC-1 %$ %ETH_PORT_CHANNEL-5-PORT_DOWN: port-channel2: Ethernet1/7
  is down
04:44:04 NX-1 %$ VDC-1 %$ %ETHPORT-5-IF_DOWN_INITIALIZING: Interface Ethernet1/7 is
  down (Initializing)
```

```
04:44:04 NX-1 %$ VDC-1 %$ %ETH_PORT_CHANNEL-5-PORT_DOWN: port-channel2: Ethernet1/8
  is down
04:44:04 NX-1 %$ VDC-1 %$ %ETHPORT-5-IF_DOWN_INITIALIZING: Interface Ethernet1/8 is
  down (Initializing)
04:44:06 NX-1 %$ VDC-1 %$ %ETH_PORT_CHANNEL-5-PORT_HOT_STANDBY: port-channel2: Ethernet1/
  7 goes to hot-standby
04:44:06 NX-1 %$ VDC-1 %$ %ETH_PORT_CHANNEL-5-PORT_HOT_STANDBY: port-channel2: Ethernet1/
  8 goes to hot-standby
NX-1# show port-channel summary
Flags:   D - Down          P - Up in port-channel (members)
         I - Individual    H - Hot-standby (LACP only)
         s - Suspended     r - Module-removed
         S - Switched      R - Routed
         U - Up (port-channel)
         p - Up in delay-lacp mode (member)
         M - Not in use. Min-links not met
--------------------------------------------------------------------------------
Group Port-         Type     Protocol  Member Ports
      Channel
--------------------------------------------------------------------------------
1     Po1(SU)       Eth      LACP      Eth1/1(P)    Eth1/2(P)
2     Po2(SU)       Eth      LACP      Eth1/3(P)    Eth1/4(P)    Eth1/5(P)
                                       Eth1/6(P)    Eth1/7(H)    Eth1/8(H)
```

虽然可以仅在端口通道的一台交换机上配置端口通道成员接口的最大数量，但是，为了加快故障排查速度并为操作人员提供帮助，建议同时在两台交换机配置该选项。交换机的接口显示为"suspended"（挂起）状态。

端口通道的主交换机可以通过检查 LACP 端口优先级来控制哪些成员接口（及相关联的链路）处于活动状态。优选端口优先级较低的接口，如果端口优先级相同，那么就优选接口号较小的接口。

5.1.5 LACP 系统优先级

LACP 系统优先级负责确定哪台交换机是端口通道的主交换机。如果成员接口的数量大于与端口通道接口相关联的最大成员接口数量，那么端口通道的主交换机就要确定端口通道中的哪些成员接口处于活动状态，优选系统优先级较小的交换机。可以使用全局命令 **lacp system-priority** *priority* 更改 LACP 系统优先级。

例 5-10 给出了验证和更改 LACP 系统优先级的配置方式。

例 5-10 查看和更改 LACP 系统优先级

```
NX-1# show lacp system-identifier
32768,88-5a-92-de-61-7c

NX-1# configuration t
Enter configuration commands, one per line. End with CNTL/Z.
NX-1(config)# lacp system-priority 1

NX-1# show lacp system-identifier
1,88-5a-92-de-61-7c
```

1. LACP 接口优先级

如果成员接口的数量大于端口通道的最大成员接口数，那么主交换机就可以通过 LACP 端口的接口优先级选择端口通道中的哪些成员接口处于活动状态，优选端口优先级较小的端口。可以

通过接口配置命令 **lacp port-priority** *priority* 设置接口优先级。

例 5-11 更改了 NX-1 Eth1/8 的端口优先级，使其成为首选接口。由于 NX-1 是 port-channel 2 的主交换机，因而端口 Eth1/8 变为活动状态，端口 Eth1/6 和 Eth1/7 处于热备状态（因为前面将最大链路数设置为 4）。

例 5-11　更改 LACP 端口优先级

```
NX-1(config)# interface e1/8
NX-1(config-if)# lacp port-priority 1
05:00:08 NX-1 %$ VDC-1 %$ %ETH_PORT_CHANNEL-5-PORT_DOWN: port-channel2: Ethernet1/6
  is down
05:00:08 NX-1 %$ VDC-1 %$ %ETHPORT-5-IF_DOWN_INITIALIZING: Interface Ethernet1/6 is
  down (Initializing)
05:00:12 NX-1 %$ VDC-1 %$ %ETH_PORT_CHANNEL-5-PORT_UP: port-channel2: Ethernet1/8
  is up
05:00:12 NX-1 %$ VDC-1 %$ %ETHPORT-5-IF_UP: Interface Ethernet1/8 is up in mode
  trunk
```

```
NX-1# show port-channel summary
Flags:  D - Down         P - Up in port-channel (members)
        I - Individual   H - Hot-standby (LACP only)
        s - Suspended    r - Module-removed
        S - Switched     R - Routed
        U - Up (port-channel)
        p - Up in delay-lacp mode (member)
        M - Not in use. Min-links not met
--------------------------------------------------------------------------------
Group Port-       Type     Protocol  Member Ports
      Channel
--------------------------------------------------------------------------------
1     Po1(SU)     Eth      LACP      Eth1/1(P)   Eth1/2(P)
2     Po2(SU)     Eth      LACP      Eth1/3(P)   Eth1/4(P)   Eth1/5(P)
                                     Eth1/6(s)   Eth1/7(s)   Eth1/8(P)
```

2．LACP 快速模式

最早的 LACP 标准每 30s 发送一次 LACP 数据包，如果在 3 个时间间隔内都未收到 LACP 数据包，那么就认为该链路不可用。因此，从端口通道中删除成员接口之前，链路可能会出现 90s 的丢包现象。

后来对该标准做出了优化调整，即每秒钟通告一次 LACP 数据包。与原始 LACP 标准的 90s 相比，这种模式可以在 3s 内识别和删除故障链路，因而称为 LACP 快速模式。可以通过接口配置命令 **lacp rate fast** 在成员接口上启用 LACP 快速模式。

注：为了成功建立端口通道，必须同时将两台交换机的所有接口都配置为 LACP 快速或 LACP 慢速模式。

注：使用 LACP 快速模式时，请检查各平台的版本说明，以确保支持 ISSU（In-Service Software Upgrade，不中断软件升级）和平滑切换。

例 5-12 解释了如何在本地和邻居接口上识别当前 LACP 状态以及将接口转换为 LACP 快速模式的方式。

例 5-12　配置 LACP 快速模式并验证 LACP 的速度状态

```
NX-1# show lacp interface ethernet1/1 | i Timeout|Local|Neighbor
Local Port: Eth1/1   MAC Address= 88-5a-92-de-61-7c
  LACP_Timeout=Long Timeout (30s)
```

```
  Partner information refresh timeout=Long Timeout (90s)
Neighbor: 0x11c
  LACP_Timeout=Long Timeout (30s)
NX-1# conf t
Enter configuration commands, one per line. End with CNTL/Z.
NX-1(config)# interface Eth1/1
NX-1(config-if)# lacp rate fast
NX-1# show lacp interface ethernet1/1 | i Timeout|Local|Neighbor
Local Port: Eth1/1   MAC Address= 88-5a-92-de-61-7c
  LACP_Timeout=Short Timeout (1s)
  Partner information refresh timeout=Long Timeout (90s)
Neighbor: 0x11c
  LACP_Timeout=Long Timeout (30s)
```

3. 平滑收敛

Nexus 交换机默认通过端口通道接口命令 **lacp graceful-convergence** 启用 LACP 平滑收敛特性。Nexus 交换机连接非思科对等设备时，其默认的平滑故障切换特性可能会延迟关闭禁用端口的时间。

另一种场景是与不完全支持 LACP 规范的设备建立 LACP 邻接关系。例如，不兼容的 LACP 设备在收到 Sync LACP 消息（建立 LACP 邻接关系的第 2 步）之后，可能会在向对等体发送 Collecting LACP 消息之前就开始发送数据。由于本地交换机仍未达到 Collecting（收集）状态，因而将丢弃这些数据包。

该解决方案在连接不兼容的 LACP 设备时，会通过 **no lacp graceful-convergence** 命令删除端口通道接口的 LACP 平滑收敛特性。此后，Nexus 交换机向对等体发送 Sync LACP 消息之前，会等待更长的时间以进行端口初始化，这样就能确保端口在发送 Sync LACP 消息之后即可接收数据包。

4．挂起单个端口

在默认情况下，如果 Nexus 交换机未收到对等体发来的 LACP PDU，那么就会将该 LACP 端口置于挂起状态。一般来说，该行为有助于防止因交换机配置差错而引发的环路问题。但是，该行为也可能会导致某些要求在 LACP 逻辑上启动端口的服务器出现问题。

可以通过端口通道接口命令 **no lacp suspend-individual** 禁用该功能，从而改变该行为。

5.1.6　端口通道成员接口的一致性

由于端口通道是逻辑接口，因而所有成员接口都必须拥有相同的特性。成员接口的以下选项必须匹配。

- **端口类型**：必须将接口中的所有端口都一致性地配置为 L2 交换端口或 L3 路由端口。差错消息 "port not compatible [Ethernet Layer]"（端口不兼容[以太层]）表示的就是该故障状态。
- **端口模式**：必须将 L2 端口通道配置为接入端口或中继端口，不能混用。差错消息 "port not compatible [port mode]"（端口不兼容[端口模式]）表示的就是该故障状态。
- **本征 VLAN**：必须通过命令 **switchport trunk native vlan** *vlan-id* 为 L2 中继端口通道上的成员接口配置相同的本征 VLAN。否则将会显示差错消息 "port not compatible [port native VLAN]"（端口不兼容[端口本征 VLAN]）。
- **允许的 VLAN**：必须通过命令 **switchport trunk allow** *vlan-ids* 为 L2 中继端口通道上的成员接口配置相同的允许 VLAN。否则将会显示差错消息 "port not compatible [port allowed VLAN list]"（端口不兼容[端口允许的 VLAN 列表]）。
- **速率**：所有成员接口的速率必须相同。否则就会将接口置于挂起状态，同时显示系统日志消息 "%ETH_PORT_CHANNEL-5-IF_DOWN_SUSPENDED_BY_SPEED"。

- **双工模式**：所有成员接口的双工模式必须相同。否则将会显示系统日志消息"command failed: port not compatible [Duplex Mode]"（命令失败：端口不兼容[双工模式]）。这一点仅适用于 100 Mbit/s 及以下速率的接口。
- **MTU**：必须为所有 L3 成员接口配置相同的 MTU（Maximum Transmission Unit，最大传输单元）。如果某接口的 MTU 与其他成员接口不匹配，那么就无法将该接口添加到端口通道中，此时将显示系统日志消息"command failed: port not compatible [Ethernet Layer]"（命令失败：端口不兼容[以太层]）。该差错消息与端口类型的差错消息相同，需要检查成员接口的详细配置以识别不匹配的 MTU。
- **加载间隔**：必须为所有成员接口配置相同的加载间隔。否则将显示系统日志消息"command failed: port not compatible [load interval]"（命令失败：端口不兼容[加载间隔]）。
- **风暴控制**：必须为端口通道的所有成员端口配置相同的风暴控制机制。否则将出现系统日志消息"port not compatible [Storm Control]"（端口不兼容[风暴控制]）。

注：命令 **show port-channel compatible-parameters** 可以显示完整的必须匹配的兼容性参数列表。

一般来说，为 Nexus 交换机配置端口通道时，需要将成员接口设置为适当的交换端口类型（L2 或 L3），然后再将接口与端口通道关联起来，其余端口通道配置操作则通过端口通道接口来完成。

如果出现了一致性差错，那么就可以利用 **show port-channel summary** 命令定位成员接口、查看成员接口配置，并将正确配置应用于希望加入端口通道组的接口。

5.1.7 排查 LACP 接口建立故障

排查 Nexus 交换机的端口通道接口建立故障时，应检查以下内容。
- 确保链路仅位于两台设备之间。
- 确认成员端口均处于活动状态。
- 确定两端链路均静态设置为"on"或已启用 LACP，且至少有一侧设置为"active"。
- 确保所有成员接口的端口配置均一致（LACP 端口优先级除外）。
- 验证两台设备发送和接收的 LACP 数据包。

5.1.8 排查流量负载均衡故障

流经端口通道接口的流量并不是按照逐个数据包轮询的方式通过成员链路向外转发，而是根据数据包报头字段计算散列值，并根据散列结果通过链路转发数据包。负载均衡散列选项是系统范围的配置参数，可以使用全局命令 **port-channel load-balance ether** *hash* 进行配置，其中的选项 *hash* 支持以下可选关键字。
- **destination-ip**：目的 IP 地址。
- **destination-mac**：目的 MAC 地址。
- **destination-port**：目的 TCP/UDP 端口。
- **source-dest-ip**：源和目的 IP 地址（包括 L2）。
- **source-dest-ip-only**：仅源和目的 IP 地址。
- **source-dest-mac**：源和目的 MAC 地址。
- **source-dest-port**：源和目的 TCP/UDP 端口（包括 L2 和 L3）。
- **source-dest-port-only**：仅源和目的 TCP/UDP 端口。
- **source-ip**：源 IP 地址。
- **source-mac**：源 MAC 地址。

- **source-port**：源 TCP/UDP 端口。

端口通道中的某些成员链路的利用率可能高于其他链路，这种情况是可能存在的，取决于端口通道的配置以及流经端口通道的流量。

命令 **show port-channel traffic [interface port-channel** *port-channel-number*]可以显示所有成员接口以及穿越该成员接口的流量负荷，如例 5-13 所示。

例 5-13　查看成员接口的流量负荷

```
NX-1# show port-channel traffic
NOTE: Clear the port-channel member counters to get accurate statistics

ChanId      Port Rx-Ucst Tx-Ucst Rx-Mcst Tx-Mcst Rx-Bcst Tx-Bcst
------ --------- ------- ------- ------- ------- ------- -------
    1     Eth1/1  98.68%  66.66%   4.08%  85.12%  70.95%    0.0%
    1     Eth1/2   1.31%  33.33%  95.91%  14.87%  29.04% 100.00%
```

可以通过命令 **show port-channel load-balance** 查看负载均衡散列选项（见例 5-14），默认的系统散列选项是 *source-dest-ip*，即根据数据包报头中的源 IP 地址和目的 IP 地址计算散列值。

例 5-14　查看端口通道散列算法

```
NX-1# show port-channel load-balance

Port Channel Load-Balancing Configuration:
System: source-dest-ip

Port Channel Load-Balancing Addresses Used Per-Protocol:
Non-IP: source-dest-mac
IP: source-dest-ip
```

如果成员链路的流量分发不均匀，那么更改散列值可能会在成员链路之间提供不同的分配比率。例如，如果与路由器建立了端口通道，那么将 MAC 地址作为散列算法的一部分就可能会影响流量流，因为路由器的 MAC 地址保持不变（源或目的 MAC 地址始终是路由器的 MAC 地址），此时一种好的方式是使用源/目的 IP 地址或者会话端口进行散列计算。

注： 必须以 2 的 *N* 次方（2、4、8、16）方式将成员链路添加到端口通道中，以确保散列计算的一致性。

在极少数情况下，故障排查操作可能需要确定数据包穿越端口通道中的具体成员链路，处理随机丢包故障时还可能需要做进一步诊断（包括光纤、ASIC 等）。可以通过命令 **show port-channel load-balance** [**forwarding-path interface port-channel** *number* { . | **vlan** *vlan_ID* } [**dst-ip** *ipv4-addr*] [**dst-ipv6** *ipv6-addr*] [**dst-mac** *dst-mac-addr*] [**l4-dst-port** *dst-port*] [**l4-src-port** *src-port*] [**src-ip** *ipv4-addr*] [**src-ipv6** *ipv6-addr*] [**src-mac** *src-mac-addr*]]识别成员链路。

例 5-15 解释了如何在 NX-1 上识别从 192.168.2.2 到 192.168.1.1 的数据包所穿越的 port-channel 1 的成员链路。

例 5-15　识别特定网络流量穿越的成员链路

```
NX-1# show port-channel load-balance forwarding-path interface port-channel 1 srcinterface
   ethernet 1/1 dst-ip 192.168.1.1 src-ip 192.168.2.2
Missing params will be substituted by 0's.
       Outgoing port id: Ethernet1/1
Param(s) used to calculate load-balance:
       dst-ip:    192.168.1.1
       src-ip:    192.168.2.2
```

```
            dst-mac: 0000.0000.0000
            src-mac: 0000.0000.0000
            VLAN: 0
```

5.2 vPC

端口通道为网络设计带来了很多好处，但是只能使用两台设备（一台本地设备和一台远程设备）。NX-OS 提供了一项名为 vPC（virtual Port-Channel，虚拟端口通道）的功能特性，该功能特性允许两台 Nexus 交换机在 vPC 域中创建虚拟交换机，此后，vPC 对等体就可以为远程设备提供一个逻辑的二层（L2）端口通道。

图 5-4 给出了 vPC 的拓扑结构示例。NX-2 和 NX-3 是同一个 vPC 域的成员，均配置了 vPC，为连接 NX-1 提供了逻辑端口通道。从 NX-1 的角度来看，NX-1 仅连接了一台交换机。

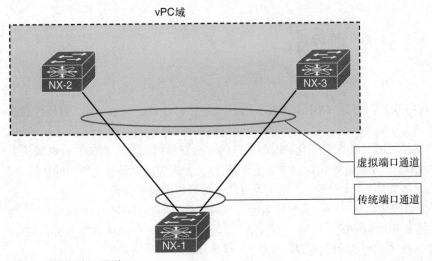

图 5-4 虚拟端口通道

> **注：** 与交换机堆栈或 VSS（Virtual Switching System，虚拟交换系统）集群技术不同，vPC 中的各个交换机端口的配置仍然保持独立，也就是说，Nexus 交换机的配置是独立的。

5.2.1 vPC 基础

vPC 域只允许两台 Nexus 交换机。vPC 的功能特性包含了 vPC 对等保活链路（vPC peer-keepalive link）、vPC 成员链路以及实际的 vPC 接口等组件（见图 5-5）。

1. vPC 域

Nexus 交换机可以同时拥有常规端口通道和 vPC 接口。端口通道与 vPC 接口之间的 LACP 通告使用不同的 LACP System-ID，两台 Nexus 对等交换机均为 vPC 成员链路使用虚拟 LACP System-ID。

vPC 域中的一台交换机是主设备（Primary device），另一台交换机是从设备（Secondary device）。Nexus 交换机选择角色优先级较低的交换机作为主设备，如果优先级相同，那么就优选 MAC 地址较小的 Nexus 交换机。由于 vPC 在确定主设备时不会出现抢占行为，因而引入了运行中的主设备（Operational primary device）和运行中的从设备（Operational secondary device）概念。

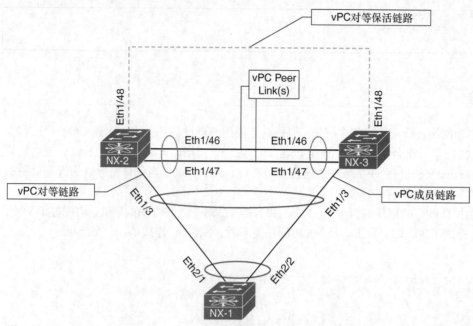

图 5-5 vPC 组件

为了更好地解释这些概念，假设 NX-2 和 NX-3 位于同一个 vPC 域中，且 NX-2 的角色优先级较低。

- **第 1 步**：两台交换机刚启动并初始化的时候，任何一台交换机都没有被选作 vPC 域的主设备。此后，NX-2 成为主设备和运行中的主设备，NX-3 成为从设备和运行中的从设备。
- **第 2 步**：重新加载 NX-2，此时 NX-3 将成为主设备和运行中的主设备。
- **第 3 步**：NX-2 完成初始化之后，虽然其角色优先级较低，但是并不会抢占 NX-3。此时的 NX-2 是主设备和运行中的从设备，而 NX-3 是从设备和运行中的从设备。仅当 NX-3 重新加载或关闭所有 vPC 接口的时候，NX-2 才成为运行中的主设备。

2. vPC 对等保活

vPC 对等保活链路负责监视 vPC 对等设备的运行状况，定期发送保活消息（系统默认周期为 1s），心跳包大小为 96 字节，使用 UDP 端口 3200。如果对等链路出现故障，那么就会检查跨 vPC 对等链路的连通性。由于跨对等保活链路的网络流量并不大，因而使用 1Gbit/s 接口。

如果没有收到任何保活消息，那么 vPC 对等设备就认为检测到了对等体故障。如果认为 vPC 对等体不可用，那么就会立即启动一个保持超时定时器（hold-timeout timer）。在保持超时期间（系统默认值为 5s），从 vPC 设备将忽略所有 vPC 保活消息，以确保在对 vPC 接口采取处理措施之前网络能够收敛。保持超时时间到期后，就会立即启动超时定时器（系统默认值为 3s）。如果在这个时间间隔内未收到 vPC 保活消息，那么就会关闭从 vPC 交换机上的 vPC 接口，该行为可以防止出现脑裂（split-blain）现象。

> 注：虽然从技术上来说将 VLAN 接口用作对等保活接口也可行，但实际上并不建议这么做，因为这样做可能会引发混乱。此外，对等保活链路应尽可能直连（管理端口除外）。

3. vPC 对等链路

vPC 对等链路负责在设备之间同步状态并转发数据。例如，假设服务器已连接 NX-1，且正与连接在 NX-2 上的主机进行通信。由于 NX-1 的端口通道执行散列处理，因而流量通过 Ethernet2/2 链路向外发送给 NX-3。NX-3 通过 vPC 对等链路将数据包转发给 NX-2，因而 NX-2

可以将流量转发给直连主机。

vPC 对等链路必须使用 10Gbit/s 或更高速率的以太网端口。为了确保有足够带宽可以将 vPC 对等体发送的流量重定向到适当的远程 vPC 对等体，通常都要使用端口通道。此外，对于模块化的 Nexus 交换机来说，应该将对等链路分散到不同的线卡/模块上，以确保对等链路在交换机硬件出现故障时仍能保持正常运行。

4．vPC 成员链路

vPC 成员链路是 vPC 域中的 Nexus 交换机上的各条独立链路。端口通道标识符与 vPC 成员端口相关联。

5．vPC 操作行为

NX-OS 通过以下 3 个方面的调整来修改 STP 的操作行为。

- vPC 对等链路永远不会进入阻塞状态。
- 仅运行中的主 Nexus 交换机生成和处理 BPDU，运行中的从 Nexus 交换机通过对等链路将收到的 BPDU 转发给运行中的主 Nexus 交换机。
- vPC 成员端口永远也不会向外宣告从 vPC 对等链路上收到的流量，这是环路预防机制的一部分。

网络设备运行了 HSRP（Hot Standby Router Protocol，热备路由器协议）之后，可以为网段上的主机提供容错型的虚拟 IP。启用 HSRP 之后，只有一台网络设备可以转发虚拟 IP 的流量。不过，对于某些部署了 vPC 的 Nexus 平台来说，两台 Nexus 交换机都能转发虚拟网关的流量，这样就能有效提高带宽并减少通过 vPC 对等链路发送的三层（L3）网络流量。

5.2.2 vPC 配置

vPC 的配置步骤如下。

- **第 1 步**：启用 vPC 功能特性。必须使用命令 **feature vpc** 启用 vPC 功能特性。
- **第 2 步**：启用 LACP 功能特性。vPC 端口通道需要使用 LACP，因而必须使用命令 **feature lacp** 启用 LACP 功能特性。
- **第 3 步**：配置对等保活链路。必须配置对等保活链路，思科建议为对等保活链路创建专用 VRF（Virtual Routing and Forwarding，虚拟路由和转发），然后使用命令 **ip address** *ip-address mask* 将 IP 地址与接口相关联。

> 注：虽然可以为对等保活链路使用管理接口，但这样做需要管理交换机在对等设备之间提供连接。如果系统拥有多块管理引擎卡（如 Nexus 7000/9000），那么每个 vPC 对等体上的主用和备用管理端口都必须连接在管理交换机上。

- **第 4 步**：配置 vPC 域。vPC 域是两个 Nexus 对等体都使用的逻辑区域，可以通过命令 **vpc domain** *domain-id* 创建 vPC 域。两台设备的域 ID 必须匹配。

  在 vPC 域上下文中，必须使用命令 **peer-keepalive destination** *remote-nexus-ip* [**hold-timeout** *secs* | **interval** *msecs* {**timeout** *secs*} | **source** *local-nexus-ip* | **vrf** *name*]标识保活接口。虽然源接口可选，但仍然建议在配置过程中静态分配源接口。可以通过可选关键字 **hold-timeout**、**interval** 和 **timeout** 分别配置保活宣告间隔、保持超时定时器和超时值。

  虽然 NX-OS 可以自动为 LACP 消息创建 vPC 系统 MAC 地址，但是也可以通过 **system-mac** *mac-address* 命令定义 MAC 地址。vPC 域的 LACP 系统优先级是 32768，但是也可以通过命令 **system-priority** *priority* 进行修改，以增大或减小 LACP 优先级。

- **第 5 步**：配置 vPC 设备优先级（可选）。vPC 设备优先级的配置命令是 **role priority**，其中，*priority* 的取值范围是 1~65535，值越小越优。设备优先级较优的节点是主 vPC 节点，另一个节点则是从节点。

- **第 6 步**：配置 vPC 系统优先级（可选）。两台交换机之间的常规端口通道协商必须标识主交换机，这个概念也同样适用于 vPC 接口。可以通过 vPC 域配置命令 **system-priority** *priority* 配置 vPC LACP 系统优先级。

- **第 7 步**：配置 vPC 自动恢复链路（可选但建议配置）。作为一种安全机制，vPC 对等体在检测到另一个 vPC 对等体之前不会启用任何 vPC 接口。在某些故障情况下（如电源故障），两台 vPC 设备可能会出现同时重启情形，致使无法检测到对方。由于此时两台设备均不转发流量，因而可能会出现流量丢失问题。

  vPC 自动恢复功能可以让其中的一个 vPC 对等体继续转发流量。初始化之后，如果 vPC 对等链路断开，且连续 3 条对等保活消息都没有收到响应消息，那么从设备就将承担运行中的主设备角色并初始化 vPC 接口，从而能够转发部分流量。有关 vPC 自动恢复特性的详细内容将在本章稍后进行讨论。

  可以通过 vPC 域配置命令 **auto-recovery [reload-delay** *delay*]启用该功能。启用该功能之前，默认时延为 240s，不过也可以通过可选关键字 *reload-delay* 更改时延值，取值范围是 240~3600。

- **第 8 步**：配置 vPC。可以通过命令 **channel-group** *port-channel-number* **mode active** 将端口分配给端口通道。通过命令 **vpc** *vpc-id* 为端口通道接口分配唯一的 vPC 标识符，*vpc-id* 需要与远程对等设备相匹配。

例 5-16 给出了图 5-5 中的 NX-2 的 vPC 配置示例。

例 5-16　vPC 配置

```
NX-2# configuration t
Enter configuration commands, one per line. End with CNTL/Z.
! Enable the vPC and LACP features
NX-2(config)# feature vpc
NX-2(config)# feature lacp
! Creation of the vPC Peer-KeepAlive VRF and association of IP address
NX-2(config)# vrf context VPC-KEEPALIVE
NX-2(config-vrf)# address-family ipv4 unicast
NX-2(config-vrf-af-ipv4)# interface Ethernet1/48
NX-2(config-if)# description vPC-KeepAlive
NX-2(config-if)# no switchport
NX-2(config-if)# vrf member VPC-KEEPALIVE
NX-2(config-if)# ip address 192.168.1.1/30
NX-2(config-if)# no shutdown
! Configuration of the vPC Domain
NX-2(config-if)# vpc domain 100
NX-2(config-vpc-domain)# peer-keepalive destination 192.168.1.2 source 192.168.1.1
  vrf VPC-KEEPALIVE
! Configuration of the vPC Peer-Link
NX-2(config)# interface Ethernet1/46-47
NX-2(config-if-range)# description vPC-PeerLink
NX-2(config-if-range)# channel-group 100 mode active
NX-2(config-if-range)# interface port-channel 100
NX-2(config-if)# switchport mode trunk
NX-2(config-if)# vpc peer-link
! Creation of the vPC Port-Channel
NX-2(config)# interface ethernet 1/1
```

```
NX-2(config-if)# channel-group 1 mode active
NX-2(config-if)# interface port-channel 1
NX-2(config-if)# vpc 1
NX-2(config-if)# switchport mode trunk
```

5.2.3 vPC 验证

目前已经完成了两台 Nexus 交换机的 vPC 配置,接下来需要检查 vPC 域的运行状况。

1. 验证 vPC 域状态

可以通过 **show vpc** 命令验证 vPC 域的运行状态,输出结果包括对等状态、vPC 保活状态、一致性检查状态、vPC 设备角色(主/从)、vPC 端口通道接口、支持的 VLAN 以及本地交换机上配置的 vPC 列表等。

例 5-17 给出了 NX-2 的 **show vpc** 命令输出结果示例。

例 5-17 查看 vPC 状态

```
NX-2# show vpc
Legend:
                (*) - local vPC is down, forwarding via vPC peer-link

vPC domain id                     : 100
Peer status                       : peer adjacency formed ok
vPC keep-alive status             : peer is alive
Per-vlan consistency status       : success
Type-2 consistency status         : success
vPC role                          : primary
Number of vPCs configured         : 1
Peer Gateway                      : Disabled
Dual-active excluded VLANs        : -
Graceful Consistency Check        : Enabled
Auto-recovery status              : Disabled
Delay-restore status              : Timer is off.(timeout = 30s)
Delay-restore SVI status          : Timer is off.(timeout = 10s)
Operational Layer3 Peer-router    : Disabled

vPC Peer-link status
---------------------------------------------------------------------
id    Port   Status  Active vlans
--    ----   ------  --------------------------------------------------
1     Po100  up      1,10,20


vPC status
----------------------------------------------------------------------------
Id    Port        Status  Consistency  Reason            Active vlans
--    ---------   ------  -----------  ------            ---------------
1     Po1         up      success      success           1
```

如前所述,对等链路应处于转发状态。这一点可以通过 **show spanning-tree** 命令的 STP 状态加以验证,例 5-18 所示的 vPC 接口(端口通道 100)处于转发状态,且被标识为网络点对点端口。

例 5-18 查看 STP 行为变化

```
NX-2# show spanning-tree vlan 1

VLAN0001
```

```
 Spanning tree enabled protocol rstp
  Root ID    Priority    28673
             Address     885a.92de.617c
             Cost        1
             Port        4096 (port-channel1)
             Hello Time  2  sec  Max Age 20 sec  Forward Delay 15 sec

  Bridge ID  Priority    32769  (priority 32768 sys-id-ext 1)
             Address     88f0.3187.3b8b
             Hello Time  2  sec Max Age 20 sec  Forward Delay 15 sec
Interface        Role Sts Cost      Prio.Nbr Type
---------------- ---- --- --------- -------- --------------------------------
Po1              Root FWD 1         128.4096 (vPC) P2p
Po100            Desg FWD 1         128.4195 (vPC peer-link) Network P2p
Eth1/45          Desg FWD 2         128.45   P2p

NX-3# show spanning-tree vlan 1
! Output omitted for brevity
Interface        Role Sts Cost      Prio.Nbr Type
---------------- ---- --- --------- -------- --------------------------------
Po1              Root FWD 1         128.4096 (vPC) P2p
Po100            Root FWD 1         128.4195 (vPC peer-link) Network P2p
Eth1/45          Altn BLK 2         128.45 P2p
```

2. 验证 vPC 对等保活

如果出现了 vPC 对等保活问题，那么就可以通过命令 **show vpc peer-keepalive** 查看对等保活链路的详细状态，包括 vPC 保活状态、对等体处于活动或关闭状态的持续时间、用于保活的接口以及保活定时器等。例 5-19 显示了 vPC 对等保活状态。

例 5-19 查看 vPC 对等保活状态

```
NX-2# show vpc peer-keepalive

vPC keep-alive status               : peer is alive
--Peer is alive for                 : (1440) seconds, (939) msec
--Send status                       : Success
--Last send at                      : 2017.11.03 03:37:49 799 ms
--Sent on interface                 : Eth1/48
--Receive status                    : Success
--Last receive at                   : 2017.11.03 03:37:49 804 ms
--Received on interface             : Eth1/48
--Last update from peer             : (0) seconds, (414) msec

vPC Keep-alive parameters
--Destination                       : 192.168.1.2
--Keepalive interval                : 1000 msec
--Keepalive timeout                 : 5 seconds
--Keepalive hold timeout            : 3 seconds
--Keepalive vrf                     : VPC-KEEPALIVE
--Keepalive udp port                : 3200
--Keepalive tos                     : 192
```

如果状态显示为"down"，那么就需要验证每台交换机是否可以从所配置的 VRF 上下文 ping 通对端交换机。如果 ping 测试失败，那么就要排查两台交换机之间的基本连接故障。

3. vPC 一致性检查器

与端口通道接口一样，vPC 域中的两台 Nexus 交换机的特定参数必须匹配。NX-OS 提供了名为一致性检查器（Consistency-Checker）的特定进程，以确保设备配置的兼容性并防止意外丢包。一致性检查器可以检测两类差错。

- 类型 1（Type 1）差错。
- 类型 2（Type 2）差错。

（1）类型 1 差错

出现类型 1 vPC 一致性检查器差错时，运行中的从 Nexus 交换机的 vPC 实例和 vPC 成员端口将进入挂起状态并停止转发网络流量，运行中的主 Nexus 交换机仍会转发网络流量。为了避免出现类型 1 一致性差错，必须确保以下设置完全匹配。

- 端口通道模式：on、off 或 active（打开、关闭或活动）。
- 每个通道的链路速率。
- 每个通道的双工模式。
- 每个通道的中继模式。
  - 本征 VLAN。
  - 中继允许的 VLAN。
  - 带标记的本征 VLAN 流量。
- STP 模式。
- 多生成树的 STP 区域配置。
- 每个 VLAN 的启用/禁用状态必须相同。
- STP 全局设置。
  - 网桥保障（BA）设置。
  - 端口类型设置（建议将所有 vPC 对等链路端口都设置为网络端口）。
  - 环路保护设置。
- STP 接口设置。
  - 端口类型设置。
  - 环路保护。
  - 根保护。
  - MTU。
  - 允许的 VLAN 比特设置。

> 注：NX-OS 版本 5.2 提供了名为平滑一致性检查器（Graceful Consistency Checker）的功能特性，该功能特性可以更改类型 1 的不一致行为。平滑一致性检查器允许运行中的主设备转发流量，如果禁用该功能特性，那么就会完全关闭 vPC。系统默认启用该功能特性。

（2）类型 2 差错

类型 2 vPC 一致性检查器差错表示可能存在非期望的潜在转发问题，如某个 VLAN 接口在这个节点上有，但是在另一个节点上却没有。

（3）确认 vPC 一致性检查器设置

从例 5-20 的 **show vpc** 命令输出结果可以看出，一致性检查器检测出一致性检查状态故障并提供了差错原因，根据不同的差错情况，可能所有 vPC 接口或仅一个接口处于故障状态。

例 5-20 vPC 状态（存在一致性检查器差错）

```
NX-2# show vpc
Legend:
              (*) - local vPC is down, forwarding via vPC peer-link
```

```
vPC domain id                       : 100
Peer status                         : peer adjacency formed ok
vPC keep-alive status               : peer is alive
Configuration consistency status    : failed
Per-vlan consistency status         : success
Configuration inconsistency reason: vPC type-1 configuration incompatible - STP
Mode inconsistent
Type-2 consistency status           : success
vPC role                            : primary
Number of vPCs configured           : 1
Peer Gateway                        : Disabled
Dual-active excluded VLANs          : -
Graceful Consistency Check          : Enabled
Auto-recovery status                : Disabled
Delay-restore status                : Timer is off.(timeout = 30s)
Delay-restore SVI status            : Timer is off.(timeout = 10s)
Operational Layer3 Peer-router      : Disabled
vPC Peer-link status
---------------------------------------------------------------------
id   Port    Status Active vlans
--   ----    ------ ------------------------------------------------
1    Po100   up     1,10,20


vPC status
---------------------------------------------------------------------
Id   Port         Status Consistency Reason               Active vlans
--   ----         ------ ----------- ------               ------------
1    Po1          up     failed      Global compat check  1,10,20
                                     failed
```

命令 **show vpc consistency-parameters** {**global** | **vlan** | **vpc** *vpc-id* | **port-channel** *port-channel-identifier*}可以提供本地与远端设备的设置对比信息以及一致性检查器的差错类型。所执行的命令参数取决于 **show vpc** 命令的输出结果。

例 5-21 给出了 **show vpc consistency-parameters global** 命令的输出结果示例。

例 5-21   show vpc consistency-parameters global 命令输出结果

```
NX-3(config)# show vpc consistency-parameters global

    Legend:
        Type 1 : vPC will be suspended in case of mismatch
Name                            Type Local Value            Peer Value
--------------                  ---- ---------------------- ----------------------
QoS (Cos)                       2    ([0-7], [], [], [],    ([0-7], [], [], [],
                                     [], [])                [], [])
Network QoS (MTU)               2    (1500, 1500, 1500,     (1500, 1500, 1500,
                                     1500, 0, 0)            1500, 0, 0)
Network Qos (Pause:             2    (F, F, F, F, F, F)     (F, F, F, F, F, F)
T->Enabled, F->Disabled)
Input Queuing (Bandwidth)       2    (0, 0, 0, 0, 0, 0)     (0, 0, 0, 0, 0, 0)
Input Queuing (Absolute         2    (F, F, F, F, F, F)     (F, F, F, F, F, F)
Priority: T->Enabled,
F->Disabled)
Output Queuing (Bandwidth       2    (100, 0, 0, 0, 0, 0)   (100, 0, 0, 0, 0, 0)
```

```
                                Remaining)
Output Queuing (Absolute         2      (F, F, F, T, F, F)    (F, F, F, T, F, F)
Priority: T->Enabled,
F->Disabled)
Vlan to Vn-segment Map           1      No Relevant Maps      No Relevant Maps
STP Mode                         1      Rapid-PVST            Rapid-PVST
STP Disabled                     1      None None
STP MST Region Name              1      ""                    ""
STP MST Region Revision          1      0                     0
STP MST Region Instance to       1
VLAN Mapping
STP Loopguard                    1      Disabled              Disabled
STP Bridge Assurance             1      Enabled               Enabled
STP Port Type, Edge              1      Normal, Disabled,     Normal, Disabled,
BPDUFilter, Edge BPDUGuard              Disabled              Disabled
STP MST Simulate PVST            1      Enabled               Enabled
Nve Admin State, Src Admin       1      None                  None
State, Secondary IP, Host
Reach Mode
Nve Vni Configuration            1      None                  None
Interface-vlan admin up          2
Interface-vlan routing           2      1                     1
capability
Allowed VLANs                    -      1,10,20               1,10,20
Local suspended VLANs            -      -                     -
```

例 5-22 给出了 **show vpc consistency-parameters vlan** 命令输出结果示例，可以看出，配置不一致会引起非期望的转发行为。

例 5-22　show vpc consistency-parameters vlan 命令输出结果

```
NX-2# show vpc consistency-parameters vlan

Name                         Type   Reason Code             Pass Vlans
-------------                ----   -----------             ----------
Vlan to Vn-segment Map         1    success                 0-4095
STP Mode                       1    success                 0-4095
STP Disabled                   1    success                 0-4095
STP MST Region Name            1    success                 0-4095
STP MST Region Revision        1    success                 0-4095
STP MST Region Instance to     1    success                 0-4095
VLAN Mapping
STP Loopguard                  1    success                 0-4095
STP Bridge Assurance           1    success                 0-4095
STP Port Type, Edge            1    success                 0-4095
BPDUFilter, Edge BPDUGuard
STP MST Simulate PVST          1    success                 0-4095
Nve Admin State, Src Admin     1    success                 0-4095
State, Secondary IP, Host
Reach Mode
Nve Vni Configuration          1    success                 0-4095
Pass Vlans                     -                            0-4095
```

可以使用 **show vpc consistency-parameters** {**vpc** *vpc-id* | **port-channel** *port-channel- identifier*} 命令显示与端口通道接口直接相关的 vPC 一致性参数，通过确定 *vpc-id*（可能与端口通道接口号不同）即可查看指定端口通道。例 5-23 显示了 **show vpc consistency- parameters vpc** *vpc-id* 命令

的输出结果。

例 5-23　show vpc consistency-parameters vpc vpc-id 命令输出结果

```
NX-2# show vpc consistency-parameters vpc 1

    Legend:
        Type 1 : vPC will be suspended in case of mismatch

Name                        Type  Local Value             Peer Value
-------------               ----  ----------------------  ----------------------
STP Port Type               1     Default                 Default
STP Port Guard              1     Default                 Default
STP MST Simulate PVST       1     Default                 Default
lag-id                      1     [(1,                    [(1,
                                  88-5a-92-de-61-7c,      88-5a-92-de-61-7c,
                                  8000, 0, 0), (7f9b,     8000, 0, 0), (7f9b,
                                  0-23-4-ee-be-64, 8001,  0-23-4-ee-be-64, 8001,
                                  0, 0)]                  0, 0)]
mode                        1     active                  active
delayed-lacp                1     disabled                disabled
Speed                       1     10 Gb/s                 10 Gb/s
Duplex                      1     full                    full
Port Mode                   1     trunk                   trunk
Native Vlan                 1     1                       1
MTU                         1     1500                    1500
LACP Mode                   1     on                      on
Interface type              1     port-channel            port-channel
Admin port mode             1     trunk                   trunk
Switchport Isolated         1     0                       0
vPC card type               1     N9K TOR                 N9K TOR
Allowed VLANs               -     1,10,20                 1,10,20
Local suspended VLANs       -     -                       -
```

5.2.4　高级 vPC 功能特性

接下来将讨论一些常见的高级 vPC 功能特性及设计方案。

1. vPC 孤立端口

vPC 孤立端口（vPC Orphan Port）是与 vPC 对等体相关联的非 vPC 端口，vPC 对等体有一个关联 VLAN，且 VLAN 位于 vPC 或 vPC 对等链路上。也就是说，如果服务器连接在 VLAN 10 的 Ethernet 1/1 上，而 vPC 接口包含 VLAN 10，那么 Ethernet 1/1 就是一个孤立端口。如果 vPC 接口不包含 VLAN 10，那么 Ethernet 1/1 就是常规接入端口。可以通过命令 **show vpc orphan-ports** 查看孤立端口信息。例 5-24 给出了定位 vPC 孤立端口的方式。

例 5-24　查看 vPC 孤立端口

```
NX-2# show vpc orphan-ports
Note:
--------::Going through port database. Please be patient..::--------

VLAN            Orphan Ports
-------         ------------------------
1               Eth1/44, Eth1/45
```

一般来说，应通过以下规则来避免孤立端口。

- 确保所有下游设备均通过 vPC 接口进行连接。
- 通过 **switchport trunk allowed vlan** 命令将 VLAN 从所有 vPC 接口和 vPC 对等链路接口剪除。如果设备只能连接一条网络链路，那么这些 VLAN 就可以与接口相关联。这一点更多的是设计方式的变化。

如果从对等体挂起了自己的 vPC 端口（因为对等链路或对等保活链路出现了故障），那么就可以将运行中的从交换机上的孤立端口挂起，以避免丢包。可以通过接口配置命令 **vpc orphan-port suspend** 完成挂起操作。

例 5-25 给出了在 NX-2 的 Ethernet 1/44 和 1/45 接口启用该功能特性的配置示例。

例 5-25　在 vPC 故障期间挂起 vPC 孤立端口

```
NX-2# configure terminal
Enter configuration commands, one per line. End with CNTL/Z.
NX-2(config-if)# interface Eth1/44-45
NX-2(config-if)# vpc orphan-port suspend
```

2．vPC 自动恢复

作为一种安全机制，vPC 对等体在检测到另一个 vPC 对等体之前不会启用任何 vPC 接口。在某些故障情况下（如电源故障），两台 vPC 设备可能都要重启，致使无法检测到对方。由于这两台设备均不转发流量，因而此时可能会出现流量丢失问题。

vPC 自动恢复功能可以让其中的一个 vPC 对等体继续转发流量。初始化之后，如果 vPC 对等链路断开，且连续 3 条对等保活消息都没有收到响应消息，那么从设备就将承担运行中的主设备角色并初始化 vPC 接口，从而能够转发部分流量。

可以通过 vPC 域配置命令 **auto-recovery [reload-delay** *delay*]启用该功能特性。启用该功能特性之前，默认时延为 240s，不过也可以通过可选关键字 *reload-delay* 更改时延值，取值范围是 240～3600。例 5-26 给出了 vPC 自动恢复特性的配置和验证示例。

例 5-26　vPC 自动恢复特性的配置和验证

```
NX-2# configure terminal
Enter configuration commands, one per line. End with CNTL/Z.
NX-2(config)# vpc domain 100
NX-2(config-vpc-domain)# auto-recovery

NX-2# show vpc
! Output omitted for brevity
Legend:
..
Auto-recovery status              : Enabled, timer is off.(timeout = 240s)
```

3．vPC 对等网关

vPC 对等网关（vPC peer-gateway）功能允许 vPC 设备将数据包路由到 vPC 对等体的路由器 MAC 地址。该功能可以克服负载均衡器或 NAS（Network Attached Storage，网络附属存储）设备（希望优化数据包转发行为）配置差错引发的故障问题。

以图 5-6 拓扑结构为例，NX-2 和 NX-3 充当 VLAN 100 和 VLAN 200 的网关，NX-2 和 NX-3 为 Web 服务器以及连接了 NAS 设备的 NX-1 配置了 vPC。NX-1 负责交换（而不是路由）来去 NAS 设备的数据包。

Web 服务器向 NAS 设备（172.32.100.22）发送数据包的时候，会计算散列值以确定应该通过哪条链路将数据包发送给 NAS 设备。假设 Web 服务器将数据包发送给 NX-2，然后 NX-2 将数据包的源 MAC 地址更改为 00c1.5c00.0011（路由进程的一部分）并将数据包转发给 NX-1，最后再由 NX-1 将数据包转发（交换）给 NAS 设备。

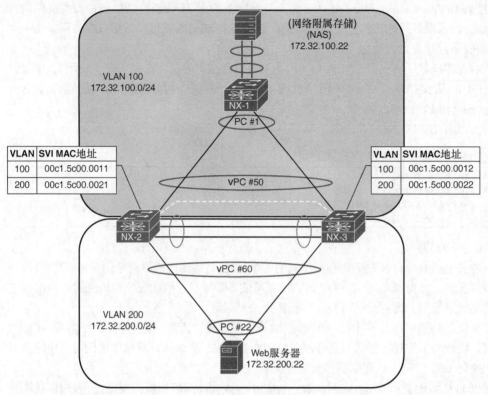

图 5-6 vPC 对等网关

此后，NAS 设备将创建应答包，在生成应答包的报头时使用 HSRP 网关的目的 MAC 地址 00c1.1234.0001，然后将数据包转发给 NX-1。NX-1 根据源 IP 地址和目的 IP 地址计算散列值，然后再将数据包转发给 NX-3。由于 NX-2 和 NX-3 都有 HSRP 网关的目的 MAC 地址，因而能够将数据包路由给网络 172.32.200.0/24 并转发回 Web 服务器。以上是正确且正常的转发行为。

不过，NAS 服务器启用了优化数据包流的功能特性之后，就会出现问题。NAS 设备收到 Web 服务器的数据包并生成应答包的报头之后，仅使用最初接收到的数据包的源和目的 MAC 地址。NX-1 收到应答包之后，将计算散列值并将数据包转发给 NX-3。此时，NX-3 并没有 MAC 地址 00c1.5c00.0011（NX-2 的 VLAN 100 接口），因而无法将数据包转发给 NX-1。出现丢包故障的原因是无法通过对等链路转发 vPC 成员端口收到的数据包（出于环路预防机制）。

在 NX-2 和 NX-3 上启用 vPC 对等网关功能之后，就可以让 NX-3 将数据包路由给 NX-2 的 MAC 地址，反之亦然。可以在 vPC 域配置模式下通过命令 **peer-gateway** 启用 vPC 对等网关功能，可以通过 **show vpc** 命令验证 vPC 对等网关功能，如例 5-27 所示。

例 5-27 配置和验证 vPC 对等网关

```
NX-2(config)# vpc domain 100
NX-2(config-vpc-domain)# peer-gateway

NX-2# show vpc
! Output omitted for brevity
Legend:
                (*) - local vPC is down, forwarding via vPC peer-link

vPC domain id                     : 100
```

```
..
vPC role                              : primary
Number of Cs configured               : 1
Peer Gateway                          : Enabled
```

> **注**：NX-OS 会在 SVI（已经在 vPC 中继链路上启用了该 VLAN）上自动禁用 IP 重定向功能。

> **注**：通过对等网关功能转发数据包时，其 TTL（Time To Live，生存时间）会被递减，因而 TTL 为 1 的数据包可能会在传输过程中丢失（因 TTL 到期）。

4. vPC ARP 同步

前面说明了如何根据拥有常规端口通道接口的设备计算出来的散列值使流量变得不对称。虽然设备可以在正常运行期间以常规方式构建 ARP（Address Resolution Protocol，地址解析协议）表（IP 到 MAC 地址的映射），但是在节点重新加载并启用之后，这种方式的速度就显得不够快了。NX-OS 提供了 ARP 同步功能，可以保持两个 vPC 对等体的 ARP 表同步，从而能够大大加快 vPC 对等体重启后的构建过程。

可以在 vPC 域配置模式下，通过 **ip arp synchronize** 命令启用 ARP 同步功能。例 5-28 给出了在 NX-2 上启用 ARP 同步功能的配置示例。

例 5-28　启用 vPC ARP 同步功能

```
NX-2# conf t
Enter configuration commands, one per line. End with CNTL/Z.
NX-2(config)# vpc domain 100
NX-2(config-vpc-domain)# ip arp synchronize
```

5. 备份三层路由

下面考虑一种特殊设计场景，假设某 Nexus 交换机在充当网关的同时还向网段上的主机提供 vPC 功能，那么就可能会出现某些故障情形，或者触发类型 1 一致性检查差错并关闭 vPC 端口。

在图 5-7 所示的简单拓扑结构中，NX-2 和 NX-3 为 VLAN 200 配置了 SVI 接口（充当 Web 服务器的网关）。NX-2、NX-3 和 R4 都运行 OSPF，因而 NX-2 和 NX-3 可以将数据包转发给 R4。NX-3 是运行中的主 Nexus 交换机。

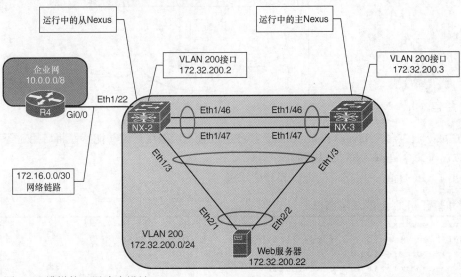

图 5-7　糟糕的三层路由设计

如果 vPC 对等链路中断（物理中断或者触发了类型 1 一致性检查器差错的意外变更），那么

NX-2 就会暂停 vPC 成员端口上的活动并关闭 VLAN 200 的 SVI。NX-3 中断与 NX-2 之间的路由协议邻接关系，从而无法为 Web 服务器提供到企业网络的连接。此后，NX-3 收到的 Web 服务器发给企业网的所有数据包都将被丢弃。

解决该故障场景的方法是在 vPC 对等体之间部署专用三层连接，可以是单独的三层链路或三层端口通道接口。

> 注：需要记住的是，vPC 对等链路无法将路由协议流量作为瞬时流量进行传输。例如，假设 NX-2 的 Eth1/22 是一个交换端口（属于 VLAN 200），且 R4 的 Gi0/0 接口配置了 IP 地址 172.32.200.5。虽然 R4 能够 ping 通 NX-3，但无法与 NX-3 建立 OSPF 邻接关系，因为 OSPF 数据包无法通过 vPC 对等链路进行传输。此时，只要部署前面所说的第二种解决方案即可解决该问题。

6. vPC 上的三层路由

从 L2 的角度来看，vPC 接口仅转发流量，而无法将 IP 地址直接分配给 vPC 接口。通过将 IP 地址分配给 vPC 设备上的 SVI，vPC 设备可以向下游设备提供网关服务。

不过，vPC 功能无法提供 L2 逻辑链路以构建路由协议邻接关系。但 NX-OS 版本 7.3 为 SVI 提供了相关功能，可以通过 vPC 接口与路由器建立路由协议邻接关系。

> 注：vPC 上的三层路由仅限于单播，不支持多播网络流量。

以图 5-8 为例，NX-2 和 NX-3 希望通过 vPC 接口利用 OSPF（Open Shortest Path First，开放最短路径优先）与 R4 交换路由。NX-2 和 NX-3 在 vPC 上启用了三层路由，从而能够与 R4 建立 OSPF 邻居邻接关系。从本质上来说，该设计方案将 NX-2、NX-3 和 R4 放在了同一个 LAN 网段上。

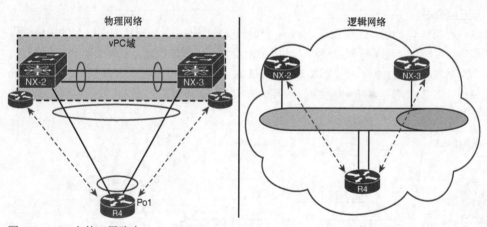

图 5-8 vPC 上的三层路由

可以在 vPC 域下通过命令 **layer3 peer-router** 为 vPC 配置三层路由。该功能特性将启用对等网关功能，可以通过命令 **show vpc** 加以验证。

例 5-29 给出了 vPC 上的三层路由的配置和验证示例。

例 5-29 配置和验证 vPC 上的三层路由

```
NX-2# configure terminal
Enter configuration commands, one per line. End with CNTL/Z.
NX-2(config)# vpc domain 100
NX-2(config-vpc-domain)# layer3 peer-router
NX-2# show vpc
! Output omitted for brevity
```

```
..
Delay-restore SVI status          : Timer is off.(timeout = 10s)
Operational Layer3 Peer-router    : Enabled
```

> **注：** 如果 vPC 对等关系未建立或 vPC 配置不一致，那么就可以收集 **show tech vpc** 命令的输出结果并联系思科技术支持中心。

5.3 FabricPath

从目前来看，几乎所有的传统 L2 网络都启用了 STP 以构建无环拓扑。但基于 STP 的 L2 网络设计方案存在一些明显的限制，其中之一就是 STP 无法充分利用多条并行转发路径。即使存在可用的物理冗余路径，STP 也会阻塞这些额外路径，强制流量仅使用其中的一条路径。其他限制还包括以下几点。

- STP 的收敛过程是中断性的。
- MAC 地址表无法扩展。
- 树形拓扑结构只能提供有限带宽。
- 树形拓扑结构存在次优路径问题。
- 主机泛洪会影响整个网络。
- 本地故障会影响整个网络，使故障排查非常困难。

为了克服上述挑战，思科在 2008 年推出了 vPC 功能，使以太网设备能够同时连接两台独立的 Nexus 交换机，并将多条链路聚合为一条逻辑端口通道。vPC 可以为用户提供双活转发路径，从而有效克服了 STP 的局限性。不过，虽然 vPC 克服了大多数的 STP 挑战，但仍然存在其他挑战。例如，vPC 无法提供第三或第四台汇聚层交换机，以进一步提高下游交换机的密度或带宽。此外，vPC 也无法克服传统 STP 设计方案存在的 VLAN 扩展问题。

思科的 FabricPath 功能为构建简洁、可扩展且支持多路径的 L2 交换矩阵提供了很好的技术基础。从控制平面的角度来看，FabricPath 采用了基于路由协议的 SPF（Shortest Path First，最短路径优先），可以在 FabricPath 域内选择去往目的端的最佳路径。FabricPath 使用 L2 IS-IS 协议，可以提供处理单播、广播和多播数据包的所有 IS-IS 功能。此时，不需要为 L2 IS-IS 启用单独进程，所有启用了 FabricPath 的接口均自动启用 L2 IS-IS。

FabricPath 为灵活的 L2 桥接以太网提供了三层路由能力，可以为路由和交换域提供如下好处。

- 路由。
  - 多路径（ECMP），两台设备之间最多允许 256 条活动链路。
  - 快速收敛。
  - 高扩展性。
- 交换。
  - 配置简单。
  - 即插即用。
  - 部署灵活。

由于 FabricPath 的核心运行在 L2 IS-IS 上，脊节点（Spine node）与叶节点（Leaf node）之间未启用 STP，因而提供了可靠的 L2 互联连接。在入站边缘设备上执行单个 MAC 地址查找操作可以识别交换矩阵的出端口，此后就可以使用可用的最短路径交换流量。

FabricPath 设计方案允许主机充分利用多个活动三层默认网关的优势，如图 5-9 所示，图中的主机看到的只是一个默认网关。交换矩阵将流量透明且同时转发给所有活动默认网关，从而将多路径能力从交换矩阵内部扩展到交换矩阵外部的三层域。

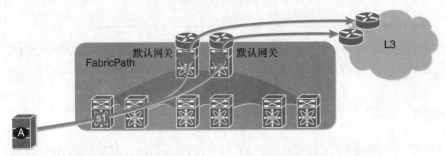

图 5-9　访问多个活动默认网关

该交换矩阵还可以扩展三层网络，可以在边缘或交换矩阵内部创建任意数量的路由接口，所连接的三层设备与这些路由接口对等，从而提供无缝的三层网络集成。

5.3.1　FabricPath 术语和组件

在理解启用 FabricPath（FP）的网络中的数据包流之前，必须首先理解 FabricPath 体系架构的各种术语和组件。图 5-10 给出了一个标准的启用了 FabricPath 的 Spine-leaf 拓扑结构，也称为 Clos 交换架构。拓扑结构中的叶交换机或边缘交换机拥有两类不同的接口。

- FP（FabricPath）核心端口。
- CE（Classical Ethernet，传统以太网）边缘端口。

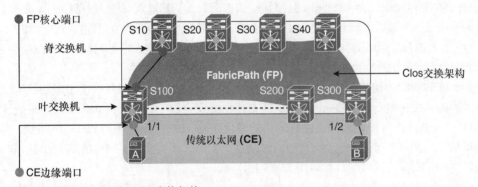

图 5-10　FabricPath 和 Clos 交换架构

FP 核心端口负责提供到脊交换机的连接，是启用了 FabricPath 的接口。FP 核心网络负责执行以下功能。

- 发送和接收 FP 帧。
- 避免使用 STP，不需要进行 MAC 地址学习，也不需要由 FP 核心端口维护 MAC 地址表。
- 使用由 IS-IS 计算的路由表来确定最佳路径。

CE 边缘端口是常规的中继端口或接入端口，可以提供与主机或其他传统交换机的连接。CE 端口负责执行以下功能。

- 发送和接收常规以太网帧。
- 运行 STP，执行 MAC 地址学习并维护 MAC 地址表。

FP 边缘设备负责维护 MAC 地址与 switch-ID（由 IS-IS 自动分配给所有交换机）的关联关系。此外，FP 还引入了一个新的数据平面封装，实现方式是在传统以太网报头顶部增加一个 16 字节 FP 帧。图 5-11 给出了 FP 封装报头格式，也称为 MAC-in-MAC 报头。外部 FP 报头包括外层目的地址、外层源地址和 FP 标签。FP 报头中的外层源地址或目的地址字段包含了以下重要字段信息。

- **SID（Switch-ID，交换机 ID）**：通过唯一编号标识每台 FP 交换机。
- **sSID（sub-Switch ID，子交换机 ID）**：标识通过 vPC+ 连接的设备和主机。
- **LID（Local ID，本地 ID）**：标识数据帧来自或去往的确切端口。出站 FP 交换机使用 LID 来确定出接口，因而不需要在 FP 核心端口上执行 MAC 地址学习操作。LID 具有本地意义。

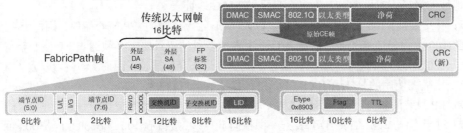

图 5-11 FabricPath 数据包结构

FP 标签主要包括以下 3 个字段。

- **Etype**：以太类型设置为 0x8903。
- **Ftag**：Ftag 是转发标签，是一个唯一的 10 比特数字，用于标识拓扑结构和/或分发树。对于单播数据包来说，Ftag 可以标识所要使用的 FP IS-IS 拓扑。对于多播数据包（广播、未知单播和多播数据包）来说，Ftag 可以标识所要使用的分发树。FTAG 1 用于 BUM（Broadcast，Unknown unicast，and Multicast，广播、未知单播和多播）流量，FTAG 2 用于多播流量。Ftag 是一个 10 比特字段，可以支持 1024 个转发树或拓扑结构。
- **TTL**：每经过一台交换机，就要递减一次 TTL 值，目的是防止数据帧无限循环。

注：如果需要的拓扑结构超过了 1024，那么就可以将 Ftag 值设置为 0，用 VLAN 来标识多目的树的拓扑结构。

5.3.2 FabricPath 数据包流

为了更好地理解 FabricPath 域中的数据包转发过程，下面将以图 5-12 所示拓扑结构和处理步骤为例加以说明。图中包含 4 台脊交换机（S10、S20、S30 和 S40）和 3 台叶/边缘交换机（S100、S200 和 S300）。每台叶交换机都有去往 4 台脊交换机的连接。主机 A（MAC 地址 A）连接在 S100 的 CE 端口上，主机 B（MAC 地址 B）连接在 S300 的 CE 端口上。

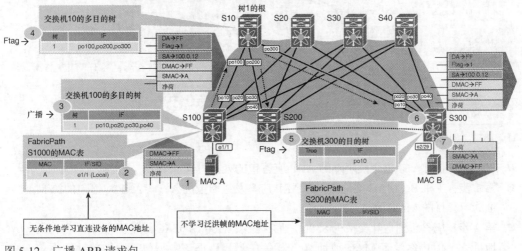

图 5-12 广播 ARP 请求包

在主机 A 和主机 B 都不知道对方 MAC 地址的时候，发送的第一个数据包就是广播 ARP 请求包。主机 A 发送给主机 B 的广播帧的数据包处理过程如下。

- **第 1 步**：主机 A 向主机 B 发送 ARP 请求。由于 ARP 请求是广播包，因而源 MAC 为 A，目的 MAC 为 FF（广播地址）。
- **第 2 步**：数据包到达叶交换机 S100 的 CE 边缘端口时，S100 根据 MAC 地址 A 和学到该 MAC 地址的接口（本例为 e1/1）更新自己的 MAC 地址表。
- **第 3 步**：叶交换机 S100 用 FP 报头封装该以太网帧。由于启用了 FP 核心端口，因而 IS-IS 已经在 S100 上预先计算了多目的树。Tree 1 表示多目的树且要求必须通过 FP 核心端口（po10、po20、po30 和 po40）发送数据包。由于这是一个广播帧，因而 Ftag 被设置为 1。请注意，广播图使用的是 Tree ID 1（Ftag 1）且将数据包转发给 S10。
- **第 4 步**：FP 封装的数据帧到达 S10 之后，将接收 Ftag 1 并查找 Tree ID。
- **第 5 步**：对于本例来说，Tree ID 1 表示一棵指向 FP 核心接口 po100、po200 和 po300（面向叶交换机 S100、S200 和 S300）的多目的树。此外，节点 S10 还要执行 RPF（Reverse Path Forwarding，反向路径转发）检查，以验证 L2 接口收到的数据包，并通过多目的 FP 封装的数据帧将数据包发送给 S200 和 S300。
- **第 6 步**：出站 FP 交换机（S300）在链路 po10 上收到该数据包之后，执行 RPF 检查以验证数据包的接收情况，并将数据包泛洪给 CE 端口。
- **第 7 步**：S300 删除 FP 报头，并根据广播帧在 VLAN 内泛洪该数据包。请注意，出站 FP 交换机（本例为 S300）不会使用 MAC 地址 A 更新其 MAC 地址表，因为边缘设备不会从 FP 核心端口收到的泛洪帧（MAC 地址被设置为 FF）学习 MAC 地址。此后，将原始广播包发送给主机 B。

ARP 请求到达主机 B 之后，生成的 ARP 应答包是单播数据包。图 5-13 给出了主机 B 发送给主机 A 的 ARP 应答包的处理过程。

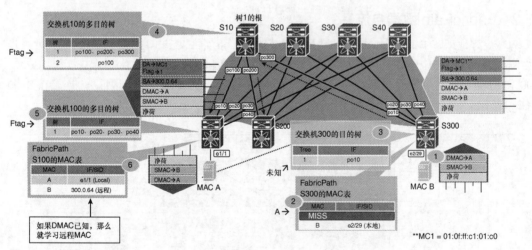

图 5-13　单播 ARP 应答包

从主机 B 到主机 A 穿越交换矩阵的 ARP 应答包的处理过程如下。

- **第 1 步**：主机 B（MAC 地址 B）将 ARP 应答发送回主机 A。该 ARP 应答包的源 MAC 地址为 B、目的 MAC 为 A。
- **第 2 步**：应答包到达叶交换机 S300 之后，虽然 S300 用 MAC 地址 B 更新了自己的 MAC 地址表，但仍然没有 MAC 地址 A 的相关信息，因而该数据包是未知单播数据包。

- **第3步**：S300（入站 FP 交换机）确定所要使用的树。未知单播通常使用第一个 Tree ID（Ftag 1）。Tree ID 1 指向交换机 S300 上的所有 FP 核心接口（po10、po20、po30 和 po40）。入站 FP 交换机将外层目的 MAC 地址设置为周知"Flood-to-Fabric"（泛洪到交换矩阵）多播地址，即 MC1—01:0F:FF:C1:01:C0。
- **第4步**：将 FP 封装的未知单播包发送给所有脊交换机，其他 FP 交换机都接受入站交换机选择的 Tree ID（本例为 Tree 1）。数据包到达 Tree 1 的根（S10）之后，将使用相同的 Ftag 1，并通过 po100 和 po200 接口向外转发该数据包（不通过 po300 向外转发该数据包，因为该接口是收到该数据帧的接口）。
- **第5步**：数据包到达 S100 之后，在 FP 树上执行查找操作，并使用 Tree ID 1（设置为 po10）。由于来自 S10 的数据包是在 S100 交换机的 po10 上收到的，因而不会将该数据包再次转发回交换矩阵。
- **第6步**：解封装 FP 报头，并将 ARP 应答转发给 MAC 地址为 A 的主机。此时，S100 使用 MAC 地址 B（IF/SID 指向 S300）更新了自己的 MAC 地址表，因为该目的 MAC 在帧内是已知的。

此后，主机 A 再向主机 B 发送数据包的时候，从主机 A 发送出来的数据包的源地址和目的地址就分别被设置为 MAC A 和 MAC B。交换机 S100 在 CE 端口上收到数据包之后，就已经知道了该目的 MAC，且指向启用了 FP 功能的网络中的 S300。由于存在到多条去往 S300 的路径，因而使用基于流的散列表来查找 FP 路由表以找出去往 S300 的最短路径。接着使用 FP 报头封装数据包，源 Switch-ID（SWID）为 S100，目的 Switch-ID（SWID）为 S300，Ftag 为 1。脊交换机收到数据包之后，执行去往 S300 的 FP 路由查找操作，并将数据包发送给去往 S300 的出接口。数据包到达 S300 之后，将使用主机的 MAC 地址（IF/SID 指向 S100）更新其 MAC 地址表。

5.3.3 FabricPath 配置

为了更好地理解 FabricPath 的配置过程并验证启用了 FabricPath 功能特性的网络，接下来将以图 5-14 所示拓扑结构为例加以说明。图中包含了两个脊节点（NX-10 和 NX-20）和 3 个叶节点（NX-1、NX-2 和 NX-3），终端主机节点（主机 A 和主机 B）连接在叶节点 NX-1 和 NX-3 上。

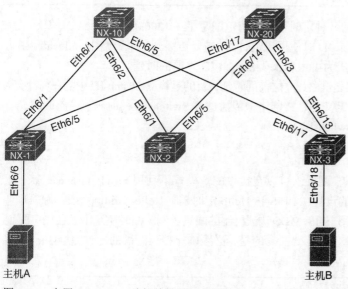

图 5-14 启用 FabricPath 功能特性的拓扑结构

启用 FabricPath 功能特性的方法与启用其他功能特性有所不同。首先需要安装 FabricPath 功能特性集，然后启用 FabricPath 功能特性集，接着再启用 FabricPath 功能特性。例 5-30 给出了启用 FabricPath 功能特性的配置示例。FabricPath 使用 DRAP（Dynamic Resource Allocation Protocol，动态资源分配协议）来分配 Switch-ID，也可以在 Nexus 交换机上使用命令 **fabricpath switch-id [1-4094]** 手工配置 Switch-ID。需要为 FabricPath 域中的每台交换机都配置一个唯一的 Switch-ID。

例 5-30　启用 FabricPath 功能特性

```
NX-1# configure terminal
NX-1(config)# install feature-set fabricpath
NX-1(config)# feature-set fabricpath
NX-1(config)# fabricpath switch-id 1
```

启用了 **feature-set fabricpath** 之后，需要配置 FP VLAN，包括两类 VLAN。
- CE（Classical Ethernet，传统以太网）VLAN。
- FP（FabricPath）VLAN。

FP VLAN 是启用了 FP 功能的链路承载的 VLAN，CE VLAN 是常规以太网链路（如中继或接入端口）承载的常规 VLAN。如果要将 VLAN 启用为 FP VLAN，那么就需要在 VLAN 配置模式下使用命令 **mode fabricpath**。配置了 FP VLAN 之后，需要使用命令 **switchport mode fabricpath** 配置 FP 核心链路。最后，将 CE 链路配置为中继或接入端口。例 5-31 给出了拓扑结构中的 NX-1 的 FP VLAN、FP 端口及 CE 端口的配置情况。

例 5-31　启用 FP VLAN、FP 核心端口及 CE 边缘端口

```
NX-1(config)# vlan 100, 200, 300, 400, 500
NX-1(config-vlan)# mode fabricpath
NX-1(config)# interface eth 6/1, eth6/5
NX-1(config-if-range)# switchport
NX-1(config-if-range)# switchport mode fabricpath
NX-1(config-if-range)# no shut
NX-1(config)# interface eth6/6
NX-1(config-if)# switchport
NX-1(config-if)# switchport access vlan 100
NX-1(config-if)# no shut
```

此外，还需要配置 FabricPath 的各种定时器，取值范围是 1s～1200s。
- **allocate-delay**：分配新 Switch-ID 时使用该定时器，需要传播到整个网络中。allocate-delay 定义了传播新 Switch-ID 且成为可用和永久 Switch-ID 之前的时延。
- **linkup-delay**：该定时器定义是链路建立之前的时延，目的是检测 Switch-ID 中的任何冲突。
- **transition-delay**：该定时器定义的是在网络中传播过渡 Switch-ID 的时延。在此期间，网络中存在所有旧的和新的 Switch-ID 值。

注：这些定时器的默认值均为 10s。

FabricPath 无须进行特定的 IS-IS 配置。可以在接口配置模式下利用 **fabricpath isis** 命令配置认证机制以及其他与 IS-IS 相关的设置（如 IS-IS Hello 定时器、Hello-padding 及度量等）。

例 5-32 给出了为 FabricPath 启用 IS-IS MD5 认证机制的配置示例，同时还给出了在接口配置模式下定义各种 IS-IS 设置的配置示例。需要注意的是，必须确保接口两端的配置完全匹配。

例 5-32　在 FP 端口上启用身份认证

```
NX-1(config)# interface ethernet6/1
NX-1(config-if)# fabricpath isis ?
```

```
  authentication                  Set hello authentication keychain
  authentication-check            Check authentication on received hellos
  authentication-type             Set hello authentication type
  csnp-interval                   Set CSNP interval in seconds
  hello-interval                  Set Hello interval in seconds
  hello-multiplier                Set multiplier for Hello holding time
  hello-padding                   Pad IS-IS hello PDUs to full MTU
  lsp-interval                    Set LSP transmission interval
  mesh-group                      Set IS-IS mesh group
  metric                          Configure the metric for interface
  mtu-check                       Check mtu on received hellos, if its padded
  retransmit-interval             Set per-LSP retransmission interval
  retransmit-throttle-interval    Set interface LSP retransmission interval

NX-1(config-if)# fabricpath isis authentication key-chain cisco
NX-1(config-if)# fabricpath isis authentication-type md5
NX-1(config-if)# exit
NX-1(config)# key chain cisco
NX-1(config-keychain)# key 1
NX-1(config-keychain-key)# key-string cisco
```

5.3.4 FabricPath 验证与故障排查

通过 FabricPath 启用了交换矩阵之后，需要验证网络中的 Switch-ID。此时可以利用 **show fabricpath switch-id** 命令完成该操作，该命令可以显示交换矩阵学到的所有 Switch-ID 及其当前状态。如果不是以静态方式分配 Switch-ID，那么 IS-IS 进程就会动态分配 Switch-ID。检查 **show fabricpath switch-id** 命令的输出结果即可验证交换矩阵中的所有节点信息，如例 5-33 所示。

例 5-33　show fabricpath switch-id 命令输出结果

```
NX-1# show fabricpath switch-id
                    FABRICPATH SWITCH-ID TABLE
Legend: '*' - this system
        '[E]' - local Emulated Switch-id
        '[A]' - local Anycast Switch-id
Total Switch-ids: 5
================================================================================
    SWITCH-ID      SYSTEM-ID        FLAGS       STATE       STATIC    EMULATED/
                                                                      ANYCAST
---------------+----------------+-----------+-----------+----------+-----------
      10         6c9c.ed4e.c141   Primary     Confirmed   Yes        No
      20         6c9c.ed4e.c143   Primary     Confirmed   Yes        No
*    100         6c9c.ed4e.4b41   Primary     Confirmed   Yes        No
     200         6c9c.ed4e.4b42   Primary     Confirmed   Yes        No
     300         6c9c.ed4e.4b43   Primary     Confirmed   Yes        No
```

除验证 FP Switch-ID 之外，验证哪些接口参与了 FP 核心链路和 CE 侧链路也非常重要。可以利用命令 **show fabricpath isis interface [brief]** 轻松识别叶节点或脊节点上的 FP 核心链路，使用关键字 **brief** 后的输出结果可以显示核心接口的概要信息，如通过命令 **switchport mode fabricpath** 启用的接口、当前状态、电路类型（通常为 0x1 或 L1）、MTU、度量、优先级、邻接关系以及邻接关系可用状态等信息。如果希望获得更多详细信息，那么就不要在命令中使用关键字 **brief**，命令 **show fabricpath isis interface** 可以显示认证机制（如果启用）、邻接信息和拓扑结构信息，同时还能显示前面摘要输出结果中的信息。例 5-34 同时显示了叶节点 NX-1 上的 FP 核心接口的摘要信息和详细信息。

例 5-34 验证 FP 核心接口

```
NX-1# show fabricpath isis interface brief
Fabricpath IS-IS domain: default
Interface    Type  Idx State      Circuit   MTU  Metric Priority Adjs/AdjsUp
-----------------------------------------------------------------------------
Ethernet6/1  P2P   1   Up/Ready   0x01/L1   1500 40     64       1/1
Ethernet6/5  P2P   3   Up/Ready   0x01/L1   1500 40     64       1/1

NX-1# show fabricpath isis interface
Fabricpath IS-IS domain: default
Interface: Ethernet6/1
  Status: protocol-up/link-up/admin-up
  Index: 0x0001, Local Circuit ID: 0x01, Circuit Type: L1
  Authentication type MD5
  Authentication keychain is cisco
  Authentication check specified
  Extended Local Circuit ID: 0x1A280000, P2P Circuit ID: 0000.0000.0000.00
  Retx interval: 5, Retx throttle interval: 66 ms
  LSP interval: 33 ms, MTU: 1500
  P2P Adjs: 1, AdjsUp: 1, Priority 64
  Hello Interval: 10, Multi: 3, Next IIH: 00:00:01
  Level  Adjs   AdjsUp  Metric   CSNP   Next CSNP   Last LSP ID
  1      1      1       40       60     Inactive    ffff.ffff.ffff.ff-ff
  Topologies enabled:
    Level Topology Metric  MetricConfig Forwarding
    0     0        40      no           UP
    1     0        40      no           UP

Interface: Ethernet6/5
  Status: protocol-up/link-up/admin-up
  Index: 0x0003, Local Circuit ID: 0x01, Circuit Type: L1
  No authentication type/keychain configured
  Authentication check specified
  Extended Local Circuit ID: 0x1A284000, P2P Circuit ID: 0000.0000.0000.00
  Retx interval: 5, Retx throttle interval: 66 ms
  LSP interval: 33 ms, MTU: 1500
  P2P Adjs: 1, AdjsUp: 1, Priority 64
  Hello Interval: 10, Multi: 3, Next IIH: 00:00:03
  Level  Adjs   AdjsUp  Metric   CSNP   Next CSNP   Last LSP ID
  1      1      1       40       60     Inactive    ffff.ffff.ffff.ff-ff
  Topologies enabled:
    Level Topology Metric  MetricConfig Forwarding
    0     0        40      no           UP
    1     0        40      no           UP
```

此外，还可以利用命令 **show fabricpath isis adjacency [detail]** 验证叶节点与脊节点之间的 IS-IS 邻接关系。例 5-35 显示了 NX-1 和 NX-10 的 IS-IS 邻接关系信息，该命令显示了 System-ID、电路类型、加入 FabricPath IS-IS 邻接关系的接口、拓扑结构 ID 以及转发状态等信息。此外，该命令还可以显示 FabricPath 过渡到当前状态的最后时间（邻接关系翻动的最后时间）。

例 5-35 验证 IS-IS 邻接关系

```
NX-1# show fabricpath isis adjacency detail
Fabricpath IS-IS domain: default Fabricpath IS-IS adjacency database:
System ID       SNPA          Level State Hold Time  Interface
NX-10           N/A           1     UP    00:00:29   Ethernet6/1
  Up/Down transitions: 1, Last transition: 00:07:36 ago
```

```
    Circuit Type: L1
    Topo-id: 0, Forwarding-State: UP

NX-20            N/A           1       UP       00:00:22      Ethernet6/5
    Up/Down transitions: 1, Last transition: 00:12:13 ago
    Circuit Type: L1
    Topo-id: 0, Forwarding-State: UP
```

接下来，验证是否在边缘/叶交换机上配置了必要的 FabricPath VLAN。可以使用命令 **show fabricpath isis vlan-range** 加以验证。配置了 FP VLAN 及面向 CE 的接口之后，边缘设备就可以学到连接在边缘节点上的主机的 MAC 地址，可以使用传统命令 **show mac address-table vlan** *vlan-id* 加以验证。例 5-36 验证了 FP VLAN 以及从连接在 FP VLAN 100 上的主机学到的 MAC 地址。

例 5-36　验证 FabricPath VLAN 和 MAC 地址表

```
NX-1# show fabricpath isis vlan-range
Fabricpath IS-IS domain: default
MT-0
Vlans configured:
100, 200, 300, 400, 500, 4040-4041
```

```
NX-1# show mac address-table vlan 100
Note: MAC table entries displayed are getting read from software.
Use the 'hardware-age' keyword to get information related to 'Age'

Legend:
        * - primary entry, G - Gateway MAC, (R) - Routed MAC, O - Overlay MAC
        age - seconds since last seen,+ - primary entry using vPC Peer-Link,
        (T) - True, (F) - False ,  ~~~ - use 'hardware-age' keyword to retrieve
age info
   VLAN     MAC Address       Type        age       Secure NTFY Ports/SWID.SSID.LID
---------+-----------------+--------+---------+------+----+------------------
* 100      30e4.db97.e8bf    dynamic    ~~~       F     F   Eth6/6
  100      30e4.db98.0e7f    dynamic    ~~~       F     F   300.0.97
```

```
NX-3# show mac address-table vlan 100
Note: MAC table entries displayed are getting read from software.
Use the 'hardware-age' keyword to get information related to 'Age'

Legend:
        * - primary entry, G - Gateway MAC, (R) - Routed MAC, O - Overlay MAC
        age - seconds since last seen,+ - primary entry using vPC Peer-Link,
        (T) - True, (F) - False ,  ~~~ - use 'hardware-age' keyword to retrieve
age info
   VLAN     MAC Address       Type        age       Secure NTFY Ports/SWID.SSID.LID
---------+-----------------+--------+---------+------+----+------------------
  100      30e4.db97.e8bf    dynamic    ~~~       F     F   100.0.85
* 100      30e4.db98.0e7f    dynamic    ~~~       F     F   Eth6/18
```

与三层 IS-IS 相似，二层 IS-IS 也要在网络中维护多个拓扑，每个拓扑在 FabricPath 域中均表示为 Tree ID，这些树都是交换矩阵中的多目的树。如果要查看 FabricPath 域中的 IS-IS 拓扑，那么就可以使用命令 **show fabricpath isis topology [summary]**。例 5-37 显示了当前拓扑结构中的不同 IS-IS 拓扑。

例 5-37　FabricPath 拓扑信息

```
NX-1# show fabricpath isis topology summary
FabricPath IS-IS Topology Summary
```

```
Fabricpath IS-IS domain: default
MT-0
  Configured interfaces:  Ethernet6/1  Ethernet6/5
 Max number of trees: 2  Number of trees supported: 2
    Tree id: 1, ftag: 1, root system: 6c9c.ed4e.c143, 20
    Tree id: 2, ftag: 2, root system: 6c9c.ed4e.c141, 10
Ftag Proxy Root: 6c9c.ed4e.c143

NX-1# show fabricpath isis topology
FabricPath IS-IS Topology
Fabricpath IS-IS domain: default
MT-0
Fabricpath IS-IS Graph 0 Level-1 for MT-0 IS routing table
NX-3.00, Instance 0x0000005C
    *via NX-10, Ethernet6/1, metric 80
    *via NX-20, Ethernet6/5, metric 80
NX-10.00, Instance 0x0000005C
    *via NX-10, Ethernet6/1, metric 40
NX-20.00, Instance 0x0000005C
    *via NX-20, Ethernet6/5, metric 40
! Output omitted for brevity
```

如果出现了流量转发问题或者无法学习 MAC 地址，那么就可以检查 FP IS-IS 邻接关系是否已建立且 FP IS-IS 路由是否在 URIB（Unicast Routing Information Base，单播路由信息库）中。可以通过命令 **show fabricpath route** [**detail** | [**switch-id** *switch-id*]加以验证，该命令可以显示远程节点（叶节点或脊节点）的路由，路由以 ftag/switch-id/subswitch-id 的形式呈现。从例 5-38 可以看出，远程边缘设备 NX-3 的路由携带 FTAG 1、Switch-ID 300 和 Subswitch-ID 0（因为未启用 vPC+）。

例 5-38　验证 URIB 中的 FabricPath 路由

```
NX-1# show fabricpath route detail
FabricPath Unicast Route Table
'a/b/c' denotes ftag/switch-id/subswitch-id
'[x/y]' denotes [admin distance/metric]
ftag 0 is local ftag
subswitch-id 0 is default subswitch-id


FabricPath Unicast Route Table for Topology-Default

0/100/0, number of next-hops: 0
        via ---- , [60/0], 0 day/s 00:16:18, local
1/10/0, number of next-hops: 1
        via Eth6/1, [115/40], 0 day/s 00:16:01, isis_fabricpath-default
1/20/0, number of next-hops: 1
        via Eth6/5, [115/40], 0 day/s 00:15:52, isis_fabricpath-default
1/200/0, number of next-hops: 2
        via Eth6/1, [115/80], 0 day/s 00:16:01, isis_fabricpath-default
        via Eth6/5, [115/80], 0 day/s 00:15:52, isis_fabricpath-default
1/300/0, number of next-hops: 2
        via Eth6/1, [115/80], 0 day/s 00:09:14, isis_fabricpath-default
        via Eth6/5, [115/80], 0 day/s 00:09:14, isis_fabricpath-default
NX-1# show fabricpath route switchid 300
! Output omitted for brevity
ftag 0 is local ftag
subswitch-id 0 is default subswitch-id
```

```
FabricPath Unicast Route Table for Topology-Default

1/300/0, number of next-hops: 2
        via Eth6/1, [115/80], 0 day/s 00:34:49, isis_fabricpath-default
        via Eth6/7, [115/80], 0 day/s 00:34:49, isis_fabricpath-default
```

从前面的输出结果可以很清楚地看出，NX-3 的路由的 Ftag 为 1。在 URIB 中验证了该路由之后，还要接着验证该路由是否已经安装到了 FIB（Forwarding Information Base，转发信息库）中。为了验证该路由是否位于 FIB 中，可以使用线卡命令 **show fabricpath unicast routes vdc** *vdc-number* [**ftag** *ftag*] [**switchid** *switch-id*]，该命令可以显示线卡上的软件表中的硬件路由信息及 RPF 接口。此外，作为与平台相关信息的一部分，该命令还能显示硬件表地址，利用该地址可以进一步验证该路由的硬件转发信息。从例 5-39 可以看出，软件表显示该路由是一条远程路由，RPF 接口是 Eth6/5，硬件表地址是 0x18c0。

例 5-39　验证硬件中特定 FP 路由的软件表

```
NX-1# attach module 6
Attaching to module 6 ...
To exit type 'exit', to abort type '$.'

module-6# show fabricpath unicast routes vdc 1 ftag 1 switchid 300
Route in VDC 1
---------------
--------------------------------------------------------------------------------
FTAG | SwitchID | SubSwitchID | Loc/Rem | RPF | RPF Intf | Num Paths | Merge V
--------------------------------------------------------------------------------
0001 |   0300   |    0000     | Remote  | Yes |  Eth6/5  |     2     |    1
--------------------------------------------------------------------------------
PD Information for Prefix:

FE num |   ADDR TYPE   | HTBL ADDR | TCAM ADDR | SWSI
-----------------------------------------------------------
     0 |  HASH TABLE   | 000018c0  | 000000ff  | 000011c1
     1 |  HASH TABLE   | 000018c0  | 000000ff  | 000011c1
-----------------------------------------------------------
PD Information for ECMP:
       Common Info
-----------------------------------
AMM key : 0x24
-----------------------------------
Next Hop |   Interface   |   LID
-----------------------------------
       0 |        Eth6/1 | 00000050
       1 |        Eth6/5 | 00000054
-----------------------------------
         Per FE Info
-----------------------------------
FE num | MP_base  | ref_count
-----------------------------------
     0 | 00000024 |     1
     1 | 00000024 |     1
-----------------------------------
```

注：例 5-39 中的命令与 Nexus 7000/7700 系列交换机上的 F2 和 F3 线卡模块有关，具体的验证命令因线卡和平台的不同而有所不同（如 Nexus 5500）。

使用软件表中的硬件地址执行命令 **show hardware internal forwarding inst** *instance-id* **table sw start** *hw-entry-addr* **end** *hw-entry-addr*，其中，*instance-id* 值位于前面示例中的 FE num 字段，*hw-entry-addr* 地址是前面示例中高亮显示的地址。该命令的输出结果可以显示 Switch-ID（swid）、Subswitch-ID（swid）及其他各种字段。需要特别注意的一个重要字段就是 ssw_ctrl，如果 ssw_ctrl 字段为 0x0 或 0x3，那么就表明该交换机没有 Subswitch-ID（仅用于 vPC+场景）；如果配置了 vPC+，那么该字段值通常为 0x1。需要注意的另一个字段就是 local 字段。如果 local 字段被设置为 n，那么就表明该路由存在多路径，从而需要对多路径表进行验证，如例 5-40 所示。

例 5-40 验证 URIB 中的 FabricPath 路由

```
module-6# show hardware internal forwarding inst 0 table sw start 0x18c0 end 0x18c0
-------------------------------------------------------------
---------------------- SW Table ------------------------
                      (INST# 0)
-------------------------------------------------------------
[18c0]| KEY
[18c0]| vdc                   :         0    sswid             :         0
[18c0]| swid                  :       12c    ftag              :         1

[18c0]| DATA
[18c0]| valid                 :         y    mp_mod            :         1
[18c0]| mp_base               :        24    local             :         n
[18c0]| cp_to_sup1            :         n    cp_to_sup2        :         n
[18c0]| drop                  :         n    dc3_si            :       11c1
[18c0]| data_tbl_ptr          :         0    ssw_ctrl          :         0
[18c0]| iic_port_idx          :        54
[18c0]| l2tunnel_remote (CR only) :        0
```

Switch-ID 重复可能会导致启用了 FabricPath 的网络出现转发故障和不稳定。如果要检查网络是否存在重复或有冲突的 Switch-ID，那么就可以使用命令 **show fabricpath conflict all**。如果出现了与 FabricPath 相关的任何错误，那么就可以考虑使用命令 **show system internal fabricpath switch-id event-history errors** 来验证特定 Switch-ID 的事件历史日志，也可以利用 **show tech-support fabricpath** 命令收集更多信息，进行进一步调查。

注：出现了 FabricPath 故障之后，可以考虑在故障期间收集以下 **show tech-support** 输出结果：
```
show tech u2rib
show tech pixm
show tech eltm
show tech l2fm
show tech fabricpath isis
show tech fabricpath topology
show tech fabricpath switch-id
```
除这些 **show tech** 命令之外，**show tech details** 的输出结果对于排查 FabricPath 环境故障来说也非常有用。

5.3.5 FabricPath 设备

Nexus 7000/7700 和 Nexus 5500 系列交换机均支持 FabricPath，如果希望了解与扩展性及支持该功能特性的交换模块相关的详细信息，可以查阅 FabricPath Configuration Guide（FabricPath 配置指南）。

5.4 仿真交换机和 vPC+

在现代数据中心设计场景中，服务器通常要连接多台边缘设备以提供冗余机制。对于启用了 FabricPath 功能特性的网络来说，仅边缘交换机学习 L2 MAC 地址，包括 MAC 地址到 Switch-ID 的映射信息（在数据平面是一对一的关联关系），如果主机的位置从一台交换机变更到另一台交换机，那么该关联或映射关系就会检测出 MAC 地址出现了移动。vPC 可以将物理连接到两台 Nexus 交换机的链路视为单个端口通道，并连接服务器或其他交换机，从而提供无环拓扑、消除生成树阻塞端口，并最大程度地提高带宽利用率。

对于启用了 FabricPath 的网络来说，如果主机或以太网交换机通过端口通道连接了两台 FabricPath 边缘交换机，那么支持相同配置就显得至关重要。由于双连接主机可以通过两台边缘交换机将数据包发送给 FabricPath 网络，因而会产生 MAC 振荡问题。图 5-15 给出了双连接主机引发的 MAC 振荡问题，图中的主机 A 双归上连边缘交换机 S1 和 S2，且正在与主机 B 进行通信。从交换机 S1 发送帧时，MAC-A 与交换机 S1 相关联；从交换机 S2 发送帧时，MAC-A 与交换机 S2 相关联，这样就产生了 MAC 振荡问题。为了解决这个问题，FabricPath 提供了仿真交换机（emulated switch）功能特性。

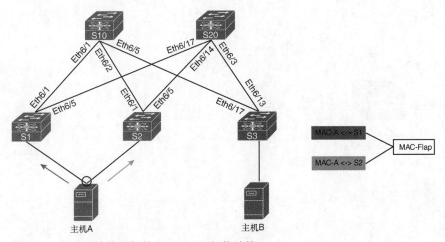

图 5-15 拥有双连接主机的 FabricPath 拓扑结构

仿真交换机架构允许两台或多台 FabricPath 交换机以单台交换机的方式为其余 FabricPath 网络提供服务。数据包离开被仿真为单台交换机的 FabricPath 交换机时，会通过分配给仿真交换机的仿真 switch-ID 映射源 MAC 地址。如果在 FabricPath 中实现了仿真交换机特性，由两台 FabricPath 边缘交换机为第三台设备提供 vPC，那么就称为 vPC+。因此，两台仿真交换机必须通过对等链路进行直连，且这两台交换机之间还应该存在对等保活路径。

使用仿真交换机时，必须了解多目的数据包的转发机制。FabricPath 网络通过计算以共享节点为根的多目的树来消除多目的帧的重复问题（确保到所有交换机都是无环路径），这就意味着对于特定多目的树来说，应该只有一台仿真交换机宣告去往仿真交换机的连接，且负责将数据包转发给仿真交换机。同样，仿真交换机发出的流量也可以通过其中的一台仿真交换机沿着图路径（包含去往仿真交换机的可达性）进入 FabricPath 网络。否则，数据包就会被 IIC（Ingress Interface Check，入站接口检查）丢弃。因此，对于每个多目的树来说，仅使用其中的一台仿真交换机处理入站和出站流量。

5.4.1 vPC+配置

配置 vPC+时,必须在 Nexus 交换机上启用以下两项主要功能特性。
- FabricPath。
- vPC。

为了更好地理解 vPC+的工作原理,下面将以图 5-16 所示拓扑结构为例加以说明。该拓扑结构中的 NX-1 和 NX-2 与 SW-12 构成了一个 vPC,NX-3 和 NX-4 与 SW-34 构成了一个 vPC。4 台 Nexus 交换机之间的所有链路都是启用了 FabricPath 的链路(包括 vPC 对等链路)。

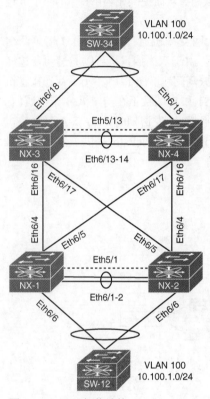

图 5-16　vPC+拓扑结构

首先分析例 5-41 中的 NX-1 和 NX-3 的 vPC 和 FabricPath 配置,其中的大多数配置与前面的 vPC 和 FabricPath 部分相似,主要区别在于 vPC 配置模式下的 **fabricpath switch-id** *switch-id* 命令,该命令为两台仿真交换机 NX-1 和 NX-2 分配了相同的 Switch-ID(分配的 Switch-ID 为 100),同时也为 NX-3 和 NX-4 分配了相同的 Switch-ID(分配的 Switch-ID 为 200)。

例 5-41　NX-1 和 NX-3 的 vPC+配置

```
NX-1
install feature-set fabricpath
feature-set fabricpath
feature vpc
vlan 100,200,300,400,500
  mode fabricpath
!
fabricpath switch-id 100
!
vpc domain 10
```

```
  peer-keepalive destination 10.12.1.2 source 10.12.1.1 vrf default
  fabricpath switch-id 100
!
interface port-channel1
  switchport mode fabricpath
  vpc peer-link
!
interface Ethernet6/4
  switchport mode fabricpath
!
interface Ethernet6/5
  switchport mode fabricpath
!
interface port-channel10
  switchport
  switchport mode trunk
  vpc 10
```

```
NX-3
install feature-set fabricpath
feature-set fabricpath
feature vpc
vlan 100,200,300,400,500
  mode fabricpath
!
fabricpath switch-id 200
!
vpc domain 20
  peer-keepalive destination 10.34.1.4 source 10.34.1.3 vrf default
  fabricpath switch-id 200
!
interface port-channel1
  switchport mode fabricpath
  vpc peer-link
!
interface Ethernet6/16
  switchport mode fabricpath
!
interface Ethernet6/17
  switchport mode fabricpath
!
interface port-channel20
  switchport
  switchport mode trunk
  vpc 20
```

5.4.2 vPC+验证和故障排查

配置了 FabricPath 和 vPC 之后，还要验证 vPC 对等邻接关系的建立情况。如果配置了 vPC+，那么命令 **show vpc** 不但能够显示与 vPC 相关的信息，而且还能显示 vPC+ Switch-ID（仿真 switch-ID）、DF（Designated Forwarder，指派转发器）标志（表示该交换机是否充当 DF）以及 FP MAC 路由（由 vPC 管理器分配）。例 5-42 给出了 **show vpc** 命令输出结果示例，以高亮方式显示了这些字段信息。

例 5-42 show vpc 命令输出结果

```
NX-1# show vpc
Legend:
```

```
                                  (*) - local vPC is down, forwarding via vPC peer-link
vPC domain id                             : 10
vPC+ switch id                            : 100
Peer status                               : peer adjacency formed ok
vPC keep-alive status                     : peer is alive
vPC fabricpath status                     : peer is reachable through fabricpath
Configuration consistency status          : success
Per-vlan consistency status               : success
Type-2 consistency status                 : success
vPC role                                  : primary
Number of vPCs configured                 : 1
Peer Gateway                              : Disabled
Dual-active excluded VLANs and BDs        : -
Graceful Consistency Check                : Enabled
Auto-recovery status                      : Enabled (timeout = 240 seconds)
Fabricpath load balancing                 : Disabled
Operational Layer3 Peer-router            : Disabled
Port Channel Limit                        : limit to 244
Self-isolation                            : Disabled
vPC Peer-link status
---------------------------------------------------------------------------
id Port Status Active vlans Active BDs
-- ---- ------ ------------------------------------------------------------
1  Po1  up     100,200,300,400,500 -
vPC status
Id : 10
  Port : Po10
  Status : up
  Consistency : success
  Reason : success
  Active Vlans : 100,200,300,400,500
  VPC+ Attributes: DF: Yes, FP MAC: 100.11.65535
```

验证 FabricPath 的仿真 Switch-ID 时，命令 **show fabricpath switch-id** 不但显示了静态 Switch-ID，而且显示了仿真 Switch-ID。此外，代表本地仿真交换机的节点旁边还会设置标志 E。例 5-43 通过 **show fabricpath switch-id** 命令显示了仿真 Switch-ID。

例 5-43 验证仿真 Switch-ID

```
NX-1# show fabricpath switch-id
                    FABRICPATH SWITCH-ID TABLE
Legend: '*' - this system
        '[E]' - local Emulated Switch-id
        '[A]' - local Anycast Switch-id
Total Switch-ids: 8
===========================================================================
    SWITCH-ID     SYSTEM-ID       FLAGS       STATE       STATIC   EMULATED/
                                                                   ANYCAST
    --------------+---------------+-----------+-----------+--------------------
*     10          6c9c.ed4f.28c2  Primary     Confirmed   Yes      No
      20          d867.d97f.fc42  Primary     Confirmed   Yes      No
      30          6c9c.ed4f.28c3  Primary     Confirmed   Yes      No
      40          d867.d97f.fc43  Primary     Confirmed   Yes      No
[E]   100         6c9c.ed4f.28c2  Primary     Confirmed   No       Yes
      100         d867.d97f.fc42  Primary     Confirmed   No       Yes
      200         6c9c.ed4f.28c3  Primary     Confirmed   No       Yes
      200         d867.d97f.fc43  Primary     Confirmed   No       Yes
```

如果运行仿真交换机特性的两台边缘设备都从远程边缘节点学到了 MAC 地址，那么除显示接口地址之外，还会显示远程边缘节点在 vPC 接口上分配的 MAC 路由。例 5-44 显示了 NX-1 和 NX-3 节点的 MAC 地址表信息。可以看出，NX-1 通过 FP MAC 路由 200.11.65535 分配的接口学到了连接在 NX-3/NX-4 vPC 链路上的远程主机的 MAC 地址，而连接在 NX-1 和 NX-2 上的主机则通过 FP MAC 路由 100.11.65535 分配的接口学到了 vPC 链路。

例 5-44　MAC 地址表

```
NX-1# show mac address-table vlan 100
Note: MAC table entries displayed are getting read from software.
Use the 'hardware-age' keyword to get information related to 'Age'

Legend:
        * - primary entry, G - Gateway MAC, (R) - Routed MAC, O - Overlay MAC
        age - seconds since last seen,+ - primary entry using vPC Peer-Link, E -
EVPN entry
        (T) - True, (F) - False ,  ~~~ - use 'hardware-age' keyword to retrieve
age info
   VLAN/BD    MAC Address      Type      age      Secure NTFY Ports/SWID.SSID.LID
---------+-----------------+--------+---------+------+----+------------------
   100       0022.56b9.007f   dynamic   ~~~        F    F   200.11.65535
*  100       24e9.b3b1.8cff   dynamic   ~~~        F    F   Po10

NX-3# show mac address-table vlan 100
Note: MAC table entries displayed are getting read from software.
Use the 'hardware-age' keyword to get information related to 'Age'

Legend:
        * - primary entry, G - Gateway MAC, (R) - Routed MAC, O - Overlay MAC
        age - seconds since last seen,+ - primary entry using vPC Peer-Link, E -
EVPN entry
        (T) - True, (F) - False ,  ~~~ - use 'hardware-age' keyword to retrieve
age info
   VLAN/BD    MAC Address      Type      age      Secure NTFY Ports/SWID.SSID.LID
---------+-----------------+--------+---------+------+----+------------------
*  100       0022.56b9.007f   dynamic   ~~~        F    F   Po20
   100       24e9.b3b1.8cff   dynamic   ~~~        F    F   100.11.65535
```

如果 MAC 学习出现了问题，那么就要检查设备之间是否存在 IS-IS 邻接关系。设备需要根据连接关系与 vPC 对等体以及其他脊设备或边缘节点建立 IS-IS 邻接关系。建立了 IS-IS 邻接关系之后，就可以通过 IS-IS 学习 FP 路由。通过 vPCM（vPC Manager，vPC 管理器）从 vPC 对等体学到的仿真交换机路由，在 URIB 中显示为通过 vPCM 学到的路由，如例 5-45 所示。进一步分析 URIB 可以看出，从远程仿真交换机学到的路由则带有标志或属性 E。

例 5-45　验证 URIB 中的本地和远程 FP 路由

```
NX-1# show fabricpath route detail
FabricPath Unicast Route Table
'a/b/c' denotes ftag/switch-id/subswitch-id
'[x/y]' denotes [admin distance/metric]
ftag 0 is local ftag
subswitch-id 0 is default subswitch-id


FabricPath Unicast Route Table for Topology-Default
```

```
0/10/0, number of next-hops: 0
        via ---- , [60/0], 0 day/s 02:16:33, local
0/10/3, number of next-hops: 1
        via sup-eth1, [81/0], 0 day/s 02:16:33, fpoam
0/100/1, number of next-hops: 0
0/100/3, number of next-hops: 1
        via sup-eth1, [81/0], 0 day/s 01:54:49, fpoam
0/100/11, number of next-hops: 1
        via Po10, [80/0], 0 day/s 01:47:44, vpcm
1/20/0, number of next-hops: 1
        via Po1, [115/20], 0 day/s 01:33:20, isis_fabricpath-default
1/30/0, number of next-hops: 1
        via Eth6/4, [115/40], 0 day/s 02:13:59, isis_fabricpath-default
1/40/0, number of next-hops: 1
        via Eth6/5, [115/40], 0 day/s 02:13:14, isis_fabricpath-default
1/100/0, number of next-hops: 0
        via ---- , [60/0], 0 day/s 01:54:49, local
1/100/0, number of next-hops: 1
        via Po1, [115/20], 0 day 00:00:00, isis_fabricpath-default
1/200/0, number of next-hops: 2
        via Eth6/4, [115/40], 0 day/s 01:41:01, isis_fabricpath-default
        via Eth6/5, [115/40], 0 day/s 01:41:01, isis_fabricpath-default
2/100/0, number of next-hops: 0
        via ---- , [60/0], 0 day/s 01:54:49, local
NX-1# show fabricpath isis route detail
Fabricpath IS-IS domain: default MT-0
Topology 0, Tree 0, Swid routing table
20, L1
Attribute:P* Instance 0x0000002B
 via port-channel1, metric 20*
! Output omitted for brevity
100, L1
Attribute:E* Instance 0x0000002B
 via port-channel1, metric 20*
200, L1
Attribute:E* Instance 0x0000002B
 via Ethernet6/4, metric 40*
 via Ethernet6/5, metric 40*
Legend E:Emulated, A:Anycast P:Physical, * - directly connected
```

由于 vPC+ 设计方案中的边缘端口通常是 vPC 端口，因而验证与 vPC 链路相关的 vPCM 信息至关重要。可以通过命令 **show system internal vpcm info interface** *interface-id* 加以验证，该命令可以显示端口通道接口的外层 FP MAC 地址、VLAN、vPC 对等信息以及 PSS 中存储的信息。请注意，PSS 中的信息可以在链路出现震荡或 VDC/交换机重新加载之后恢复信息。例 5-46 显示了 NX-1 节点 port-channel 10 的 vPCM 信息，并以高亮方式显示了 FP MAC 地址以及来自 vPC 对等体的相关信息。

例 5-46 验证 vPCM 中的 FP MAC 信息

```
NX-1# show system internal vpcm info interface po10
! Output omitted for brevity
port-channel10 - if_index: 0x16000009
--------------------------------------------------------------------------------

IF Elem Information:

        IF Index: 0x16000009
```

```
            MCEC NUM: 10 Is MCEC
            Allowed/Config VLANs    : 6 - [1,100,200,300,400,500]
            Allowed/Config BDs      : 0 - []

MCECM DB Information:
            IF Index        : 0x16000009
            vPC number      : 10
            Num members     : 0
            vPC state       : Up
            Internal vPC state: Up
            Compat Status   :
                    Old Compat Status       : Pass
                    Current Compat Status: Pass
                    Reason Code             : SUCCESS
                    Param compat reason code:0(SUCCESS)
            Individual mode: N
            Flags : 0x0
            Is AutoCfg Enabled : N
            Is AutoCfg Sync Complete : N
            Number of members: 0
            FEX Parameters:
                    vPC is a non internal-vpc
                    vPC is a non fabric-vpc
                    FPC bringup: FALSE
                    Parent vPC number: 0
            Card type               : F2
            Hardware prog state     : No R2 prog
            Fabricpath outer MAC address info: 100.11.65535
            Designated forwarder state: Allow
            Assoc flag: Disassociated
            Is switchport: Yes  Is shared port: FALSE

            Up VLANs                : 5 - [100,200,300,400,500]
            Suspended VLANs         : 1 - [1]
            Compat check pass VLANs: 4096 - [0-4095]
            Compat check fail VLANs : 0 - []

            Up BDs                  : 0 - []
            Suspended BDs           : 0 - []
            Compat check pass BDs   : 0 - []
            Compat check fail BDs   : 0 - []
            Compat check pass VNIs  : 0 - []
            Compat check fail VNIs  : 0 - []

            vPC Peer Information:

            Peer Number : 10
            Peer IF Index: 0x16000009
            Peer state      : Up
            Card type               : F2
            Fabricpath outer MAC address info of peer: 100.11.0
            Peer configured VLANs   : 6 - [1,100,200,300,400,500]
            Peer Up VLANs           : 5 - [100,200,300,400,500]
            Peer configured VNIs    : 0 - []
            Peer Up BDs             : 0 - []
PSS Information:
```

```
                IF Index       : 0x16000009
                vPC number: 10
                vPC state: Up
                Internal vPC state: Up
                Old Compat Status: Pass
                Compat Status: Pass
                Card type      : F2
                Fabricpath outer MAC address info: 100.11.65535
                Designated forwarder state: Allow
                Up VLANs                    : 5 - [100,200,300,400,500]
                Suspended VLANs             : 1 - [1]
                Up BDs                      : 0 - []
                Suspended BDs               : 0 - []

                vPC Peer Information:
                Peer number: 10
                Peer if_index: 0x16000009
                Peer state: Up
                Card type      : F2
                Fabricpath outer MAC address info of peer: 100.11.0
                Peer configured VLANs       : 6 - [1,100,200,300,400,500]
                Peer Up VLANs               : 5 - [100,200,300,400,500]
            ..
```

其他的平台级验证操作已经在前面的 FabricPath 章节讨论过了，目的是验证 FabricPath 路由在线卡上为的硬件编程情况。

> 注：与平台相关的命令不但与具体的平台相关，而且与 Nexus 7000/7700 机箱上的线卡相关。如果出现了 vPC+故障，那么就可以采集以下 **show tech-support** 命令的输出结果：
>
> show tech-support fabricpath
> show tech-support vpc
>
> 此外，5.3 节还给出了其他有用的 **show tech-support** 命令。

5.5 本章小结

本章从 L2 转发角度介绍了在交换机之间提供弹性机制并增加容量的相关技术和功能特性。端口通道和虚拟端口通道允许交换机利用物理成员端口创建逻辑接口，排查相关故障问题时，确保所有成员端口的配置一致性是非常重要的排查内容。本章详细介绍了这类故障问题的排查技术及差错消息。

FabricPath 提供了一种灵活的、无须生成树的协议机制，能够有效提高链路吞吐量和扩展性，同时还能最大程度地减少与生成树相关的广播问题。FabricPath 的处理过程非常简单，即在 L2 域中以封装状态路由数据包，转发给主机之前解封装数据包。排查 FabricPath 拓扑结构中的数据包转发故障时，不但要用到一些基本的 STP 和端口转发故障排查概念，而且还要结合一些 IS-IS 网络故障排查技术。

第三部分

三层路由故障排查

第 6 章　IP 和 IPv6 服务故障排查
第 7 章　EIGRP 故障排查
第 8 章　OSPF 故障排查
第 9 章　IS-IS 故障排查
第 10 章　Nexus 路由映射故障排查
第 11 章　BGP 故障排查

第 6 章

IP 和 IPv6 服务故障排查

本章主要讨论如下主题。
- IP SLA（Service Level Agreement，服务等级协议）。
- 对象跟踪。
- IPv4 服务。
- IPv6 服务。
- FHRP（First-Hop Redundancy Protocol，第一跳冗余协议）。

6.1 IP SLA

IP SLA 是一种网络性能监控应用，允许用户执行服务等级监控、故障排查和资源规划。IP SLA 是一种能够感知应用程序的综合操作代理，可以通过测量响应时间、网络可靠性、资源可用性、应用程序性能、抖动、连接时间和丢包等参数来监控网络性能。该功能特性收集的统计信息有助于实现 SLA 监控、故障排查、问题分析以及网络拓扑结构设计等操作。IP SLA 主要包括以下两个实体。

- **IP SLA 发送端**：IP SLA 发送端根据用户配置的操作类型生成有效的测量流量并报告度量指标。除报告度量指标之外，IP SLA 发送端还负责检测阈值违规事件并发送通知。图 6-1 给出了面向不同操作类型的 IP SLA 测量选项。

图 6-1 面向不同操作类型的 IP SLA 测量选项

- **IP SLA 响应端**：响应端与发送端运行在不同的交换机上，负责响应 UDP/TCP（User Datagram Protocol/Transmission Control Protocol，用户数据报协议/传输控制协议）探针并对控制包做出响应。响应端检查发送端发送的数据包，通过控制包和 CLI（Command-Line Interface，命令行接口）来确定 TCP/UDP 端口和地址。

系统默认不启用 IP SLA 功能特性。如果要启用 IP SLA 功能特性，则需要使用命令 **feature sla [responder | sender]**。除非 IP SLA 发送端同时充当远程设备的响应端，否则不需要在同一台设备上同时启用 SLA 发送端和响应端功能。

可以在 Nexus 上配置不同的 IP SLA 探针。
- ICMP 回显（ICMP Echo）。
- ICMP 抖动（ICMP Jitter）。

- UDP 回显（UDP Echo）。
- UDP 抖动（UDP Jitter）。
- TCP 连接（TCP Connect）。

6.1.1 ICMP 回显探针

发送端将 ICMP（Internet Control Message Protocol，Internet 控制报文协议）ping 包（回显数据包）发送给目的端（响应端）以测量可达性和 RTT（Round-Trip Time，往返时间）。发送端会创建一个 ICMP ping 包并添加时间戳，然后发送给另一个节点（响应端）。响应端收到探针之后，添加自己的时间戳并响应 ICMP ping 包。发送端收到该数据包之后，计算 RTT 并存储统计信息。例 6-1 说明了两台 Nexus 交换机 NX-1 与 NX-2 之间的 ICMP 回显探针的配置情况。可以使用命令 **ip sla** *number* 配置 IP SLA 探针，在此配置模式下可以进一步利用命令 **icmp-echo** *dest-ip-address* [**source-interface** *interface-id* | **source-ip** *src-ip-address*]定义 ICMP 回显探针，利用命令选项 **request-data-size** *size* 以自定义数据大小发送探针。

例 6-1　IP SLA ICMP 回显探针配置

```
feature sla sender
!
ip sla 10
  icmp-echo 192.168.2.2 source-interface loopback0
    request-data-size 1400
    frequency 5
ip sla schedule 10 start-time now
```

配置完成后，探针并不会自行启动。可以通过全局配置命令 **ip sla schedule** *number* **start-time** [**now** | **after** *time* | *time*]立即启动或者在特定时间段后启动探针，其中，参数 *time* 的格式是 hh:mm:ss。

配置 ICMP 回显探针时，不需要在探针的目的远程设备上配置 IP SLA 响应端。探针启动后，可以使用命令 **show ip sla statistics** [*number*] [**aggregated** |**details**]验证探针的统计信息，选项 **aggregated** 可以显示汇总统计信息，选项 **details** 可以显示详细统计信息。例 6-2 显示了例 6-1 配置的 ICMP 回显探针的统计信息。请注意，对于 **show ip sla statistics** 命令的输出结果来说，应仔细验证 RTT 值、返回代码、成功次数和失败次数等字段值。对于 **show ip sla statistics aggregated** 命令的输出结果来说，此时的 RTT 值显示为汇总值（如该探针 RTT 最小值/平均值/最大值）。

例 6-2　IP SLA 统计信息

```
NX-1# show ip sla statistics 10

IPSLAs Latest Operation Statistics

IPSLA operation id: 10
        Latest RTT: 2 milliseconds
Latest operation start time: 06:16:13 UTC Sat Sep 30 2017
Latest operation return code: OK
Number of successes: 678
Number of failures: 0
Operation time to live: 213 sec
NX-1# show ip sla statistics aggregated

IPSLAs aggregated statistics

IPSLA operation id: 10
```

```
Type of operation: icmp-echo
Start Time Index: 05:19:48 UTC Sat Sep 30 2017

RTT Values:
        Number Of RTT: 694              RTT Min/Avg/Max: 2/2/7 milliseconds
Number of successes: 694
Number of failures: 0
```

> **注：** 可以使用命令 **show running-config sla sender** 或命令 **show ip sla configuration** *number* 查看 IP SLA 探针的配置信息。

6.1.2 UDP 回显探针

发送端将用户自定义大小的单个 UDP 包发送给目的端（响应端）并测量 RTT。发送端将创建 UDP 包，按照用户自定义参数打开套接字，然后将数据包发送给另一个节点（响应端）并记下时间。响应端收到探针之后，将使用相同的套接字参数将相同的数据包发送回发送端。发送端收到数据包之后，记录接收时间、计算 RTT，并存储统计信息。

如果要定义 UDP 回显 IP SLA 探针，那么就可以使用命令 **udp-echo** [*dest-ip-address* | *dest-hostname*] *dest-port-number* **source-ip** [*src-ip-address* | *src-hostname*] **source-port** *src-port-number* [**control** [**enable** | **disable**]]。例 6-3 给出了交换机 NX-1 的 UDP 回显探针和交换机 NX-2 的响应端配置示例，同时还显示了探针启动后的统计信息。请注意，必须在远端配置响应端，否则探测将失败。如果要配置 IP SLA 响应端，那么就可以使用命令 **ip sla responder**。如果要配置 UDP 回显探针响应端，那么就可以使用命令 **ip sla responseer udp-echo ipaddress** *ip-address* **port** *port-number*。

例 6-3 IP SLA UDP 回显探测及统计信息

```
NX-1 (Sender)
ip sla 11
  udp-echo 192.168.2.2 5000 source-ip 192.168.1.1 source-port 65000
    tos 180
    frequency 10
ip sla schedule 11 start-time now

NX-2 (Responder)
ip sla responder
ip sla responder udp-echo ipaddress 192.168.2.2 port 5000

NX-1
NX-1# show ip sla statistics 11 details

IPSLAs Latest Operation Statistics

IPSLA operation id: 11
        Latest RTT: 2 milliseconds
Latest operation start time: 04:45:38 UTC Sun Oct 01 2017
Latest operation return code: OK
Number of successes: 3
Number of failures: 6
Operation time to live: 3459 sec
Operational state of entry: Active
Last time this entry was reset: Never
```

> **注：** 配置了 UDP 回显探针响应端之后，响应端设备将持续监听响应端节点上指定的 UDP 端口。

6.1.3 UDP 抖动探针

UDP 抖动探针以稳定的时间间隔将一串 UDP 数据包发送给目的端以测量数据包抖动情况。此外，还可以指定一个编解码器，以确定净荷大小、包间隔以及所要发送的数据包数量，从而模拟特定的语音编解码器。发送端将数据包发送给响应端之前，会将自己的发送时间戳添加到数据包中。响应端收到数据包之后，会将数据包副本发送回发送端，并携带接收端的接收时间和处理时间。发送端收到该副本之后，将使用该数据包的接收时间加上前 3 个值（发送端发送时间、响应端接收时间和响应端处理时间）来计算 RTT 值并更新抖动统计信息。

UDP Plus 操作是 UDP 回显操作的一个超集，除测量 UDP RTT 之外，UDP Plus 操作还可以测量各个方向的丢包和抖动情况。抖动是数据包之间的时延差异，抖动统计信息对于分析 VoIP（Voice over IP，IP 语音）网络中的流量非常有用。

UDP 抖动探针的配置命令是 **udp-jitter** [*dest-ip-address* | *dest-hostname*] *dest-port-number* **codec** *codec-type* [**codec-numpackets** *number-of-packets*] [**codec-size** *number-of-bytes*] [**codec-interval** *milliseconds*] [**advantage-factor** *value*] **source-ip** [*src-ip-address* | *src-hostname*] **source-port** *src-port-number* [**control** [**enable** | **disable**]]。IP SLA UDP 抖动操作的默认请求包大小为 32 字节，可以在 **ip sla** 命令下通过选项 **request-data-size** 修改该数值。

表 6-1 给出了 **udp-jitter** 配置中的部分选项信息。

表 6-1 udp-jitter 配置选项

选项	描述
control {**enable** | **disable**}	（可选）启动/禁止向 IP SLA 响应端发送 IP SLA 控制消息
codec *codec-type*	以 ICPIF（Calculated Planning Impairment Factor，计算计划损伤因子）和 MOS（Mean Opinion Score，平均意见得分）值的形式生成预估的语音质量得分。可以利用 UDP 抖动探针配置以下 3 类代码 ■ g711alaw：G.711 A 律（64 kbit/s 传输） ■ g711ulaw：G.711 U 律（64 kbit/s 传输） ■ g729a：G.729（8 kbit/s 传输）
codec-numpackets *number-of-packets*	（可选）指定每次操作发送的数据包数量，取值范围是 1~60000，默认值为 1000
codec-size *number-of-bytes*	（可选）指定每个发送的数据包的字节数（也称为净荷大小或请求包大小），取值范围是 16~1500，默认值取决于具体的编解码器
codec-interval *milliseconds*	指定操作所使用的数据包之间的间隔（时延），取值范围是 1~60000，默认值为 20
advantage-factor *value*	指定用于 ICPIF 计算的期望因子。根据测量得到的减损减去该值即可得到最终的 ICPIF 值（以及相应的 MOS 值）

例 6-4 给出了利用 g729a 编解码器配置 UDP 抖动探针的相关信息，该编解码器的 ToS（Type of Service，服务类型）值被设置为 180。通过命令 **ip sla schedule** 及选项 **life** [*time-in-seconds* | **forever**] 指定探针的生存时间。有关 UDP 抖动探针的更详细信息可以查看统计信息，包括 UDP 抖动探针的单向时延、抖动时间、丢包及语音得分等统计信息。

例 6-4 IP SLA UDP 抖动探针及统计信息

```
NX-1
ip sla 15
  udp-jitter 192.168.2.2 5000 codec g729a codec-numpackets 50 codec-interval 100
    tos 180
    verify-data
    frequency 10
ip sla schedule 15 life forever start-time now
```

```
NX-1# show ip sla statistics 15 details

IPSLAs Latest Operation Statistics

IPSLA operation id: 15
Type of operation: udp-jitter
        Latest RTT: 20 milliseconds
Latest operation start time: 06:53:31 UTC Sun Oct 01 2017
Latest operation return code: OK
RTT Values:
        Number Of RTT: 47                RTT Min/Avg/Max: 9/20/34 milliseconds
Latency one-way time:
        Number of Latency one-way Samples: 0
        Source to Destination Latency one way Min/Avg/Max: 0/0/0 milliseconds
        Destination to Source Latency one way Min/Avg/Max: 0/0/0 milliseconds
        Source to Destination Latency one way Sum/Sum2: 0/0
        Destination to Source Latency one way Sum/Sum2: 0/0
Jitter Time:
        Number of SD Jitter Samples: 0
        Number of DS Jitter Samples: 0
        Source to Destination Jitter Min/Avg/Max: 0/0/0 milliseconds
        Destination to Source Jitter Min/Avg/Max: 0/0/0 milliseconds
        Source to destination positive jitter Min/Avg/Max: 0/0/0 milliseconds
        Source to destination positive jitter Number/Sum/Sum2: 0/0/0
        Source to destination negative jitter Min/Avg/Max: 0/0/0 milliseconds
        Source to destination negative jitter Number/Sum/Sum2: 0/0/0
        Destination to Source positive jitter Min/Avg/Max: 0/0/0 milliseconds
        Destination to Source positive jitter Number/Sum/Sum2: 0/0/0
        Destination to Source negative jitter Min/Avg/Max: 0/0/0 milliseconds
        Destination to Source negative jitter Number/Sum/Sum2: 0/0/0
        Interarrival jitterout: 0        Interarrival jitterin: 0
        Jitter AVG: 0
        Over thresholds occured: FALSE
Packet Loss Values:
        Loss Source to Destination: 0
        Source to Destination Loss Periods Number: 0
        Source to Destination Loss Period Length Min/Max: 0/0
        Source to Destination Inter Loss Period Length Min/Max: 0/0
        Loss Destination to Source: 0
        Destination to Source Loss Periods Number: 0
        Destination to Source Loss Period Length Min/Max: 0/0
        Destination to Source Inter Loss Period Length Min/Max: 0/0
        Out Of Sequence: 0       Tail Drop: 50
        Packet Late Arrival: 0   Packet Skipped: 0
Voice Score Values:
        Calculated Planning Impairment Factor (ICPIF): 11
        MOS score: 4.06
Number of successes: 23
Number of failures: 1
Operation time to live: forever
Operational state of entry: Active
Last time this entry was reset: Never
```

注：如果希望 IP SLA 检查每个应答包的数据损坏情况，那么就可以在 **ip sla** 配置模式下使用 **verify-data** 命令。

随着时延和抖动的增加，MOS 得分也将下降。这类统计信息可以帮助网络设计和实施团队为应用程序优化网络。

6.1.4 TCP 连接探针

TCP 连接探针针对给定的源和目的 IP 地址及端口进行无阻塞的 TCP 连接，将当前时间标记为操作时间。如果立即收到响应端的成功响应，那么发送端就会使用当前时间和操作时间的差值来更新 RTT 及其统计组件。如果发送端没有立即收到成功的响应消息，那么就会设置一个定时器并等待由目的 IP 地址返回的响应消息所触发的回呼操作，如果收到了回呼消息，那么发送端就可以像以前一样更新 RTT 和统计组件。

TCP 连接操作负责测量连接目的设备所需的时间，可以测试虚拟电路的可用性或应用程序的可用性。如果目的设备是思科路由器，那么 IP SLA 探针就会与用户指定的端口号建立 TCP 连接。如果目的设备是非思科 IP 主机，那么就必须指定一个周知目的端口号（如 FTP[File Transfer Protocol，文件传输协议]的 21、Telnet 的 23、HTTP[Hypertext Transfer Protocol，超文本传输协议]服务器的 80）。该操作对于测试 Telnet 或 HTTP 的连接时间来说非常有用。

如果要定义 IP SLA TCP 连接探针，那么就可以使用命令 **tcp-connect** [*dest-ip-address* | *dest-hostname*] *dest-port-number* **source-ip** [*src-ip-address* | *src-hostname*] **source-port** *srcport-number* [**control** [**enable** | **disable**]]。对于 TCP 连接探针来说，必须使用命令 **ip sla responder tcp-connect ipaddress** *ip-address* **port** *port-number* 在目的路由器/交换机上配置响应端。例 6-5 给出了 IP SLA TCP 连接探针的配置示例，目的是探测交换机 NX-1 与 NX-2 之间的 TCP 连接。

例 6-5 IP SLA TCP 连接探针的配置及统计信息

```
NX-1 (Sender)
ip sla 20
  tcp-connect 192.168.2.2 10000 source-ip 192.168.1.1
ip sla schedule 20 life forever start-time now

NX-2 (Responder)
ip sla responder tcp-connect ipaddress 192.168.2.2 port 10000

NX-1
NX-1# show ip sla statistics 20 details

IPSLAs Latest Operation Statistics

IPSLA operation id: 20
        Latest RTT: 304 milliseconds
Latest operation start time: 16:23:03 UTC Sun Oct 01 2017
Latest operation return code: OK
Number of successes: 2
Number of failures: 0
Operation time to live: forever
Operational state of entry: Active
Last time this entry was reset: Never
```

注：如果希望了解 IP SLA 的其他命令选项信息，可以参考 Nexus CCO（Cisco Connection Online，思科在线）文档。

6.2 对象跟踪

为了确保负载均衡与故障切换能力，需要在网络中部署多种 IP 和 IPv6 服务（如 FHRP[First-Hop Redundancy Protocol，第一跳冗余协议]）以实现高可靠性和高可用性。不过，在 WAN 链路中断等故障情况下（这类故障在网络中可能比路由器故障更常见），即便部署了这些功能，也无法保证网络的正常运行，因而此时可能会导致链路出现严重中断。

对象跟踪特性为影响和控制网络中的故障切换操作提供了一种非常灵活且可自定义的操作机制。该功能特性可以跟踪网络中的指定对象，并在对象状态发生变化进而影响网络流量时采取必要的操作。对象跟踪特性的主要目的是允许路由器系统中的进程和协议监控同一系统中其他不相关进程和协议的属性，从而实现以下目的。

- 提供最佳服务等级。
- 提高网络的可用性和恢复速度。
- 减少网络中断及中断持续时间。

HSRP（Hot Standby Router Protocol，热备份路由器协议）、VRRP（Virtual Router Redundancy Protocol，虚拟路由器冗余协议）和 GLBP（Gateway Load Balancing Protocol，网关负载均衡协议）等客户端可以在被跟踪对象中注册其所要跟踪的对象，并在对象状态发生变化时采取必要的响应措施。除这些协议之外，可以使用该功能特性的客户端还有以下设备。

- EEM（Embedded Event Manager，嵌入式事件管理器）。
- vPC（virtual Port-Channel，虚拟端口通道）。

对象跟踪特性可以跟踪以下对象。

- 线路协议状态变更。
- 线路可达性。
- 对象跟踪列表。

对象跟踪的配置命令是 **track** *number* <*object-type*> <*object-instance*> <*object-parameter*>，其中，*number* 的取值范围是 1～1000，*object-type* 指的是支持的被跟踪对象（如接口、IP 路由或跟踪列表），*object-instance* 指的是被跟踪对象的实例（如接口名、路由前缀、掩码等），*object-parameter* 指的是与对象类型相关的参数。

6.2.1 接口的对象跟踪

命令 **track** *number* **interface** *interface-id* [**line protocol** | **ip routing** | **ipv6 routing**]的作用是创建跟踪对象，目的是跟踪接口的状态以及接口的线路协议状态或 IP/IPv6 路由。例 6-6 给出了跟踪接口的线路协议和 IP 路由状态的配置示例。从第一次测试可以看出，如果接口被关闭，那么其线路协议和 IP 路由的跟踪状态都将显示为关闭状态。从第二次测试可以看出，如果将接口设置为 L2 端口（通过 **switchport** 命令配置该接口），那么只有 track 2（IP 路由跟踪）的跟踪状态显示为关闭（因为端口 Eth2/5 当前已禁用 IP 路由）。可以通过命令 **show track** 检查已配置跟踪对象的状态信息。

例 6-6 接口状态的对象跟踪

```
NX-1
NX-1(config)# track 1 interface ethernet 2/5 line-protocol
NX-1(config)# track 2 interface ethernet 2/5 ip routing
NX-1(config)# interface ethernet2/5
NX-1(config-if)# shut

NX-1# show track
Track 1
  Interface Ethernet2/5 Line Protocol
  Line Protocol is DOWN
  2 changes, last change 00:00:08

Track 2
  Interface Ethernet2/5 IP Routing
  IP Routing is DOWN
```

```
  2 changes, last change 00:00:08
NX-1(config)# interface ethernet2/5
NX-1(config-if)# no shut
NX-1(config-if)# switchport
```

```
NX-1# show track
Track 1
  Interface Ethernet2/5 Line Protocol
  Line Protocol is UP
  5 changes, last change 00:00:41

Track 2
  Interface Ethernet2/5 IP Routing
  IP Routing is DOWN
  4 changes, last change 00:00:42
```

6.2.2 路由状态的对象跟踪

如果要配置路由状态跟踪对象，那么就可以使用命令 **track** *number* **ip route** *ip-address/mask* **reachability**（适用于 IPv4 路由）或命令 **track** *number* **ipv6 route** *ipv6-address/mask* **reachability**（适用于 IPv6 路由），该命令可以创建一个路由对象以跟踪路由的可达性。例 6-7 给出了 IPv4 和 IPv6 路由状态跟踪对象的配置示例，如果被跟踪路由的可达性丢失（如数据包丢失或路由协议震荡等原因），那么对象跟踪就会中断。此外，还可以配置对象跟踪的启动时延和中断时延。可以通过命令 **delay [down | up]** *time-in-seconds* 设置对象跟踪的启动和中断时延（以秒为单位），**delay** 命令选项的作用是防止瞬态或非持久事件导致的跟踪中断。

例 6-7 路由状态的对象跟踪

```
NX-1
NX-1(config)# track 5 ip route 192.168.2.2/32 reachability
NX-1(config-track)# delay down 3
NX-1(config-track)# delay up 1
NX-1# show track 5
Track 5
  IP Route 192.168.2.2/32 Reachability
  Reachability is UP
  3 changes, last change 00:02:07
  Delay up 1 secs, down 3 secs
```

此外，还可以通过命令 **track** *number* **ip sla [reachability | status]** 为 IP SLA 探针配置跟踪对象，因而跟踪对象可以间接验证远端前缀的可达性。使用 IP SLA 探针的好处在于，网络运营商不仅能够使用 IP SLA 来验证可达性，还能跟踪 UDP 回显、UDP 抖动和 TCP 连接等其他探针的状态。

例 6-8 给出了 IP SLA 探针的对象跟踪配置示例。请注意，**show track** 命令的输出结果不但显示了状态信息，而且显示了操作代码和 RTT 信息，这些信息实际上都是 **show ip sla statistics** 命令输出结果的一部分。

例 6-8 IP SLA 探针的对象跟踪

```
NX-1
ip sla 10
  icmp-echo 192.168.2.2 source-interface loopback0
    request-data-size 1400
    frequency 5
ip sla schedule 10 start-time now
```

```
!
track 10 ip sla 10 state
NX-1# show track 10
Track 10
  IP SLA 10 State
  State is UP
  1 changes, last change 00:01:01
  Latest operation return code: OK
  Latest RTT (millisecs): 3
```

6.2.3 跟踪列表状态的对象跟踪

可以根据多个对象的状态创建条件式跟踪对象，实现方式是跟踪列表（Track-List）。可以在跟踪列表中添加多个对象，通过 **and** 和 **or** 函数组成的布尔表达式得到这些对象的跟踪列表状态。跟踪列表对象的配置命令是 **track** *number* **list boolean [and | or]**，在跟踪列表配置模式下，可以通过命令 **object** *object-number* **[not]** 列出多个对象。请注意，如果对象的条件不应该为 True，那么就应该使用选项 **not**。

例 6-9 给出了通过 **and** 和 **or** 布尔表达式在跟踪列表上配置对象跟踪的示例。请注意示例的第二部分，跟踪列表对象 20 的 **show track** 命令输出结果显示其状态为 DOWN，这是因为对象 2 的状态不是 UP（因为接口 Eth2/5 启用了 IP 路由）。

例 6-9 跟踪列表的对象跟踪

```
NX-1
! Previous track configurations
NX-1# show run track
track 1 interface Ethernet2/5 line-protocol
track 2 interface Ethernet2/5 ip routing
track 5 ip route 192.168.2.2/32 reachability
 delay up 1 down 3
track 10 ip sla 10

! Track List with Boolean AND for matching track 1 and not matching track 2.
NX-1(config)# track 20 list boolean and
NX-1(config-track)# object 1
NX-1(config-track)# object 2 not

! Track List with Boolean OR for matching track 1 or matching track 5.
NX-1(config)# track 30 list boolean or
NX-1(config-track)# object 1
NX-1(config-track)# object 5
NX-1# show track 20
Track 20
  List  Boolean and
  Boolean and is DOWN
  2 changes, last change 00:03:14
  Track List Members:
    object 2 not UP
    object 1 UP

NX-1# show track 30
Track 30
  List Boolean or
  Boolean or is UP
  1 changes, last change 00:01:23
```

```
Track List Members:
   object 5 UP
   object 1 UP
```

此外,还可以指定用于维护跟踪列表状态的阈值。此时,可以采用以下两种方式定义阈值。
- 百分比。
- 权重。

这两种方法都可以用于跟踪列表,方法是使用命令 **track** *number* **list threshold [percentage | weight]**。

可以使用命令 **threshold percentage up** *value* **down** *value* 配置百分比阈值。如果处于 UP 状态的已配置跟踪列表对象的百分比超出了配置阈值,那么跟踪状态就保持为 UP 状态,否则就进入 DOWN 状态。

与此相似,可以使用命令 **threshold weight up** *value* **down** *value* 配置权重阈值。如果处于 UP 状态的对象的组合权重超出了配置阈值,那么跟踪状态将保持在 UP 状态。

例 6-10 给出了通过百分比和权重方式配置跟踪列表对象阈值的配置示例。对于第一个配置示例(百分比阈值)来说,由于 UP 百分比被配置为 60,因而至少有两个对象处于 UP 状态。对于第二个配置示例(权重阈值)来说,由于 UP 状态的权重被配置为 45,因而仅当对象 1 以及其他两个对象(对象 2 或对象 5)中的一个处于 UP 状态时,跟踪对象才保持在 UP 状态。

例 6-10 跟踪列表对象的阈值

```
NX-1(config)# track 100 list threshold percentage
NX-1(config-track)# threshold percentage up 60 down 40
NX-1(config-track)# object 1
NX-1(config-track)# object 2
NX-1(config-track)# object 5
NX-1(config)# track 200 list threshold weight
NX-1(config-track)# threshold weight up 45 down 30
NX-1(config-track)# object 1 weight 30
NX-1(config-track)# object 2 weight 15
NX-1(config-track)# object 5 weight 15
```

6.2.4 将跟踪对象与静态路由相结合

可以将对象跟踪与静态路由集成在一起,从而在对象跟踪出现中断后对流量转发行为施加影响。只要跟踪状态为 UP,那么跟踪对象就可以与静态路由相关联(将路由安装到路由表,即 URIB[Unicast Routing Information Base,单播路由信息库]中)。Netstack/IP 组件会持续监听来自对象跟踪组件的消息(通过 MTS[Message and Transactional Service,消息和事务服务]进行交换)。如果跟踪状态转为 DOWN,那么 NX-OS 上的跟踪组件就会向 Netstack 组件发送 MTS 消息,并删除 URIB 中与跟踪对象相关联的路由。如果希望跟踪状态出现中断后能够切换流量,那么就需要预先配置另一条指向相同目的端的静态路由,该静态路由的 AD(Administrative Distance,管理距离)值相对较大且拥有不同的下一跳。例 6-11 给出了跟踪对象与静态路由集成使用的示例,可以在故障事件发生后实现快速恢复。对于本例来说,只要 Track 1 的状态为 UP,那么就会在路由表中安装下一跳为 10.12.1.2 的 192.168.2.2/32 路由。

例 6-11 结合使用跟踪对象和静态路由

```
NX-1
NX-1(config)# ip route 192.168.2.2/32 10.12.1.2 track 1
NX-1(config)# ip route 192.168.2.2/32 10.13.1.3 254
```

```
NX-1# show ip route 192.168.2.2/32
192.168.2.2/32, ubest/mbest: 1/0
    *via 10.12.1.2, [1/0], 00:00:48, static

! Shutting down the interface E2/5 having subnet 10.12.1.0/24

NX-1(config)# interface e2/5
NX-1(config-if)# shutdown

NX-1# show track 1
Track 1
  Interface Ethernet2/5 Line Protocol
  Line Protocol is DOWN
  8 changes, last change 00:00:03

! Change in track status leading to failover of traffic to different next-hop
NX-1# show ip route 192.168.2.2/32
192.168.2.2/32, ubest/mbest: 1/0
    *via 10.13.1.3, [254/0], 00:00:40, static
```

> **注：** 如果网络出现了与对象跟踪有关的故障问题，那么就可以收集 **show tech track** 命令的输出结果并联系思科 TAC（Technical Assistance Center，技术支持中心）。

6.3 IPv4 服务

NX-OS 提供了大量关键网络服务，可以实现网络的灵活性、可扩展性、可靠性和安全性，从而解决企业网或数据中心面临的诸多关键问题。本节将讨论以下 IP 服务。

- DHCP 中继。
- DHCP 监听。
- 动态 ARP 检查。
- IP 源防护。
- 单播 RPF。

6.3.1 DHCP 中继

与传统的思科 IOS 或思科 IOS XE 软件不同，NX-OS 不支持 DHCP（Dynamic Host Configuration Protocol，动态主机配置协议）服务器功能，不过，可以让 NX-OS 设备充当 DHCP 中继代理。DHCP 中继代理是一种可以在 DHCP 客户端与 DHCP 服务器（两者位于不同的子网中）之间中继 DHCP 请求/应答的设备。中继代理负责监听客户端的请求并增加一些重要数据，如客户端的链路信息（服务器需要利用该信息为客户端分配地址）。DHCP 服务器应答时，由中继代理将信息转发回客户端。

虽然 DHCP 中继代理是一项非常有用的功能特性，但同时也带来了一些安全问题。

- 端口上的主机无法看到其他端口上的主机流量。
- 连接在城域端口上的主机不再可信，需要使用某种机制来安全地识别这些主机。
- 防范恶意主机发起的网络欺骗攻击（IP 耗尽、IP 欺骗、DoS[Denial of Service，拒绝服务]攻击等）。

DHCP 选项 82 有助于解决这些问题。DHCP 选项 82 定义在 RFC 3046 中，是一种新型容器选项，包含了中继代理收集的子选项信息。图 6-2 给出了 DHCP 中继代理的数据包格式。

```
+---------+---------+---------+---------+---------+
|  代码   |  长度   |  代理   |  信息   |  字段   |
+---------+---------+---------+---------+---------+
|   82    |    N    |   i1    |   ...   |   iN    |
+---------+---------+---------+---------+---------+
```

图 6-2 中继代理的数据包格式

长度 N 表示的是代理信息字段的字节总数，该字段包含了多个由 SubOpt/Length/Value 三元组构成的子选项。

下面将以启用了 DHCP 选项 82 功能特性的接入交换机为例，说明 DHCP 消息流的交互过程。

- 客户端从交换机端口向外广播 DHCPDISCOVER 消息（UDP 目的端口 67 和 UDP 源端口 68）。
- 交换机上的中继代理拦截该广播请求并插入选项 82 数据（电路 ID 和远程 ID），将中继代理的 IP 地址放到 DHCP 数据包的 giaddr 字段中，将 UDP 源端口 68 替换为中继代理服务器端口 67，然后以单播方式将客户端请求发送给同一 UDP 目的端口上的 DHCP 服务器。中继代理接口启用了选项 82 功能之后，就会将 DHCP 服务器的 IP 地址配置为 IP Helper 地址。
- DHCP 服务器收到并处理中继来的 DHCP 客户端请求，支持选项 82 数据的 DHCP 服务器将以 DHCPOFFER 消息作为响应，响应消息的 yiaddr 字段包含了可用网络地址以及所有的选项 82 数据。响应消息是 UDP 单播消息，将直接路由给 UDP 目的端口 67 的中继代理。
- 中继代理收到服务器的应答消息之后，将删除消息中的选项 82 数据，然后以单播或广播方式将 DHCPOFFER 消息发送回客户端。
- 客户端可能会从多个 DHCP 服务器收到多条 DHCPOFFER 消息。客户端决定接收特定 DHCP 服务器发送的 DHCPOFFER 消息之后，就以广播方式向 DHCP 服务器发送一条 DHCPREQUEST 消息（UDP 目的端口 67），在消息中通过服务器标识符选项来标明所选择的服务器。
- 同样，中继代理会拦截广播请求并插入选项 82 数据，然后将请求消息中继给 DHCP 服务器。
- 选定的 DHCP 服务器通过提交已分配的 IP 地址来确认该请求，并以单播方式向中继代理发送 DHCPACK 消息（UDP 目的端口 67）。
- 中继代理收到应答消息之后，将删除消息中的选项 82 数据，然后将消息中继回客户端。
- 客户端收到携带配置参数的 DHCPACK 消息之后，将对所分配的 IP 地址执行 ARP（Address Resolution Protocol，地址解析协议）检查，以确保该地址未被其他主机占用。如果检测到该 IP 地址已被占用，那么就会向服务器发送 DHCPDECLINE 消息并重新启动配置进程。同样，中继代理将拦截该消息并进行中继。
- 如果客户端决定放弃 IP 地址租约，那么就会向服务器发送 DHCPRELEASE 消息。由于 DHCPRELEASE 消息是发送给服务器的单播消息，因而此时无须使用中继代理。
- 收到 DHCPRELEASE 消息后，服务器就会将该 IP 地址标记为未分配地址。

如果要在 NX-OS 上启用 DHCP 中继代理，那么就需要使用命令 **feature dhcp** 在系统上启用 DHCP 功能特性。如果要让设备充当 DHCP 中继代理，那么就要配置全局命令 **ip dhcp relay**。可以使用命令 **ip dhcp relay address** *ip-address* 在接口上配置 DHCP 中继，其中，变量 *ip-address* 是 DHCP 服务器地址。如果要启用 DHCP 选项 82，那么就要配置全局命令 **ip dhcp relay information option**。

为了更好地理解 DHCP 中继功能，下面将以图 6-3 所示拓扑结构为例加以说明，拓扑结构中的 NX-1 充当中继代理。

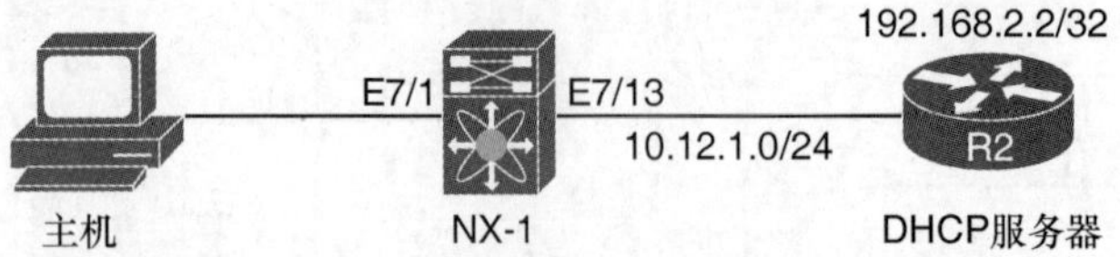

```
           E7/1    E7/13      192.168.2.2/32
  主机       NX-1  10.12.1.0/24   R2
                              DHCP服务器
```

图 6-3 配置了 DHCP 中继代理的拓扑结构

例 6-12 给出了 NX-1 的 DHCP 中继代理配置示例。请注意，配置了 DHCP 中继的接口拥有去往 DHCP 服务器的可达性。

例 6-12　DHCP 中继配置

```
NX-1
NX-1(config)# feature dhcp
NX-1(config)# ip dhcp relay
NX-1(config)# interface e7/1
NX-1(config-if)# ip dhcp relay address 192.168.2.2
```

配置完成后，如果客户端希望请求 IP 地址，那么 DHCP 中继代理就会在客户端与服务器之间帮助进行消息交换。可以使用命令 **show ip dhcp relay** 验证接口是否已经启用了 DHCP 中继功能。完成消息交换之后，可以使用命令 **show ip dhcp relay statistics** 验证中继代理在服务器与客户端之间双向接收和转发的所有消息的统计信息。例 6-13 给出了 NX-1 的 DHCP 中继配置和统计信息示例。命令 **show ip dhcp relay statistics** 的输出结果显示了中继代理接收、转发和丢弃的各种 DHCP 数据包的全部统计信息。此外，该输出结果还显示了中继代理丢弃数据包的原因及其统计信息。

例 6-13　验证 DHCP 中继信息和统计信息

```
NX-1# show ip dhcp relay
DHCP relay service is enabled
Insertion of option 82 is disabled
Insertion of VPN suboptions is disabled
Insertion of cisco suboptions is disabled
Global smart-relay is disabled
Relay Trusted functionality is disabled
Relay Trusted Port is Globally disabled
V4 Relay Source Address HSRP is Globally disabled
! Output omitted for brevity
Helper addresses are configured on the following interfaces:
 Interface          Relay Address     VRF Name
 -------------      -------------     --------
 Ethernet7/1        192.168.2.2
NX-1# show ip dhcp relay statistics
-----------------------------------------------------------------
Message Type            Rx              Tx              Drops
-----------------------------------------------------------------
Discover                22              22              0
Offer                   1               1               0
Request(*)              1               1               0
Ack                     1               1               0
Release(*)              0               0               0
Decline                 0               0               0
Inform(*)               0               0               0
Nack                    0               0               0
-----------------------------------------------------------------
Total                   25              25              0
-----------------------------------------------------------------
```

```
DHCP L3 FWD:
Total Packets Received                          :       0
Total Packets Forwarded                         :       0
Total Packets Dropped                           :       0
Non DHCP:
Total Packets Received                          :       0
Total Packets Forwarded                         :       0
Total Packets Dropped                           :       0
DROP:
DHCP Relay not enabled                          :       0
Invalid DHCP message type                       :       0
Interface error                                 :       0
Tx failure towards server                       :       0
Tx failure towards client                       :       0
Unknown output interface                        :       0
Unknown vrf or interface for server             :       0
Max hops exceeded                               :       0
Option 82 validation failed                     :       0
Packet Malformed                                :       0
Relay Trusted port not configured               :       0
* - These counters will show correct value when switch
receives DHCP request packet with destination ip as broadcast
address. If request is unicast it will be HW switched
```

配置了 DHCP 中继地址之后，就可以在 Nexus 交换机上执行 ACL（Access Control List，访问控制列表）编程操作。

- 如果 L3 接口是物理/子接口，那么就对 RACL（Router ACL，路由器 ACL）进行编程，将所有 DHCP 数据包都重定向到管理引擎上的 DHCP 监听进程。
- 如果 L3 接口是 SVI（Switched Virtual Interface，交换式虚接口），那么就在硬件上对 VACL（VLAN ACL）进行编程。
- 如果 L3 接口是端口通道，那么就在端口通道的所有成员接口上对 RACL 进行编程。

编程后的 RACL 或 VACL 包含了以下两类主要参数。

- 过滤器。
  - 允许源端口 67 和 68 去往任意目的端口。
  - 允许任意端口去往目的端口 67 和 68。
- 操作。
  - 重定向到管理引擎上的 DHCP 监听进程。

由于 DHCP 进程向 Netstack 注册了上述例外原因，因而 LC 抓取的所有 DHCP 请求/应答消息都将通过 Netstack 快速 MTS 队列进入 DHCP 监听进程。

在接口上配置了 DHCP 中继特性之后，就可以通过命令 **show system internal access-list interface** *interface-id* [**module** *slot*] 检查是否已在硬件中进行了 ACL 编程，如例 6-14 所示。请注意，输出结果中的策略类型是 DHCP，策略名称是 Relay。该命令的输出结果显示了 ACL 保留的 TCAM（Ternary Content Addressable Memory，三元内容可寻址存储器）表项数以及邻接关系数。

例 6-14 验证线卡上的 DHCP 中继的 ACL

```
NX-1# show system internal access-list interface ethernet 7/1 module 7
Policies in ingress direction:
       Policy type                    Policy Id       Policy name
-----------------------------------------------------------------
```

```
                DHCP                    4           Relay
No Netflow profiles in ingress direction


INSTANCE 0x0
---------------

  Tcam 1 resource usage:
  ----------------------
  Label_b = 0x201
  Bank 0
  ------
    IPv4 Class
      Policies: DHCP(Relay) [Merged]
      Netflow profile: 0
      Netflow deny profile: 0
      5 tcam entries

    0 l4 protocol cam entries
    0 mac etype/proto cam entries
    0 lous
    0 tcp flags table entries
    1 adjacency entries

No egress policies
No Netflow profiles in egress direction
```

在线卡上对 ACL 进行编程之后，就可以使用命令 **show system internal access-list input statistics [module** *slot*]查看该 ACL 的硬件统计信息。例 6-15 显示了 DHCP 中继 ACL 的统计信息，可以看出，与来自源端口 67 的流量相匹配的次数为 5 次。如果在正常网络运行条件下，DHCP 的操作有问题，那么就可以利用 **show system internal access-list input statistics [module** *slot*] 命令以及前一个示例中的命令，来确定是否已在硬件中对 DHCP 中继 ACL 进行了编程，且统计信息计数器一直都处于递增状态。

例 6-15 验证线卡上的 DHCP 中继的 ACL 统计信息

```
NX-1# show system internal access-list input statistics module 7
                VDC-2 Ethernet7/1 :
                =====================


INSTANCE 0x0
---------------

  Tcam 1 resource usage:
  ----------------------
  Label_b = 0x201
  Bank 0
  ------
    IPv4 Class
      Policies: DHCP(Relay) [Merged]
      Netflow profile: 0
      Netflow deny profile: 0
      Entries:
        [Index] Entry [Stats]
        ---------------------
```

```
    [0058:000e:000e] prec 1 redirect(0x0) udp 0.0.0.0/0 255.255.255.255/32 eq 68
flow-label 68 [0]
    [0059:000f:000f] prec 1 redirect(0x0) udp 0.0.0.0/0 255.255.255.255/32 eq 67
flow-label 67 [5]
    [005a:0010:0010] prec 1 redirect(0x0) udp 0.0.0.0/0 eq 68 255.255.255.255/32
flow-label 262144 [0]
    [005b:0011:0011] prec 1 redirect(0x0) udp 0.0.0.0/0 eq 67 255.255.255.255/32
flow-label 196608 [0]
    [005c:0012:0012] prec 1 permit ip 0.0.0.0/0 0.0.0.0/0 [0]
```

6.3.2 DHCP 监听

DHCP 监听是一种 L2 安全功能特性，可以解决通过 DHCP 消息发起的某些 DoS 攻击，而且能够有效防范 IP 欺骗（指的是恶意主机尝试使用其他主机的 IP 地址）。DHCP 监听特性的操作分为以下两个层面。

- 发现。
- 执行。

发现操作包括拦截 DHCP 消息和建立{IP 地址,MAC 地址,端口,VLAN}记录数据库等功能，通常将该数据库称为绑定表。执行操作包括 DHCP 消息验证、限速以及 DHCP 广播到单播的转换功能。

DHCP 监听提供了以下安全功能。

- 通过 DHCP 消息防范 DoS 攻击。
- DHCP 消息验证。
- 创建有助于验证 DHCP 消息的 DHCP 绑定表。
- 插入/删除选项 82。
- 对接口上 DHCP 消息的数量进行限速。

注：DHCP 监听功能与 DHCP 中继代理相关联，可以在 DHCP 客户端与服务器处于不同子网时，实现相同的安全功能特性。

为了更好地理解 DHCP 监听特性的工作原理，下面将以图 6-4 所示拓扑结构为例加以说明。该拓扑结构中的 DHCP 服务器与客户端位于同一个 VLAN 100 中，Nexus 交换机 NX-1 为 DHCP 服务器与客户端主机之间提供了二层连接。

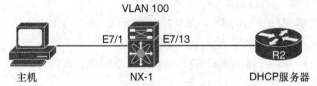

图 6-4　包含 DHCP 服务器和客户端的拓扑结构

如果要启用 DHCP 监听功能，那么就要首先在 Nexus 交换机上运行全局配置命令 **ip dhcp snooping**，然后通过命令 **ip dhcp snooping vlan** *vlan-id* 为指定 VLAN 启用 DHCP 监听功能。通常将连接 DHCP 服务器的端口配置为可信端口，将连接客户端的端口配置为非可信端口。为了将连接服务器的端口配置为可信端口，需要启用接口配置命令 **ip dhcp snooping trust**。例 6-16 给出了 NX-1 的 DHCP 监听配置示例。启用了 DHCP 监听功能之后，可以使用命令 **show ip dhcp snooping** 验证交换机的 DHCP 监听状态，从 **show ip dhcp snooping** 命令的输出结果可以看出，NX-1 已经为 VLAN 100 启用了 DHCP 监听功能。

例 6-16 配置和验证 DHCP 监听功能

```
NX-1
NX-1(config)# ip dhcp snooping
NX-1(config)# ip dhcp snooping vlan 100
NX-1(config)# interface e7/13
NX-1(config-if)# ip dhcp snooping trust
NX-1# show ip dhcp snooping
Switch DHCP snooping is enabled
DHCP snooping is configured on the following VLANs:
100
DHCP snooping is operational on the following VLANs:
100
Insertion of Option 82 is disabled
Verification of MAC address is enabled
DHCP snooping trust is configured on the following interfaces:
Interface              Trusted
------------           -------
Ethernet7/13           Yes
```

客户端与服务器之间交换了请求/应答消息之后，就会在为非可信端口配置了 DHCP 监听功能的设备建立一条绑定表项。IPSG（IP Source Guard，IP 源防护）和 DAI（Dynamic ARP Inspection，动态 ARP 检查）功能也使用绑定表。如果要查看绑定表信息，可以使用命令 **show ip dhcp snooping binding**，如例 6-17 所示。从输出结果可以看出，交换机已经为非可信端口 Eth7/1 构建了绑定表项，同时还显示了利用所列 MAC 地址为主机分配的 IP 地址。

例 6-17 DHCP 监听绑定数据库

```
NX-1# show ip dhcp snooping binding
MacAddress          IpAddress       Lease(Sec)    Type           VLAN    Interface
-----------------   -------------   ----------    ----------     ----    -------------
d4:2c:44:fa:cf:47   10.12.1.2       84741         dhcp-snoop     100     Ethernet7/1
```

在非可信端口上收到数据包之后，将执行如下验证操作。

- 允许从客户端（源端口 68）向服务器（目的端口 67）发送请求消息（BOOTREQUEST）。
- 利用绑定表项验证 DHCPRELEASE/DHCPDECLINE 消息，以防止主机释放或拒绝其他主机的地址。

在可信端口上收到数据包之后，将执行如下验证操作。

- 对于从服务器（源端口 67）发送给客户端（目的端口 68）的应答消息（BOOTREPLY）来说，需要执行绑定表更新和删除选项 82 操作，然后转发数据包。
- 对于从客户端（源端口 68）发送给服务器（目的端口 67）的请求消息来说，直接转发请求消息即可，无须执行任何验证操作。

因此，为了执行前面提到的验证操作，需要在线卡上安装 DHCP 监听功能的 ACL，从而能够查看不同表项的已编程 ACL 的统计信息。例 6-18 显示了硬件编程的 DHCP 监听 ACL 及其统计信息。

例 6-18 DHCP 监听 ACL 编程信息

```
NX-1# show system internal access-list module 7
              VLAN 100 :
              ==========
Policies in ingress direction:
        Policy type                 Policy Id      Policy name
```

```
--------------------------------------------------------------
      DHCP                            4         Snooping

No Netflow profiles in ingress direction

INSTANCE 0x0
---------------

  Tcam 1 resource usage:
  ----------------------
  Label_b = 0x201
   Bank 0
   ------
     IPv4 Class
       Policies: DHCP(Snooping)   [Merged]
       Netflow profile: 0
       Netflow deny profile: 0
       5 tcam entries

   0 l4 protocol cam entries
   0 mac etype/proto cam entries
   0 lous
   0 tcp flags table entries
   1 adjacency entries
! Output omitted for brevity
NX-1# show system internal access-list input statistics module 7
              VLAN 100  :
              =========

INSTANCE 0x0
---------------

  Tcam 1 resource usage:
  ----------------------
  Label_b = 0x201
   Bank 0
   ------
     IPv4 Class
       Policies: DHCP(Snooping)   [Merged]
       Netflow profile: 0
       Netflow deny profile: 0
       Entries:
         [Index] Entry [Stats]
         --------------------
  [0058:000e:000e] prec 1 redirect(0x0) udp 0.0.0.0/0 0.0.0.0/0 eq 68 flow-label
 68  [0]
  [0059:000f:000f] prec 1 redirect(0x0) udp 0.0.0.0/0 0.0.0.0/0 eq 67 flow-label
 67  [5]
  [005a:0010:0010] prec 1 redirect(0x0) udp 0.0.0.0/0 eq 68 0.0.0.0/0 flow-label
 262144  [0]
  [005b:0011:0011] prec 1 redirect(0x0) udp 0.0.0.0/0 eq 67 0.0.0.0/0 flow-label
 196608  [0]
  [005c:0012:0012] prec 1 permit ip 0.0.0.0/0 0.0.0.0/0   [0]
INSTANCE 0x3
---------------

  Tcam 1 resource usage:
```

```
                   ----------------------
                   Label_b = 0x201
                     Bank 0
                     ------
                      IPv4 Class
                        Policies: DHCP(Snooping)    [Merged]
                        Netflow profile: 0
                        Netflow deny profile: 0
                        Entries:
                          [Index] Entry [Stats]
                          --------------------
                   [0058:000e:000e] prec 1 redirect(0x0) udp 0.0.0.0/0 0.0.0.0/0 eq 68 flow-label
                    68   [2]
                   [0059:000f:000f] prec 1 redirect(0x0) udp 0.0.0.0/0 0.0.0.0/0 eq 67 flow-label
                    67   [0]
                   [005a:0010:0010] prec 1 redirect(0x0) udp 0.0.0.0/0 eq 68 0.0.0.0/0 flow-label
                    262144   [0]
                   [005b:0011:0011] prec 1 redirect(0x0) udp 0.0.0.0/0 eq 67 0.0.0.0/0 flow-label
                    196608   [0]
                   [005c:0012:0012] prec 1 permit ip 0.0.0.0/0 0.0.0.0/0   [0]
```

6.3.3 动态 ARP 检查

安全性对于任何网络环境来说都是非常重要的问题。恶意主机可以毒化主机以及路由器和交换机的 ARP 缓存。DAI（Dynamic ARP Inspection，动态 ARP 检查）功能可以保护主机和其他网络设备免受 ARP 缓存中毒的影响。DAI 可以验证连接在交换机上的主机发送的 ARP 请求和响应消息的完整性，根据 DHCP 监听功能创建的绑定表来检查每个 ARP 数据包的 MAC-IP 绑定是否正确，如果检查失败，那么就不转发该 ARP 数据包。

DAI 需要以逐个 VLAN 的方式进行启用，且支持 src-MAC、dst-MAC 和 IP 地址验证功能。DAI 根据有效单播 IP 地址的监听绑定表项来验证 ARP 数据包的[源,目的]和[MAC,IP]地址。如果设备上没有相应的绑定表项，那么就要在设备执行 ARP 检查之前，为入站接口配置 DAI 可信端口。

例 6-19 显示了 VLAN 100 的 DAI 配置示例，并通过命令 **show ip arp inspection statistics vlan** *vlan-id* 显示了 ARP 请求/应答的统计信息以及转发的数据包数量。可以使用命令 **ip arp inspection vlan** *vlan-id* 为 VLAN 配置 DAI。由于面向服务器的端口被配置为可信端口，因而可以通过接口级命令 **ip arp inspection trust** 启用 DAI。在 DAI 可信端口上，不检查 rx 和 tx 数据包。

例 6-19 配置和验证 DAI

```
NX-1(config)# ip arp inspection vlan 100
NX-1(config)# interface e7/13
NX-1(config-if)# ip arp inspection trust
NX-1# show ip arp inspection statistics vlan 100

Vlan : 100
-----------
ARP Req Forwarded      = 2
ARP Res Forwarded      = 3
ARP Req Dropped        = 0
ARP Res Dropped        = 0
DHCP Drops             = 0
DHCP Permits           = 5
SMAC Fails-ARP Req     = 0
SMAC Fails-ARP Res     = 0
```

```
DMAC Fails-ARP Res = 0
IP Fails-ARP Req   = 0
IP Fails-ARP Res   = 0
```

对于 DAI 来说,需要在线卡上编程 ARP 监听 ACL(VACL)。需要注意的是,由于 DAI 功能与 DHCP 监听功能一起启用,因而会在线卡上同时看到两个 ACL。例 6-20 显示了线卡上编程的 ACL 及其统计信息。

例 6-20 DAI 的 ACL 编程和统计信息

```
NX-1# show system internal access-list vlan 100 module 7
Policies in ingress direction:
        Policy type                  Policy Id       Policy name
---------------------------------------------------------------
      DHCP                               4           Snooping
      ARP                                5           Snooping

No Netflow profiles in ingress direction

INSTANCE 0x0
---------------
  Tcam 1 resource usage:
  ----------------------
  Label_b = 0x202
  Bank 0
  ------
    IPv4 Class
      Policies: DHCP(Snooping)   [Merged]
      Netflow profile: 0
      Netflow deny profile: 0
      5 tcam entries
    ARP Class
      Policies: ARP(Snooping)    [Merged]
      Netflow profile: 0
      Netflow deny profile: 0
      3 tcam entries

  0 l4 protocol cam entries
  0 mac etype/proto cam entries
  0 lous
  0 tcp flags table entries
  2 adjacency entries

! Output omitted for brevity
NX-1# show system internal access-list input statistics module 7
              VLAN 100   :
              =========


INSTANCE 0x0
---------------

  Tcam 1 resource usage:
  ----------------------
  Label_b = 0x202
  Bank 0
  ------
    IPv4 Class
```

```
      Policies: DHCP(Snooping)    [Merged]

! Output omitted for brevity

    ARP Class
      Policies: ARP(Snooping)    [Merged]
      Netflow profile: 0
      Netflow deny profile: 0
      Entries:
        [Index] Entry [Stats]
        --------------------
  [0062:0018:0018] prec 1 redirect(0x0) arp/response ip 0.0.0.0/0 0.0.0.0/0 0000
.0000.0000 0000.0000.0000    [2]
  [0063:0019:0019] prec 1 redirect(0x0) arp/request ip 0.0.0.0/0 0.0.0.0/0 0000.
0000.0000 0000.0000.0000    [1]
  [0064:001a:001a] prec 1 permit arp-rarp/all ip 0.0.0.0/0 0.0.0.0/0 0000.0000.0
000 0000.0000.0000    [0]
```

ARP ACL

对于非 DHCP（未启用 DHCP 监听功能）场景来说，可以定义 ARP ACL 来过滤恶意 ARP 请求和响应，没有任何数据包会被重定向给管理引擎。ARP 数据包到达线卡之后，会根据用户为 ARP 检查过滤器配置的 ACL 列表，在线卡上转发或丢弃该 ARP 数据包。需要以每个 VLAN 为基础配置 ARP ACL 过滤器，ARP ACL 的配置命令是 **arp access-list** *acl-name*，接收[**permit** | **deny**] [**request** | **response**] **ip** *ip-address subnet-mask* **mac** [*mac-address mac-address-range*]形式的表项。例 6-21 给出了一个 ARP ACL 示例，该 ARP ACL 是 VLAN 100 的 ARP 检查过滤器。配置完成后，就在硬件中对该 ARP ACL 进行编程，同时，还可以使用相同的 **show system internal access-list input statistics** [**module** *slot*]命令验证相应的统计信息。

例 6-21 配置和验证 ARP ACL

```
NX-1(config)# arp access-list ARP-ACL
NX-1(config-arp-acl)# permit request ip 10.12.1.0 255.255.255.0 mac d42c.44fa.cf47
  d42c.44fa.efff log
NX-1(config-arp-acl)# deny ip any mac any log
NX-1(config-arp-acl)# exit
NX-1(config)# ip arp inspection filter ARP-ACL vlan 100
NX-1# show arp access-lists ARP-ACL

ARP access list ARP-ACL
10 permit request ip 10.12.1.0 255.255.255.0 mac d42c.44fa.cf47 d42c.44fa.efff log
20 deny ip any mac any log
NX-1# show system internal access-list input statistics module 7
            VLAN 100   :
            =========
INSTANCE 0x0
---------------

  Tcam 1 resource usage:
  ----------------------
  Label_b = 0x202
   Bank 0
   ------
    ARP Class
      Policies: VACL(ARP-ACL) ARP(Snooping)    [Merged]
      Netflow profile: 0
```

```
            Netflow deny profile: 0
            Entries:
                [Index] Entry [Stats]
                ---------------------
     [005d:0013:0013] prec 1 redirect(0x0) arp/request ip 10.12.1.0/24 0.0.0.0/0 d4
2c.44fa.cf47 d42c.44fa.efff      [1]
     [005e:0014:0014] prec 2 deny arp-rarp/all ip 0.0.0.0/0 0.0.0.0/0 0000.0000.000
0 0000.0000.0000  log    [2]
! Output omitted for brevity
```

6.3.4 IP 源防护

IPSG（IP Source Guard，IP 源保护）提供了 IP 和 MAC 过滤器，可以限制 DHCP 监听非可信端口上的 IP 流量。仅允许源 IP 和 MAC 地址与有效 IP 源绑定（静态 IP 和 DHCP 绑定）一致的 IP 流量，其他 IP 流量（DHCP 除外）均被丢弃。也就是说，IPSG 防范 IP 欺骗的方式是仅允许通过特定端口上的 DHCP 监听功能获得的 IP 地址。

在 DHCP 监听非可信二层端口上启用 IPSG 功能之后，刚开始 IPSG 会阻塞该端口上的所有 IP 通信（DHCP 监听进程抓取的 DHCP 数据包除外）。客户端从 DHCP 服务器收到有效 IP 地址之后，对于连接在交换机上的主机的 IP 流量来说，仅当 MAC-IP 地址与 IPSG 模块编程的地址相匹配时，IPSG 才允许该流量。IPSG 功能会从绑定表中提取 MAC-IP 绑定表项，并在线卡的 RPF（Reverse Path Forwarding，反向路径转发）表中对 SMAC（源 MAC）-IP 绑定检查进行编程，从而在硬件中提供每个端口的 IP 流量过滤器。

可以通过命令 **ip verify source dhcp-snooping-vlan** 在连接主机的交换机的 L2 端口上启用 IPSG。使用命令 **show ip verify source interface** *interface-id* 为主机分配了 IP 地址之后，就可以验证 IPSG 表。DHCP 服务器分配了 DHCP 地址之后，例 6-22 在端口 Ethernet7/1 上启用了 IPSG 并显示 IPSG 表信息，其中，端口 Ethernet7/1 是面向主机的端口（非可信端口）。

例 6-22　IPSG 配置

```
NX-1(config)# interface e7/1
NX-1(config-if)# ip verify source dhcp-snooping-vlan
NX-1# show ip verify source interface e7/1
IP source guard is enabled on this interface.

Interface        Filter-mode     IP-address       Mac-address        Vlan
-----------      -----------     ----------       --------------     ----
Ethernet7/1      active          10.12.1.3        d4:2c:44:fa:cf:47  100
```

启用了 IPSG 并通过 FIB（Forwarding Information Base，转发信息库）对 IP-MAC 表项进行编程之后，就要检查所有流量的 IP-MAC 绑定信息。例如，所有来自该客户端（拥有有效 IP-MAC 绑定信息）的 ping 操作都可以正常工作。如果删除了 IPSG 绑定表项，那么 ping 操作将失败，FIB 将丢弃所有这类无效流量。RPF 表中的 SMAC-IP 绑定信息是通过 SAL（Security Abstraction Layer，安全抽象层）进程进行编程的，其中，SAL 是一种运行在 Nexus 系统上的 VDC（Virtual Device Context，虚拟设备上下文）本地强制进程。SAL 通过 NX-OS 基础设施提供系统启动、重启、HA（High Availability，高可用性）功能以及进程间通信，因而 SAL 在管理引擎中被视为硬件抽象层，用于对 FIB 中的 IPSG 绑定数据库进行编程，从而确保数据包转发阶段的安全性。

可以使用命令 **show system internal sal info database vlan** *vlan-id* 查看 SAL 数据库信息，该命令提供了 IPv4 和 IPv6 表 ID 信息。有了表 ID 信息之后，就可以通过 **show system internal forwarding table** *table-id* **route** *ip-address/mask* [**module** *slot*]命令验证 FIB 信息，其中，*table-id* 是

从 SAL 数据库输出结果得到的字段。例 6-23 给出了利用 SAL 数据库信息验证 IPSG FIB 编程情况的示例。

例 6-23　查看 SAL 数据库信息并验证 IPSG FIB

```
NX-1# show system internal sal info database vlan 100

VLAN ID: 100
Security Features Enabled: IP Source
Table-id Information:
    V4 Table-id: 17
    V6 Table-id: 0x80000011
! Table 17 in hex is 0x11
NX-1# show system internal forwarding table 0x11 route 10.12.1.3/32 module 7

Routes for table 17

----+---------------------+----------+----------+----------
Dev | Prefix              | PfxIndex | AdjIndex | LIF
----+---------------------+----------+----------+----------
 0   10.12.1.3/32           0x406      0x4d       0xfff
```

6.3.5　单播 RPF

URPF（Unicast Reverse Path Forwarding，单播反向路径转发）是一种在网络边缘匹配源 IP 地址以丢弃特定流量的技术。也就是说，URPF 可以防止网络遭受源 IP 欺骗攻击，允许合法源端将流量发送给目的服务器。URPF 有两种不同的实现模式。

- **宽松模式（Loose mode）**：如果在 FIB 中执行的数据包源地址查找操作返回的结果是匹配，且 FIB 查找结果表明至少可以通过一个真实接口访问该源地址，那么就表明宽松模式检查成功。该模式下，收到数据包的入站接口不需要匹配 FIB 查找结果中的任何接口。
- **严格模式（Strict mode）**：如果 URFP 在 FIB 中找到了数据包源地址的匹配项且收到数据包的入站接口与 FIB 匹配项中的某个单播 RPF 接口相匹配，那么就表明严格模式检查成功。如果检查失败，那么就丢弃数据包。如果预计数据包流是对称的，那么就执行严格模式的单播 RPF 检查。

最多可以在 8 个 ECMP 接口使用严格模式 URPF，如果接口数量超出了 8 个，那么就要进入宽松模式。宽松模式 URPF 最多支持 16 个 ECMP 接口。URPF 适用于 L3 接口、SVI、L3 端口通道以及子接口。需要注意的是，严格模式 URPF 不兼容/32 ECMP 路由，因而不建议在连接核心层设备的上行链路上使用严格模式 URPF，因为这样做可能会丢弃/32 路由。

URPF 的配置命令是 **ip verify unicast source reachable-via [any [allow-default] | rx]**，其中，选项 **rx** 表示启用严格模式 URPF，选项 **any** 表示启用宽松模式 URPF，选项 **allow-default** 与宽松模式一起使用，作用是包括路由表中未明确包含的 IP 地址。例 6-24 给出了在 L3 接口上启用 URPF 严格模式的配置示例。配置完成后，可以通过命令 **show ip interface** *interface-id* 查看接口是否启用了 URPF，可以看出，本例在接口 Eth7/1 上启用了严格模式的 URPF。

例 6-24　配置和验证 URPF

```
NX-1(config)# interface e7/1
NX-1(config-if)# ip verify unicast source reachable-via rx
NX-1# show ip interface e7/1
IP Interface Status for VRF "default"(1)
Ethernet7/1, Interface status: protocol-up/link-up/admin-up, iod: 175,
```

```
  IP address: 10.13.1.1, IP subnet: 10.13.1.0/24 route-preference: 0, tag: 0
  IP broadcast address: 255.255.255.255
  IP multicast groups locally joined: none
  IP MTU: 1500 bytes (using link MTU)
  IP primary address route-preference: 0, tag: 0
  IP proxy ARP : disabled
  IP Local Proxy ARP : disabled
  IP multicast routing: disabled
  IP icmp redirects: enabled
  IP directed-broadcast: disabled
  IP Forwarding: disabled
  IP icmp unreachables (except port): disabled
  IP icmp port-unreachable: enabled
  IP unicast reverse path forwarding: strict
  IP load sharing: none
! Output omitted for brevity
```

6.4 IPv6 服务

随着数据中心的快速发展，IPv6 在网络中的重要性也日益显著，用于克服当前面临的各种编址和安全挑战。NX-OS 提供了丰富的 IPv6 服务，这些服务可以为大规模数据中心环境提供有效的可靠性和安全性。本节将讨论以下 IPv6 服务。

- 邻居发现。
- IPv6 地址分配。
- IPv6 第一跳安全。

本节将详细介绍这些功能，并解释如何在 NX-OS 交换机上执行相应的故障排查操作。

6.4.1 邻居发现

IPv6 ND（Neighbor Discovery，邻居发现）定义在 RFC 4861 中，是一组确定两个 IPv6 邻居节点之间关系的消息和进程。IPv6 ND 建立在 RFC 2463 定义的 ICMPv6 之上，IPv6 ND 替换了 IPv4 使用的 ARP、ICMP 重定向以及 ICMP 路由器发现消息等协议。IPv6 ND 和 ICMPv6 对于 IPv6 的运行来说至关重要。

IPv6 ND 定义了 5 种 ICMPv6 数据包，节点在建立通信之前，可以通过这些数据包了解必须掌握的信息。

- 路由器请求消息（Router Solicitation）（ICMPv6 类型 133，代码 0）。
- 路由器宣告消息（Router Advertisement）（ICMPv6 类型 134，代码 0）。
- 邻居请求消息（Neighbor Solicitation）（ICMPv6 类型 135，代码 0）。
- 邻居宣告消息（Neighbor Advertisement）（ICMPv6 类型 136，代码 0）。
- 重定向消息（Redirect Message）（ICMPv6 类型 137，代码 0）。

启用了接口之后，主机就可以向外发送 RS（Router Solicitation，路由器请求），请求路由器立即（而不是在下一个计划时间）生成 RA（Router Advertisement，路由器宣告）。发送了 RS 消息之后，源地址字段将被设置为发送 NIC（Network Interface Card，网络接口卡）的 MAC 地址，以太网报头中的目的地址字段将被设置为 33:33:00:00:00:02。IPv6 报头中的源地址字段将被设置为分配给发送接口的链路本地 IPv6 地址或 IPv6 未指定地址（::），目的地址字段则被设置为具有链路本地范围的全部路由器（All-Routers）多播地址（FF02:2），且跳数限制被设置为 255。

路由器会定期或者以响应 RS 消息的方式来通告其存在性以及各种链路和 Internet 参数。RA

消息包含用于同一链路判定（on-link determination）和/或地址配置的前缀、建议的跳数限制值以及 MTU（Maximum Transmission Unit，最大传输单元）等信息。在 RA 消息的以太网报头中，源地址字段被设置为发送 NIC，目的地址字段被设置为 33:33:00:00:00:01 或者从单播地址发送 RS 消息的主机的单播 MAC 地址。与 RS 消息类似，IPv6 报头中的源地址字段被设置为分配给发送接口的链路本地地址，目的地址字段被设置为具有本地链路范围的全部节点（All-Nodes）多播地址（FF02:1）或者发送 RS 消息的主机的单播 IPv6 地址，且跳数限制字段被设置为 255。

节点发送 NS（Neighbor Solicitation，邻居请求）消息以确定邻居的链路层地址或者验证仍然可以通过缓存的链路层地址访问邻居。此外，还可以利用 RS 执行 DAD（Duplicate Address Detection，重复地址检测）操作。在 NS 消息的以太网报头中，目的 MAC 地址对应于目的端的请求节点地址（solicited-node address）。在单播 NS 消息中，目的地址字段被设置为邻居的单播 MAC 地址。在 IPv6 报头中，源地址字段被设置为发送接口的 IPv6 地址或未指定地址（::)(在 DAD 期间）。在多播 NS 消息中，目的地址被设置为请求节点地址。在单播 NS 消息中，目的地址被设置为目的端 IPv6 单播地址。

NA（Neighbor Advertisement，邻居宣告）消息是 NS 消息的响应消息。节点也可以发送非请求的邻居宣告消息，以宣告链路层地址的变更信息。在请求的 NA 消息的以太网报头中，目的 MAC 被设置为初始 NS 发送端的单播 MAC 地址。在非请求的 NA 消息中，目的 MAC 被设置为 33:33:00:00:00:01，这是链路本地范围全部节点多播地址。在 IPv6 报头中，源地址被设置为分配给发送接口的 IPv6 单播地址。在请求 NA 消息中，目的 IPv6 地址被设置为初始 NS 消息的发送端的 IPv6 单播地址。在非请求 NA 消息中，目的地址字段被设置为本地链路范围的全部节点多播地址（FF02::1）。

路由器利用 RM（Redirect Message，重定向消息）通知主机有更好的第一跳去往目的端。以太网报头中的目的 MAC 被设置为初始发送端的单播 MAC。IPv6 报头中的源地址字段被设置为发送接口的单播 IPv6 地址，目的地址字段被设置为初始主机的单播地址。

如果要启用邻居发现功能，首先需要在接口上启用 IPv6 或配置 IPv6 地址。可以使用命令 **ipv6 address** *ipv6-address* [**eui64**]或命令 **ipv6 address use-link-local-only** 配置 IPv6 地址，命令选项 **eui64** 表示以 EUI64 格式配置 IPv6 地址，命令选项 **use-link-local-only** 表示在接口上手动配置链路本地地址，而不使用自动分配的链路本地地址。图 6-5 所示拓扑结构中的两台交换机 NX-1 与 NX-2 之间启用了 IPv6 链路，该链路配置了子网为 2002:10:12:1:://64 的 IPv6 地址。

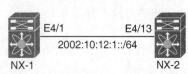

图 6-5 启用 IPv6 的拓扑结构

如果链路两侧均配置了 IPv6 地址，且从其中一侧发起了 ping 操作，那么将启动 ND 进程并建立 IPv6 邻居关系。可以使用命令 **show ipv6 neighbor [detail]**查看 IPv6 邻居。例 6-25 显示了两台交换机之间的 IPv6 邻居关系。可以看出，配置了 IPv6 地址且用户对 IPv6 单播地址或远程对等体的链路本地地址发起 ping 操作之后，就会启动 IPv6 ND 进程、进行消息交换并建立 IPv6 邻居关系。

例 6-25 IPv6 邻居发现

```
NX-1
NX-1(config)# interface Eth4/1
NX-1(config-if)# ipv6 address 2002:10:12:1::1/64
```

```
NX-2
NX-2(config)# interface Eth4/13
NX-2(config-if)# ipv6 address 2002:10:12:1::2/64
```

```
NX-1
! IPv6 neighbor output after initiating ipv6 ping
NX-1# show ipv6 neighbor

Flags: # - Adjacencies Throttled for Glean
       G - Adjacencies of vPC peer with G/W bit

IPv6 Adjacency Table for VRF default
Total number of entries: 2
Address                Age         MAC Address      Pref  Source    Interface
2002:10:12:1::2        00:11:51    0002.0002.0012   50    icmpv6    Ethernet4/1
fe80::202:ff:fe02:12
                       00:00:04    0002.0002.0012   50    icmpv6    Ethernet4/1
```

为了更好地理解邻居发现进程，可以使用 Ethanalyzer 工具。Ethanalyzer 不但能够识别 IPv6 ND 进程，而且还能协助解决各种 ND 故障。NX-1 向对等设备发起 ICMPv6 Ping 操作之后，Ethanalyzer 的输出结果如例 6-26 所示。可以看出，NX-1 启动 ping 操作时，会向 NX-2 发送一条 NS 消息，应答包是从 NX-2 发送给 NX-1 的 NA 消息。请注意，作为 NA 消息的一部分，此时设置了 rtr（Router，路由器）、sol（Solicited，请求）和 ovr（Override，覆盖）标志，目的地址被设置为 2002:10:12:1::2 且通过 MAC 地址 0002.0002.0012 可达。

例 6-26 IPv6 邻居发现：Ethanalyzer 抓取的信息

```
NX-1
NX-1# ethanalyzer local interface inband display-filter "ipv6" limit-captured-frames 0
Capturing on inband
2017-10-14 21:25:51.314297 2002:10:12:1::1 -> ff02::1:ff00:2 ICMPv6 86 Neighbor
  Solicitation for 2002:10:12:1::2 from 00:01:00:01:00:12
4 2017-10-14 21:25:51.315476 2002:10:12:1::2 -> 2002:10:12:1::1 ICMPv6 86 Neighbor
  Advertisement 2002:10:12:1::2 (rtr, sol, ovr) is at 00:02:00:02:00:12
2017-10-14 21:25:53.319291 2002:10:12:1::1 -> 2002:10:12:1::2 ICMPv6 118 Echo (ping)
  request id=0x1eaf, seq=1, hop limit=255
2017-10-14 21:25:53.319620 2002:10:12:1::2 -> 2002:10:12:1::1 ICMPv6 118 Echo (ping)
  reply id=0x1eaf, seq=1, hop limit=2 (request in 2580)
! Output omitted for brevity
```

排查 IPv6 ND 故障时，有必要检查全局和接口级 IPv6 ND 信息。此时，命令 **show ipv6 nd [global traffic | interface** *interface-id*] 就显得非常有用，如例 6-27 所示。可以看出，接口级 IPv6 ND 信息显示了 ND 进程发送各种 ICMPv6 消息的时间以及这些 ND 消息的参数等详细信息。命令选项 **global traffic** 可以显示交换机上的全局统计信息，这些统计信息是所有启用了 IPv6 且参与 IPv6 ND 进程的接口的累积统计信息。

例 6-27 IPv6 ND 接口信息

```
NX-1
NX-1# show ipv6 nd interface ethernet 4/1
ICMPv6 ND Interfaces for VRF "default"
Ethernet4/1, Interface status: protocol-up/link-up/admin-up
  IPv6 address:
    2002:10:12:1::1/64 [VALID]
  IPv6 link-local address: fe80::201:ff:fe01:12 [VALID]
```

```
    ND mac-extract : Disabled
  ICMPv6 active timers:
      Last Neighbor-Solicitation sent: 00:07:38
      Last Neighbor-Advertisement sent: 00:06:26
      Last Router-Advertisement sent: 00:01:34
      Next Router-Advertisement sent in: 00:06:50
  Router-Advertisement parameters:
      Periodic interval: 200 to 600 seconds
      Send "Managed Address Configuration" flag: false
      Send "Other Stateful Configuration" flag: false
      Send "Current Hop Limit" field: 2
      Send "MTU" option value: 1500
      Send "Router Lifetime" field: 1800 secs
      Send "Reachable Time" field: 0 ms
      Send "Retrans Timer" field: 0 ms
      Suppress RA: Disabled
      Suppress MTU in RA: Disabled
  Neighbor-Solicitation parameters:
      NS retransmit interval: 1000 ms
      ND NUD retry base: 1
      ND NUD retry interval: 1000
      ND NUD retry attempts: 3
  ICMPv6 error message parameters:
      Send redirects: true (0)
      Send unreachables: false
  ICMPv6 DAD parameters:
      Maximum DAD attempts: 1
      Current DAD attempt : 1
NX-1# show ipv6 nd global traffic
  ICMPv6 packet Statisitcs (sent/received):
  Total Messages           :        260/274
  Error Messages           :        0/0
  Interface down drop count :       0/0
  Adjacency not recovered from AM aft HA:      0/0
! Output omitted for brevity
  Echo Request             :        10/4
  Echo Replies             :        4/21
  Redirects                :        0/0
  Packet Too Big           :        0/0
  Router Advertisements    :        115/125
  Router Solicitations     :        0/0
  Neighbor Advertisements  :        65/65
  Neighbor Solicitations   :        66/59
  Fastpath Packets         :        0
  Ignored Fastpath Packets :        0
   Duplicate router RA received:     0/0
  ICMPv6 MLD Statistics (sent/received):
  V1 Queries:         0/0
  V2 Queries:         0/0
  V1 Reports:         0/0
  V2 Reports:         0/0
  V1 Leaves :         0/0
```

6.4.2 IPv6 地址分配

IPv6 网络中的主机通过以下地址分配方式获取 IPv6 地址。

- **手动配置**：可以使用 CLI 或 GUI（Graphical User Interface，图形用户界面）手动配置 IPv6 地址。
- **SLAAC（Stateless Auto-Address Configuration，无状态自动地址配置）**：SLAAC 自动执行 IPv6 地址分配进程。IPv6 主机使用从路由器接收到的 RA 消息中的 IPv6 前缀及其链路本地地址组合来形成唯一的 IPv6 地址。SLAAC 不会选择性地分配地址，也不会在客户端所允许的地址上实施分配策略。SLAAC 仅要求 DHCP 服务器提供其他必要配置信息，如 DNS 服务器和域搜索列表。目前，NX-OS 不支持该地址分配方式。
- **有状态地址配置**：有状态地址配置由 DHCPv6 服务器分配 IPv6 地址。由于客户端与服务器之间的 DHCPv6 基本操作使用链路本地地址进行通信，因而同一链路中必须有 DHCP 服务器实体，该实体可以是 DHCPv6 服务器，也可以是 DHCPv6 中继代理（如果没有服务器）。可以使用被称为 All_DHCP_Relay_Agents_and_Servers（全部 DHCP 中继代理和服务器）的预定义的链路范围多播地址（FF02::1:2）与邻接的中继代理和服务器进行通信，所有服务器和中继代理都是该多播组的成员。在很多情况下，出于实际需求和网络管理等原因，DHCPv6 服务器并不总是与每条链路都直连，因而需要使用中继代理，由中继代理拦截本地链路的 DHCPv6 消息，然后再转发给管理员配置的 DHCPv6 服务器。可以为中继代理配置目的地址列表，包括单播地址、All_DHCP_Servers（全部 DHCP 服务器）多播地址或网络管理员选择的其他地址。

1. DHCPv6 中继代理

DHCPv6 中继代理与 IPv4 DHCP 中继代理相似，DHCPv6 中继代理负责拦截客户端发送的 DHCPv6 请求数据包、执行基本验证操作、构造 DHCPv6 RELAY-FORWARD 消息，然后再转发给在中继代理上配置的所有 DHCPv6 服务器。与此相似，DHCPv6 中继代理会拦截服务器发送的所有 DHCPv6 RELAY-REPLY 数据包并执行基本的验证操作，然后再根据接收到的数据包的中继消息选项构造实际的 DHCPv6 RELAY-REPLY 消息，并转发给目的客户端。此外，NX-OS 还支持中继链功能，即 DHCPv6 中继代理从其他中继代理获取 RELAY-FORWARD 消息，并转发给相应的 DHCPv6 服务器，同时还在 RELAY-FORWARD 消息中插入 Remote-ID 和 Interface-ID 选项，并在转发给客户端/中继代理之前从 RELAY-REPLY 消息中删除这些信息。

如果要启用 DHCPv6 中继代理，那么就可以使用全局配置命令 **ipv6 dhcp relay**。在全局范围内启用了 DHCPv6 中继代理之后，还必须在面向客户端的接口（可以是 L3 接口、SVI、L3 端口通道或子接口）上启用中继代理，配置命令是 **ipv6 dhcp relay address** *ipv6-address* [**use-vrf** *vrf-name* **interface** *interface-id*]（如果可以通过不同的 VRF 访问 DHCP 服务器，那么就使用 **use-vrf** 选项）。可以通过全局配置命令 **ipv6 dhcp relay option vpn** 跨 VRF 启用 DHCPv6 中继代理。NX-OS 的 DHCPv6 中继代理还提供了另一种可选项，可以通过全局配置命令 **ipv6 dhcp relay source-interface** *interface-id* 指定源接口。

配置了 DHCPv6 中继并在客户端与服务器之间交换消息之后，可以使用 **show ipv6 dhcp relay statistics** 命令查看中继代理的统计信息。例 6-28 显示了 module 7（该模块连接客户端）的 DHCPv6 中继代理统计信息。需要记住的是，客户端发送 DHCPv6 请求时，会向与其相连的路由器或交换机发送 DHCPv6 请求消息。第一跳或中继代理收到这些请求消息之后，会将其作为 RELAY-FORWARD 消息中继给 DHCPv6 服务器。如果将来自客户端的原始消息以 RELAY-FORWARD 消息中继给服务器，那么服务器就会发送 RELAY-REPLY 消息，从而将响应消息回送给客户端。如果 DHCPv6 数据包被丢弃了，那么输出结果还会显示这些丢包计数器。

例 6-28　DHCPv6 中继统计信息

```
NX-1# show ipv6 dhcp relay statistics interface Eth7/1
-------------------------------------------------------------------------------
Message Type                        Rx              Tx              Drops
-------------------------------------------------------------------------------
SOLICIT                             5               0               0
ADVERTISE                           0               2               0
REQUEST                             1               0               0
CONFIRM                             0               0               0
RENEW                               0               0               0
REBIND                              0               0               0
REPLY                               0               2               0
RELEASE                             1               0               0
DECLINE                             0               0               0
RECONFIGURE                         0               0               0
INFORMATION_REQUEST                 0               0               0
RELAY_FWD                           0               7               0
RELAY_REPLY                         4               0               0
UNKNOWN                             0               0               0
-------------------------------------------------------------------------------
Total                               11              11              0
-------------------------------------------------------------------------------

DHCPv6 Server stats:
-------------------------------------------------------------------------------
Relay Address       VRF name        Dest. Interface     Request     Response
-------------------------------------------------------------------------------
2001:10:12:1::1     ---             ---                 7           4
DROPS:
------
DHCPv6 Relay is disabled                                :   0
Max hops exceeded                                       :   0
Packet validation fails                                 :   0
Unknown output interface                                :   0
Invalid VRF                                             :   0
Option insertion failed                                 :   0
Direct Replies (Recnfg/Adv/Reply) from server:              0
IPv6 addr not configured                                :   0
Interface error                                         :   0
VPN Option Disabled                                     :   0
IPv6 extn headers present                               :   0
```

与 IPv4 DHCP 中继代理相似，DHCPv6 中继也在硬件中进行 ACL 编程。可以通过命令 **show system internal access-list input statistics [module** *slot*] 查看相应的统计信息，如例 6-29 所示。

例 6-29　DHCPv6 中继 ACL 线卡统计信息

```
NX-1# show system internal access-list input statistics module 7
             VDC-4 Ethernet7/1 :
             =====================

INSTANCE 0x0
---------------

  Tcam 1 resource usage:
  ----------------------
  Label_b = 0x201
```

```
      Bank 0
      ------
        IPv6 Class
          Policies: DHCP(DHCPV6 Relay)   [Merged]
          Netflow profile: 0
          Netflow deny profile: 0
          Entries:
            [Index] Entry [Stats]
            --------------------
 [0058:000e:000e] prec 1 redirect(0x0) udp 0x0/0 0xffffffb8/32 eq 547
 flow-label 547 [6]
 [0059:000f:000f] prec 1 permit ip 0x0/0 0x0/0 [8]
```

2. DHCPv6 中继 LDRA

DHCPv6 主要运行在二层（使用链路范围多播地址），这是因为客户端在没有 IPv6 地址或完成 DHCPv6 处理之前不知道 DHCPv6 服务器的位置。如果网络中的 DHCPv6 服务器与客户端不在同一条链路中，那么就可以使用 DHCPv6 中继代理，使 DHCPv6 服务器能够处理本地链路之外的客户端请求。

DHCPv6 中继代理在上游 DHCPv6 消息（从客户端到服务器）中增加了 Interface-ID 选项，以标识客户端所连接的接口，DHCPv6 中继代理在将下游 DHCPv6 消息转发给 DHCPv6 客户端时使用该信息。

如果终端主机与 DHCPv6 中继代理直连，那么将毫无疑问能够正常工作。但是，对于某些网络配置来说，DHCPv6 客户端与中继代理之间可能存在一台或多台二层设备。此时，就很难使用 DHCPv6 中继代理的 Interface-ID 选项识别客户端，由于这些二层设备靠近终端主机，因而需要二层设备在 DHCPv6 消息中附加 Interface-ID 选项。通常将这类设备称为 LDRA（Lightweight DHCPv6 Relay Agent，轻量级 DHCPv6 中继代理）。

如果客户端没有 IPv6 地址或者不知道 DHCPv6 服务器的位置，那么 DHCPv6 就会向保留的链路范围多播地址 ff02::1:2 发送 Information-Request 消息。客户端在 UDP 端口 546 上监听 DHCPv6 消息，服务器和中继代理在 UDP 端口 547 上监听 DHCPv6 消息。LDRA 会检查入站接口是 L3 接口还是 L2 接口，如果是 L3 接口，那么就将数据包发送给 DHCPv6 中继代理；如果是 L2 接口，那么就执行以下检查。

- LDRA 检查入站接口或 VLAN 是否启用或禁用了 LDRA。如果禁用了 LDRA，那么就正常交换数据包。
- 启用了 LDRA 功能的入站接口可以分为以下类别（如果不属于这些类别，那么 LDRA 就会丢弃该数据包）。
    - 客户端侧可信接口。
    - 客户端侧非可信接口。
    - 服务器侧非可信接口。
- 如果在客户端侧非可信接口上收到了 RELAY-FORWARD 消息，那么就丢弃该数据包。
- 如果数据包属于以下类型的消息，那么就丢弃以下数据包。
    - ADVERTISE 消息。
    - REPLY 消息。
    - RECONFIGURE RELAY-REPLY 消息。
- 如果跳数大于最大允许值，那么就丢弃该数据包。
- 如果数据包通过了所有的校验检查，那么就会创建一个新帧并中继给服务器。

- msg-type：RELAY-FORWARD。
- hop-count：
  - 如果收到的消息不是 RELAY-FORWARD，那么就将跳数设置为 0；
  - 否则，将跳数递增 1。
- link-address：未指定地址（::）。
- peer-address：客户端的链路本地地址（入站帧的 IP 报头中收到的源 IP 地址）。
- Interface-ID 选项：填充 Interface-ID 详细信息以标识收到该数据包的接口。
- Relay-Message 选项：原始收到的消息。

如前所示，link-address 参数必须设置为 0。LDRA 在 RELAY-FORWARD 消息中包含了 Interface-ID 选项和 Relay-Message 选项，其他选项均是可选项。LDRA 利用 Interface-ID 来标识交换机和收到该数据包的接口。Interface-ID 是一个非传导值，服务器不会解析 Interface-ID 选项的信息。LDRA 利用交换机的 MAC 地址以及接口 ifindex 创建一个字符串，并将其用作 Interface-ID。如果入站消息是 RELAY-FORWARD 消息且在客户侧可信接口上收到了该消息，那么就表明网络中的二层或三层代理已可用，且位于本地中继代理之前。

如果启用了 LDRA 的设备收到了服务器发送的响应消息，那么设备将执行以下操作。

- 如果在服务器侧的可信接口上未收到 RELAY-REPLY 消息，那么就丢弃该数据包。如果是其他类型的 DHCP 消息，那么也丢弃该数据包。
- 如果数据包没有 Interface-ID 选项，那么就丢弃该数据包。
- LDRA 检查该数据包是否是发送给自己的数据包。
- LDRA 会在转发请求时添加 Interface-ID 选项，其中包含了交换机和接口的详细信息。二层中继代理可以提取该信息。
- 如果 Interface-ID 选项包含了与交换机相关的信息，且与 L2 中继代理的标识符不匹配，那么就简单地转发该数据包，而不进行任何更改，因为该消息是发送给网络中的其他中继代理的。

如果要在全局范围内启用 LDRA，那么就要配置 **ipv6 dhcp-ldra** 命令，然后利用命令 **ipv6 dhcp-ldra** 在 L2 接口上启用 LDRA。此外，LDRA 还允许通过命令 **ipv6 dhcp-ldra attach-policy [client-facing-untrusted | client-facing-trusted | client-facing-disabled | server-facing-trusted]** 指定接口策略，支持以下策略选项。

- **client-facing-untrusted**：接口上连接的是常规客户端。
- **client-facing-trusted**：网络中位于该设备之前的其他 L2 或 L3 中继代理都连接在该接口上。连接在该接口上的中继代理应该位于实际的客户端与该 L2 中继代理之间（位于上游网络）。
- **server-facing-trusted**：位于该设备之后且去往服务器端的所有 DHCP 服务器或 L3 中继代理都应该连接在该接口上。如果该接口连接了中继代理，那么中继代理就应该位于该设备与实际 DHCP 服务器之间的网络路径上。

注：如果出现了与 DHCPv6 中继代理相关的故障问题，那么就可以收集 **show tech dhcp** 命令的输出结果，并联系思科 TAC（Technical Assistance Center，技术支持中心）。

6.4.3 IPv6 第一跳安全

IPv4 FHS（First-Hop Security，第一跳安全）已经在数据中心应用了很多年，特别是思科 Catalyst 交换机等设备，最近也开始用于 Nexus 交换机。FHS 功能解决的安全漏洞与园区网面临的安全威胁相似。对于数据中心环境来说，主机通常是专用于物理服务器的应用程序或者是共享同一物理服务器的 VM（Virtual Machine，虚拟机）。与其他部署 FHS 的环境不同，服务器和 VM 不可信的主要原因是它们可能是攻击者通过某种允许协议（如 HTTP）安装的恶意软件的受害者。

IPv6 FHS 中的某些功能与 IPv4 相似。例如，安装在 VM 上的某些恶意软件可能会发送 RA（Router Advertisement，路由器宣告）消息，从而伪装成链路上的其他 VM 的默认网关。需要注意的是，虽然这种场景貌似合理，但实际上，企业和园区网络中的恶意 RA 消息主要来自粗心的用户。这种情况对于数据中心环境来说并不是很严重，因为 VM 和服务器通常都是受管设备。当然，某些 VM 也可能处于非受管状态，此时用户的粗心问题就会演变为非常重要的问题。表 6-2 列出了常见的 IPv6 网络攻击行为以及相应的 FHS 防范技术。

表 6-2　　　　　　　　　　　　　IPv6 攻击和防范技术

攻击	攻击防范功能	能力
非法 VM 发起的 MAC 欺骗	端口安全	限制端口上的 MAC 地址
被感染 VM 发起的 IPv6 地址欺骗	IPv6 源保护	以端口/MAC/VLAN 为基础验证 IP 源地址
其他 VM、主机和网络设备上的 ND 缓存中毒	IPv6 监听	监控 ND 和 DHCP 流量并收集地址分配信息，为源和目的地址保护提供一个有效的源和目的地址列表
非法 DHCP 服务器	DHCPv6 保护	防止不可信实体充当 DHCP 服务器
非法路由器	RA 保护	防止不可信实体充当路由器

本节将详细讨论 RA 保护、IPv6 监听以及 DHCPv6 保护等安全功能。

注：虽然还有很多其他 FHS 技术，但本章并不讨论这些技术。有关更多详细信息，请参阅 Cisco 官网文档。

1. RA 保护

RA 保护（RA Guard）功能允许二层交换机用户将指定交换机端口配置为面向路由器的端口，其他端口收到 RA（Router Advertisement，路由器宣告）消息之后都会丢弃这些消息，因而永远也不会到达该链路的终端主机。RA 保护功能会执行更深入的深度包检查操作，以验证 RA 的源端、前缀列表、优先级以及所携带的其他信息。RA 保护功能定义在 RFC 6105 中，目的是检查路由器的 ND（Neighbor Discovery，邻居发现）流量（如 RS、RA 和重定向消息）并丢弃非法消息。RA 保护功能可以基于策略配置（如主机端口不允许 RA 消息）来阻塞未授权消息。

如果要启用 IPv6 RA 保护功能，需要首先定义 RA 保护策略，然后在接口上应用该策略。RA 保护策略的配置命令是 **ipv6 nd raguard policy** *policy-name*。表 6-3 列出了 RA 保护策略的所有可用选项。

表 6-3　　　　　　　　　　　　　RA 保护策略子配置选项

选项	描述
device-role [host | router | monitor | switch]	定义连接在端口上的设备的角色，可以是主机、路由器、监控器或交换机
hop-limit [maximum | minimum limit]	验证指定的跳数限制。如果未配置，那么就忽略该检查
managed-config-flag [on | off]	检查所宣告的 managed-config 标志处于 on 还是 off 状态。如果未配置，那么就忽略该检查
other-config-flag [on | off]	验证所宣告的其他配置参数
router-preference maximum [high | low | medium]	检查所宣告的默认路由优先级参数小于还是等于指定极限值
trusted-port	指定将策略应用于可信端口

定义了 RA 保护策略之后，就可以通过接口级配置命令 **ipv6 nd raguard attach-policy**

policy-name 将其应用到接口上。例 6-30 给出了 RA 保护配置示例,命令 **show ipv6 nd raguard policy** *policy-name* 显示了关联在不同接口上的 RA 保护策略。

例 6-30 配置 IPv6 RA 保护功能

```
NX-1(config)# ipv6 nd raguard policy RAGUARD
NX-1(config-raguard-policy)# device-role router
NX-1(config-raguard-policy)# trusted-port
NX-1(config-raguard-policy)# exit
NX-1(config)# interface e7/1
NX-1(config-if)# ipv6 nd raguard attach-policy RAGUARD
NX-1# show ipv6 nd raguard policy RAGUARD

Policy RAGUARD configuration:
  trusted-port
  device-role router
Policy RAGUARD is applied on the following targets:
Target           Type    Policy            Feature       Target range
Eth7/1           PORT    RAGUARD           RA guard      vlan all
```

> **注:** 排查 IPv6 RA 保护故障时,可以使用调试命令 **debug ipv6 snooping raguard**,输出结果将显示在调试日志文件中。

2. IPv6 监听

IPv6 监听是 ND 监听和 DHCPv6 监听两种功能特性的组合。IPv6 ND 监听功能可以分析 IPv6 邻居发现流量并确定该流量对链路上的节点是否有害,检查期间,ND 监听功能会收集地址绑定信息(IP、MAC、端口)(如果有)并存储到绑定表中。此后,如果两个客户端之间发生争用情况,那么就可以利用绑定表项确定地址所有权。IPv6 DHCP 监听功能在客户端与服务器之间抓取 DHCPv6 数据包,然后从监听到的数据包中学习已分配的地址并存储到绑定表中。此外,IPv6 监听功能还能限制链路上的节点所能索要的地址数量,从而有助于保护交换机绑定表免受 DoS 泛洪攻击。图 6-6 解释了 IPv6 监听功能的角色及其防止设备免受无效或非法主机攻击的保护方式。

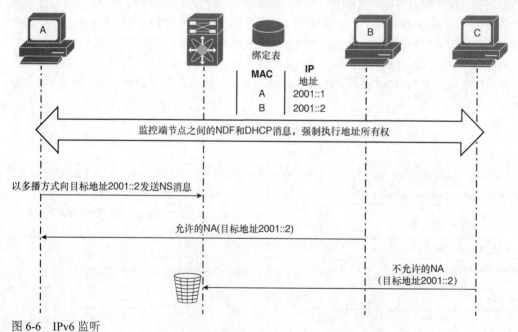

图 6-6 IPv6 监听

可以通过下面两个简单步骤配置 IPv6 监听功能。
- 定义监听策略。
- 将监听策略附加到 VLAN 上。

IPv6 监听策略的配置命令是 **ipv6 snooping policy** *policy-name*。可以在 IPv6 监听策略中指定多种选项（见表 6-4）。

表 6-4　　　　　　　　　　　　　　IPv6 监听策略子选项

子选项	描述
device-role [node | switch]	定义连接在端口上的设备的角色，默认设备角色是 node。设备角色（配合 **trusted-port** 命令）对于该策略所应用的接口学到的表项的优先级水平有直接影响 **device-role node** 拥有接入端口的优先级，**device-role node** 拥有中继端口的优先级
tracking [enable [reachable-lifetime *value*] | **disable** [**stalelifetime** *value*]]	覆盖该策略所应用的端口上的默认跟踪策略 如果跟踪表项不是期望表项且表项应该位于绑定表中以防止被窃，那么该选项对于可信端口来说就非常有用。在这种情况下，可以配置命令 **tracking disable stale-lifetime** *infinite*
trusted port	使用该策略在端口上接收消息时，将不执行任何验证操作。不过，为了防止地址欺骗，仍然会进行消息分析，这样就可以利用这些消息携带的绑定信息维护绑定表。从这些端口发现的绑定信息比从非可信端口收到的绑定信息更可信
validate source-mac	收到包含链路层地址选项的 NDP（Neighbor Discovery Protocol，邻居发现协议）消息后，将根据链路层地址选项检查源 MAC 地址，如果不同，则丢弃该数据包
protocol [dhcp | ndp]	指定应该将哪种协议重定向到监听组件以进行分析检查
security-level [glean | inspect | guard]	指定 IPv6 监听功能实现的安全级别，默认值为 **guard** **glean**：学习绑定信息，但不丢弃数据包 **inspect**：学习绑定信息，并在检测到问题时丢弃数据包 **guard**：与 **inspect** 相似，发现威胁时也丢弃 IPv6、ND、RA 和 IPv6 DHCP 服务器数据包

定义了监听策略之后，就可以在 **vlan configuration** *vlan-id* 子配置模式下使用命令 **ipv6 snooping attach-policy** *policy-name* 附加该策略。

例 6-31 给出了 VLAN 100 的 IPv6 监听策略的配置示例。

例 6-31　配置 IPv6 监听

```
NX-1(config)# ipv6 snooping policy V6-SNOOP-VLAN-100
NX-1(config-snoop-policy)# device-role switch
NX-1(config-snoop-policy)# protocol ndp
NX-1(config-snoop-policy)# trusted-port
NX-1(config-snoop-policy)# exit
NX-1(config)# vlan configuration 100
NX-1(config-vlan-config)# ipv6 snooping attach-policy V6-SNOOP-VLAN-100
NX-1# show ipv6 snooping policy

Policy V6-SNOOP-VLAN-100 configuration:
  trusted-port
  security-level guard
  device-role switch
  gleaning from Neighbor Discovery
  gleaning from DHCP
  NOT gleaning from protocol unkn
Policy V6-SNOOP-VLAN-100 is applied on the following targets:
Target              Type  Policy                Feature       Target range
vlan 100            VLAN  V6-SNOOP-VLAN-100     Snooping      vlan all
```

与其他 FHS 功能相似，IPv6 监听功能也在硬件中进行 ACL 编程，可以通过命令 **show system internal access-list interface** *interface-id* **[module** *slot*]加以验证，可以通过命令 **show system internal access-list input statistics [module** *slot*]显示相应的统计信息，如例 6-32 所示。

例 6-32　验证 IPv6 监听的硬件统计信息

```
NX-1# show system internal access-list input statistics module 7
              VLAN 100 :
              =========

INSTANCE 0x0
--------------

  Tcam 1 resource usage:
  ----------------------
  Label_b = 0x201
   Bank 0
   ------
    IPv6 Class
      Policies: DHCP_FHS(DHCP SISF)    [Merged]
      Netflow profile: 0
      Netflow deny profile: 0
      Entries:
        [Index] Entry [Stats]
        --------------------
[0058:000e:000e] prec 1 redirect(0x0) icmp 0x0/0 0x0/0 137 0 flow-label 35072     [0]
[0059:000f:000f] prec 1 redirect(0x0) icmp 0x0/0 0x0/0 136 0 flow-label 34816     [0]
[005a:0010:0010] prec 1 redirect(0x0) icmp 0x0/0 0x0/0 135 0 flow-label 34560     [0]
[005b:0011:0011] prec 1 redirect(0x0) icmp 0x0/0 0x0/0 134 0 flow-label 34304     [0]
[005c:0012:0012] prec 1 redirect(0x0) icmp 0x0/0 0x0/0 133 0 flow-label 34048     [0]
[005d:0013:0013] prec 1 redirect(0x0) udp 0x0/0 0x0/0 eq 547 flow-label 547       [0]
[005e:0014:0014] prec 1 redirect(0x0) udp 0x0/0 0x0/0 eq 546 flow-label 546       [0]
[005f:0015:0015] prec 1 redirect(0x0) udp 0x0/0 eq 547 0x0/0 flow-label 196608    [0]
[0060:0016:0016] prec 1 permit ip 0x0/0 0x0/0    [0]
```

3．DHCPv6 保护

DHCPv6 保护（DHCPv6 Guard）功能的主要目的是阻塞来自非法 DHCP 服务器或中继代理的 DHCP 应答或宣告消息。根据实际部署情况，DHCPv6 保护将决定桥接、交换或阻塞这些消息。此外，DHCPv6 保护功能还会验证消息中的相关信息，如消息中的地址和前缀是否位于指定范围。可以将设备配置为客户端或服务器模式，从而保护客户端免于接收非法 DHCP 服务器的应答消息。

由于设备的默认安全模式是 **guard**，因而在默认情况下，所有配置了 DHCPv6 保护功能的端口均处于客户端模式，也就是说，所有端口在默认情况下均丢弃所有 DHCPv6 服务器消息。因此，对于有实际应用意义的 DHCPv6 部署方案来说，至少应该将一个端口的角色配置为 **dhcp-server**，使该端口允许 DHCPv6 服务器消息。这也是实现合理安全级别的最简单配置方式。

在接口或 VLAN 上启用了 DHCP 保护功能之后，DHCP 保护功能将在硬件中对 ACL 进行编程，通过 ACL 将 DHCP 数据包导出给管理引擎。ACL 过滤器如下。

- 匹配 UDP 协议。
- 源端口应该是 DHCP 客户端端口（端口 546）或服务器端口（端口 547）。
- 目的端口应该是 DHCP 客户端端口（端口 546）或服务器端口（端口 547）。

可以通过命令 **show system internal access-list input interface** *interface-id* **module** *slot* 验证线卡或硬件的 ACL 配置。

如果在同一设备上同时配置了 DHCPv6 保护功能和 DHCPv6 中继功能，那么将首先由 DHCPv6 保护功能处理 FHS 收到的 DHCPv6 请求数据包。DHCPv6 保护进程完成所有处理操作之后，再将数据包提供给 DHCPv6 中继功能，由其将数据包中继给指定服务器。

与此类似，DHCP 中继代理首先要处理其他中继代理或 DHCP 服务器发来的 DHCP 应答数据包。如果收到的数据包不是中继数据包（RELAY-FORWARD 或 RELAY-REPLY），那么就传递给 DHCP 保护功能进行处理。

> **注：**事实上，DHCP 保护和 DHCP 中继代理功能仅在第一跳进行协同工作。在以后的各跳中，DHCP 中继代理的优先级高于 DHCP 保护功能。系统为这两项功能维护独立的统计信息。

如果要配置 DHCPv6 保护策略，那么就可以使用命令 **ipv6 dhcp guard policy** *policy-name*。在策略配置下，首先要定义设备角色，即客户端、服务器或监控器。然后再定义已宣告的最小和最大所允许的服务器优先级。此外，还可以使用 **trusted port** 命令选项指定设备是否连接在可信端口上。完成策略配置之后，可以使用命令 **ipv6 dhcp guard attach-policy** *policy-name* 命令将策略附加到指定端口或 VLAN 上。例 6-33 给出了 DHCPv6 保护功能的配置示例。如果要检查策略配置情况，可以使用命令 **show ipv6 dhcp guard policy** 或 **show ipv6 snooping policies** 来验证 IPv6 监听和 DHCPv6 保护策略。请注意，DHCPv6 保护功能与 IPv6 监听功能协同使用。

例 6-33　配置 DHCPv6 保护功能并验证保护策略

```
NX-1(config)# ipv6 dhcp guard policy DHCPv6-Guard
NX-1(config-dhcpg-policy)# device-role ?
  client    Attached device is a client (default)
  monitor   Attached device is a monitor/sniffer
  server    Attached device is a dhcp server

NX-1(config-dhcpg-policy)# device-role server
NX-1(config-dhcpg-policy)# preference max 255
NX-1(config-dhcpg-policy)# preference min 0
NX-1(config-dhcpg-policy)# trusted port
NX-1(config-dhcpg-policy)# exit
NX-1(config-if)# interface e7/13
NX-1(config-if)# switchport
NX-1(config-if)# switchport access vlan 200
NX-1(config-if)# ipv6 dhcp guard attach-policy DHCPv6-Guard
NX-1(config-if)# exit
NX-1(config)# vlan configuration 200
NX-1(config-vlan-config)# ipv6 dhcp guard attach-policy DHCPv6-Guard
NX-1(config-vlan-config)# end
NX-1# show ipv6 dhcp guard policy
Dhcp guard policy: DHCPv6-Guard
        Trusted Port
        Target: Eth7/13 vlan 200
NX-1# show ipv6 snooping policies


Target              Type  Policy              Feature       Target range
Eth7/13             PORT  DHCPv6-Guard        DHCP Guard    vlan all
vlan 100            VLAN  V6-SNOOP-VLAN-100   Snooping      vlan all
vlan 200            VLAN  DHCPv6-Guard        DHCP Guard    vlan all
```

6.5 FHRP

FHRP(First-Hop Redundancy Protocol,第一跳冗余协议)可以为网络提供路由冗余能力。FHRP 旨在为网络提供第一跳 IP 网关的透明故障切换。可以利用以下功能特性在交换机或路由器上实现 FHRP。

- HSRP(Hot Standby Routing Protocol,热备路由协议)。
- VRRP(Virtual Router Redundancy Protocol,虚拟路由器冗余协议)。
- GLBP(Gateway Load-Balancing Protocol,网关负载均衡协议)。

接下来将详细解释这些 FHRP 协议的工作方式和故障排查方式。

6.5.1 HSRP

HSRP 定义在 RFC 2281 中,可以提供第一跳设备(通常充当主机的网关设备)的透明故障切换。HSRP 能够为配置了默认网关 IP 地址的以太网中的 IP 主机提供路由冗余机制。最少需要两台设备才能启用 HSRP 功能,其中的一台设备充当主用设备并负责转发数据包;另一台设备充当备用设备,可以在故障情况下接管主用设备的角色。

在一个网段当中,需要为属于同一个 HSRP 组的每个启用了 HSRP 的接口配置一个虚拟 IP。HSRP 选择其中的某个接口充当 HSRP 活动路由器。除虚拟 IP 之外,还要为 HSRP 组分配一个虚拟 MAC 地址。主用路由器接收数据包并将数据包路由给该 HSRP 组的虚拟 MAC 地址。HSRP 主用设备出现故障后,HSRP 备用设备将控制该组的虚拟 IP 和 MAC 地址。如果 HSRP 组中有两台以上的设备,那么就会选择一台新的 HSRP 备用设备。网络运营商通过定义接口优先级(默认值为 100)来控制应该由哪台设备应充当 HSRP 主用设备,优先级较高的设备将充当 HSRP 主用设备。

启用了 HSRP 功能的接口通过发送和接收基于 UDP 的多播 Hello 消息,来检测故障状况并指定主用和备用路由器。如果备用设备未收到 Hello 消息或者主用设备无法发送 Hello 消息,那么拥有次高优先级的备用设备将成为 HSRP 主用设备。请注意,设备之间的 HSRP 主用状态的切换操作对于网段上的所有主机来说都是透明的。

HSRP 支持两种版本:版本 1 和版本 2,表 6-5 列出了这两种 HSRP 版本之间的主要区别。

表 6-5　　　　　　　　　　HSRP 版本 1 与版本 2 对比

	HSRP 版本 1	HSRP 版本 2
定时器	不支持毫秒定时器	支持毫秒定时器
组范围	0~255	0~4095
多播地址	224.0.0.2	224.0.0.102
MAC 地址范围	0000.0C07.AC*xy*,其中的 *xy* 是 HSRP 组号的十六进制数值	0000.0C9F.F000~0000.0C9F.FFFF
认证	不支持认证	支持 MD5 认证

注:HSRP 从版本 1 到版本 2 是中断性的升级,因为这两种版本的 MAC 地址发生了变化。

如果在网段上配置了 HSRP,且选定了主用设备和备用设备,那么 HSRP 控制包将包含以下字段。

- **源 MAC**:活动设备的虚拟 MAC 或备用/侦听设备的接口 MAC。
- **目的 MAC**:版本 1 为 0100.5e00.0002,版本 2 为 0100.5e00.0066。
- **源 IP**:接口 IP。

- **目的 IP**：版本 1 为 224.0.0.2，版本 2 为 224.0.0.102。
- **UDP 端口**：1985。

为了更好地理解 HSRP 功能，下面将以图 6-7 所示拓扑结构为例加以说明，例中的 VLAN 10 运行了 HSRP。

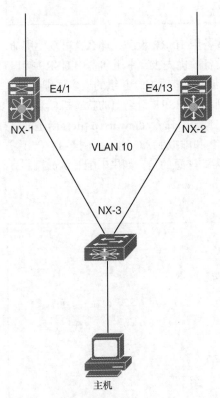

图 6-7　HSRP 拓扑结构

如果要启用 HSRP，那么就需要使用命令 **feature hsrp**。配置完成后，HSRP 默认以 HSRP 版本 1 方式进行运行。如果希望手动更改 HSRP 的版本，那么就需要在 HSRP 配置模式下使用命令 **hsrp version [1 | 2]**进行配置。

例 6-34 给出了 VLAN 10 的 HSRP 配置实例。可以看出，本例为 HSRP 配置了组号 10 和 VIP（Virtual IP，虚拟 IP）10.12.1.1，NX-1 的优先级被设置为 110，意味着 NX-1 将充当主用 HSRP 网关。此外，本例还为 HSRP 配置了抢占模式，即 NX-1 出现故障且 HSRP 主用网关切换为 NX-2 之后，如果 NX-1 恢复正常再次可用，那么 NX-1 将重新接管主用设备的角色。

例 6-34　配置 HSRP

```
NX-1
interface Vlan10
  no shutdown
  no ip redirects
  ip address 10.12.1.2/24
  hsrp version 2
  hsrp 10
   preempt
    priority 110
    ip 10.12.1.1
NX-2
```

```
interface Vlan10
  no shutdown
  no ip redirects
  ip address 10.12.1.3/24
  hsrp version 2
  hsrp 10
    ip 10.12.1.1
```

如果要查看 HSRP 组的状态并确定哪台设备充当主用或备用 HSRP 设备，那么就可以使用命令 **show hsrp brief**，该命令可以显示 HSRP 组信息、本地设备的优先级、主用和备用 HSRP 接口地址以及组地址（HSRP VIP）。也可以使用命令 **show hsrp [detail]** 查看 HSRP 组的更多详细信息，该命令不但列出了 HSRP 组的详细信息，而且按时间序列列出了 HSRP 组经历的状态机信息。该命令对于排查与 HSRP 有限状态机有关的故障时非常有用。此外，命令 **show hsrp [detail]** 还可以显示为该 HSRP 组配置的认证机制以及该 HSRP 组的虚拟 IP 和虚拟 MAC 地址。例 6-35 给出了 **show hsrp brief** 和 **show hsrp detail** 命令的输出结果示例，需要注意的是，如果未配置认证机制，那么 **show hsrp detail** 命令的认证部分将显示 *Authentication text "cisco"*。

例 6-35　HSRP 验证及详细信息

```
NX-1
NX-1# show hsrp brief
*:IPv6 group   #:group belongs to a bundle
              P indicates configured to preempt.
              |
 Interface  Grp  Prio P State   Active addr   Standby addr   Group addr
  Vlan10    10   110  P Active  local         10.12.1.3      10.12.1.1
    (conf)
NX-1# show hsrp detail
Vlan10 - Group 10 (HSRP-V2) (IPv4)
  Local state is Active, priority 110 (Cfged 110), may preempt
    Forwarding threshold(for vPC), lower: 1 upper: 110
  Hellotime 3 sec, holdtime 10 sec
  Next hello sent in 0.951000 sec(s)
  Virtual IP address is 10.12.1.1 (Cfged)
  Active router is local
  Standby router is 10.12.1.3 , priority 100 expires in 9.721000 sec(s)
  Authentication text "cisco"
  Virtual mac address is 0000.0c9f.f00a (Default MAC)
  2 state changes, last state change 00:03:07
  IP redundancy name is hsrp-Vlan10-10 (default)

----- Detailed information -----
State History
---------------------------------------------------------------
 Time                  Prev State      State        Event
---------------------------------------------------------------
 (20)-20:04:55 Active                  Active       Sby Timer Expired

 (20)-19:38:58 Standby                 Active       Act Timer Expired.

 (20)-19:38:47 Speak                   Standby      Sby Timer Expired.

 (20)-19:38:47 Listen                  Speak        Act Timer Expired.

 (20)-19:38:37 Initial                 Listen       If Enabled-VIP.
```

在接口上配置了 HSRP 之后,接口会根据 HSRP 版本自动加入相应的 HSRP 多播组。可以通过命令 **show ip interface** *interface-id* 查看该信息,该命令不提供接口 HSRP 虚拟 IP 信息。如果希望查看虚拟 IP 和 HSRP 多播组信息,那么就需要使用命令 **show ip interface** *interface-id* **vaddr**。例 6-36 给出了这两条命令的输出结果示例。

例 6-36　HSRP 多播组和 VIP

```
NX-1# show ip interface vlan 10
IP Interface Status for VRF "default"(1)
Vlan10, Interface status: protocol-up/link-up/admin-up, iod: 148,
  IP address: 10.12.1.2, IP subnet: 10.12.1.0/24 route-preference: 0, tag: 0
  IP broadcast address: 255.255.255.255
  IP multicast groups locally joined:
      224.0.0.102
  IP MTU: 1500 bytes (using link MTU)
! Output omitted for brevity
NX-1# show ip interface vlan 10 vaddr
IP Interface Status for VRF "default"(1)
Vlan10, Interface status: protocol-up/link-up/admin-up, iod: 148,
  IP address: 10.12.1.2, IP subnet: 10.12.1.0/24 route-preference: 0, tag: 0
  Virtual IP address(406): 10.12.1.1, IP subnet: 10.12.1.1/32
  IP broadcast address: 255.255.255.255
  IP multicast groups locally joined:
      224.0.0.102
  IP MTU: 1500 bytes (using link MTU)
```

请注意,主用 HSRP 网关设备会用虚拟 IP 地址和虚拟 MAC 地址填充 ARP 表,如例 6-37 所示。可以看出,虚拟 IP 地址 10.12.1.1 映射到 MAC 地址 0000.0c9f.f00a(组 10 的虚拟 MAC 地址)。

例 6-37　ARP 表中的 HSRP 虚拟 MAC 地址与虚拟 IP 地址

```
NX-1# show ip arp vlan 10

Flags: * - Adjacencies learnt on non-active FHRP router
       + - Adjacencies synced via CFSoE
       # - Adjacencies Throttled for Glean
       D - Static Adjacencies attached to down interface

IP ARP Table
Total number of entries: 1
Address         Age          MAC Address        Interface
10.12.1.1       -            0000.0c9f.f00a     Vlan10
```

如果 HSRP 处于关闭状态或者在两台设备之间来回震荡,或者 HSRP 尚未在两台设备之间建立正确的状态(如两台设备均显示为主用/主用状态),那么就可能需要进行抓包或调试操作,以检查 HSRP Hello 包是否正在到达对端或者是否正在交换机本地生成 Hello 包。由于 HSRP 控制包发往 CPU,因而需要使用 Ethanalyzer 来抓取这些数据包。可以通过 **hsrp** 的显示过滤器(display-filter)来抓取 HSRP 控制包并确定是否未收到任何数据包。

除 Ethanalyzer 之外,还可以同时启用 HSRP 调试功能,以查看是否已收到 Hello 包。可以通过命令 **debug hsrp engine packet hello interface** *interface-id* **group** *group-number* 启用 Hello 包的 HSRP 调试功能,该命令可以显示设备与对等体之间往来的 Hello 包及其他信息,如认证、Hello 和保持定时器等。

例 6-38 给出了抓取 Hello 包的 Ethanalyzer 和 HSRP 调试示例。请注意,HSRP 版本 2 分配了一个 6 字节 ID 来标识 HSRP Hello 包的发送端（通常是接口 MAC 地址）。

例 6-38　Ethanalyzer 和 HSRP Hello 调试

```
NX-2
NX-2# ethanalyzer local interface inband display-filter hsrp limit-captured-frames 0
Capturing on inband
1 2017-10-21 07:45:18.646334    10.12.1.2 -> 224.0.0.102   HSRPv2 94 Hello (state
  Active)
2 2017-10-21 07:45:18.915261    10.12.1.3 -> 224.0.0.102   HSRPv2 94 Hello (state
  Standby)
2 2017-10-21 07:45:21.503535    10.12.1.2 -> 224.0.0.102   HSRPv2 94 Hello (state
  Active)
4 2017-10-21 07:45:21.602261    10.12.1.3 -> 224.0.0.102   HSRPv2 94 Hello (state
  Standby)
NX-1
NX-1# debug logfile hsrp
NX-1# debug hsrp engine packet hello interface vlan 10 group 10
NX-1#
NX-1# show debug logfile hsrp
! Below hello packet is received by remote peer
2017 Oct 20 19:49:45.351470 hsrp: Vlan10[10/V4]: Hello in from 10.12.1.3 Peer/My
 State Standby/Active pri 100 ip 10.12.1.1
2017 Oct 20 19:49:45.351516 hsrp: Vlan10[10/V4]: hel 3000 hol 10000 auth cisco
2017 Oct 20 19:49:46.739041 hsrp: Vlan10[10/V4]: Hello out Active pri 110 ip
  10.12.1.1

! Below packet is the hello packet generated locally by the switch

2017 Oct 20 19:49:46.739063 hsrp: Vlan10[10/V4]: hel 3000 hol 10000 id
  5087.8940.2042

2017 Oct 20 19:49:48.039809 hsrp: Vlan10[10/V4]: Hello in from 10.12.1.3 Peer/My
 State Standby/Active pri 100 ip 10.12.1.1
2017 Oct 20 19:49:48.039829 hsrp: Vlan10[10/V4]: hel 3000 hol 10000 auth cisco
2017 Oct 20 19:49:49.595505 hsrp: Vlan10[10/V4]: Hello out Active pri 110 ip
  10.12.1.1
2017 Oct 20 19:49:49.595526 hsrp: Vlan10[10/V4]: hel 3000 hol 10000 id
  5087.8940.2042
```

常见的 HSRP 故障就是 HSRP 组始终处于关闭状态，出现这种情况的主要原因有以下几种。
- 未配置虚拟 IP。
- 接口处于关闭状态。
- 未配置接口 IP。

因此，在排查 HSRP 组处于关闭状态的故障时，应着重检查上述信息。

HSRPv6

HSRPv6（HSRP for IPv6，用于 IPv6 的 HSRP）可以为 IPv6 主机提供与 IPv4 HSRP 相同的功能。HSRP IPv6 组拥有一个从 HSRP 组号派生出来的虚拟 MAC 地址，同时还有一个虚拟 IPv6 链路本地地址（该地址默认由 HSRP 虚拟 MAC 地址派生得到）。HSRPv6 组处于活动状态时，会定期向 HSRP 虚拟 IPv6 链路本地地址发送 RA 消息；如果 HSRPv6 组变为非活动状态（迁移到待机状态），那么在发送完最后一条 RA 消息之后将停止发送 RA。

HSRPv6 使用与 IPv4 HSRP 不同的 MAC 地址范围和 UDP 端口，主要参数如下。
- HSRP 版本 2。

- UDP 端口：2029。
- MAC 地址范围：0005.73A0.0000 ～ 0005.73A0.0FFF。
- Hello 多播地址：FF02::66（链路本地范围多播地址）。
- 跳数限制：255。

启用 HSRPv6 不需要单独的功能特性，可以通过命令 **feature hsrp** 为 IPv4 和 IPv6 地址簇启用 HSRP。例 6-39 在 VLAN 10 的 NX-1 和 NX-2 之间配置了 HSRPv6，本例将 NX-2 的优先级设置为 110，意味着 NX-2 将充当主用交换机，NX-1 将充当备用交换机。请注意，本例使用命令 **ip** *ipv6-address* 定义了一个虚拟 IPv6 地址，但是该虚拟 IPv6 地址是备用虚拟 IP 地址，主用虚拟 IPv6 地址由系统自动为该 HSRPv6 组分配。

例 6-39　配置 HSRPv6

```
NX-1
interface Vlan10
  no shutdown
  no ipv6 redirects
  ipv6 address 2001:db8::2/48
  hsrp version 2
  hsrp 20 ipv6
    ip 2001:db8::1
NX-2
interface Vlan10
  no shutdown
  no ipv6 redirects
  ipv6 address 2001:db8::3/48
  hsrp version 2
  hsrp 20 ipv6
    preempt
    priority 110
    ip 2001:db8::1
```

与 IPv4 相似，可以使用命令 **show hsrp [group** *group-number*] [**detail**]查看 HSRPv6 组信息，该命令可以显示设备的当前状态、优先级、主用和备用虚拟 IPv6 地址、虚拟 MAC 地址以及组的状态历史等信息。例 6-40 列出了在 VLAN 10 上配置的 HSRP 组 20 的详细输出结果，请注意，虚拟 IPv6 地址是基于为该组分配的虚拟 MAC 地址计算得到的，已配置的虚拟 IPv6 地址位于备用 VIP 列表下。

例 6-40　HSPRv6 组详细信息

```
NX-2
NX-2# show hsrp group 20 detail
Vlan10 - Group 20 (HSRP-V2) (IPv6)
  Local state is Active, priority 110 (Cfged 110), may preempt
    Forwarding threshold(for vPC), lower: 1 upper: 110
  Hellotime 3 sec, holdtime 10 sec
  Next hello sent in 1.621000 sec(s)
  Virtual IP address is fe80::5:73ff:fea0:14 (Implicit)
  Active router is local
  Standby router is fe80::5287:89ff:fe40:2042 , priority 100 expires in 9.060000
sec(s)
  Authentication text "cisco"
  Virtual mac address is 0005.73a0.0014 (Default MAC)
  2 state changes, last state change 00:02:40
  IP redundancy name is hsrp-Vlan10-20-V6 (default)
```

```
    Secondary VIP(s):
                    2001:db8::1

    ----- Detailed information -----
    State History
    ----------------------------------------------------------------
     Time                Prev State        State         Event
    ----------------------------------------------------------------
     (21)-20:22:39 Standby           Active        Act Timer Expired.

     (21)-20:22:28 Speak             Standby       Sby Timer Expired.

     (21)-20:22:28 Listen            Speak         Act Timer Expired.

     (21)-20:22:18 Initial           Listen        If Enabled-VIP.

     (21)-20:22:18 No Trans          Initial       N/A.
```

在接口上配置并分配虚拟 IPv6 地址，并不会启动 HSRPv6，可以使用命令 **show ipv6 interface** *interface-id* 加以验证。还必须将虚拟 IPv6 地址和虚拟 MAC 地址添加到 ICMPv6 中，可以通过命令 **show ipv6 icmp vaddr [link-local | global]** 查看该信息，其中，关键字 **link-local** 的作用是显示主用虚拟 IPv6 地址，该地址是通过虚拟 MAC 自动计算得到的；关键字 **global** 的作用是显示手动配置的虚拟 IPv6 地址。例 6-41 给出了这两条命令的输出结果示例。

例 6-41　验证 HSRPv6 虚拟地址

```
NX-2
NX-2# show ipv6 interface vlan 10
IPv6 Interface Status for VRF "default"(1)
Vlan10, Interface status: protocol-up/link-up/admin-up, iod: 121
  IPv6 address:
    2001:db8::3/48 [VALID]
  IPv6 subnet:  2001:db8::/48
  IPv6 link-local address: fe80::e6c7:22ff:fe1e:9642 (default) [VALID]
  IPv6 virtual addresses configured:
        fe80::5:73ff:fea0:14   2001:db8::1
  IPv6 multicast routing: disabled
! Output omitted for brevity
NX-2# show ipv6 icmp vaddr link-local
  Virtual IPv6 addresses exists:
  Interface: Vlan10, context_name: default (1)
    Group id: 20, Protocol: HSRP, Client UUID: 0x196, Active: Yes (1) client_state:1
      Virtual IPv6 address: fe80::5:73ff:fea0:14
      Virtual MAC: 0005.73a0.0014, context_name: default (1)

NX-2# show ipv6 icmp vaddr global
    Group id: 20, Protocol: HSRP, Client UUID: 0x196, Active: Yes
      Interface: Vlan10, Virtual IPv6 address: 2001:db8::1
      Virtual MAC: 0005.73a0.0014, context_name: default (1) flags:3
```

如果 HSRPv6 邻居处于震荡状态，那么就可以使用与 IPv4 相同的 Ethanalyzer 进行分析。例 6-42 显示了 HSRPv6 控制包的 Ethanalyzer 分析结果，包含了 HSRP 主用和备用交换机的数据包信息。

例 6-42　用于 HSRPv6 的 Ethanalyzer

```
NX-2
NX-2# ethanalyzer local interface inband display-filter hsrp limit-captured-frames 0
```

```
Capturing on inband
20:32:29.596977  fe80::5287:89ff:fe40:2042 -> ff02::66     HSRPv2 114 Hello (state
  Standby)
20:32:29.673860  fe80::e6c7:22ff:fe1e:9642 -> ff02::66     HSRPv2 114 Hello (state
  Active)
20:32:32.307507  fe80::5287:89ff:fe40:2042 -> ff02::66     HSRPv2 114 Hello (state
  Standby)
20:32:32.333125  fe80::e6c7:22ff:fe1e:9642 -> ff02::66     HSRPv2 114 Hello (state
  Active)
```

注：如果出现了与 HSRP 或 HSRPv6 相关的故障问题，那么就可以在故障状态下收集 **show tech hsrp** 命令的输出结果。

6.5.2 VRRP

VRRP（Virtual Router Redundancy Protocol，虚拟路由器冗余协议）最初定义在 RFC 2338 中，定义的是版本 1。RFC 3768 和 RFC 5798 分别定义了 VRRP 版本 2 和版本 3。NX-OS 目前仅支持 VRRP 版本 2 和版本 3。VRRP 的工作原理与 HSRP 类似。VRRP 在多台设备中选举一个成员作为 VRRP 主设备，承担默认网关角色，从而提供设备级冗余，消除了单点故障问题。其他非 VRRP 主设备的成员组成一个 VRRP 组，担任备用角色。如果 VRRP 主用设备出现故障，那么就由 VRRP 备用设备承担 VRRP 主用设备角色，并充当默认网关。

可以通过命令 **feature vrrp** 启用 VRRP 功能特性。VRRP 的配置与 HSRP 类似，可以使用命令 **vrrp** *group-number* 配置 VRRP。在接口 VRRP 配置模式下，网络运营商可以定义虚拟 IP、优先级、认证机制等。在 VRRP 配置模式下，需要配置 **no shutdown** 命令以启用 vrrp 组。例 6-43 给出了 NX-1 与 NX-2 之间的 VRRP 配置示例。

例 6-43 配置 VRRP

```
NX-1
interface Vlan10
  no shutdown
  no ip redirects
  ip address 10.12.1.2/24
  vrrp 10
    priority 110
    authentication text cisco
    address 10.12.1.1
    no shutdown
NX-2
interface Vlan10
  no shutdown
  no ip redirects
  ip address 10.12.1.3/24
  vrrp 10
    authentication text cisco
    address 10.12.1.1
    no shutdown
```

如果要验证 VRRP 的状态，那么就可以使用命令 **show vrrp [master | backup]**，选项 **master** 和 **backup** 可以分别显示相应设备节点的信息。此外，还可以通过命令 **show vrrp [detail]** 收集与 VRRP 相关的更多详细信息。例 6-44 给出了详细的 VRRP 输出结果及 VRRP 状态信息。请注意，命令 **show vrrp detail** 的输出结果显示了虚拟 IP 地址和虚拟 MAC 地址，VRRP 虚拟 MAC 地址

的格式为0000.5e00.01*xy*，其中，*xy*是VRRP组号的十六进制表示。

例6-44　VRRP状态及详细信息

```
NX-1
NX-1# show vrrp master
     Interface  VR IpVersion Pri   Time Pre State    VR IP addr
     ----------------------------------------------------------
        Vlan10  10    IPV4    110   1 s  Y  Master    10.12.1.1

NX-1# show vrrp detail

Vlan10 - Group 10 (IPV4)
    State is Master
    Virtual IP address is 10.12.1.1
    Priority 110, Configured 110
    Forwarding threshold(for VPC), lower: 1 upper: 110
    Advertisement interval 1
    Preemption enabled
    Authentication text "cisco"
    Virtual MAC address is 0000.5e00.010a
    Master router is Local
NX-2
NX-2# show vrrp backup
     Interface  VR IpVersion Pri   Time Pre State    VR IP addr
     ----------------------------------------------------------
        Vlan10  10    IPV4    100   1 s  Y  Backup    10.12.1.1
```

如果出现了VRRP震荡问题，那么就可以通过命令**show vrrp statistics**确定震荡原因是出现了某种错误还是错误地接收了某个数据包，该命令可以显示设备成为主用设备的次数以及其他错误统计信息，如TTL差错、无效数据包长度和地址列表不匹配等。例6-45给出了**show vrrp statistics**命令的输出结果示例，可以看出，NX-1收到了组10的5次认证失败统计信息。

例6-45　VRRP统计信息

```
NX-1
NX-1# show vrrp statistics

Vlan10 - Group 10 (IPV4) statistics

Number of times we have become Master : 1
Number of advertisement packets received : 0
Number of advertisement interval mismatch : 0
Authentication failure cases : 5
TTL Errors : 0
Zero priority advertisements received : 0
Zero priotiy advertisements sent : 0
Invalid type field received : 0
Mismatch in address list between ours & received packets : 0
Invalid packet length : 0
```

VRRP版本2仅支持IPv4地址簇，而VRRP版本3（VRRPv3）则同时支持IPv4和IPv6地址簇。对于NX-OS来说，不能在同一设备上同时启用VRRP和VRRPv3。如果已在Nexus交换机上启用了VRRP功能特性，那么再启用VRRPv3的时候就会显示一条错误消息，声称当前已启用VRRPv2。因此，有必要将VRRP迁移到VRRPv3，迁移过程对服务的影响很小。从VRRP版本2迁移到版本3的过程如下。

- **第 1 步**：使用命令 **no feature vrrp** 禁用 VRRP 功能。
- **第 2 步**：使用命令 **feature vrrpv3** 启用 VRRPv3 功能。
- **第 3 步**：在接口配置模式下使用命令 **vrrpv3** *group-number* **address-family [ipv4 | ipv6]** 配置 VRRPv3 组。
- **第 4 步**：使用地址配置命令定义 VRRPv3 主用和备用虚拟 IP 地址。
- **第 5 步**：使用命令 **vrrpv2** 启用与 VRRP 版本 2 的后向兼容性。这样做有助于与其他启用 VRRP 版本 2 的设备交换状态信息。
- **第 6 步**：在 VRRPv3 组上运行 **no shutdown** 命令。

例 6-46 给出了 NX-1 交换机从 VRRPv2 到 VRRPv3 的迁移配置示例。

例 6-46　VRRPv3 迁移配置

```
NX-1
NX-1(config)# feature vrrpv3
Cannot enable VRRPv3: VRRPv2 is already enabled

NX-1(config)# no feature vrrp
NX-1(config)# feature vrrpv3
NX-1(config)# interface vlan 10
NX-1(config-if)# vrrpv3 10 address-family ipv4
NX-1(config-if-vrrpv3-group)# address 10.12.1.1 primary
NX-1(config-if-vrrpv3-group)# address 10.12.1.5 secondary
NX-1(config-if-vrrpv3-group)# vrrpv2
NX-1(config-if-vrrpv3-group)# preempt
NX-1(config-if-vrrpv3-group)# no shutdown
NX-1(config-if-vrrpv3-group)# end
```

可以通过命令 **show vrrpv3 [brief | detail]** 验证 VRRPv3 组的相关信息，其中，**show vrrpv3 brief** 命令可以显示与组相关的摘要信息，如组号、地址簇、优先级、抢占、状态、主用地址和组地址（是虚拟组 IP）；**show vrrpv3 detail** 命令可以显示其他详细信息，如为 VRRPv2 和 VRRPv3 收发的宣告消息、虚拟 MAC 地址以及与错误和状态迁移有关的其他统计信息。例 6-47 显示了 **show vrrpv3** 命令的摘要和详细输出结果。

例 6-47　show vrrpv3 命令输出结果

```
NX-1
NX-1# show vrrpv3 brief

 Interface         Grp  A-F  Pri  Time  Own  Pre  State   Master addr/Group addr
 Vlan10            10   IPv4 100  0     N    Y    MASTER  10.12.1.2(local)  10.12.1.1

NX-1# show vrrpv3 detail

Vlan10 - Group 10 - Address-Family IPv4
  State is MASTER
  State duration 1 mins 3.400 secs
  Virtual IP address is 10.12.1.1
  Virtual secondary IP addresses:
    10.12.1.5
  Virtual MAC address is 0000.5e00.010a
  Advertisement interval is 1000 msec
  Preemption enabled
  Priority is 100
  Master Router is 10.12.1.2 (local), priority is 100
```

```
    Master Advertisement interval is 1000 msec (expires in 594 msec)
    Master Down interval is unknown
    VRRPv3 Advertisements: sent 72 (errors 0) - rcvd 0
    VRRPv2 Advertisements: sent 32 (errors 0) - rcvd 15
    Group Discarded Packets: 0
      VRRPv2 incompatibility: 0
      IP Address Owner conflicts: 0
      Invalid address count: 0
      IP address configuration mismatch : 0
      Invalid Advert Interval: 0
      Adverts received in Init state: 0
      Invalid group other reason: 0
    Group State transition:
      Init to master: 0
      Init to backup: 1 (Last change Sat Oct 21 16:16:39.737 UTC)
      Backup to master: 1 (Last change Sat Oct 21 16:16:43.347 UTC)
      Master to backup: 0
      Master to init: 0
      Backup to init: 0
```

此外，也可以通过 **show vrrpv3 statistics** 命令查看差错统计信息。该命令可以显示已丢弃的数据包计数器以及各种丢包原因（如 TTL 无效、校验和无效或消息类型无效），输出结果的后半部分与 **show vrrpv3 detail** 命令相似。例 6-48 显示了命令 **show vrrpv3 statistics** 的输出结果示例。

例 6-48　show vrrpv3 statistics 命令输出结果

```
NX-1
NX-1# show vrrpv3 statistics

VRRP Global Statistics:
  Dropped Packets : 0
VRRP Statistics for Vlan10
  Header Discarded Packets: 0
    Invalid TTL/Hop Limit: 0
    Invalid Checksum: 0
    Invalid Version: 0
    Invalid Msg Type: 0
    Invalid length/Incomplete packet: 0
    Invalid group no: 0
    Invalid packet other reason: 0

VRRP Statistics for Vlan10 - Group 10 - Address-Family IPv4
  State is MASTER
  State duration 40.332 secs
  VRRPv3 Advertisements: sent 560 (errors 0) - rcvd 0
  VRRPv2 Advertisements: sent 520 (errors 0) - rcvd 89
  Group Discarded Packets: 0
    VRRPv2 incompatibility: 0
    IP Address Owner conflicts: 0
    Invalid address count: 0
    IP address configuration mismatch : 89
    Invalid Advert Interval: 0
    Adverts received in Init state: 0
    Invalid group other reason: 0
  Group State transition:
    Init to master: 0
    Init to backup: 1 (Last change Sat Oct 21 16:16:39.737 UTC)
```

```
Backup to master: 2 (Last change Sat Oct 21 16:25:56.905 UTC)
Master to backup: 1 (Last change Sat Oct 21 16:24:28.198 UTC)
Master to init: 0
Backup to init: 0
```

注：如果出现了与 VRRP 相关的故障问题，那么就可以在故障状态下收集命令 **show tech vrrp [brief]** 或 **show tech vrrpv3 [brief]** 的输出结果，并联系思科 TAC（Technical Assistance Center，技术支持中心）。

6.5.3 GLBP

顾名思义，GLBP（Gateway Load-Balancing Protocol，网关负载均衡协议）可以为网段提供网关冗余和负载均衡能力。GLBP 通过确保 GLBP 组的每个成员都能将流量转发给适当的网关，来确保主用/备用网关的冗余性以及负载平衡能力。可以通过命令 **feature glbp** 在 NX-OS 上启用 GLBP。定义 GLBP 组时，可以配置以下参数。

- 组号、主用和备用 IP 地址。
- 选择 AVG（Active Virtual Gateway，活动虚拟网关）所用的优先级值。
- 抢占时间和抢占延迟时间。
- 虚拟转发器的优先级和抢占延迟时间。
- 初始权重以及备用网关成为 AVG 的上下阈值。
- 网关负载均衡方法。
- MD5 和明文认证属性。
- GLBP 定时器值。
- 接口跟踪。

GLBP 提供了 3 种负载均衡机制。

- **无**：此时的功能类似于 HSRP。
- **主机相关机制**：利用主机 MAC 地址来确定将数据包重定向到哪个虚拟转发器 MAC。该方法可以确保主机使用相同的虚拟 MAC 地址（只要组中的虚拟转发器数量保持不变）。
- **循环机制**：每个虚拟转发器的 MAC 地址依次轮流应答虚拟 IP 地址。
- **加权机制**：为 GLBP 组中的每台设备配置不同的权重，从而定义设备之间的负载均衡比率。

例 6-49 显示了 NX-1 与 NX-2 之间的 GLBP 配置情况。

例 6-49 配置 GLBP

```
NX-1
NX-1(config)# interface vlan 10
NX-1(config-if)# glbp 10
NX-1(config-if-glbp)# timers 1 4
NX-1(config-if-glbp)# priority 110
NX-1(config-if-glbp)# preempt
NX-1(config-if-glbp)# load-balancing ?
  host-dependent  Load balance equally, source MAC determines forwarder choice
  round-robin     Load balance equally using each forwarder in turn
  weighted        Load balance in proportion to forwarder weighting

NX-1(config-if-glbp)# load-balancing host-dependent
NX-1(config-if-glbp)# forwarder preempt ?
  <CR>
  delay  Wait before preempting

NX-1(config-if-glbp)# forwarder preempt
```

```
NX-1(config-if-glbp)# ip 10.12.1.1
NX-1(config-if-glbp)# end
NX-2
NX-2(config-if-glbp)# interface vlan 10
NX-2(config-if)# glbp 10
NX-2(config-if-glbp)# ip 10.12.1.1
NX-2(config-if-glbp)# timers 1 4
NX-2(config-if-glbp)# load-balancing host-dependent
NX-2(config-if-glbp)# forwarder preempt
NX-2(config-if-glbp)# end
```

与 HSRP 版本 2 相似，GLBP 通过多播地址 224.0.0.102 传送 Hello 包，但使用的 UDP 源和目的端口号是 3222。

可以通过命令 **show glbp [brief]** 查看 GLBP 组的详细信息，该命令可以显示已配置的虚拟 IP、组状态以及与该组相关的其他信息。此外，该命令的输出结果还会显示与转发器相关的信息，包括转发器的 MAC 地址和 IP 地址等信息。例 6-50 给出了命令 **show glbp** 和 **show glbp brief** 的输出结果示例，显示了与 GLBP 组 10 相关的信息以及转发器及其状态信息。

例 6-50　show glbp 和 show glbp brief 命令输出结果

```
NX-1
NX-1# show glbp

Extended-hold (NSF) is Disabled

Vlan10 - Group 10
  State is Active
    4 state change(s), last state change(s) 00:01:54
  Virtual IP address is 10.12.1.1
  Hello time 1 sec, hold time 4 sec
    Next hello sent in 990 msec
  Redirect time 600 sec, forwarder time-out 14400 sec
  Preemption enabled, min delay 0 sec
  Active is local
  Standby is 10.12.1.3, priority 100 (expires in 3.905 sec)
  Priority 110 (configured)
  Weighting 100 (default 100), thresholds: lower 1, upper 100
  Load balancing: host-dependent
  Group members:
    5087.8940.2042 (10.12.1.2) local
    E4C7.221E.9642 (10.12.1.3)
  There are 2 forwarders (1 active)
  Forwarder 1
   State is Active
     2 state change(s), last state change 00:01:50
    MAC address is 0007.B400.0A01 (default)
    Owner ID is 5087.8940.2042
    Preemption enabled, min delay 30 sec
    Active is local, weighting 100

  Forwarder 2
   State is Listen
     1 state change(s), last state change 00:00:40
    MAC address is 0007.B400.0A02 (learnt)
    Owner ID is E4C7.221E.9642
    Redirection enabled, 599.905 sec remaining (maximum 600 sec)
```

```
        Time to live: 14399.905 sec (maximum 14400 sec)
        Preemption enabled, min delay 30 sec
        Active is 10.12.1.3 (primary), weighting 100 (expires in 3.905 sec)
NX-1# show glbp brief
Interface        Grp Fwd Pri State      Address          Active rtr       Standby rtr
Vlan10           10  -   110 Active     10.12.1.1        local            10.12.1.3

! Below is the list of forwarders
Vlan10           10  1   7   Active     0007.B400.0A01   local            -
Vlan10           10  2   7   Listen     0007.B400.0A02   10.12.1.3        -
```

如果要排查 GLBP 故障问题，那么就可以使用 Ethanalyzer 等工具来抓取 GLBP 控制包。Ethanalyzer 可以提供正在收发的作为 GLBP 控制包的一部分的详细信息。例 6-51 给出了 GLBP 数据包的 Ethanalyzer 输出结果示例。

例 6-51 GLBP 数据包的 Ethanalyzer 输出结果

```
NX-2
NX-2# ethanalyzer local interface inband display-filter glbp limit-captured-frames 0
Capturing on inband
2017-10-22 20:33:43.857524    10.12.1.2 -> 224.0.0.102  GLBP 102 G: 10, Hello, I
Pv4, Request/Response?
2017-10-22 20:33:43.857934    10.12.1.3 -> 224.0.0.102  GLBP 102 G: 10, Hello, I
Pv4, Request/Response?
2 2017-10-22 20:33:44.858861   10.12.1.2 -> 224.0.0.102  GLBP 102 G: 10, Hello, I
Pv4, Request/Response?
4 2017-10-22 20:33:44.859474   10.12.1.3 -> 224.0.0.102  GLBP 102 G: 10, Hello, I
Pv4, Request/Response?
NX-2# ethanalyzer local interface inband display-filter glbp limit-captured-frames 1
  detail
Capturing on inband
1
Frame 1: 102 bytes on wire (816 bits), 102 bytes captured (816 bits) on interface
0
    Interface id: 0
    Encapsulation type: Ethernet (1)
    Arrival Time: Oct 22, 2017 20:33:54.873326000 UTC
    [Time shift for this packet: 0.000000000 seconds]
    Epoch Time: 1508704434.873326000 seconds
    [Time delta from previous captured frame: 0.000000000 seconds]
    [Time delta from previous displayed frame: 0.000000000 seconds]
    [Time since reference or first frame: 0.000000000 seconds]
    Frame Number: 1
    Frame Length: 102 bytes (816 bits)
    Capture Length: 102 bytes (816 bits)
    [Frame is marked: False]
    [Frame is ignored: False]
    [Protocols in frame: eth:ip:udp:glbp]
Ethernet II, Src: Cisco_00:0a:02 (00:07:b4:00:0a:02), Dst: IPv4mcast_00:00:66
  (01:00:5e:00:00:66)
    Destination: IPv4mcast_00:00:66 (01:00:5e:00:00:66)
        Address: IPv4mcast_00:00:66 (01:00:5e:00:00:66)
        .... ..0. .... .... .... .... = LG bit: Globally unique address (factory
 default)
        .... ...1 .... .... .... .... = IG bit: Group address (multicast/broadcast)
    Source: Cisco_00:0a:02 (00:07:b4:00:0a:02)
```

```
              Address: Cisco_00:0a:02 (00:07:b4:00:0a:02)
              .... ..0. .... .... .... .... = LG bit: Globally unique address (factory
default)
              .... ...0 .... .... .... .... = IG bit: Individual address (unicast)
          Type: IP (0x0800)
   Internet Protocol Version 4, Src: 10.12.1.3 (10.12.1.3), Dst: 224.0.0.102
   (224.0.0.102)
          Version: 4
          Header length: 20 bytes
          Differentiated Services Field: 0xc0 (DSCP 0x30: Class Selector 6; ECN: 0x00:
   Not-ECT (Not ECN-Capable Transport))
              1100 00.. = Differentiated Services Codepoint: Class Selector 6 (0x30)
              .... ..00 = Explicit Congestion Notification: Not-ECT (Not ECN-Capable
   Transport) (0x00)
          Total Length: 88
          Identification: 0xbd59 (48473)
          Flags: 0x00
              0... .... = Reserved bit: Not set
              .0.. .... = Don't fragment: Not set
              ..0. .... = More fragments: Not set
          Fragment offset: 0
          Time to live: 255
          Protocol: UDP (17)
          Header checksum: 0x1206 [correct]
              [Good: True]
              [Bad: False]
          Source: 10.12.1.3 (10.12.1.3)
          Destination: 224.0.0.102 (224.0.0.102)
   User Datagram Protocol, Src Port: glbp (3222), Dst Port: glbp (3222)
          Source port: glbp (3222)
          Destination port: glbp (3222)
          Length: 68
         Checksum: 0x9f53 [validation disabled]
              [Good Checksum: False]
              [Bad Checksum: False]
   Gateway Load Balancing Protocol
          Version?: 1
          Unknown1: 0
          Group: 10
          Unknown2: 0000
          Owner ID: e4:c7:22:1e:96:42 (e4:c7:22:1e:96:42)
          TLV l=28, t=Hello
              Type: Hello (1)
              Length: 28
              Unknown1-0: 00
              VG state?: Standby (16)
              Unknown1-1: 00
              Priority: 100
              Unknown1-2: 0000
              Helloint: 1000
              Holdint: 4000
              Redirect: 600
              Timeout: 14400
              Unknown1-3: 0000
              Address type: IPv4 (1)
              Address length: 4
              Virtual IPv4: 10.12.1.1 (10.12.1.1)
```

```
    TLV l=20, t=Request/Response?
    Type: Request/Response? (2)
    Length: 20
    Forwarder?: 2
    VF state?: Active (32)
    Unknown2-1: 00
    Priority: 167
    Weight: 100
    Unknown2-2: 00384002580000
    Virtualmac: Cisco_00:0a:02 (00:07:b4:00:0a:02)
```

注：如果出现了与 GLBP 相关的故障问题，那么就可以收集 **show tech glbp** 命令的输出结果，并联系思科 TAC（Technical Assistance Center，技术支持中心）。

6.6 本章小结

  NX-OS 为 Nexus 平台提供了大量有用的 IP 和 IPv6 服务以及面向数据中心的路由和交换功能，使 Nexus 交换机能够满足不同网络层次的功能需求。本章首先详细介绍了 IP SLA 功能，不但能够跟踪可达性、限定指定源端与目的端之间的抖动值，而且支持基于 UDP 和 TCP 的测量探针。除 IP SLA 之外，还可以利用对象跟踪功能在系统中执行条件式跟踪操作，对象跟踪功能不但能够跟踪接口、IP 或 IPv6 路由以及跟踪列表，而且能与静态路由一起集成使用。

  作为 IPv4 服务的一部分，NX-OS 支持 DHCP 中继、监听以及其他与 IPv4 安全相关的功能特性。本章详细介绍了如何在数据中心环境中使用 DHCP 中继和 DHCP 监听功能来扩展 DHCP 服务器的能力，同时保护网络免受攻击。如果 DHCP 服务器与主机位于不同的 VLAN 或子网中，那么就可以使用 DHCP 中继功能。此外，本章还介绍了如何使用 DAI、IP 源保护和 URPF 等安全功能特性，启用了这些服务之后，NX-OS 会在硬件中配置 ACL 以允许指定流量。

  对于 IPv6 服务来说，本章详细讨论了 IPv6 邻居发现进程和 IPv6 第一跳安全功能，如 RA 保护、IPv6 监听和 DHCPv6 保护等。除此以外，本章还讨论了多种 FHRP 协议，如用于 IPv4 和 IPv6 的 HSRP、VRRP 以及 GLBP。FHRP 协议可以为主机提供网关冗余机制。最后，本章还讨论了这些 FHRP 协议的工作方式以及配置和故障排查方式。

第 7 章

EIGRP 故障排查

本章主要讨论如下主题。
- EIGRP 基础知识。
- EIGRP 邻居邻接关系故障排查。
- EIGRP 路径选择和路由丢失故障排查。
- 收敛问题。

EIGRP（Enhanced Interior Gateway Routing Protocol，增强型内部网关路由协议）是企业网常见的增强型距离矢量路由协议。最初的 EIGRP 是思科专有协议，后来思科在 2013 年通过信息性的 RFC（Request for Comment，征求意见书）向 IETF（Internet Engineering Task Force，互联网工程任务组）通告了 EIGRP，目前已成为 RFC 7868 标准。

本章主要讨论与 EIGRP 邻居邻接关系建立、路径选择、路由丢失以及收敛问题等有关的故障识别与排查技术。

7.1 EIGRP 基础知识

Nexus 交换机可以运行多个 EIGRP 进程，每个进程都与同一公共路由域（也称为 AS[Autonomous System，自治系统]）中的其他路由器或 NX-OS 交换机建立邻接关系。同一 AS 中的 EIGRP 设备仅与同一 AS 的成员交换路由，并使用相同的度量计算公式。

在路由选择方面，EIGRP 采用了跳数之外的计算因子，而且在路由选择算法中增加了新的计算逻辑。EIGRP 采用 DUAL（Diffusing Update Algorithm，扩散更新算法）来识别网络路径，并使用预先计算的无环备份路径提供快速收敛能力。

接下来将以图 7-1 所示拓扑结构为例，说明 NX-1 计算去往网络 10.4.4.0/24 的最佳路径的过程。

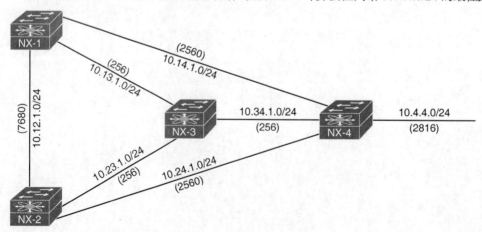

图 7-1　EIGRP 参考拓扑

表 7-1 列出了 EIGRP 的关键术语、定义以及与图 7-1 所示拓扑结构之间的关系。

表 7-1　　　　　　　　　　　　　　EIGRP 术语

术语	定义
后继路由	去往目的端的拥有最小路径度量的路由 NX-1 去往 NX-4 上的 10.4.4.0/24 的后继路由是 NX-1→NX-3→NX-4
后继路由器	后继路由的第一跳路由器 10.4.4.0/24 的后继路由器是 NX-3
FD（Feasible Distance，可行距离）	到达目的端的度量值最小的路径的度量值。可行距离是在本地进行计算的，计算公式将在本章后面的 7.1.2 节进行描述 NX-1 去往 10.4.4.0/24 网络的 FD 为 3328（256 + 256 + 2816）
RD（Reported Distance，报告距离）	由路由器报告的去往指定前缀的距离，报告距离是宣告路由器的可行距离 NX-3 以 RD 值 3072（256+2816）宣告前缀 10.4.4.0/24 NX-4 以 RD 值 2816 宣告去往 NX-1 和 NX-2 的前缀 10.4.4.0/24
可行性条件	如果要将某条路由视为备用路由，那么接收到该路由的报告距离必须小于本地计算的可行距离，这种逻辑可以确保无环路径
可行后继路由	该路由满足可行性条件，被维护为备用路由（可行后继路由），可行性条件可以确保备用路由是无环路由 NX-1→NX-4 是可行后继路由，因为 RD 值 2816 小于路径 NX-1→NX-3→NX-4 的 FD 值 3328

7.1.1　拓扑表

EIGRP 维护了一个拓扑表，该表是 DUAL 的一个重要组成部分，包含了标识无环备用路由的相关信息。拓扑表包含了 EIGRP AS 中宣告的所有网络前缀，表中的每条表项都包含以下信息。

- 网络前缀。
- 宣告该前缀的临近 EIGRP 邻居。
- 距离每个邻居的度量（报告距离、跳数）。
- 用于度量计算的数值（负荷、可靠性、总时延、最小带宽）。

命令 **show ip eigrp topology** [*network-prefix/prefix-length*] [**active** | **all-links**]可以显示 EIGRP 拓扑表，可选关键字 **active** 可以显示处于主动状态的前缀，关键字 **all-links** 可以显示所有路径（包括非后继路由和可行后继路由）。图 7-2 显示了图 7-1 中的 NX-1 的拓扑表信息。

```
NX01# show ip eigrp topology
! Output omitted for brevity
IPv4-EIGRP Topology Table for AS(100) /
ID(192.168.1.1)

Codes: P - Passive, A - Active, U - Update, Q -
Query,  R - Reply,
        r - reply Status, s - sia Status

P 10.12.1.0/24, 1 successors, FD is 2816
        via Connected, Ethernet1/1
P 10.13.1.0/24, 1 successors, FD is 2816
        via Connected, Ethernet2/2
P 10.14.1.0/24, 1 successors, FD is 5120
        via Connected, Ethernet1/3
```

图 7-2　EIGRP 拓扑表

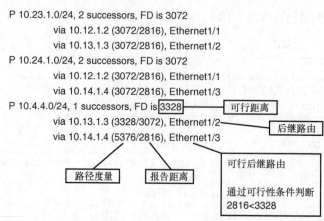

图 7-2　EIGRP 拓扑表（续）

分析网络 10.4.4.0/24 的时候可以看出，NX-1 计算出的后继路由的 FD 为 3328，后继路由器（上游路由器）宣告的后继路由的报告距离（RD）为 3072。第二条路径的度量值为 5376、RD 值为 2816。由于 2816 小于 3072，因而第二条路径表项满足可行性条件，从而将该表项标识为该前缀的可行后继路由。

路由 10.4.4.0/24 显示为 P（Passive，被动），表示该拓扑结构稳定。如果拓扑结构处于变化状态，那么在计算新路径时，路由将处于 A（Active，主动）状态。

7.1.2　路径度量计算

对于所有路由协议来说，度量计算都是极为重要的组成部分。EIGRP 采用多种因子来计算路径的度量值。度量计算默认使用带宽和时延参数，不过也可以包含接口负荷和可靠性等参数。图 7-3 给出的路径度量公式定义在 RFC 7868（该 RFC 解释了 EIGRP 协议）中。

$$度量 = \left[\left(K_1 \times 带宽 + \frac{K_2 \times 带宽}{256 - Load} + K_3 \times 时延\right) \times \frac{K_5}{K_4 + 可靠性}\right]$$

图 7-3　EIGRP 度量计算公式

EIGRP 使用 K 值来定义该计算公式使用的系数以及这些系数在计算度量值时的关联影响。常见的一种误解是将 K 值直接应用于带宽、负荷、时延或可靠性，这是不准确的。例如，K_1 和 K_2 引用的都是 BW（BandWidth，带宽）。

BW 表示路径中最慢的链路，以 10Gbit/s（10^7）为单位。链路速率是从接口上配置的接口带宽采集到的，时延是路径中测量的总时延，以十分之一微秒（μs）为单位。

EIGRP 的度量计算公式基于 IGRP 度量计算公式，但是将输出结果乘以 256，从而将度量值从 24 位调整为 32 位。考虑到以上因素，EIGRP 的度量计算公式如图 7-4 所示。

$$度量 = 256 \times \left[\left(K_1 \times \frac{10^7}{最小带宽} + \frac{K_2 \times 最小带宽}{256 - 负荷} + \frac{K_3 \times 总时延}{10}\right) \times \frac{K_5}{K_4 + 可靠性}\right]$$

图 7-4　RFC 定义的 EIGRP 度量计算公式

默认情况下，K_1 和 K_3 的值为 1，K_2、K_4 和 K_5 的值为 0。将这些默认 K 值代到公式中，即可得到简化后的计算公式，如图 7-5 所示。

$$度量 = 256 \times \left[\left(1 \times \frac{10^7}{最小带宽} + \frac{0 \times 最小带宽}{256 - 负荷} + \frac{1 \times 总时延}{10} \right) \times \frac{\frac{10^7}{0}}{0 + 可靠性} \right]$$

简化后

$$度量 = 256 \times \left(\frac{10^7}{最小带宽} + \frac{总时延}{10} \right)$$

图 7-5 使用默认 K 值之后的 EIGRP 度量计算公式

注： EIGRP 还提供了一个专门处理高速接口的度量计算公式（称为 EIGRP 宽度量），该公式增加了第六个 K 值。有关 EIGRP 宽度量的详细内容将在本章后面进行讨论。

EIGRP 的更新包中包含了与每个前缀相关联的路径属性，EIGRP 的路径属性可以包括跳数、累计时延、最小带宽链路速率以及报告距离。在更新包的传播过程中，EIGRP 会在每一跳都更新这些路径属性，从而允许每台路由器都能独立识别最短路径。

表 7-2 列出了常见的网络类型、链路速率、时延以及使用图 7-5 所示简化公式得到的 EIGRP 度量值。

表 7-2　　　　　　　　　　　　　EIGRP 接口默认度量

接口类型	链路速率（kbit/s）	时延	度量
串行接口	64	20000μs	40512000
T1	1544	20000μs	2170031
以太网	10000	1000μs	281600
快速以太网	100000	100μs	28160
千兆以太网	1000000	10μs	2816
万兆以太网	10000000	10μs	512

注： 串行接口与 T1 接口的时延完全相同，因而唯一的区别参数就是链路速率。此外，千兆以太网接口和万兆以太网接口的时延也完全相同。

以图 7-1 的拓扑结构为例，利用图 7-5 的公式来计算 NX-1 到网络 10.4.4.0/24 的度量。两台 Nexus 交换机之间的链路速率为 1Gbit/s，总时延为 30μs（10.4.4.0/24 链路时延为 10μs，10.34.1.0/24 链路时延为 10μs，10.13.1.0/24 时延为 10μs）。

可以使用命令 **show ip eigrp topology** *network/prefix-length* 直接从 EIGRP 拓扑表中查询指定前缀的 EIGRP 度量。例 7-1 显示了 NX-1 关于网络 10.4.4.0/24 的拓扑表输出结果。可以看出，输出结果包含了后继路由、可行后继路径以及该前缀的 EIGRP 状态。每条路径都包含了 EIGRP 属性的最小带宽、总时延、接口可靠性、负荷及跳数。

例 7-1 特定前缀的 EIGRP 拓扑表

```
NX-1# show ip eigrp topology 10.4.4.0/24
IP-EIGRP (AS 1): Topology entry for 10.4.4.0/24
  State is Passive, Query origin flag is 1, 1 Successor(s), FD is 3328
  Routing Descriptor Blocks:
  10.13.1.3 (Ethernet1/2), from 10.13.1.3, Send flag is 0x0
```

```
          Composite metric is (3328/3072), Route is Internal
          Vector metric:
          Minimum bandwidth is 1000000 Kbit
          Total delay is 30 microseconds
             Reliability is 255/255
             Load is 1/255
             Minimum MTU is 1500
             Hop count is 2
             Internal tag is 0
   10.14.1.4 (Ethernet1/3), from 10.14.1.4, Send flag is 0x0
          Composite metric is (5376/2816), Route is Internal
          Vector metric:
          Minimum bandwidth is 1000000 Kbit
          Total delay is 110 microseconds
             Reliability is 255/255
             Load is 1/255
             Minimum MTU is 1500
             Hop count is 1
             Internal tag is 0
```

注： 除后继路由和可行后继路由之外，EIGRP 拓扑表还维护了其他路径，可以通过命令 **show ip eigrp topology all-links** 显示其他路径信息。

7.1.3 EIGRP 通信

EIGRP 通过 5 种数据包与其他路由器进行通信（见表 7-3）。EIGRP 使用自己的 IP 协议号（88），并在可能的情况下使用多播数据包，也可以根据需要使用单播数据包。在可能的情况下，EIGRP 设备之间的通信基于多播组地址 224.0.0.10 或 MAC 地址 01:00:5e:00:00:0a。

表 7-3　　　　　　　　　　　　　　　EIGRP 数据包类型

类型	数据包名称	功能
1	Hello	该数据包负责发现 EIGRP 邻居，而且还可以在邻居不可用的时候检测邻居
2	确认（Acknowledgement，ACK）	该数据包发送给发端路由器，为其他包含非零序号的 EIGRP 数据包提供确认
3	更新（Update）	负责与其他 EIGRP 路由器传送路由和可达性信息的数据包
4	查询（Query）	在路由收敛期间，发送该数据包以查找其他路径
5	应答（Reply）	该数据包是查询数据包的响应数据包

EIGRP 使用 RTP（Reliable Transport Protocol，可靠传输协议）来确保按序分发数据包，并确保路由器能够收到特定数据包。所有的 EIGRP 数据包都包含一个序列号，如果序列号为零，那么就不需要从收端 EIGRP 路由器接收响应消息；如果序列号为其他值，那么就需要接收包含原始序列号的确认包。

为了确保收到数据包，必须采用可靠的传输方法。所有的更新、查询和应答包都必须进行可靠传输，而 Hello 包和确认包不需要确认，因而可以采取不可靠传输方式。

如果发端路由器在重传超时到期时，仍未从邻居路由器收到确认包，那么就会通知无响应路由器停止处理其多播数据包。发端路由器将通过单播发送所有流量，直到邻居实现完全同步。实现同步之后，发端路由器将通知目的路由器再次处理多播数据包。所有的单播数据包都需要进行确认。对于每个需要确认的数据包来说，EIGRP 最多重试 16 次，如果邻居路由器的重试次数达到了 16 次，那么就会重置该邻居关系。

7.1.4 EIGRP 基本配置

NX-OS 交换机要求在 EIGRP 进程及接口配置子模式下配置 EIGRP。在 NX-OS 设备上配置 EIGRP 的步骤如下。

- **第 1 步**：启用 EIGRP 功能特性。必须通过全局配置命令 **feature eigrp** 启用 EIGRP。
- **第 2 步**：定义 EIGRP 进程标签。必须使用全局配置命令 **router eigrp** *process-tag* 定义 EIGRP 进程，其中，*process-tag* 最长为 20 个字母数字字符。
- **第 3 步**：定义路由器 ID（可选）。RID（Router-ID，路由器 ID）是一个 32 比特的唯一编号，用于标识 EIGRP 路由器。EIGRP 使用 RID 作为环路预防机制。可以手工或动态设置 RID，但是要求每个 EIGRP 进程的 RID 都必须唯一。可以通过命令 **router-id** *router-id* 静态设置 RID。

  如果没有手工配置 RID，那么 NX-OS 始终优选 Loopback 0 的 IP 地址。如果没有 Loopback 0，那么 NX-OS 就选择配置的第一个环回接口的 IP 地址。如果没有环回接口，那么 NX-OS 就选择配置中的第一个物理接口的 IP 地址。
- **第 4 步**：定义地址簇。EIGRP 可以在同一个 EIGRP 进程下同时支持 IPv4 和 IPv6 地址簇，因而需要通过命令 **address-family [ipv4 | ipv6] unicast** 定义地址簇。

  如果 Nexus 交换机仅处理 IPv4 地址，那么该步骤可选。
- **第 5 步**：为 EIGRP 进程定义 ASN（Autonomous System Number，自治系统号）。必须使用命令 **autonomous-system** *as-number* 为 EIGRP 进程定义自治系统。

  如果 EIGRP 进程标签只是数字且与 EIGRP 进程使用的 ASN 相匹配，那么该步骤可选。
- **第 6 步**：在接口上启用 EIGRP。可以通过命令 **interface** *interface-id* 选择将要启用 EIGRP 的接口，然后再利用命令 **ip router eigrp** *process-tag* 在该接口上启用 EIGRP 进程。

> 注：与 IOS 设备不同，在接口上启用 EIGRP 会将所有备用直连网络都通告到拓扑表中。

例 7-2 所示配置仅在接口 Ethernet1/1、VLAN 10 和 Loopback 0 上启用了 EIGRP。

例 7-2　EIGRP 基本配置

```
NX-1# configure terminal
NX-1(config)# feature eigrp
NX-1(config)# router eigrp NXOS
17:10:19 NX-1 %$ VDC-1 %$ eigrp[27525]: EIGRP-5-HA_INFO: EIGRP HA info msg -
  SYSMGR_SUPSTATE_ACTIVE
NX-1(config-router)# autonomous-system 12
NX-1(config-router)# address-family ipv4 unicast
NX-1(config-router-af)# interface Ethernet1/1
NX-1(config-if)# ip router eigrp NXOS
NX-1(config-if)# interface vlan10
NX-1(config-if)# ip router eigrp NXOS
NX-1(config-if)# interface loopback0
NX-1(config-if)# ip router eigrp NXOS
17:10:19 NX-1 %$ VDC-1 %$ %EIGRP-5-NBRCHANGE_DUAL: eigrp-NXOS [27525] (defaultbase)
  IP-EIGRP(0) 123: Neighbor 10.12.1.2 (Ethernet1/2) is up: new adjacency
```

7.2　EIGRP 邻居邻接关系故障排查

EIGRP 在处理路由并将路由添加到 RIB（Routing Information Base，路由信息库）（又称为路由表）之前，需要建立邻居关系。邻居邻接表对于跟踪邻居状态以及发送给每个邻居的更新消息来说至关重要。本节将详细说明排查 NX-OS 交换机 EIGRP 邻居邻接关系故障的过程。

图 7-6 所示简单拓扑结构包含了两台 Nexus 交换机，下面将以此为例解释 EIGRP 邻接关系故障的排查过程。

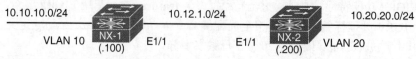

图 7-6　包含两台 NX-OS 交换机的拓扑结构

首先使用命令 **show ip eigrp neighbors [detail]** [*interface-id* | *neighbor-ip-address* | **vrf** {*vrf-name* | **all**}]验证已成功建立 EIGRP 邻接关系的设备。例 7-3 给出了在 NX-1 上运行该命令之后的输出结果示例。

例 7-3　显示 EIGRP 邻居

```
NX-1# show ip eigrp neighbors
IP-EIGRP neighbors for process 12 VRF default
H   Address                  Interface        Hold    Uptime    SRTT    RTO    Q     Seq
                                              (sec)             (ms)           Cnt   Num
0   10.12.1.200              Eth1/1           14      00:09:45  1       50     0     17
```

表 7-4 简要说明了例 7-3 显示的一些关键字段信息。

表 7-4　　　　　　　　　　　EIGRP 邻居的字段信息

字段	描述
Address	EIGRP 邻居的 IP 地址
Interface	检测到邻居的接口
Holdtime	收到邻居的数据包以确保其仍然有效的剩余时间
SRTT	数据包发送到邻居且从邻居收到应答消息的时间（以毫秒为单位）
RTO	重传超时（等待 ACK）
Q Cnt	队列中等待发送的数据包（更新/查询/应答包）数量
Seq Num	最后从该路由器收到的序列号

除要在 NX-OS 设备的网络接口上启用 EIGRP 之外，还要确保以下参数匹配，才能保证两台路由器成为邻居。

- 接口必须处于主动状态（Active）。
- 设备之间必须存在使用主用子网的连接。
- 自治系统号（ASN）必须匹配。
- 度量计算公式的 K 值。
- Hello 和保持定时器。
- 认证参数。

7.2.1　验证主动接口

配置 EIGRP 的时候，一种最佳实践是验证路由器仅在期望接口上运行 EIGRP。命令 **show ip eigrp interface** [**brief** | *interface-id*] [**vrf** {*vrf-name* | **all**}]可以显示所有处于主动状态的 EIGRP 接口。如果使用了可选关键字 **brief**，那么就可以显示例 7-4 所示的摘要信息。

例 7-4 显示主动 EIGRP 接口

```
NX-1# show ip eigrp interfaces brief
IP-EIGRP interfaces for process 123 VRF default

                 Xmit Queue    Mean   Pacing Time   Multicast    Pending
Interface  Peers Un/Reliable   SRTT   Un/Reliable   Flow Timer   Routes
Eth1/1     1     0/0           6      0/0           50           0
Lo0        0     0/0           0      0/0           0            0
Vlan10     0     0/0           0      0/0           0            0
```

表 7-5 简要说明了例 7-4 显示的一些关键字段信息。

表 7-5　　　　　　　　　　　EIGRP 接口字段信息

字段	描述
Interface	运行 EIGRP 的接口
Peers	在该接口上检测到的对等体数量
Xmt Queue Un/Reliable	传输队列中剩余的不可靠/可靠数据包数量，值为零时表示网络稳定
Mean SRTT	数据包发送给邻居且从邻居收到应答消息的平均时间（以毫秒为单位）
Pacing Time Un/Reliable	用于确定应何时将 EIGRP 数据包从接口发送出去（适用于不可靠和可靠数据包）
Multicast Flow Timer	路由器发送多播包的最大时间（以秒为单位）
Pending Routes	传输队列中需要发送的路由数量

7.2.2　被动接口

有些网络拓扑需要将网段宣告给 EIGRP，但又要防止邻居在该网段建立邻接关系。常见应用场景就是在园区拓扑中宣告接入层网络。

为了更好地解释这样做可能产生的问题（NX-1 与 NX-2 之间无法建立 EIGRP 邻接关系），下面就以例 7-5 为例，来分析这两台交换机的 EIGRP 接口状态以及对等链路 E1/1 的状态，可以看出，显示结果与期望并不一致。

例 7-5 找出主动 EIGRP 接口

```
NX-1# show ip eigrp interfaces brief
IP-EIGRP interfaces for process 12 VRF default

                 Xmit Queue    Mean   Pacing Time   Multicast    Pending
Interface  Peers Un/Reliable   SRTT   Un/Reliable   Flow Timer   Routes
Vlan10     0     0/0           0      0/0           0            0

NX-2# show ip eigrp interfaces brief
IP-EIGRP interfaces for process 12 VRF default

                 Xmit Queue    Mean   Pacing Time   Multicast    Pending
Interface  Peers Un/Reliable   SRTT   Un/Reliable   Flow Timer   Routes
Lo0        0     0/0           0      0/0           0            0
Vlan20     0     0/0           0      0/0           0            0
```

如前所述，显示 EIGRP 接口时并不显示被动接口。如果使用命令 **show ip eigrp** [*process-tag*]，那么就可以显示指定 EIGRP 进程的主动接口和被动接口的数量，如例 7-6 所示。

例 7-6 查看 EIGRP 被动接口

```
NX-1# show ip eigrp
! Output omitted for brevity
IP-EIGRP AS 12 ID 192.168.100.100 VRF default
```

```
Number of EIGRP interfaces: 1 (0 loopbacks)
Number of EIGRP passive interfaces: 2
Number of EIGRP peers: 0
```

接下来需要检查以下配置信息。

- 接口参数命令 **ip passive-interface eigrp** *process-tag*，该命令仅将指定接口设置为被动接口。
- 全局 EIGRP 配置命令 **passive-interface default**，该命令将该 EIGRP 进程下的所有接口均设置为被动状态。接口参数命令 **no ip passive-interface eigrp** *process-tag* 的优先级高于全局命令，可以将接口设置为主动状态。

例 7-7 给出了 NX-1 和 NX-2 配置示例，此时这两台 Nexus 交换机无法建立 EIGRP 邻接关系。为了确保这两台交换机能够建立邻接关系，要求这两台交换机的 Ethernet1/1 接口都必须处于主动状态。因而需要将命令 **no ip passive-interface eigrp NXOS** 移到 NX-1 的接口 E1/1，将命令 **ip passive interface eigrp NXOS** 从 NX-2 的 E1/1 移到 VLAN20。

例 7-7 带有被动接口的 EIGRP 配置

```
NX-1# show run eigrp
! Output omitted for brevity
router eigrp NXOS
  autonomous-system 12
  passive-interface default
  address-family ipv4 unicast

interface Vlan10
  ip router eigrp NXOS
  no ip passive-interface eigrp NXOS

interface loopback0
  ip router eigrp NXOS

interface Ethernet1/1
  ip router eigrp NXOS

NX-2# show run eigrp
! Output omitted for brevity
router eigrp NXOS
  autonomous-system 12
  address-family ipv4 unicast

interface Vlan20
  ip router eigrp NXOS

interface loopback0
  ip router eigrp NXOS

interface Ethernet1/1
  ip router eigrp NXOS
  ip passive-interface eigrp NXOS
```

注：除将接口设置为被动状态之外，还可以通过命令 **ip eigrp** *process-tag* **shutdown** 临时关闭接口的 EIGRP，这样做的好处是既能在接口上禁用 EIGRP，又能将 EIGRP 配置保留在接口上。

7.2.3 验证 EIGRP 数据包

排查 EIGRP 邻接关系故障的一个关键步骤，就是要确保设备正在发送或接收 EIGRP 网络流量。

命令 **show ip eigrp traffic** 可以从高层视角以摘要方式显示该设备收发的数据包类型，如例 7-8 所示。

例 7-8 EIGRP 流量统计

```
NX-1# show ip eigrp traffic
IP-EIGRP Traffic Statistics for AS 12 VRF default
  Hellos sent/received: 1486/623
  Updates sent/received: 13/8
  Queries sent/received: 0/0
  Replies sent/received: 0/0
  Acks sent/received: 6/8
  Input queue high water mark 1, 0 drops
  SIA-Queries sent/received: 0/0
  SIA-Replies sent/received: 0/0
  Hello Process ID: (no process)
  PDM Process ID: (no process)
```

如果希望获取设备收发数据包的详细信息，那么就可以使用调试功能。命令 **debug ip eigrp packets [siaquery | siareply | hello | query | reply | request | update | verbose]** 可以为指定数据包类型启用调试功能，如例 7-9 所示。

例 7-9 调试 EIGRP 数据包

```
NX-1# debug ip eigrp packets hello
! Output omitted for brevity
03:58:18.813041 eigrp: NXOS [26942] EIGRP: Received HELLO on Vlan10
03:58:18.814582 eigrp: NXOS [26942] nbr 10.10.10.10
03:58:18.814613 eigrp: NXOS [26942] AS 12, Flags 0x0, Seq 0/0idbQ 0/0
03:58:18.814618 eigrp: NXOS [26942] iidbQ un/rely 0/0
03:58:18.814623 eigrp: NXOS [26942] peerQ un/rely 0/0
03:58:18.965326 eigrp: NXOS [26942] EIGRP: Received HELLO on Ethernet1/1
03:58:18.965415 eigrp: NXOS [26942] nbr 10.12.1.200
03:58:18.965424 eigrp: NXOS [26942] AS 12, Flags 0x0, Seq 0/0idbQ 0/0
03:58:18.965430 eigrp: NXOS [26942] iidbQ un/rely 0/0
03:58:18.965435 eigrp: NXOS [26942] peerQ un/rely 0/0
03:58:20.244286 eigrp: NXOS [26942] EIGRP: Sending HELLO on Vlan10
03:58:20.244304 eigrp: NXOS [26942] AS 12, Flags 0x0, Seq 0/0idbQ 0/0
03:58:20.244310 eigrp: NXOS [26942] iidbQ un/rely 0/0
03:58:21.273574 eigrp: NXOS [26942] EIGRP: Sending HELLO on Ethernet1/1
03:58:21.273651 eigrp: NXOS [26942] AS 12, Flags 0x0, Seq 0/0idbQ 0/0
03:58:21.273660 eigrp: NXOS [26942] iidbQ un/rely 0/0
```

执行 EIGRP 调试操作时，仅显示到达管理引擎 CPU 的数据包。如果调试输出结果未显示任何数据包，那么就需要进一步检查交换机的 QoS（Quality of Service，服务质量）策略、ACL（Access Control List，访问控制列表）以及 CoPP（Control Plane Policing，控制平面策略），也可以仅验证离开或进入接口的数据包。

接口可能部署也可能未部署 QoS 策略。如果部署了 QoS 策略，那么就必须检查策略映射（policy-map）是否存在丢包问题，然后再将其引用到与 EIGRP 路由协议相匹配的分类映射。由于 CoPP 策略基于 QoS 设置，因而上述逻辑也同样适用于 CoPP 策略。

例 7-10 给出了 CoPP 策略的检查过程。

- 使用命令 **show run copp all** 检查 CoPP 策略。该命令将显示相关的策略映射名称、已定义的类别以及每个类别的警管速率。
- 验证分类映射信息，以确定该分类映射的条件匹配设置。
- 验证了分类映射之后，使用命令 **show policy-map interface control-plane** 检查该类别的策

略映射。

例 7-10 验证 EIGRP 的 CoPP 策略

```
NX-1# show run copp all
! Output omitted for brevity
class-map type control-plane match-any copp-system-p-class-critical
  match access-group name copp-system-p-acl-bgp
  match access-group name copp-system-p-acl-rip
  match access-group name copp-system-p-acl-vpc
  match access-group name copp-system-p-acl-bgp6
  match access-group name copp-system-p-acl-lisp
  match access-group name copp-system-p-acl-ospf
  match access-group name copp-system-p-acl-rip6
  match access-group name copp-system-p-acl-rise
  match access-group name copp-system-p-acl-eigrp
  match access-group name copp-system-p-acl-lisp6
  match access-group name copp-system-p-acl-ospf6
  match access-group name copp-system-p-acl-rise6
  match access-group name copp-system-p-acl-eigrp6
  match access-group name copp-system-p-acl-otv-as
  match access-group name copp-system-p-acl-mac-l2pt
  match access-group name copp-system-p-acl-mpls-ldp
  match access-group name copp-system-p-acl-mpls-rsvp
  match access-group name copp-system-p-acl-mac-l3-isis
  match access-group name copp-system-p-acl-mac-otv-isis
  match access-group name copp-system-p-acl-mac-fabricpath-isis
..
policy-map type control-plane copp-system-p-policy-strict
  class copp-system-p-class-critical
    set cos 7
    police cir 36000 kbps bc 250 ms conform transmit violate drop

NX-1# show run aclmgr all | section copp-system-p-acl-eigrp
ip access-list copp-system-p-acl-eigrp
  10 permit eigrp any any

NX-1# show policy-map interface control-plane class copp-system-p-class-critical
! Output omitted for brevity
Control Plane
  service-policy input copp-system-p-policy-strict

    class-map copp-system-p-class-critical (match-any)
     ..
     module 1:
        conformed 1623702 bytes,
          5-min offered rate 995 bytes/sec
          peak rate 1008 bytes/sec at Tues 16:08:39 2018
        violated 0 bytes,
          5-min violate rate 0 bytes/sec
          peak rate 0 bytes/sec
```

注：该 CoPP 策略源自 Nexus 7000 交换机，实际的策略名称和分类映射可能会因不同的平台而异。

接下来需要确定是否在接口上发送或接收了数据包，相应的操作包括为 EIGRP 协议创建特定 ACE（Access Control Entity，访问控制实体）。为了确保正确计数，EIGRP 的 ACE 应该位于所有不明确的 ACE 表项之前。可以通过 ACL 配置命令 **statistics per-entry** 显示每个 ACE 的命中情况。

例 7-11 给出了在 Ethernet1/1 接口上检测 EIGRP 流量的 ACL 配置示例。请注意，该 ACL 包含了一条 **permit ip any** 命令，作用是允许所有流量通过该接口。否则，可能会导致流量丢失。

例 7-11 利用 ACL 验证 EIGRP 数据包

```
NX-1# configure terminal
NX-1(config)# ip access-list EIGRP
NX-1(config-acl)# permit eigrp any any
NX-1(config-acl)# permit icmp any any
NX-1(config-acl)# permit ip any any
NX-1(config-acl)# statistics per-entry
NX-1(config-acl)# interface e1/1
NX-1(config-if)# ip access-group EIGRP in

NX-1# show ip access-list
IP access list EIGRP
        statistics per-entry
        10 permit eigrp any any [match=108]
        20 permit icmp any any [match=5]
        30 permit ip any any [match=1055]
```

例 7-11 使用的是以太网接口（通常表示的是一种一对一的通信关系），但是对于 SVI（Switched Virtual Interfaces，交换式虚接口）（也称为接口 VLAN）等多路接入接口来说，可能需要在特定 ACE 中指定邻居。

例 7-12 给出了应用在 SVI 上的 ACL 配置情况，该 ACL 为邻居 10.12.100.200 配置了一条 ACE 表项，其他邻居的 EIGRP 数据包则通过第二条表项进行收集。

例 7-12 利用 ACL 对 EIGRP 数据包进行精细化验证

```
ip access-list EIGRP
  statistics per-entry
    permit eigrp 10.12.100.200/32 any any
    permit eigrp any any
    permit icmp any any
    permit ip any any
interface vlan 10
  ip access-group EIGRP in

NX-1# show ip access-list
IP access list EIGRP
        statistics per-entry
        10 permit eigrp 10.12.100.200/32 any log [match=100]
        20 permit eigrp any any [match=200]
        30 permit icmp any any [match=0]
        40 permit ip any any [match=5]
```

另一种替代 ACL 的方法就是使用内置的 NX-OS Ethanalyzer 来抓取 EIGRP 数据包。例 7-13 给出了该命令的语法形式，可以通过可选关键字 **detail** 查看数据包的详细信息。

例 7-13 利用 Ethanalyzer 验证 EIGRP 数据包

```
NX-1# ethanalyzer local interface inband capture-filter "proto eigrp"
Capturing on inband
2017-09-03 04:21:12.688751    10.12.1.2 -> 224.0.0.10    EIGRP Hello
2017-09-03 04:21:12.690573    10.12.1.1 -> 224.0.0.10    EIGRP Hello
2017-09-03 04:21:12.701393    10.12.1.2 -> 224.0.0.10    EIGRP Hello
2017-09-03 04:21:12.705344    10.12.1.2 -> 10.12.1.1     EIGRP Update
2017-09-03 04:21:12.705344    10.12.1.2 -> 10.12.1.1     EIGRP Update
```

7.2.4 必须存在使用主用子网的连接

EIGRP 路由器必须能够通过与主 IP 地址相关联的网络与对等路由器进行通信。EIGRP 仅使用主 IP 地址建立邻接关系，无法使用从 IP 地址建立邻接关系。本节将 NX-2 的子网掩码从 10.12.1.200/24 更改为 10.12.1.200/25，这样就可以将 NX-2 从 NX-1 所处的网络（10.12.1.100）调整到网络 10.12.1.128/25 中。

例 7-14 中的 NX-1 检测到 NX-2 并将其注册为邻居，但 NX-2 并未检测到 NX-1。

例 7-14　NX-1 检测到 NX-2 并将其注册为邻居

```
NX-1# show ip eigrp neighbor
IP-EIGRP neighbors for process 12 VRF default
H   Address                 Interface       Hold  Uptime    SRTT  RTO   Q    Seq
                                            (sec)           (ms)        Cnt  Num
0   10.12.1.200             Eth1/1          13    00:00:10  1     5000  1    0
NX-2# show ip eigrp neighbor
IP-EIGRP neighbors for process 12 VRF default
```

此外，NX-1 会在超出重试限制之后不断改变 NX-2（10.12.1.200）的邻居状态，如例 7-15 所示。

例 7-15　由于重试限制导致 EIGRP 邻接关系中断

```
NX-1
13:28:06 NX-1 %$ VDC-1 %$ %EIGRP-5-NBRCHANGE_DUAL: eigrp-NXOS [26809] (default-base)
    IP-EIGRP(0) 12: Neighbor 10.12.1.200 (Ethernet1/1) is down: retry limit exceeded
13:28:09 NX-1 %$ VDC-1 %$ %EIGRP-5-NBRCHANGE_DUAL: eigrp-NXOS [26809] (default-base)
    IP-EIGRP(0) 12: Neighbor 10.12.1.200 (Ethernet1/1) is up: new adjacency
21:19:00 NX-1 %$ VDC-1 %$ %EIGRP-5-NBRCHANGE_DUAL: eigrp-NXOS [26809] (default-base)
    IP-EIGRP(0) 123: Neighbor 10.12.1.200 (Ethernet1/1) is down: retry limit exceeded
21:19:00 NX-1 %$ VDC-1 %$ %EIGRP-5-NBRCHANGE_DUAL: eigrp-NXOS [26809] (default-base)
    IP-EIGRP(0) 123: Neighbor 10.12.1.200 (Ethernet1/1) is up: new adjacency
```

注： 与 IOS 路由器不同，NX-OS 不提供系统日志消息 "*is blocked: not on common subnet*"。

请注意，EIGRP 最多为需要确认的数据包重试 16 次，一旦邻居达到了重试限制 16 次，EIGRP 就会重置邻居关系。可以通过命令 **show ip eigrp neighbors detail** 检测 NX-OS 的实际重试次数，如例 7-16 所示。

例 7-16　查看邻居的 EIGRP 重试次数

```
NX-1# show ip eigrp neighbors detail
IP-EIGRP neighbors for process 12 VRF default
H   Address                 Interface       Hold  Uptime    SRTT  RTO   Q    Seq
                                            (sec)           (ms)        Cnt  Num
0   10.12.1.200             Eth1/1          14    00:01:12  1     5000  1    0
    Version 8.0/1.2, Retrans: 15, Retries: 15, BFD state: N/A, Waiting for Init,
    Waiting for Init Ack
     UPDATE seq 13 ser 0-0 Sent 78084 Init Sequenced
NX-1# show ip eigrp neighbors detail
IP-EIGRP neighbors for process 12 VRF default
H   Address                 Interface       Hold  Uptime    SRTT  RTO   Q    Seq
                                            (sec)           (ms)        Cnt  Num
0   10.12.1.200             Eth1/1          13    00:01:19  1     5000  1    0
    Version 8.0/1.2, Retrans: 16, Retries: 16, BFD state: N/A, Waiting for Init,
    Waiting for Init Ack
     UPDATE seq 13 ser 0-0 Sent 79295 Init Sequenced
```

接下来需要在两个节点之间 ping 主 IP 地址，以验证两者之间的连通性，如例 7-17 所示。

例 7-17　验证两个主用子网之间的连接性

```
NX-1# ping 10.12.1.200
PING 10.12.1.200 (10.12.1.200): 56 data bytes
Request 0 timed out
Request 1 timed out
Request 2 timed out
Request 3 timed out
Request 4 timed out

--- 10.12.1.200 ping statistics ---
5 packets transmitted, 0 packets received, 100.00% packet loss

NX-2# ping 10.12.1.100
PING 10.12.1.100 (10.12.1.100): 56 data bytes
ping: sendto 10.12.1.100 64 chars, No route to host
Request 0 timed out
ping: sendto 10.12.1.100 64 chars, No route to host
Request 1 timed out
ping: sendto 10.12.1.100 64 chars, No route to host
Request 2 timed out
ping: sendto 10.12.1.100 64 chars, No route to host
Request 3 timed out
ping: sendto 10.12.1.100 64 chars, No route to host
Request 4 timed out
```

可以看出，NX-1 无法 ping 通 NX-2，NX-2 也无法 ping 通 NX-1，因为没有去往主机的路由。这也意味着 NX-1 可能能够将数据包发送给 NX-2，但 NX-2 却没有发送 ICMP 响应消息的路由。

例 7-18 显示了 NX-1 和 NX-2 的路由表信息，可以帮助确定故障原因。

例 7-18　NX-1 和 NX-2 邻接关系路由表

```
NX-1# show ip route 10.12.1.200
IP Route Table for VRF "default"
'*' denotes best ucast next-hop
'**' denotes best mcast next-hop
'[x/y]' denotes [preference/metric]
'%<string>' in via output denotes VRF <string>

10.12.1.200/32, ubest/mbest: 1/0, attached
    *via 10.12.1.200, Eth1/1, [250/0], 00:30:29, am

NX-2# show ip route 10.12.1.100
IP Route Table for VRF "default"
'*' denotes best ucast next-hop
'**' denotes best mcast next-hop
'[x/y]' denotes [preference/metric]
'%<string>' in via output denotes VRF <string>

Route not found
```

此时需要检查两台设备的 IP 地址配置情况，可以看出，当前配置的子网掩码的前缀长度不匹配。解决该问题之后，EIGRP 设备即可进行正常通信。

7.2.5　EIGRP ASN 不匹配

EIGRP 要求 EIGRP Hello 包中的 ASN 必须匹配，才能建立邻接关系。但是，由于 EIGRP 默认使用进程标签中指定的数字作为 ASN，因而对于 Nexus 交换机来说可能会出现问题。该方式与 IOS 路由器的传统 EIGRP 配置方式相同。

如果在 EIGRP 配置中指定了 ASN，那么就使用该值代替标识 EIGRP 进程的数字。如果 EIGRP 的进程标记是字母数字，且未指定 ASN，那么就假定 ASN 为 0，表示该实例处于关闭状态。

例 7-19 显示的配置可能会让初级网络工程师感到困惑，不知道此时的 EIGRP ASN 是 12 还是 1234？

例 7-19 令人困惑的 EIGRP ASN 配置

```
NX-1# show run eigrp
! Output omitted for brevity

router eigrp 12
  autonomous-system 1234

interface Ethernet1/1
  ip router eigrp 12
```

不幸的是，即使 EIGRP ASN 不匹配，系统也不会提供任何调试或日志消息。因此，必须仔细检查双方的 EIGRP ASN 以确保完全相同。

使用命令 **show ip eigrp** 检查 EIGRP 协议时可以看到 EIGRP 实例的 ASN，ASN 列在 router-id 的旁边。使用命令 **show ip eigrp interfaces brief** 查看 EIGRP 接口时，也能看到 ASN 信息。例 7-20 给出了在 Nexus 交换机上查看 ASN 的配置示例。

例 7-20 查看 EIGRP ASN

```
NX-1# show ip eigrp
! Output omitted for brevity
IP-EIGRP AS 1234 ID 192.168.100.100 VRF default
  Process-tag: 12
  Instance Number: 1
  Status: running

NX-1# show ip eigrp interfaces brief
IP-EIGRP interfaces for process 1234 VRF default

                Xmit Queue   Mean   Pacing Time   Multicast    Pending
Interface  Peers Un/Reliable SRTT   Un/Reliable   Flow Timer   Routes
Eth1/1      0     0/0         0      0/0           0            0
Lo0         0     0/0         0      0/0           0            0
Vlan10      0     0/0         0      0/0           0            0
```

注： 在 EIGRP 配置中明确指定 AS，有助于避免给水平有限的网络工程师带来困惑，这是一种非常好的最佳实践。

7.2.6 K 值不匹配

EIGRP 通过 K 值来定义最佳路径计算公式所使用的因子。为了确保路由逻辑的一致性并防止出现路由环路，所有的 EIGRP 邻居都必须使用相同的 K 值。K 值包含在 EIGRP 的 Hello 包中。

例 7-21 显示了指示 K 值不匹配的 syslog 消息。通过 **show ip eigrp** 命令查看 EIGRP 进程，即可确定本地路由器的 K 值。

例 7-21 EIGRP K 值不匹配

```
04:11:19 NX-1 %$ VDC-1 %$ %EIGRP-5-NBRCHANGE_DUAL: eigrp-NXOS [30489] (defaultbase)
  IP-EIGRP(0) 12: Neighbor 10.12.1.200 (Ethernet1/1) is down: K-value mismatch
04:11:37 NX-1 %$ VDC-1 %$ last message repeated 3 times
04:11:37 NX-1 %$ VDC-1 %$ %EIGRP-5-NBRCHANGE_DUAL: eigrp-NXOS [30489]
  (default-base) IP-EIGRP(0) 12: Neighbor 10.12.1.200 (Ethernet1/1) is down:
  Interface Goodbye received
```
```
NX-1# show ip eigrp
! Output omitted for brevity
IP-EIGRP AS 12 ID 192.168.100.100 VRF default
  Process-tag: NXOS
  Instance Number: 1
  Status: running
  none
  Metric weights: K1=1 K2=1 K3=1 K4=1 K5=1
```

可以在 EIGRP 进程中，通过命令 **metric weights** *TOS* K_1 K_2 K_3 K_4 K_5 [K_6] 配置 Nexus 交换机的 K 值。除非配置了 EIGRP 宽度量，否则 K_6 可选。此处未使用 TOS，应将其设置为 0。例 7-22 给出了带有自定义 K 值的 EIGRP 配置示例。

例 7-22 带有自定义 K 值的 EIGRP 配置

```
NX-1# show run eigrp
! Output omitted for brevity
router eigrp NXOS
  autonomous-system 12
  metric weights 0 1 1 1 1 1
  address-family ipv4 unicast
```

7.2.7 Hello 和保持定时器问题

EIGRP Hello 包的辅助功能是确保 EIGRP 邻居的健康性和可用性。EIGRP Hello 包以一定的间隔进行发送，该间隔被称为 Hello 定时器。Nexus 交换机的默认 EIGRP Hello 定时器为 5s。

EIGRP 使用的第二种定时器是保持定时器，表示 EIGRP 认为路由器可达且运行正常的时间。保持时间的默认值是 Hello 间隔的 3 倍，默认为 15s，如果是低速接口，那么默认为 180s。保持时间处于递减状态，收到 Hello 包之后，保持定时器就会重置并开始倒计时。如果保持定时器达到零，那么 EIGRP 就宣称邻居不可达，并将拓扑变更信息通告给 DUAL 算法。

如果 EIGRP Hello 定时器大于其他 EIGRP 邻居的保持定时器，那么会话就会处于持续震荡状态。例 7-23 中的 NX-1 一直都在周期性地重置与 NX-2 之间的邻接关系（因为 NX-1 的保持定时器到期）。

例 7-23 由于保持定时器到期而导致 EIGRP 出现邻接关系故障

```
NX-1
03:11:35 NX-1 %$ VDC-1 %$ %EIGRP-5-NBRCHANGE_DUAL: eigrp-NXOS [30489] (defaultbase)
  IP-EIGRP(0) 12: Neighbor 10.12.1.200 (Ethernet1/1) is down: holding time
  expired
03:11:39 NX-1 %$ VDC-1 %$ %EIGRP-5-NBRCHANGE_DUAL: eigrp-NXOS [30489] (defaultbase)
  IP-EIGRP(0) 12: Neighbor 10.12.1.200 (Ethernet1/1) is up: new adjacency
03:11:54 NX-1 %$ VDC-1 %$ %EIGRP-5-NBRCHANGE_DUAL: eigrp-NXOS [30489] (defaultbase)
  IP-EIGRP(0) 12: Neighbor 10.12.1.200 (Ethernet1/1) is down: holding time
  expired
03:11:59 NX-1 %$ VDC-1 %$ %EIGRP-5-NBRCHANGE_DUAL: eigrp-NXOS [30489] (defaultbase)
```

```
      IP-EIGRP(0) 12: Neighbor 10.12.1.200 (Ethernet1/1) is up: new adjacency
NX-2
03:11:35 NX-2 %$ VDC-1 %$ %EIGRP-5-NBRCHANGE_DUAL: eigrp-NXOS [26807] (defaultbase)
      IP-EIGRP(0) 12: Neighbor 10.12.1.100 (Ethernet1/1) is down: Interface
      Goodbye received
03:11:39 NX-2 %$ VDC-1 %$ %EIGRP-5-NBRCHANGE_DUAL: eigrp-NXOS [26807] (defaultbase)
      IP-EIGRP(0) 12: Neighbor 10.12.1.100 (Ethernet1/1) is up: new adjacency
03:11:54 NX-2 %$ VDC-1 %$ %EIGRP-5-NBRCHANGE_DUAL: eigrp-NXOS [26807] (defaultbase)
      IP-EIGRP(0) 12: Neighbor 10.12.1.100 (Ethernet1/1) is down: Interface
      Goodbye received
03:11:59 NX-2 %$ VDC-1 %$ %EIGRP-5-NBRCHANGE_DUAL: eigrp-NXOS [26807] (defaultbase)
      IP-EIGRP(0) 12: Neighbor 10.12.1.100 (Ethernet1/1) is up: new adjacency
```

可以通过命令 **show ip eigrp interface** [*interface-id*] [**vrf** {*vrf-name* | **all**}]查看接口的 EIGRP Hello 定时器和保持定时器状态。如果使用了可选关键字 **brief**，那么就无法查看定时器信息。例 7-24 给出了 NX-1 和 NX-2 的输出结果示例。

例 7-24 验证 EIGRP Hello 和保持定时器

```
NX-1# show ip eigrp interfaces
! Output omitted for brevity
IP-EIGRP interfaces for process 12 VRF default

                Xmit Queue    Mean   Pacing Time   Multicast    Pending
Interface  Peers Un/Reliable  SRTT   Un/Reliable   Flow Timer   Routes
Eth1/1       1     0/0          1      0/0            50           0
  Hello interval is 5 sec
  Holdtime interval is 15 sec
  Next xmit serial <none>

NX-2# show ip eigrp interfaces
! Output omitted for brevity
IP-EIGRP interfaces for process 12 VRF default

                Xmit Queue    Mean   Pacing Time   Multicast    Pending
Interface  Peers Un/Reliable  SRTT   Un/Reliable   Flow Timer   Routes
Eth1/1       1     0/0          3      0/0            50           0
  Hello interval is 120 sec
  Holdtime interval is 15 sec
  Next xmit serial <none>
```

可以看出，NX-2 的 Hello 定时器为 120s，超出了 NX-1 的保持定时器（15s），这就是 NX-1 不断拆除 EIGRP 邻接关系的原因。

例 7-25 利用接口命令 **ip hello-interval eigrp** *process-tag hello-time* 修改了 Hello 间隔。只要将 Hello 间隔恢复为默认值或小于 15s（NX-1 的保持定时器），就可以让交换机建立邻接关系。

例 7-25 修改了 Hello 定时器之后的 EIGRP 配置信息

```
NX-2# show run eigrp
! Output omitted for brevity
router eigrp NXOS
  autonomous-system 12
  address-family ipv4 unicast

interface Ethernet1/1
  ip router eigrp NXOS
  ip hello-interval eigrp NXOS 120
```

注：可以使用命令 **ip hold-time eigrp** *process-tag hold-time* 修改 EIGRP 接口的保持定时器。

7.2.8 EIGRP 认证问题

认证是一种确保只有经授权的 EIGRP 设备才能成为 EIGRP 邻居的安全机制。所有的 EIGRP 数据包都包含了预先计算的密码散列值，接收路由器会对散列值进行解密。如果密码不匹配，那么路由器就会丢弃数据包，从而防止形成邻接关系。

但不幸的是，如果启用的认证机制不匹配或密码不一致，则系统不会提供相应的调试或日志消息。需要在两台设备上分别验证 EIGRP 认证机制，以确保双方都启用了完全相同的 EIGRP 认证机制。

如果在接口上显式配置了认证机制，那么就会在 EIGRP 接口的下方显示相应的状态信息（见例 7-26），可以看出，NX-2 仅在 Ethernet1/1 上启用了认证机制。

例 7-26 查看接口的 EIGRP 认证机制

```
NX-2# show ip eigrp interfaces
! Output omitted for brevity
Eth1/1            1         0/0         2         0/0         50          0
  Hello interval is 5 sec
  Holdtime interval is 15 sec
  Next xmit serial <none>
  Un/reliable mcasts: 0/55  Un/reliable ucasts: 85/65
  Mcast exceptions: 0  CR packets: 0  ACKs suppressed: 20
  Retransmissions sent: 8  Out-of-sequence rcvd: 0
  Authentication mode is md5,  key-chain is "EIGRP"
Lo0               0         0/0         0         0/0         0           0
  ..
  Authentication mode is not set
Vlan20            0         0/0         0         0/0         0           0
  ..
  Authentication mode is not set
```

EIGRP 使用密钥链（Keychain）功能，采取 MD5 算法对密码进行加密。密钥链允许配置多个密码和序列号，而且还能设置有效期，从而轮换使用这些密码。如果使用基于时间的密钥链，那么就必须保证 Nexus 交换机的时间与 NTP 同步，而且要在密钥迭代之间提供一定的重叠时间。

散列由密钥号和密码组成，EIGRP 认证仅加密密码，而不对整个 EIGRP 数据包进行加密。可以通过命令 **show key chain [mode decrypt]** 查看密码信息，可选关键字 **mode decrypt** 可以在一对双引号之间以纯文本方式显示密码，有助于检测非期望字符（如空格）。例 7-27 显示了密钥链密码的验证方式。

例 7-27 验证密钥链

```
NX-1# show key chain
Key-Chain EIGRP
  Key 1 -- text 7 "0802657d2a36"
    accept lifetime (always valid) [active]
    send lifetime (always valid) [active]

NX-1# show key chain mode decrypt
Key-Chain EIGRP
  Key 1 -- text 0 "CISCO"
    accept lifetime (always valid) [active]
    send lifetime (always valid) [active]
```

> **注**：如果密钥号不同，则即使密码相同，两台 EIGRP 设备的散列值也不匹配，因而必须确保密钥号和密码都匹配。

1. 基于接口的 EIGRP 认证

解决与认证相关的故障问题时，需要确保在两个邻居的接口上都启用了认证机制，且密码完全相同。在 Nexus 交换机上启用 EIGRP 认证机制的过程如下。

- **第 1 步**：创建密钥链。可以通过命令 **key chain** *key-chain-name* 创建本地密钥链。
- **第 2 步**：标识密钥序列号。可以通过命令 **key** *key-number* 指定密钥序列号，其中，*key-number* 的取值范围是 0～2147483647。
- **第 3 步**：指定密码。可以通过命令 **key-string** *text* 输入预共享密码。可以根据需要重复步骤 2 和 3，以配置多个密钥串。
- **第 4 步**：标识接口的密钥链。必须通过命令 **ip authentication key-chain eigrp** *process-tag key-chain-name* 指定接口使用的密钥链。
- **第 5 步**：为接口启用认证机制。通过用命令 **ip authentication key-chain eigrp mode eigrp** *process-tag* **md5** 在接口上启用认证机制。

例 7-28 给出了在 Ethernet1/1 上启用 EIGRP 认证机制的配置示例。

例 7-28　EIGRP 接口级认证

```
key chain EIGRP
  key 1
    key-string CISCO

router eigrp NXOS
  autonomous-system 12
  authentication mode md5

interface Ethernet1/1
  ip router eigrp NXOS
  ip authentication key-chain eigrp NXOS EIGRP
  ip authentication key-chain eigrp mode eigrp NXOS md5
```

2. 全局 EIGRP 认证

截至本书写作之时，如果在全局范围内启用了 EIGRP 认证，那么执行 **show ip eigrp interfaces** 命令的时候就不会在任何接口下显示 EIGRP 认证。排查全局 EIGRP 认证启用故障时，必须仔细核查配置信息，此时的主要区别在于，创建了密钥链之后，必须在 EIGRP 进程中通过以下命令启用认证机制。

- **authentication mode md5**。
- **authentication key-chain** *key-chain-name*。

例 7-29 显示了全局启用 EIGRP 认证机制的配置示例。

例 7-29　EIGRP 进程级认证

```
key chain EIGRP
  key 1
    key-string CISCO

router eigrp NXOS
  autonomous-system 12
  authentication mode md5
  authentication key-chain EIGRP
  address-family ipv4 unicast
```

注：基于接口的 EIGRP 认证会覆盖全局 EIGRP 认证设置。

7.3 EIGRP 路径选择和路由丢失故障排查

接下来以图 7-7 拓扑结构为例，解释 EIGRP 协议的故障排查方法。图中的所有路由器均使用 10Gbit/s 链路进行互联，NX-1 正在宣告两个网络 10.1.1.0/24 和 10.11.11.0/24，NX-6 正在宣告网络 10.6.6.0/24。

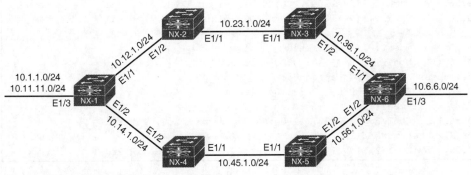

图 7-7 路径选择示例拓扑结构

例 7-30 显示了 NX-1 和 NX-6 的部分路由表信息。请注意，对于所宣告的网络前缀来说，NX-1 与 NX-6 之间的双方向均存在两条路径。

例 7-30 NX-1 和 NX-6 的路由表

```
NX-1# show ip route
! Output omitted for brevity
10.6.6.0/24, ubest/mbest: 2/0
    *via 10.12.1.2, Eth1/1, [90/129088], 00:00:02, eigrp-NXOS, internal
    *via 10.14.1.4, Eth1/2, [90/129088], 00:11:27, eigrp-NXOS, internal
NX-6# show ip route
! Output omitted for brevity
10.1.1.0/24, ubest/mbest: 2/0
    *via 10.36.1.3, Eth1/1, [90/1280], 00:00:07, eigrp-NXOS, internal
    *via 10.56.1.5, Eth1/2, [90/1280], 00:00:07, eigrp-NXOS, internal
10.11.11.0/24, ubest/mbest: 2/0
    *via 10.36.1.3, Eth1/1, [90/1280], 00:00:07, eigrp-NXOS, internal
    *via 10.56.1.5, Eth1/2, [90/1280], 00:00:07, eigrp-NXOS, internal
```

使用 **show ip route [eigrp]** 命令可以查看安装在 RIB 中的 EIGRP 路由信息，可选关键字 **eigrp** 的作用是仅显示 EIGRP 学到的路由。EIGRP 路由通过 eigrp-*process-tag* 进行标识。

源自 AS 内部的 EIGRP 路由的 AD（Administrative Distance，管理距离）值为 90，且在进程标签（process-tag）的后面标记 *internal*。源自 AS 外部的路由被称为外部 EIGRP 路由，外部 EIGRP 路由的 AD 值为 170，且在进程标签的后面标记 *external*。将 AD 值较高的外部 EIGRP 路由安装到 RIB 中可以起到预防环路的作用。

例 7-31 显示了图 7-7 所示拓扑结构的 EIGRP 路由信息，方括号中的第二个数字就是选定路由的度量。

例 7-31 查看 NX-1 的 EIGRP 路由

```
NXOS6# show ip route
! Output omitted for brevity
IP Route Table for VRF "default"
'*' denotes best ucast next-hop
'**' denotes best mcast next-hop
'[x/y]' denotes [preference/metric]
10.1.1.0/24, ubest/mbest: 2/0
    *via 10.36.1.3, Eth1/1, [90/1280], 00:00:07, eigrp-NXOS, internal
    *via 10.56.1.5, Eth1/2, [90/1280], 00:00:07, eigrp-NXOS, internal
10.11.11.0/24, ubest/mbest: 2/0
    *via 10.36.1.3, Eth1/1, [90/1280], 00:00:07, eigrp-NXOS, internal
    *via 10.56.1.5, Eth1/2, [90/1280], 00:00:07, eigrp-NXOS, internal
..
10.36.1.0/24, ubest/mbest: 1/0, attached
    *via 10.36.1.6, Eth1/1, [0/0], 06:08:04, direct
10.56.1.1/32, ubest/mbest: 1/0, attached
    *via 10.56.1.6, Eth1/2, [0/0], 06:08:04, local
172.16.0.0/16, ubest/mbest: 1/0
    *via 10.56.1.5, Eth1/2, [170/3072], 00:00:02, eigrp-NXOS, external
192.168.2.2/32, ubest/mbest: 1/0
    *via 10.56.1.5, Eth1/2, [90/130816], 00:01:43, eigrp-NXOS, internal
```

7.3.1 负载均衡

EIGRP 允许将多条后续路由（度量相同）安装到 RIB 中。将去往同一前缀的多条路径安装到 RIB 中，就称为 ECMP（Equal-Cost MultiPath，等价多路径）路由。截至本书写作之时，Nexus 节点默认的最大 ECMP 路径为 8。

可以在 EIGRP 进程中通过命令 **maximum-paths** 修改默认的 ECMP 设置，最多可以将 ECMP 默认值增加到 16。

NX-OS 不支持 EIGRP 的非等价负载均衡（允许将后继路由和可行后继同时安装到 EIGRP RIB 中），可以在其他思科操作系统中通过 **variance** 命令启用非等价负载均衡机制。

7.3.2 末梢

EIGRP 的末梢（Stub）功能可以为 EIGRP 路由器节省宝贵的路由器资源。EIGRP 末梢路由器可以在 EIGRP Hello 包中将自己宣告为末梢路由器，相邻路由器可以检测末梢字段并更新 EIGRP 邻居表，以反映路由器的末梢状态。

如果路由处于主动状态，那么 EIGRP 就不会向 EIGRP 末梢路由器发送 EIGRP 查询消息，这样就可以加快 EIGRP AS 内部的路由收敛速度，因为这样做可以缩小前缀的查询域范围。

EIGRP 末梢路由器不会宣告从其他 EIGRP 对等体学到的路由。在默认情况下，EIGRP 末梢路由器仅宣告直连路由和汇总路由。不过，也可以进行定制化配置，让末梢路由器仅接收路由，或者宣告重分发路由、直连路由或汇总路由的任意组合。

例 7-32 显示了 NX-1 和 NX-6 的路由表信息，可以看出，与例 7-30 显示的基线路由表有所不同。

例 7-32 定制化配置之后的路由表信息

```
NX-1# show ip route eigrp-NXOS
! Output omitted for brevity
10.6.6.0/24, ubest/mbest: 1/0
    *via 10.14.1.4, Eth1/2, [90/129088], 00:14:42, eigrp-NXOS, internal
```

```
NX-6# show ip route eigrp-NXOS
! Output omitted for brevity
10.1.1.0/24, ubest/mbest: 1/0
    *via 10.56.1.5, Eth1/2, [90/1280], 00:15:24, eigrp-NXOS, internal
10.11.11.0/24, ubest/mbest: 1/0
    *via 10.56.1.5, Eth1/2, [90/1280], 00:15:24, eigrp-NXOS, internal
```

对于从 NX-1 到 NX-6 的路由来说，看起来似乎只有下面那条路径可用（NX-1→NX-4→NX-5→NX-6），上面那条路径（NX-1→NX2→NX-3→NX-6）有问题吗？首先需要检查 EIGRP 邻接关系，如例 7-33 所示。

例 7-33　故障排查：验证 EIGRP 邻居邻接关系

```
NX-1# show ip eigrp neighbors
IP-EIGRP neighbors for process 100 VRF default
H   Address              Interface       Hold  Uptime    SRTT  RTO   Q    Seq
                                         (sec)           (ms)        Cnt  Num
0   10.12.1.2            Eth1/1          14    00:01:15  1     50    0    72
1   10.14.1.4            Eth1/2          14    00:24:07  1     50    0    70

NX-2# show ip eigrp neighbors
IP-EIGRP neighbors for process 100 VRF default
H   Address              Interface       Hold  Uptime    SRTT  RTO   Q    Seq
                                         (sec)           (ms)        Cnt  Num
1   10.12.1.1            Eth1/2          13    00:01:39  1     50    0    75
0   10.23.1.3            Eth1/1          13    00:01:43  1     50    0    62

NX-3# show ip eigrp neighbors
IP-EIGRP neighbors for process 100 VRF default
H   Address              Interface       Hold  Uptime    SRTT  RTO   Q    Seq
                                         (sec)           (ms)        Cnt  Num
0   10.23.1.2            Eth1/1          13    00:02:07  1     50    0    73
1   10.36.1.6            Eth1/2          12    00:19:42  1     50    0    86

NX-6# show ip eigrp neighbors
IP-EIGRP neighbors for process 100 VRF default
H   Address              Interface       Hold  Uptime    SRTT  RTO   Q    Seq
                                         (sec)           (ms)        Cnt  Num
1   10.36.1.3            Eth1/1          11    00:19:03  1     50    0    61
0   10.56.1.5            Eth1/2          13    00:19:03  1     50    0    34
```

可以看出，所有路由器都建立了邻接关系。使用可选关键字 **detail** 可以为故障问题提供更多有用信息，例 7-34 显示了命令 **show ip eigrp neighbors detail** 的输出结果。

例 7-34　进一步验证 EIGRP 邻居邻接关系

```
NX-1# show ip eigrp neighbors detail
IP-EIGRP neighbors for process 100 VRF default
H   Address              Interface       Hold  Uptime    SRTT  RTO   Q    Seq
                                         (sec)           (ms)        Cnt  Num
0   10.12.1.2            Eth1/1          14    00:00:10  1     50    0    89
    Version 8.0/1.2, Retrans: 0, Retries: 0, BFD state: N/A, Prefixes: 1
    Stub Peer Advertising ( CONNECTED/DIRECT SUMMARY ) Routes
    Suppressing queries
1   10.14.1.4            Eth1/2          14    00:28:26  1     50    0    86
    Version 8.0/1.2, Retrans: 1, Retries: 0, BFD state: N/A, Prefixes: 4

NX-6# show ip eigrp neighbors detail
IP-EIGRP neighbors for process 100 VRF default
H   Address              Interface       Hold  Uptime    SRTT  RTO   Q    Seq
```

```
                                     (sec)         (ms)          Cnt Num
1   10.36.1.3              Eth1/1    13   00:23:18   1     50    0    76
    Version 8.0/1.2, Retrans: 0, Retries: 0, BFD state: N/A, Prefixes: 2
0   10.56.1.5              Eth1/2    13   00:23:18   1     50    0    38
    Version 8.0/1.2, Retrans: 0, Retries: 0, BFD state: N/A, Prefixes: 5
```

可以看出，NX-1 检测到 10.12.1.2 对等体（NX-2）已经配置了 EIGRP 末梢功能。末梢功能将阻止 NX-2 将 E1/2 接口学到的路由宣告给 E1/1 接口，反之亦然。

接下来需要验证并删除 EIGRP 配置。EIGRP 命令 **eigrp stub** {**direct** | **leak-map** *leak-map-name* | **receive-only** | **redistributed** | **static** | **summary**}可以在交换机上配置末梢功能，如例 7-35 所示。只要删除了末梢配置，就可以让路由穿过 NX-2 进行宣告。

> **注**：选项 **receive-only** 不能与其他 EIGRP 末梢选项结合使用。如果 EIGRP 路由器配置了 **receive-only** 末梢选项，那么在网络设计时一定要注意连接在 EIGRP 路由器上的网络的双向连接，确保路由器知道如何发送回程流量。

例 7-35 EIGRP 末梢配置

```
NX-2# show run eigrp
router eigrp NXOS
  autonomous-system 100
  stub

interface Ethernet1/1
  ip router eigrp NXOS

interface Ethernet1/2
  ip router eigrp NXOS
```

> **注**：截至本书写作之时，NX-OS 仅在 Enterprise Services 许可中全面支持 EIGRP 的所有功能，而面向特定平台的 LAN Base 许可仅支持 EIGRP 末梢功能。在实际使用时，必须仔细检查当前的许可选项，否则很可能会出现故障问题。

7.3.3 最大跳数

EIGRP 是一种混合型距离矢量路由协议，可以跟踪跳数。

除按前缀进行过滤之外，EIGRP 还支持按跳数进行过滤。在默认情况下，EIGRP 路由器仅允许将最多 100 跳的路由安装到 EIGRP 拓扑表中。如果路由的 EIGRP 跳数路径属性大于 100，那么就不会将这类路由安装到 EIGRP 拓扑表中。可以通过 EIGRP 配置命令 **metric maximum-hops** *hop-count* 更改跳数限制。

与前面一样，NX-1 的路由表标记了变更信息，看起来有些路径已经消失了。例 7-36 给出了 NX-1 和 NX-6 的路由表信息。

例 7-36 NX-1 和 NX-6 的路由表

```
NX-1# show ip route eigrp
! Output omitted for brevity
10.6.6.0/24, ubest/mbest: 1/0
    *via 10.14.1.4, Eth1/2, [90/1280], 00:29:28, eigrp-NXOS, internal
10.23.1.0/24, ubest/mbest: 1/0
    *via 10.12.1.2, Eth1/1, [90/768], 00:00:50, eigrp-NXOS, internal
10.36.1.0/24, ubest/mbest: 1/0
```

```
         *via 10.12.1.2, Eth1/1, [90/1024], 00:00:50, eigrp-NXOS, internal
10.45.1.0/24, ubest/mbest: 1/0
         *via 10.14.1.4, Eth1/2, [90/768], 00:34:40, eigrp-NXOS, internal
10.56.1.0/24, ubest/mbest: 1/0
         *via 10.14.1.4, Eth1/2, [90/1024], 00:34:37, eigrp-NXOS, internal
NX-6# show ip route eigrp
10.1.1.0/24, ubest/mbest: 2/0
         *via 10.36.1.3, Eth1/1, [90/1280], 00:01:20, eigrp-NXOS, internal
         *via 10.56.1.5, Eth1/2, [90/1280], 00:29:59, eigrp-NXOS, internal
10.11.11.0/24, ubest/mbest: 2/0
         *via 10.36.1.3, Eth1/1, [90/1280], 00:01:20, eigrp-NXOS, internal
         *via 10.56.1.5, Eth1/2, [90/1280], 00:29:59, eigrp-NXOS, internal
10.12.1.0/24, ubest/mbest: 1/0
         *via 10.36.1.3, Eth1/1, [90/1024], 00:01:20, eigrp-NXOS, internal
10.14.1.0/24, ubest/mbest: 1/0
         *via 10.56.1.5, Eth1/2, [90/1024], 00:29:59, eigrp-NXOS, internal
10.23.1.0/24, ubest/mbest: 1/0
         *via 10.36.1.3, Eth1/1, [90/768], 00:29:59, eigrp-NXOS, internal
10.45.1.0/24, ubest/mbest: 1/0
         *via 10.56.1.5, Eth1/2, [90/768], 00:29:59, eigrp-NXOS, internal
```

可以看出，NX-1 缺失了图中上面那条去往网络 10.6.6.0/24 的路径（NX-1→NX-2→NX-3→NX-6），而 NX-6 拥有去往网络 10.1.1.0/24 和 10.11.11.0/24 的全部路径。这就意味着存在双向连接且未部署 EIGRP 末梢功能。此外，上述信息还表明所有路径均存在 EIGRP 邻接关系，因而执行了某种形式的过滤或路径控制操作。

检查 NX-1、NX-2、NX-3 和 NX-6 的 EIGRP 配置有助于确定故障根源。可以看出，NX-2 配置了最大跳数功能并将其设置为 1（见例 7-37），这样一来，会平等看待所有路由（从 NX-6 的角度来看）。因此，删除 **metric maximum-hops** 命令或者将最大跳数值更改为正常值，就能让路由表恢复正常。

例 7-37 配置了最大跳数功能的配置信息

```
NX-2# show run eigrp
! Output omitted for brevity
router eigrp NXOS
  autonomous-system 100
  metric maximum-hops 1

interface Ethernet1/1
  ip router eigrp NXOS

interface Ethernet1/2
  ip router eigrp NXOS
```

7.3.4 分发列表

EIGRP 可以为单个接口应用分发列表（distribute list）来过滤路由。分发列表的配置命令是 **ip distribute-list eigrp** *process-tag* {**route-map** *route-map-name* | **prefix-list** *prefix-list-name* {**in** | **out**}，相应的配置规则如下。

- 如果将方向设置为 **in**，那么入站过滤机制就会在执行 DUAL 处理之前删除路由，因而不会将路由安装到 RIB 中。
- 如果将方向设置为 **out**，那么就会在出站路由宣告期间执行过滤操作，因而 DUAL 会处理路由并将路由安装到接收路由器的本地 RIB 中。

- 宣告或接收所有通过前缀列表的路由,过滤所有未通过前缀列表的路由。
- 除了指定前缀列表,还可以指定路由映射来修改路径属性(还有过滤功能)。

网络工程师检查 NX-6 的路由表之后发现,路由 10.1.1.0/24 少了一条路径,而路由 10.11.11.0/24 则拥有两条路径。例 7-38 显示了 NX-6 的当前路由表信息,与例 7-30 显示的原始路由表有所不同。

例 7-38　仅一条路由缺失路径

```
NX-6# show ip route eigrp-NXOS
IP Route Table for VRF "default"
! Output omitted for brevity
10.1.1.0/24, ubest/mbest: 1/0
    *via 10.56.1.5, Eth1/2, [90/1280], 00:05:41, eigrp-NXOS, internal
10.11.11.0/24, ubest/mbest: 2/0
    *via 10.36.1.3, Eth1/1, [90/1280], 00:41:15, eigrp-NXOS, internal
    *via 10.56.1.5, Eth1/2, [90/1280], 00:05:41, eigrp-NXOS, internal
..
10.45.1.0/24, ubest/mbest: 1/0
    *via 10.56.1.5, Eth1/2, [90/768], 00:05:43, eigrp-NXOS, internal
```

网络 10.11.11.0/24 有两条路径,且连接在同一台 Nexus 交换机(NX-1)上,这表明启用了某种形式的路径控制机制。沿缺失路径检查路由表信息,就应该能够找到产生该行为的路由器。

例 7-39 给出了 NX-2 的路由表信息。在源自 NX-1 的路径看起来更优的情况下,显示了来自 NX-3 的 10.1.1.0/24 的路径。

例 7-39　路由 10.1.1.0/24 的路径已变更

```
NX-2# show ip rout eigrp-NXOS
! Output omitted for brevity

10.1.1.0/24, ubest/mbest: 1/0
    *via 10.23.1.3, Eth1/1, [90/1792], 00:00:03, eigrp-NXOS, internal
10.11.11.0/24, ubest/mbest: 1/0
    *via 10.12.1.1, Eth1/2, [90/768], 00:40:28, eigrp-NXOS, internal
..
10.56.1.0/24, ubest/mbest: 1/0
    *via 10.23.1.3, Eth1/1, [90/1024], 23:45:07, eigrp-NXOS, internal
```

这就意味着过滤操作是在 NX-1(出站方向)或 NX-2(入站方向)上进行的。例 7-40 显示了 NX-2 的配置情况,可以看出,NX-2 过滤了 10.1.1.0/24 的入站路径。请注意,Seq 5 拒绝路由 10.1.1.0/24,而 Seq 10 则允许所有其他路由。

例 7-40　分发列表配置示例

```
NX-2
interface Ethernet1/2
  description To NX-1
  ip router eigrp NXOS
  ip distribute-list eigrp NXOS prefix-list DISTRIBUTE out

ip prefix-list DISTRIBUTE seq 5 deny 10.1.1.0/24
ip prefix-list DISTRIBUTE seq 10 permit 0.0.0.0/0 le 32
```

7.3.5　偏移列表

修改 EIGRP 的路径度量可以为 EIGRP 提供流量工程能力,修改接口的时延参数可以改变路

由器接口接收和宣告的所有路由。偏移列表（offset list）允许根据更新的方向、特定前缀或者方向与前缀的组合来修改路由属性。可以在接口配置模式下通过命令 **ip offset-list eigrp** *process-tag* {**route-map** *route-map-name* | **prefix-list** *prefix-list-name* {**in** | **out**} *off-set value* 应用偏移列表，相应的配置规则如下。

- 如果将方向设置为 **in**，那么将路由添加到 EIGRP 拓扑表中时增加偏移值。
- 如果将方向设置为 **out**，那么将路由宣告给 EIGRP 邻居时为路径度量增加偏移值（偏移值由偏移列表指定）。
- 所有通过路由映射或前缀列表的路由都会将度量添加到路径属性中。

偏移值是根据附加时延计算得到的，附加时延值会被添加到 EIGRP 路径属性的现有时延中。包含偏移时延之后的路径度量计算公式如图 7-8 所示。

$$度量 + 偏移值 = 256 \times \left(\left(\frac{10^7}{最小带宽} + \frac{总时延}{10} \right) + 偏移时延 \right)$$

简化后

$$偏移值 = 256 \times 偏移时延$$

图 7-8　EIGRP 偏移值计算公式

例 7-41 显示了 NX-2 的偏移列表配置情况，可以看出，该配置将从 NX-1 收到的前缀 10.1.1.0/24 的路径度量增加了 256。

例 7-41　偏移列表配置示例

```
NX-2
interface Ethernet1/2
  description To NX-1
  ip router eigrp NXOS
  ip offset-list eigrp NXOS prefix-list OFFSET in 256

ip prefix-list OFFSET seq 5 permit 10.1.1.0/24
```

例 7-42 显示了前缀 10.1.1.0/24（由图 7-8 中的 NX-1 宣告给 NX-2）的拓扑表项，可以看出，此时的路径度量已从 768 增加到了 1024，时延也增加了 10μs。

例 7-42　10.1.1.0/24 的 EIGRP 路径属性

```
Before Offset List is Applied on NX-2
NX-2# show ip eigrp topology 10.1.1.0/24
! Output omitted for brevity
IP-EIGRP (AS 100): Topology entry for 10.1.1.0/24
  10.12.1.1 (Ethernet1/2), from 10.12.1.1, Send flag is 0x0
      Composite metric is (768/512), Route is Internal
      Vector metric:
        Minimum bandwidth is 10000000 Kbit
        Total delay is 20 microseconds
        Reliability is 255/255
After Offset List is Applied on NX-2
NX-2# show ip eigrp topology 10.1.1.0/24
! Output omitted for brevity
IP-EIGRP (AS 100): Topology entry for 10.1.1.0/24
  10.12.1.1 (Ethernet1/2), from 10.12.1.1, Send flag is 0x0
```

```
      Composite metric is (1024/768), Route is Internal
      Vector metric:
        Minimum bandwidth is 10000000 Kbit
        Total delay is 30 microseconds
```

例 7-41 增加的度量值可以通过 EIGRP 路径度量计算公式计算得到，也就是增加了 10ns 时延值。在路径上的某个节点增加度量值之后，可能并不会让后面的节点也增加相同的度量值，具体取决于该路径的带宽是否在下游方向出现了变化。

例 7-43 解释了为何度量值增加（256）仅影响来自 10.1.1.0/24 的路径，而不影响来自 10.11.11.0/24 的路径。

例 7-43　NX-6 的路径调整

```
NX-6# show ip route
! Output omitted for brevity
10.1.1.0/24, ubest/mbest: 1/0
    *via 10.56.1.5, Eth1/2, [90/1280], 00:11:15, eigrp-NXOS, internal
10.11.11.0/24, ubest/mbest: 2/0
    *via 10.36.1.3, Eth1/1, [90/1280], 00:11:12, eigrp-NXOS, internal
    *via 10.56.1.5, Eth1/2, [90/1280], 00:11:15, eigrp-NXOS, internal

NX-6# show ip eigrp topology 10.1.1.0/24
! Output omitted for brevity
IP-EIGRP (AS 100): Topology entry for 10.1.1.0/24
  10.56.1.5 (Ethernet1/2), from 10.56.1.5, Send flag is 0x0
      Composite metric is (1280/1024), Route is Internal
      Vector metric:
        Minimum bandwidth is 10000000 Kbit
        Total delay is 40 microseconds
      ..
  10.36.1.3 (Ethernet1/1), from 10.36.1.3, Send flag is 0x0
      Composite metric is (1536/1280), Route is Internal
      Vector metric:
        Minimum bandwidth is 10000000 Kbit
        Total delay is 50 microseconds

NX-6# show ip eigrp topology 10.11.11.0/24
! Output omitted for brevity
IP-EIGRP (AS 100): Topology entry for 10.11.11.0/24
  10.36.1.3 (Ethernet1/1), from 10.36.1.3, Send flag is 0x0
      Composite metric is (1280/1024), Route is Internal
      Vector metric:
        Minimum bandwidth is 10000000 Kbit
        Total delay is 40 microseconds
  10.56.1.5 (Ethernet1/2), from 10.56.1.5, Send flag is 0x0
      Composite metric is (1280/1024), Route is Internal
      Vector metric:
        Minimum bandwidth is 10000000 Kbit
        Total delay is 40 microseconds
```

7.3.6　基于接口的设置

EIGRP 根据接口协商的连接速率自动为接口分配时延和带宽。在某些情况下，EIGRP 也会根据流量工程要求调整这些数值。如果实际的流量流与预期不符，那么就可以检查如下 EIGRP 配置命令。

- **ip bandwidth eigrp** *process-tag bandwidth*，该命令可以在计算最小带宽路径属性时更改 EIGRP 进程使用的带宽值。
- **ip delay eigrp** *process-tag delay-value* [*picoseconds*]，该命令可以在增加总时延路径属性时改变 EIGRP 进程所使用的接口时延。

如果使用了这些命令，那么从该关联接口收到或通告的所有前缀都会受到影响，而偏移列表则可以有针对性地选择特定前缀。

> 注：如前所述，使用路由映射时，可以通过 EIGRP 偏移列表或分发列表来控制路径度量。在这两种情况下，EIGRP 都会通过总时延路径属性来修改度量值。但是，如果为 EIGRP 度量增加或减小的数值较小，那么 IOS 路由器可能会出现精度无效问题，因为 IOS 路由器使用的是整数函数，这些设备可能无法登记数值 4007 与 4008 之间的差异，但 Nexus 交换机可以做到。

一般来说，如果数值凑整后对路径决策没有产生影响，那么就需要使用较大的数值，以确保能够有效影响远离变更地点的路径决策。

7.3.7 重分发

每种路由协议都有各自不同的方法来计算路由的最佳路径。例如，EIGRP 可以使用带宽、时延、负载和可靠性等参数来计算最佳路径，而 OSPF 则主要通过路径度量来计算 SPT（Shortest Path First Tree，最短路径优先树），OSPF 无法使用 EIGRP 的路径属性来计算 SPF 树，EIGRP 也无法仅使用总路径量度来运行 DUAL（Diffusing Update Algorithm，扩散更新算法）。目的协议必须提供与目的协议相关的度量，以便目的协议可以为重分发路由计算最佳路径。

可以通过命令 **redistribute** [**bgp** *asn* | **direct** | **eigrp** *process-tag* | **isis** *process-tag* | **ospf** *process-tag* | **rip** *process-tag* | **static**] **route-map** *route-map-name* 将路由重分发到 EIGRP 中。对于 Nexus 交换机来说，路由映射是重分发进程的一部分。

每种协议在重分发时都会提供一个种子度量，以允许目的协议计算最佳路径。EIGRP 在设置种子度量时使用的逻辑规则如下。

- Nexus 交换机的默认种子度量：最小带宽为 100000 kbit/s，时延为 1000μs，可靠性为 255，负载为 1，MTU（Maximum Transmission Unit，最大传输单元）为 1492。
- 默认种子度量并不是必需的，而且在 EIGRP 进程之间进行重分发时会保留路径属性。

> 注：Nexus 交换机的默认种子度量与 IOS 和 IOS XR 路由器不同，它们使用的默认种子度量值为无穷大，将种子度量设置为无穷大可以阻止将路由安装到拓扑表中。

在需要的情况下，可以按需修改默认种子度量的带宽、负载、时延、可靠性和 MTU 数值。此时，可以通过 EIGRP 进程命令 **metric weights** *tos bandwidth delay reliability load mtu* 修改重分发到进程中的所有路由的相关参数，也可以通过命令 **set metric weights** *bandwidth delay reliability load mtu* 在路由映射中进行选择性修改。

例 7-44 给出了重分发进程的配置示例。此时的 NX-1 重分发了 10.1.1.0/24 和 10.11.11.0/24 的直连路由，而不是通过 EIGRP 路由协议宣告这些路由。请注意，该路由映射可以是一条非常简单的不包含任何条件匹配项的 **permit** 语句。

例 7-44 配置 NX-1 重分发

```
router eigrp NXOS
 autonomous-system 100
 redistribute direct route-map REDIST
```

```
!
route-map REDIST permit 10
```

例 7-45 显示了 NX-2 的路由表信息。路由 10.1.1.0/24 和 10.11.11.0/24 被标记为外部路由，AD 被设置为 170。此外，本例还显示了拓扑表信息，显示了 EIGRP 的路径度量。请注意，EIGRP 包含了源端协议（Connected）的属性信息，该信息是 NX-1 发送的路由宣告的一部分。

例 7-45　NX-2 的外部路由

```
NX-2# show ip route eigrp-NXOS
! Output omitted for brevity
10.1.1.0/24, ubest/mbest: 1/0
    *via 10.12.1.1, Eth1/2, [170/51456], 00:00:07, eigrp-NXOS, external
10.11.11.0/24, ubest/mbest: 1/0
    *via 10.12.1.1, Eth1/2, [170/51456], 00:00:07, eigrp-NXOS, external
10.14.1.0/24, ubest/mbest: 1/0
    *via 10.12.1.1, Eth1/2, [90/768], 00:33:45, eigrp-NXOS, internal
```

```
NX-2# show ip eigrp topology 10.1.1.0/24
IP-EIGRP (AS 100): Topology entry for 10.1.1.0/24
  State is Passive, Query origin flag is 1, 1 Successor(s), FD is 51456
  Routing Descriptor Blocks:
  10.12.1.1 (Ethernet1/2), from 10.12.1.1, Send flag is 0x0
      Composite metric is (51456/51200), Route is External
      Vector metric:
        Minimum bandwidth is 100000 Kbit
        Total delay is 1010 microseconds
        Reliability is 255/255
        Load is 1/255
        Minimum MTU is 1492
        Hop count is 1
        Internal tag is 0
      External data:
        Originating router is 10.1.1.1
        AS number of route is 0
        External protocol is Connected, external metric is 0
        Administrator tag is 0 (0x00000000)
```

注： EIGRP 利用 router-id 来预防外部路由环路，EIGRP 路由器不会安装包含与自身 router-id 相同的外部路由。为了避免出现外部 EIGRP 路由问题，必须确保 EIGRP AS 中的所有设备的 router-id 的唯一性。

7.3.8 传统度量与宽度量

最初的 EIGRP 规范以 10μs 为单位测量时延，以每秒千字节为单位测量带宽，因而对于高速接口来说扩展性很差。从前面的表 7-2 可以看出，千兆以太网和万兆以太网接口的时延完全相同。

例 7-46 给出了常见 LAN 接口速率的度量计算过程。请注意，11Gbit/s 接口和 20Gbit/s 接口的度量值没有任何区别，虽然两个接口的带宽速率不同，但最后计算得到的度量值却都是 256。

例 7-46　常见 LAN 接口速率的度量计算

```
GigabitEthernet:
Scaled Bandwidth = 10,000,000 / 1000000
Scaled Delay = 10 / 10
Composite Metric = 10 + 1 * 256 = 2816

10 GigabitEthernet:
Scaled Bandwidth = 10,000,000 / 10000000
```

```
Scaled Delay = 10 / 10
Composite Metric = 1 + 1 * 256 = 512
11 GigabitEthernet:
Scaled Bandwidth = 10,000,000 / 11000000
Scaled Delay = 10 / 10
Composite Metric = 0 + 1 * 256 = 256
20 GigabitEthernet:
Scaled Bandwidth = 10,000,000 / 20000000
Scaled Delay = 10 / 10
Composite Metric = 0 + 1 * 256 = 256
```

EIGRP 还支持另一种度量模式，称为宽度量（wide metric）。该解决方案解决了大容量接口的扩展性问题。NX-OS 支持 EIGRP 宽度量模式，需要进行必要的配置才能启用该度量模式。

注：IOS 路由器仅在命名配置模式下支持 EIGRP 宽度量模式，而 IOS-XR 路由器则默认使用宽度量模式。

图 7-9 给出了 EIGRP 宽度量计算公式。可以看出，该计算公式引入了一个额外的 K 值（K_6），从而为抖动或未来其他属性的测量增加了一个扩展属性。

$$\text{宽度量} = \left[\left(K_1 \times \text{带宽} + \frac{K_2 \times \text{带宽}}{256 - \text{负荷}} + K_3 \times \text{时延} + K_6 \times \text{扩展属性} \right) \times \frac{K_5}{K_4 + \text{可靠性}} \right]$$

图 7-9 EIGRP 宽度量计算公式

正如 EIGRP 扩展到 256 以适应 IGRP 一样，EIGRP 的宽度量也被扩展到 65535 以适应高速链路。宽度量模式最大可支持高达 655Tbit/s（65535×10^7）的接口速率，而不会出现任何扩展性问题。时延指的是接口总时延，宽度量模式以皮秒（10^{-12}）为单位，而不是以微秒（10^{-6}）为单位，同样很好地支持了高速接口。图 7-10 给出了更新后的宽度量计算公式（考虑了时延和扩展性的转换）。

$$\text{宽度量} = 65535 \times \left[\left(\frac{K_1 \times 10^7}{\text{最小带宽}} + \frac{\frac{K_2 \times 10^7}{\text{最小带宽}}}{256 - \text{负荷}} + \frac{K_3 \times \text{时延}}{10^{-6}} + K_6 \times \text{扩展属性} \right) \times \frac{K_5}{K_4 + \text{可靠性}} \right]$$

图 7-10 更新后的 EIGRP 宽度量公式

EIGRP 的宽度量在设计之初就考虑了后向兼容性。EIGRP 的宽度量将 K_1 和 K_3 设置为 1，将 K_2、K_4、K_5 和 K_6 设置为 0，即可实现后向兼容（因为此时的 K 值度量计算公式与传统度量计算公式完全匹配）。只要 $K_1 - K_5$ 相同且不设置 K_6，那么这两种度量模式就可以在路由器之间建立邻接关系。

注：可以通过命令 **show ip eigrp** 显示 Nexus 交换机使用的度量模式。如果存在 K_6 度量，那么路由器使用的就是宽度量模式。

如果对等路由器使用的是传统度量模式，那么 EIGRP 能够检测到该情况，且无缩放的度量计算公式如图 7-11 所示。

$$\text{无缩放的带宽} = \left(\frac{\text{EIGRP带宽} \times \text{EIGRP传统度量}}{\text{缩放带宽}} \right)$$

图 7-11 无缩放的 EIGRP 度量计算公式

如果路由穿越了同时部署了传统度量和宽度量的网络设备，那么这种度量转换就会导致路由明细度的下降。最终结果就是，通过宽度量对等体学到的路径始终优于通过传统度量学到的路径，从而出现次优路由问题。

再次回到图 7-7 所示的拓扑结构，看一看将 Nexus 交换机调整为 EIGRP 宽度量模式之后的效果。例 7-47 显示了网络 10.1.1.0/24（通过接口 Ethernet1/3 宣告）的路径度量变化前后的情况，请注意，虽然最小带宽没有发生变化，但此时的时延已改成以皮秒为单位。

例 7-47　NX-1 传统度量与宽度量对比

```
Classic Metrics on all other Nexus switches
! Output omitted for brevity
NX-1# show ip eigrp topology 10.1.1.0/24
IP-EIGRP (AS 100): Topology entry for 10.1.1.0/24
  State is Passive, Query origin flag is 1, 1 Successor(s), FD is 512
  Routing Descriptor Blocks:
  0.0.0.0 (loopback0), from Connected, Send flag is 0x0
      Composite metric is (512/0), Route is Internal
      Vector metric:
        Minimum bandwidth is 10000000 Kbit
        Total delay is 10 microseconds
        Reliability is 255/255
        Load is 1/255
        Minimum MTU is 1500
```

```
Wide Metrics on NX-1. Classic Metrics on all other Nexus switches
NX-1# show ip eigrp topology 10.1.1.0/24
! Output omitted for brevity
IP-EIGRP (AS 100): Topology entry for 10.1.1.0/24
  State is Passive, Query origin flag is 1, 1 Successor(s), FD is 131072
  Routing Descriptor Blocks:
  0.0.0.0 (loopback0), from Connected, Send flag is 0x0
      Composite metric is (131072/0), Route is Internal
      Vector metric:
        Minimum bandwidth is 10000000
        Total delay is 1000000
        Reliability is 255/255
        Load is 1/255
        Minimum MTU is 1500
        Hop count is 0
        Internal tag is 0
```

注： 请注意微秒（10^{-6}）与皮秒（10^{-12}）之间的差异，10 微秒等于 10000000 皮秒。从 NX-1 度量时延值可以看出，已经删除了小数位（删除了一个 0）。

例 7-48 显示了 NX-6 关于网络 10.1.1.0/24 的 EIGRP 拓扑表，此时全网配置的都是传统度量。请注意，例中的两条路径的 FD 均为 1280。

例 7-48　所有 Nexus 交换机的传统度量

```
NX-6# show ip eigrp topology 10.1.1.0/24
! Output omitted for brevity
IP-EIGRP (AS 100): Topology entry for 10.1.1.0/24
  State is Passive, Query origin flag is 1, 2 Successor(s), FD is 1280
  Routing Descriptor Blocks:
  10.36.1.3 (Ethernet1/1), from 10.36.1.3, Send flag is 0x0
      Composite metric is (1280/1024), Route is Internal
      Vector metric:
```

```
      Minimum bandwidth is 10000000 Kbit
      Total delay is 40 microseconds
      ..
      Hop count is 3
   10.56.1.5 (Ethernet1/2), from 10.56.1.5, Send flag is 0x0
      Composite metric is (1280/1024), Route is Internal
      Vector metric:
      Minimum bandwidth is 10000000 Kbit
      Total delay is 40 microseconds
      ..
      Hop count is 3
```

例 7-49 显示了 NX-6 关于网络 10.1.1.0/24 的 EIGRP 拓扑表，此时 NX-1 和 NX-2 启用了宽度量模式，其余交换机配置的都是 EIGRP 传统度量。请注意，路径 NX-1→NX-2→NX-3→NX-6 的总时延已更改为 30μs，这是因为该路径的前两跳采用的是皮秒而不是微秒来计算时延，因而减少了 10 μs。NX-6 仅将该路径用来转发流量。

例 7-49　NX-1 和 NX-2 启用了宽度量模式

```
NX-6# show ip eigrp topology 10.1.1.0/24
! Output omitted for brevity
IP-EIGRP (AS 100): Topology entry for 10.1.1.0/24
  State is Passive, Query origin flag is 1, 1 Successor(s), FD is 1024
  Routing Descriptor Blocks:
  10.36.1.3 (Ethernet1/1), from 10.36.1.3, Send flag is 0x0
      Composite metric is (1024/768), Route is Internal
      Vector metric:
      Minimum bandwidth is 10000000 Kbit
      Total delay is 30 microseconds
      ..
      Hop count is 3
  10.56.1.5 (Ethernet1/2), from 10.56.1.5, Send flag is 0x0
      Composite metric is (1280/1024), Route is Internal
      Vector metric:
      Minimum bandwidth is 10000000 Kbit
      Total delay is 40 microseconds
      ..
      Hop count is 3
```

例 7-50 显示了 NX-5 和 NX-6 关于网络 10.1.1.0/24 的 EIGRP 拓扑表，此时的 NX-1、NX-2 和 NX-3 启用了宽度量模式。请注意，路径 NX-1→NX-2→NX-3→NX-6 的总时延已经减小为 20μs，而路径 NX-1→NX-4→NX-5→NX-6 则不再满足 NX-6 的可行后继条件，因而没有显示在拓扑表中。

需要注意的是，此时 NX-5 计算出路径 NX-1→NX-2→NX-3→NX-6→NX-5 的总时延等于路径 NX-1→NX-4→NX-5。因此，如果部署了负载均衡机制，那么将有部分流量沿较长路径进行次优转发。

例 7-50　NX-1、NX-2 和 NX-3 启用了宽度量模式

```
NX-6# show ip eigrp topology 10.1.1.0/24
! Output omitted for brevity
IP-EIGRP (AS 100): Topology entry for 10.1.1.0/24
  State is Passive, Query origin flag is 1, 1 Successor(s), FD is 768
  Routing Descriptor Blocks:
  10.36.1.3 (Ethernet1/1), from 10.36.1.3, Send flag is 0x0
      Composite metric is (768/512), Route is Internal
      Vector metric:
```

```
              Minimum bandwidth is 10000000 Kbit
              Total delay is 20 microseconds
              ..
              Hop count is 3
NX-5# show ip eigrp topology 10.1.1.0/24
! Output omitted for brevity
IP-EIGRP (AS 100): Topology entry for 10.1.1.0/24
  State is Passive, Query origin flag is 1, 2 Successor(s), FD is 1024
  Routing Descriptor Blocks:
  10.45.1.4 (Ethernet1/1), from 10.45.1.4, Send flag is 0x0
      Composite metric is (1024/768), Route is Internal
      Vector metric:
          Minimum bandwidth is 10000000 Kbit
          Total delay is 30 microseconds
          ..
          Hop count is 2
  10.56.1.6 (Ethernet1/2), from 10.56.1.6, Send flag is 0x0
      Composite metric is (1024/768), Route is Internal
      Vector metric:
          Minimum bandwidth is 10000000 Kbit
          Total delay is 30 microseconds
          ..
          Hop count is 4
```

如果以非摘要方式查看 EIGRP 的接口信息,那么就可以确定传统度量或宽度量 EIGRP 邻居的数量。例 7-51 以 NX-6 为例显示了相关命令和输出结果。

例 7-51 查看传统度量 EIGRP 邻居和宽度量 EIGRP 邻居的数量

```
NX-6# show ip eigrp interfaces
! Output omitted for brevity
IP-EIGRP interfaces for process 100 VRF default

                 Xmit Queue   Mean   Pacing Time   Multicast    Pending
Interface  Peers Un/Reliable  SRTT   Un/Reliable   Flow Timer   Routes
Eth1/1       1    0/0          1      0/0            50           0
  Hello interval is 5 sec
  ..
  Classic/wide metric peers: 0/1
Eth1/2       1    0/0          1      0/0            50           0
  Hello interval is 5 sec
  ..
  Classic/wide metric peers: 1/0
```

例 7-52 显示了 NX-6 和 NX-5 关于网络 10.1.1.0/24 的 EIGRP 拓扑表信息,此时的 NX-1、NX-2、NX-3 和 NX-6 启用了宽度量模式。NX-6 仅包含宽度量路径,且时延仅以皮秒为单位进行显示。

例 7-52 NX-1、NX-2、NX-3 和 NX-6 启用了宽度量模式

```
NX-6# show ip eigrp topology 10.1.1.0/24
! Output omitted for brevity
IP-EIGRP (AS 100): Topology entry for 10.1.1.0/24
  State is Passive, Query origin flag is 1, 1 Successor(s), FD is 327680
  Routing Descriptor Blocks:
  10.36.1.3 (Ethernet1/1), from 10.36.1.3, Send flag is 0x0
      Composite metric is (327680/262144), Route is Internal
      Vector metric:
```

```
              Minimum bandwidth is 10000000
              Total delay is 4000000
              ..
              Hop count is 3
NX-5# show ip eigrp topology 10.1.1.0/24
! Output omitted for brevity
IP-EIGRP (AS 100): Topology entry for 10.1.1.0/24
  State is Passive, Query origin flag is 1, 1 Successor(s), FD is 768
  Routing Descriptor Blocks:
  10.56.1.6 (Ethernet1/2), from 10.56.1.6, Send flag is 0x0
      Composite metric is (768/512), Route is Internal
      Vector metric:
        Minimum bandwidth is 10000000 Kbit
        Total delay is 20 microseconds
        ..
        Hop count is 4
  10.45.1.4 (Ethernet1/1), from 10.45.1.4, Send flag is 0x0
      Composite metric is (1024/768), Route is Internal
      Vector metric:
        Minimum bandwidth is 10000000 Kbit
        Total delay is 30 microseconds
        ..
        Hop count is 2
```

此时，NX-5 按照无缩放的 EIGRP 度量计算公式，将路径 NX-1→NX-2→NX-3→NX-6→NX-5 计算为最佳路径，致使去往网络 10.1.1.0/24 的所有流量都采用了较长的路径。

请注意，启用 EIGRP 宽度量模式时必须进行认真规划。启用了宽度量模式之后，为了确保最佳路由，最好在整个区域或者沿同一路径去往目的端的所有设备上都启用宽度量模式。

7.4 收敛问题

如果链路出现故障且接口协议进入了关闭状态，那么连接在接口上的所有邻居也都将进入关闭状态。EIGRP 邻居进入关闭状态之后，对于该 EIGRP 邻居来说是后继路由器（上游路由器）的所有前缀都必须重新计算路径。

如果 EIGRP 检测到失去了某条路径的后继路由器，那么可行后继路由器将立即成为提供备用路由的后继路由器。Nexus 交换机则因为 EIGRP 路径度量出现了变化，向外部发送该路径的更新（Update）消息。下游交换机则为所有受影响的前缀运行各自的 DUAL 算法，以考虑该新 EIGRP 度量的影响。如果从后继交换机收到某前缀的新 EIGRP 度量，那么就可能会导致后继路由或可行后继路由出现变更。

图 7-12 给出了 NX-1 与 NX-3 之间出现链接故障后的场景示例。

此时将发生以下操作。

- NX-3 将 NX-2 通告的可行后继路径安装为后继路由。
- NX-3 向 NX-5 发送更新包，说明前缀 10.1.1.0/24 的新 RD 为 19。
- NX-5 收到 NX-3 发送的更新包，计算出去往 10.1.1.0/24 的路径 NX-3→NX-2→NX-1 的 FD 为 29。
- NX-5 将该路径与从 NX-4 收到的路径进行对比，从后者收到的路径度量为 25。
- NX-5 选择经由 NX-4 的路径作为后继路由。

例 7-53 显示了 NX-1-NX-3 链接出现故障后，NX-5 关于前缀 10.1.1.0/24 的 EIGRP 拓扑表输

出结果。

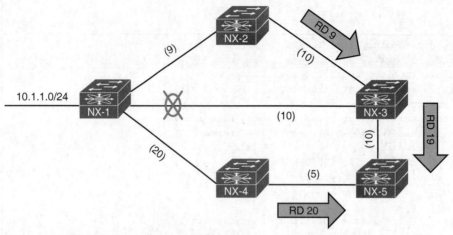

图 7-12 出现链路故障之后的 EIGRP 拓扑结构

例 7-53 网络 10.1.1.0/24 的 EIGRP 拓扑表信息

```
NX-5# show ip eigrp topology 10.1.1.0/24
IP-EIGRP (AS 100): Topology entry for 10.1.1.0/24
  State is Passive, Query origin flag is 1, 1 Successor(s), FD is 25
  Routing Descriptor Blocks:
  10.45.1.4 (Ethernet1/2), from 10.45.1.4, Send flag is 0x0
      Composite metric is (25/20), Route is Internal
      Vector metric:
      ..
      Hop count is 2
      Originating router is 192.168.1.1
  10.35.1.3 (Ethernet1/1), from 10.35.1.3, Send flag is 0x0
      Composite metric is (29/19), Route is Internal
      Vector metric:
      ..
      Hop count is 3
      Originating router is 192.168.1.1
```

如果该前缀没有可用的可行后继路由，那么 DUAL 就必须执行新的路由计算操作。EIGRP 拓扑表中的路由状态从 Passive（被动）状态更改为 Active（主动）状态。

7.4.1 主动状态查询

检测到拓扑结构发生变化的路由器将向 EIGRP 邻居发送查询消息以查找该路由的相关信息，查询消息包含了将时延设置为无穷大的网络前缀，以便其他路由器知道该前缀已经处于主动状态。路由器发送 EIGRP 查询包之后，会以逐个前缀的方式为每个邻居设置应答状态标志。

收到查询包之后，EIGRP 路由器将执行以下操作之一。

- 应答该查询包，说明路由器没有去往该前缀的路由。
- 如果查询包不是来自该路由的后继路由器，那么接收端路由器就会检测到设置为无穷大的时延，但是由于不是来自后继路由器而加以忽略，接收端路由器将使用该路由的 EIGRP 属性进行应答。
- 如果查询包来自该路由的后继路由器，那么接收端路由器就会检测到设置为无穷大的时延，在 EIGRP 拓扑结构中将该前缀设置为主动状态，并向所有下游 EIGRP 邻居发送该路由的查询包。

该查询过程将在路由器之间不断持续，直至路由器建立查询边界。如果某路由器不再将该前缀标记为主动状态，那么就表明建立了查询边界，意味着该路由器将以如下方式响应查询。

- 没有去往该前缀的路由。
- 以 EIGRP 属性进行应答（因为该查询不是来自后继路由器）。

如果路由器收到了所有下游查询的应答消息，那么就表明已经完成了 DUAL 算法，从而将路由更改为被动状态，并向所有向其发送查询包的上游路由器发送应答包。收到特定前缀的应答包之后，就会为该邻居和前缀标记应答包。此后，将继续沿上游方向执行应答进程，直至第一台发送查询包的路由器收到应答为止。

图 7-13 给出了 NX-1 与 NX-2 之间出现链路故障之后的拓扑结构示例。

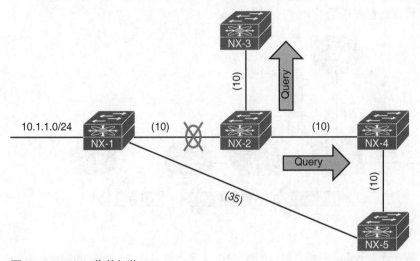

图 7-13　EIGRP 收敛拓扑

从 NX-2 的角度来看，NX-2 计算去往网络 10.1.1.0/24 的新路由的步骤如下。

- **第 1 步**：NX-2 检测到链路故障。NX-2 没有该路由的可行后继路由，从而将前缀 10.1.1.0/24 设置为主动状态，并向 NX-3 和 NX-4 发送查询包。
- **第 2 步**：NX-3 收到 NX-2 发送的查询包并处理设置为无穷大的时延字段。NX-3 没有其他 EIGRP 邻居，因而向 NX-2 发送应答消息，说明不存在其他路由。NX-4 收到 NX-2 发送的查询包并处理设置为无穷大的时延字段，由于收到查询包的是后继路由器，且不存在该前缀的可行后继路由，因而 NX-4 将该路由标记为主动路由，并向 NX-5 发送查询包。
- **第 3 步**：NX-5 收到 NX-4 发送的查询包并检测到时延字段被设置为无穷大。由于收到查询包的是非后继路由器，且后继路由器位于其他接口上，因而 NX-5 通过适当的 EIGRP 属性向 NX-4 发送关于网络 10.4.4.0/24 的应答消息。
- **第 4 步**：NX-4 收到 NX-5 发送的应答消息之后，确认该应答包并计算新路径。由于这是 NX-4 上最后一个未完成的查询包，因而 NX-4 将该前缀设置为被动状态。在满足所有查询包的情况下，NX-4 以新的 EIGRP 度量响应 NX-2 的查询。
- **第 5 步**：NX-2 收到 NX-4 发送的应答消息之后，确认该应答包并计算新路径。由于这是 NX-2 最后一个未完成的查询数据包，因而 NX-2 将该前缀设置为被动状态。

7.4.2　SIA

DUAL 在快速查找无环路径方面非常有效，通常在几秒之内就能找到备用路径。有时 EIGRP

查询可能会因丢包、邻居缓慢或跳数过多等问题出现延时。EIGRP 最多可以为应答消息等待主动定时器的一半时间（默认为 90s），如果路由器在 90s 内未收到响应消息，那么发送端路由器就会向未响应的 EIGRP 邻居发送 SIA-Query（SIA[Stuck-in-Active，卡在主动状态]查询）消息。

收到 SIA-Query 消息之后，路由器应该在 90s 之内响应 SIA-Reply(SIA 应答) 消息，SIA-Reply 消息包含了路由信息或提供与该查询过程相关的信息。如果路由器在主动定时器超时之前都未能响应 SIA-Query 消息，那么 EIGRP 就会认为该路由器处于 SIA 状态。如果将某邻居声明为 SIA 状态，那么 DUAL 就会删除来自该邻居的所有路由，并将这种情况视为该邻居对所有路由均返回了不可达消息。可以通过命令 **show ip eigrp topology active** 查看所有主动状态查询情况。

图 7-14 所示拓扑结构中的 NX-1 与 NX-2 之间出现了链路故障，NX-2 向 NX-4 和 NX-3 发送关于网络 10.1.1.0/24 和 10.12.1.0/24 的查询包，NX-4 向 NX-2 发送了应答包，而 NX-3 则将查询包发送给了 R5，R5 又将查询包发送给了 R6。

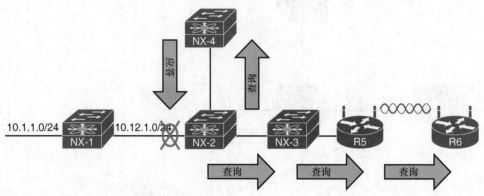

图 7-14　EIGRP SIA 拓扑结构

网络工程师发现链路故障的 syslog 消息之后，立即在 NX-2 上运行 **show ip eigrp topology active** 命令，得到例 7-54 所示的输出结果。

例 7-54　SIA 定时器的输出结果

```
NX-2# show ip eigrp topology active
IP-EIGRP Topology Table for AS(100)/ID(10.23.1.2) VRF default

Codes: P - Passive, A - Active, U - Update, Q - Query, R - Reply,
       r - reply Status, s - sia Status

A 10.1.1.0/24, 0 successors, FD is 768
    1 replies, active 00:00:08, query-origin: Local origin
        via 10.12.1.1 (Infinity/Infinity), Ethernet1/2
    Remaining replies:
        via 10.23.1.3, r, Ethernet1/1
A 10.12.1.0/24, 1 successors, FD is Inaccessible
    1 replies, active 00:00:08, query-origin: Local origin
        via Connected (Infinity/Infinity), Ethernet1/2
    Remaining replies:
        via 10.23.1.3, r, Ethernet1/1
```

10.23.1.3 旁边的 "r" 表示 NX-2 仍在等待 NX-3 的应答消息。NX-1 被注册为中断状态且路径被设置为无穷大。接下来在 NX-3 上执行 **show ip eigrp topology** 命令，可以看出 NX-3 正在等待 R5 的响应消息。接下来在 R5 上运行该命令，可以看出 R5 正在等待 R6 的响应消息。在 R6 上执行该命令后没有显示任何主动前缀，表明 R6 从未收到 R5 发送的查询消息。因此，R5 发送的查询消息可能被无线网络连接丢弃了。

过了 90s 的时间窗口之后，交换机 NX-2 就会发出 SIA-Query 消息（查看 EIGRP 流量计数器即可发现）。例 7-55 显示了 90s 时间窗口前后的流量计数器信息。

例 7-55 通过 EIGRP 流量计数器查看 SIA 查询和应答情况

```
Before 90 second window
NX-2# show ip eigrp traffic
IP-EIGRP Traffic Statistics for AS 100 VRF default
  Hellos sent/received: 65/64
  Updates sent/received: 0/0
  Queries sent/received: 2/0
  Replies sent/received: 0/1
  Acks sent/received: 1/2
  Input queue high water mark 3, 0 drops
  SIA-Queries sent/received: 0/0
  SIA-Replies sent/received: 0/0
  Hello Process ID: (no process)
  PDM Process ID: (no process)

After 90 second window
NX-2# show ip eigrp traffic
IP-EIGRP Traffic Statistics for AS 100 VRF default
  Hellos sent/received: 115/115
  Updates sent/received: 7/6
  Queries sent/received: 2/0
  Replies sent/received: 0/1
  Acks sent/received: 7/9
  Input queue high water mark 3, 0 drops
  SIA-Queries sent/received: 2/0
  SIA-Replies sent/received: 0/2
  Hello Process ID: (no process)
  PDM Process ID: (no process)
```

例 7-56 显示了收到 SIA-Reply 消息之后的 EIGRP 拓扑表信息。此后，SIA 消息就出现在系统日志中，且 EIGRP 对等状态被重置。

例 7-56 收到 SIA-Reply 消息之后的拓扑表

```
NX-2# show ip eigrp topology active
IP-EIGRP Topology Table for AS(100)/ID(10.23.1.2) VRF default

Codes: P - Passive, A - Active, U - Update, Q - Query, R - Reply,
       r - reply Status, s - sia Status

A 10.1.1.0/24, 0 successors, FD is 768
    1 replies, active 00:04:40, query-origin: Local origin, retries(3)
        via 10.12.1.1 (Infinity/Infinity), Ethernet1/2
        via 10.23.1.3 (Infinity/Infinity), r, Ethernet1/1, serno 112
A 10.12.1. 0/24, 1 successors, FD is Inaccessible
    1 replies, active 00:04:40, query-origin: Local origin, retries(3)
        via Connected (Infinity/Infinity), Ethernet1/2
        via 10.23.1.3 (Infinity/Infinity), r, Ethernet1/1, serno 111

NX-2
03:57:41 NX-2 %EIGRP-3-SIA_DUAL:  eigrp-NXOS [8394] (default-base) Route
  10.12.1.0/24 stuck-in-active state in IP-EIGRP(0) 100. Cleaning up
03:57:41 NX-2 %EIGRP-5-NBRCHANGE_DUAL:  eigrp-NXOS [8394] (default-base) IP-EIGRP(0)
```

```
100: Neighbor 10.23.1.3 (Ethernet1/1) is down: stuck in active
03:57:42 NX-2 %EIGRP-5-NBRCHANGE_DUAL: eigrp-NXOS [8394] (default-base) IP-EIGRP(0)
  100: Neighbor 10.23.1.3 (Ethernet1/1) is up: new adjacency
```

路由器因繁忙而使无效路由卡在路由表中，可能会让人感到很沮丧。解决这个问题的办法有两种。

- 在 EIGRP 进程下，使用命令 **timers active-time** {*disabled* | *1-65535_minutes*}将主动定时器调整为其他值。
- 在网络设计方案中使用网络汇总机制。EIGRP 汇总机制对于创建查询边界以减少查询操作的执行范围来说非常有用。

通过 **show ip eigrp** 命令查看 EIGRP 进程即可显示主动定时器的相关信息，SIA 定时器信息显示在 *Active interval* 字段中，例 7-57 显示主动定时器的间隔为 3min。

例 7-57　SIA 定时器的输出结果

```
NX-2# show ip eigrp
! Output omitted for brevity
IP-EIGRP AS 100 ID 10.23.1.2 VRF default
 ..
 Max paths: 8
 Active Interval: 3 minute(s)
```

7.5　本章小结

本章讨论了 EIGRP 的常见故障识别与排查技术。

排查 EIGRP 与其他设备的邻接关系故障时，必须确保以下参数匹配。

- 接口必须处于主动状态。
- 设备之间必须存在使用主用子网的连接。
- 自治系统号。
- 度量计算公式的 *K* 值。
- 认证参数。

EIGRP 是一种距离矢量路由协议，根据从下游邻居收到的信息创建拓扑结构图。在排查 EIGRP 的次优路径选择或路由丢失故障时，最好从目的端开始，逐步朝路由源端进行排查。在排查每一跳的故障过程中应检查以下信息，以确定是否对路径信息进行了显式修改。

- 与面向源端的设备之间的 EIGRP 邻接关系。
- EIGRP 末梢功能特性的启用情况。
- 基于跳数或分发列表的过滤机制。
- 度量控制机制，包括通过偏移列表来增加路径度量或者显式配置接口的带宽或时延。
- 路由器是否使用了两种不同的进程来包含上游和下游路由接口。此时，需要在不同的进程之间相互重分发路由。
- EIGRP 宽度量规划不当，未考虑高速接口的比例系数问题。

EIGRP 的 DUAL 算法非常智能，克服了大多数矢量路由协议所面临的难题。DUAL 提供了极好的快速收敛能力，但是如果远程路由器无响应，那么有时也可能会出现收敛问题。可以通过减小 SIA 定时器间隔或路由汇总机制来缩短路由收敛时间。

第 8 章

OSPF 故障排查

本章主要讨论如下主题。
- OSPF 基础知识。
- OSPF 邻居邻接关系故障排查。
- OSPF 路由丢失故障排查。
- OSPF 路径选择故障排查。

OSPF（Open Shortest Path First，开放最短路径优先）是一种链路状态路由协议，可以为每台路由器提供所有目的网络的完整视图。网络中的每台路由器都使用完整的网络视图计算最佳、最短、无环路径。

本章将重点讨论 OSPF 邻居邻接关系、路径选择和路由丢失等故障的识别与排查问题。

8.1 OSPF 基础知识

OSPF 将包含链路状态和度量信息的 LSA（Link-State Advertisement，链路状态宣告）宣告给邻居路由器，路由器将收到的 LSA 存储在被称为 LSDB（Link-State DataBase，链路状态数据库）的本地数据库中，然后将收到的 LSA 宣告给邻居路由器。这些相同的 LSA 会在整个 OSPF 域中泛洪（就像宣告路由器所宣告的那样）。LSDB 可以提供全网拓扑结构，实质上就是为路由器提供了完整网络视图。

所有的路由器都运行 Dijkstra SPF（Shortest Path First，最短路径优先）算法，以构造基于最短路径的无环拓扑。每台路由器将自己视为树的顶部，且树中包含了 OSPF 域内的所有网络目的端。虽然每台 OSPF 路由器的 SPT（SPF Tree，SPF 树）都各不相同，但区域中的所有 OSPF 路由器用于计算 SPT 的 LSDB 都相同。

8.1.1 路由器间通信

OSPF 运行在自己的协议（89）上，并尽可能通过多播方式减少不必要的流量。OSPF 的两个组播地址如下。
- AllSPFRouters：IPv4 地址 224.0.0.5 或 MAC 地址 01:00:5E:00:00:05。
  所有运行 OSPF 的路由器都应该能够接收这些数据包。
- AllDRouters：IPv4 地址 224.0.0.6 或 MAC 地址 01:00:5E:00:00:06。
  与指派路由器（Designated Router）进行通信时使用该地址。

OSPF 协议定义了 5 类数据包。表 8-1 列出了这些 OSPF 数据包的类型及功能说明。

表 8-1　　　　　　　　　　　　　　　OSPF 数据包类型

类型	数据包名称	功能说明
1	Hello	**发现和维护邻居** 所有 OSPF 接口都周期性地向外发送该数据包，以发现新邻居并确保其他邻居仍然在线
2	DBD 或 DDP（Database Description，数据库描述）	**汇总数据库的内容** 首次建立 OSPF 邻接关系时交换该数据包，这些数据包可以描述 LSDB 的内容
3	LSR（Link State Request，链路状态请求）	**数据库下载** 如果路由器认为其部分 LSDB 已经过时，那么就会利用该数据包请求邻居的部分数据库
4	LSU（Link State Update，链路状态更新）	**数据库更新** 是特定网络链路的显式 LSA，通常在直接响应 LSR 时发送该数据包
5	链路状态确认（Link State Ack）	**泛洪确认** 响应 LSA 泛洪时发送这些数据包，从而为 LSA 泛洪提供一种可靠的传输特性

8.1.2　OSPF Hello 包

OSPF Hello 包负责发现和维护邻居。在大多数情况下，路由器会将 Hello 包发送给 AllSPFRouters 地址（224.0.0.5）。表 8-2 列出了 OSPF Hello 包的部分数据信息。

表 8-2　　　　　　　　　　　　　　　OSPF Hello 包字段

数据字段	描述
RID（Router-ID，路由器 ID）	OSPF 区域中唯一的 32 比特 ID
认证选项	允许 OSPF 路由器之间进行安全通信，以防范恶意行为。支持的认证选项包括 None（无）、Clear Text（明文）或 MD5
Area-ID（区域 ID）	OSPF 接口所属的 OSPF 区域，Area-ID 是一个 32 比特数字，可以写成点分十进制方式（0.0.1.0）或十进制方式（256）
接口地址掩码	发送 Hello 包的接口的主 IP 的网络掩码
接口优先级	为指派路由器选举进程的路由器接口优先级
Hello 间隔	路由器在接口上发送 Hello 包的时间间隔（以秒为单位）
失效（Dead）间隔	路由器在宣布邻居路由器失效之前等待邻居路由器向其发送 Hello 包的时间间隔（以秒为单位）
指派路由器和备用指派路由器	该网络链路的指派路由器和备用指派路由器的 IP 地址
主动邻居	网段上看到的 OSPF 邻居列表（必须在失效间隔内收到了该邻居的 Hello 包）

8.1.3　邻居状态

OSPF 邻居指的是共享公共的已启用 OSPF 的网络链路的路由器。OSPF 路由器通过 OSPF Hello 包发现其他邻居。相邻的 OSPF 邻居指的是在两个邻居之间共同同步的 OSPF 数据库的 OSPF 邻居。

每个 OSPF 进程都会为相邻的 OSPF 邻居以及每台路由器的状态维护一张表格。表 8-3 列出了 OSPF 邻居的状态信息。

表 8-3　　　　　　　　　　　　　　　OSPF 邻居状态

状态	描述
Down（中断）	邻居关系的初始状态，表示没有从该路由器收到任何 LSA
Attempt（尝试）	该状态与不支持广播且需要显式邻居配置的 NBMA（Nonbroadcast Multiple Access，非广播多路接入）网络相关，该状态表示未收到最近信息，但路由器仍在尝试进行通信
Init（初始）	虽然从其他路由器收到了 Hello 包，但仍未建立双向通信

续表

状态	描述
2-Way（双向）	已经建立双向通信，如果需要指派路由器或备用指派路由器，那么就在该状态进行选举
ExStart（启动前）	该状态是建立邻接关系的首个状态，路由器会为 LSDB 的同步操作选择主从设备
Exchange（交换）	该状态下的路由器会通过 DBD 包交换链路状态信息
Loading（加载）	该状态下的路由器会向邻居发送 LSR 包，以请求更多在 Exchange 状态下发现（但未收到）的最新的 LSA
Full（完全邻接）	邻居路由器已经完全邻接

8.1.4 指派路由器

以太网（LAN）等多路接入网络允许在同一个网段上存在两台以上路由器。随着网段上的路由器数量的增加，可能会导致 OSPF 出现扩展性问题。额外的路由器会在网段上泛洪更多的 LSA，而且随着 OSPF 邻居邻接关系的增多，OSPF 流量也将变得过量。如果 6 台路由器同时共享同一个多路接入网络，那么就会在该网络上形成 15 个 OSPF 邻接关系以及 15 次数据库泛洪。

为了克服这种效率低下问题，多路接入网络使用了 DR（Designated Router，指派路由器）的概念。DR 可以大大减少多路接入网段上的 OSPF 邻接关系数量，此时的路由器只要与 DR 建立完全的 OSPF 邻接关系即可，彼此之间无须建立邻接关系。此后，由 DR 负责在出现更新后将更新消息泛洪给网段上的所有 OSPF 路由器。

如果 DR 出现了故障，那么 OSPF 就必须建立新的邻接关系，调用所有新的 LSA，而且可能会导致潜在的路由短暂丢失问题。如果 DR 出现了故障，那么 BDR（Backup Designated Router，备用指派路由器）就会成为新的 DR，此后还要选举新的 BDR。为了实现切换时间的最小化，BDR 也要与网段上的所有 OSPF 路由器建立完全的 OSPF 邻接关系。

DROther 是已启用 DR 的网段上既不是 DR 也不是 BDR 的路由器。

注：OSPF 优先级最高的邻居将被选为 DR 和 BDR，如果优先级相同，那么就选择 RID（Router ID，路由器 ID）较大的邻居。可以通过命令 **ip ospf priority** *0-255* 在接口上设置 OSPF 优先级，如果将优先级设置为零，那么就可以防止该路由器成为网段的 DR。

8.1.5 区域

OSPF 通过在路由域中使用多个 OSPF 区域（Area）来实现路由表的可扩展性。每个 OSPF 区域都包含一组连接在一起的网络和主机。OSPF 采用两层分层结构，Area 0 是一个特殊区域，称为骨干区域，其他 OSPF 区域都必须连接到 Area 0 上，也就是说，Area 0 提供了非骨干区域之间的传输连接。非骨干区域将路由发布到骨干区域中，然后由骨干区域将路由发布到其他非骨干区域中。

虽然实际的区域拓扑对于区域外部来说不可见，但完全能够提供到区域外部路由器的连接性。这就意味着区域外部的路由器没有该区域的完整拓扑结构图，这样就能大大减少区域内的 OSPF 网络流量。将 OSPF 路由域划分成多个区域之后，并非所有的 OSPF 路由器都拥有相同的 LSDB，但同一区域内的所有路由器都拥有相同的区域 LSDB。路由流量的减少会大大降低对路由器内存及其他资源的消耗，从而实现良好的扩展能力。

ABR（Area Border Router，区域边界路由器）是连接 Area 0 与其他 OSPF 区域的 OSPF 路由器。ABR 负责将一个 OSPF 区域中的路由宣告并注入另一个 OSPF 区域中。每台 ABR 都需要加入 Area 0，否则无法将路由宣告到其他区域中。

如果某路由器负责将外部路由重分发到 OSPF 域中，那么就将该路由器称为 ASBR（Autonomous System Boundary Router，自治系统边界路由器）。ASBR 可以是任意 OSPF 路由器，且 ASBR 功能与 ABR 功能相独立。

8.1.6 LSA

理解 OSPF 拓扑表的构建方式以及各类 LSA（Link State Advertisement，链路状态宣告）的作用方式，有助于识别和排查 OSPF 的路由丢失问题。表 8-4 列出 OSPF LSA 的相关信息。

表 8-4　　　　　　　　　　　　　　　　OSPF LSA 类型

LSA 类型	描述
Type-1（1 类）	路由器链路——每台 OSPF 路由器都会宣告 Type-1 LSA，Type-1 LSA 是 LSDB 中最基本的 LSA，所有启用了 OSPF 的链路（接口）都有 Type-1 LSA 表项，反映实际的网络 路由器链路分为末梢路由器链路和转接路由器链路，末梢路由器链路包含网络掩码，而转接路由器链路不包含网络掩码
Type-2（2 类）	网络链路——Type-2 LSA 表示使用 DR 的多路接入网段。DR 始终宣告 Type-2 LSA，并将 Type-1 转接链路 LSA 连接在一起。此外，Type-2 LSA 还为 Type-1 转接链路提供网络掩码 如果还未选举 DR，那么 LSDB 中就没有 Type-2 LSA，因为相应的 Type-1 转接路由器链路 LSA 是末梢链路。Type-2 LSA 不会以 Type-1 LSA 相同的方式泛洪到源端 OSPF 区域之外
Type-3（3 类）	汇总链路——ABR 的角色是参与多个 OSPF 区域并确保与 Type-1 LSA 相关联的网络在非源端 OSPF 区域内可达，ABR 不会将 Type-1 或 Type-2 LSA 转发到其他区域中 ABR 收到 Type-1 LSA 之后，会创建一条引用原始 Type-1 LSA 中的网络的 Type-3 LSA，使用 Type-2 LSA 来确定多路接入网络的网络掩码。然后 ABR 将该 Type-3 LSA 宣告到其他区域中 如果 ABR 从 Area 0（骨干区域）收到了 Type-3 LSA，那么就会为非骨干区域生成一条 Type-3 LSA，但是将自己列为宣告路由器
Type-4（4 类）	ASBR 汇总——Type-4 LSA 为 Type-5 LSA 定位 ASBR，Type-5 LSA 在整个 OSPF 域中不加修改地进行泛洪，识别 ASBR 的唯一机制就是 RID，路由器会检查 Type-5 LSA 以确定 RID 是否位于本地区域，如果不是，那么就需要通过某种机制来定位 ASBR 仅 Type-1 或 Type-2 LSA 提供了在区域内定位 RID 的方法。如果路由器与 ASBR 处于不同的区域中，那么路由器就可以通过 Type-4 LSA 定位 ASBR
Type-5（5 类）	AS 外部——ASBR 将路由重分发到 OSPF 之后，该外部路由将以 Type-5 LSA 的方式在整个 OSPF 域中进行泛洪。Type-5 LSA 不与某个特定区域相关联，而且会在所有 ABR 上泛洪，泛洪期间仅修改 LSA 老化值
Type-7（7 类）	NSSA 外部——NSSA（Not So Stubby Area，非完全末梢区域）区域是一种减小区域内 LSDB 的方法，其做法是阻止 Type-4 和 Type-5 LSA，同时又允许将网络重分发到区域中。Type-7 LSA 仅存在于进行路由重分发的 NSSA 区域中 ASBR 将外部路由作为 Type-7 LSA 注入 NSSA 区域中。ABR 不会将 Type-7 LSA 宣告到源端 NSSA 区域之外，但是会将 Type-5 LSA 宣告到其他 OSPF 区域中。如果 Type-5 LSA 穿过了 Area 0，那么第二台 ABR 就会为 Type-5 LSA 创建一条 Type-4 LSA

注：每条 LSA 都包含了宣告路由器的 RID，其中，路由器 RID 代表路由器以及链路的互联方式。

图 8-1 显示了一个多区域 OSPF 拓扑，且外部路由重分发到了 Area 56 中。图的左侧是该拓扑结构的网络前缀，并在要宣告的网段下方显示了相应类型的 LSA。本例标出了每种 LSA 的位置。请注意，Area 1234 是一个广播区域，包含一个 DR，该 DR 生成了一条 Type-2 LSA。NX-6 将网络 100.65.0.0/16 重分发到了 OSPF 中，NX-5 则为 ASBR（NX-6）宣告了第一条 Type-4 LSA。

注：Cisco Press 出版的 *IP Routing on Cisco IOS, IOS XE and IOS XR* 详细介绍了 OSPF LSA，同时还以可视化方式解释了路由器利用 LSA 构建拓扑表的方式。

OSPF 将路由分为以下 3 类。

- **区域内路由**：指的是位于 OSPF 区域内的网络路由，路由表中的区域内路由的旁边有 *intra* 标记。
- **区域间路由**：指的是位于其他 OSPF 区域内的网络路由，路由表中的区域间路由的旁边有 *inter* 标记。
- **外部路由**：指的是重分发到 OSPF 域中的路由，路由表中的外部路由的旁边有 *type-1* 或 *type-2* 标记。

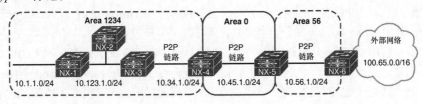

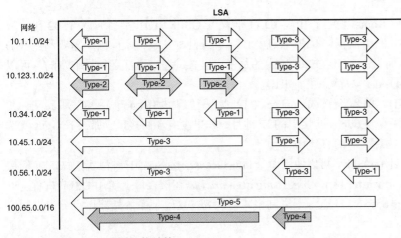

图 8-1 OSPF LSA 示例拓扑结构

例 8-1 显示了图 8-1 中的 NX-1 的路由表信息，包括了区域内、区域间和外部 OSPF 路由。

例 8-1 OSPF 路由表示例

```
NX-1# show ip route ospf
IP Route Table for VRF "default"
'*' denotes best ucast next-hop
'**' denotes best mcast next-hop
'[x/y]' denotes [preference/metric]
'%<string>' in via output denotes VRF <string>

10.34.1.0/24, ubest/mbest: 1/0
    *via 10.123.1.3, Eth1/1, [110/80], 00:10:30, ospf-NXOS, intra
10.45.1.0/24, ubest/mbest: 1/0
    *via 10.123.1.3, Eth1/1, [110/120], 00:10:15, ospf-NXOS, inter
10.56.1.0/24, ubest/mbest: 1/0
    *via 10.123.1.3, Eth1/1, [110/160], 00:10:15, ospf-NXOS, inter
10.65.0.0/16, ubest/mbest: 1/0
    *via 10.123.1.3, Eth1/1, [110/20], 00:08:02, ospf-NXOS, type-2
```

8.2 OSPF 邻居邻接关系故障排查

讨论完 OSPF 协议的基本内容之后，接下来将解释 OSPF 的配置方式以及 NX-OS OSPF 邻居邻接关系的故障排查方式。

8.2.1 OSPF 基本配置

NX-OS 交换机要求在 OSPF 进程及接口配置子模式下配置 OSPF。在 NX-OS 设备上配置 OSPF 的步骤如下。

- **第 1 步**：启用 OSPF 功能特性。必须使用全局配置命令 **feature ospf** 启用 OSPF 功能特性。
- **第 2 步**：定义 OSPF 进程标签。必须使用全局配置命令 **router ospf** *process-tag* 定义 OSPF 进程，其中，*process-tag* 最长支持 20 个字母数字字符。
- **第 3 步**：启用 OSPF 邻居日志记录功能（推荐）。NX-OS 默认并不记录 OSPF 邻居邻接关系的建立或解除操作。可以通过 OSPF 配置命令 **log-adjacency-changes** [**detail**]启用日志记录功能（也建议这么做），可选关键字 **detail** 可以列出表 8-3 所列的 OSPF 邻居状态。
- **第 4 步**：定义 RID（Router-ID，路由器 ID）（推荐）。OSPF RID 是一个唯一的 32 比特数字，用于标识 OSPF 路由器，与术语"邻居 ID"（Neighbor ID）同义。由于需要利用 RID 来构建拓扑表，因而 RID 对于 OSPF 域中的每个 OSPF 进程来说都必须是唯一的。可以通过命令 **router-id** *router-id* 静态设置 RID。

    如果未手动配置 RID，那么将首选 Loopback 0 IP 地址。如果没有 Loopback 0，那么 NX-OS 就选择配置中的第一个 Loopback 接口的 IP 地址；如果没有环回接口，那么 NX-OS 就选择配置中的第一个物理接口的 IP 地址。

- **第 5 步**：在接口上启用 OSPF。可以使用命令 **interface** *interface-id* 选择将要启用 OSPF 的接口，然后利用命令 **ip router ospf** *process-tag* **area** *area-id* 在该接口上启用 OSPF 进程。可以以十进制格式（1-65536）或点分十进制格式输入 *area-id*，但系统始终以点分十进制格式存储 *area-id*。

在接口上启用了 OSPF 之后，系统会默认宣告从网络，可以通过命令 **ip router ospf** *process-tag* **area** *area-id* **secondaryaries none** 禁用该行为。

无论实际的子网掩码是什么，OSPF 都将环回接口宣告为/32。可以通过 **ip ospf advertise-subnet** 命令更改该行为，从而可以通过 LSA 宣告子网掩码。

> **注**：通常情况下，每个接口都只能位于一个区域内。但目前可以通过命令 **ip router ospf** *process-tag* **multi-area** *area-id* 让接口通过点对点 OSPF 链路存在于多个区域中。

例 8-2 在接口 Ethernet1/1、VLAN 10 和 Loopback 0 上启用了 OSPF。

例 8-2　OSPF 基本配置

```
NX-1# configure terminal
Enter configuration commands, one per line. End with CNTL/Z.
NX-1(config)# feature ospf
NX-1(config)# router ospf NXOS
NX-1(config-router)# log-adjacency-changes detail
NX-1(config-router)# router-id 192.168.100.100
NX-1(config-if)# ip router ospf NXOS area 0
NX-1(config-if)# interface vlan 10
NX-1(config-if)# ip router ospf NXOS area 0
NX-1(config-if)# interface e1/1
NX-1(config-if)# ip router ospf NXOS area 0
12:58:33 NX-1 %$ VDC-1 %$ %OSPF-5-NBRSTATE: ospf-NXOS [13016] Process NXOS, Nbr
10.12.1.200 on Ethernet1/1 from DOWN to INIT, HELLORCVD
```

```
12:58:42 NX-1 %$ VDC-1 %$ %OSPF-5-NBRSTATE: ospf-NXOS [13016] Process NXOS, Nbr
  10.12.1.200 on Ethernet1/1 from INIT to EXSTART, ADJOK
12:58:42 NX-1 %$ VDC-1 %$ %OSPF-5-NBRSTATE: ospf-NXOS [13016] Process NXOS, Nbr
  10.12.1.200 on Ethernet1/1 from EXSTART to EXCHANGE, NEGDONE
12:58:42 NX-1 %$ VDC-1 %$ %OSPF-5-NBRSTATE: ospf-NXOS [13016] Process NXOS, Nbr
  10.12.1.200 on Ethernet1/1 from EXCHANGE to LOADING, EXCHDONE
12:58:42 NX-1 %$ VDC-1 %$ %OSPF-5-NBRSTATE: ospf-NXOS [13016] Process NXOS, Nbr
  10.12.1.200 on Ethernet1/1 from LOADING to FULL, LDDONE
```

OSPF 在处理并将路由添加到 RIB 之前，需要建立邻居关系。邻居邻接表对于跟踪邻居状态以及发送给每个邻居的更新来说至关重要。本节将解释 NX-OS 交换机上的 OSPF 邻居邻接关系故障排查过程。

图 8-2 以包含两台 Nexus 交换机的简单拓扑结构为例，说明了 OSPF 邻接关系故障的排查过程。

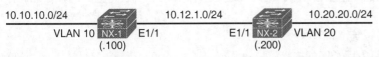

图 8-2 包含两台 NX-OS 交换机的简单拓扑结构

8.2.2 OSPF 邻居验证

首先，使用命令 **show ip ospf neighbors** [*interface-id* [**detail** | **summary**] | *neighbor-id* [**detail**] | **vrf** {*vrf-name* | **all**}] [**summary**]验证已成功建立 OSPF 邻接关系的设备，关键字 **summary** 的作用是显示 OSPF 邻居的数量以及与这些邻居相关联的接口。

例 8-3 在 NX-1 上以不同的形式运行了该命令，请注意关键字 **detail** 所包含的一些额外信息，如 *Dead timer* 和 *last change*。

例 8-3 显示 OSPF 邻居

```
NX-1# show ip ospf neighbors
 OSPF Process ID NXOS VRF default
 Total number of neighbors: 2
 Neighbor ID     Pri State            Up Time  Address        Interface
 192.168.200.200   1 FULL/DR          00:02:04 10.12.1.200    Eth1/1
NX-1# show ip ospf neighbors summary
 OSPF Process ID NXOS VRF default, Neighbor Summary
 Interface Down Attempt Init TwoWay ExStart Exchange Loading Full Total
    Total    0     0     0    0       0       0       0      1    1
    Eth1/1   0     0     0    0       0       0       0      1    1
NX-1# show ip ospf neighbors 192.168.200.200 detail
 Neighbor 192.168.200.200, interface address 10.12.1.200
    Process ID NXOS VRF default, in area 0.0.0.0 via interface Ethernet1/1
    State is FULL, 6 state changes, last change 00:03:21
    Neighbor priority is 1
    DR is 10.12.1.200 BDR is 10.12.1.100
    Hello options 0x2, dbd options 0x42
    Last non-hello packet received never
    Dead timer due in 00:00:34
```

表 8-5 简要列出了例 8-3 显示的字段信息。

表 8-5　　　　　　　　　　　　　　OSPF 邻居状态字段

字段	描述
邻居 ID（Neighbor ID）	邻居路由器的路由器 ID（RID）
PRI	邻居的接口优先级，用于 DR/BDR 选举进程
状态（State）	第一个字段是表 8-3 描述的邻居状态 第二个字段是 DR、BDR 或 DROther 角色（如果接口需要 DR）。如果是非 DR 网络链路，那么第二个字段将显示为 "-" 排查邻居关系故障时查看 OSPF 邻居的状态非常有用，实际的状态切换速度可能比看到的要快（取决于 LSDB 的大小）
失效时间（Dead Time）	在路由器被宣告为不可达之前剩下的失效时间
地址（Address）	OPSF 邻居的主 IP 地址
接口（Interface）	OSPF 邻居所连接的本地接口

除在 NX-OS 设备的网络接口上启用 OSPF 之外，还必须保证以下参数完全匹配，才能确保两台路由器成为邻居。

- 接口必须处于 *Active*（主动）状态。
- 设备之间必须存在使用主用子网的连接。
- 设备之间的 MTU（Maximum Transmission Unit，最大传输单位）必须匹配。
- 路由器 ID 必须唯一。
- 接口区域必须匹配。
- OSPF 末梢区域标志必须匹配。
- DR 必须匹配（基于 OSPF 网络类型）。
- OSPF Hello 和失效定时器必须匹配。
- 认证参数必须匹配。

8.2.3　OSPF 接口确认

如果邻居邻接关系丢失，那么验证正确的接口是否正在运行 OSPF 将至关重要。命令 **show ip ospf interface [brief]** 可以显示所有已启用 OSPF 的接口。例 8-4 给出了该命令简要模式的输出示例。

例 8-4　以简要模式显示 OSPF 接口信息

```
NX-1# show ip ospf interface brief
 OSPF Process ID NXOS VRF default
 Total number of interface: 3
 Interface          ID    Area       Cost   State       Neighbors Status
 Eth1/1             2     0.0.0.0    4      BDR         1         up
 VLAN10             3     0.0.0.0    4      DR          0         up
 Lo0                1     0.0.0.0    1      LOOPBACK    0         up
```

表 8-6 列出了例 8-3 显示的字段信息。

表 8-6　　　　　　　　　　　　　　OSPF 接口字段

字段	描述
接口（Interface）	启用了 OSPF 的接口
区域（Area）	该接口所关联的区域。区域始终以点分十进制格式显示
开销（Cost）	SPF 算法利用开销来计算路径的度量
状态（State）	当前的接口状态 DR、BDR、DROTHER、LOOP 或 Down
邻居（Neighbor）	已建立邻接关系的网段的邻居 OSPF 路由器数量
状态（Status）	该接口的协议线路状态，值为 Down 表示该接口不可达

例 8-5 以非简要模式显示了 **show ip ospf interface** 命令的输出结果，请注意，输出结果包含了主 IP 地址、接口网络类型、DR、BDR 和 OSPF 接口定时器等信息。

例 8-5　OSPF 接口输出结果

```
NX-1# show ip ospf interface
! Output omitted for brevity
 Ethernet1/1 is up, line protocol is up
    IP address 10.12.1.100/24
    Process ID NXOS VRF default, area 0.0.0.0
    Enabled by interface configuration
    State BDR, Network type BROADCAST, cost 4
    Index 2, Transmit delay 1 sec, Router Priority 1
    Designated Router ID: 192.168.200.200, address: 10.12.1.200
    Backup Designated Router ID: 192.168.100.100, address: 10.12.1.100
    1 Neighbors, flooding to 1, adjacent with 1
    Timer intervals: Hello 10, Dead 40, Wait 40, Retransmit 5
      Hello timer due in 00:00:00
    No authentication
    Number of opaque link LSAs: 0, checksum sum 0
```

8.2.4　被动接口

有些网络拓扑结构需要将某个网络段宣告给 OSPF，但同时又要防止邻居在该网段上建立邻接关系。由于显示 OSPF 接口时会显示被动接口，因而查看是否配置了被动接口的最快方法就是使用 **show ip ospf** [*process-tag*] 命令检查 OSPF 进程。例 8-6 给出了该命令的输出结果示例以及被动接口统计信息。

例 8-6　确定是否配置了被动 OSPF 接口

```
NX-1# show ip ospf
 Routing Process NXOS with ID 192.168.100.100 VRF default
..
   Area BACKBONE(0.0.0.0)
        Area has existed for 00:22:02
        Interfaces in this area: 3 Active interfaces: 3
        Passive interfaces: 1  Loopback interfaces: 1
```

找到被动接口之后，接下来需要检查以下配置信息。

- 接口参数命令 **ip ospf passive-interface**，该命令仅将该接口设置为被动接口。
- 全局 OSPF 配置命令 **passive-interface default**，该命令将 OSPF 进程下的所有接口都设置为被动状态。接口参数命令 **no ip ospf passive-interface** 的优先级高于全局命令，可以将该接口设置为主动状态。

例 8-7 显示了 NX-1 和 NX-2 的配置信息，此时两台 Nexus 交换机无法建立 OSPF 邻接关系。两台交换机上的 Ethernet1/1 接口都必须处于主动状态才能建立邻接关系，因而将命令 **ip ospf passive-interface** 从 Eth1/1 接口移到 NX-1 的 VLAN 10，将命令 **no ip ospf passive-interface** 从 VLAN 20 移到 NX-2 的 Eth1/1 接口，这样就可以建立邻接关系。

例 8-7　含有被动接口的 OSPF 配置

```
NX-1# show run ospf
! Output omitted for brevity
feature ospf

router ospf NXOS
```

```
  router-id 192.168.100.100
  log-adjacency-changes detail

interface loopback0
  ip router ospf NXOS area 0.0.0.0

interface Ethernet1/1
  ip ospf passive-interface
  ip router ospf NXOS area 0.0.0.0

interface VLAN10
  ip router ospf NXOS area 0.0.0.0

NX-2# show run ospf
! Output omitted for brevity
router ospf NXOS
  router-id 192.168.200.200
  log-adjacency-changes detail
  passive-interface default

interface loopback0
  ip router ospf NXOS area 0.0.0.0

interface Ethernet1/1
  ip router ospf NXOS area 0.0.0.0

interface VLAN20
  no ip ospf passive-interface
  ip router ospf NXOS area 0.0.0.0
```

8.2.5 OSPF 数据包验证

排查 OSPF 邻接关系故障的重要步骤就是确保设备正在发送或接收 OSPF 网络流量。命令 **show ip ospf traffic** [*interface-id*] [**detail**]可以从高层视角显示设备收发的各类数据包汇总信息。

例 8-8 显示了该命令的使用示例。请注意，输出结果中的错误包和有效包是分开的。如果执行命令时指定了特定接口，那么就可以更详细地了解该接口收发的数据包信息。

例 8-8　OSPF 流量统计

```
NX-1# show ip ospf traffic
 OSPF Process ID NXOS VRF default, Packet Counters (cleared 00:32:34 ago)
 Total: 319 in, 342 out
 LSU transmissions: first 23, rxmit 4, for req 5 nbr xmit 0
 Flooding packets output throttled (IP/tokens): 0 (0/0)
 Ignored LSAs: 0, LSAs dropped during SPF: 0
 LSAs dropped during graceful restart: 0
 Errors: drops in      0, drops out     0, errors in     0,
         errors out    0, hellos in     0, dbds in       0,
         lsreq in      0, lsu in        0, lsacks in     0,
         unknown in    0, unknown out   0, no ospf       0,
         bad version   0, bad crc       0, dup rid       0,
         dup src       0, invalid src   0, invalid dst   0,
         no nbr        0, passive       1, wrong area    14,
         pkt length    0, nbr changed rid/ip addr        0
         bad auth      0, no vrf        0
         bad reserved  0, no vrf        0
```

```
              hellos        dbds       lsreqs        lsus        acks
   In:          253           18          5           25          18
   Out:         275           19          5           32          11
```

可以通过调试功能了解路由器上各种进程的详细信息,具体来说,命令 **debug ip ospf [adjacency | hello | packet]** 可以显示到达交换机管理引擎的数据包的处理情况,用户可以据此验证路由器是否收到或宣告了数据包。

例 8-9 显示了 OSPF Hello 包以及数据包调试操作示例。

例 8-9   OSPF Hello 包及数据包调试

```
NX-1# debug ip ospf packet
NX-1# debug ip ospf hello
13:59:28.140175 ospf: NXOS [16748] (default) LAN hello out, ivl 10/40, options 0x02,
   mask /24, prio 1, dr 0.0.0.0, bdr 0.0.0.0 nbrs 0 on Ethernet1/1 (area 0.0.0.0)
13:59:28.140631 ospf: NXOS [16748] (default) sent: prty:6 HELLO to 224.0.0.5/
   Ethernet1/1
13:59:29.165361 ospf: NXOS [16748] (default) rcvd: prty:0 ver:2 t:HELLO len:44
   rid:192.168.200.200 area:0.0.0.0 crc:0x732d aut:0 aukid:0 from 10.12.1.200/
   Ethernet1/1
13:59:29.165460 ospf: NXOS [16748] (default) LAN hello in, ivl 10/40, options 0x02,
   mask /24, prio 1, dr 0.0.0.0, bdr 0.0.0.0 on Ethernet1/1 from 10.12.1.200
```

注: 可以将调试输出结果重定向到日志文件中,具体可参见第 2 章。

表 8-7 列出了例 8-9 所示调试输出结果中的字段信息。

表 8-7   OSPF 调试输出结果中的相关字段

字段	描述
ivl	在 Hello 包中提供 Hello 和失效定时器
options	确定与该接口相关联的区域是常规 OSPF 区域、OSPF 末梢区域或 OSPF NSSA 区域,这些值均以十六进制格式显示,本章将在后面解释这些选项的验证方式
mask	该接口上的主 IP 地址的子网掩码
prio	用于 DR/BDR 选举进程的接口优先级
dr	DR 的路由器 ID
bdr	BDR 的路由器 ID
nbrs	该网段上检测到的邻居数量

一般来说,使用调试命令来查找故障根源是最不推荐的方法,因为启用调试操作可能会生成大量数据。NX-OS 在后台提供了事件历史记录(event-history)功能,不会影响交换机的性能,为用户提供了另一种有效的故障排查方法。命令 **show ip ospf event-history [hello | adjacency | event]** 可以为 OSPF 邻接关系故障提供非常有用的信息,关键字 **hello** 提供的信息与例 8-9 的调试命令相同。

例 8-10 给出了 **show ip ospf event-history hello** 命令的输出结果示例,请注意 NX-1 输出结果的差异。

例 8-10   通过 OSPF 事件历史记录查看 Hello 包信息

```
NX-1# show ip ospf event-history hello
OSPF HELLO events for Process "ospf-NXOS"
18:20:03.890150 ospf NXOS [16748]: LAN hello out, ivl 10/40, options 0x02, mask /24,
   prio 1, dr 10.12.1.200, bdr 0.0.0.0 nbrs 1 on Ethernet1/1 (area 0.0.0.0)
18:19:59.777890 ospf NXOS [16748]: LAN hello in, ivl 10/40, options 0x02, mask /24,
   prio 1, dr 10.12.1.200, bdr 0.0.0.0 on Ethernet1/1 from 10.12.1.200
```

```
18:19:56.320192 ospf NXOS [16748]: LAN hello out, ivl 10/40, options 0x02, mask /24,
  prio 1, dr 10.12.1.200, bdr 0.0.0.0 nbrs 1 on Ethernet1/1 (area 0.0.0.0)
18:19:52.101250 ospf NXOS [16748]: LAN hello in, ivl 10/40, options 0x02, mask /24,
  prio 1, dr 10.12.1.200, bdr 0.0.0.0 on Ethernet1/1 from 10.12.1.200
```

在交换机上执行 OSPF 调试操作只能显示到达管理引擎的数据包。如果数据包未显示在调试或事件历史记录中，那么就需要检查 QoS（Quality of Service，服务质量）策略、CoPP（Control Plane Policing，控制平面策略），或者验证进入或离开接口的数据包。

接口可能部署，也可能未部署 QoS 策略。如果已部署 QoS 策略，那么就必须检查策略映射是否存在任何丢包，然后将其引用到与 OSPF 路由协议相匹配的分类映射包。由于 CoPP 策略也基于 QoS 设置，因而上述排查过程也同样适用于 CoPP 策略。

例 8-11 给出了检查交换机 CoPP 策略的逻辑过程。

- 使用命令 **show running-config copp all** 检查 CoPP 策略，可以显示相关的策略映射名称、定义的类别和警管速率。
- 分析分类映射，以确定该分类映射的条件匹配设置。
- 验证了分类映射之后，使用命令 **show policy-map interface control-plane** 检查该类别的策略映射。

例 8-11 验证 OSPF 的 CoPP 策略

```
NX-1# show run copp all
! Output omitted for brevity

class-map type control-plane match-any copp-system-p-class-critical
  match access-group name copp-system-p-acl-bgp
  match access-group name copp-system-p-acl-rip
  match access-group name copp-system-p-acl-vpc
  match access-group name copp-system-p-acl-bgp6
  match access-group name copp-system-p-acl-lisp
  match access-group name copp-system-p-acl-ospf
..
policy-map type control-plane copp-system-p-policy-strict
  class copp-system-p-class-critical
    set cos 7
    police cir 36000 kbps bc 250 ms conform transmit violate drop
..

NX-1# show run aclmgr all | section copp-system-p-acl-ospf
ip access-list copp-system-p-acl-ospf
  10 permit ospf any any
NX-1# show policy-map interface control-plane class copp-system-p-class-critical
! Output omitted for brevity
Control Plane
  service-policy input copp-system-p-policy-strict

    class-map copp-system-p-class-critical (match-any)
      ..
      module 1:
        conformed 1429554 bytes,
          5-min offered rate 1008 bytes/sec
          peak rate 1008 bytes/sec at Mon 19:03:31
        violated 0 bytes,
          5-min violate rate 0 bytes/sec
          peak rate 0 bytes/sec
```

注： 该 CoPP 策略源自 Nexus 7000 交换机，实际的策略名称和分类映射可能会因不同的平台而异。

由于 CoPP 运行在 RP 级别，因而有可能接口收到了数据包却没有转发给 RP。下一阶段就是确定接口是否发送或接收了数据包，相应的操作包括为 OSPF 协议创建特定的 ACE（Access Control Entity，访问控制实体）。为了确保计数正确，OSPF 的 ACE 应该位于所有不明确的 ACE 表项之前。可以通过 ACL 配置命令 **statistics per-entry** 显示每条 ACE 的命中情况。

例 8-12 的 ACL 配置负责检测 Ethernet1/1 接口上的 OSPF 流量。请注意，该 ACL 包含了一条 **permit ip any any** 命令以允许所有流量通过该接口，否则可能会导致流量丢失。

例 8-12 利用 ACL 验证 OSPF 数据包

```
NX-1# configure terminal
Enter configuration commands, one per line. End with CNTL/Z.
NX-1(config)# ip access-list OSPF
NX-1(config-acl)# permit ospf any 224.0.0.5/32
NX-1(config-acl)# permit ospf any 224.0.0.6/32
NX-1(config-acl)# permit ospf any any
NX-1(config-acl)# permit ip any any
NX-1(config-acl)# statistics per-entry
NX-1(config-acl)# int Eth1/1
NX-1(config-if)# ip access-group OSPF in
```

```
NX-1# show ip access-list
IP access list OSPF
        statistics per-entry
        10 permit ospf any 224.0.0.5/32 [match=12]
        20 permit ospf any 224.0.0.6/32 [match=2]
        30 permit ospf any any [match=7]
        40 permit ip any any [match=5]
```

注： 例中配置了 3 条 OSPF ACE 表项，前两条 ACE 与 DR 和 BDR 通信的多播组相关联，第三条 ACE 则用于初始 Hello 包。

注： 例 8-12 使用的是以太网接口，这类接口通常表示一对一的关系，但是对于 SVI（Switched Virtual Interface，交换式虚接口）（也称为接口 VLAN）等多路接入接口来说，可能需要在特定接口中指定邻居。

作为 ACL 替代方案，可以使用 NX-OS 内置的 Ethanalyzer 来抓取 OSPF 数据包。例 8-13 给出了该命令的语法示例，可以通过关键字 **detail** 查看数据包的详细信息。

例 8-13 利用 Ethanalyzer 验证 OSPF 数据包

```
NX-1# ethanalyzer local interface inband capture-filter "proto ospf"
Capturing on inband
2017-09-09 18:45:59.419456    10.12.1.1 -> 224.0.0.5    OSPF Hello Packet
2017-09-09 18:46:01.826241    10.12.1.2 -> 224.0.0.5    OSPF Hello Packet
2017-09-09 18:46:08.566112    10.12.1.1 -> 224.0.0.5    OSPF Hello Packet
2017-09-09 18:46:11.119443    10.12.1.2 -> 224.0.0.5    OSPF Hello Packet
2017-09-09 18:46:16.456222    10.12.1.1 -> 224.0.0.5    OSPF Hello Packet
```

8.2.6 必须存在使用主用子网的连接

OSPF 路由器必须能够通过使用与主 IP 地址相关联的网络与对等路由器进行通信，不能通过从 IP 地址建立邻接关系。OSPF Hello 包包含来自宣告接口的子网掩码，然后检查这些子网掩码及数据包的源 IP，以验证路由器是否位于同一子网中。

本节将 NX-2 的子网掩码从 10.12.1.200/24 更改为 10.12.1.200/25。从而将 NX-2 调整到网络 10.12.1.128/25 中，该网络与 NX-1 所处的网络（10.12.1.100）不同。

检查 OSPF 的邻居表没有看到这两台交换机的任何表项。下面将利用命令 **show ip ospf event-history** 检查 OSPF Hello 包（见例 8-14），可以看出，OSPF 能够检测到路由器之间存在错误子网掩码。

例 8-14　NX-1 和 NX-2 检测到错误子网掩码

```
NX-1# show ip ospf event-history hello
OSPF HELLO events for Process "ospf-NXOS"
00:28:57.260176 ospf NXOS [16748]: LAN hello out, ivl 10/40, options 0x02, mask /24,
  prio 0, dr 0.0.0.0, bdr 0.0.0.0 nbrs 0 on Ethernet1/1 (area 0.0.0.0)
00:28:50.465118 ospf NXOS [16748]: Bad mask
00:28:49.620142 ospf NXOS [16748]: LAN hello out, ivl 10/40, options 0x02, mask /24,
  prio 0, dr 0.0.0.0, bdr 0.0.0.0 nbrs 0 on Ethernet1/1 (area 0.0.0.0)
00:28:42.178993 ospf NXOS [16748]: Bad mask
```

```
NX-2# show ip ospf event-history hello
OSPF HELLO events for Process "ospf-NXOS"
00:28:03.330191 ospf NXOS [7223]: LAN hello out, ivl 10/40, options 0x02, mask /25,
  prio 1, dr 10.12.1.200, bdr 0.0.0.0 nbrs 0 on Ethernet1/1 (area 0.0.0.0)
00:27:58.216842 ospf NXOS [7223]: Bad mask
00:27:54.860129 ospf NXOS [7223]: LAN hello out, ivl 10/40, options 0x02, mask /25,
  prio 1, dr 10.12.1.200, bdr 0.0.0.0 nbrs 0 on Ethernet1/1 (area 0.0.0.0)
00:27:31.491788 ospf NXOS [16748]: Bad mask
```

如果故障问题是由子网不匹配引起的，那么就无法在 OSPF 调试消息或事件历史记录识别 Hello 包。通过 **ping** *neighbor-ipaddress* 或 **show ip route** *neighbor-ipaddress* 命令验证连接性即可看出该网络不在匹配网络上。只有确保 OSPF 路由器的主接口位于同一子网中，才能保证通信的正常进行。

> **注：** OSPF RFC 2328 规定，只有在点对点 OSPF 网络上使用了 **ip unnumbered** 命令之后，邻居才能通过非相连网络建立邻接关系。由于 NX-OS 不支持无编号 IP 寻址方案，因而该规则对于本用例来说并不适用。

8.2.7　MTU 要求

DBD 数据包的 OSPF 报头包含了接口 MTU。OSPF 在 ExStart 和 Exchange 邻居状态下需要交换 DBD 数据包，路由器会检查 DBD 数据包中包含的接口 MTU，以确保匹配性。如果 MTU 不匹配，那么 OSPF 设备之间就无法建立邻接关系。

例 8-15 显示 NX-1 和 NX-2 在 3min 前开始建立邻居邻接关系，目前卡在 ExStart 状态。

例 8-15　OSPF 邻居卡在 EXSTART 邻居状态

```
NX-1# show ip ospf neighbors
 OSPF Process ID NXOS VRF default
 Total number of neighbors: 1
 Neighbor ID     Pri State            Up Time  Address        Interface
 192.168.200.200   1 EXSTART/DR       00:03:47 10.12.1.200    Eth1/1
```

```
NX-2# show ip ospf neighbors
 OSPF Process ID NXOS VRF default
 Total number of neighbors: 1
 Neighbor ID     Pri State            Up Time  Address        Interface
 192.168.100.100   0 EXSTART/DROTHER  00:03:49 10.12.1.100    Eth1/1
```

检查 OSPF 的事件历史记录以确定交换机卡在 ExStart 状态的原因。例 8-16 显示了 NX-1 的 OSPF 邻接关系事件历史记录，可以看出，NX-2 的 MTU 大于 NX-1 接口的 MTU。

例 8-16　NX-1 的 OSPF 邻接关系事件历史记录显示 MTU 不匹配

```
NX-1# show ip ospf event-history adjacency
Adjacency events for OSPF Process "ospf-NXOS"
07:04:01.681927 ospf NXOS [16748]:    DBD from 10.12.1.200, mtu too large
07:04:01.681925 ospf NXOS [16748]: seqnr 0x40196423, dbdbits 0x7, mtu 9216, options
    0x42
07:04:01.681923 ospf NXOS [16748]: Got DBD from 10.12.1.200 with 0 entries
07:04:01.680135 ospf NXOS [16748]: mtu 1500, opts: 0x42, ddbits: 0x7, seq:
    0x11f2da90
07:04:01.680133 ospf NXOS [16748]: Sent DBD with 0 entries to 10.12.1.200 on
    Ethernet1/1
07:04:01.680131 ospf NXOS [16748]: Sending DBD to 10.12.1.200 on Ethernet1/1
07:04:01.381284 ospf NXOS [16748]:    DBD from 10.12.1.200, mtu too large
07:04:01.381282 ospf NXOS [16748]: seqnr 0x40196423, dbdbits 0x7, mtu 9216, options
    0x42
07:04:01.381280 ospf NXOS [16748]: Got DBD from 10.12.1.200 with 0 entries
07:04:01.201829 ospf NXOS [16748]: Nbr 10.12.1.200: EXSTART --> EXSTART, event
    TWOWAYRCVD
```

注：MTU 消息仅出现在 MTU 较小的设备上。

可以利用命令 **show interface** *interface-id* 显示这两台交换机的 MTU 值（见例 8-17），可以看出，NX-2 的 MTU 大于 NX-1。

例 8-17　检查接口的 MTU 值

```
NX-1# show interface E1/1 | i MTU
  MTU 1500 bytes, BW 10000000 Kbit, DLY 10 usec
NX-2# show int E1/1 | i MTU
  MTU 9216 bytes, BW 1000000 Kbit, DLY 10 usec
```

OSPF 协议本身并不知道如何处理分段。如果数据包大于接口 MTU，那么就需要使用 IP 分段机制。如果在 MTU 较小的交换机上配置接口参数命令 **ip ospf mtu-ignore**，那么就可以忽略 MTU 的安全检查。例 8-18 在 NX-1 上配置了该命令，从而允许 NX-1 忽略 NX-2 较大的 MTU。

例 8-18　配置 OSPF 忽略接口 MTU

```
NX-1# show run ospf
! Output omitted for brevity
router ospf NXOS
  router-id 192.168.100.100
  log-adjacency-changes

interface Ethernet1/1
  ip ospf mtu-ignore
  ip router ospf NXOS area 0.0.0.0

interface VLAN10
  ip ospf passive-interface
  ip router ospf NXOS area 0.0.0.0
```

虽然该技术能够让交换机建立邻接关系，但后续仍有可能出现问题。因此，最简单的解决方案就是更改 MTU，以保证所有设备的 MTU 均匹配。

> **注**：如果 OSPF 接口是 VLAN 接口（SVI），那么就必须确保所有二层（L2）端口都支持在 SVI 上配置 MTU。例如，如果 VLAN 10 的 MTU 为 9000，那么将应该配置所有中继端口均支持 MTU 值 9000。

8.2.8 唯一的 RID

RID（Router-ID，路由器 ID）为 OSPF 路由器提供了唯一的标识符。作为安全机制的一部分，Nexus 交换机会丢弃与自己 RID 相同的数据包，此时将显示 syslog 消息 "*using our routerid, packet dropped*" 以及其他设备的接口和 RID。例 8-19 给出了 NX-1 的 syslog 消息示例。

例 8-19 路由器 ID 重复

```
07:15:51 NX-1 %OSPF-4-DUPRID: ospf-NXOS [16748] (default) Router 10.12.1.200 on
  interface Ethernet1/1 is using our routerid, packet dropped
07:16:01 NX-1 %OSPF-4-SYSLOG_SL_MSG_WARNING: OSPF-4-DUPRID: message repeated 1 times
  in last 16 sec
```

可以通过 **show ip ospf** 命令查看 OSPF 进程来检查 RID，如例 8-20 所示。

例 8-20 查看 OSPF RID

```
NX-1# show ip ospf
! Output omitted for brevity
 Routing Process NXOS with ID 192.168.12.12 VRF default
 Routing Process Instance Number 1
```

目前的一种最佳实践就是在 OSPF 进程中利用命令 **router-id** *router-id* 静态设置 RID。因此，更改了其中一台 Nexus 交换机的 RID 之后，就应该能够建立邻接关系。

> **注**：RID 是 OSPF 拓扑表（通过 LSDB 构建的）的关键组件，所有 OSPF 设备都要维护唯一的 RID。有关 OSPF 拓扑表的更多信息，请参阅 Cisco Press 出版的 *IP Routing on Cisco IOS, IOS-XE, and IOS XR* 一书的第 7 章。

8.2.9 接口区域号必须匹配

OSPF 要求 OSPF Hello 包中的 Area-ID 必须匹配才能建立邻接关系，否则将显示 syslog 消息 "*received for wrong area*" 以及另一台设备的接口和 Area-ID。

例 8-21 显示了 NX-1 和 NX-2 的 syslog 消息示例。

例 8-21 邻居配置了不同区域的 syslog 消息

```
06:47:52 NX-1 %OSPF-4-AREA_ERR: ospf-NXOS [16748] (default) Packet from 10.12.1.200
  on Ethernet1/1 received for wrong area 0.0.0.1
06:48:02 NX-1 %OSPF-4-SYSLOG_SL_MSG_WARNING: OSPF-4-AREA_ERR: message repeated 1
  times in last 151289 sec
06:48:10 NX-1 %OSPF-4-AREA_ERR: ospf-NXOS [16748] (default) Packet from 10.12.1.200
  on Ethernet1/1 received for wrong area 0.0.0.1
06:48:20 NX-1 %OSPF-4-SYSLOG_SL_MSG_WARNING: OSPF-4-AREA_ERR: message repeated 1
  times in last 17 sec
06:49:19 NX-2 %OSPF-4-AREA_ERR: ospf-NXOS [7223] (default) Packet from 10.12.1.100
  on Ethernet1/1 received for wrong area 0.0.0.0
06:49:29 NX-2 %OSPF-4-SYSLOG_SL_MSG_WARNING: OSPF-4-AREA_ERR: message repeated 1
  times in last 18 sec
06:49:35 NX-2 %OSPF-4-AREA_ERR: ospf-NXOS [7223] (default) Packet from 10.12.1.100
  on Ethernet1/1 received for wrong area 0.0.0.0
06:49:45 NX-2 %OSPF-4-SYSLOG_SL_MSG_WARNING: OSPF-4-AREA_ERR: message repeated 1
  times in last 16 sec
```

出现这种情况时，可以利用 **show ip ospf interface brief** 命令检查 OSPF 接口以确定 Area-ID 的配置情况。例 8-22 给出了 NX-1 和 NX-2 的输出结果示例，可以看出，NX-1 和 NX-2 的 Ethernet1/1 接口的区域号并不相同。

例 8-22　Ethernet1/1 接口的 OSPF 区域不同

```
NX-1# show ip ospf interface brief
OSPF Process ID NXOS VRF default
Total number of interface: 3
Interface              ID       Area            Cost      State         Neighbors Status
Eth1/1                 2        0.0.0.0         4         DROTHER       0         up
VLAN10                 3        0.0.0.0         4         DR            0         up
Lo0                    1        0.0.0.0         1         LOOPBACK      0         up

NX-2# show ip ospf interface brief
OSPF Process ID NXOS VRF default
Total number of interface: 3
Interface              ID       Area            Cost      State         Neighbors Status
Eth1/1                 2        0.0.0.1         40        DR            0         up
VLAN20                 3        0.0.0.0         40        DR            0         up
Lo0                    1        0.0.0.0         1         LOOPBACK      0         up
```

因此，将 NX-1 和 NX-2 的接口区域号更改为相同值之后，就可以在两者之间建立邻接关系。

> 注：由于 Area-ID 始终以点分十进制格式存储在 Nexus 交换机上，因而与其他以十进制格式存储 Area-ID 的设备一起使用时，可能会引起混乱。将十进制格式转换为点分十进制格式的步骤如下。
>
> - **第 1 步**：将十进制数值转换为二进制。
> - **第 2 步**：将二进制数值从最右边的数字开始分成 4 个八比特组。
> - **第 3 步**：根据需要添加零以补齐每个八比特组。
> - **第 4 步**：将每个八比特组转换为十进制格式（提供点分十进制格式）。

8.2.10　OSPF 末梢（区域标志）设置必须匹配

OSPF Hello 包包含一个选项字段（E 比特），该字段反映了该区域包含 Type-5 LSA（末梢功能）设置的能力。区域中的接口必须是以下类型才能建立邻接关系。

- **常规**：该区域允许外部路由（Type-5 LSA）。
- **末梢/完全末梢**：该区域不允许外部 LSA（Type-5 LSA），也不允许路由重分发。
- **NSSA（Not So Stubby Area，非完全末梢区域）/完全 NSSA**：该区域不允许外部 LSA（Type-5 LSA），但是允许路由重分发。

OSPF Hello 事件历史记录检测到 OSPF 的区域设置不匹配，从例 8-23 可以看出，NX-1 检测到区域标志与接口配置不一致。

例 8-23　OSPF 事件历史记录显示区域标志不匹配

```
NX-1# show ip ospf event-history hello
OSPF HELLO events for Process "ospf-NXOS"
07:27:01.940673 ospf NXOS [10809]: LAN hello out, ivl 10/40, options 0x00, mask /24,
   prio 1, dr 10.12.1.100, bdr 0.0.0.0 nbrs 0 on Ethernet1/1 (area 0.0.0.1)
07:27:00.422461 ospf NXOS [10809]: Hello packet options mismatch ours: 0, theirs
   0x2
07:26:52.750167 ospf NXOS [10809]: LAN hello out, ivl 10/40, options 0x00, mask /24,
   prio 1, dr 10.12.1.100, bdr 0.0.0.0 nbrs 0 on Ethernet1/1 (area 0.0.0.1)
07:26:51.446550 ospf NXOS [10809]: Hello packet options mismatch ours: 0, theirs
   0x2
```

接下来验证无法建立邻接关系的两台路由器的区域设置情况。从例 8-24 可以看出，NX-1 将 Area 1 配置为末梢区域，而 NX-2 却没有进行相应配置。

例 8-24 验证 OSPF 区域设置

```
NX-1# show running-config ospf
! Output omitted for brevity
router ospf NXOS
  router-id 192.168.100.100
  area 0.0.0.1 stub
  log-adjacency-changes

NX-2(config-if)# show running-config ospf
! Output omitted for brevity
  router-id 192.168.200.200
  log-adjacency-changes
```

因此，将这两台路由器的区域都设置为相同的末梢区域，就可以通过区域标志检查，从而建立邻接关系。

8.2.11 DR 要求

不同类型的介质可以提供不同的特性，也可能会限制网段所允许的节点数量。表 8-8 列出了 5 种 OSPF 网络类型，包括可以在 NX-OS 上配置的网络类型以及能够建立对等连接的网络类型。

表 8-8　　　　　　　　　　NX-OS 支持的 OSPF 网络类型

接口类型	可以在 NX-OS 上配置	OSPF Hello 包的 DR/BDR 字段	能够建立对等关系
广播型（Broadcast）	是	是	广播型，无须更改非广播接口，需要调整 OSPF 定时器
非广播型（Non-Broadcast）	否	是	非广播型，无须更改广播接口，需要调整 OSPF 定时器
点到点型（Point-to-Point）	是	否	点到点型，无须更改点到多点，需要调整 OSPF 定时器
点到多点型（Point-to-Multipoint）	否	否	点到多点型，无须更改点到点，需要调整 OSPF 定时器
环回接口（Loopback）	否	不适用	不适用

由于以太网可以为网段上的两台以上 OSPF 设备提供网络连接，因而需要 DR。Nexus 交换机的默认 OSPF 网络类型是广播 OSPF 网络类型，因为其所有接口都是以太网接口，且广播网络类型还提供了 DR。

> **注：** 在需要 DR（广播/非广播型）的 OSPF 网段上，如果因为将所有接口的 OSPF 优先级都设置为零而不能将路由器选为 DR，那么就无法建立邻接关系，此时邻居关系将卡在 2-Way 状态。

有时 Nexus 交换机只要为接口建立一个 OSPF 邻接关系即可。如通过直连电缆互连的两个以太网端口，两个以太网端口均配置为三层端口（L3），在这种情况下，将 OSPF 网络类型设置为 P2P（Point-to-Point，点对点），不但能够更快地建立邻接关系（不需要选举 DR），而且不会因为 DR 功能而浪费宝贵的 CPU 周期。

仅当 DR 和 BDR Hello 选项相匹配的情况下，OSPF 才能建立邻接关系。例 8-25 显示 NX-1 与 NX-2 卡在 Init 状态，NX-2 未将 NX-1 视为 OSPF 邻居，表明 OSPF 网络类型不兼容。

例 8-25 OSPF 邻接关系故障

```
NX-1# show ip ospf neighbors
 OSPF Process ID NXOS VRF default
 Total number of neighbors: 4
 Neighbor ID     Pri State         Up Time   Address       Interface
 192.168.200.200   1 INIT/DROTHER  00:03:47  10.12.1.200   Eth1/1
NX-2# show ip ospf neighbors
```

可以通过命令 **show ip ospf interface** 确认 Ethernet1/1 接口的 OSPF 网络类型。可以看出，NX-1 被配置为广播型（需要 DR），而 NX-2 则被配置为点对点型（不需要 DR），因而 DR 要求不匹配是邻接关系失败的原因。例 8-26 显示了 NX-1 和 NX-2 的 OSPF 网络类型差异情况。

例 8-26 验证接口的 OSPF 网络类型

```
NX-1# show ip ospf interface | i line|Network
 Ethernet1/1 is up, line protocol is up
    State DR, Network type BROADCAST, cost 4
 VLAN10 is up, line protocol is up
    State DR, Network type BROADCAST, cost 4
 loopback0 is up, line protocol is up
    State LOOPBACK, Network type LOOPBACK, cost 1
NX-2# show ip ospf interface | i line|Network
 Ethernet1/1 is up, line protocol is up
    State P2P, Network type P2P, cost 40
 VLAN20 is down, line protocol is down
    State DOWN, Network type BROADCAST, cost 40
 loopback0 is up, line protocol is up
    State LOOPBACK, Network type LOOPBACK, cost 1
```

由于这两台 Nexus 交换机使用的都是 L3 以太网端口，因而需要更改其中一台交换机的 OSPF 网络类型。此处建议将两台交换机都配置为 OSPF 点对点网络类型，因而可以通过命令 **ip ospf network point-to-point** 将 NX-1 的 Ethernet1/1 接口配置为 OSPF 点对点网络类型，从而允许两台交换机建立邻接关系。例 8-27 给出了 NX-1 和 NX-2 的配置示例，该配置允许两者建立邻接关系。

例 8-27 配置 OSPF 网络类型

```
NX-1# show running-config ospf
! Output omitted for brevity
interface Ethernet1/1
  ip ospf network point-to-point
  ip router ospf NXOS area 0.0.0.0

NX-2# show running-config ospf
! Output omitted for brevity
interface Ethernet1/1
  ip ospf network point-to-point
  ip router ospf NXOS area 0.0.0.0
```

8.2.12 定时器

OSPF Hello 包的辅助功能是确保相邻的 OSPF 邻居运行正常且可用，OSPF 可以按照设定的时间间隔（称为 Hello 定时器）发送 Hello 包。OSPF 使用的另一个定时器是 OSPF 失效间隔定时器（Dead Interval Timer），其默认值为 Hello 定时器的 4 倍。从邻居路由器收到 Hello 包之后，OSPF 失效定时器就会重置为初始值并重新开始递减。

> 注：OSPF Hello 定时器的默认间隔取决于具体的 OSPF 网络类型，更改 Hello 定时器间隔也会同时修改默认的失效间隔。

如果路由器在 OSPF 失效间隔定时器达到零之前仍未收到 Hello 包，那么邻居状态将会切换为 Down 状态，OSPF 路由器则会立即发出反映拓扑结构变化情况的适当 LSA，区域内的所有路由器都将运行相应的 SPF 算法。

建立邻接关系时，必须保证 OSPF Hello 定时器和 OSPF 失效间隔定时器匹配。如果定时器不匹配，那么就会在 OSPF Hello 包事件历史记录中显示定时器信息。从例 8-28 可以看出，NX-1 收到的 Hello 包显示 OSPF 定时器不兼容。

例 8-28　OSPF 定时器不兼容

```
NX-1# show ip ospf event-history hello
OSPF HELLO events for Process "ospf-NXOS"
14:09:47.542331 ospf NXOS [12469]: :  LAN hello out, ivl 10/40, options 0x02, mask
  /24, prio 1, dr 10.10.10.12, bdr 10.10.10.11 nbrs 3 on VLAN10 (area 0.0.0.0)
14:09:45.881230 ospf NXOS [12469]: :  LAN hello in, ivl 10/40, options 0x12, mask
  /24, prio 1, dr 10.10.10.12, bdr 10.10.10.11 on VLAN10 from 10.10.10.11
14:09:45.873642 ospf NXOS [12469]: :  LAN hello in, ivl 10/40, options 0x12, mask
  /24, prio 1, dr 10.10.10.12, bdr 10.10.10.11 on VLAN10 from 10.10.10.12
14:09:45.140175 ospf NXOS [12469]: :  LAN hello out, ivl 10/40, options 0x02, mask
  /24, prio 1, dr 10.12.1.100, bdr 0.0.0.0 nbrs 0 on Ethernet1/1 (area 0.0.0.0)
14:09:42.522692 ospf NXOS [12469]: :    Mismatch in configured hello interval
14:09:39.910300 ospf NXOS [12469]: :  LAN hello out, ivl 10/40, options 0x02, mask
  /24, prio 1, dr 10.10.10.12, bdr 10.10.10.11 nbrs 3 on VLAN10 (area 0.0.0.0)
14:09:39.725303 ospf NXOS [12469]: :  LAN hello in, ivl 10/40, options 0x12, mask
  /24, prio 1, dr 10.10.10.12, bdr 10.10.10.11 on VLAN10 from 1
0.10.10.10
```

此时需要通过命令 **show ip ospf interface** 检查这两台交换机的 OSPF 接口，以查看 Hello 定时器和失效定时器。例 8-29 显示了 NX-1 和 NX-2 的 Ethernet1/1 接口的 OSPF 定时器信息，可以看出，这两台交换机的 Hello 定时器和失效定时器并不相同。

例 8-29　OSPF Hello 定时器不相同

```
NX-1# show ip ospf interface | i line|Timer
 Ethernet1/1 is up, line protocol is up
    Timer intervals: Hello 10, Dead 40, Wait 40, Retransmit 5
 VLAN10 is up, line protocol is up
    Timer intervals: Hello 10, Dead 40, Wait 40, Retransmit 5
 loopback0 is up, line protocol is up

NX-2# show ip ospf interface | i line|Timer
 Ethernet1/1 is up, line protocol is up
    Timer intervals: Hello 15, Dead 60, Wait 60, Retransmit 5
 VLAN20 is down, line protocol is down
    Timer intervals: Hello 10, Dead 40, Wait 40, Retransmit 5
 loopback0 is up, line protocol is up
```

为了排查配置差错，例 8-30 显示了这两台交换机的配置信息。可以看出，NX-2 在 Ethernet1/1 接口上通过命令 **ip ospf hello-interval 15** 修改了 Hello 间隔，因而可以在 NX-2 上删除 **ip ospf hello-interval** 命令，也可以在 NX-1 上设置相同的定时器，使交换机形成邻接关系。

例 8-30　OSPF Hello 定时器不匹配

```
NX-1# show run ospf
! Output omitted for brevity
```

```
  interface Ethernet1/1
    ip router ospf NXOS area 0.0.0.0
NX-2# show run ospf
  interface Ethernet1/1
    ip ospf hello-interval 15
    ip router ospf NXOS area 0.0.0.0
```

> **注**：IOS 路由器支持 OSPF 快速包（fast-packet）Hello，可以实现亚秒级的邻居故障检测。但 Nexus 和 IOS XR 不支持 OSPF 快速包 Hello，因而可以使用 BFD（Bidirectional Forwarding Detection，双向转发检测）在 IOS、IOS XR 和 Nexus 设备之间实现快速收敛，BFD 是首选的亚秒级故障检测方法。

8.2.13 认证

OSPF 支持两种认证方式：明文和 MD5 加密散列。明文模式几乎没有任何安全能力，因为任何有权访问链路的人都能通过网络嗅探器看到密码。MD5 加密散列模式使用散列函数，因而永远也不会向外发送密码，是目前公认为较为安全的认证模式。

OSPF 认证以逐个接口为基础，或者运行在区域内的所有接口上。由于密码只能以接口参数的方式进行设置，因而必须逐个接口设置密码。如果未指定接口，那么就会将默认密码设置为空。

命令 **area** *area-id* **authentication** 的作用是为 OSPF 区域启用明文认证，而接口参数命令 **ip ospf authentication** 则仅在接口上启用明文认证。可以通过接口参数命令 **ip ospf authentication-key** *password* 设置明文密码。

例 8-31 通过上述两个命令在 NX-1 的 Ethernet1/1 接口和 NX-2 的所有 Area 0 接口上显示了明文认证信息。

例 8-31 OSPF 明文认证

```
NX-1# conf t
Enter configuration commands, one per line. End with CNTL/Z.
NX-1(config)# int eth1/1
NX-1(config-if)# ip ospf authentication
NX-1(config-if)# ip ospf authentication-key CISCO
NX-1 %OSPF-4-AUTH_ERR:  ospf-NXOS [8792] (default) Received packet from 10.12.1.200
  on Ethernet1/1 with bad authentication 0

NX-2# conf t
Enter configuration commands, one per line. End with CNTL/Z.
NX-2(config)# router ospf NXOS
NX-2(config-router)# area 0 authentication
NX-2(config-router)# int eth1/1
NX-2(config-if)# ip ospf authentication-key CISCO
```

请注意 NX-1 在启用认证机制后产生的认证差错。如果 OSPF 认证参数不匹配，那么 Nexus 交换机就会生成包含"*bad authentication*"的 syslog 消息，要求进一步验证认证配置。

查看 OSPF 接口并分析验证选项即可验证认证设置。例 8-32 验证了 NX-1 和 NX-2 接口使用的是 OSPF 明文密码。

例 8-32 验证 OSPF 明文认证

```
NX-1# show ip ospf interface
 Ethernet1/1 is up, line protocol is up
    IP address 10.12.1.100/24
    Process ID NXOS VRF default, area 0.0.0.0
    Enabled by interface configuration
```

```
      State P2P, Network type P2P, cost 4
      Index 2, Transmit delay 1 sec
      0 Neighbors, flooding to 0, adjacent with 0
      Timer intervals: Hello 10, Dead 40, Wait 40, Retransmit 5
        Hello timer due in 00:00:06
      Simple authentication
      Number of opaque link LSAs: 0, checksum sum 0
NX-2# show ip ospf interface | i protocol|authent
 Ethernet1/1 is up, line protocol is up
    Simple authentication
 VLAN20 is down, line protocol is down
    Simple authentication
 loopback0 is up, line protocol is up
```

请注意,密码是以加密格式进行存储的。因此,在接口上显式配置密码时,重新配置密码可能会更加容易。例 8-33 显示了查看密码的方式。

例 8-33 查看简单认证的 OSPF 密码

```
NX-2# sho run ospf
! Output omitted for brevity
router ospf NXOS
  router-id 192.168.200.200
  area 0.0.0.0 authentication

interface loopback0
  ip router ospf NXOS area 0.0.0.0

interface Ethernet1/1
  ip ospf authentication-key 3 bdd0c1a345e1c285
  ip router ospf NXOS area 0.0.0.0
```

命令 **area** *area-id* **authentication message-digest** 的作用是为 OSPF 区域启用 MD5 认证,而接口参数命令 **ip ospf authentication message-digest** 则为该接口启用 MD5 认证。可以通过接口参数命令 **ip ospf message-digest-key** *key #* **md5** *password* 配置 MD5 密码,或者通过命令 **ip ospf authentication key-chain** *key-chain-name* 使用密钥链。MD5 认证是密钥 ID 和密码的散列值。如果密钥不匹配,那么节点之间的散列也不同。

注: 有关密钥链创建的详细内容请参考第 7 章。

例 8-34 使用上述两条命令在 NX-1 的 Ethernet1/1 接口和 NX-2 的所有 Area 0 接口上显示了 OSPF 加密认证信息,可以看出,NX-2 正在使用密钥链来维护密码。

例 8-34 OSPF 加密认证

```
NX-1# conf t
Enter configuration commands, one per line. End with CNTL/Z.
NX-1(config)# int eth1/1
NX-1(config-if)# ip ospf authentication message-digest
NX-1(config-if)# ip ospf message-digest-key 2 md5 CISCO

NX-2# conf t
NX-2(config)# key chain OSPF-AUTH
NX-2(config-keychain)# key 2
NX-2(config-keychain-key)# key-string CISCO
NX-2(config-keychain-key)# router ospf NXOS
NX-2(config-router)# area 0 authentication message-digest
```

```
NX-2(config-router)# int eth1/1
NX-2(config-if)# ip ospf authentication key-chain OSPF-AUTH
```

例 8-35 显示了加密密码认证的验证方式。NX-2 直接声明了认证使用的密钥链名称。请注意，虽然 NX-2 的 VLAN 20 启用了加密认证，但无法识别密码，这是因为 VLAN 20 使用的是默认密钥 ID 0。

例 8-35　验证 OSPF 加密认证

```
NX-1# show ip ospf interface | i protocol|auth
 Ethernet1/1 is up, line protocol is up
   Message-digest authentication, using key id 2
 VLAN10 is up, line protocol is up
   No authentication
 loopback0 is up, line protocol is up

NX-2# show ip ospf interface | i protocol|auth
 Ethernet1/1 is up, line protocol is up
   Message-digest authentication, using keychain OSPF-AUTH (ready)
 VLAN20 is down, line protocol is down
   Message-digest authentication, using default key id 0
 loopback0 is up, line protocol is up
```

使用密钥链的好处是可以验证密码（见例 8-36），网络工程师能够检查密码，而不是强制他们重新输入密码。

例 8-36　查看密钥链密码

```
NX-2# show key chain OSPF-AUTH
Key-Chain OSPF-AUTH
  Key 2 -- text 7 "072c087f6d26"
    accept lifetime (always valid) [active]
    send lifetime (always valid) [active]

NX-2# show key chain OSPF-AUTH mode decrypt
Key-Chain OSPF-AUTH
  Key 2 -- text 0 "CISCO"
    accept lifetime (always valid) [active]
    send lifetime (always valid) [active]
```

启用了认证机制之后，有必要检查系统日志是否存在"*bad authentication*"差错消息。如果出现了该差错消息，那么就需要验证该网络链路上的所有对等体的认证选项和密码信息。

8.3　OSPF 路由丢失故障排查

讨论了 OSPF 邻接关系的故障排查方式之后，接下来将讨论 OSPF 路由丢失故障的排查方式。

8.3.1　不连续网络

未完全理解 OSPF 设计要求的网络工程师可能会创建图 8-3 所示的拓扑结构，虽然 NX-2 和 NX-3 有 OSPF 接口位于 Area 0 中，但 Area 12 的流量必须穿越 Area 23 才能到达 Area 34。采用这种设计方式的 OSPF 网络就是不连续网络，因为区域间的流量试图穿越非骨干区域。

从例 8-37 可以看出，NX-2 和 NX-3 似乎与 OSPF 域中的所有网络都完全连接，NX-2 连接了网络 10.34.1.0/24 和网络 192.168.4.4/32，NX-3 连接了网络 10.12.1.0/24 和网络 192.168.1.1/32。

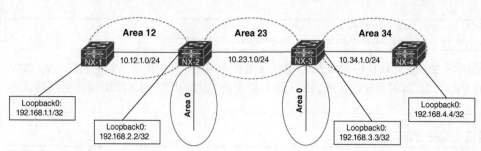

图 8-3 不连续网络

例 8-37 验证 NX-2 和 NX-3 的远程区域路由

```
NX-2# show ip route ospf-NXOS
! Output omitted for brevity

10.34.1.0/24, ubest/mbest: 1/0
    *via 10.23.1.3, Eth1/2, [110/80], 00:02:56, ospf-NXOS, inter
192.168.1.1/32, ubest/mbest: 1/0
    *via 10.12.1.1, Eth1/1, [110/41], 00:04:37, ospf-NXOS, intra
192.168.3.3/32, ubest/mbest: 1/0
    *via 10.23.1.3, Eth1/2, [110/41], 00:06:14, ospf-NXOS, inter
192.168.4.4/32, ubest/mbest: 1/0
    *via 10.23.1.3, Eth1/2, [110/81], 00:02:35, ospf-NXOS, inter

NX-3# show ip route ospf
! Output omitted for brevity

10.12.1.0/24, ubest/mbest: 1/0
    *via 10.23.1.2, Eth1/2, [110/80], 00:07:29, ospf-NXOS, inter
192.168.1.1/32, ubest/mbest: 1/0
    *via 10.23.1.2, Eth1/2, [110/81], 00:06:10, ospf-NXOS, inter
192.168.2.2/32, ubest/mbest: 1/0
    *via 10.23.1.2, Eth1/2, [110/41], 00:07:29, ospf-NXOS, inter
192.168.4.4/32, ubest/mbest: 1/0
    *via 10.34.1.4, Eth1/1, [110/41], 00:04:14, ospf-NXOS, intra
```

例 8-38 列出了 NX-1 和 NX-4 的路由表信息，可以看出，NX-1 缺少 Area 34 的路由表项，NX-4 缺少 Area 12 的路由表项。Area 12 的 Type-1 LSA 到达 NX-2 之后，NX-2 会生成一条 Type-3 LSA 并发送给 Area 23 和 Area 0。NX-3 收到 Type-3 LSA 之后，将其插入 Area 23 的 LSDB 中，但 NX-3 并不会为 Area 0 或 Area 34 创建新的 Type-3 LSA。

例 8-38 验证 NX-1 和 NX-4 的远程区域路由

```
NX-1# show ip route ospf
! Output omitted for brevity

10.23.1.0/24, ubest/mbest: 1/0
    *via 10.12.1.2, Eth1/1, [110/80], 00:13:12, ospf-NXOS, inter
192.168.2.2/32, ubest/mbest: 1/0
    *via 10.12.1.2, Eth1/1, [110/41], 00:13:12, ospf-NXOS, inter

NX-4# show ip route ospf
! Output omitted for brevity

10.23.1.0/24, ubest/mbest: 1/0
    *via 10.34.1.3, Eth1/1, [110/80], 00:11:54, ospf-NXOS, inter
192.168.3.3/32, ubest/mbest: 1/0
    *via 10.34.1.3, Eth1/1, [110/41], 00:11:54, ospf-NXOS, inter
```

Type-3 LSA 进入其他 OSPF 区域时，OSPF ABR 的处理逻辑如下。

- 从非骨干区域收到 Type-1 LSA 之后，会创建进入骨干区域和非骨干区域的 Type-3 LSA。
- 从 Area 0 收到 Type-3 LSA 之后，会为非骨干区域创建 Type-3 LSA。
- 从非骨干区域收到 Type-3 LSA 之后，ABR 仅将其插入源端区域的 LSDB 中，而不为其他非骨干区域创建 Type-3 LSA。

解决这种不连续网络的一种简单方法就是在 NX-2 与 NX-3 之间安装虚链路。虚链路将 Area 0 扩展到非骨干区域，从而解决了 ABR 的限制问题，相当于在 ABR 与另一台多区域 OSPF 路由器之间为 OSPF 运行了一条虚拟隧道。虚链路将 Area 0 扩展到 Area 23，使 Area 0 成为连续的 OSPF 区域。

可以通过命令 **area** *area-id* **virtual-link** *endpoint-rid* 为 OSPF 路由进程配置虚链路。例 8-39 在两端设备上都配置了该命令。

例 8-39 配置虚链路

```
NX-2
router ospf NXOS
  area 0.0.0.23 virtual-link 192.168.3.3

NX-3
router ospf NXOS
  area 0.0.0.23 virtual-link 192.168.2.2
```

在 NX-2 与 NX-3 之间配置了虚链路之后，NX-1 的路由表如例 8-40 所示，可以看出，此时已经存在网络 192.168.4.4，且虚链路显示为 OSPF 接口。

例 8-40 配置了虚链路之后验证网络连接

```
NX-1# show ip route ospf
! Output omitted for brevity

10.23.1.0/24, ubest/mbest: 1/0
    *via 10.12.1.2, Eth1/1, [110/80], 00:22:47, ospf-NXOS, inter
10.34.1.0/24, ubest/mbest: 1/0
    *via 10.12.1.2, Eth1/1, [110/120], 00:00:13, ospf-NXOS, inter
192.168.2.2/32, ubest/mbest: 1/0
    *via 10.12.1.2, Eth1/1, [110/41], 00:22:47, ospf-NXOS, inter
192.168.3.3/32, ubest/mbest: 1/0
    *via 10.12.1.2, Eth1/1, [110/81], 00:00:13, ospf-NXOS, inter
192.168.4.4/32, ubest/mbest: 1/0
    *via 10.12.1.2, Eth1/1, [110/121], 00:00:13, ospf-NXOS, inter

NX-2# show ip ospf interface brief
 OSPF Process ID NXOS VRF default
 Total number of interface: 4
 Interface        ID    Area         Cost   State      Neighbors Status
 VL1              4     0.0.0.0      40     P2P        1         up
 Eth1/1           1     0.0.0.12     40     P2P        1         up
 Eth1/2           2     0.0.0.23     40     P2P        1         up
 Lo0              3     0.0.0.0      1      LOOPBACK   0         up
```

8.3.2 RID 重复

RID（Router-ID，路由器 ID）对于拓扑结构的创建来说至关重要。如果两台相邻路由器的 RID 相同，那么就无法建立前面所说的邻接关系。如果两台拥有相同 RID 的路由器之间有一台中间路由器，那么就无法将这些路由安装到拓扑结构中。

RID 在 OSPF LSA 中充当唯一的标识符。如果两台路由器宣告了拥有相同 RID 的 LSA，那么就会导致 OSPF 拓扑出现混乱，从而导致无法安装路由或者将数据包转发给错误的路由器。此外，还可能会妨碍 LSA 的传播，因而收端路由器可能会认为存在环路。

图 8-4 给出了一个示例拓扑结构，所有的 Nexus 交换机都在网络空间 192.168.0.0/16 中宣告对等网络和环回地址。NX-2 和 NX-4 配置了相同的 RID 192.168.4.4。NX-3 位于 NX-2 与 NX-4 之间且拥有不同的 RID，因而允许 NX-2 和 NX-4 与它们的对等体建立完全的邻居邻接关系。

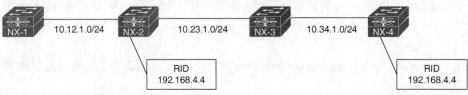

图 8-4 路由器 ID 重复的拓扑结构

从 NX-1 的角度来看（NX-1 的路由表如例 8-41 所示），一个很明显的问题就是缺少了 NX-4 的环回接口（192.168.4.4/32）。

例 8-41　NX-1 的路由表缺少了 NX-4 环回接口

```
NX-1# show ip route ospf
! Output omitted for brevity

10.23.1.0/24, ubest/mbest: 1/0
    *via 10.12.1.2, Eth1/1, [110/80], 2d08h, ospf-NXOS, intra
10.34.1.0/24, ubest/mbest: 1/0
    *via 10.12.1.2, Eth1/1, [110/120], 2d08h, ospf-NXOS, intra
192.168.2.2/32, ubest/mbest: 1/0
    *via 10.12.1.2, Eth1/1, [110/41], 2d08h, ospf-NXOS, intra
192.168.3.3/32, ubest/mbest: 1/0
    *via 10.12.1.2, Eth1/1, [110/81], 2d08h, ospf-NXOS, intra
```

NX-2 和 NX-4 则存在 LSA 故障且显示了 syslog 消息 "*Possible router-id collision*"，如例 8-42 所示。

例 8-42　syslog 消息显示存在 LSA 及重复 RID 问题

```
05:15:23 NX-2 %OSPF-4-SELF_LSA:    ospf-NXOS [9225]    context default: Received updated
  self-originated router LSA. Possible router-id collision
05:16:55 NX-4 %OSPF-4-SELF_LSA:    ospf-NXOS [8486]    context default: Received updated
  self-originated router LSA. Possible router-id collision
```

例 8-43 显示了这两台 Nexus 交换机的路由表信息以及 syslog 消息 "*Possible router-id collision*"，可以看出，NX-2 缺少了 NX-1 的环回接口（192.168.1.1/32）和 NX-4 的环回接口（192.168.4.4/32），而 NX-4 缺少了 10.12.1.0/24 和 NX-2 的环回接口（192.168.2.2/32）。

例 8-43　NX-2 和 NX-4 的路由表

```
NX-2# show ip route ospf
! Output omitted for brevity

10.34.1.0/24, ubest/mbest: 1/0
    *via 10.23.1.3, Eth1/2, [110/80], 2d08h, ospf-NXOS, intra
192.168.1.1/32, ubest/mbest: 1/0
    *via 10.12.1.1, Eth1/1, [110/41], 2d08h, ospf-NXOS, intra
192.168.3.3/32, ubest/mbest: 1/0
```

```
    *via 10.23.1.3, Eth1/2, [110/41], 2d08h, ospf-NXOS, intra
NX-4# show ip route ospf
! Output omitted for brevity

10.23.1.0/24, ubest/mbest: 1/0
    *via 10.34.1.3, Eth1/1, [110/80], 2d08h, ospf-NXOS, intra
192.168.3.3/32, ubest/mbest: 1/0
    *via 10.34.1.3, Eth1/1, [110/41], 2d08h, ospf-NXOS, intra
```

可以通过 show ip ospf 命令检查两台报告了 "*Possible router-id collision*" 问题的 Nexus 交换机的 OSPF 进程，从而快速检查 RID 问题。从例 8-44 可以看出，NX-2 和 NX-4 拥有相同的 RID。

例 8-44　syslog 消息显示存在 LSA 及重复 RID 问题

```
NX-2# show ip ospf | i ID
Routing Process NXOS with ID 192.168.4.4 VRF default
NX-4# show ip ospf | i ID
Routing Process NXOS with ID 192.168.4.4 VRF default
```

需要记住的是，RID 可以动态设置，也可以静态设置。通常来说，出现 RID 重复问题的主要原因是复制另一台路由器的配置时没有更改 RID。可以在 OSPF 进程下使用命令 **router-id** *router-id* 更改 RID。更改了 Nexus 交换机的 RID 之后，OSPF 进程会立即重启。

8.3.3　过滤路由

NX-OS 提供了多种方法，可以在网络进入 OSPF 数据库之后过滤这些网络。ABR 的路由过滤操作面向内部 OSPF 网络，而 ASBR 的路由过滤操作则面向外部 OSPF 网络。如果发现某些路由在某个区域中有但是在其他区域中没有，那么就需要检查以下配置。

- **区域过滤**：通过进程级配置命令 **area** *area-id* **filter-list route-map** *routemap-name* {**in**|**out**}，让 ABR 收到路由或者在路由宣告给 ABR 之后进行路由过滤。
- **路由汇总**：通过命令 **area** *area-id* **range** *summary-network* [**not-advertise**]，让 ABR 汇总内部路由。如果配置了关键字 **not-advertise**，那么就不会为任何组件路由生成 Type-3 LSA，从而将它们隐藏在源端区域。

命令 **summary-address** *summary-network* [**not-advertise**]的作用是在 ASBR 上汇总外部路由，关键字 **not-advertise** 的作用是不为汇总网络内的组件路由生成任何 Type-5/Type-7 LSA。

> **注：** 如果 Type 7 LSA 被转换为 Type 5 LSA，那么 NSSA 区域的 ABR 将充当 ASBR。出现这种场景时，将仅在 ABR 上执行外部路由汇总。

8.3.4　路由重分发

可以通过命令 **redistribute** [**bgp** *asn* | **direct** | **eigrp** *process-tag* | **isis** *process-tag* | **ospf** *process-tag* | **rip** *process-tag* | **static**] **route-map** *route-map-name* 将路由重分发到 OSPF 中。Nexus 交换机的路由重分发进程需要用到路由映射。

每种协议在重分发时都提供了一个种子度量，允许目的协议计算最佳路径。OSPF 为种子度量提供了如下默认设置。

- 网络被配置为 OSPF Type-2 外部网络。
- 默认重分发度量值为 20（源协议为 BGP 时除外，此时的默认种子度量为 1）。

可以根据需要将 OSPF 外部网络类型（类型 1 和类型 2）、重分发度量以及路由标记的默认种子度量更改为其他值。

例 8-45 给出了路由重分发的配置示例,可以看出,NX-1 重分发了直连路由 10.1.1.0/24 和 10.11.11.0/24,而不是通过 OSPF 路由协议宣告这些路由。请注意,路由映射可以是只有一条无任何匹配条件的简单 **permit** 语句。

例 8-45 配置 NX-1 重分发

```
router ospf NXOS
  redistribute direct route-map REDIST
!
route-map REDIST permit 10
```

8.3.5 OSPF 转发地址

OSPF Type-5 LSA 包含一个被称为 FA(Forwarding Address,转发地址)的字段,可以在源端使用共享网段时优化转发流量。RFC 2328 定义的转发地址场景并不常见,图 8-5 给出了一个示例拓扑。

- 在 Area 0 的所有链路上(10.13.1.0/24、10.24.1.0/24 和 10.34.1.0/24)启用 OSPF。
- 用户试图连接防火墙之外的 DMZ(172.16.1.1)中的代理服务器。
- NX-1 有一条指向防火墙(10.120.1.10)的去往网络 172.16.1.0/24 的静态路由。
- NX-1 以 Type-1 外部路由的方式将静态路由重分发到 OSPF 中。
- NX-1 和 NX-2 通过 VLAN 120(10.120.1.0/24)与防火墙直连。

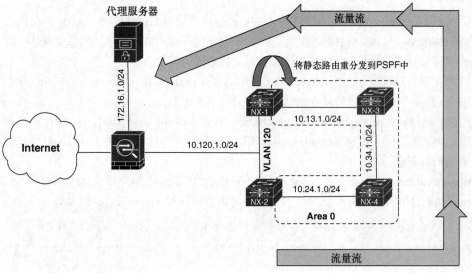

图 8-5 默认 OSPF 转发地址

例 8-46 显示的 NX-1 配置将网络 172.16.1.0/24 宣告到 OSPF 域中。此外,NX-1 的静态路由经验证可安装到 OSPF 数据库中,然后在 NX-3 上进行检查。

例 8-46 NX-1 将 172.16.1.0/24 重分发到 OSPF 中

```
NX-1
ip route 172.16.1.0/24 10.120.1.10
!
route-map REDIST permit 10
  set metric-type type-1
!
router ospf NXOS
```

```
   redistribute static route-map REDIST
   log-adjacency-changes
!
interface Ethernet1/1
   ip router ospf NXOS area 0.0.0.0
```

```
NX-1# show ip route
! Output omitted for brevity
10.13.1.0/24, ubest/mbest: 1/0, attached
    *via 10.13.1.1, Eth2/1, [0/0], 00:09:19, direct
..
10.120.1.0/24, ubest/mbest: 1/0, attached
    *via 10.120.1.1, Eth2/9, [0/0], 00:09:15, direct
..
172.16.1.0/24, ubest/mbest: 1/0
    *via 10.120.1.10, [1/0], 00:09:15, static
```

```
NX-3# show ip route ospf-NXOS
! Output omitted for brevity
10.24.1.0/24, ubest/mbest: 1/0
    *via 10.34.1.4, Eth2/2, [110/80], 00:09:57, ospf-NXOS, intra
172.16.1.0/24, ubest/mbest: 1/0
    *via 10.13.1.1, Eth2/1, [110/60], 00:02:30, ospf-NXOS, type-1
```

例 8-47 显示了外部路由的 Type-5 LSA，该路由去往网络 172.16.1.0/24 的代理服务器。此处的 ASBR 是 NX-1（192.168.1.1），所有的 Nexus 交换机都将去往网络 172.16.1.0/24 的数据包转发给该设备。请注意，此处的转发地址是默认值 0.0.0.0。

例 8-47　OSPF Type-5 LSA 中的默认 FA

```
NX-4# show ip ospf database external detail
        OSPF Router with ID (192.168.4.4) (Process ID NXOS VRF default)

                Type-5 AS External Link States

   LS age: 199
   Options: 0x2 (No TOS-capability, No DC)
   LS Type: Type-5 AS-External
   Link State ID: 172.16.1.0 (Network address)
   Advertising Router: 192.168.1.1
   LS Seq Number: 0x80000002
   Checksum: 0x7c98
   Length: 36
   Network Mask: /24
        Metric Type: 1 (Same units as link state path)
        TOS: 0
        Metric: 20
        Forward Address: 0.0.0.0
        External Route Tag: 0
NX-1# show ip ospf | i ID
 Routing Process NXOS with ID 192.168.1.1 VRF default
```

从例 8-48 可以看出，来自 NX-2（和 NX-4）的流量使用的是次优路由（NX-2→NX-4→NX-3→NX-1→FW），最佳路由允许 NX-2 通过直连网络 10.120.1.0/24 去往防火墙。

例 8-48　验证次优路由

```
NX-2# trace 172.16.1.1
traceroute to 172.16.1.1 (172.16.1.1), 30 hops max, 40 byte packets
```

```
 1  10.24.1.4 (10.24.1.4)      1.402 ms   1.369 ms   1.104 ms
 2  10.34.1.3 (10.34.1.3)      2.886 ms   2.846 ms
 3  10.13.1.1 (10.13.1.1)      4.052 ms   3.527 ms   3.659 ms
 4  10.120.1.10 (10.120.1.10)  5.221 ms *
NX-4# trace 172.16.1.1
traceroute to 172.16.1.1 (172.16.1.1), 30 hops max, 40 byte packets
 1  10.34.1.3 (10.34.1.3)      1.485 ms   1.29 ms    1.18 ms
 2  10.13.1.1 (10.13.1.1)      2.385 ms   2.34 ms    2.478 ms
 3  10.120.1.10 (10.120.1.10)  3.856 ms * *
```

RFC 2328 针对这类场景在 OSPF Type-5 LSA 中指定了转发地址。如果转发地址为 0.0.0.0，那么所有路由器都将数据包转发给 ASBR，从而导致潜在的次优路由。

出现下列情况时，OSPF 的转发地址将从 0.0.0.0 更改为源路由协议中的下一跳 IP 地址。

- ASBR 指向下一跳 IP 地址的接口启用了 OSPF。对于本例来说，NX-1 的 VLAN 120 接口已经启用了 OSPF（与静态路由 172.16.1.0/24 的下一跳地址 10.120.1.10 相关联）。
- 接口未设置为被动状态。
- 接口是广播或非广播 OSPF 网络类型。

目前已在 NX-1 和 NX-2 的 VLAN 120 接口（与 Area 120 相关联）上启用了 OSPF。图 8-6 给出了当前拓扑情况，由于 VLAN 接口默认为广播 OSPF 网络类型，因而满足上述所有条件，从而将 FA 设置为显式 IP 地址。

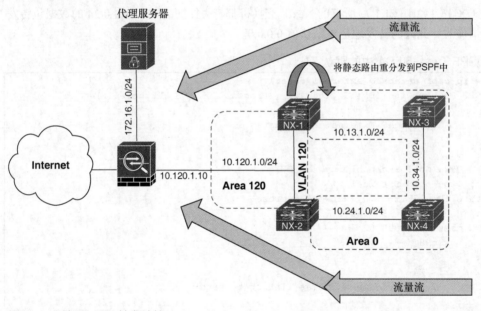

图 8-6 非默认 OSPF 转发地址

例 8-49 显示了网络 172.16.1.0/24 的 Type-5 LSA。由于 NX-1 的 10.120.1.1 接口已经启用了 OSPF，且该接口是广播网络类型，因而转发地址从 0.0.0.0 更改为 10.120.1.10。

例 8-49 查看非默认的 OSPF 转发地址

```
NX-2# show ip ospf database external detail
! Output omitted for brevity

        OSPF Router with ID (192.168.2.2) (Process ID NXOS VRF default)

                Type-5 AS External Link States
```

```
    LS Type: Type-5 AS-External
    Link State ID: 172.16.1.0 (Network address)
    Advertising Router: 192.168.1.1
    Network Mask: /24
        Metric Type: 1 (Same units as link state path)
        TOS: 0
        Metric: 20
        Forward Address: 10.120.1.10
        External Route Tag: 0
```

例 8-50 验证了 NX-2 与 NX-4 之间的连接性，可以看出，由于转发地址已更改为 10.120.1.10，因而两者之间的连接已经采用了最佳路径。

例 8-50　验证最佳路由

```
NX-2# traceroute 172.16.1.1
traceroute to 172.16.1.1 (172.16.1.1), 30 hops max, 40 byte packets
 1  10.120.1.10 (10.120.1.10)  2.845 ms *   3.618 ms

NX-4# traceroute 172.16.1.1
traceroute to 172.16.1.1 (172.16.1.1), 30 hops max, 40 byte packets
 1  10.24.1.2 (10.24.1.2)   1.539 ms  1.288 ms  1.071 ms
 2  10.120.1.10 (10.120.1.10)  3.4 ms *   3.727 ms
```

一名初级网络工程师发现不再需要网络 10.120.1.0/24，因而对 Area 120 的 LSA 实施过滤操作，阻止 Area 120 的 LSA 宣告到 Area 0 中，如例 8-51 所示。

例 8-51　配置 OSPF 过滤网络 10.120.1.10

```
NX-1
router ospf NXOS
  redistribute static route-map REDIST
  area 0.0.0.120 range 10.0.0.0/8 not-advertise
  log-adjacency-changes

NX-2
router ospf NXOS
  area 0.0.0.120 range 10.0.0.0/8 not-advertise
  log-adjacency-changes
```

该初级网络工程师更改配置后，Area 0 中的所有路由器都没有了网络 172.16.1.0/24，仅存在另一个对等网络，如例 8-52 所示。

例 8-52　验证缺少网络 172.16.1.0/24

```
NX-3# show ip route ospf
! Output omitted for brevity

10.24.1.0/24, ubest/mbest: 1/0
    *via 10.34.1.4, Eth1/2, [110/80], 00:23:31, ospf-NXOS, intra
NX-4# show ip route ospf
! Output omitted for brevity

10.13.1.0/24, ubest/mbest: 1/0
    *via 10.34.1.3, Eth1/2, [110/80], 00:23:42, ospf-NXOS, intra
```

如果 Type-5 LSA 的转发地址不是默认值，那么该地址必须是区域内或区域间 OSPF 路由。如果无法解析该 FA，那么就会忽略该 LSA，而且不会安装到 RIB 中。FA 为外部下一跳地址引入多条路径提供了一种有效解决方案，否则，完全没有必要在 LSA 中包含 FA。为了解决上述故障问题，需要删除 NX-1 和 NX-2 的过滤器，从而恢复正常连接。

> 注：对于上述场景来说，如果 NX-1 出现了故障，那么将没有任何冗余机制可以提供连接。因此，通常需要在其他路由器上也重复上述配置，从而提供有效的弹性能力。此外，在 ABR 上应用路由过滤机制时，请仔细考虑外部网络。

8.4 OSPF 路径选择故障排查

OSPF 利用 Dijkstra 的 SPF（Shortest Path First，最短路径优先）算法来创建最短路径的无环拓扑，所有路由器都使用相同逻辑来计算每个网络的最短路径。路径选择进程按照以下路径选择顺序对路径进行优先级排序。

- 区域内路由。
- 区域间路由。
- 外部 Type-1 路由。
- 外部 Type-2 路由。

接下来将详细解释上述内容。

8.4.1 区域内路由

通过 Type-1 LSA 宣告的路由始终优于 Type-3 和 Type-5 LSA。如果存在多条区域内路由，那么就在 RIB 中安装总路径度量最小的路径。如果度量相同，那么就将度量相同的路由都安装到 RIB 中。

> 注：即使区域内路由的路径度量大于区域间路由的路径度量，也仍然选择区域内路径。

8.4.2 区域间路由

区域间路由使用总路径度量最小的路径去往目的端。如果两条路由的度量相同，那么就将这两条路由都将安装到 RIB 中。区域间路由的所有路径都必须经过 Area 0。

以图 8-7 为例，NX-1 正在计算去往 NX-6 的路径。NX-1 将使用路径 NX-1→NX-3→NX-5→NX-6，因为该路径的总路径度量为 35，而 NX-1→NX-2→NX-4→NX-6 路径的总路径度量为 40。

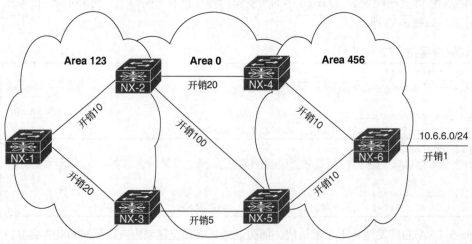

图 8-7　区域间路由选择

8.4.3 外部路由选择

本章前面将 OSPF 外部路由简单地解释为 Type-1 或 Type-2，Type-1 和 Type-2 外部 OSPF 路由的主要区别如下。

- Type-1 路由始终优于 Type-2 路由。
- Type-1 路径度量等于：重分发度量+到达 ASBR 的总路径度量。
- Type-2 路径度量等于：重分发度量。

另一个关键判断因素就是确定设备运行在 RFC 1583 模式还是 RFC 2328 模式。思科 NX-OS 交换机默认运行在 2328 模式下，而思科 IOS、IOS XE 和 IOS XR 则仅支持 1583 模式。接下来将根据设备运行在 RFC 1583 或 RFC 2328 模式下来解释相应的路径选择逻辑。

8.4.4 E1 和 N1 外部路由

OSPF 外部 Type-1 路由的路径选择逻辑如下。

- **RFC 1583 模式**：外部 OSPF Type-1 路由计算公式是重分发度量+到达宣告网络的 ASBR 的最小路径度量。越靠近始发端 ASBR 的路由器的 Type-1 路径度量越小，但距离 ASBR 十跳之外的路由器的路径度量则通常较高。

  如果路径度量相同，那么就将两条路由都将安装到 RIB 中。如果 ASBR 位于不同的区域中，那么流量路径就必须经过 Area 0。ABR 路由器不会将 O E1 和 O N1 路由同时安装到 RIB 中，O N1 对于典型的 NSSA 区域来说优先级较高，因而不会在 ABR 上安装 O E1 路由。

- **RFC 2328 模式**：如果 ASBR 与计算路由器位于同一区域，那么优选首先到达 ASBR 的路由。如果 ASBR 与计算路由器不在同一区域中，那么最佳路径计算规则就遵循上面的 RFC 1583 模式。

> **注**：NSSA 区域提供了一种可选项，可以禁止将重分发路由宣告到 NSSA 区域之外（将 P 比特设置为零），这样做可能会改变路由选择结果。该内容不在本书写作范围之内，感兴趣的读者可参考 RFC 2328 和 3101。

图 8-8 拓扑结构中的 NX-1 和 NX-3 需要计算去往外部网络（100.65.0.0/16）的路径，该外部网络由 NX-6 和 NX7 进行重分发。

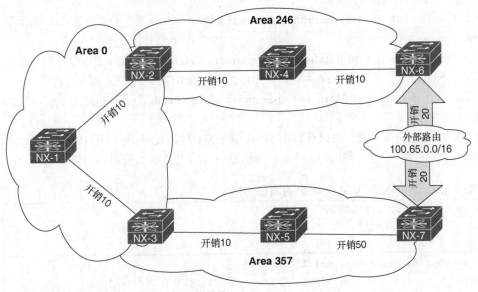

图 8-8　外部 Type-2 路由选择示例拓扑结构

由于路径 NX-1→NX-2→NX-4→NX-6 的度量值为 50,小于路径 NX-1→NX-3→NX-5→NX-7 的度量值 90,因而 NX-1 选择路径 NX-1→NX-2→NX-4→NX-6 去往网络 100.65.0.0/16,而 NX-3 选择的则是路径度量较大的路径 NX-3→NX-5→NX-7。

上述路径选择基于 RFC 2328 逻辑规则,因为 NX-1 与 ASBR 不在同一区域内,而 NX-3 与 NX-7 位于相同区域内。例 8-53 分别从 NX-1 和 NX-3 的角度显示了路由表和路径度量信息。

例 8-53 Type-1 网络的外部 OSPF 路径选择

```
NX-1# show ip route ospf | b 100.65
100.65.0.0/16, ubest/mbest: 1/0
    *via 10.12.1.2, Eth1/1, [110/50], 00:33:55, ospf-NXOS, type-1
NX-3# show ip route ospf | b 100.65
100.65.0.0/16, ubest/mbest: 1/0
    *via 10.35.1.5, Eth1/1, [110/80], 00:00:15, ospf-NXOS, type-1
```

8.4.5 E2 和 N2 外部路由

OSPF 外部 Type-2 路由的路径选择逻辑如下。

- **RFC 1583 模式**:OSPF 外部 Type-2 路由的度量不加上到达 ASBR 的路径度量。如果重分发度量相同,那么路由器就会对比转发开销。转发开销指的是去往宣告该网络的 ASBR 的度量,转发开销越小越优。如果转发开销相同,那么就将这两条路由都安装到路由表中。ABR 路由器不会将 O E2 和 O N2 路由同时安装到 RIB 中,O N2 对于典型的 NSSA 区域来说较优,因而不会在 ABR 上安装 O E2 路由。
- **RFC 2328 模式**:如果 ASBR 与计算路由器位于同一区域,那么就优选首先到达 ASBR 的路由。如果 ASBR 与计算路由器不在同一区域,那么最佳路径计算规则就遵循上面的 RFC 1583 模式。

仍然以图 8-6 的拓扑结构为例,所有路径的度量均为 20。路径决策的第一步就是检查网络 100.65.0.0/16 的 ASBR 是否与计算路由器处于同一区域。由于 NX-1 与 ASBR 不在同一区域,因而根据转发开销选择路径。转发开销是在 NX-OS 上计算得到的。

例 8-54 给出了转发开销的计算步骤。

- **第 1 步**:通过 **show ip ospf database external** *network* 命令查看 OSPF LSDB 来识别 ASBR。
- **第 2 步**:通过 **show ip ospf database asbr-summary detail** 命令检查 ABR 报告的 ASBR 地址(Type-4 LSA)的度量(提供了从 ASBR 到该区域 ABR 的路径度量。)
- **第 3 步**:通过 **show ip ospf database router** *abr-ip-address* **detail** 命令找到去往 Type-4 LSA 的 ABR 的度量。
- **第 4 步**:结合这两个度量,可以计算出 NX-1 到 NX-6 的转发开销为 30,NX-1 到 NX-7 的转发开销为 70。由于去往 NX-6 的路径开销最小,因而 NX-1 选择该路径。

例 8-54 NX-1 的 OSPF 外部 Type-2 路由的路径选择

```
NX-1# show ip route ospf | b 100.65
100.65.0.0/16, ubest/mbest: 1/0
    *via 10.12.1.2, Eth1/1, [110/20], 00:04:33, ospf-NXOS, type-2
NX-1# show ip ospf database external 100.65.0.0
        OSPF Router with ID (192.168.1.1) (Process ID NXOS VRF default)
```

```
                  Type-5 AS External Link States
Link ID         ADV Router       Age      Seq#       Checksum Tag
100.65.0.0      192.168.6.6      31       0x80000002 0x277b   0
100.65.0.0      192.168.7.7      375      0x80000002 0x1a86   0

NX-1# show ip ospf database asbr-summary detail | i ID|Metric
        OSPF Router with ID (192.168.1.1) (Process ID NXOS VRF default)
  Link State ID: 192.168.6.6 (AS Boundary Router address)
    TOS:    0 Metric: 20
  Link State ID: 192.168.7.7 (AS Boundary Router address)
    TOS:    0 Metric: 60

NX-1# show ip ospf database router 192.168.2.2 detail | i Router|Metric
        OSPF Router with ID (192.168.1.1) (Process ID NXOS VRF default)
                Router Link States (Area 0.0.0.0)
  LS Type: Router Links
  Advertising Router: 192.168.2.2
    Link connected to: a Router (point-to-point)
    (Link ID) Neighboring Router ID: 192.168.1.1
    (Link Data) Router Interface address: 10.12.1.2
      TOS    0 Metric: 10
      TOS    0 Metric: 10

NX-1# show ip ospf database router 192.168.3.3 detail | i Router|Metric
        OSPF Router with ID (192.168.1.1) (Process ID NXOS VRF default)
                Router Link States (Area 0.0.0.0)
  LS Type: Router Links
  Advertising Router: 192.168.3.3
    Link connected to: a Router (point-to-point)
    (Link ID) Neighboring Router ID: 192.168.1.1
    (Link Data) Router Interface address: 10.13.1.3
      TOS    0 Metric: 10
      TOS    0 Metric: 10
```

由于 NX-3 与 NX-7 位于同一区域，因而 NX-3 的路径选择基于 RFC 2328 规则。从例 8-55 可以看出，NX-3 选择的路径是 NX-3→NX-5→NX-7。

例 8-55 NX-3 的 OSPF 外部 Type-2 路由的路径选择

```
NX-3# show ip route ospf | b 100.65
100.65.0.0/16, ubest/mbest: 1/0
    *via 10.35.1.5, Eth1/1, [110/80], 00:00:15, ospf-NXOS, type-1
```

8.4.6 RFC 1583 和 RFC 2328 混合设备问题

通常来说，利用 RFC 2328 逻辑就足以找到外部路由的下一跳，但是，如果网络中还存在不支持 RFC 2328 逻辑的设备，那么就可能会出现次优路径（如上节所述）或路由环路。图 8-9 给出的示例拓扑结构就出现了路由环路，因为 IOS 路由器使用的是 RFC 1583 逻辑。

NX-3 使用 RFC 2328 标准，选择 R7 作为网络 100.65.0.0/16 的 ASBR，并将数据包转发给 R5。R5 使用 RFC 1583 标准，将数据包转发回 NX-3，从而出现了路由环路。例 8-56 通过简单的 traceroute 测试（从 NX-3 到网络 100.65.0.0/16）验证了该路由环路。

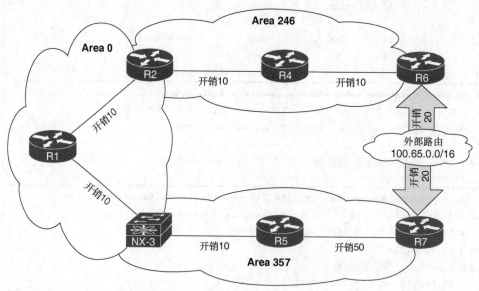

图 8-9 Nexus 和 IOS 设备的外部 Type-2 路由选择

例 8-56　因网络中存在混合的 OSPF 设备而产生路由环路

```
NX-3# trace 100.65.1.1
traceroute to 100.65.1.1 (100.65.1.1), 30 hops max, 40 byte packets
 1  10.35.1.5 (10.35.1.5)  1.819 ms  1.124 ms  0.982 ms
 2  10.35.1.3 (10.35.1.3)  1.9 ms  1.459 ms  1.534 ms
 3  10.35.1.5 (10.35.1.5)  2.427 ms  2.214 ms  2.111 ms
```

解决该问题的方法是通过 OSPF 命令 **rfc1583compatibility**，让 Nexus 交换机运行在 RFC 1583 模式下。例 8-57 给出了解决这类路由环路问题的配置示例。

例 8-57　验证 RFC1583 兼容性

```
NX-3# show run ospf
! Output omitted for brevity
router ospf NXOS
  rfc1583compatibility
```

> **注**：RFC 1583 与 RFC 2328 之间的另一个重要差异就是汇总度量。对于 RFC 1583 来说，ABR 将所有组件路由中的最小度量作为汇总路由的度量，而 RFC 2328 则将所有组件路由中的最大度量作为汇总路由的度量。在 ABR 上配置 **rfc1583compatibility** 命令会改变上述操作行为。

8.4.7　接口链路开销

接口开销是 Dijkstra SPF 计算过程不可缺少的组成部分，因为最短路径度量是从路由器到目的端的接口开销（度量）累计值。OSPF 通过下列公式为接口分配 OSPF 链路开销（度量）。

开销=接口带宽/参考带宽

NX-OS 的默认参考带宽为 40Gbit/s，其他思科 OS（IOS 和 IOS XR）的默认参考带宽为 100Mbit/s。表 8-9 列出了常见网络接口类型的 OSPF 开销（基于默认参考带宽）。

表 8-9　使用默认设置的 OSPF 接口开销

接口类型	默认 NX-OS OSPF 开销	默认 IOS OSPF 开销
T1	无	64
以太网	4000	10
快速以太网	400	1
GE	40	1
40GE	4	1
100GE	1	1

请注意，从表 8-9 可以看出，IOS 路由器与快速以太网接口和 100GE 接口相关联的链路开销并没有任何区别，这样就可能会出现次优路径选择问题，如果路径中插入了 NX-OS 交换机，那么这个问题将会更加明显。

图 8-10 给出了因参考带宽设置错误而导致故障的示例拓扑，图中的两个 WAN 服务提供商之间的连接应该使用 10GE 路径（R1→NX-3→NX-4→R2），而 R1 与 R2 之间的 1GE 链路则仅作为备份路径，因为流量很可能会被仅支持关键业务流量的 QoS 策略丢弃。

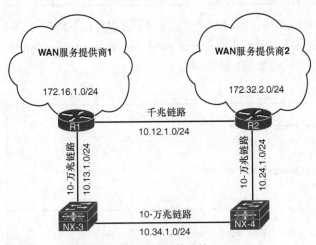

图 8-10　参考带宽问题的示例拓扑结构

例 8-58 显示了基于默认参考带宽的 R1 路由表，可以看出，172.16.1.0/24 与 172.32.2.0/24 之间的流量流经的是备用 1GE 链路（10.12.1.0/24），与期望流量模式不符。请注意，R1 使用 1GE 链路去往网络 172.32.2.0/24 的 OSPF 路径的度量值为 2。

例 8-58　R1 路由表（基于默认的 OSPF 自动开销带宽）

```
R1# show ip route ospf | b Gate
Gateway of last resort is not set

    10.0.0.0/8 is variably subnetted, 6 subnets, 2 masks
O      10.24.1.0/24 [110/2] via 10.12.1.2, 00:10:24, GigabitEthernet0/1
O      10.34.1.0/24 [110/5] via 10.13.1.3, 00:10:09, TenGigabitEthernet1/2
    172.32.0.0/24 is subnetted, 1 subnets
O      172.32.2.0 [110/2] via 10.12.1.2, 00:03:20, GigabitEthernet0/1
```

接下来首先关闭 1GE 链路，然后使用 10GE 路径来检查 OSPF 度量，如例 8-59 所示。请注意，R1 使用 10GE 链路去往网络 172.32.2.0/24 的 OSPF 路径的度量值为 10。

例 8-59　关闭了 1GE 链路之后的 R1 路由表

```
R1# conf t
Enter configuration commands, one per line. End with CNTL/Z.
R1(config)# int gi0/1
R1(config-if)# shut
16:04:43.107: %OSPF-5-ADJCHG: Process 1, Nbr 192.168.2.2 on GigabitEthernet0/1 from
  FULL to DOWN, Neighbor Down: Interface down or detached
16:04:45.077: %LINK-5-CHANGED: Interface GigabitEthernet0/1, changed state to
  administratively down
16:04:46.077: %LINEPROTO-5-UPDOWN: Line protocol on Interface GigabitEthernet0/1,
  changed state to down
R1(config-if)# do show ip route ospf | b Gatewa
Gateway of last resort is not set

     10.0.0.0/8 is variably subnetted, 5 subnets, 2 masks
O       10.12.1.0/24 [110/10] via 10.13.1.3, 00:00:09, TenGigabitEthernet1/2
O       10.24.1.0/24 [110/9] via 10.13.1.3, 00:00:09, TenGigabitEthernet1/2
O       10.34.1.0/24 [110/5] via 10.13.1.3, 00:11:40, TenGigabitEthernet1/2
     172.32.0.0/24 is subnetted, 1 subnets
O       172.32.2.0 [110/10] via 10.13.1.3, 00:00:09, TenGigabitEthernet1/2
```

R1 和 R2 选择次优路径的原因是参考带宽不同，因而需要更改参考带宽以匹配 NX-OS 的默认参考带宽 40Gbit/s。设置 IOS 和 NX-OS 设备参考带宽的命令是 **auto-cost reference-bandwidth** *speed-in-megabits*，例 8-60 给出了更改 R1 和 R2 参考带宽的配置示例。

例 8-60　更改 R1 和 R2 的 OSPF 参考带宽

```
R1(config-if)# router ospf 1
R1(config-router)# auto-cost reference-bandwidth ?
  <1-4294967>  The reference bandwidth in terms of Mbits per second

R1(config-router)# auto-cost reference-bandwidth 40000
% OSPF: Reference bandwidth is changed.
       Please ensure reference bandwidth is consistent across all routers.

R2# conf t
Enter configuration commands, one per line. End with CNTL/Z.
R2(config)# router ospf 1
R2(config-router)# auto-cost reference-bandwidth 40000
% OSPF: Reference bandwidth is changed.
       Please ensure reference bandwidth is consistent across all routers.
```

接下来利用 10GE 路径检查新的 OSPF 度量开销，然后在 R1 上重新激活 1GE 链路。例 8-61 显示了上述操作情况，同时验证了当前连接网络 172.16.1.0/24 与 172.32.2.0/24 的路径情况。

例 8-61　在 R1 和 R2 上配置新的 OSPF 参考带宽后验证新路径

```
R1(config-router)# do show ip route ospf | b Gate
Gateway of last resort is not set

     10.0.0.0/8 is variably subnetted, 5 subnets, 2 masks
O       10.12.1.0/24 [110/13] via 10.13.1.3, 00:01:55, TenGigabitEthernet1/2
O       10.24.1.0/24 [110/12] via 10.13.1.3, 00:01:55, TenGigabitEthernet1/2
O       10.34.1.0/24 [110/8] via 10.13.1.3, 00:01:55, TenGigabitEthernet1/2
     172.32.0.0/24 is subnetted, 1 subnets
O       172.32.2.0 [110/13] via 10.13.1.3, 00:01:55, TenGigabitEthernet1/2

R1# conf t
```

```
Enter configuration commands, one per line. End with CNTL/Z.
R1(config)# int gi0/1
R1(config-if)# no shut
16:09:10.887: %LINK-3-UPDOWN: Interface GigabitEthernet0/1, changed state to up
16:09:11.887: %LINEPROTO-5-UPDOWN: Line protocol on Interface GigabitEthernet0/1,
  changed state to up
16:09:16.623: %OSPF-5-ADJCHG: Process 1, Nbr 192.168.2.2 on GigabitEthernet0/1 from
  LOADING to FULL, Loading Done
R1(config-if)# do show ip route ospf | b Gate
Gateway of last resort is not set
      10.0.0.0/8 is variably subnetted, 6 subnets, 2 masks
         10.24.1.0/24 [110/12] via 10.13.1.3, 00:02:46, TenGigabitEthernet1/2
         10.34.1.0/24 [110/8] via 10.13.1.3, 00:02:46, TenGigabitEthernet1/2
      172.32.0.0/24 is subnetted, 1 subnets
         172.32.2.0 [110/13] via 10.13.1.3, 00:02:46, TenGigabitEthernet1/2
```

可以看出，此时网络 172.16.1.0/24 与 172.32.2.0/24 之间的路径使用的仍然是 10GE 路径，因为 1GE 路径的路径度量开销为 41，即（40000/1000）+ 1（环回）。

注：另一种解决方案是通过命令 **ip ospf cost** *1-65535* 在 NX-OS 和 IOS 设备的接口上静态设置 OSPF 开销。

8.5 本章小结

本章概述了 OSPF 路由协议之后，详细讨论了设备间的邻接关系故障、路由丢失以及路径选择等故障排查方法。

两台路由器必须确保以下参数完全兼容，才能成为邻居。

- 接口必须处于主动状态（Active）。
- 设备之间必须存在使用主用子网的连接。
- 设备之间的 MTU 必须匹配。
- 路由器 ID 必须唯一。
- 接口区域必须匹配。
- DR 必须匹配（基于 OSPF 网络类型）。
- OSPF 末梢区域标志必须匹配。

OSPF 是一种链路状态路由协议，基于 LSA 构建完整的网络拓扑视图。OSPF 路由域出现路由缺失的主要原因通常是网络设计不佳或跨区域边界宣告路由时实施了错误的路由过滤策略。本章列出了一些常见的 OSPF 错误设计方案，这些错误设计会导致路径信息丢失。

OSPF 会建立从计算路由器到全部目的网络的无环拓扑，所有路由器都使用相同的逻辑来计算每个网络的最短路径。路径选择进程按照以下路径选择顺序对路径进行优先级排序。

- 区域内路由。
- 区域间路由。
- 外部 Type-1 路由。
- 外部 Type-2 路由。

如果重分发度量相同，那么 Nexus 交换机将默认基于 RFC 2328 规则选择外部路径（如果存在多个 ABSR，那么区域内连接将优于区域间连接）。思科 IOS 和 IOS XR 路由器基于 RFC 1583 规则选择外部路径，按照最小转发开销来选择 ABSR。如果网络混合部署了 Nexus 交换机和 IOS 或 IOS XR 路由器，那么就可能会产生路由环路问题，此时可以将 Nexus 交换机配置为 RFC 1583 兼容模式。

第 9 章

IS-IS 故障排查

本章主要讨论如下主题。
- IS-IS 基础知识。
- IS-IS 邻居邻接关系故障排查。
- IS-IS 路由丢失故障排查。
- IS-IS 路径选择故障排查。

IS-IS（Intermediate System-to-Intermediate System，中间系统到中间系统）是一种链路状态路由协议，常用于服务提供商和部分企业网络中。IS-IS 提供了快速收敛特性，支持大规模网络，且支持多种协议。思科在很多底层技术中使用了 IS-IS，如 OTV（Overlay Transport Virtualization，叠加传输虚拟化）、ACI（Application Centric Infrastructure，以应用为中心的基础设施）和 SD-Access（Software Defined Access，软件定义接入网）。

本章将重点讨论 IS-IS 邻居邻接关系、路径选择和路由丢失等故障的识别与排查技术。

9.1 IS-IS 基础知识

IS-IS 采用两级分层结构，由 Level 1（L1）和 Level 2（L2）连接组成。IS-IS 通信发生在 L1、L2 或两者（L1-L2）区域，L2 路由器仅与其他 L2 路由器通信，L1 路由器仅与其他 L1 路由器通信，L1-L2 路由器负责在 L1 与 L2 之间提供连接。L2 路由器可以与相同区域或不同区域中的 L2 路由器进行通信，而 L1 路由器只能与相同区域中的其他 L1 路由器进行通信。IS-IS 路由器之间可以建立以下类型的邻接关系。

- L1⟵⟶L1。
- L2⟵⟶L2。
- L1-L2⟵⟶L1。
- L1-L2⟵⟶L2。
- L1-L2⟵⟶L1-L2。

注：本章将频繁使用术语 L1 和 L2，这里仅指 IS-IS 的层级，不要与 OSI 模型相混淆。

IS-IS 利用 LSP（Link-State Packet，链路状态包）构建类似于 OSPF LSDB（Link-State DataBase，链路状态数据库）的 LSPDB（Link-State Packet DataBase，链路状态包数据库），然后运行 Dijkstra SPF（Shortest Path First，最短路径优先）算法构造无环路的最短路径拓扑。

9.1.1 区域

虽然 OSPF 和 IS-IS 使用的都是两层结构，但两种协议的工作方式并不相同。OSPF 提供区域间连接的方式是允许路由器参与多个区域，而 IS-IS 则是将所有路由器及其全部接口都放到特定区域中。OSPF 分层结构中的区域将前缀宣告到骨干区域中，然后通过骨干区域将这些前缀宣告给非骨干区域。L2 是 IS-IS 的骨干区域，与 OSPF 不同的是，只要 L2 的邻接关系连续，L2 就能跨越多个区域。

图 9-1 解释了 OSPF 与 IS-IS 之间的基本区别。可以看出，IS-IS 的骨干域跨越了 4 个区域，而 OSPF 的骨干域仅限于 Area 0。

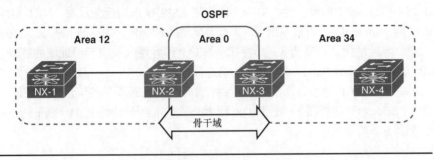

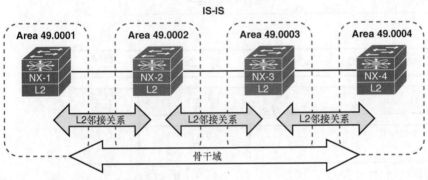

图 9-1 OSPF 与 IS-IS 的区域对比

图 9-2 中的 NX-1 与 NX-2 之间建立了 L1 邻接关系，NX-4 与 NX-5 之间建立了 L1 邻接关系。虽然 NX-2 和 NX-4 是 L1-L2 路由器，但 NX-1 和 NX-5 仅支持 L1 连接，区域地址相同才能建立 L1 邻接关系。NX-2 与 NX-3 之间建立了 L2 邻接关系，NX-3 与 NX-4 之间建立了 L2 邻接关系。NX-2 和 NX-4 是 L1-L2 路由器，可以与其他路由器建立 L1 和 L2 邻接关系。

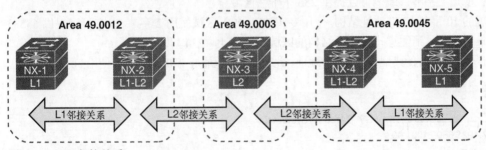

图 9-2 IS-IS 邻接关系

位于相同层级的所有 L1 IS-IS 路由器都维护一个相同的 LSPDB 副本，且所有 L1 路由器都不知道其层级（区域）之外的其他路由器或网络。与此相似，L2 路由器也维护了一个与其他 L2 路由器相同的独立 LSPDB，L2 路由器仅知道 L2 LSPDB 中的其他 L2 路由器及网络。

L1-L2 路由器负责将 L1 前缀注入 L2 拓扑中，L1-L2 路由器不会将 L2 路由宣告到 L1 区域，但是会在 L1 LSP 中设置附加比特，表示该路由器连接在 IS-IS 骨干网络上。如果 L1 路由器没有去往特定网络的路由，那么就会在 LSPDB 中搜索设置了附加比特的最近路由器，由该路由器充当默认网关。

9.1.2 NET 编址

IS-IS 路由器通过 LSP（Link-State Packet，链路状态包）共享区域的拓扑结构，路由器可以据此构建 LSPDB。IS-IS 使用 NET 地址来构建 LSPDB 拓扑，NET 地址包含在所有 LSP 的 IS 报头中。确保路由器在 IS-IS 路由域中的唯一性，对于正确构建 LSPDB 来说至关重要。NET 编址基于 OSI（Open System Interconnection，开放系统互联）参考模型的 NSAP（Network Service Access Point，网络服务接入点）地址结构，长度为 8～20 字节，NSAP 编址形式取决于编址域的逻辑。

NET 地址的 IDP（Inter-Domain Part，域间部分）动态长度可能会产生不必要的混淆。大多数网络工程师是从右到左读取 NET 地址，而不是从左到右读取。对于最简单的编址形式来说，第一个字节始终是 SEL（Selector，选择器）（数值为 00），接下来的 6 字节是系统 ID，剩下的 1～13 字节是区域地址，如图 9-3 所示。

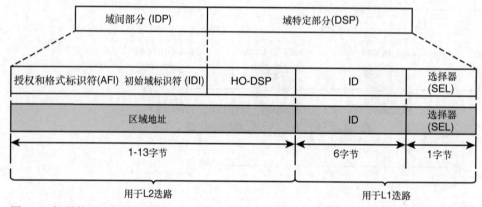

图 9-3 扩展的 NSAP 地址结构

图 9-4 列出了 3 种不同的 NET 编址形式。

- 简单的 8 字节 NET 地址结构。该结构不需要 AFI（Authority and Format Identifier，授权和格式标识符），因为长度不是 NSAP 地址 IDP 的一部分。请注意，区域地址（Area Address）的长度是 1 字节，最多可以提供 256 个唯一区域。
- 常规的 10 字节 NET 地址结构。使用专用 AFI（49），区域使用 2 字节，最多可以提供 65535 个唯一区域。请注意，该地址结构的区域地址是 49.1234。
- 包含了区域地址的典型 OSI NSAP 地址。此时的区域地址是 49.0456.1234，使用专用 AFI（49）。

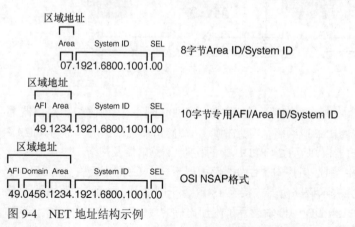

图 9-4 NET 地址结构示例

> 注：从本质上来说，路由器的系统 ID 等同于 EIGRP 或 OSPF 的路由器 ID。NET 地址用于构造网络拓扑结构，因而必须唯一。

9.1.3 路由器间通信

与其他路由协议不同，由于路由器间通信并没有封装在 OSI 模型的第三层（网络层），因而 IS（Intermediate System，中间系统）的通信与协议无关。IS 通信使用 OSI 模型的第二层，IP、IPv6 及其他协议都使用 OSI 模型的第三层编址。

IS PDU（Protocol Data Unit，协议数据单元）（数据包）遵循相同的报头结构，可以通过报头结构来标识 PDU 的类型。每类 PDU 的数据都位于报头之后，最后一个字段使用可选的可变长度字段，包含与该 IS PDU 类型相关的特性信息。

可以将 IS 数据包分为 3 类 PDU，每类 PDU 都可以区分 L1 和 L2 路由信息。

- **IIH（IS-IS Hello）包**：IIH 包负责发现和维护邻居关系。
- **LSP（Link-State Packet，链路状态包）**：LSP 负责提供路由器及相关网络的信息。LSP 与 OSPF LSA 类似，区别在于 OSPF 使用多种 LSA。
- **SNP（Sequence Number Packet，系列号包）**：SNP 负责控制路由器之间的 LSP 同步进程。
  - CSNP（Complete Sequence Number Packet，完全序列号包）为宣告路由器的 LSPDB 提供 LSP 报头，以确保 LSPDB 同步。
  - PSNP（Partial Sequence Numbers Packet，部分序列号包）负责在点到点网络上确认收到 LSP，并在 LSPDB 失步时请求丢失的链路状态信息。

9.1.4 IS 协议报头

每个 IS 数据包都包含一个描述 PDU 的公共报头，8 个字段的长度均为 1 字节，且位于所有数据包中。表 9-1 列出了 IS 协议报头的字段信息。

表 9-1　　　　　　　　　　　　　IS-IS 数据包类型

字段	描述
域内路由协议鉴别符（协议标识符）	网络层标识符由 ISO 分配 IS-IS 通信使用 0x83 ES-IS 通信使用 0x81
PDU 报头长度	PDU 报头的长度，因为它在本质上是动态的
版本	协议版本标识符
系统 ID 长度	系统 ID 在 1~8 字节，网络设备商将其标准化为 6 字节。该字段表示系统 ID 的长度。值为 0 时表示默认长度为 6 字节
PDU 类型	PDU 类型的 1 字节表示形式 用于指示该 PDU 是 Hello、LSP 还是 SNP
保留	表示数据包的层级。值为 1 时表示仅 L1，值为 2 时表示手工模式的 L2
最大区域	取值介于 1~254 之间，表示路由器支持的区域数，默认值为 3

ISO 10589 指出，必须对 IS 数据包报头字段中的 0 值采取特殊处理方式，包括 IS 类型、LSPF 数据库过载比特和最大区域地址字段。0 值表示表 9-1 中的默认设置。

9.1.5 TLV

IS PDU 的一部分使用包含路由信息的可变模块。每个模块都指定了信息类型、数据长度以及值本身，通常称为 TLV（Type, Length, and Value，类型—长度—值）三元组。每个 TLV 都维护一个 1 字节的数字标签，以标识数据的类型（功能）和长度。TLV 支持嵌套功能，因而子 TLV 可以位于另一个 TLV 中。

TLV 为 IS 协议提供了丰富的功能特性和扩展能力。为 IS 协议开发新的功能特征之后，就可以在现有结构中增加新 TLV。例如，增加了 TLV#232（IPv6 接口地址）和#236（IPv6 可达性）之后，就可以为 IS 协议增加 IPv6 支持能力。

9.1.6 IS PDU 编址

IS 设备利用二层地址进行通信，源地址始终是网络接口的二层地址，目的地址则因网络类型的不同而有所变化。Nexus 交换机基于以太网，因而使用二层 MAC 地址进行 IS-IS 通信。

ISO 标准将网络介质分为两类：广播网络（Broadcast network）和通用拓扑网络（General topology network）。

广播网络通过单个数据包向多台设备提供通信。广播接口以多播方式（使用周知二层地址）进行通信，因而只有运行 IS-IS 的节点才会处理这些流量。IS-IS 不会在广播网络上发送单播流量，因为网段上的所有路由器都应该知道网络中的状态。

表 9-2 列出了 IS 通信使用的目的 MAC 地址信息。

表 9-2　　　　　　　　　　　　IS-IS 目标 MAC 地址

名称	目的 MAC 地址
全部 L1 IS（All L1 IS）	0180.c200.0014
全部 L2 IS（All L2 IS）	0180.c200.0015
全部中间系统	0900.2b00.0005
全部终端系统	0900.2b00.0004

通用拓扑网络与网络介质相关，如果设备向外发送了单个数据包，那么只能与一台设备进行通信。IS-IS 文档通常将通用拓扑网络称为点对点网络。点到点网络与定向目的地址进行通信，该目的地址与远程设备的二层地址相匹配。帧中继等 NBMA 技术可能无法保证通过单个数据包与所有设备进行通信，一个常见最佳实践是在 NMBA 技术上使用点对点子接口，以确保 IS-IS 节点之间的正确通信。

9.1.7 IIH（IS-IS Hello）包

IIH（IS-IS Hello）包负责发现和维护邻居。IS-IS 通信过程包含了 5 种 Hello 包（如表 9-3 所示），前 3 种 Hello 包与 IS-IS 邻居邻接关系有关，另外两种 Hello 包与 ES-IS 通信相关。

与其他 IS-IS 路由器建立 L1-L2 邻接关系的路由器会在广播链路上同时发送 L1 和 L2 IIH。为了节省广域网链路带宽，点到点链路使用点到点 Hello 包，可以同时为 L1 和 L2 邻接关系提供服务。

表 9-3 列出了 5 种 IS Hello 包类型的描述信息。

表 9-3　　　　　　　　　　　　IS-IS Hello 类型

类型	描述
L1 IS-IS Hello (IIH) PDU Type 15	发现、建立和维护 L1 IS-IS 邻居
L2 IS-IS Hello (IIH) PDU Type 16	发现、建立和维护 L2 IS-IS 邻居
点到点 Hello (IIH) IS-IS PDU Type 17	发现、建立和维护点到点 IS-IS 邻居
ESH（End System Hello，终端系统 Hello）	用于 ES（End System，终端系统）发现 IS（Intermediate System，中间系统），反之亦然。与 ICMP 类似
ISH（Intermediate System Hello，中间系统 Hello）	用于 ES（End System，终端系统）为路由器选择进程发现 IS（Intermediate System，中间系统），反之亦然

表 9-4 列出了 IIH Hello 包的描述信息。

表 9-4　　　　　　　　　　　IIH 数据包字段

类型	描述
电路类型	0×1 仅 Level 1 0×2 仅 Level 2 0×3 Level 1 和 Level 2
系统 ID	发送 IIH 的路由器的系统 ID
保持定时器	用于该中间系统的保持定时器
PDU 长度	PDU 的总长度
优先级	用来选举 DIS（Designated Intermediate System，指派中间系统）的路由器接口优先级（点到点 IIH 不包含该字段）
DIS 的系统 ID	广播网段的 DIS 的系统 ID（点到点 IIH 不包含该字段）

表 9-5 列出了 IIH PDU 中的常见 TLV。

表 9-5　　　　　　　　　　　IIH PDU 中的常见 TLV

TLV 编号	名称	描述
1	区域地址	宣告路由器的区域地址列表
6	IS 邻居	邻居 IS 路由器的 SNPA（Subnetwork Pint of Attachment，子网连接点）列表，SNPA 是 IS-IS 路由器的二层硬件地址 （点到点 Hello 包没有）
8	填充	负责将数据包扩展到整个 MTU（Maximum Transmission Unit，最大传输单元），忽略该 TLV 中的数据
10	认证	标识认证类型，包含明文密码或 MD5 散列
132	IP 接口地址	传输接口的 IP 地址列表（包含从 IP 地址）
240	邻接状态	用于点到点链路，目的是确保三向 IS-IS 握手

9.1.8　LSP

LSP（Link-State Packet，链路状态包）与 OSPF LSA 相似，负责宣告邻居和附属网络，但 IS-IS 仅使用两种类型的 LSP。IS-IS 为每个层级都定义了一种 LSP，L1 LSP 在源端区域进行泛洪，L2 LSP 则在整个 L2 网络中进行泛洪。

1. LSP ID

LSP ID 是 8 字节定长字段，为 LSP 发起端提供唯一标识。LSP ID 包括以下内容。

- **系统 ID（System ID）（6 字节）**：系统 ID 从路由器配置的 NET 地址中提取。
- **伪节点 ID（Pseudonode ID）（1 字节）**：伪节点 ID 标识特定伪节点（虚拟路由器）或物理路由器的 LSP。伪节点 ID 为 0 的 LSP 描述来自系统的链路，称为非伪节点 LSP。伪节点 ID 不为 0 的 LSP 表示该 LSP 是伪节点 LSP。伪节点 ID 与执行 DIS（Designated Intermediate System，指派中间系统）功能的路由器接口的电路 ID 相关。如果该层级的 DIS 是同一台路由器，那么伪节点 ID 在所有广播网段中都是唯一的。有关伪节点和 DIS 的详细内容将在本章后面进行讨论。
- **分段 ID（Fragment ID）（1 字节）**：如果 LSP 大于向外发送该 LSP 的接口的最大 MTU 值，那么就必须对 LSP 进行分段。IS-IS 在创建 LSP 的时候对 LSP 进行分段，分段 ID 允许接收端路由器处理分段后的 LSP。

图 9-5 给出了两个 LSP ID 示例。左边的 LSP ID 表示这是一台 IS 路由器，右边的 LSP ID 表示这是一台 DIS 路由器（因为伪节点 ID 不为 0）。

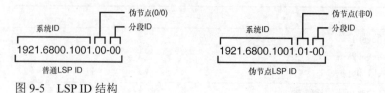

图 9-5　LSP ID 结构

2. 属性字段

LSP 报头的最后一部分是一个 8 比特段，规定了 IS-IS 的 4 个组件。

- **分区比特（Partition Bit）**：分区比特标识路由器是否支持分区修复功能。分区修复功能允许与 L1 路由器属于相同区域的 L2 路由器修复损坏的 L1 区域。思科和大多数网络设备商不支持分区修复功能。
- **连接比特（Attached Bit）**：接下来的 4 比特是连接比特，由通过 L2 骨干网连接其他区域的 L1-L2 路由器进行设置，连接比特位于 L1 LSP 中。
- **过载比特（Overload Bit）**：过载比特指示路由器处于过载状态。计算 SPF 的时候，其他路由器应避免通过该路由器发送流量。恢复正常后，路由器会宣告一条无过载比特的新 LSP，SPF 计算过程也将恢复正常，路由器无须再避开前面曾经过载的节点。
- **路由器类型（Router Type）**：最后两个比特表示该 LSP 来自 L1 还是 L2 路由器。

3. LSP 包和 TLV

表 9-6 列出了 LSP 中的常见 TLV，可以利用这些 TLV 构建拓扑表并将路由安装到 RIB（Routing Information Base，路由信息库）中。

表 9-6　LSP PDU 中的常见 TLV

TLV 编号	名称	描述
1	区域地址（Area Address）	被配置路由器上的区域地址列表
2	IS 邻居（IS Neighbor）	来自邻接 IS 路由器的 SNPA（Subnetwork Point of Address，子网地址点）列表也将相关的接口度量。SNPA 是 IS-IS 路由器的二层硬件地址
10	认证（Authentication）	标识认证类型，包含明文密码或 MD5 散列
128	IP 内部可达性信息（IP Internal Reachability Information）	宣告路由器的内部 IS-IS 网络及接口度量列表
130	IP 外部可达性信息（IP External Reachability Information）	外部路由重分发到 IS-IS 中时，该 TLV 列出了外部（重分发）网络及相关度量的列表，度量可以是内部或外部
132	IP 接口地址（IP Interface Address）	发送接口的 IP 地址列表（包括从 IP 地址）（TLV 中的 IP 地址数量限制为 63 个）
137	主机名（Hostname）	用来标识路由器的路由器主机名，可以代替系统 ID

9.1.9　DIS

广播网络允许网段上存在两台以上的路由器，随着网段上的路由器数量的不断增多，可能会产生 IS-IS 的扩展性问题。路由器数量越多，在网段上泛洪的 LSP 就越多，确保数据库同步所需的资源也就越多。

IS-IS 克服此类低效问题的方法是创建伪节点（虚拟路由器）来管理广播网段的同步问题。广播网段上的路由器被称为 DIS（Designated Intermediate System，指派中间系统），负责承担伪节点的角色。如果 DIS 路由器出现故障，那么就由其他路由器成为新的 DIS 并承担相应的职责。每个 IS-IS 层级（L1 和 L2）都有一个伪节点和 DIS，因而广播网段可以拥有两个伪节点和两个 DIS。

将逻辑伪节点插入广播网段之后，就可以在 LSPDB 中将多路接入网段转换为多个点到点网络。

> 注：虽然人们很自然地将 IS-IS DIS 的行为与 OSPF DR（Designated Router，指派路由器）的行为关联在一起，但两者的操作方式并不相同。所有的 IS-IS 路由器都会相互建立一个完整的邻居邻接关系，所有路由器都可以向网段上的其他 IS-IS 路由器宣告非伪节点 LSP，而 OSPF 则规定将 LSA 发送给 DR，再由 DR 宣告到网段上。

DIS 宣告伪节点 LSP，告诉其他路由器其连接在伪节点上。伪节点 LSP 的作用类似于 OSPF Type-2 LSA，因为其作用是指示相连接的邻居并通知节点充当 DIS 的路由器。连接伪节点的路由器的系统 ID 列在 IS 可达性 TLV 中（接口度量设置为 0），因为 SPF 使用非伪节点 LSP 的度量来计算 SPF 树。

伪节点每 10s 宣告一次 CSNP（Complete Sequence Number Packet，完全序列号包）。IS-IS 路由器会检查其 LSPDB 以验证 CSNP 中列出的所有 LSP 是否存在，且序列号是否与 CSNP 中的版本相匹配。

- 如果 LSP 丢失或者路由器的 LSP 比 CSNP 中的 LSP 过时（序列号较小），那么路由器就会宣告一个 PSNP（Partial Sequence Numbers Packet，部分序列号包），以请求正确或丢失的 LSP。所有 IS-IS 路由器都接收 PSNP，但只有 DIS 发送正确的 LSP，这样就能大大减少网段上的流量。
- 如果路由器检测到 CSNP 中的序列号小于本地存储在 LSPDB 中的 LSP 的序列号，那么就用较大的序列号来宣告本地 LSP。所有 IS-IS 路由器都接收 LSP 并进行相应地处理。DIS 应该发送更新后的 CSNP，其中包含所宣告 LSP 的更新序列号。

9.1.10 路径选择

了解了以下关键定义之后，IS-IS 的路径选择就显得非常简单。
- 区域内路由指的是从相同层级和区域地址中的其他路由器学到的路由。
- 区域间路由指的是从 L2 路由器学到的路由（这些路由来自 L1 路由器或不同区域地址的 L2 路由器）。
- 外部路由指的是重分发到 IS-IS 域的路由。外部路由可以选择两种度量类型。
    - 内部度量，可以与 IS-IS 路径度量进行直接比较，Nexus 交换机默认选择该度量类型。IS-IS 对待这些路由的优先级与通过 TLV#128 宣告的路由相同。
    - 外部度量，不能与内部路径度量进行比较。

IS-IS 最佳路径选择步骤如下，每个阶段都要识别拥有最小路径度量的路由。
- 第 1 步：L1 区域内路由。
  使用内部度量的 L1 外部路由。
- 第 2 步：L2 区域内路由。
  使用内部度量的 L2 外部路由。
  L1→L2 区域间路由。
  使用内部度量的 L1→L2 区域间外部路由。
- 第 3 步：使用内部度量的泄露路由（L2→L1）。
- 第 4 步：使用外部度量的 L1 外部路由。
- 第 5 步：使用外部度量的 L2 外部路由。
  L1→L2 带外部度量的区域间外部路由。

- **第 6 步**：使用外部度量的泄露路由（L2→L1）。

> **注**：正常的 IS-IS 配置仅使用前 3 个步骤。使用外部度量的外部路由要求在重分发时在路由映射中显式指定外部度量类型。

9.2 IS-IS 邻居邻接关系故障排查

讨论了 IS-IS 协议的基本内容之后，本节将简要回顾 NX-OS 的 IS-IS 配置方式并分析邻居邻接关系故障的排查问题。

9.2.1 IS-IS 基本配置

NX-OS 交换机要求在 IS-IS 进程及接口配置子模式下配置 IS-IS。Nexus 交换机的 IS-IS 配置步骤如下。

- **第 1 步**：启用 IS-IS 功能特性。必须通过全局配置命令 **feature isis** 启用 IS-IS 功能特性。
- **第 2 步**：定义 IS-IS 进程标签。必须使用全局配置命令 **router isis** *instance-tag* 定义 IS-IS 进程，其中，*instance-tag* 最长可支持 20 个字母数字字符。
- **第 3 步**：定义 IS-IS NET 地址。必须使用命令 **net** *net-address* 配置 NET 地址。
- **第 4 步**：定义 IS-IS 类型（可选）。Nexus 交换机默认工作在 L1-L2 IS-IS 类型下，意味着与 L1 邻居建立 L1 邻接关系，与 L2 邻居建立 L2 邻接关系，并与其他 L1-L2 IS-IS 对等体建立两条会话（L1 和 L2）。

  可以通过命令 **is-type**{**level-1** | **level-1-2** | **level-2**}更改 IS-IS 路由器类型。
- **第 5 步**：允许将 L1 路由传播给 L2 网络（可选）。Nexus 交换机不会自动将 L1 网络传播给 L2 网络，需要使用命令 **distribute level-1 into level-2** {**all** | **route-map** *route-map-name*}。
- **第 6 步**：在接口上启用 IS-IS。首先利用命令 **interface** *interface-id* 选择需要启用 IS-IS 的接口，然后通过命令 **ip router isis** *instance-tag* 在该接口上启用 IS-IS 进程。

例 9-1 在接口 Ethernet1/1、VLAN 10 和 Loopback 0 上启用了 IS-IS。

例 9-1　IS-IS 基本配置

```
NX-1# configure terminal
Enter configuration commands, one per line.  End with CNTL/Z.
NX-1(config)# feature isis
NX-1(config)# router isis NXOS
13:27:13 NX-1 isis[11140]: ISIS-6-START: Process start. Reason - configuration
NX-1(config-router)# net 49.0012.0000.0000.0001.00
NX-1(config-router)# interface lo0
NX-1(config-if)# ip router isis NXOS
NX-1(config-if)# interface ethernet1/1
NX-1(config-if)# ip router isis NXOS
NX-1(config-if)# interface VLAN10
NX-1(config-if)# ip router isis NXOS
```

```
NX-2# conf t
Enter configuration commands, one per line.  End with CNTL/Z.
NX-2(config)# feature isis
NX-2(config)# router isis NXOS
13:32:22 NX-1 isis[11140]: ISIS-6-START: Process start. Reason - configuration
NX-2(config-router)# net 49.0012.0000.0000.0002.00
NX-2(config-router)# interface lo0
```

```
NX-2(config-if)# ip router isis NXOS
NX-2(config-if)# interface ethernet1/1
NX-2(config-if)# ip router isis NXOS
NX-2(config-if)# interface VLAN20
NX-2(config-if)# ip router isis NXOS
13:33:11 NX-2 %ISIS-5-ADJCHANGE:  isis-NXOS [24168]  LAN adj L2 0000.0000.0001
    over Ethernet1/1 - INIT (New) on MT--1
13:33:11 NX-2 %ISIS-5-ADJCHANGE:  isis-NXOS [24168]  LAN adj L2 0000.0000.0001
    over Ethernet1/1 - UP on MT-0
13:33:11 NX-2 %ISIS-5-ADJCHANGE:  isis-NXOS [24168]  LAN adj L1 0000.0000.0001
    over Ethernet1/1 - INIT (New) on MT--1
13:33:11 NX-2 %ISIS-5-ADJCHANGE:  isis-NXOS [24168]  LAN adj L1 0000.0000.0001
    over Ethernet1/1 - UP on MT-0
```

IS-IS 要求邻接路由器在处理 LSP 之前建立邻接关系。IS-IS 邻居邻接进程包括 3 个状态：Down（关闭）、Initialization（初始化）和 Up（启用）。本节将详细讨论 Nexus 交换机的 IS-IS 邻居邻接关系故障排查过程。

图 9-6 的简单拓扑结构中包含了两台 Nexus 交换机，下面将以此为例讨论 IS-IS 邻接关系故障的排查问题。

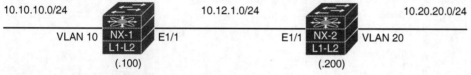

图 9-6　包含两台 NX-OS 交换机的简单拓扑结构

9.2.2　IS-IS 邻居验证

首先通过命令 **show isis adjacency**[**interface** *interface-id*][**detail** | **summary**][**vrf**{*vrf-name*}]验证已成功建立 IS-IS 邻接关系的设备，关键字 **detail** 可以显示邻居的正常运行时间以及邻居节点配置的所有从 IP 地址。

例 9-2 显示了 NX-1 非详细版本的命令输出结果，可以看出，L1 邻接关系有一条表项，L2 邻接关系也有一条单独表项，这是与其他路由器建立 L1-L2 邻接关系的预期结果。

例 9-2　显示 IS-IS 邻居

```
NX-1# show isis adjacency
IS-IS process: NXOS VRF: default
IS-IS adjacency database:
Legend: '!': No AF level connectivity in given topology
System ID       SNPA             Level  State  Hold Time  Interface
NX-2            0021.21ae.c123   1      UP     00:00:07   Ethernet1/1
NX-2            0021.21ae.c123   2      UP     00:00:07   Ethernet1/1
```

表 9-7 列出了例 9-2 显示的字段信息。请注意，NX-2 的保持时间相对较低，因为 NX-2 是网络 10.12.1.0/24 的 DIS。

表 9-7　　　　　　　　　　　　　　IS-IS 邻居状态字段

字段	描述
系统 ID	系统 ID（SEL）从 NET 地址中提取出来
SNPA（Subnetwork Point of Addresses，子网地址点）	IS-IS 路由器的二层硬件地址，Nexus 交换机始终显示 MAC 地址（因为基于以太网）
层级	与邻居建立的邻接关系类型：L1、L2 或 L1-L2

续表

字段	描述
状态	显示邻居处于 Up 或 Down 状态
保持时间	接收其他 IIH 以维护 IS-IS 邻接关系所需的时间
接口	与邻居路由器建立对等关系的接口

> **注**：事实上，系统 ID 引用的是路由器的主机名，而不是 6 字节的系统 ID。IS-IS 在可选的 TLV#137（属于 LSP 的一部分）中提供了一个名称到系统 ID 的映射。可以在 IS-IS 路由器配置下通过命令 **no hostname dynamic** 禁用该功能特性。

例 9-3 显示了使用关键字 **summary** 和 **detail** 的 **show isis adjacency** 命令输出结果。请注意，可选关键字 **detail** 可以显示特定邻居状态切换的精确定时器。

例 9-3　显示 IS-IS 邻居（使用关键字 summary 和 detail）

```
NX-1# show isis adjacency summary
IS-IS process: NXOS VRF: default
IS-IS adjacency database summary:
Legend: '!': No AF level connectivity in given topology
P2P          UP        INIT      DOWN      All
   L1         0         0         0         0
   L2         0         0         0         0
   L1-2       0         0         0         0
   SubTotal   0         0         0         0

LAN          UP        INIT      DOWN      All
   L1         1         0         0         1
   L2         1         0         0         1
   SubTotal   2         0         0         2

Total                   2         0         0         2

NX-1# show isis adjacency detail
IS-IS process: NXOS VRF: default
IS-IS adjacency database:
Legend: '!': No AF level connectivity in given topology
System ID       SNPA             Level   State   Hold Time   Interface
NX-2            0021.21ae.c123   1       UP      00:00:06    Ethernet1/1
  Up/Down transitions: 1, Last transition: 00:38:30 ago
  Circuit Type: L1-2
  IPv4 Address: 10.12.1.200
  IPv6 Address: 0::
  Circuit ID: NX-2.01, Priority: 64
  BFD session for IPv4 not requested
  BFD session for IPv6 not requested
  Restart capable: 1; ack 0;
  Restart mode: 0; seen(ra 0; csnp(0; l1 0; l2 0)); suppress 0

NX-2            0021.21ae.c123   2       UP      00:00:08    Ethernet1/1
  Up/Down transitions: 1, Last transition: 00:38:30 ago
  Circuit Type: L1-2
  IPv4 Address: 10.12.1.200
  IPv6 Address: 0::
  Circuit ID: NX-2.01, Priority: 64
  BFD session for IPv4 not requested
```

```
BFD session for IPv6 not requested
Restart capable: 1; ack 0;
Restart mode: 0; seen(ra 0; csnp(0; l1 0; l2 0)); suppress 0
```

除在 Nexus 交换机的网络接口上启用 IS-IS 之外，还要确保以下参数匹配，才能保证两台交换机成为邻居。

- IS-IS 接口必须处于主动（Active）状态。
- 设备之间必须存在使用主用子网的连接。
- MTU 必须匹配。
- L1 邻接关系要求区域地址匹配对等的 L1 路由器，且系统 ID 在两个邻居之间必须唯一。
- L1 路由器可以与 L1 或 L1-L2 路由器建立邻接关系，但不能与 L2 路由器建立邻接关系。
- L2 路由器可以与 L2 或 L1-L2 路由器建立邻接关系，但不能与 L1 路由器建立邻接关系。
- DIS 要求必须匹配。
- IIH 认证类型和证书必须匹配（如有）。

9.2.3 IS-IS 接口确认

如果某个 IS-IS 层级出现了邻居邻接关系丢失问题，那么就必须验证正确的接口是否正在运行该层级的 IS-IS。命令 **show isis interface**[*interface-id* | **brief**][**level-1** | **level-2**][**vrf** *vrf-name*]可以列出所有启用了 IS-IS 的接口及其相关信息。如果指定了特定接口，那么就可以将输出结果限定为指定接口。

为简洁起见，例 9-4 省略了部分输出结果，重点显示了如下信息。

- IS-IS 接口运行为 L1-L2 接口（思科默认值）。
- 默认 MTU 是以太网的 1500。
- LAN ID（伪节点 ID）是 NX-1.01。
- L1 和 L2 度量被设置为 10（思科默认值）。
- 在 L1 和 L2 层级建立了两个 IS-IS 邻接关系。
- 接口优先级对于 L1 和 L2 来说都是 64（思科默认值）。

例 9-4 IS-IS 接口验证

```
NX-1# show isis interface
! Output omitted for brevity
IS-IS process: NXOS VRF: default
Ethernet1/1, Interface status: protocol-up/link-up/admin-up
 IP address: 10.12.1.100, IP subnet: 10.12.1.0/24
  IPv6 routing is disabled
  Level1
    No auth type and keychain
    Auth check set
  Level2
    No auth type and keychain
    Auth check set
  Index: 0x0002, Local Circuit ID: 0x01, Circuit Type: L1-2
  BFD IPv4 is locally disabled for Interface Ethernet1/1
  BFD IPv6 is locally disabled for Interface Ethernet1/1
  MTR is disabled
  LSP interval: 33 ms, MTU: 1500
  Level   Metric-0  Metric-2   CSNP  Next CSNP   Hello  Multi   Next IIH
  1       4         0          10    00:00:07    10     3       00:00:04
  2       4         0          10    00:00:08    10     3       0.384739
```

```
    Level   Adjs    AdjsUp  Pri  Circuit ID              Since
    1       1       0       64   NX-1.01                 00:57:39
    2       1       0       64   NX-1.01                 00:57:39
    Topologies enabled:
      L MT  Metric  MetricCfg  Fwdng  IPV4-MT  IPV4Cfg  IPV6-MT  IPV6Cfg
      1 0   4       no         UP     UP       yes      DN       no
      2 0   4       no         UP     UP       yes      DN       no       Metric (L1/L2):
    10/10

loopback0, Interface status: protocol-up/link-up/admin-up
    IP address: 192.168.100.100, IP subnet: 192.168.100.100/32
    IPv6 routing is disabled
    Level1
      No auth type and keychain
      Auth check set
    Level2
      No auth type and keychain
      Auth check set
    Index: 0x0001, Local Circuit ID: 0x01, Circuit Type: L1-2
    BFD IPv4 is locally disabled for Interface loopback0
    BFD IPv6 is locally disabled for Interface loopback0
    MTR is disabled
    Level      Metric
    1          1
    2          1
    Topologies enabled:
      L MT  Metric  MetricCfg  Fwdng  IPV4-MT  IPV4Cfg  IPV6-MT  IPV6Cfg
      1 0   1       no         UP     UP       yes      DN       no
      2 0   1       no         UP     UP       yes      DN       no

NX-1# show isis interface brief
IS-IS process: NXOS VRF: default
Interface    Type   Idx State       Circuit    MTU  Metric   Priority  Adjs/AdjsUp

                                                     L1 L2   L1 L2    L1    L2
-------------------------------------------------------------------------------
Topology: TopoID: 0
Vlan10       Bcast  3   Down/Ready  0x02/L1-2 1500  4  4     64 64    0/0   0/0
Topology: TopoID: 0
loopback0    Loop   1   Up/Ready    0x01/L1-2 1500  1  1     64 64    0/0   0/0
Topology: TopoID: 0
VLAN10       Bcast  2   Up/Ready    0x01/L1-2 1500  4  4     64 64    1/0   1/0
Topology: TopoID: 0
VLAN10       Bcast  4   Up/Ready    0x03/L1-2 1500  4  4     64 64    0/0   0/0
```

命令 **show isis** 可以列出所有 IS-IS 接口以及路由器的 IS-IS 配置摘要信息。例 9-5 显示了该命令的输出结果示例，显示了系统 ID、MTU、度量类型、区域地址和拓扑结构模式等信息。

例 9-5 IS-IS 协议验证

```
NX-1# show isis
! Output omitted for brevity
ISIS process : NXOS
 Instance number : 1
VRF: default
 System ID : 0000.0000.0001   IS-Type : L1-L2
 SAP : 412   Queue Handle : 15
 Maximum LSP MTU: 1492
```

```
    Stateful HA enabled
    Graceful Restart enabled. State: Inactive
    Last graceful restart status : none
    Start-Mode Complete
    BFD IPv4 is globally disabled for ISIS process: NXOS
    BFD IPv6 is globally disabled for ISIS process: NXOS
    Topology-mode is base
    Metric-style : advertise(wide), accept(narrow, wide)
    Area address(es) :
      49.0012
    Process is up and running
    VRF ID: 1
    Stale routes during non-graceful controlled restart
    Interfaces supported by IS-IS :
      loopback0
      Ethernet1/1
      Vlan10
    Topology : 0
    Address family IPv4 unicast :
      Number of interface : 3
      Distance : 115
    Address family IPv6 unicast :
      Number of interface : 0
      Distance : 115
    Topology : 2
    Address family IPv4 unicast :
      Number of interface : 0
      Distance : 115
    Address family IPv6 unicast :
      Number of interface : 0
      Distance : 115
    Level1
    No auth type and keychain
    Auth check set
    Level2
    No auth type and keychain
    Auth check set
```

9.2.4 被动接口

有些网络拓扑结构需要将某个网络段宣告给 IS-IS，但又要防止邻居在该网段上建立邻接关系。显示 IS-IS 接口时，被动接口会被显示为 *Inactive*。**show isis interface** 命令可以显示所有 IS-IS 接口及其当前状态，例 9-6 显示了该命令的输出结果，可以看出，接口 Ethernet1/1 对于 L1 来说处于被动状态，但是对于 L2 来说却处于主动状态。

例 9-6 确定是否配置了被动 IS-IS 接口

```
NX-1# show isis interface
! Output omitted for brevity

Ethernet1/1, Interface status: protocol-up/link-up/admin-up
  IP address: 10.12.1.100, IP subnet: 10.12.1.0/24
  IPv6 routing is disabled
  Level1
    No auth type and keychain
    Auth check set
```

```
    Level2
      No auth type and keychain
      Auth check set
    Index: 0x0002, Local Circuit ID: 0x01, Circuit Type: L1-2
    BFD IPv4 is locally disabled for Interface Ethernet1/1
    BFD IPv6 is locally disabled for Interface Ethernet1/1
    MTR is disabled
    Passive level: level-1
    LSP interval: 33 ms, MTU: 1500
    Level-2 Designated IS: NX-2
    Level   Metric-0  Metric-2   CSNP    Next CSNP   Hello   Multi   Next IIH
    1         4         0         10     Inactive     10      3      Inactive
    2         4         0         10     00:00:06     10      3      00:00:03
    Level   Adjs    AdjsUp  Pri  Circuit ID          Since
    1         0        0    64   0000.0000.0000.00   00:01:55
    2         1        1    64   NX-2.01             00:01:57

    Topologies enabled:
      L  MT  Metric  MetricCfg  Fwdng  IPV4-MT  IPV4Cfg  IPV6-MT  IPV6Cfg
      1   0     4       no       UP      DN       yes      DN       no
      2   0     4       no       UP      UP       yes      DN       no
```

找到被动接口之后，接下来需要检查以下配置信息。

- 接口参数命令 **isis passive-interface** {**level-1** | **level-2** | **level-1-2**}，该命令仅将该接口设置为指定 IS-IS 层级的被动状态。
- 全局 IS-IS 配置命令 **passive-interface default** {**level-1** | **level-2** | **level-1-2**}，该命令将 IS-IS 进程下的所有接口都设置为被动状态。接口参数命令 **no isis passive-interface** {**level-1** | **level-2** | **level-1-2**}的优先级高于全局配置命令，因而使该接口处于主动状态。

例 9-7 显示了 NX-1 和 NX-2 的配置信息，此时两台 Nexus 交换机无法在 L1 或 L2 建立 IS-IS 邻接关系。两台交换机的 Ethernet1/1 接口在各个 IS-IS 层级都必须处于主动状态，才能建立邻接关系。让这些接口处于主动状态的方式是删除 NX-1 Ethernet1/1 接口的命令 **isis passive interface level-1**，并在 NX-2 的 Ethernet1/1 上配置命令 **no isis passive interface level-1-2**，从而允许两台交换机在 L1 和 L2 建立邻接关系。

例 9-7　含有被动接口的 IS-IS 配置

```
NX-1# show run isis
! Output omitted for brevity
router isis NXOS
  net 49.0012.0000.0000.0001.00

interface loopback0
  ip router isis NXOS

interface Ethernet1/1
  ip router isis NXOS
  isis passive-interface level-1

interface VLAN10
  ip router isis NXOS
NX-2# show run isis
! Output omitted for brevity
router isis NXOS
  net 49.0012.0000.0000.0002.00
```

```
  passive-interface default level-1-2

interface loopback0
  ip router isis NXOS

interface Ethernet1/1
  ip router isis NXOS
  no isis passive-interface level-1

interface VLAN20
  ip router isis NXOS
```

9.2.5 IS-IS 数据包验证

排查 IS-IS 邻接关系故障的一个重要步骤就是确保设备正在发送或接收 IS-IS 网络流量。**show isis traffic**[**interface** *interface-id*]命令可以从高层视角显示设备收发的各类数据包汇总信息。

例 9-8 显示了该命令的输出结果。请注意，输出结果中的认证差错与其他差错是分开的。如果运行该命令时指定了特定接口，那么就可以更详细地了解该接口收发的数据包信息。

例 9-8 IS-IS 流量统计

```
NX-1# show isis traffic
IS-IS process: NXOS
VRF: default
IS-IS Traffic:
PDU           Received        Sent  RcvAuthErr  OtherRcvErr  ReTransmit
LAN-IIH          30087       11023           0          506         n/a
P2P-IIH              0           0           0            0         n/a
CSNP              4387        4630           0            0         n/a
PSNP                 0           0           0            0         n/a
LSP                353         187           0            0           0
```

可以通过调试功能了解路由器的各种进程信息，具体来说，命令 **debug isis** {**adjacency** | **iih** | **lsp** {**flooding** | **generation**}}可以显示到达交换机管理引擎的数据包的处理情况，用户可以据此验证路由器是否收到或宣告了数据包。

例 9-9 显示了 L1 和 L2 IIH 包的收发情况。

例 9-9 IS-IS Hello 调试

```
NX-1# debug isis iih
NX-1# conf t
Enter configuration commands, one per line. End with CNTL/Z.
NX-1(config)# int Ethernet1/1
NX-1(config-if)# no shut
03:25:37 NX-1 %ETHPORT-5-SPEED: Interface Ethernet1/1, operational speed changed to
  1 Gbps
03:25:37 NX-1 %ETHPORT-5-IF_DUPLEX: Interface Ethernet1/1, operational duplex mode
  changed to Full
03:25:37 NX-1 %ETHPORT-5-IF_RX_FLOW_CONTROL: Interface Ethernet1/1, operational
  Receive Flow Control state changed to off
03:25:37 NX-1 %ETHPORT-5-IF_TX_FLOW_CONTROL: Interface Ethernet1/1, operational
  Transmit Flow Control state changed to off
03:25:37 NX-1 %ETHPORT-5-IF_UP: Interface Ethernet1/1 is up in Layer3
03:25:37.567524 isis: NXOS L2 IIH timer expired for interface Ethernet1/1
03:25:37.567620 isis: NXOS Sending normal restart tlv
03:25:37.567642 isis: NXOS Build L2 LAN IIH for Ethernet1/1 len 1497
```

```
03:25:37.567664 isis: NXOS Send L2 LAN IIH over Ethernet1/1 len 1497 prio 6,dmac
  0180.c200.0015
03:25:37.580195 isis: NXOS L1 IIH timer expired for interface Ethernet1/1
03:25:37.580286 isis: NXOS Sending normal restart tlv
03:25:37.580303 isis: NXOS Build L1 LAN IIH for Ethernet1/1 len 1497
03:25:37.580324 isis: NXOS Send L1 LAN IIH over Ethernet1/1 len 1497 prio 6,dmac
  0180.c200.0014
03:25:37.583037 isis: NXOS Receive L1 LAN IIH over Ethernet1/1 from NX-2 (0021.21ae.
  c123) len 1497 prio 0
03:25:37.583102 isis: NXOS Failed to find IPv6 address TLV MT-0
03:25:37 NX-1 %ISIS-5-ADJCHANGE: isis-NXOS LAN adj L1 NX-2 over Ethernet1/1 - INIT
  (New) on MT--1
03:25:37.583158 isis: NXOS isis_iih_find_bfd_enable: MT 0 : isis_topo_bfd_required =
  FALSE
03:25:37.583176 isis: NXOS isis_iih_find_bfd_enable: MT 0 : isis_topo_usable = TRUE
03:25:37.583193 isis: NXOS isis_receive_lan_iih: isis_bfd_required = 0, isis_
  neighbor_useable 1
03:25:37.583229 isis: NXOS Set adjacency NX-2 over Ethernet1/1 IPv4 address to
  10.12.1.200
03:25:37.583271 isis: NXOS isis_receive_lan_iih BFD TLV: Bring UP adjacency
03:25:37.583295 isis: NXOS 2Way Advt pseudo-lsp : LAN adj L1 NX-2 over Ethernet1/1
03:25:37 NX-1 %ISIS-5-ADJCHANGE: isis-NXOS LAN adj L1 NX-2 over Ethernet1/1 - UP
  on MT-0
03:25:37.583365 isis: NXOS Obtained Restart TLV RR=0, RA=0, SA=0
03:25:37.583383 isis: NXOS Process restart tlv for adjacency NX-2 over Ethernet1/1
  address 10.12.1.200
03:25:37.583397 isis: NXOS Process restart info for NX-2 on Ethernet1/1: RR=no,
  RA=no SA=no
03:25:37.583410 isis: NXOS Restart TLV present SA did not change SA state unsuppress
  adj changed
03:25:37.583467 isis: NXOS    Timer started with holding time 30 sec
03:25:37.583484 isis: NXOS Sending triggered LAN IIH on Ethernet1/1
03:25:37.583501 isis: NXOS Sending triggered LAN IIH on Ethernet1/1
03:25:37.583516 isis: NXOS isis_receive_lan_iih: Triggering DIS election
03:25:37.583571 isis: NXOS LAN IIH parse complete
03:25:37.604100 isis: NXOS Receive L2 LAN IIH over Ethernet1/1 from NX-2 (0021.21ae.
  c123) len 1497 prio 0
```

一般来说，使用调试命令来查找故障根源是最不推荐的方法，因为启用调试操作可能会生成大量数据。NX-OS 在后台提供了事件历史记录（event-history）功能，不影响交换机的性能，为用户提供了另一种有效的故障排查方法。命令 **show isis event-history [adjacency | dis | iih | lsp-flood | lsp-gen]** 可以为 OSPF 邻接关系故障提供非常有用的信息，关键字 **iih** 提供的信息与例 9-9 的调试命令相同。

例 9-10 显示了 **show isis even history iih** 命令的输出结果示例，请注意 NX-1 输出结果的差异，可以看出，显示的信息没有太大差别。

例 9-10　通过 IS-IS 事件历史记录查看 Hello 包信息

```
NX-1# show isis event-history iih
ISIS NXOS process

 iih Events for ISIS process
03:33:27.593010 isis NXOS [11140]: [11145]: 2Way Advt pseudo-lsp : LAN adj L1 NX-2
  over Ethernet1/1
03:33:27.592977 isis NXOS [11140]: [11145]: Set adjacency NX-2 over Ethernet1/1 IPv4
```

```
   address to 10.12.1.200
03:33:27.592957 isis NXOS [11140]: [11145]: isis_receive_lan_iih: isis_bfd_required
   = 0, isis_neighbor_useable 1
03:33:27.592904 isis NXOS [11140]: [11145]: Failed to find IPv6 address TLV MT-0
03:33:27.592869 isis NXOS [11140]: [11145]: Receive L1 LAN IIH over Ethernet1/1 from
   NX-2 (0021.21ae.c123) len 1497 prio 0
03:33:27.590316 isis NXOS [11140]: [11141]: isis_elect_dis(): Sending triggered LAN
   IIH on Ethernet1/1
03:33:27.590253 isis NXOS [11140]: [11141]: Advertising MT-0 adj 0000.0000.0000.00
   for if Ethernet1/1
03:33:27.590241 isis NXOS [11140]: [11141]: Advertising MT-0 adj NX-2.01 for if
   Ethernet1/1
03:33:27.590181 isis NXOS [11140]: [11141]: Send L1 LAN IIH over Ethernet1/1 len
   1497 prio 6,dmac 0180.c200.0014
03:33:27.582343 isis NXOS [11140]: [11145]: Sending triggered LAN IIH on Ethernet1/1
03:33:27.582339 isis NXOS [11140]: [11145]: Sending triggered LAN IIH on Ethernet1/1
03:33:27.582307 isis NXOS [11140]: [11145]: Process restart tlv for adjacency NX-2
   over Ethernet1/1 address 10.12.1.200
03:33:27.582242 isis NXOS [11140]: [11145]: 2Way Advt pseudo-lsp : LAN adj L2 NX-2
   over Ethernet1/1
03:33:27.582207 isis NXOS [11140]: [11145]: Set adjacency NX-2 over Ethernet1/1 IPv4
   address to 10.12.1.200
03:33:27.582154 isis NXOS [11140]: [11145]: isis_receive_lan_iih: isis_bfd_required
   = 0, isis_neighbor_useable 1
03:33:27.582101 isis NXOS [11140]: [11145]: Failed to find IPv6 addr
   ess TLV MT-0
03:33:27.582066 isis NXOS [11140]: [11145]: Receive L2 LAN IIH over Ethernet1/1 from
   NX-2 (0021.21ae.c123) len 1497 prio 0
03:33:27.579283 isis NXOS [11140]: [11141]: Send L2 LAN IIH over Ethernet1/1 len
   1497
prio 6,dmac 0180.c200.0015
```

在交换机上执行 IS-IS 调试操作仅显示到达管理引擎 CPU 的数据包。如果数据包未显示在调试或事件历史记录中，那么就需要检查 QoS（Quality of Service，服务质量）策略、CoPP（Control Plane Policing，控制平面策略），或者验证进入或离开接口的数据包。

接口可能部署也可能未部署 QoS 策略。如果已部署 QoS 策略，则必须检查策略映射是否存在任何丢包，然后将其引用到与 IS-IS 路由协议相匹配的分类映射。由于 CoPP 策略也基于 QoS 设置，因而上述排查过程也同样适用于 CoPP 策略。

例 9-11 给出了检查交换机 CoPP 策略的逻辑过程。

- 使用命令 **show running-config copp all** 检查 CoPP 策略，可以显示相关的策略映射名称、定义的类别和警管速率。
- 验证分类映射，以确定该分类映射的条件匹配设置。
- 验证了分类映射之后，使用命令 **show policy-map interface control-plane** 检查该类别的策略映射。如果发现丢包，那么就需要修改 CoPP 策略以适应更高的 IS-IS 数据包流。

例 9-11 验证 IS-IS 的 CoPP 策略

```
NX-1# show run copp all
! Output omitted for brevity

class-map type control-plane match-any copp-system-p-class-critical
  ..
  match access-group name copp-system-p-acl-mac-l2pt
  match access-group name copp-system-p-acl-mpls-ldp
```

```
    match access-group name copp-system-p-acl-mpls-rsvp
    match access-group name copp-system-p-acl-mac-l3-isis
    match access-group name copp-system-p-acl-mac-otv-isis
    match access-group name copp-system-p-acl-mac-fabricpath-isis
..
policy-map type control-plane copp-system-p-policy-strict
  class copp-system-p-class-critical
    set cos 7
    police cir 36000 kbps bc 250 ms conform transmit violate drop
```

```
NX-1# show run aclmgr all | section copp-system-p-acl-mac-l3-isis
mac access-list copp-system-p-acl-mac-l3-isis
  10 permit any 0180.c200.0015 0000.0000.0000
  20 permit any 0180.c200.0014 0000.0000.0000
  30 permit any 0900.2b00.0005 0000.0000.0000
```

```
NX-1# show policy-map interface control-plane class copp-system-p-class-critical
! Output omitted for brevity
Control Plane
  service-policy input copp-system-p-policy-strict

    class-map copp-system-p-class-critical (match-any)
      ..
      module 1:
        conformed 816984 bytes,
          5-min offered rate 999 bytes/sec
          peak rate 1008 bytes/sec at Wed 16:08:39
        violated 0 bytes,
          5-min violate rate 0 bytes/sec
          peak rate 0 bytes/sec
```

注：该 CoPP 策略源自 Nexus 7000 交换机，实际的策略名称和分类映射可能会因不同的平台而异。

查看数据包是否到达 Nexus 交换机的另一种可选技术就是交换机内置的 Ethanalyzer。使用 Ethanalyzer 的原因是 IS-IS 使用二层编址，可以限制三层端口的抓包。此时需要使用命令 **ethanalyzer local interface inband [capture-filter "ether host** *isis-mac-address*"**] [detail]**，其中的 **capture-filter** 可以将流量限制为特定类型流量，过滤器 **ether host** *isis-mac-address* 可以根据表 9-2 中的数值将流量限制为 IS-IS。可选关键字 **detail** 可以提供匹配流量的数据包级视图。例 9-12 给出了利用 Ethanalyzer 识别 L2 IIH 包的应用示例。

例 9-12 利用 Ethanalyzer 验证 IS-IS 数据包

```
NX-1# ethanalyzer local interface inband capture-filter "ether host
  01:80:c2:00:00:15"

Capturing on inband
09:08:42.979127 88:5a:92:de:61:7c -> 01:80:c2:00:00:15 ISIS L2 HELLO,
 System-ID: 0000.0000.0001
09:08:46.055807 88:5a:92:de:61:7c -> 01:80:c2:00:00:15 ISIS L2 HELLO,
 System-ID: 0000.0000.0001
09:08:47.489024 88:5a:92:de:61:7c -> 01:80:c2:00:00:15 ISIS L2 CSNP,
 Source-ID: 0000.0000.0001.00, Start LSP-ID: 0000.0000.0000.00-00, End LSP-ID: ff
 ff.ffff.ffff.ff-ff
09:08:48.570401 00:2a:10:03:f2:80 -> 01:80:c2:00:00:15 ISIS L2 HELLO,
 System-ID: 0000.0000.0002
09:08:49.215861 88:5a:92:de:61:7c -> 01:80:c2:00:00:15 ISIS L2 HELLO,
 System-ID: 0000.0000.0001
```

```
09:08:52.219001 88:5a:92:de:61:7c -> 01:80:c2:00:00:15 ISIS L2 HELLO,
 System-ID: 0000.0000.0001

NX-1# ethanalyzer local interface inband capture-filter "ether host
  01:80:c2:00:00:15" detail

Capturing on inband
Frame 1 (1014 bytes on wire, 1014 bytes captured)
    Arrival Time: May 22, 2017 09:07:16.082561000
    [Time delta from previous captured frame: 0.000000000 seconds]
    [Time delta from previous displayed frame: 0.000000000 seconds]
    [Time since reference or first frame: 0.000000000 seconds]
    Frame Number: 1
    Frame Length: 1014 bytes
    Capture Length: 1014 bytes
    [Frame is marked: False]
    [Protocols in frame: eth:llc:osi:isis]
IEEE 802.3 Ethernet
    Destination: 01:80:c2:00:00:15 (01:80:c2:00:00:15)
        Address: 01:80:c2:00:00:15 (01:80:c2:00:00:15)
        .... ...1 .... .... .... .... = IG bit: Group address (multicast/broadcast)
        .... ..0. .... .... .... .... = LG bit: Globally unique address (factory
 default)
    Source: 88:5a:92:de:61:7c (88:5a:92:de:61:7c)
        Address: 88:5a:92:de:61:7c (88:5a:92:de:61:7c)
        .... ...0 .... .... .... .... = IG bit: Individual address (unicast)
        .... ..0. .... .... .... .... = LG bit: Globally unique address (factory
 default)
    Length: 1000
Logical-Link Control
    DSAP: ISO Network Layer (0xfe)
    IG Bit: Individual
    SSAP: ISO Network Layer (0xfe)
    CR Bit: Command
    Control field: U, func=UI (0x03)
        000. 00.. = Command: Unnumbered Information (0x00)
        .... ..11 = Frame type: Unnumbered frame (0x03)
ISO 10589 ISIS InTRA Domain Routeing Information Exchange Protocol
    Intra Domain Routing Protocol Discriminator: ISIS (0x83)
    PDU Header Length   : 27
    Version (==1)       : 1
    System ID Length    : 0
    PDU Type            : L2 HELLO (R:000)
    Version2 (==1)      : 1
    Reserved (==0)      : 0
    Max.AREAs: (0==3)   : 0
    ISIS HELLO
        Circuit type            : Level 2 only, reserved(0x00 == 0)
        System-ID {Sender of PDU} : 0000.0000.0001
        Holding timer           : 9
        PDU length              : 997
        Priority                : 64, reserved(0x00 == 0)
        System-ID {Designated IS} : 0000.0000.0001.01
        Area address(es) (4)
            Area address (3): 49.0012
        Protocols Supported (1)
            NLPID(s): IP (0xcc)
```

```
            IP Interface address(es) (4)
              IPv4 interface address   : 10.12.1.100 (10.12.1.100)
            IS Neighbor(s) (6)
              IS Neighbor: 00:2a:10:03:f2:80
            Restart Signaling (1)
              Restart Signaling Flags  : 0x00
                  .... .0.. = Suppress Adjacency: False
                  .... ..0. = Restart Acknowledgment: False
                  .... ...0 = Restart Request: False
            Padding (255)
            Padding (255)
            Padding (255)
            Padding (171)
```

9.2.6 必须存在使用主用子网的连接

虽然 IS-IS 运行在 OSI 模型的第二层，但主 IP 地址必须与对等 IS-IS 路由器位于相同的网络上。IS-IS IIH 数据包包含接口 IP 地址，接收端路由器必须能够解析去往收到 IIH 包的接口的直连路由，从而添加到 IIH IS 邻居表项中。如果路由器没有在 IIH IS 邻居表项中看到自己，那么就会将会话保持在 INIT 状态，而不会进入 UP 状态。

本节将 NX-2 的子网掩码从 10.12.1.200/24 更改为 10.12.1.200/25。从而将 NX-2 调整到网络 10.12.1.128/25 中，该网络与 NX-1 所处的网络（10.12.1.100）不同。

检查 IS-IS 邻居表后可以看出，NX-1 与 NX-2 处于 INIT 状态，但 NX-2 没有检测到 NX-1，如例 9-13 所示。

例 9-13　NX-1 与 NX-2 卡在 INIT 状态

```
NX-1# show isis adjacency
IS-IS process: NXOS VRF: default
IS-IS adjacency database:
Legend: '!': No AF level connectivity in given topology
System ID       SNPA            Level  State  Hold Time  Interface
NX-2            0021.21ae.c123  1      INIT   00:00:29   Ethernet1/1
NX-2            0021.21ae.c123  2      INIT   00:00:23   Ethernet1/1

NX-2# show isis adjacency
IS-IS process: NXOS VRF: default
IS-IS adjacency database:
Legend: '!': No AF level connectivity in given topology
System ID       SNPA            Level  State  Hold Time  Interface
```

接下来需要检查 NX-1 和 NX-2 的 IS-IS 事件历史记录，以了解邻接关系及 IIH 包信息。可以看出，NX-1 有 NX-2 的邻接表项，但 NX-2 没有任何邻接表项。检查完 IIH 事件历史记录之后可以看出，NX-2 显示无法找到可用 IP 地址，如例 9-14 所示。

例 9-14　NX-1 和 NX-2 事件历史记录

```
NX-1# show isis event-history adjacency
ISIS NXOS process

 adjacency Events for ISIS process
04:33:36.052173 isis NXOS [11140]: : Set adjacency NX-2 over Ethernet1/1 IPv4
   address to 10.12.1.200
04:33:36.052112 isis NXOS [11140]: : LAN adj L2 NX-2 over Ethernet1/1 - INIT (New)
   T -1
04:33:36.052105 isis NXOS [11140]: : isis_init_topo_adj LAN adj 2 NX-2 over
```

```
                          Ethernet1/1 - LAN MT-0
04:33:30.612053 isis NXOS [11140]: : Set adjacency NX-2 over Ethernet1/1 IPv4
  address to 10.12.1.200
04:33:30.611992 isis NXOS [11140]: : LAN adj L1 NX-2 over Ethernet1/1 - INIT (New)
  T -1
04:33:30.611986 isis NXOS [11140]: : isis_init_topo_adj LAN adj 1 NX-2 over
  Ethernet1/1 - LAN MT-0

NX-1# show isis event-history iih
ISIS NXOS process

 iih Events for ISIS process
04:40:30.890260 isis NXOS [11140]: [11141]: Send L1 LAN IIH over Ethernet1/1 len
  1497 prio 6,dmac 0180.c200.0014
04:40:28.712993 isis NXOS [11140]: [11145]: Process restart tlv for adjacency
  0000.0000.0002 over Ethernet1/1 address 10.12.1.200
04:40:28.712988 isis NXOS [11140]: [11145]: Neighbor TLV missing in hello from
  0000.0000.0002 , hence adjacency in INIT state
04:40:28.712986 isis NXOS [11140]: [11145]: Fail to find iih nbr tlv
04:40:28.712946 isis NXOS [11140]: [11145]: isis_receive_lan_iih: isis_bfd_required
  = 0, isis_neighbor_useable 1
04:40:28.712941 isis NXOS [11140]: [11145]: Failed to find IPv6 address TLV MT-0
04:40:28.712896 isis NXOS [11140]: [11145]: Receive L2 LAN IIH over Ethernet1/1 from
  0000.0000.0002 (0021.21ae.c123) len 1497 prio 0
04:40:27.023004 isis NXOS [11140]: [11145]: Process restart tlv for adjacency
  0000.0000.0002 over Ethernet1/1 address 10.12.1.200
04:40:27.022997 isis NXOS [11140]: [11145]: Neighbor TLV missing in hello from
  0000.0000.0002 , hence adjacency in INIT state

NX-2# show isis event-history adjacency
ISIS NXOS process

 adjacency Events for ISIS process

NX-2# show isis event-history iih
ISIS NXOS process

 iih Events for ISIS process
04:39:22.419356 isis NXOS [24168]: [24185]: Receive L1 LAN IIH over
Ethernet1/1 from 0000.0000.0001 (0012.34ed.82a8) len 1497 prio 0
04:39:18.419396 isis NXOS [24168]: [24185]: Failed to find IPv6 address TLV MT-0
04:39:18.419394 isis NXOS [24168]: [24185]: isis_iih_find_ipv4_addr: Unable to find
  IPv4 address for Ethernet1/1
04:39:18.419385 isis NXOS [24168]: [24185]: Fail to find usable IPv4 address
04:39:18.419356 isis NXOS [24168]: [24185]: Receive L2 LAN IIH over
Ethernet1/1 from 0000.0000.0001 (0012.34ed.82a8) len 1497 prio 0
04:39:15.939106 isis NXOS [24168]: [24185]: Failed to find IPv6 address TLV MT-0
04:39:15.939104 isis NXOS [24168]: [24185]: isis_iih_find_ipv4_addr: Unable to find
  IPv4 address for Ethernet1/1
04:39:15.939095 isis NXOS [24168]: [24185]: Fail to find usable IPv4 address
```

接下来检查并修正两台 IS-IS 路由器接口上的 IP 地址/子网掩码，以便建立连接。

9.2.7 MTU 要求

IS-IS Hello（IIH）包用 TLV#8 进行填充，以充满网络接口的 MTU（Maximum Transmission Unit，最大传输单元）。填充 IIH 的好处是能够检测远程接口上的超大帧或 MTU 不匹配差错。如果两个接口的 MTU 相同，那么广播接口传送 L1 和 L2 IIH 就会浪费带宽。

为了解释 MTU 不匹配的故障排查过程，将 NX-1 的 MTU 设置为 1000，而 NX-2 的 MTU 则保持为 1500。

首先检查 IS-IS 的邻接状态（见例 9-15），可以看出，NX-1 未检测到 NX-2，而 NX-2 检测到了 NX-1。

例 9-15　NX-1 未检测到 NX-2

```
NX-1# show isis adjacency
! Output omitted for brevity
System ID        SNPA              Level  State   Hold Time   Interface

NX-2# show isis adjacency
! Output omitted for brevity
System ID        SNPA              Level  State   Hold Time   Interface
NX-1             0012.34ed.82a8    1      INIT    00:00:29    Ethernet1/1
NX-1             0012.34ed.82a8    2      INIT    00:00:29    Ethernet1/1
```

接下来检查 IS-IS IIH 事件历史记录以确定问题原因。从例 9-16 可以看出，NX-1 正在发送长度为 997 的 IIH 包，且 NX-2 收到了这些数据包；NX-2 正在向 NX-1 发送长度为 1497 的 IIH 包，且 NX-1 收到了这些数据包。IIH 的包长表明出现了 MTU 问题。

例 9-16　MTU 不匹配时的 NX-1 IS-IS 邻接关系事件历史记录

```
NX-3# show isis event-history iih
ISIS NXOS process

 iih Events for ISIS process
15:25:30.583389 isis NXOS [13932]: [13933]: Send L1 LAN IIH over Ethernet1/1 len 997
    prio 6,dmac 0180.c200.0014
15:25:29.536721 isis NXOS [13932]: [13933]: Send L2 LAN IIH over Ethernet1/1 len 997
    prio 6,dmac 0180.c200.0015
15:25:25.824258 isis NXOS [13932]: [13937]: Process restart tlv for adjacency NX-2
    over Ethernet1/1 address 10.12.1.200
15:25:25.824168 isis NXOS [13932]: [13937]: Failed to find IPv6 address TLV MT-0
15:25:25.824094 isis NXOS [13932]: [13937]: Receive L1 LAN IIH over Ethernet1/1 from
    NX-2 (002a.1003.f280) len 1497 prio 0
15:25:25.281611 isis NXOS [13932]: [13937]: Process restart tlv for adjacency NX-2
    over Ethernet1/1 address 10.12.1.200
15:25:25.281521 isis NXOS [13932]: [13937]: Failed to find IPv6 address TLV MT-0
15:25:25.281446 isis NXOS [13932]: [13937]: Receive L2 LAN IIH over Ethernet1/1 from
    NX-2 (002a.1003.f280) len 1497 prio 0
15:25:18.019441 isis NXOS [13932]: [13937]: Receive L1 LAN IIH over Ethernet1/1 from
    NX-2 (002a.1003.f280) len 1497 prio 0
15:25:17.456734 isis NXOS [13932]: [13933]: Send L2 LAN IIH over Ethernet1/1 len 997
    prio 6,dmac 0180.c200.0015
15:25:15.166714 isis NXOS [13932]: [13933]: Send L1 LAN IIH over Ethernet1/1 len 997
    prio 6,dmac 0180.c200.0014
```

此时可以使用命令 **show interface** *interface-id* 检查两台交换机的 MTU 值（见例 9-17），可以看出，NX-2 的 MTU 大于 NX-1。

例 9-17　检查接口 MTU

```
NX-1# show interface e1/1 | i MTU
  MTU 1000 bytes, BW 10000000 Kbit, DLY 10 usec

NX-2# show interface e1/1 | i MTU
  MTU 1500 bytes, BW 1000000 Kbit, DLY 10 usec
```

思科引入了一项功能特性，可以让路由器从接口向外发送前 5 个 IIH 包之后禁用 MTU 填充功能。这样做不但解决了带宽浪费问题，而且提供了一种检查路由器之间 MTU 值的有效机制。Nexus 交换机使用接口参数命令 **no isis hello padding [always]** 禁用 IIH 填充功能，关键字 **always** 表示不填充任何 IIH 包，虽然这样做能够让 NX-1 建立邻接关系，但后续仍有可能出现问题。最佳解决方案是将接口 MTU 修改为两台设备接口可以接受的最大 MTU。

> 注：如果 IS-IS 接口是 VLAN 接口（SVI），那么就必须确保所有二层（L2）端口都支持在 SVI 上配置 MTU。例如，如果 VLAN 10 的 MTU 为 9000，那么就应该配置所有中继端口均支持 MTU 值 9000。

9.2.8 唯一的系统 ID

系统 ID 为同一区域中的 IS-IS 路由器提供了唯一的标识符。作为安全机制的一部分，Nexus 交换机会丢弃与自己系统 ID 相同的数据包，此时将显示系统日志消息 "*Duplicate system ID*" 以及其他设备的接口和系统 ID。例 9-18 给出了 NX-2 的系统日志消息示例。

例 9-18 系统 ID 重复

```
05:48:56 NX-2 %ISIS-4-LAN_DUP_SYSID:  isis-NXOS [24168]  L1 LAN IIH - Duplicate
  system ID 0000.0000.0001 detected over Ethernet1/1 from 0012.34ed.82
05:48:57 NX-2 %ISIS-4-SYSLOG_SL_MSG_WARNING: ISIS-4-LAN_DUP_SYSID: message repeated
  12 times in last 237176 sec
```

通常情况下，复制另一台交换机的 IS-IS 配置时可能会出现系统 ID 重复问题。必须更改 NET 地址的系统 ID 部分，才能建立邻接关系。

9.2.9 L1 邻接关系要求区域必须匹配

如前所述，IS-IS 路由器可以工作在 L1、L2 或 L1-L2 层级。L1 邻接关系位于区域内，只能与其他 L1 或 L1-L2 路由器建立邻接关系。L1 邻接关系要求必须与将要建立 L1 邻接关系的路由器的区域 ID 匹配。

例 9-19 显示了 NX-1 和 NX-2 的 IS-IS 邻接表，可以看出，两台 Nexus 交换机均已建立 L2 邻接关系，但是并没有建立本章前面所说的 L1 邻接关系。

例 9-19 只有 L2 IS-IS 邻接关系

```
NX-1# show isis adjacency
! Output omitted for brevity
System ID         SNPA              Level  State  Hold Time   Interface
NX-2              002a.1003.f280    2      UP     00:00:28    Ethernet1/1

NX-2# show isis adjacency
! Output omitted for brevity
System ID         SNPA              Level  State  Hold Time   Interface
NX-1              885a.92de.617c    2      UP     00:00:09    Ethernet1/1
```

经过逻辑推导，由于 NX-1 和 NX-2 已经建立了 L2 邻接关系，因而两者可以建立并维持 IS-IS 数据包的双向传输，从而表明认证参数不正确、定时器无效或区域 ID 不匹配。

例 9-20 显示了 NX-1 和 NX-2 的 IS-IS 事件历史记录。请注意，在指示收到 L1 IIH 消息的前面出现了差错消息 "*No common area*"。

例 9-20 IS-IS 事件历史记录表明区域不匹配

```
NX-1# show isis event-history iih
ISIS NXOS process
```

```
 iih Events for ISIS process
03:30:01.385298 isis NXOS [27230]: [27235]: Failed to find IPv6 address TLV MT-0
03:30:01.385260 isis NXOS [27230]: [27235]: Receive L2 LAN IIH over Ethernet1/1 from
  NX-2 (002a.1003.f280) len 1497 prio 0
03:30:00.470215 isis NXOS [27230]: [27231]: Send L2 LAN IIH over Ethernet1/1 len
  1497 prio 6,dmac 0180.c200.0015
03:29:57.250206 isis NXOS [27230]: [27231]: Send L2 LAN IIH over Ethernet1/1 len
  1497 prio 6,dmac 0180.c200.0015
03:29:57.095233 isis NXOS [27230]: [27235]: No common area
03:29:57.095231 isis NXOS [27230]: [27235]: Failed to find IPv6 address TLV MT-0
03:29:57.095199 isis NXOS [27230]: [27235]: Receive L1 LAN IIH over Ethernet1/1 from
  NX-2 (002a.1003.f280) len 1497 prio 0
```

```
NX-2# show isis event-history iih
ISIS NXOS process

 iih Events for ISIS process
03:29:52.986467 isis NXOS [12392]: [12442]: Receive L2 LAN IIH over Ethernet1/1 from
  NX-1 (885a.92de.617c) len 1497 prio 0
03:29:520.780227 isis NXOS [12392]: [12404]: Send L2 LAN IIH over Ethernet1/1 len
  1497 prio 6,dmac 0180.c200.0015

03:29:51.966543 isis NXOS [12392]: [12442]: No common area
03:29:51.966542 isis NXOS [12392]: [12442]: Failed to find IPv6 address TLV MT-0
03:29:51.966510 isis NXOS [12392]: [12442]: Receive L1 LAN IIH over Ethernet1/1 from
  NX-1 (885a.92de.617c) len 1497 prio 0
```

最后一步就是验证配置并检查 NET 编址。例 9-21 显示了 NX-1 和 NX-2 的 NET 表项，可以看出，NX-1 的区域号是 49.0012，NX-2 的区域号是 49.0002。

例 9-21 验证 NET 编址

```
NX-1# show run isis
! Output omitted for brevity
router isis NXOS
  net 49.0012.0000.0000.0001.00
```
```
NX-2# show run isis
! Output omitted for brevity
router isis NXOS
  net 49.0002.0000.0000.0002.00
```

因此，更改 NET 地址的区域部分，使其与任意一台 Nexus 交换机相匹配，即可建立 L1 邻接关系。

9.2.10 检查 IS-IS 邻接能力

IS-IS 路由器没有相应的机制来检测其区域是否位于 L2 骨干网的末端（边缘）或中间（转接），只有操作人员才能将区域识别为转接区域，所以思科默认所有路由器都是 L1-L2。虽然该默认行为能够确保所有路由器都能路由转接流量，但同时也限制了协议的可扩展性。

可以通过 IS-IS 配置命令 **is-type** {level-1 | level-1-2 | level-2-only} 设置 Nexus 交换机运行的 IS-IS 层级。

查看 IS-IS 进程即可验证上述设置，如例 9-22 所示。

例 9-22 验证 IS-IS 进程的层级类型

```
NX-1# show isis
ISIS process : NXOS
```

```
  Instance number : 1
  UUID: 1090519320
  Process ID 27230
 VRF: default
  System ID : 0000.0000.0001 IS-Type : L2
NX-1# show run isis
! Output omitted for brevity
router isis NXOS
  net 49.0012.0000.0000.0001.00
  is-type level-2
```

某些拓扑结构设计可能要求特定接口应该仅建立特定 IS-IS 层级的邻接关系，此时就可以使用接口参数命令 **isis circuit-type {level-1 | level-1-2 | level-2-only}**。

查看 IS-IS 进程即可验证上述设置（见例 9-23），可以看出，Ethernet1/1 被设置为仅允许 L1 连接。

例 9-23 验证 IS-IS 接口的层级类型

```
NX-1# show isis interface | i protocol|Type
loopback0, Interface status: protocol-up/link-up/admin-up
  Index: 0x0001, Local Circuit ID: 0x01, Circuit Type: L1-2
Ethernet1/1, Interface status: protocol-up/link-up/admin-up
  Index: 0x0002, Local Circuit ID: 0x01, Circuit Type: L1
EthernetVlan10, Interface status: protocol-down/link-down/admin-down
  Index: 0x0003, Local Circuit ID: 0x02, Circuit Type: L1-2
NX-1# show run isis
! Output omitted for brevity
router isis NXOS
  net 49.0012.0000.0000.0001.00
  is-type level-1-2

interface loopback0
  ip router isis NXOS

interface Ethernet1/1
  isis circuit-type level-1
  ip router isis NXOS

interface EthernetVlan10
  ip router isis NXOS
```

可以将 Nexus 交换机设置为特定的 IS-IS 层级，且电路设置与全局 IS-IS 设置不同。如果设置不同，那么 Nexus 交换机就会在建立邻接关系时使用最严格的层级。表 9-8 列出了路由器能够建立的邻接关系类型（仅基于 IS-IS 路由器类型和 IS-IS 电路类型）。

表 9-8　　　　　　　　　　IS-IS 邻居邻接能力

	路由器 IS 类型 L1	路由器 IS 类型 L2	路由器 IS 类型 L1-L2
电路类型 L1	L1	不配置（默认值）	L1
电路类型 L2	不配置（默认值）	L2	L2
电路类型 L1-L2	L1	L2	L1 和 L2

如果事件历史记录缺失 IIH 包，那么就需要在两台路由器上验证 IS-IS 路由器和接口层级的设置情况。

9.2.11　DIS 要求

Nexus 交换机的默认 IS-IS 接口是广播接口，因而需要设置 DIS。但是，只与两台 IS-IS 路由器直连的广播接口不仅无法获得伪节点的好处，还要耗费资源来选择 DIS，不但要将 CSNP 持续泛洪到网段中，而且该层级的所有路由器的 LSPDB 都必须包含不必要的伪节点 LSP。IS-IS 可以通过接口命令 **isis network point-to-point** 让通用拓扑接口按照点到点接口方式运行。

如果两台 IS-IS Nexus 交换机分别拥有广播接口和 IS-IS 点对点接口，那么就无法建立邻接关系，两台设备均不显示 IS-IS 邻接关系，而且通用拓扑交换机还会在 IS-IS IIH 事件历史记录中报告消息 "*Fail: Receiving P2P IIH over LAN interface xx*"，IS-IS 事件历史记录可以显示宣告 P2P 接口的邻居。如果检测到这些消息，那么就要在其中的某个节点上更改接口类型，以确保接口的一致性。

例 9-24 显示了 NX-2 的 IS-IS 事件历史记录以及 NX-1 和 NX-2 的相关配置。

例 9-24　IS-IS 接口类型不匹配

```
NX-2# show isis event-history iih
ISIS NXOS process

 iih Events for ISIS process
02:50:33.000228 isis NXOS [24168]: [24169]: Send L2 LAN IIH over Ethernet1/1 len
   1497 prio 6,dmac 0180.c200.0015
02:50:30.200875 isis NXOS [24168]: [24185]: P2P IIH parse failed!
02:50:30.200873 isis NXOS [24168]: [24185]: Fail: Receiving P2P IIH over LAN
   interface Ethernet1/1
02:50:30.200870 isis NXOS [24168]: [24185]: Receive P2P IIH over Ethernet1/1 from
   NX-1 len 1497 prio 0
02:50:25.390172 isis NXOS [24168]: [24169]: Send L1 LAN IIH over Ethernet1/1 len
   1497 prio 6,dmac 0180.c200.0014
NX-1# show run isis
! Output omitted for brevity
router isis NXOS
  net 49.0012.0000.0000.0001.00

interface loopback0
  ip router isis NXOS
interface Ethernet1/1
  isis network point-to-point
  ip router isis NXOS

interface EthernetVlan10
  ip router isis NXOS

NX-2# show run isis
! Output omitted for brevity
router isis NXOS
  net 49.0012.0000.0000.0002.00

interface loopback0
  ip router isis NXOS

interface Ethernet1/1
  ip router isis NXOS

interface EthernetVlan20
  ip router isis NXOS
```

在 NX-2 的 Ethernet1/1 接口上增加命令 **isis network point-to-point** 之后，即可将两个接口设置为同一类型，从而允许建立邻接关系。

9.2.12　IIH 认证

IS-IS 允许对需要建立邻接关系的 IIH 数据包进行认证。IIH 认证需要对逐个接口进行配置，IIH 认证机制为每个 IS-IS 层级使用不同的设置。对于大多数设计方案来说，认证一种 PDU 就足够了。

IS-IS 提供了两种认证机制：明文和 MD5 加密散列。明文模式的安全性很低，因为任何可以访问链路的人都能通过网络嗅探器看到密码。MD5 加密散列使用散列算法，因而永远也不会在 PDU 中包含密码，目前人们已广泛接受这种更为安全的认证模式。所有的 IS-IS 认证参数都存储在 TLV#10 中（属于 IIH 的一部分）。

Nexus 交换机使用接口参数命令 **isis authentication key-chain** *key-chain-name*{**level-1** | **level-2**}启用 IIH 认证机制，通过命令 **isis authentication-type** {**md5** | **cleartext**}{**level-1** | **level-2**}选择认证类型。

例 9-25 显示 NX-1 的 Ethernet1/1 接口启用了 MD5 认证。

例 9-25　NX-1 的 L1 IIH 认证模式

```
NX-1# conf t
Enter configuration commands, one per line. End with CNTL/Z.
NX-1(config)# key chain IIH-AUTH
NX-1(config-keychain)# key 2
NX-1(config-keychain-key)# key-string CISCO
NX-1(config-keychain-key)# interface Ethernet1/1
NX-1(config-if)# isis authentication key-chain CISCO level-1
NX-1(config-if)# isis authentication-type md5 level-1
```

在 NX-1 的 Ethernet1/1 接口上配置了 L1 IIH 认证之后，NX-1 与 NX-2 之间的 L1 邻接关系将中断。事实上，NX-2 一直都在试图启动与 NX-1 之间的会话，但始终被卡在 INIT 状态，如例 9-26 所示。由于 L2 认证模式未做任何变更，因而 L2 邻接关系始终保持正常。

例 9-26　NX-1 的 L1-IIH 认证模式影响了 L1 邻接关系

```
NX-1# show isis adjacency
! Output omitted for brevity
System ID       SNPA              Level   State   Hold Time   Interface
NX-2            002a.1003.f280    2       UP      00:00:29    Ethernet1/1

NX-2# show isis adjacency
! Output omitted for brevity
System ID       SNPA              Level   State   Hold Time   Interface
NX-1            885a.92de.617c    1       INIT    00:00:29    Ethernet1/1
NX-1            885a.92de.617c    2       UP      00:00:07    Ethernet1/1
```

除检查配置信息之外，还可以通过命令 **show isis interface** [*interface-id*]显示接口的认证参数。例 9-27 给出了 NX-1 的输出结果，可以看出，L1 启用了认证机制，而 L2 没有启用认证机制。

例 9-27　查看 IIH 认证

```
NX-1# show isis interface  Ethernet1/1
! Output omitted for brevity
IS-IS process: NXOS VRF: default
Ethernet1/1, Interface status: protocol-up/link-up/admin-up
  IP address: 10.12.1.100, IP subnet: 10.12.1.0/24
```

```
    IPv6 routing is disabled
    Level1
      Auth type:MD5
      Auth keychain: IIH-AUTH
       Auth check set
    Level2
      No auth type and keychain
       Auth check set
```

可以通过命令 **show key chain key- chainname [mode decrypt]** 查看密钥链中的密码，可选关键字 **mode decrypt** 表示以明文方式显示密码，如例 9-28 所示。

例 9-28 查看密钥链密码

```
NX-1# show key chain IIH-AUTH
Key-Chain IIH-AUTH
  Key 2 -- text 7 "072c087f6d26"
    accept lifetime (always valid) [active]
    send lifetime (always valid) [active]
NX-1# show key chain IIH-AUTH mode decrypt
Key-Chain IIH-AUTH
  Key 2 -- text 0 "CISCO"
    accept lifetime (always valid) [active]
    send lifetime (always valid) [active]
```

启用了认证机制之后，必须检查系统日志是否存在差错消息 "*bad authentication*"，如果存在认证差错，那么就需要验证该网络链路上的所有对等体的认证选项和密码。

9.3 IS-IS 路由丢失故障排查

讨论了 IS-IS 邻居邻接关系的故障排查方法之后，本节将解释路由丢失故障的排查方式，同时还将讨论运行 IS-IS 的 Nexus 交换机与各种设备（如 IOS、IOS XR）混合部署时可能存在的各种问题。

9.3.1 系统 ID 重复

IS-IS 系统 ID 在拓扑结构的创建过程中起着非常重要的作用。如果两台相邻路由器在同一 L1 区域中具有相同的系统 ID，那么就无法建立前面所说的邻接关系。如果两台路由器在同一 L1 区域中具有相同的系统 ID，且存在一台中间路由器，那么将无法在拓扑结构中安装这些路由。

图 9-7 给出了一个示例拓扑，所有 Nexus 交换机都位于同一区域且仅存在 L1 邻接关系。NX-2 和 NX-4 配置了相同的系统 ID 0000.0000.0002，NX-3 位于 NX-2 与 NX-4 之间且具有不同的系统 ID，因而允许 NX-2 与 NX-4 建立完全的邻居邻接关系。

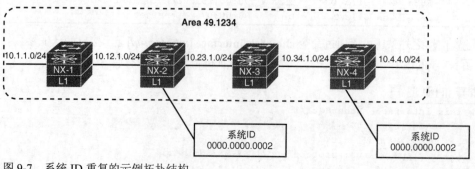

图 9-7 系统 ID 重复的示例拓扑结构

从 NX-1 的角度来看,第一个明显故障就是 NX-4 的网络 10.4.4.0/24 丢失了,如例 9-29 所示。

例 9-29 NX-1 的路由表缺少 NX-4 的网络 10.4.4.0/24

```
NX-1# show ip route isis
! Output omitted for brevity

10.23.1.0/24, ubest/mbest: 1/0
    *via 10.12.1.2, Eth1/1, [115/8], 00:16:56, isis-NXOS, L1
10.34.1.0/24, ubest/mbest: 1/0
    *via 10.12.1.2, Eth1/1, [115/12], 00:16:49, isis-NXOS, L1
```

NX-2 和 NX-4 出现与系统 ID 重复的故障消息:*L1 LSP—Possible duplicate system ID*,如例 9-30 所示。

例 9-30 系统日志消息显示 LSP 存在系统 ID 重复问题

```
15:45:26 NX-2 %ISIS-4-LSP_DUP_SYSID:  isis-NXOS [15772]  L1 LSP - Possible duplicate
  system ID 0000.0000.0002 detected
15:41:47 NX-4 %ISIS-4-LSP_DUP_SYSID:  isis-NXOS [23550]  L1 LSP - Possible duplicate
  system ID 0000.0000.0002 detected
```

例 9-31 显示的两台 Nexus 交换机路由表均出现了系统日志消息 "*Possible duplicate system ID*"。请注意,NX-2 仅缺少 NX-4 的接口(10.4.4.0/24),而 NX-4 却缺少 10.12.1.0/24 和 NX-1 的以太网接口(10.1.1.0/24)。检查 IS-IS 数据库就会看到一个指示 NX-2 存在问题的标志(*)。

例 9-31 NX-2 和 NX-4 的路由表

```
NX-2# show ip route is-is
! Output omitted for brevity
10.1.1.1/32, ubest/mbest: 1/0
    *via 10.12.1.1, Eth2/1, [115/8], 00:04:00, isis-NXOS, L1
10.34.1.0/24, ubest/mbest: 1/0
    *via 10.23.1.3, Eth2/2, [115/8], 00:04:03, isis-NXOS, L1

NX-2(config-router)# do show isis database
IS-IS Process: NXOS LSP database VRF: default
IS-IS Level-1 Link State Database
  LSPID                 Seq Number   Checksum  Lifetime  A/P/O/T
  NX-1.00-00            0x00000004   0x42EC    939       0/0/0/1
  NX-1.01-00            0x00000003   0x804A    960       0/0/0/1
  NX-2.00-00         *  0x00000134   0xDC3E    1199      0/0/0/1
  NX-2.01-00         ?  0x00000003   0xA027    974       0/0/0/1
  NX-3.00-00            0x00000021   0x9D74    1196      0/0/0/1
  NX-3.02-00            0x0000001D   0x5D4E    1110      0/0/0/1

NX-4# show ip route is-is
! Output omitted for brevity
10.23.1.0/24, ubest/mbest: 1/0
    *via 10.34.1.3, Eth2/1, [115/8], 00:04:02, isis-NXOS, L1

NX-4(config-router)# do show isis database
IS-IS Process: NXOS LSP database VRF: default
IS-IS Level-1 Link State Database
  LSPID                 Seq Number   Checksum  Lifetime  A/P/O/T
  NX-1.00-00            0x00000004   0x42EC    914       0/0/0/1
  NX-1.01-00            0x00000003   0x804A    936       0/0/0/1
  NX-4.00-00         *  0x00000139   0xAC16    1194      0/0/0/1
  NX-4.01-00         *  0x00000003   0xA027    954       0/0/0/1
  NX-3.00-00            0x00000021   0x9D74    1173      0/0/0/1
  NX-3.02-00            0x0000001D   0x5D4E    1087      0/0/0/1
```

可以通过命令 show isis | i system 检查这两台报告了"Possible duplicate system ID"差错消息的 Nexus 交换机的 IS-IS 进程，从而快速检查路由器的系统 ID。从例 9-32 可以看出，NX-2 和 NX-4 的系统 ID 相同。

例 9-32　验证 IS-IS 系统 ID

```
NX-2# show isis | i System
  System ID : 0000.0000.0002    IS-Type : L1
NX-4# show isis | i System
  System ID : 0000.0000.0002    IS-Type : L1
```

9.3.2　接口链路开销

IS-IS 接口度量是 Dijkstra SPF 计算过程的重要组成部分，因为最短路径度量是从源路由器到目的路由器的累积接口度量。Nexus 交换机根据以下公式分配 IS-IS 接口度量。

接口度量=接口带宽/参考带宽

NX-OS 的默认参考带宽为 40 Gbit/s，其他思科 OS（IOS 和 IOS XR）则将接口链路度量静态设置为 10（不考虑接口速率）。表 9-9 列出了常见网络接口类型的默认 IS-IS 度量（基于默认参考带宽）。

表 9-9　　　　　　　　　　　使用默认设置的 IS-IS 接口开销

接口类型	默认 NX-OS IS-IS 开销	默认 IOS IS-IS 开销
快速以太网	400	10
GE	40	10
10GE	4	10
40GE	1	10

从表 9-9 可以看出，对于 IOS 路由器来说，与快速以太网接口和 40GE 接口相关联的链路开销没有任何区别。从本质上来说，如果 Nexus 交换机与 IOS 设备在 IS-IS 拓扑结构中进行交互，那么就存在次优路由问题。以图 9-8 拓扑结构为例，R1 与 R2 应该使用 10GE 路径（R1→NX-3→NX-4→R2），因为 R1 与 R2 之间的 GE 链路应该仅被用作备用路径。

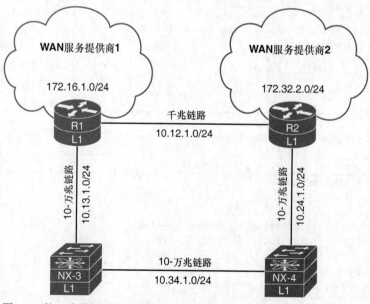

图 9-8　接口度量问题的示例拓扑结构

例 9-33 显示了 R1 的路由表信息，所有设备均使用默认接口度量。可以看出，172.16.1.0/24 与 172.32.2.0/24 之间的流量使用的是备用 GE 链路（10.12.1.0/24），与期望通信模式不符。请注意，使用 GE 链路去往网络 172.32.2.0/24 的 IS-IS 路径度量为 20。

例 9-33　基于默认接口度量带宽的 R1 路由表

```
R1# show ip route isis | begin Gateway
Gateway of last resort is not set

     10.0.0.0/8 is variably subnetted, 6 subnets, 2 masks
i L1    10.24.1.0/24 [115/18] via 10.13.1.3, 00:04:51, TenGigabitEthernet2/2
i L1    10.34.1.0/24 [115/14] via 10.13.1.3, 00:04:51, TenGigabitEthernet2/2
     172.16.0.0/16 is variably subnetted, 3 subnets, 2 masks
i L1    172.32.2.0/24 [115/20] via 10.12.1.2, 00:00:08, GigabitEthernet0/1
```

IS-IS 的一个优势就在于可以在路由器之上构造网络。查看 IS-IS 拓扑表（而不是查看路由表）的命令是 **show isis topology**，IS-IS 拓扑表列出了到达目的路由器、下一跳节点和出站接口的总路径度量。例 9-34 从 R1 和 NX-3 的角度显示了拓扑表信息，可以看出，R1 选择经 Gi0/1 的直连链路去往 R2。

例 9-34　基于默认度量的 R1 和 NX-3 的 IS-IS 拓扑表

```
R1# show isis topology

IS-IS TID 0 paths to level-1 routers
System Id       Metric      Next-Hop         Interface     SNPA
R1              --
R2              10          R2               Gi0/1         fa16.3e10.00b6
NX-3            10          NX-3             Te2/2         0012.1298.1231
NX-4            14          NX-3             Te2/2         0012.1298.1231

NX-3# show isis topology
IS-IS process: NXOS
VRF: default
IS-IS Level-1 IS routing table
R1.00, Instance 0x0000001D
   *via R1, Ethernet1/2, metric 4
R1.03, Instance 0x0000001D
   *via R1, Ethernet1/2, metric 14
R2.00, Instance 0x0000001D
   *via NX-4, Ethernet1/1, metric 8
R2.02, Instance 0x0000001D
   *via NX-4, Ethernet1/1, metric 8
NX-4.00, Instance 0x0000001D
   *via NX-4, Ethernet1/1, metric 4

IS-IS Level-2 IS routing table
```

请注意 R1 和 NX-3 相互指向对方时，因度量值冲突而产生的影响。为了确保路由选择最佳路径，可以采取以下 3 种选项。

- 在非 Nexus 交换机的 IS-IS 设备上静态设置 IS-IS 度量，IOS 设备可以使用接口参数命令 **isis metric** *metric-value*。
- 在 Nexus 交换机接口上使用接口参数命令 **isis metric** *metric-value* {**level-1** | **level-2**}，静态设置 IS-IS 度量以反映更优的网络链路。
- 将 Nexus 交换机的参考带宽调整为更大值，从而使这些链路更优。可以使用 IS-IS 进程配置命令 **reference-bandwidth** *reference-bw* {**gbps** | **mbps**} 设置参考带宽。

由于 R1 与 R2 之间没有任何中间路由器，因而唯一可行的选项就是修改 R1 和 R2 的 IS-IS 度量。从例 9-35 可以看出，链路 10.12.1.0/24 的度量被静态设置为 40，10Gbit/s 接口的度量被设置为 4，该值基于参考带宽 40Gbit/s。

例 9-35　在 R1 和 R2 上设置静态 IS-IS 度量

```
R1# conf t
Enter configuration commands, one per line. End with CNTL/Z.
R1(config)# interface GigabitEthernet0/1
R1(config-if)# isis metric ?
  <1-16777214>  Default metric
  maximum       Maximum metric. All routers will exclude this link from their
                SPF

R1(config-if)# isis metric 40
R1(config-if)# interface TenGigabitEthernet2/2
R1(config-if)# isis metric 4

R2(config)# int GigabitEthernet0/1
R2(config-if)# isis metric 40
R2(config)# int TenGigabitEthernet2/2
R2(config-if)# isis metric 4
```

调整了度量之后，接着检查 R1 和 NX-3 的 IS-IS 路由表和拓扑表，如例 9-36 所示。可以看出，此时网络 10.13.1.0/24 与 10.24.1.0/24 的接口度量已经匹配，而且 R1 已经选择了 10Gbit/s 路径作为去往 R2 的首选路径。

例 9-36　静态设置度量之后的 IS-IS 路由和拓扑表

```
R1# show ip route isis | beg Gate
Gateway of last resort is not set

      10.0.0.0/8 is variably subnetted, 6 subnets, 2 masks
i L1     10.24.1.0/24 [115/12] via 10.13.1.3, 00:01:08, TenGigabitEthernet2/2
i L1     10.34.1.0/24 [115/8] via 10.13.1.3, 00:01:08, TenGigabitEthernet2/2
      172.16.0.0/16 is variably subnetted, 3 subnets, 2 masks
i L1     172.32.2.0/24 [115/22] via 10.13.1.3, 00:01:08, TenGigabitEthernet2/2

R1# show isis topology

IS-IS TID 0 paths to level-1 routers
System Id           Metric    Next-Hop        Interface    SNPA
R1                  --
R2                  12        NX-3            Gi0/2        0012.1298.1231
NX-3                4         NX-3            Te2/2        0012.1298.1231
NX-4                8         NX-3            Te2/2        0012.1298.1231

NX-3# show isis topology
IS-IS process: NXOS
VRF: default
IS-IS Level-1 IS routing table
R1.00, Instance 0x00000023
   *via R1, Ethernet1/2, metric 4
R2.00, Instance 0x00000023
   *via NX-4, Ethernet1/1, metric 8
R2.02, Instance 0x00000023
   *via NX-4, Ethernet1/1, metric 8
NX-4.00, Instance 0x00000023
   *via NX-4, Ethernet1/1, metric 4
```

9.3.3 度量模式不匹配

IS-IS 接口度量是 LSP 的一个重要组件。RFC 1195 将接口度量指定为 6 比特字段（取值范围是 1～63），且包含在 IS 邻居 TLV（2）和 IP 可达性 TLV（128 和 130）中。将接口度量限制为 63 可能会产生一些问题，因为 IS-IS 拓扑结构中的网络带宽可能差异很大，链路带宽范围是 10 Mbit/s ~ 100Gbit/s。通常将这类度量称为窄度量（narrow metric）。

RFC 3784 提供了一种 24 比特的接口度量方式，允许将度量设置为 1～16777214。24 比特度量位于扩展 IS 可达性 TLV（22）和扩展 IP 可达性 TLV（135）中。通常将这类度量称为宽度量（wide metric）。

Nexus 交换机默认同时接受窄度量或宽度量方式，但是仅宣告宽度量，而 IOS 和 IOS XR 默认仅接受和宣告窄度量，因而在拓扑结构中集成非 Nexus 交换机时可能会出现问题。图 9-9 显示了拥有多种设备类型的简单 L1 拓扑，图中的所有设备和接口都启用了 IS-IS。

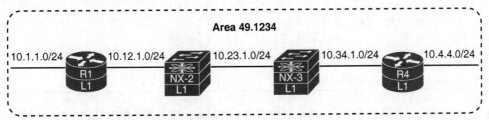

图 9-9 拥有多种设备类型的简单 IS-IS L1 拓扑

例 9-37 显示了 R1 和 NX-2 的 IS-IS 路由表项，可以看出，R1 的路由表中没有任何 IS-IS 路由，而 NX-2 则拥有去往拓扑结构中的所有网络的路由。

例 9-37 R1 和 NX-2 的 IS-IS 路由表项

```
R1# show ip route isis
Codes: L - local, C - connected, S - static, R - RIP, M - mobile, B - BGP
       D - EIGRP, EX - EIGRP external, O - OSPF, IA - OSPF inter area
       N1 - OSPF NSSA external type 1, N2 - OSPF NSSA external type 2
       E1 - OSPF external type 1, E2 - OSPF external type 2
       i - IS-IS, su - IS-IS summary, L1 - IS-IS level-1, L2 - IS-IS level-2
       ia - IS-IS inter area, * - candidate default, U - per-user static route
       o - ODR, P - periodic downloaded static route, H - NHRP, l - LISP
       a - application route
       + - replicated route, % - next hop override, p - overrides from PfR

Gateway of last resort is not set

NX-2# show ip route isis
! Output omitted for brevity

10.1.1.0/24, ubest/mbest: 1/0
    *via 10.12.1.1, Eth1/1, [115/14], 00:02:30, isis-NXOS, L1
10.4.4.0/24, ubest/mbest: 1/0
    *via 10.23.1.3, Eth1/2, [115/18], 00:02:14, isis-NXOS, L1
10.34.1.0/24, ubest/mbest: 1/0
    *via 10.23.1.3, Eth1/2, [115/8], 00:17:28, isis-NXOS, L1
```

排查路由丢失故障的第一步就是验证邻居邻接关系并检查 IS-IS 拓扑表。例 9-38 显示了 R1 和 NX-2 的拓扑表信息，可以看出，R1 去往其他路由器的所有度量均显示为双星号（**），而 NX-2 则显示了正常度量值。这是因为 R1 仅配置了窄度量，所使用的 TLV 与 NX-2 宣告的宽度量 TLV 不同。

例 9-38　度量类型不匹配的 IS-IS 拓扑表

```
R1# show isis topology

IS-IS TID 0 paths to level-1 routers
System Id            Metric          Next-Hop            Interface      SNPA
R1                   --
NX-2                 **
NX-3                 **
R4                   **
```

```
NX-2# show isis topology
IS-IS process: NXOS
VRF: default
IS-IS Level-1 IS routing table
R1.00, Instance 0x0000000C
    *via R1, Ethernet1/1, metric 4
NX-3.00, Instance 0x0000000C
    *via NX-3, Ethernet1/2, metric 4
R4.00, Instance 0x0000000C
    *via NX-3, Ethernet1/2, metric 8
R4.01, Instance 0x0000000C
    *via NX-3, Ethernet1/2, metric 8
```

为了证实上述推断，可以查看 IS-IS 协议以检查 R1 和 NX-2 的度量类型（见例 9-39），可以看出，R1 被设置为仅接受和生成窄度量，而 NX-OS 则同时接受窄度量和宽度量，但是仅宣告宽度量。

例 9-39　检查 IS-IS 度量配置

```
R1# show isis protocol | i metric
  Generate narrow metrics:    level-1-2
  Accept narrow metrics:      level-1-2
  Generate wide metrics:      none
  Accept wide metrics:        none
```

```
NX-2# show isis | i Metric
  Metric-style : advertise(wide), accept(narrow, wide)
```

可以通过命令 **metric-style transition** 将 Nexus 交换机置于度量转换模式，让 Nexus 交换机使用窄度量和宽度量 TLV 来填充 LSP，这样就允许运行在窄度量模式下的其他路由器计算拓扑结构的总路径度量。例 9-40 显示了 NX-2 IS-IS 度量转换模式的配置和验证过程。

例 9-40　IS-IS 度量转换模式的配置和验证

```
NX-2# show run isis
! Output omitted for brevity

router isis NXOS
  net 49.1234.0000.0000.0002.00
  is-type level-1
  metric-style transition
```

```
NX-2# show isis | i Metric
  Metric-style : advertise(narrow, wide), accept(narrow, wide)
```

将 Nexus 交换机置于 IS-IS 度量转换模式后，例 9-41 显示了此时 IOS 路由器的 IS-IS 拓扑表和路由表。

例 9-41 将 NX-OS 设置为度量转换模式之后验证 IOS 设备

```
R1# show isis topology

IS-IS TID 0 paths to level-1 routers
System Id        Metric      Next-Hop           Interface       SNPA
R1               --
NX-2             10          NX-2               Gi0/1           0022.2222.2222
NX-3             14          NX-2               Gi0/1           0023.3333.3333
R4               18          NX-2               Gi0/1           0022.2222.2222

R1# show ip route isis
! Output omitted for brevity

Gateway of last resort is not set

      10.0.0.0/8 is variably subnetted, 7 subnets, 2 masks
i L1     10.4.4.0/24 [115/28] via 10.12.1.2, 00:01:30, GigabitEthernet0/1
i L1     10.23.1.0/24 [115/14] via 10.12.1.2, 00:02:55, GigabitEthernet0/1
i L1     10.34.1.0/24 [115/18] via 10.12.1.2, 00:01:30, GigabitEthernet0/1

R4# show ip route isis
! Output omitted for brevity

Gateway of last resort is not set

      10.0.0.0/8 is variably subnetted, 7 subnets, 2 masks
i L1     10.1.1.0/24 [115/28] via 10.34.1.3, 00:01:54, GigabitEthernet0/1
i L1     10.12.1.0/24 [115/18] via 10.34.1.3, 00:01:54, GigabitEthernet0/1
i L1     10.23.1.0/24 [115/14] via 10.34.1.3, 00:01:54, GigabitEthernet0/1
```

9.3.4 L1 到 L2 的路由传播

IS-IS 运行在两级层次结构上，L1-L2 路由器的主要功能是充当 L1 路由器到 L2 IS-IS 骨干网的网关。图 9-10 给出了一个简单示例拓扑，NX-1 和 NX-2 位于 Area 49.0012，而 NX-3 和 NX-4 位于 Area 49.0034。NX-2 应该将 NX-1 的网络 10.1.1.0/24 宣告给 Area 49.0034，NX-3 应该将 NX-4 的网络 10.4.4.0/24 宣告给 Area 49.0012。

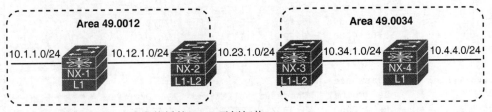

图 9-10 IS-IS L1 到 L2 路由传播的 IS-IS 示例拓扑

例 9-42 显示了 4 台 Nexus 交换机的路由表。可以看出，NX-3 缺少网络 10.1.1.0/24，该网络在 NX-2 路由表中以 IS-IS L1 路由的方式存在。NX-4 的网络 4.4.4.0/24 也存在同样情况，该网络出现在 NX-3 上。

例 9-42 NX-1、NX-2、NX-3 和 NX-4 的路由表

```
NX-1# show ip route isis
! Output omitted for brevity

0.0.0.0/0, ubest/mbest: 1/0
```

```
        *via 10.12.1.2, Eth1/1, [115/4], 00:02:43, isis-NXOS, L1
10.23.1.0/24, ubest/mbest: 1/0
        *via 10.12.1.2, Eth1/1, [115/8], 00:02:33, isis-NXOS, L1
NX-2# show ip route isis
! Output omitted for brevity

10.1.1.1/32, ubest/mbest: 1/0
        *via 10.12.1.1, Eth1/1, [115/8], 00:01:42, isis-NXOS, L1
10.34.1.0/24, ubest/mbest: 1/0
        *via 10.23.1.3, Eth1/2, [115/8], 00:01:38, isis-NXOS, L2
NX-3# show ip route isis
! Output omitted for brevity

10.4.4.4/32, ubest/mbest: 1/0
        *via 10.34.1.4, Eth1/1, [115/8], 00:02:45, isis-NXOS, L1
10.12.1.0/24, ubest/mbest: 1/0
        *via 10.23.1.2, Eth1/2, [115/8], 00:02:45, isis-NXOS, L2
NX-4# show ip route isis
! Output omitted for brevity

0.0.0.0/0, ubest/mbest: 1/0
        *via 10.34.1.3, Eth1/1, [115/4], 00:00:28, isis-NXOS, L1
10.23.1.0/24, ubest/mbest: 1/0
        *via 10.34.1.3, Eth1/1, [115/8], 00:03:48, isis-NXOS, L1
```

接下来利用命令 **show isis database [level-1 | level-2] [detail]** [*lsp-id*]检查 IS-IS 数据库，以验证 LSPDB 拥有适当的 LSP。只要为宣告路由器指定 IS-IS 层级或 LSP ID，就可以对 LSP 进行限制。

例 9-43 显示了 NX-2 LSPDB 中的所有 L1 和 L2 LSP，可以看出，NX-2 已经收到了 NX-1 的 L1 LSP 和 NX-3 的 L2 LSP。

例 9-43 NX-2 的 LSPDB

```
NX-2# show isis database
IS-IS Process: NXOS LSP database VRF: default
IS-IS Level-1 Link State Database
    LSPID                   Seq Number     Checksum   Lifetime   A/P/O/T
    NX-1.00-00              0x00000006     0x9FC1     743        0/0/0/1
    NX-1.01-00              0x00000002     0x8249     1137       0/0/0/1
    NX-2.00-00            * 0x0000000A     0x9AE9     1179       1/0/0/3

IS-IS Level-2 Link State Database
    LSPID                   Seq Number     Checksum   Lifetime   A/P/O/T
    NX-2.00-00            * 0x00000003     0x0E82     1179       0/0/0/3
    NX-3.00-00              0x00000003     0x5CF5     1153       0/0/0/3
    NX-3.02-00              0x00000002     0x952F     1152       0/0/0/3
NX-2# show isis database level-2
IS-IS Process: NXOS LSP database VRF: default
IS-IS Level-2 Link State Database
    LSPID                   Seq Number     Checksum   Lifetime   A/P/O/T
    NX-2.00-00            * 0x00000003     0x0E82     1155       0/0/0/3
    NX-3.00-00              0x00000003     0x5CF5     1130       0/0/0/3
    NX-3.02-00              0x00000002     0x952F     1129       0/0/0/3
```

表 9-10 列出了例 9-43 显示的部分关键字段信息。

表 9-10　通用 IS-IS 数据库字段

字段	描述
LSP ID	LSP ID 是一个定长 8 字节字段，为 LSP 发起端提供唯一的标识。LSP ID 的系统 ID 部分包括交换机的主机名（而不是数字形式）。IS-IS 在可选 TLV#137 下提供了名称到系统 ID 的映射，该 TLV#137 是 LSP 的一部分，可以简化故障排查。检查 LSPDB 时，一定要确保所有设备都拥有唯一的主机名
生存期	被宣告 LSP 在超时并被清除之前保持有效的时间
A（Attached Bit，连接比特）	标识该路由器是否为 L1-L2 路由器，并提供与 IS-IS L2 骨干网的连接
P（Partition Bit，分区比特）	表示是否在该 LSP 上设置分区修复比特
O（Overload Bit，过载比特）	表示是否在宣告路由器上设置了过载比特。过载比特表示正在执行系统维护或路由器刚刚启动并等待完全收敛。过载比特作为流量工程的一种形式，可以在允许的情况下通过其他路径引导流量，本质上与在所有链路上分摊开销具有相同的效果（设置高接口开销） Nexus 交换机使用命令 **set overload-bit** 设置过载比特
T（Topology Bit，拓扑比特）	指示路由器的功能。值 1 表示路由器是 IS-IS L1 路由器，值 3 表示路由器是 L1 或 L1-L2 路由器，取决于 LSP ID 是否位于两个 IS-IS 层级中

使用可选关键字 **detail** 可以在查看 LSPDB 时显示所有网络、度量和 TLV 类型。例 9-44 显示了 NX-2 的所有 L2 IS-IS LSP 信息，包含了 NX-2 向其他 L2 邻居宣告的所有网络。请注意，NX-2 的 LSP 没有网络表项 10.1.1.0/24，NX-3 的 LSP 也没有网络表项 10.4.4.0/24。

例 9-44　检查 NX-2 的 L2 LSPDB 详细信息

```
NX-2# show isis database level-2 detail
IS-IS Process: NXOS LSP database VRF: default
IS-IS Level-2 Link State Database
  LSPID                   Seq Number    Checksum  Lifetime  A/P/O/T
  NX-2.00-00            * 0x00000003    0x0E82    1135      0/0/0/3
    Instance        :  0x00000003
    Area Address    :  49.0012
    NLPID           :  0xCC
    Router ID       :  192.168.2.2
    IP Address      :  192.168.2.2
    Hostname        :  NX-2             Length : 4
    Extended IS     :  NX-3.02          Metric : 4
    Extended IP     :       10.23.1.0/24 Metric : 4           (U)
    Extended IP     :       10.12.1.0/24 Metric : 4           (U)
    Digest Offset   :  0
  NX-3.00-00              0x00000003    0x5CF5    1109      0/0/0/3
    Instance        :  0x00000001
    Area Address    :  49.0034
    NLPID           :  0xCC
    Router ID       :  192.168.3.3
    IP Address      :  192.168.3.3
    Hostname        :  NX-3             Length : 4
    Extended IS     :  NX-3.02          Metric : 4
    Extended IP     :       10.23.1.0/24 Metric : 4           (U)
    Extended IP     :       10.34.1.0/24 Metric : 4           (U)
    Digest Offset   :  0
  NX-3.02-00              0x00000002    0x952F    1108      0/0/0/3
    Instance        :  0x00000001
    Extended IS     :  NX-2.00          Metric : 0
    Extended IS     :  NX-3.00          Metric : 0
    Digest Offset   :  0
```

> 注：需要记住的是，实际路由器的 LSP ID 的伪节点部分为 0，包含所有链路。如果 LSP ID 的伪节点部分不为 0，那么反映的就是网段的 DIS，并列出与其相连的路由器的 LSP ID。LSP ID NX-3.02-00 是 NX-2 到 NX-3 网络链路的 DIS。

IS-IS LSPDB 表名 NX-1 的 L1 路由没有传播到 NX-2 的 L2 数据库，NX-4 和 NX-3 也出现了相似的路由传播行为。出现这种情况的原因是 NX-OS 与其他思科操作系统（IOS、IOS XR 等）的操作方式不同，Nexus 交换机需要在 L1-L2 路由器上通过命令 **distribute level-1 into level-2 {all | route-map** *route-map-name*} 进行显式配置，将 L1 路由注入 L2 拓扑中。

例 9-45 显示了 NX-2 和 NX-3 的相关 IS-IS 配置，此时已经可以将 L1 路由传播到 L2 LSPDB 中。

例 9-45 配置 L1 路由传播

```
NX-2# show run isis
! Output omitted for brevity
router isis NXOS
  net 49.0012.0000.0000.0002.00
  distribute level-1 into level-2 all
  log-adjacency-changes

NX-3# show run isis
! Output omitted for brevity
router isis NXOS
  net 49.0034.0000.0000.0003.00
  distribute level-1 into level-2 all
  log-adjacency-changes
```

配置 NX-2 和 NX-3 传播 L1 路由之后，例 9-46 显示了 NX-3 宣告给 NX2 的 LSP。可以看出，目前已包含 L1 路由 10.4.4.0/24。

例 9-46 启用 L1 路由传播之后的 NX-3 LSP

```
NX-2# show isis database level-2 detail NX-3.00-00
IS-IS Process: NXOS LSP database VRF: default
IS-IS Level-2 Link State Database
  LSPID                 Seq Number    Checksum  Lifetime   A/P/O/T
  NX-3.00-00            0x00000004    0x7495    1069       0/0/0/3
    Instance      : 0x00000002
    Area Address  : 49.0034
    NLPID         : 0xCC
    Router ID     : 192.168.3.3
    IP Address    : 192.168.3.3
    Hostname      : NX-3           Length : 4
    Extended IS   : NX-3.02        Metric : 4
    Extended IP   :       10.4.4.0/24  Metric : 8      (U)
    Extended IP   :       10.23.1.0/24 Metric : 4      (U)
    Extended IP   :       10.34.1.0/24 Metric : 4      (U)
    Digest Offset : 0
```

配置 NX-2 和 NX-3 传播 L1 路由之后，例 9-47 显示了 NX-2 和 NX-4 的路由表。可以看出，网络 10.1.1.0/24 和 10.4.4.0/24 对于两台 L1-L2 交换机来说均可达。

例 9-47 启用 L1 路由传播之后的 NX-2 和 NX-4 路由表

```
NX-2# show ip route isis
! Output omitted for brevity

10.1.1.0/24, ubest/mbest: 1/0
```

```
    *via 10.12.1.1, Eth1/1, [115/8], 00:11:52, isis-NXOS, L1
10.4.4.0/24, ubest/mbest: 1/0
    *via 10.23.1.3, Eth1/2, [115/12], 00:00:40, isis-NXOS, L2
10.34.1.0/24, ubest/mbest: 1/0
    *via 10.23.1.3, Eth1/2, [115/8], 00:11:48, isis-NXOS, L2
NX-3# show ip route isis
! Output omitted for brevity

10.1.1.0/24, ubest/mbest: 1/0
    *via 10.23.1.2, Eth1/2, [115/12], 00:01:44, isis-NXOS, L2
10.4.4.0/24, ubest/mbest: 1/0
    *via 10.34.1.4, Eth1/1, [115/8], 00:12:13, isis-NXOS, L1
10.12.1.0/24, ubest/mbest: 1/0
    *via 10.23.1.2, Eth1/2, [115/8], 00:05:12, isis-NXOS, L2
```

9.3.5 次优路由

如前所述，L1-L2 路由器充当 L1 路由器去往 L2 IS-IS 骨干网的网关。虽然 L1-L2 路由器不向 L1 区域宣告 L2 路由，但是会在 L1 LSP 中设置连接比特，表示该路由器连接了 IS-IS 骨干网。如果 L1 路由器缺少特定网络的路由，那么就会在 LSPDB 中搜索携带连接比特的最近的路由器，将该路由器作为网关设备。

图 9-11 中的 Area 49.1234 连接了 Area 49.0005 和 Area 49.0006，NX-1 和 NX-3 是 L1 路由器，NX-2 和 NX-4 是 L1-L2 路由器。

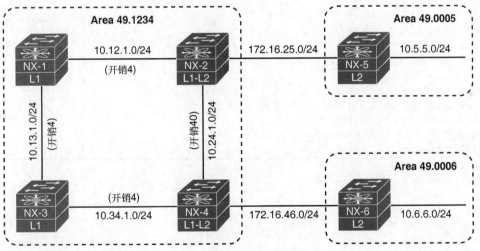

图 9-11 IS-IS 区域间拓扑

问题在于 NX-1 试图连接网络 10.6.6.0/24 时出现了次优路由，使用了开销较大的 10.24.1.0/24 网络链路。NX-3 连接网络 10.5.5.0/24 时也出现了同样问题。例 9-48 显示了 NX-1 和 NX-3 选择的次优路径。

例 9-48 次优路径选择

```
NX-1# trace 10.6.6.6 so lo0
traceroute to 10.6.6.6 (10.6.6.6) from 192.168.1.1 (192.168.1.1), 30 hops max, 40
  byte packets
 1  10.12.1.2 (10.12.1.2)  1.95 ms  1.36 ms  1.397 ms
 2  10.24.1.4 (10.24.1.4)  2.758 ms  2.498 ms  2.423 ms
 3  172.16.46.6 (172.16.46.6)  4.037 ms *  4.103 ms
NX-3# trace 10.5.5.5 so lo0
traceroute to 10.5.5.5 (10.5.5.5) from 192.168.3.3 (192.168.3.3), 30 hops max, 40
```

```
                 byte packets
  1  10.34.1.4 (10.34.1.4)    1.826 ms   1.127 ms   1.249 ms
  2  10.24.1.2 (10.24.1.2)    2.434 ms   2.461 ms   2.19 ms
  3  172.16.25.5 (172.16.25.5)  5.262 ms
```

例 9-49 显示了 NX-1 和 NX-3 的 IS-IS 数据库，发现 NX-2 和 NX-4 都设置了连接比特 "A"。从本质上来说，连接比特提供了一条指向 L1-L2 宣告路由器的 L1 默认路由。

例 9-49　Area 49.1234 的 IS-IS 数据库

```
NX-1# show isis database
IS-IS Process: NXOS LSP database VRF: default
IS-IS Level-1 Link State Database
  LSPID                  Seq Number    Checksum   Lifetime   A/P/O/T
  NX-1.00-00           * 0x0000001D    0xC67A     1038       0/0/0/1
  NX-2.00-00             0x00000021    0xAF03     1120       1/0/0/3
  NX-3.00-00             0x0000001F    0xD222     1055       0/0/0/1
  NX-4.00-00             0x00000021    0x94B5     1154       1/0/0/3
```

NX-1 和 NX-2 必须找出设置了连接比特的最近的路由器。一般来说，这是一项手动操作（交叉引用 IS-IS 拓扑表和 IS-IS 数据库），但 NX-OS 可以实现自动化操作。例 9-50 显示了 NX-1 和 NX-3 的 IS-IS 拓扑表。

例 9-50　Area 49.1234 的 IS-IS 拓扑

```
NX-1# show isis topology
IS-IS process: NXOS
VRF: default
IS-IS Level-1 IS routing table
NX-2.00, Instance 0x00000022
   *via NX-2, Ethernet1/2, metric 4
NX-3.00, Instance 0x00000022
   *via NX-3, Ethernet1/1, metric 4
NX-4.00, Instance 0x00000022
   *via NX-3, Ethernet1/1, metric 8
0000.0000.0000.00, Instance 0x00000022, Default
   *via NX-2, Ethernet1/2, metric 4

NX-3# show isis topology
IS-IS process: NXOS
VRF: default
IS-IS Level-1 IS routing table
NX-1.00, Instance 0x0000001F
   *via NX-1, Ethernet1/1, metric 4
NX-2.00, Instance 0x0000001F
   *via NX-1, Ethernet1/1, metric 8
NX-4.00, Instance 0x0000001F
   *via NX-4, Ethernet1/2, metric 4
0000.0000.0000.00, Instance 0x0000001F, Default
   *via NX-4, Ethernet1/2, metric 4
```

例 9-51 显示了 NX-1 和 NX-3 的路由表。可以看出，路由表中没有网络 105.5.0/24 或 10.64.0/24 的路由表项，因而使用了默认路由。请注意，默认路由与例 9-50 的 IS-IS 拓扑表项有关。

例 9-51　NX-1 和 NX-3 路由表

```
NX-1# show ip route isis
! Output omitted for brevity
```

```
    0.0.0.0/0, ubest/mbest: 1/0
        *via 10.12.1.2, Eth1/2, [115/4], 00:07:05, isis-NXOS, L1
    10.24.1.0/24, ubest/mbest: 1/0
        *via 10.12.1.2, Eth1/2, [115/44], 00:07:05, isis-NXOS, L1
    10.34.1.0/24, ubest/mbest: 1/0
        *via 10.13.1.3, Eth1/1, [115/8], 00:04:39, isis-NXOS, L1
    172.16.25.0/24, ubest/mbest: 1/0
        *via 10.12.1.2, Eth1/2, [115/8], 00:07:05, isis-NXOS, L1
    172.16.46.0/24, ubest/mbest: 1/0
        *via 10.13.1.3, Eth1/1, [115/12], 00:04:39, isis-NXOS, L1
NX-3# show ip route isis
! Output omitted for brevity

    0.0.0.0/0, ubest/mbest: 1/0
        *via 10.34.1.4, Eth1/2, [115/4], 00:07:32, isis-NXOS, L1
    10.12.1.0/24, ubest/mbest: 1/0
        *via 10.13.1.1, Eth1/1, [115/8], 00:05:11, isis-NXOS, L1
    10.24.1.0/24, ubest/mbest: 1/0
        *via 10.34.1.4, Eth1/2, [115/44], 00:07:32, isis-NXOS, L1
    172.16.25.0/24, ubest/mbest: 1/0
        *via 10.13.1.1, Eth1/1, [115/12], 00:05:11, isis-NXOS, L1
    172.16.46.0/24, ubest/mbest: 1/0
        *via 10.34.1.4, Eth1/2, [115/8], 00:07:32, isis-NXOS, L1
```

可以通过路由泄露（Route leaking）机制解决次优路由问题，路由泄露是一种将 L2 路由重分发到 L1 的技术。IS-IS 路由泄露的配置命令是 **distribute level-2 into level-1** {**all** | **route-policy** *route-policy-name*}。例 9-52 以高亮方式显示了 L2 路由泄露的配置示例。

例 9-52 配置 IS-IS L2 路由泄露

```
NX-2# show run isis
! Output omitted for brevity

router isis NXOS
 net 49.1234.0000.0000.0002.00
   distribute level-1 into level-2 all
   distribute level-2 into level-1 all
   metric-style transition
NX-3# show run isis
! Output omitted for brevity

router isis NXOS
 net 49.1234.0000.0000.0003.00
   distribute level-1 into level-2 all
   distribute level-2 into level-1 all
   metric-style transition
```

注：路由泄露通常使用限制性的路由映射来控制被泄露路由；否则，以 L2 模式运行所有区域路由器或许更有用。

接下来检查 IS-IS 数据库以验证配置变更情况，查看 NX-2 和 NX-4 是否已将网络 10.5.5.0/24 和 10.6.6.0/24 宣告给了 Area 49.1234 的 IS-IS L1。完成验证操作之后，需要检查路由表以验证这些表项是否已经添加到 RIB 中。例 9-53 显示了带有 L2 路由泄露的 IS-IS 数据库。

例 9-53 带有 L2 路由泄露的 IS-IS 数据库

```
NX-1# show isis database detail NX-2.00-00
! Output omitted for brevity
```

```
IS-IS Level-1 Link State Database
    Extended IP    :    10.6.6.0/24      Metric : 54      (D)
    Extended IP    :    10.5.5.0/24      Metric : 14      (D)
..
    Extended IP    :    172.16.25.0/24   Metric : 4       (U)
NX-1# show isis database detail NX-4.00-00
! Output omitted for brevity
IS-IS Level-1 Link State Database
    Extended IP    :    10.6.6.0/24      Metric : 14      (D)
    Extended IP    :    10.5.5.0/24      Metric : 54      (D)
..
    Extended IP    :    172.16.46.0/24   Metric : 4       (U)
NX-1# show ip route isis
! Output omitted for brevity

0.0.0.0/0, ubest/mbest: 1/0
    *via 10.12.1.2, Eth1/2, [115/4], 06:41:03, isis-NXOS, L1
10.5.5.0/24, ubest/mbest: 1/0
    *via 10.12.1.2, Eth1/2, [115/18], 00:01:02, isis-NXOS, L1
10.6.6.0/24, ubest/mbest: 1/0
    *via 10.13.1.3, Eth1/1, [115/22], 00:01:20, isis-NXOS, L1
10.24.1.0/24, ubest/mbest: 1/0
    *via 10.12.1.2, Eth1/2, [115/44], 06:41:03, isis-NXOS, L1
10.34.1.0/24, ubest/mbest: 1/0
    *via 10.13.1.3, Eth1/1, [115/8], 06:38:37, isis-NXOS, L1
172.16.25.0/24, ubest/mbest: 1/0
    *via 10.12.1.2, Eth1/2, [115/8], 06:41:03, isis-NXOS, L1
172.16.46.0/24, ubest/mbest: 1/0
    *via 10.13.1.3, Eth1/1, [115/12], 06:38:37, isis-NXOS, L1
```

例 9-54 验证了 NX-1 和 NX-3 正在使用最佳路径转发流量。

例 9-54　L2 路由泄露后的路径检查

```
NX-1# trace 10.6.6.6
traceroute to 10.6.6.6 (10.6.6.6), 30 hops max, 40 byte packets
 1  10.13.1.3 (10.13.1.3)   1.41 ms   1.202 ms   1.223 ms
 2  10.34.1.4 (10.34.1.4)   2.454 ms  2.46 ms    2.588 ms
 3  172.16.46.6 (172.16.46.6)  4.368 ms
NX-3# trace 10.5.5.5
traceroute to 10.5.5.5 (10.5.5.5), 30 hops max, 40 byte packets
 1  10.13.1.1 (10.13.1.1)   1.73 ms   1.387 ms   1.409 ms
 2  10.12.1.2 (10.12.1.2)   2.376 ms  2.814 ms   2.48 ms
 3  172.16.25.5 (172.16.25.5)  4.38 ms *  4.702 ms
```

9.3.6　路由重分发

可以通过命令 **redistribute [bgp** *asn* | **direct** | **eigrp** *process-tag* | **isis** *process-tag* | **ospf** *process-tag* | **rip** *process-tag* | **static] route-map** *route- map-name* 将路由重分发到 IS-IS 中，Nexus 交换机的重分发进程需要用到路由映射，每种协议在重分发时都会提供一个种子度量，从而允许目的协议计算最佳路径。IS-IS 提供的默认重分发度量是 10。

例 9-55 提供了路由重分发进程所需的配置示例，可以看出，NX-1 重分发了直连路由 10.1.1.0/24 和 10.11.11.0/24，而没有通过 IS-IS 路由协议宣告这些路由。请注意，例中的路由映射是一条简单的 **permit** 语句，没有任何匹配条件。

例 9-55 配置 NX-1 重分发

```
router isis NXOS
  redistribute direct route-map REDIST
!
route-map REDIST permit 10
```

NX-1 启用了路由重分发并将前缀 10.1.1.0/24 和 10.11.11.0/24 注入 IS-IS 数据库中。通过查看其他设备（如 NX-2）的 LSPDB 即可验证重分发的前缀信息，如例 9-56 所示。

例 9-56 验证重分发网络

```
NX-2# show isis database detail NX-1.00-00
! Output omitted for brevity
IS-IS Process: NXOS LSP database VRF: default
IS-IS Level-1 Link State Database
  LSPID                 Seq Number     Checksum  Lifetime   A/P/O/T
  NX-1.00-00            0x00000008     0x2064    1161       0/0/0/3
    Instance         : 0x00000005
    Area Address     : 49.0012
    NLPID            : 0xCC
    Router ID        : 10.12.1.100
    IP Address       : 10.12.1.100
    Hostname         : NX-1            Length : 4
    Extended IS      : NX-1.01         Metric : 4
    Extended IP      :      10.1.1.0/24 Metric : 10       (U)
    Extended IP      :    10.11.11.0/24 Metric : 10       (U)
    Extended IP      :     10.12.1.0/24 Metric : 4        (U)
    Digest Offset    : 0
```

9.4 本章小结

本章概述了 IS-IS 路由协议之后，详细讨论了设备之间的邻接关系故障、路由丢失故障以及路径选择故障的排查方法。

为了确保两台路由器能够成为邻居，以下参数必须匹配。

- IS-IS 接口必须处于主动（Active）状态。
- 设备之间必须存在使用主用子网的连接。
- MTU 必须匹配。
- L1 邻接关系要求区域地址匹配对等的 L1 路由器，且系统 ID 在两个邻居之间必须唯一。
- L1 路由器可以与 L1 或 L1-L2 路由器建立邻接关系，但不能与 L2 路由器建立邻接关系。
- L2 路由器可以与 L2 或 L1-L2 路由器建立邻接关系，但不能与 L1 路由器建立邻接关系。
- DIS 要求必须匹配。
- IIH 认证类型和证书必须匹配（如有）。

IS-IS 是一种链路状态路由协议，基于 LSP 创建全网视图。路由数据库中出现路由丢失的主要原因是网络设计错误、度量类型不匹配或者设备配置不支持 L1 到 L2 的路由传播。本章提供了一些常见的 IS-IS 错误设计方式以及防止路径信息丢失的解决方案。

第 10 章

Nexus 路由映射故障排查

本章主要讨论如下主题。
- 基于 ACL、前缀列表和正则表达式的条件匹配。
- 路由映射（Route-Map）。
- RPM（Route Policy Manager，路由策略管理器）故障排查。
- 重分发。
- PBR（Policy-Based Routing，策略路由）。

NX-OS（Nexus Operating System，Nexus 操作系统）的路由映射提供了很多有用功能，包括过滤路由、修改路由属性和路由行为等。这些技术都基于路由特性通过条件匹配规则来实现相关操作。

在讨论路由映射之前，需要首先解释条件匹配所涉及的 ACL（Access Control List，访问控制列表）、前缀列表和 BGP 团体（Community）属性等概念。

10.1 条件匹配

路由映射通常使用某种形式的条件匹配来阻止、接受或修改某些特定前缀。虽然可以通过多种路由协议属性实现网络前缀的条件匹配，但接下来的章节主要讨论常见的前缀条件匹配技术。

10.1.1 ACL

ACL（Access Control List，访问控制列表）的最初目的是对流入或流出网络接口的数据包进行过滤，类似于基本防火墙的功能。目前的 ACL 提供了一种在路由协议中识别网络的方法，而且 ACL 对于隔离故障方向或者在排查复杂网络环境故障时确定丢包位置也非常有用。

NX-OS 的 ACL 是基于二层、三层或四层信息实现流量过滤的通用表达式。ACL 由 ACE（Access Control Entry，访问控制表项）组成，ACE 是 ACL 中负责标识所要采取的操作（允许或拒绝）以及相关数据包分类的表项。数据包分类按照自上（最小序列）而下（较大序列）的方式，直至找到匹配模式。找到匹配项之后，就可以采取相应的操作（允许或拒绝），处理操作到此结束。每个 ACL 的末尾都有一条隐式 deny ACE，作用是拒绝所有与该 ACL 早期都不匹配的数据包。

可以将 ACL 分为两类。
- **标准 ACL**：仅根据源网络定义数据包。
- **扩展 ACL**：可以根据源端、目的端、协议、端口或其他数据包属性的组合定义数据包。

标准 ACL 使用编号为 1–99、1300–1999 的表项或命名式 ACL。扩展 ACL 使用编号为 100–199、2000–2699 的表项或命名式 ACL。命名式 ACL 与标准 ACL 或扩展 ACL 一起使用，提供与 ACL 相关的功能，通常是首选方式。

利用扩展 ACL 选择网络前缀的行为取决于协议是 IGP（如 EIGRP、OSPF、IS-IS）还是 BGP

（Border Gateway Protocol，边界网关协议）。

10.1.2 ACL 和 ACL 管理器组件

NX-OS 通过 ACLMGR（ACL Manager，ACL 管理器）组件来管理 ACL，ACLMGR 是一个与平台独立的模块，是管理 IP、IPv6 和 MAC ACL 定义以及策略对象的中心位置。ACLMGR 负责处理从用户接收的配置，并将安全 ACL 与已启用接口相关联。

使用 ACLMGR 之后，ACL 将具备如下功能。
- 对象组（匹配 IP 地址、TCP 或 UDP 端口）。
- 时间范围。
- IPv6 通配符匹配。
- 基于数据包长度的匹配。
- 状态化重启。
- 逐个表项统计。

除在接口上应用 ACL 或者将 ACL 与路由映射协同使用（路由协议利用路由映射实现路由过滤功能）之外，ACL 还可以用于以下协议。
- CoPP（Control Plane Policing，控制平面策略）。
- DHCP（Dynamic Host Configuration Protocol，动态主机配置协议）。
- PBR（Policy-Based Routing，策略路由）。
- WCCP（Web Cache Communication Protocol，网络缓存通信协议）。

NX-OS 支持以下 ACL 格式。
- IPv4/IPv6 ACL。
- MAC（Media Access Control，介质访问控制）ACL。
- ARP（Address Resolution Protocol，地址解析协议）ACL。
- VLAN（Virtual LAN，虚拟局域网）访问映射（access-map）。

NX-OS 的 ACL 应用于目标对象时，会创建一个策略。NX-OS 支持以下 ACL 策略类型。
- RACL（Router ACL，路由器 ACL）。
- PACL（Port ACL，端口 ACL）。
- VACL（VLAN ACL）。
- VTY（Virtual Terminal Line，虚拟终端线路）ACL。

注：PACL 只能用于 L2/L3 物理以太网接口（包括 L2 端口通道接口）的入站数据包。

例 10-1 列出了 NX-OS 支持的各种 ACL 配置。请注意，例中配置了命令 **statistics per-entry**，目的是为 ACL 下配置的 ACE 启用统计功能。如果未配置命令 **statistics per-entry**，那么命令 **show ip access-list** 将无法显示命中特定 ACE 的数据包的任何统计信息。

例 10-1 ACL 格式

```
IP ACL
NX-1(config)# ip access-list TEST
NX-1(config-acl)# permit ip host 192.168.33.33 host 192.168.3.3
NX-1(config-acl)# permit ip any any
NX-1(config-acl)# statistics per-entry
IPv6 ACL
NX-1(config)# ipv6 access-list TESTv6
NX-1(config-ipv6-acl)# permit icmp host 2001::33 host 2001::3
NX-1(config-ipv6-acl)# permit ipv6 any any
```

```
NX-1(config-ipv6-acl)# statistics per-entry
```

MAC ACL
```
NX-1(config)# mac access-list TEST-MAC
NX-1(config-mac-acl)# permit 00c0.cf00.0000 0000.00ff.ffff any
NX-1(config-mac-acl)# permit any any
NX-1(config-mac-acl)# statistics per-entry
```

ARP ACL
```
NX-1(config)# arp access-list TEST-ARP
NX-1(config-arp-acl)# deny ip host 192.168.10.11 mac 00c0.cf00.0000 ffff.ff00.0000
NX-1(config-arp-acl)# permit ip any mac any
```

VLAN Access-map
```
NX-1(config)# vlan access-map TEST-VLAN-MAP
NX-1(config-access-map)# match ip address TEST
NX-1(config-access-map)# action drop
NX-1(config-access-map)# statistics per-entry
```

注： 可以通过命令 **show run aclmgr** 验证与 ACL 相关的配置，该命令可以显示 ACL 配置和 ACL 关联点。

例 10-2 给出了在配置和未配置 **statistics per-entry** 命令的情况下，命令 **show ip access-list** 的输出结果对比情况。可以看出，在配置了 **statistics per-entry** 命令的 ACL 配置中，显示了已确认命中的统计信息。

例 10-2 ACL 统计信息

```
Output when statistics per-entry command is configured
NX-1# show ip access-list TEST
IP access list TEST
        statistics per-entry
        10 permit ip 192.168.33.33/32 192.168.3.3/32 [match=5]
        20 permit ip any any [match=1]
Output when statistics per-entry command is not configured
NX-1# show ip access-list TEST
IP access list TEST
        10 permit ip 192.168.33.33/32 192.168.3.3/32
        20 permit ip any any
```

将 ACL 关联到接口或其他组件之后，ACL 就被编程到了 TCAM（Ternary Content Addressable Memory，三态内容寻址存储器）中。可以通过命令 **show system internal access-list interface** *interface-id* **input statistics [module** *slot*] 验证访问列表的 TCAM 编程情况，该命令可以显示 ACL 被编程到哪个寄存器组中，以及 ACL 关联到关联点之后创建的策略信息。除此之外，该命令还可以显示 ACL 中每条 ACE 表项的统计信息。

如果未配置 **statistics per-entry** 命令，那么 TCAM 计数器仅统计所有流量表项的递增情况，即 **permit ip 0.0.0.0/0 0.0.0.0/0**。在未配置 **statistics per-entry** 命令的情况下，例 10-3 显示了 TCAM 上的 ACL 表项及 TCAM 统计信息。

例 10-3 验证 TCAM 中的访问列表计数器

```
Per-Entry Statistics is Configured
NX-1# show system internal access-list interface e4/2 input statistics module 4
INSTANCE 0x0
---------------
  Tcam 1 resource usage:
  ----------------------
  Label_b = 0x2
   Bank 0
```

```
      ------
      IPv4 Class
         Policies: RACL(TEST)
         Netflow profile: 0
         Netflow deny profile: 0
         Entries:
            [Index] Entry [Stats]
            --------------------
  [0018:14242:0004] prec 1 permit-routed ip 192.168.33.33/32 192.168.3.3/32   [5]
  [0019:14c42:0005] prec 1 permit-routed ip 0.0.0.0/0 0.0.0.0/0   [2]
  [001a:15442:0006] prec 1 deny-routed ip 0.0.0.0/0 0.0.0.0/0   [0]
Per-Entry Statistics is Not Configured
NX-1# show system internal access-list interface e4/2 input statistics module 4
INSTANCE 0x0
---------------
  Tcam 1 resource usage:
  ----------------------
  Label_b = 0x3
   Bank 0
   ------
      IPv4 Class
         Policies: RACL(TEST) [Merged]
         Netflow profile: 0
         Netflow deny profile: 0
         Entries:
            [Index] Entry [Stats]
            --------------------
  [001b:15262:0007] prec 1 permit-routed ip 0.0.0.0/0 0.0.0.0/0   [33]
! Output after 5 packets are sent between the host 192.168.3.3 and 192.168.33.33
NX-1# show system internal access-list interface e4/2 input statistics module 4
INSTANCE 0x0
---------------
  Tcam 1 resource usage:
  ----------------------
  Label_b = 0x3
   Bank 0
   ------
      IPv4 Class
         Policies: RACL(TEST) [Merged]
         Netflow profile: 0
         Netflow deny profile: 0
         Entries:
            [Index] Entry [Stats]
            --------------------
  [001b:15262:0007] prec 1 permit-routed ip 0.0.0.0/0 0.0.0.0/0   [38]
```

如前所述，ACL 关联到关联点之后，ACLMGR 将创建策略，可以通过命令 **show system internal aclmgr access-lists policies** *interface-id* 验证 ACLMGR 创建的策略信息，包括策略类型和接口索引（指向关联该 ACL 的接口），如例 10-4 所示。

例 10-4 验证硬件中的访问列表计数器

```
NX-1# show system internal aclmgr access-lists policies ethernet 4/2
{
   0x11498cfc
   type = SC_TYPE_PORT; mode = SC_MODE_L3;
   flags =
```

```
            ifindex = 0x1a181000 (Ethernet4/2); vdc = 0; vlan = 0;
        2 policies: {
            ACLMGR_POLICY_INBOUND_IPV4_GHOST_RACL: 0x4400282
            ACLMGR_POLICY_INBOUND_IPV4_RACL: 0x4400283
        }
        no links
}

Policy node 0x04400282, name TEST, policy type 0x00400004
Destination: vdc = 1; vlan = 0; ifindex = 0x1a181000;
Effective destinations: no-destination
  Ifelse node 0x04400265
    TRUE action node 0x04400266, action type 0x00200002
    FALSE action node 0x04400267, action type 0x00200001
    NOMATCH action node 0x04400267, action type 0x00200001

Policy node 0x04400283, name TEST, policy type 0x00400010
Destination: vdc = 1; vlan = 0; ifindex = 0x1a181000;
Effective destinations: no-destination
  Ifelse node 0x04400265
    TRUE action node 0x04400266, action type 0x00200002
    FALSE action node 0x04400267, action type 0x00200001
    NOMATCH action node 0x04400267, action type 0x00200001
```

NX-OS 有一个 PPF（Packet Processing Filter，包处理过滤器）API，可以为客户端过滤 ACLMGR 收到和处理的安全规则，这里的客户端可以是接口、端口通道、VLAN 管理器、VSH 等。需要记住的是，ACLMGR 以 PPF 数据库的形式存储所有数据，其中的每个网元都是一个节点。根据从上一条命令收到的节点 ID，在该节点上查找 PPF 数据库即可验证策略的更多详细信息。

例 10-5 给出了 **show system internal aclmgr ppf node** *node-id* 命令的使用示例，将策略关联到关联点之后，可以通过该命令在 ACLMGR 的 PPF 数据库上查找所创建的策略节点。该命令对于排查与 ACL/过滤相关的故障问题非常有用，如 ACL 未正确过滤流量或者根本就没有匹配 NX-OS 平台上的任何 ACL 表项。

例 10-5 验证 PPF 数据库

```
NX-1# show system internal aclmgr ppf node 0x04400283
 ACLMGR PRIVATE DATA VERSION IN USE : 1
========= PPF Node: 0x4400283 ========
  .nlinks = 1
      0x4400265
  .noptlinks = 0
  .nrefs = 0
  .id = 0x4400283
  .group = 0x0
  .flags = 0x0
  .priv_data_size = 0
  .type = Policy Instance
  .dest.vdc = 1
  .dest.vrf = 0
  .dest.vlan = 0
  .dest.ifindex = 0x1a181000
  .dir = IN
  .u.pinst.type = 0x400010 (racl_ipv4)
  .u.pinst.policy.head = 0x4400265
  .u.pinst.policy.tail = 0x0
```

```
.u.pinst.policy.size = 0x0
.u.pinst.policy.el_field = 0
.u.pinst.policy.el_field = 0
```

> **注**：排查与 ACL 相关的故障问题时，建议收集故障期间的 **show tech aclmgr [detail]** 命令或 **show tech aclqos [detail]** 命令输出结果。线卡上的 ACLQOS 组件能够以线卡为基础提供 ACL 统计信息，对于排查 ACL 故障来说非常重要。

1. IGP 网络选择

如果在路由重分发期间，通过 ACL 来影响 IGP（Interior Gateway Protocol，内部网关协议）网络选择，那么就可以用 ACL 的源字段来标识网络，用目标字段来标识所允许的网络区间的最小前缀长度。表 10-1 提供了一个在 ACL 配置模式下配置的 ACL 表项示例，指定了与扩展 ACL 相匹配的网络。从第 2 行和第 3 行可以看出，目的网络 172.16.0.0 通配符的细微差异对于实际所允许的网络区间有较大影响。

表 10-1　　　　　　　　　用于 IGP 路由选择的扩展 ACL

ACL 表项	匹配的网络
permit ip any any	允许全部网络
permit ip host 172.16.0.0 host 255.240.0.0	允许 172.16.0.0/12 范围内的所有网络
permit ip host 172.16.0.0 host 255.255.0.0	允许 172.16.0.0/16 范围内的所有网络
permit host 192.168.1.1	仅允许网络 192.168.1.1/32

> **注**：用于分发列表的扩展 ACL 使用源字段来标识网络宣告源端，用目标字段来标识网络前缀。

2. BGP 网络选择

利用扩展 ACL 匹配 BGP 路由时的操作与匹配 IGP 路由完全不同，此时源字段匹配的是路由的网络部分，目标字段匹配的是网络掩码，如图 10-1 所示。在引入前缀列表之前，IOS 只能通过扩展 ACL 来匹配 BGP。

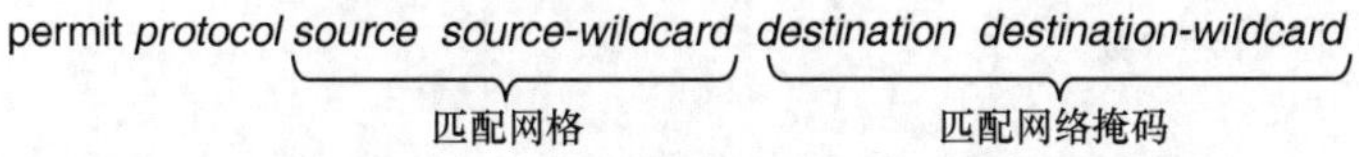

```
permit protocol source  source-wildcard  destination  destination-wildcard
               _____/     _____/
                    匹配网格                       匹配网络掩码
```

图 10-1　BGP 扩展 ACL 匹配规则

表 10-2 以具体实例解释了网络和子网掩码通配符的概念。

表 10-2　　　　　　　　　用于 BGP 路由选择的扩展 ACL

扩展 ACL	匹配的网络
permit ip 10.0.0.0 0.0.0.0 255.255.0.0 0.0.0.0	仅允许网络 10.0.0.0/16
permit ip 10.0.0.0 0.0.255.0 255.255.255.0 0.0.0.0	允许前缀长度为/24 的所有 10.0.x.0 网络
permit ip 172.16.0.0 0.0.255.255 255.255.255.0 0.0.0.255	允许前缀长度为/24～/32 的所有 172.16.x.x 网络
permit ip 172.16.0.0 0.0.255.255 255.255.255.128 0.0.0.127	允许前缀长度为/25～/32 的所有 172.16.x.x 网络

10.1.3　前缀匹配和前缀列表

前缀列表提供了另一种在路由协议中识别网络的方法，可以识别特定 IP 地址、网络或网络区间，可以通过多种前缀长度（子网掩码）选择多个网络（使用前缀匹配规范）。前缀匹配技术在网络选择方面比 ACL 更受欢迎，因为大多数网络工程师发现前缀匹配技术更容易理解。

1. 前缀匹配

前缀匹配规范的结构包括两部分：高阶比特模式和高阶比特数（确定所要匹配的比特模式中的高阶比特）。有些文档也将高阶比特模式称为地址或网络，将高阶比特数称为长度或掩码长度。

图 10-2 给出了基本的前缀匹配规范示例，其中的高阶比特模式是 192.168.0.0，高阶比特数是 16，为了更好地解释高阶比特数的位置，图中将高阶比特模式转换成了二进制模式。由于不包含额外的匹配长度参数，因而此处的高阶比特数是完全匹配参数。

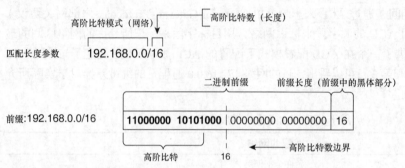

图 10-2　基本前缀匹配模式

虽然前缀匹配规范逻辑看起来似乎与访问列表的功能相同，但前缀匹配技术真正强大且灵活的地方在于，可以利用匹配长度参数在一条语句中通过指定前缀长度来识别多个网络。匹配长度参数选项如下。

- le（小于或等于）。
- ge（大于或等于）。

图 10-3 给出了一个前缀匹配规范示例，其中的高阶比特模式是 10.168.0.0，高阶比特数是 13，前缀匹配长度必须大于或等于 24。

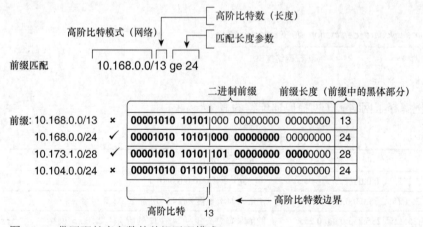

图 10-3　带匹配长度参数的前缀匹配模式

前缀 10.168.0.0/13 不满足匹配要求，因为前缀长度小于 24 比特的最小值，前缀 10.168.0.0/24 就满足匹配长度参数。前缀 10.173.1.0/28 满足匹配要求，因为前 13 比特与高阶比特模式匹配，且前缀长度位于匹配长度参数范围。前缀 10.104.0.0/24 不满足匹配要求，因为高阶比特数范围内的高阶比特模式不匹配。

图 10-4 的前缀匹配规范中的高阶比特模式为 10.0.0.0，高阶比特数为 8，前缀匹配长度必须

介于 22～26。

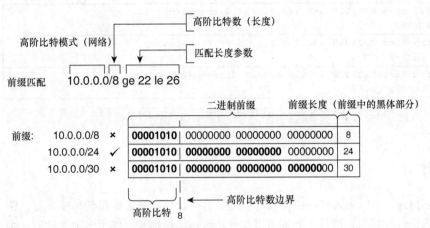

图 10-4 带有不合格匹配前缀的前缀匹配

前缀 10.0.0.0/8 不匹配的原因是前缀长度太短。前缀 10.0.0.0/24 匹配的原因是比特模式匹配且前缀长度范围是 22～26。前缀 10.0.0.0/30 不匹配的原因是比特太长。所有第 1 个八位组以 10 开头且前缀长度范围是 22～26 的前缀均将匹配该前缀匹配规范。

2. 前缀列表

前缀列表包含多个前缀匹配规范表项，每个表项都包含 **permit** 或 **deny** 操作。系统按照自上而下的顺序依次处理前缀列表，并对第一个匹配前缀应用适当的 **permit** 或 **deny** 操作。

配置 NX-OS 前缀列表的命令是全局配置命令 **ip prefix-list** *prefix-list-name* [**seq** *sequence-number*] {**permit** | **deny**} *high-order-bit-pattern/highorder-bit-count* [{**eq** *match-length-value* | **le** *le-value* | **ge** *ge-value* [**le** *le-value*]}]。

如果未提供序列号，那么就根据最高序列号自动递增 5，第一条表项是 5。这种序列方式允许删除特定表项。由于无法对前缀列表进行重新排序，因而建议预留足够的空间以便后续插入新的序列号。

例 10-6 为 RFC 1918 地址范围内的所有网络都定义了一个名为 RFC1918 的前缀列表示例，该前缀列表仅允许 192.168.0.0 网络区间存在/32 前缀，而不允许其他网络区间存在/32 前缀。

例 10-6 前缀列表示例

```
NX-1(config)# ip prefix-list RFC1918 seq 5 permit 192.168.0.0/13 ge 32
NX-1(config)# ip prefix-list RFC1918 seq 10 deny 0.0.0.0/0 ge 32
NX-1(config)# ip prefix-list RFC1918 seq 15 permit 10.0.0.0/7 ge 8
NX-1(config)# ip prefix-list RFC1918 seq 20 permit 172.16.0.0/11 ge 12
NX-1(config)# ip prefix-list RFC1918 seq 25 permit 192.168.0.0/15 ge 16
```

可以看出，序列 5 允许 192.168.0.0/13 比特模式中的全部/32 前缀，序列 10 拒绝所有比特模式中的全部/32 前缀，序列 15、20、25 则允许适当网络区间中的路由。请注意，序列顺序对于前两条表项来说非常重要，因为只有这样才能确保该前缀列表只有 192.168.0.0 存在/32 前缀。

命令 **show ip prefix-list** *prefix-list-name high-order-bit-pattern/high-orderbit-count* **first-match** 可以根据前缀列表来检查特定网络前缀，从而找到匹配序列（如果有）。

例 10-7 根据前面创建的前缀列表 RFC1918 对 3 种网络前缀模式运行上述命令。第一条命令使用高阶比特数 32（匹配序列 5），第二条命令使用高阶比特数 16（匹配序列 25），最后一条命令匹配序列 10（有一条 **deny** 操作）。

例 10-7 找到特定前缀模式的匹配序列

```
NX-1# show ip prefix-list RFC1918 192.168.1.1/32 first-match
   seq 5 permit 192.168.0.0/13 ge 32
NX-1# show ip prefix-list RFC1918 192.168.1.1/16 first-match
   seq 25 permit 192.168.0.0/15 ge 16
NX-1# show ip prefix-list RFC1918 172.16.1.1/32 first-match
   seq 10 deny 0.0.0.0/0 ge 32
```

注：可以通过例 10-7 命令验证网络前缀与前缀列表中的期望序列是否相匹配。

10.2 路由映射

路由映射（Route-Map）可以为各种路由协议提供大量有用特性。从最简单的角度来看，路由映射可以像 ACL 那样过滤网络，而且还能通过添加或修改网络属性来提供更多附加功能。必须在路由协议中引用路由映射才能影响路由协议的行为。路由映射对于 BGP 来说是非常关键的组件，因为路由映射是修改每个邻居独特路由策略的主要组件。

路由映射包括以下组件。

- **序列号**：指示路由映射的处理顺序。
- **条件匹配规则**：识别特定序列的前缀特性（网络、BGP 路径属性、下一跳等）。
- **处理操作**：允许或拒绝前缀。
- **可选操作**：取决于路由器引用路由映射的方式，可选操作包括修改、添加或删除路由特性。

路由映射的命令语法是 **route-map** *route-map-name* [**permit** | **deny**] [*sequence-number*]，规则如下。

- 如果未提供处理操作，那么就使用默认的 **permit** 操作。
- 如果未提供序列号，那么序列号默认为 10。
- 如果未包含条件匹配语句，那么隐含所有前缀均与该语句相关联。
- 匹配了条件匹配规则之后，路由映射仅在处理完所有可选操作（如已配置）之后才停止。
- 如果路由未被条件匹配，那么就为该路由执行隐含的 **deny** 操作。

例 10-8 给出了一个路由映射示例，包含了前面所说的 4 个组件。条件匹配规则基于 ACL 指定的网络区间，例中增加了必要的注释以方便了解每个序列中的路由映射行为。

例 10-8 路由映射示例

```
route-map EXAMPLE permit 10
 match ip address ACL-ONE
! Prefixes that match ACL-ONE are permitted. Route-map completes processing upon
  a match

route-map EXAMPLE deny 20
 match ip address ACL-TWO
! Prefixes that match ACL-TWO are denied. Route-map completes processing upon a match

route-map EXAMPLE permit 30
 match ip address ACL-THREE
 set metric 20
! Prefixes that match ACL-THREE are permitted and modify the metric. Route-map
  completes
! processing upon a match
```

```
route-map EXAMPLE permit 40
! Because a matching criteria was not specified, all other prefixes are permitted
! If this sequence was not configured, all other prefixes would drop because of the
! implicit deny for all route-maps
```

注： 删除特定路由映射语句时需要包含序列号，以免删除整个路由映射。

10.2.1 条件匹配

解释了路由映射的组件和处理顺序之后，本节将讨论匹配路由的方式。例 10-9 列出了 NX-OS 提供的各种可用选项。

例 10-9　路由映射示例

```
NX-1(config)# route-map TEST permit 10
NX-1(config-route-map)# match ?
  as-number        Match BGP peer AS number
  as-path          Match BGP AS path list
  community        Match BGP community list
  extcommunity     Match BGP community list
  interface        Match first hop interface of route
  ip               Configure IP features
  ipv6             Configure IPv6 features
  length           Packet length
  mac-list         Match entries of mac-lists
  metric           Match metric of route
  route-type       Match route-type of route
  source-protocol  Match source protocol
  tag              Match tag of route
  vlan             Vlan ID
```

可以看出，NX-OS 提供了多种条件匹配选项，有些选项（如 **vlan** 和 **mac-list**）仅适用于策略路由。表 10-3 列出了常用的前缀匹配方法的命令语法及其描述信息。

表 10-3　条件匹配选项

匹配命令	描述
match as-number { *number* [, *number* ...] | **as-path-accesslist name** *acl-name* }	匹配来自对等体（拥有匹配 ASN）的路由
match as-path *acl-number*	基于正则表达式查询操作选择前缀，以隔离 BGP PA（Path Attribute，路径属性）AS_PATH 中的 ASN *允许多个匹配变量
match community *communitylist-name*	基于 BGP 团体属性选择前缀 *允许多个匹配变量
match interface *interface-id*	基于扩展 BGP 团体属性选择前缀 *允许多个匹配变量
match ip address {*acl-number* | *acl-name*}	基于所关联的接口选择前缀 *允许多个匹配变量
match ip address prefix-list *prefix-list-name*	基于 ACL 定义的网络选择规则选择前缀 *允许多个匹配变量
match ip route-source prefix-list *prefix-list-name*	基于前缀选择规则选择前缀 *允许多个匹配变量
match local-preference	基于 BGP 属性本地优先级（Local Preference）选择前缀 *允许多个匹配变量

匹配命令	描述
match metric {*1-4294967295* | **external** *1-4294967295*} [*+- deviation*]	基于度量（可以是精确度量、度量区间或者在可接受的偏差范围内）选择前缀 *允许多个匹配变量
match route-type *protocolspecific-flag*	基于源路由协议中的源协议或路由的子分类选择前缀 *允许多个匹配变量
match tag *tag-value*	基于其他路由器设置的数字标记（0-4294967295）选择前缀 *允许多个匹配变量

1. 多个条件匹配规则

如果为特定路由映射序列配置了同一类型的多个变量（ACL、前缀列表、标记等），那么只要有任一个变量匹配即认为该前缀匹配，此时需要在配置中使用布尔逻辑运算符 *or*。

例 10-10 中的序列 10 要求前缀匹配 ACL-ONE 或 ACL-TWO。请注意，序列 20 未配置匹配语句，因而未匹配序列 10 的所有前缀均满足该条件并被拒绝。

例 10-10 包含多个匹配变量的路由映射示例

```
route-map EXAMPLE permit 10
 match ip address ACL-ONE ACL-TWO
!
route-map EXAMPLE deny 20
```

> **注：** 序列 20 是冗余配置，因为未匹配序列 10 的所有前缀均包含了隐式 **deny** 语句，这样配置的目的是为初级网络工程师提供更加清晰的配置信息。

如果为特定路由映射序列配置了多个匹配选项，那么前缀必须同时满足这些匹配选项才能符合该序列匹配要求，此时需要在该类配置中使用布尔逻辑运算符 *and*。

例 10-11 中的序列 10 要求前缀与 ACL-ONE 匹配，且度量值范围是 500～600。如果前缀不同时满足这两个匹配选项，那么就不符合序列 10 的要求且被拒绝（因为其他序列没有 **permit** 操作）。

例 10-11 包含多个匹配选项的路由映射示例

```
route-map EXAMPLE permit 10
 match ip address ACL-ONE
 match metric 550 +- 50
```

2. 复杂匹配

某些网络工程师发现，如果条件匹配规则使用 ACL、AS-Path ACL 或包含 **deny** 语句的前缀列表，那么路由映射就会显得过于复杂。例如，例 10-12 中的 ACL 对网络区间 172.16.1.0/24 应用了 **deny** 语句。

例 10-12 复杂匹配路由映射

```
ip access-list standard ACL-ONE
 deny    172.16.1.0 0.0.0.255
 permit 172.16.0.0 0.0.255.255
!
route-map EXAMPLE permit 10
 match ip address ACL-ONE
!
route-map EXAMPLE deny 20
 match ip address ACL-ONE
```

```
!
route-map EXAMPLE permit 30
 set metric 20
```

查看这类配置时必须遵循序列次序，然后关注条件匹配规则，应该仅在匹配后才使用处理操作和可选操作。如果匹配了条件匹配规则中的 **deny** 语句，那么该路由就会被路由映射中的序列拒绝。

由于前缀 172.16.1.0/24 被 ACL-ONE 拒绝，因而可以推断序列 10 和 20 无匹配项目，这样一来就不需要处理操作（**permit** 或 **deny**）。序列 30 未配置匹配语句，允许剩余的所有路由，因而前缀 172.16.1.0/24 将被传递给序列 30，度量值将被设置为 20。前缀 172.16.2.0/24 匹配 ACL-ONE，并传递给序列 10。

> **注：** 路由映射按照序列顺序、条件匹配规则、处理操作和可选操作依次进行处理，匹配组件中的 **deny** 语句与路由映射的序列操作相隔离。

10.2.2 可选操作

除能够传递前缀之外，路由映射还能修改路由属性。表 10-4 列出了常见的属性修改操作。

表 10-4　　　　　　　　　　　　　　　　路由映射 set 操作

Set 操作	描述
set as-path prepend {*as-number-pattern* | **last-as** *1-10*}	使用指定模式或者邻接 AS 的多次迭代为网络前缀追加 AS_PATH
set ip next-hop { *ip-address* | **peer-address** | **self** }	为所有匹配前缀设置下一跳 IP 地址，BGP 动态控制使用关键字 **peer-address** 或 **self**
set local-preference *0-4294967295*	设置 BGP PA 本地优先级（Local Preference）
set metric {+*value* | -*value* | *value*} *参数 value 的取值范围是 0-4294967295	修改现有度量或者为特定路由设置度量
set origin { **igp** | **incomplete** }	设置 BGP PA 路由来源（Origin）
set tag *tag-value*	设置数字标记（0-4294967295）以通过其他路由器识别网络
set weight *0-65535*	设置 BGP PA 权重（Weight）

10.2.3 路由策略配置不完整

路由策略的另一个常见问题就是路由映射配置不完整，具体来说，指的是引用 ACL、前缀列表、AS-Path ACL、BGP 团体列表等的条件匹配定义不完整。有时可能会接受所有路径，有时则完全不接受任何路径。无论路由映射是直接用于 BGP 邻居，还是用于路由重分发，都存在这个问题。

因此，验证路由映射的配置是否完整对于路由映射的故障排查来说非常重要。

10.2.4 诊断路由策略管理器

NX-OS 的 RPM（Route Policy Manager，路由策略管理器）进程可以为路由映射功能特性提供以下特性的路由映射功能。

- 运行为状态化、可重启、多线程进程。
- 在数据库（位于专用内存）中维护状态信息。
- 支持 IPv4 和 IPv6 地址簇。
- 为其他进程（如 BGP 或 OSPF）提供 API 交互功能。
- 与 URIB（Unicast RIB，单播 RIB）、ARP 表进行交互。

NX-OS 的 RPM 为验证 Nexus 交换机的路由映射交互情况提供了另一种方法。图 10-5 给出了一个示例拓扑，图中的 NX-2 将学自 OSPF 的路由重分发到了 EIGRP 中。

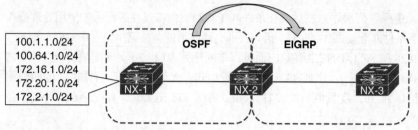

图 10-5 解释路由重分发的示例拓扑

例 10-13 显示了相关的 EIGRP 配置信息，将 OSPF 和直连路由重分发到了 EIGRP 中，通过路由映射来选择重分发到 EIGRP 的路由。

例 10-13 NX-2 重分发配置

```
NX-2
ip prefix-list PRE1 seq 5 permit 100.1.1.0/24
ip prefix-list PRE2 seq 5 permit 100.64.1.0/24
!
route-map REDIST-CONNECTED-2-EIGRP permit 10
  match interface Vlan10
route-map REDIST-OSPF-2-EIGRP permit 10
  match ip address prefix-list PRE1 PRE2
  set metric 10000 1 255 1 1500
!
router eigrp NXOS
  router-id 192.168.2.2
  address-family ipv4 unicast
    autonomous-system 100
    redistribute direct route-map REDIST-CONNECTED-2-EIGRP
    redistribute ospf NXOS route-map REDIST-OSPF-2-EIGRP
```

如果路由未按预期安装到 EIGRP 中，那么需要首先检查目的路由协议是否绑定了相关策略。此时可以通过命令 **show system internal rpm event-history rsw** 命令显示 RPM 处理的底层事件信息。

从例 10-14 的输出结果可以看出，EIGRP 应用了两个路由映射：分别用于 OSPF 和直连接口（路由重分发的源协议）。

例 10-14 查看 RPM 事件历史记录

```
NX-2# show system internal rpm event-history rsw

Routing software interaction logs of RPM
1) Event:E_DEBUG, length:98, at 211881 usecs after 02:01:35 [120] [4933]:
   Bind ack sent - client eigrp-NXOS uuid 0x41000130 for policy
   REDIST-CONNECTED-2-EIGRP
2) Event:E_DEBUG, length:93, at 211857 usecs after 02:01:35 [120] [4933]:
   Bind request - client eigrp-NXOS uuid 0x41000130 policy
   REDIST-CONNECTED-2-EIGRP
3) Event:E_DEBUG, length:114, at 17980 usecs after 02:01:21 [120] [4933]:
   Notify of clients aborted for policy REDIST-CONNECTED-2-EIGRP - change
   status: 0 - config refcount: 0
4) Event:E_DEBUG, length:85, at 95007 usecs after 02:01:06 [120] [4933]:
   Trying to notify for policy REDIST-CONNECTED-2-EIGRP - config refcount 1
```

```
5) Event:E_DEBUG, length:85, at 815063 usecs after 02:01:02 [120] [4933]:
   Trying to notify for policy REDIST-CONNECTED-2-EIGRP - config refcount 1
6) Event:E_DEBUG, length:93, at 381829 usecs after 01:59:30 [120] [4933]:
   Bind ack sent - client eigrp-NXOS uuid 0x41000130 for policy
   REDIST-OSPF-2-EIGRP
7) Event:E_DEBUG, length:88, at 381614 usecs after 01:59:30 [120] [4933]:
   Bind request - client eigrp-NXOS uuid 0x41000130 policy REDIST-OSPF-2-EIGRP
```

可以通过命令 **show system internal rpm clients** 获得与特定协议相关联的 RPM 进程数（见例 10-15），这是一种更简单的操作方法。如果 *Bind-count* 显示的数量与协议使用的期望路由映射期数量不匹配，那么就需要查看 RPM 事件历史记录以了解出错原因。

例 10-15　查看每种协议的 RPM 客户端数量

```
NX-2# show system internal rpm clients
PBR: policy based routing RF/RD: route filtering/redistribution

Client     Bind-count   Client-name        [VRF-name/Status/Id]
RF/RD      0            bgp-100
RF/RD      2            eigrp-NXOS
RF/RD      0            ospf-NXOS
RF/RD      0            icmpv6
RF/RD      0            igmp
RF/RD      0            u6rib
RF/RD      0            tcp
RF/RD      0            urib
```

除像例 10-2 那样查看前缀列表的精确信息之外，了解前缀列表的内部编程信息也非常有用。可以通过命令 **show system internal rpm ip-prefix-list** 显示 Nexus 交换机为例 10-13 中的路由映射配置的所有前缀列表。请注意，客户端显示了引用前缀列表的路由映射信息。此外，每个前缀列表表项都显示了该前缀列表的序列数量以及版本历史。例 10-16 显示了 NX-2 的 **show system internal rpm ip-prefix-list** 命令输出结果。

例 10-16　从 RPM 的角度查看前缀列表

```
NX-2# show system internal rpm ip-prefix-list
Policy name: PRE1              Type: ip prefix-list
Version: 2                     State: Ready
Ref. count: 1                  PBR refcount: 0
Stmt count: 1                  Last stmt seq: 5
Set nhop cmd count: 0          Set vrf cmd count: 0
Set intf cmd count: 0          Flags: 0x00000003
PPF nodeid: 0x00000000         Config refcount: 0
PBR Stats: No
Clients:
    REDIST-OSPF-2-EIGRP (internal - route-map)

Policy name: PRE2              Type: ip prefix-list
Version: 2                     State: Ready
Ref. count: 1                  PBR refcount: 0
Stmt count: 1                  Last stmt seq: 5
Set nhop cmd count: 0          Set vrf cmd count: 0
Set intf cmd count: 0          Flags: 0x00000003
PPF nodeid: 0x00000000         Config refcount: 0
PBR Stats: No
Clients:
    REDIST-OSPF-2-EIGRP (internal - route-map)
```

最后一种方法是从调试角度查看相关变更信息。习惯上需要从目标协议启用路由映射选项。例 10-17 显示了与将路由重分发给 EIGRP 相关联的路由映射的调试信息。

例 10-17 查看重分发的调试信息

```
NX-2# debug ip eigrp route-map
NX-2# config t
NX-2(config)# router eigrp NXOS
NX-2(config-router)# address-family ipv4 unicast
NX-2(config-router-af)# redistribute ospf NXOS route-map REDIST-OSPF-2-EIGRP
! Output omitted for brevity
02:59:50.071881 eigrp: librpm NXOS [9600] Setting up referee policy PRE1 -
   referer handle 0x820d5e4 referee_plcy 0x5f835614
02:59:50.071917 eigrp: librpm NXOS [9600] rpm_build_prefix_trie_pfl() - List:
   PRE1, No. of entries: 1
02:59:50.071954 eigrp: librpm NXOS [9600] rpm_create_prefix_in_trie()
   - ip addr: 100.1.1.0 mask_len: 24
02:59:50.071973 eigrp: librpm NXOS [9600] Successfully cloned the policy PRE1
   with version 2 rules_flag 0x00000001
02:59:50.071987 eigrp: librpm NXOS [9600] Stats AVL initialized with keysize 12
   and offset 16
02:59:50.072001 eigrp: librpm NXOS [9600] Setting up referee policy PRE2 -
   referer handle 0x820d5e4 referee_plcy 0x5f83556c
02:59:50.072016 eigrp: librpm NXOS [9600] rpm_build_prefix_trie_pfl() - List:
   PRE2, No. of entries: 1
02:59:50.072030 eigrp: librpm NXOS [9600] rpm_create_prefix_in_trie() - ip addr:
   100.64.1.0 mask_len: 24
02:59:50.072043 eigrp: librpm NXOS [9600] Successfully cloned the policy PRE2
   with version 2 rules_flag 0x00000001
02:59:50.072055 eigrp: librpm NXOS [9600] Stats AVL initialized with keysize 12
   and offset 16
02:59:50.072070 eigrp: librpm NXOS [9600] Successfully cloned the policy
   REDIST-OSPF-2-EIGRP with version 5 rules_flag 0x00000409
02:59:50.072086 eigrp: librpm NXOS [9600] Stats AVL initialized with keysize 12
   and offset 16
02:59:50.072099 eigrp: librpm NXOS [9600] Stats successfully init for policy
   REDIST-OSPF-2-EIGRP and context 0x0820d894
02:59:50.072125 eigrp: librpm NXOS [9600] rpm_bind request sent with uuid
   0x41000130 client name eigrp-NXOS policy <REDIST-OSPF-2-EIGRP> type <route-map>
02:59:50.072688 eigrp: librpm NXOS [9600] Recvd msg-type <BIND_ACK> pname
   <REDIST-OSPF-2-EIGRP> ptype <route-map> pversion <0> data <0x0820d894>
02:59:50.076902 eigrp: librpm NXOS [9600] ========== RPM Evaluation starting for
   policy REDIST-OSPF-2-EIGRP ==========
..
```

10.3 策略路由

路由器根据 IP 包的目的地址做出转发决策，但有时也可能需要考虑其他因素（如包长或源地址）来确定应该由哪台路由器转发数据包。

PBR（Policy-Based Routing，策略路由）可以根据以下数据包特征来实现数据包的条件式转发。

- 按协议类型（如 ICMP[Internet Control Message Protocol，互联网控制报文协议]、TCP[Transmission Control Protocol，传输控制协议]、UDP[User Datagram Protocol，用户数据报协议]等）进行路由。

- 按源 IP 地址、目的 IP 地址或两者进行路由。
- 根据特定转接流量对时延、链路速率或利用率的容忍度，手动为去往同一目的端的不同流量分配不同的网络路径。

路由器接口收到数据包之后，会检查这些数据包以执行 PBR 处理。PBR 会验证是否存在下一跳 IP 地址并利用指定的下一跳地址转发数据包。为了解决第一个下一跳 IP 地址可能不在 RIB 中的情况，需要再额外配置一个下一跳地址作为备用下一跳地址。如果路由表中没有指定的下一跳地址，那么就不对数据包执行条件式转发。

> 注：PBR 策略并不修改 RIB，因为这些策略不是面向所有数据包的通用策略。这一点常常使故障排查过程复杂化，因为路由表虽然显示了从路由协议学到的下一跳地址，但是并不能满足条件式转发流量对不同的下一跳地址的需求。

NX-OS 的 PBR 使用带有 **match** 和 **set** 语句的路由映射，然后将路由映射应用到入站接口。具体步骤如下。

- **第 1 步**：启用 PBR 功能。通过全局配置命令 **feature pbr** 启用 PBR 功能。
- **第 2 步**：定义路由映射。通过全局配置命令配置路由映射 **route-map** *route-map-name* [**permit** | **deny**] [*sequencenumber*]。
- **第 3 步**：确定条件匹配规则。条件匹配规则可以基于数据包长度（使用命令 **match length** *minimumlength maximum-length*）或者通过 ACL 使用数据包的 IP 地址字段（使用命令 **match ip address** {*access-list-number* | *acl-name*}）。
- **第 4 步**：指定下一跳。使用路由映射配置命令 **set ip** [**default**] **next-hop** *ip-address* [... *ip-address*] 为符合规则的数据包指定一个或多个下一跳。可选关键字 **default** 可以改变处理行为，仅在 RIP 无目的地址时才使用由路由映射指定的下一跳地址。如果 RIB 有可行路径，那么该路径就是转发数据包的下一跳地址。
- **第 5 步**：将路由映射应用于入站接口。使用接口参数命令 **ip policy route-map** *route-map-name* 应用路由映射。
- **第 6 步**：启用 PBR 统计功能（可选）。使用命令 **route-map** *route-map-name* **pbr-statistics** 启用 PBR 转发统计功能。

图 10-6 给出了 PBR 示例拓扑结构，图中的 NX-1 与 NX-6 之间的默认路径是 NX-1→NX-2→NX-3→NX-5→NX-6，这是因为链路 10.23.1.0/24 的开销小于链路 10.24.1.0/24。但本例要求从 NX-1 的 Loopback 0（192.168.1.1）到 NX-6 的 Loopback 0（192.168.6.6）的流量不能经由 NX-3 进行转发，必须经由 NX-4 进行转发（虽然该路径的开销更大）。

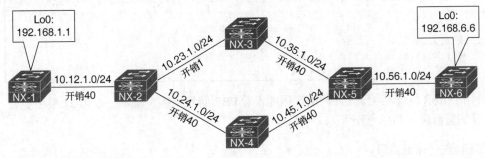

图 10-6 策略路由示例拓扑

例 10-18 显示了 NX-2 的 PBR 配置。

例 10-18　NX-2 的 PBR 配置

```
NX-2
feature pbr
!
ip access-list R1-TO-R6
  10 permit ip 192.168.1.1/32 192.168.6.6/32
!
route-map PBR pbr-statistics
route-map PBR permit 10
  match ip address R1-TO-R6
  set ip next-hop 10.24.1.4
!
interface Ethernet2/1
  description to NX-1
  ip address 10.12.1.2/24
  ip router ospf NXOS area 0.0.0.0
  ip policy route-map PBR
```

例 10-19 显示了 NX-1 发起的 traceroute 操作（显示了流经 NX-3 的流量）。由于未指定源接口，因而流量来自 IP 地址 10.12.1.1，这一点可以从 NX-2 关于网络前缀 192.168.6.6 的路由表加以确认。

例 10-19　去往 NX-6 Loopback0 接口的正常流量流

```
NX-1# traceroute 192.168.6.6
traceroute to 192.168.6.6 (192.168.6.6), 30 hops max, 40 byte packets
 1  10.12.1.2 (10.12.1.2)  2.016 ms  1.382 ms  1.251 ms
 2  10.23.1.3 (10.23.1.3)  3.24 ms  3.372 ms  3.364 ms
 3  10.35.1.5 (10.35.1.5)  5.819 ms  5.497 ms  3.844 ms
 4  10.56.1.6 (10.56.1.6)  5.577 ms *  5.359 ms

NX-2# show ip route 192.168.6.6
! Output omitted for brevity

192.168.6.6/32, ubest/mbest: 1/0
    *via 10.23.1.3, Eth2/2, [110/82], 3d01h, ospf-NXOS, intra
```

从例 10-20 可以看出，从 NX-1 Loopback 0 接口发起的 traceroute 操作在 NX-2 上被重定向到了 NX-4，traceroute 结果显示 PBR 工作正常。

例 10-20　验证 PBR 流量

```
NX-1# traceroute 192.168.6.6 source 192.168.1.1
traceroute to 192.168.6.6 (192.168.6.6) from 192.168.1.1 (192.168.1.1), 30 hops max,
  40 byte packets
 1  10.12.1.2 (10.12.1.2)  1.33 ms  1.199 ms  1.064 ms
 2  10.24.1.4 (10.24.1.4)  3.354 ms  2.978 ms  2.825 ms
 3  10.45.1.5 (10.45.1.5)  3.818 ms  3.707 ms  3.487 ms
 4  10.56.1.6 (10.56.1.6)  4.542 ms *  4.907 ms
```

NX-2 启用了 PBR 统计功能，允许网络工程师查看 PBR 转发了多少流量。例 10-21 显示了启用条件式转发流量前后的 PBR 统计信息。

例 10-21　PBR 统计输出结果

```
PBR Statistics after Example 10-32
NX-2# show route-map PBR pbr-statistics
route-map PBR, permit, sequence 10
```

```
    Policy routing matches: 0 packets

Default routing: 12 packets
PBR Statistics after Example 10-33
NX-2# show route-map PBR pbr-statistics
route-map PBR, permit, sequence 10
    Policy routing matches: 12 packets

Default routing: 12 packets
```

注：本例配置的 PBR 用于转接流量。如果要对本地生成的流量应用 PBR，那么就要使用命令 **ip local policy route-map** *route-map-name*。

10.4 本章小结

本章讨论了 NX-OS 路由映射用到的条件匹配进程所包含的一些重要功能模块。

- 访问控制列表提供了网络识别方法，扩展 ACL 则为 IGP 协议提供了网络选择和路由器宣告能力，而且为 BGP 路由提供了使用网络和子网掩码通配符的能力。
- 前缀列表可以根据高阶比特模式、高阶比特数以及前缀长度要求识别网络。
- 正则表达式能够以系统化方式解析输出结果。虽然正则表达式通常用于 BGP 过滤操作，但有时也可以用于 CLI 解析输出结果。

路由映射的路由过滤功能与 ACL 相似，同时还提供了路由属性的修改功能。路由映射由序列号、匹配条件、处理操作以及可选的修改操作组成，相应的逻辑如下。

- 如果未指定匹配规则，那么所有路由都符合该路由映射序列。
- 相同类型的多个条件匹配规则需要使用布尔逻辑运算符 *or*，不同类型的多个条件匹配规则需要使用布尔逻辑运算符 *and*。
- NX-OS 使用 RPM（RPM 与实际协议使用相互独立的进程和内存空间）。RPM 为排查协议的非期望行为提供了另一种方法。

策略路由可以将特定网络流量引导到与路由协议选择的路径不同的路径上，也就意味着默认包转发决策可以绕过路由协议。

第 11 章

BGP 故障排查

本章主要讨论如下主题。
- BGP 基础知识。
- BGP 对等故障排查。
- BGP 路由更新和路由传播。
- BGP 路由过滤。

11.1 BGP 基础知识

RFC 1654 定义的 BGP（Border Gateway Protocol，边界网关协议）是一种路径向量路由协议，具有可扩展性、灵活性和网络稳定性等特点。最初开发 BGP 的时候，主要设计考虑是面向跨公网（如 Internet）或专网的 IPv4 组织机构间的路由信息交换。人们常常将 BGP 称为 Internet 协议，因为 BGP 是唯一能够保存 Internet 路由表的协议，目前的 IPv4 路由数量已超过 600000 条，IPv6 路由数也超过了 42000 条，而且一直都在快速增长当中。

从 BGP 的角度来看，AS（Autonomous System，自治系统）是由单个组织机构控制的路由器集合。需要连接 Internet 的组织机构必须获得 ASN（Autonomous System Number，自治系统号）。最初的 ASN 长度为 2 字节（16 比特），可以提供 65535 个 ASN。后来由于资源耗尽，RFC 4893 又将 ASN 字段扩展为 4 字节（32 比特），从而能够提供 4294967295 个唯一的 ASN，与原来的 65535 个 ASN 相比有了巨大提升。IANA（Assigned Numbers Authority，互联网数字分配机构）负责分配所有的公有 ASN，以确保全球唯一性。

只要不在 Internet 上进行公开交换，所有组织机构就都能使用两段专用 ASN。ASN 64512～65535 是 16 比特 ASN 区间内的专用 ASN，4200000000～4294967294 是扩展后的 32 比特区间内的专用 ASN。

> **注**：必须使用 IANA 分配的 ASN、服务提供商分配的 ASN 或专用 ASN。不仅如此，公有前缀还与组织机构的 ASN 号相关联。因此，错误或恶意使用错误的 ASN 宣告前缀可能会导致流量丢失，甚至可能会给 Internet 造成严重破坏。

11.1.1 地址簇

虽然最初的 BGP 主要负责在组织机构之间路由 IPv4 前缀，但 RFC 2858 通过 AFI（Address-Family Identifier，地址簇标识符）扩展能力增加了 MP-BGP（Multi-Protocol BGP，多协议 BGP）功能。地址簇与特定的网络协议（如 IPv4、IPv6 等）相关，可以通过 SAFI（Subsequent Address-Family Identifier，子地址簇标识符）来提供更精细化的标识能力（如单播和多播）。MBGP 利用 BGP PA（Path Attribute，路径属性）MP_REACH_NLRI 和 MP_UNREACH_NLRI 来实现这种隔离能力，这些属性都位于 BGP 的更新消息当中，用于携带不同地址簇的网络可达性信息。

> 注：有些网络工程师将多协议 BGP 称为 MP-BGP，也有些网络工程师使用术语 MBGP，两者都是一回事。

网络工程师和设备商一直都在丰富 BGP 的功能和特性。目前 BGP 已经为 MPLS VPN（Multiprotocol Label Switching Virtual Private Network，多协议标签交换虚拟专网）、IPSec SA（Security Association，安全关联）和 VxLAN（Virtual Extensible LAN，虚拟可扩展局域网）等叠加技术提供了可扩展的信令控制平面，这些叠加技术通过 MPLS L3VPN 提供三层连接，或者通过 eVPN（ethernet VPN，以太网 VPN）提供二层连接。

每个地址簇在 BGP 中都会为每种协议维护一个独立的数据库和配置（地址簇+子地址簇），这样就可以为不同的地址簇配置不同的路由策略，即使路由器与其他路由器之间使用的是同一个 BGP 会话。为了区分 AFI 和 SAFI 数据库，BGP 的每条路由宣告都会包含 AFI 和 SAFI。表 11-1 列出了 BGP 常用的 AFI 和 SAFI。

表 11-1　　　　　　　　　　　　BGP 常用的 AFI/SAFI

AFI	SAFI	网络层信息
1	1	IPv4 单播
1	2	IPv4 多播
1	4	MPLS 标签
1	128	MPLS L3VPN IPv4
2	1	IPv6 单播
2	4	MPLS 标签
2	128	MPLS L3VPN IPv6
25	65	VPLS（Virtual Private Lan Service，虚拟专用 LAN 服务）
25	70	EVPN

11.1.2　路径属性

BGP 为每条网络路径都关联 PA（Path Attribute，路径属性），PA 可以为 BGP 提供精细化的路由策略控制能力。可以将 BGP 前缀的 PA 分为以下几种。

- 周知强制属性（Well-known mandatory）。
- 周知自选属性（Well-known discretionary）。
- 可选传递属性（Optional transitive）。
- 可选非传递属性（Optional nontransitive）。

RFC 4271 规定，所有的 BGP 实现都必须支持周知属性。每个前缀宣告都必须包含周知强制属性，可以包含也可以不包含周知自选属性。

BGP 实现可以不支持可选属性。如果设置了可选传递属性，那么就可以随着路由宣告从一个 AS 传递到另一个 AS。如果是可选非传递属性，那么就不能在不同的 AS 之间共享。BGP 中的 NLRI（Network Layer Reachability Information，网络层可达性信息）是路由更新，由网络前缀、前缀长度以及特定路由的 BGP PA 组成。

11.1.3　环路预防

BGP 是一种路径向量路由协议，与链路状态路由协议不同，BGP 并不包含完整的网络拓扑结构。BGP 的行为与距离向量协议相似，以确保无环路径。

BGP 属性 AS_PATH 是一个周知强制属性，包含了前缀宣告从源端 AS 开始穿越的所有 AS 的完整 ASN 列表。BGP 协议利用 AS_PATH 来实现环路预防，如果 BGP 路由器收到前缀宣告之后，发现自己的 AS 位于 AS_PATH 列表中，那么就会丢弃该前缀，因为路由器认为该路由宣告产生了环路。

> **注：** 本章后面还会讨论其他与 IBGP 相关的环路预防机制。

11.2 BGP 会话

BGP 会话指的是两台 BGP 路由器之间建立的邻接关系，BGP 会话是点对点会话，分为以下两类。

- **IBGP（Internal BGP，内部 BGP）会话**：与 IBGP 路由器（位于相同 AS 或参与同一 BGP 联盟的路由器）建立的会话。与 EBGP 会话相比，IBGP 会话更安全，可以降低某些 BGP 安全措施。IBGP 前缀安装到路由器的 RIB（Routing Information Base，路由信息库）中之后，分配的 AD（Administrative Distance，管理距离）值为 200。
- **EBGP（External BPG，外部 BGP）会话**：与其他 AS 中的 BGP 路由器建立的会话。EBGP 前缀安装到路由器的 RIB（Routing Information Base，路由信息库）中之后，分配的 AD 值为 20。

> **注：** AD 代表路由信息源的可信度。如果一台路由器从一个以上的路由协议学到去往某目的端的路由，且拥有相同的前缀长度，那么就会比较 AD，优选 AD 值较小的路由。

BGP 使用 TCP（Transmission Control Protocol，传输控制协议）端口 179 与其他路由器进行通信，TCP 可以处理通信（控制平面）数据包的分段、排序和可靠性（确认和重传）。BGP 可以与直连邻居建立邻接关系，也可以与多跳之外的邻居建立邻接关系。多跳会话要求路由器使用安装在 RIB 中的底层路由（静态路由或来自路由协议），与远程端点建立 TCP 会话。

> **注：** 通过同一个网络相连接的 BGP 邻居使用 ARP 表来定位对等体的 IP 地址。多跳 BGP 会话需要路由表信息来查找对等体的 IP 地址。为了提供建立 BGP TCP 会话所需的拓扑路径信息，通常都要在 IBGP 节点之间配置一条静态路由或者运行某种 IGP（Interior Gateway Protocol，内部网关协议），默认路由无法建立多跳 BGP 会话。

可以将 BGP 视为一种控制平面路由协议或一种应用程序，因为 BGP 允许与多跳之外的对等体交换路由。虽然 BGP 路由器不要求必须在数据平面（路径）交换前缀，但数据路径中的所有路由器都必须知道将要通过它们转发的所有路由。

11.2.1 BGP 标识符

BGP RID（Router-ID，路由器 ID）是一个 32 比特唯一编号，负责在宣告的前缀中将 BGP 路由器标识为 BGP 标识符。此外，RID 还可以为自治系统内宣告的路由器提供环路预防机制。可以手动或动态设置 BGP RID。必须将 RID 设置为非零值，只有这样，路由器才能成为邻居。NX-OS 节点使用处于 Up 状态的最小环回接口的 IP 地址作为 RID，如果没有处于 Up 状态的环回接口，那么就在 BGP 进程初始化的时候，使用处于 Up 状态的最小接口的 IP 地址作为 RID。

RID 通常使用路由器的 IPv4 地址（如环回地址），实际上可以使用任何 IPv4 地址（包括未在路由器上配置的 IP 地址）。NX-OS 在 BGP 路由器配置模式下通过命令 **router-id** *router-id* 静态分配 BGP RID。更改了 RID 之后，必须重置并重新建立所有 BGP 会话。

> **注：** 最佳实践是静态分配 BGP RID。

11.2.2 BGP 消息

BGP 通信需要用到 4 类消息，如表 11-2 所示。

表 11-2　　　　　　　　　　　　　　BGP 数据包类型

类型	名称	功能描述
1	OPEN（打开）	设置并建立 BGP 邻接关系
2	UPDATE（更新）	宣告、更新或撤销路由
3	NOTIFICATION（通告）	向 BGP 邻居指示差错状态
4	KEEPALIVE（保持激活）	确保 BGP 邻居始终处于激活状态

1. OPEN

OPEN 消息用于建立 BGP 邻接关系。双方在建立 BGP 对等关系之前需要协商会话能力，OPEN 消息包含了 BGP 版本号、发端路由器的 ASN、保持时间（Hold Time）、BGP 标识符以及建立会话能力所需的其他可选参数。

保持时间属性为每个 BGP 邻居设置保持定时器（以秒为单位）。收到 UPDATE 或 KEEPALIVE 消息之后，保持定时器会被重置为初始值。如果保持定时器达到零，那么就会拆除该 BGP 会话、删除来自该邻居的路由，同时还会向其他 BGP 邻居发送适当的路由撤销消息以通告受影响的前缀。保持时间是确保 BGP 邻居处于正常、有效状态的心跳机制。

建立 BGP 会话时，路由器会使用包含在两台路由器 OPEN 消息中较小的保持时间值。必须将保持时间值设置为至少 3s 或者 0，思科路由器的默认保持定时器为 180s。

2. UPDATE

UPDATE 消息负责宣告可行路由、撤销以前宣告的路由或两者。UPDATE 消息包含 NLRI（Network Layer Reachability Information，网络层可达性信息），NLRI 在宣告前缀时包含了前缀以及相关联的 BGP PA。撤销 NLRI 仅包含前缀信息。为了减少不必要的流量，UPDATE 消息也可以充当 KEEPALIVE 消息。

3. NOTIFICATION

如果检测到 BGP 会话出现差错（如保持定时器到期、邻居能力变更或请求 BGP 会话重置），那么就会发送 NOTIFICATION 消息，从而导致 BGP 连接关闭。

注：有关 BGP 消息的更多内容将在故障排查部分进行讨论。

4. KEEPALIVE

BGP 不依赖 TCP 连接状态来确保邻居的激活性。KEEPALIVE 消息按照两台 BGP 路由器协商好的保持定时器的 1/3 时间进行交换，思科设备的默认保持时间为 180s，因而默认 KEEPALIVE 间隔为 60s。如果保持时间为 0，那么就不会在 BGP 邻居之间发送 KEEPALIVE 消息。

11.2.3 BGP 邻居状态

BGP 与被称为对等体的邻居路由器建立 TCP 会话，BGP 通过 FSM（Finite State Machine，有限状态机）来维护所有 BGP 对等体及其运行状态的表格。BGP 会话可能报告以下状态。

- Idle（空闲）状态。
- Connect（连接）状态。
- Active（激活）状态。
- OpenSent（打开发送）状态。
- OpenConfirm（打开确认）状态。

- Established（建立）状态。

图 11-1 显示了 BGP FSM 以及 BGP 会话建立的状态顺序。

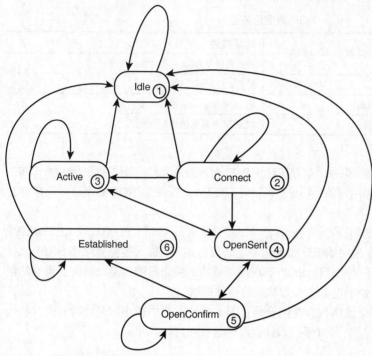

图 11-1　BGP 有限状态机

1. Idle 状态

Idle（空闲）是 BGP FSM 的第一阶段。BGP 检测到启动事件之后，会尝试初始化与 BGP 对等体之间的 TCP 连接，同时侦听来自对等路由器的新连接。

如果因差错导致 BGP 第二次返回 Idle 状态，那么就要将连接重试定时器（ConnectRetryTimer）设置为 60s，且在再次初始化连接之前该定时器必须递减为零。如果仍未离开 Idle 状态，那么连接重试定时器的时长将比上一次增加一倍。

2. Connect 状态

Connect 状态下的 BGP 将启动 TCP 连接。TCP 三向握手进程完成之后，已建立的 BGP 会话的 BGP 进程就会重置连接重试定时器并向邻居发送 OPEN 消息，同时迁移为 OpenSent 状态。

如果 Connect 阶段完成之前，连接重试定时器超时，那么就会尝试建立新的 TCP 连接、重置连接重试定时器，状态迁移到 Active 状态。如果在这个过程中收到了其他输入，那么就会迁移到 Idle 状态。

该阶段由 IP 地址较大的邻居管理连接，发起请求的路由器使用动态源端口，目的端口始终为 179。

> 注：服务提供商通常为客户网络分配的 IP 地址都比较大或比较小，这样做有助于服务提供商为 ACL 或防火墙规则创建正确指令，对故障排查来说也非常有利。

3. Active 状态

该状态下的 BGP 会启动一个新的 TCP 三向握手进程。如果建立了 TCP 连接，那么就会发送一条 OPEN 消息，将保持定时器设置为 4min，且状态迁移为 OpenSent 状态。如果 TCP 连接失败，那么状态将回到 Connect 状态并重置连接重试定时器。

4. OpenSent 状态

该状态下的发端路由器已经发送了 OPEN 消息并等待其他路由器发来的 OPEN 消息，发端路由器收到其他路由器发来的 OPEN 消息之后，会检查这两条 OPEN 消息的差错情况，将对比以下信息。

- BGP 版本必须匹配。
- OPEN 消息的源 IP 地址必须与为邻居配置的 IP 地址相匹配。
- OPEN 消息中的 AS 号必须与邻居的配置相匹配。
- BGP 标识符（RID）必须唯一。如果 RID 不存在，那么就不满足该条件。
- 安全参数（密码、TTL[Time to Live, 生存时间]等）。

如果 OPEN 消息没有任何差错，那么就协商保持时间（使用较低值），并发送 KEEPALIVE 消息（假设保持时间未设置为零）。此后，连接状态将迁移到 OpenConfirm 状态。如果在 OPEN 消息中发现了差错，那么就会发送 NOTIFICATION 消息，并将状态迁移回 Idle 状态。

如果 TCP 收到了断开连接消息，那么 BGP 将关闭连接，重置连接重试定时器，并将状态设置为 Active 状态。如果在这个过程中收到了其他输入，那么就会迁移到 Idle 状态。

5．OpenConfirm 状态

该状态下的 BGP 会等待 KEEPALIVE 消息或 NOTIFICATION 消息。收到邻居的 KEEPALIVE 消息之后，就会迁移到 Established 状态。如果保持定时器到期、出现终止事件或者收到 NOTIFICATION 消息，那么就会迁移回 Idle 状态。

6．Established 状态

该状态将建立 BGP 会话，BGP 邻居通过 UPDATE 消息交换路由。收到 UPDATE 和 KEEPALIVE 消息之后，将重置保持定时器。如果保持定时器到期，那么就会检测到差错，BGP 则将邻居迁移回 Idle 状态。

11.2.4 BGP 配置与验证

虽然 NX-OS 的 BGP 配置步骤很简单，但必须在启用了 BGP 功能特性之后才能使用 BGP 命令行，可以通过命令 **feature bgp** 在 Nexus 平台上启用 BGP 功能特性。在 NX-OS 设备上配置 BGP 的步骤如下。

- 第 1 步：创建 BGP 路由进程。利用全局配置命令 **router bgp** *as-number* 初始化 BGP 进程。
- 第 2 步：分配 BGP RID。在 BGP 路由器进程下分配唯一的 BGP RID，RID 可以是分配给物理接口或环回接口的 IP 地址。
- 第 3 步：初始化地址簇。利用 BGP 路由器配置命令 **address-family** *afi safi* 初始化地址簇，从而能够与 BGP 邻居相关联。
- 第 4 步：标识 BGP 邻居的 IP 地址和自治系统号。利用 BGP 路由器配置命令 **neighbor** *ip-address* **remote-as** *as-number* 标识 BGP 邻居的 IP 地址和自治系统号。
- 第 5 步：激活 BGP 邻居的地址簇。使用 BGP 邻居配置命令 **address-family** *afi safi* 激活 BGP 邻居的地址簇。

图 11-2 给出了一个示例拓扑（该拓扑结构也将用于下一节示例），该拓扑中的 Nexus 设备 NX-1、NX-2 和 NX-4 都属于 AS 65000，路由器 NX-6 属于 AS 65001。

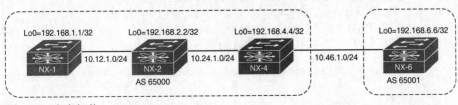

图 11-2 参考拓扑

例 11-4 显示了路由器 NX-4 的 BGP 配置示例（包括 IBGP 和 EBGP 对等配置），例中的 NX-4 试图与 NX-1 建立 IBGP 对等关系，与 NX-6 建立 EBGP 对等关系。配置 BGP 对等关系时，必须确保以下信息的正确性。

- 本地和远程 ASN。
- 源对等 IP。
- 远程对等 IP。
- 认证密码（可选）。
- EBGP 多跳（仅 EBGP）。

例 11-1 中的 NX-4 与 NX-1 建立了 IBGP 对等关系，与 NX-6 路由器建立了 EBGP 对等关系。此外，NX-4 设备还使用 **network** 命令在 IPv4 地址簇下宣告了自己的环回地址。

例 11-1　NX-OS BGP 配置

```
NX-4
feature bgp
router bgp 65000
  router-id 192.168.4.4
  address-family ipv4 unicast
    network 192.168.4.4/32
    redistribute direct route-map conn
  neighbor 10.46.1.6
    remote-as 65001
    address-family ipv4 unicast
  neighbor 192.168.1.1
    remote-as 65000
    update-source loopback0
    address-family ipv4 unicast
      next-hop-self
!
ip prefix-list connected-routes seq 5 permit 10.46.1.0/24
!
route-map conn permit 10
  match ip address prefix-list connected-routes
```

执行了 NX-4 的配置之后，NX-1 与 NX-4 之间以及 NX-4 与 NX-6 之间应该建立了对等关系。可以使用 **show bgp** *afi safi* **summary** 命令验证 BGP 对等信息，其中的 *afi* 和 *safi* 用于不同的地址簇，本例使用的是 IPv4 单播地址簇。例 11-2 显示了 NX-1 与 NX-4、NX-6 之间的 BGP 对等关系验证情况，可以看出，NX-4 交换机已经建立了 IBGP 和 EBGP 对等连接，而且从每个邻居都学到了一条前缀。

例 11-2　验证 NX-OS BGP 对等连接

```
NX-4# show bgp ipv4 unicast summary
BGP summary information for VRF default, address family IPv4 Unicast
BGP router identifier 192.168.4.4, local AS number 65000
BGP table version is 8, IPv4 Unicast config peers 2, capable peers 2
4 network entries and 4 paths using 576 bytes of memory
BGP attribute entries [3/432], BGP AS path entries [1/6]
BGP community entries [0/0], BGP clusterlist entries [0/0]

Neighbor        V    AS MsgRcvd MsgSent   TblVer  InQ OutQ Up/Down  State/PfxRcd
10.46.1.6       4 65001      24      27        8    0    0 00:16:01 1
192.168.1.1     4 65000      23      24        8    0    0 00:16:24 1
```

建立了 BGP 对等关系之后，可以使用 **show bgp** *afi safi* 命令验证 BGP 前缀，该命令可以列出各个地址簇中的所有 BGP 前缀。例 11-3 显示了 NX-4 的 BGP 前缀输出结果，可以看出，BGP 表保存了本地宣告的前缀，下一跳为 0.0.0.0，同时还有一个指示该前缀学自 IBGP 对等体（i）或 EBGP 对等体（e）的标签。

例 11-3　NX-OS BGP 表输出结果

```
NX-4# show bgp ipv4 unicast
BGP routing table information for VRF default, address family IPv4 Unicast
BGP table version is 20, local router ID is 192.168.4.4
Status: s-suppressed, x-deleted, S-stale, d-dampened, h-history, *-valid, >-best
Path type: i-internal, e-external, c-confed, l-local, a-aggregate, r-redist,
I-injected
Origin codes: i - IGP, e - EGP, ? - incomplete, | - multipath, & - backup

   Network            Next Hop         Metric     LocPrf     Weight Path
*>r10.46.1.0/24       0.0.0.0               0        100      32768 ?
*>i192.168.1.1/32     192.168.1.1                    100          0 i
*>l192.168.4.4/32     0.0.0.0                        100      32768 i
*>e192.168.6.6/32     10.46.1.6                                   0 65001 i
```

在 NX-OS 上配置了命令 **router bgp** *asn* 之后，BGP 进程就被实例化了。使用命令 **show bgp process** 可以查看 BGP 进程的详细信息和配置摘要，该命令可以显示 BGP 进程 ID、状态、已配置和处于激活状态的 BGP 对等体的数量、BGP 属性、VRF 信息、重分发以及与各种重分发语句一起使用的路由映射等信息。如果 BGP 进程出现了问题，那么就可以通过该命令来了解 BGP 的状态以及 BGP 进程的内存信息。例 11-4 显示了命令 **show bgp process** 的输出结果，突出显示了例 11-3 中的部分重要字段。

例 11-4　NX-OS BGP 进程

```
NX-4# show bgp process

BGP Process Information
BGP Process ID                 : 9618
BGP Protocol Started, reason:  : configuration
BGP Protocol Tag               : 65000
BGP Protocol State             : Running
BGP MMODE                      : Not Initialized
BGP Memory State               : OK
BGP asformat                   : asplain

BGP attributes information
Number of attribute entries    : 4
HWM of attribute entries       : 4
Bytes used by entries          : 400
Entries pending delete         : 0
HWM of entries pending delete  : 0
BGP paths per attribute HWM    : 3
BGP AS path entries            : 1
Bytes used by AS path entries  : 26

Information regarding configured VRFs:

BGP Information for VRF default
VRF Id                         : 1
```

```
VRF state                     : UP
Router-ID                     : 192.168.4.4
Configured Router-ID          : 192.168.4.4
Confed-ID                     : 0
Cluster-ID                    : 0.0.0.0
No. of configured peers       : 2
No. of pending config peers   : 0
No. of established peers      : 2
VRF RD                        : Not configured

    Information for address family IPv4 Unicast in VRF default
    Table Id                  : 1
    Table state               : UP
    Peers     Active-peers    Routes    Paths    Networks    Aggregates
    2         2               4         4        1           0

    Redistribution
        direct, route-map conn
    Wait for IGP convergence is not configured

    Nexthop trigger-delay
        critical 3000 ms
        non-critical 10000 ms

    Information for address family IPv6 Unicast in VRF default
    Table Id                  : 80000001
    Table state               : UP
    Peers     Active-peers    Routes    Paths    Networks    Aggregates
    0         0               0         0        0           0

    Redistribution
        None

    Wait for IGP convergence is not configured

    Nexthop trigger-delay
        critical 3000 ms
        non-critical 10000 ms
```

11.3 BGP 对等关系故障排查

BGP 对等关系故障主要分为两类。
- BGP 对等关系中断。
- BGP 对等关系翻动。

BGP 对等关系故障是网络运营商在生产环境中遇到的常见问题之一，虽然这是一个非常常见的问题，但 BGP 对等关系中断或者出现翻动所产生的影响可能极小（如果网络中存在冗余），也可能极大（如果与 Internet 提供商的对等关系完全中断）。本节将重点讨论这两类故障的排查问题。

11.3.1 BGP 对等关系中断故障排查

如果已配置的 BGP 会话处于非 Established 状态，那么网络工程师就将该场景称为 BGP 对等关系中断。BGP 对等关系中断是大多数 BGP 环境常见的问题之一。如果出现了以下情形，那么就可以检测出对等关系出现中断。

- 在 BGP 会话建立期间因配置差错而导致 BGP 会话中断。
- 因网络迁移或其他事件、软件或硬件升级触发 BGP 会话中断。
- 因传输问题无法维护 BGP 保持激活。

中断的 BGP 对等体处于 Idle 或 Active 状态。从对等体状态的角度来看，这些状态可能意味着以下问题。

- Idle 状态。
  - 没有连接对等体的路由。
- Active 状态。
  - 没有去往对等地址的路由（无 IP 连接）。
  - 配置差错，如更新源丢失或配置出错。
- Idle/Active 状态。
  - 建立了 TCP 会话，但 BGP 协商失败（如 AS 配置差错）。
  - 路由器不同意对等关系参数。

接下来将详细讨论 BGP 对等关系中断故障的排查步骤。

1. 验证配置

排查 BGP 对等关系故障的第一步就是验证配置并了解设计方案。很多时候 BGP 对等关系无法建立的原因是基本的配置差错，因而在配置新 BGP 会话时应注意检查以下内容。

- 本地 AS 号。
- 远程 AS 号。
- 验证网络拓扑和其他文件。

理解对等体之间的 BGP 数据包的流量流非常重要。BGP 包的源 IP 地址反映了出站接口的 IP 地址，收到 BGP 包之后，路由器将数据包的源 IP 地址关联到 BGP 邻居表。如果 BGP 包的源地址与邻居表中的表项不匹配，那么就不能将该数据包与邻居相关联，从而丢弃该数据包。

大多数部署方案中的 IBGP 对等关系是通过环回接口建立的，如果未指定更新源接口，那么就无法建立会话。如果要从接口显式获取 BGP 包，那么就需要在 **neighbor** *ip-address* 配置部分利用命令 **update-source** *interface-id* 正确配置对等体。

如果 EBGP 对等体之间存在多跳，那么就需要配置正确的跳数，确保通过命令 **ebgp- multihop** [*hop-count*] 来配置正确的跳数。如果未指定 *hop-count*，那么默认值将被设置为 255。请注意，IBGP 会话的默认 TTL 值是 255，而 EBGP 会话的默认值是 1。如果 EBGP 对等关系是在两台直连设备之间通过环回地址建立的，那么用户也可以使用 **disable- connected-check** 命令，而不是 **ebgp-multihop 2** 命令。该命令将禁用连接验证机制，默认情况下，如果 EBGP 对等体不在直连网段中，那么该机制将阻止建立会话。

另一个对于成功建立 BGP 会话来说非常重要的配置（虽然可选）就是对等体认证。认证密码配置出错或输入错误都会导致 BGP 会话失败。

2. 验证可达性和丢包

验证了配置之后，还需要验证对等 IP 之间的连接性。如果对等关系建立在环回接口之间，那么就需要执行从环回接口到环回接口的 ping 测试。如果 ping 测试未指定源接口，那么就会将出站接口的 IP 地址用作源 IP 地址（与对等 IP 地址不相关）。例 11-5 显示了 NX-1 与 NX-4 之间从环回接口到环回接口的 ping 测试情况（因为两者通过环回地址建立对等关系）。

例 11-5 以源接口作为环回接口的 ping 测试

```
NX-4# ping 192.168.1.1 source 192.168.4.4
PING 192.168.1.1 (192.168.1.1) from 192.168.4.4: 56 data bytes
```

```
64 bytes from 192.168.1.1: icmp_seq=0 ttl=253 time=4.555 ms
64 bytes from 192.168.1.1: icmp_seq=1 ttl=253 time=2.72 ms
64 bytes from 192.168.1.1: icmp_seq=2 ttl=253 time=2.587 ms
64 bytes from 192.168.1.1: icmp_seq=3 ttl=253 time=2.559 ms
64 bytes from 192.168.1.1: icmp_seq=4 ttl=253 time=2.695 ms

--- 192.168.1.1 ping statistics ---
5 packets transmitted, 5 packets received, 0.00% packet loss
round-trip min/avg/max = 2.559/3.023/4.555 ms
```

注： 有时用户在执行 ping 测试时可能会出现丢包现象，出现这个问题的最常见原因是 CoPP 策略，该策略会丢弃这类数据包。

利用前面的 ping 测试，可以验证 IBGP 和 EBGP 对等体之间的可达性。如果可达性有问题，那么就需要通过以下流程来定位故障或故障方向。

识别丢包方向。可以利用 NX-OS 的 **show ip traffic** 命令来识别丢包或丢包方向。如果从源端到目的端的 ping（ICMP）包出现完全丢包或随机丢包情况，那么就可以使用该测试方法。该命令的输出结果中包含 "*ICMP Software Processed Traffic Statistics*" 部分，该输出部分由两个子块组成：*Transmission* 和 *Reception*，这两个子块都包含了 *echo request*（回显请求）和 *echo reply*（回显应答）包统计信息。执行该测试时，需要首先确保源端设备和目的端设备的发送和接收计数器保持稳定（不递增），然后通过指定源接口或 IP 地址来启动对目的端的 ping 测试。ping 测试完成后，可以查看 **show ip traffic** 命令的输出结果以验证两端计数器的增加情况，从而得出丢包方向。例 11-6 给出了定位丢包方向的测试方法，例中的 NX-1 向 NX-4 环回接口发起了 ping 测试。第一个输出结果显示回显请求包数量为 10，回显应答包数量也是 10。NX-1 向 NX-4 环回接口发起的 ping 测试完成之后，回显请求和回显应答计数器都增加到了 15。

例 11-6　ping 测试和 show ip traffic 命令输出结果

```
NX-4
NX-4# show ip traffic | in Transmission:|Reception:|echo
Transmission:
  Redirect: 0, unreachable: 0, echo request: 33, echo reply: 10,
Reception:
  Redirect: 0, unreachable: 0, echo request: 10, echo reply: 29,
NX-1
NX-1# ping 192.168.4.4 source 192.168.1.1
PING 192.168.4.4 (192.168.4.4) from 192.168.1.1: 56 data bytes
64 bytes from 192.168.4.4: icmp_seq=0 ttl=253 time=3.901 ms
64 bytes from 192.168.4.4: icmp_seq=1 ttl=253 time=2.913 ms
64 bytes from 192.168.4.4: icmp_seq=2 ttl=253 time=2.561 ms
64 bytes from 192.168.4.4: icmp_seq=3 ttl=253 time=2.502 ms
64 bytes from 192.168.4.4: icmp_seq=4 ttl=253 time=2.571 ms

--- 192.168.4.4 ping statistics ---
5 packets transmitted, 5 packets received, 0.00% packet loss
round-trip min/avg/max = 2.502/2.889/3.901 ms
NX-4
NX-4# show ip traffic | in Transmission:|Reception:|echo
Transmission:
  Redirect: 0, unreachable: 0, echo request: 33, echo reply: 15,
Reception:
  Redirect: 0, unreachable: 0, echo request: 15, echo reply: 29,
```

与此相似，可以验证 NX-1 的回显应答接收计数器。从前面的示例可以看出，ping 测试成功，因而收到的回显请求和发送的回显应答计数器都增加了。但是，如果 ping 测试失败，那么就需要仔细检查这些计数器，并多次进行相应的迭代测试。如果向目的设备发起的 ping 测试失败，但目的设备的两个计数器仍在递增，那么就表明可能是返回路径出现了问题，用户需要接着检查该路径的返回流量。

ACL 对于解决丢包或可达性问题来说非常有效。配置与源和目的 IP 相匹配的 ACL，有助于确认数据包是否已实际到达目的路由器。唯一需要注意的是，配置 ACL 时，应该在末尾配置 **permit ip any**，否则可能会导致其他数据包被丢弃，从而给业务造成影响。

3．验证路径中的 ACL 和防火墙

大多数部署方案中的边缘路由器或 IGW（Internet Gateway，互联网网关）路由器会配置 ACL 以限制网络中的允许流量。如果在这些配置了 ACL 的链路上建立 BGP 会话，那么就必须确保 BGP 数据包（TCP 端口 179）不会被这些 ACL 丢弃。

例 11-7 显示了允许 BGP 流量通过网络链路所需的 ACL 配置示例，除 IPv4 之外，本例还显示了 IPv6 BGP 会话的 **ipv6 access-list** 配置示例。如果要在接口上应用 IPv4 ACL，那么就要在所有平台上应用 **ip access-group** *access-list-name* **{in|out}** 命令。对于 IPv6 ACL 来说，则要在 NX-OS 上使用接口命令 **ipv6 traffic-filter** *access-list-name* **{in|out}**。

例 11-7 允许 BGP 流量的 ACL

```
NX-4(config)# ip access-list v4_BGP_ACL
NX-4(config-acl)# permit tcp any eq bgp any
NX-4(config-acl)# permit tcp any any eq bgp
! Output omitted for brevity
NX-4(config)# ipv6 access-list v6_BGP_ACL
NX-4(config-ipv6-acl)# permit tcp any eq bgp any
NX-4(config-ipv6-acl)# permit tcp any any eq bgp
! Output omitted for brevity
NX-4(config)# interface Ethernet2/1
NX-4(config-if)# ip access-group v4_BGP_ACL in
NX-4(config-if)# ipv6 traffic-filter v6_BGP_ACL in
```

除在边缘设备上配置 ACL 之外，很多方案会部署防火墙设备来防止网络出现非期望的恶意通信。安装防火墙设备比在路由器和交换机上配置一个庞大的 ACL 更好。防火墙支持以下两种配置模式。

- 路由模式。
- 透明模式。

路由模式下的防火墙具备路由功能，在网络中被视为一跳路由设备。透明模式下的防火墙则不被视为去往所连接设备的一跳路由设备，仅仅相当于线路上的一个障碍物而已。因此，如果 EBGP 会话跨越了一台透明防火墙，那么就不需要 **ebgp-multihop**，即使路径当中存在多台设备而必须配置 **ebgp-multihop**，也无须将防火墙作为一跳路由设备。

防火墙可以为接口实现多种安全等级。例如，为 ASA 内部接口分配的安全等级为 100，为外部接口分配的安全等级为 0。需要配置 ACL 以允许相关流量从较不安全的接口去往较高安全等级的接口。该规则适用于路由模式防火墙和透明模式防火墙，这两种情况都需要配置 ACL。

需要在透明模式防火墙中为每个网络配置网桥组，以最大程度地减少安全环境开销。此时需要将接口配置为网桥组的一部分，同时还要为 BVI（Bridge Virtual Interface，网桥虚接口）配置管理 IP 地址。

例 11-8 给出了一个 ASA ACL 配置示例，该配置允许 ICMP 和 BGP 数据包穿越防火墙，同

时还显示了为接口应用的 ACL，所有不符合 ACL 要求的流量均被丢弃。

例 11-8 透明防火墙配置

```
interface GigabitEthernet0/0
  nameif Inside
  bridge-group 200
  security-level 100
!
interface GigabitEthernet0/1
  nameif Outside
  bridge-group 200
  security-level 0
!
! Creating BVI with Management IP and should be the same subnet
! as the connected interface subnet
interface BVI200
  ip address 10.1.13.10 255.255.255.0
!
access-list Out extended permit icmp any any
access-list Out extended permit tcp any eq bgp any
access-list Out extended permit tcp any any eq bgp
!
access-group Out in interface Outside
```

请注意，在名为 Out 的访问列表中，虽然两条允许 BGP 数据包的语句并非必需语句，但最佳实践仍然建议这么做。

用户在设备中间使用防火墙时可能遇到的其他问题主要与 ASA 防火墙的两个功能特性相关。

- 序列号随机化。
- 为 MD5 认证启用 TCP 选项 19。

ASA 防火墙默认执行序列号随机化操作，因而可能会导致 BGP 会话翻动。此外，如果利用 MD5 认证来保护 BGP 对等会话，那么就必须在防火墙策略上启用 TCP 选项 19。

4. 验证 TCP 会话

建立 BGP 对等关系之前，必须首先建立 TCP 会话，因而必须确保建立 TCP 会话且不会被两台 BGP 对等设备之间路径上的任何位置所阻塞。可以使用命令 **show sockets connection tcp** 验证 NX-OS 的 TCP 连接。从例 11-9 可以看出，TCP 处于侦听状态（端口 179），且 NX-4 已经为 IBGP 和 EBGP 对等关系建立了 TCP 连接。

例 11-9 TCP 套接字连接

```
NX-4# show sockets connection tcp
! Output omitted for brevity
Total number of tcp sockets: 6
Active connections (including servers)
Protocol State/        Recv-Q/   Local Address(port)/
         Context       Send-Q    Remote Address(port)
tcp      LISTEN        0         *(179)
         Wildcard      0         *(*)

tcp6     LISTEN        0         *(179)
         Wildcard      0         *(*)

[host]:  tcp ESTABLISHED 0       10.46.1.4(53879)
         default       0         10.46.1.6(179)
```

```
[host]: tcp     ESTABLISHED   0        192.168.4.4(179)
                default       0        192.168.1.1(21051)
```

如果未建立 BGP 对等关系，那么 TCP 表中就可能存在一条过时表项。过时表项可能显示 TCP 会话处于已建立状态，从而阻止路由器初始化另一条 TCP 连接，也就阻止路由器建立 BGP 对等连接。

作为一种有效的 BGP 对等关系中断故障排查技术，可以在 TCP 端口 179 上向目的节点 IP 执行 Telnet 测试（使用本地对等 IP 作为源地址），这样就可以验证两台 BGP 对等设备之间是否阻塞或丢弃了 TCP 会话。该测试不但能够有效验证目的路由器的各种 TCP 故障，而且能验证可能阻塞 BGP 包的 ACL 问题。

例 11-10 给出了 Telnet 测试示例，本例从 NX-1（192.168.1.1）向 NX-4（192.168.4.4）的端口 179 发起 Telnet 测试来验证 BGP 会话。从测试结果可以看出，虽然 BGP TCP 会话建立了，但很快就被关闭/断开了。

例 11-10 使用 Telnet 到端口 179

```
NX-4# show sockets connection tcp foreign 192.168.1.1 detail

Total number of tcp sockets: 4
Active connections (including servers)

NX-1# telnet 192.168.4.4 179 source 192.168.1.1
Trying 192.168.4.4...
Connected to 192.168.4.4.
Escape character is '^]'.
Connection closed by foreign host.

NX-4# show sockets connection tcp foreign 192.168.1.1 detail

Total number of tcp sockets: 5
Active connections (including servers)
[host]: Local host: 192.168.4.4 (179), Foreign host: 192.168.1.1 (40944)
  Protocol: tcp, type: stream, ttl: 64, tos: 0xc0, Id: 18
  Options: REUSEADR, pcb flags none, state:    | ISDISCONNECTED
! Output omittied for brevity
```

如果 Telnet 的源端不是在远程设备上配置的要建立 BGP 邻居关系的接口或 IP，那么该 Telnet 请求就会被拒绝。这是确认对等设备的配置是否符合文档要求的另一种方法。

排查 TCP 连接故障时，检查 Netstack 进程的事件历史记录日志也非常重要。Netstack 是 NX-OS 上的一个 2 层~4 层协议栈应用，是 NX-OS 控制平面的关键组件之一。如果 Nexus 设备出现了 TCP 会话建立故障，那么很可能是 Netstack 进程出了问题。此时，可以利用 **show sockets internal event-history event** 命令了解 BGP 对等 IP 的 TCP 状态切换情况。

例 11-11 显示了 **show sockets internal event-history event** 命令输出结果，可以看出，BGP 对等 IP 192.168.2.2 的 TCP 会话已经关闭，但是没有显示任何新的连接请求。

例 11-11 命令 show sockets internal event-history events 输出结果

```
NX-4# show sockets internal event-history event
1) Event:E_DEBUG, length:67, at 192101 usecs after Fri Sep 1 05:21:38 2017
    [138] [4226]: Marking desc 22 in mts_open for client 25394, sotype 2
! Output omitted for brevity
4) Event:E_DEBUG, length:91, at 810192 usecs after Fri Sep 1 05:17:09 2017
    [138] [4137]: PCB: Removing pcb from hash list L: 192.168.4.4.179, F: 192.16
8.1.1.21051 C: 1
5) Event:E_DEBUG, length:62, at 810184 usecs after Fri Sep 1 05:17:09 2017
```

```
          [138] [4137]: PCB: Detach L 192.168.4.4.179 F 192.168.1.1.21051
    6) Event:E_DEBUG, length:77, at 810164 usecs after Fri Sep 1 05:17:09 2017
          [138] [4137]: TCP: Closing connection L: 192.168.4.4.179, F: 192.168.1.1.21051
```

> **注：** 排查与 TCP 相关的协议故障时（如 BGP），可以抓取 **show tech netstack [detail]** 命令的输出结果，并与思科 TAC 共享这些信息。

5. OPEN 消息差错

如果 OPEN 消息中的信息有错，那么就无法建立 BGP 对等连接。此时，BGP 发话端会向对等体发送 BGP NOTIFICATION 消息，说明收到的信息有错，与路由器配置的信息不同。出现 BGP OPEN 消息差错的原因如下。

- 版本号不支持。
- 对等 AS 错误。
- BGP 路由器 ID 错误。
- 可选参数不支持。

上述差错原因中，对等 AS 错误或 BGP 路由器 ID 错误是常见的 OPEN 消息差错，这些通常是由文档或人为差错引起的。NOTIFICATION 消息对这两种错误都有明确解释，列出了差错值和期望值，如例 11-12 所示。例中的路由器期望对等 AS 位于 AS 65001 中，但收到的却是 AS 65002。

例 11-12　BGP 对等 AS 错误的 NOTIFICATION 消息

```
04:51:33 NX-4 %BGP-3-BADPEERAS:  bgp-100 [9544]  VRF default, Peer 10.46.1.6 - bad
remote-as, expecting 65001 received 65002.
```

BGP 发话端在初始 BGP 协商期间，会相互交换能力参数。如果 BGP 发话端收到自己不支持的能力，那么 BGP 就会检测到有关不支持能力（或不支持可选参数）的 OPEN 消息差错。例如，假设某 BGP 发话端具有增强型的路由刷新能力，但接收侧的 BGP 发话端运行的软件版本较旧，不支持该能力，那么就会检测到不支持该能力的 OPEN 消息差错。BGP 发话端之间协商的可选能力包括以下几种。

- 路由刷新能力。
- 4 字节 AS 能力。
- 多协议能力。
- 单会话/多会话能力。

为了解决能力不支持问题，可以在 BGP 邻居配置模式下使用命令 **dont-capability- negotiate**，该命令将禁止 BGP 对等体之间进行能力协商，从而确保建立 BGP 对等连接。

6. BGP 调试

运行调试操作应该是解决所有网络故障的最后手段，因为调试使用不当可能会给网络造成较大影响。但是，在其他故障排查技术无法理解故障原因的时候，调试就成为唯一选择。为了减轻大量调试输出产生的不利影响，用户可以使用 NX-OS 的调试日志文件。除调试日志文件之外，网络运营商还可以使用调试过滤器对调试结果进行过滤，将输出结果限定为特定邻居、前缀甚至地址簇，从而减轻对 Nexus 交换机的影响。

如果 BGP 对等关系中断，且所有故障排查操作都无法找出故障原因，那么就可以启用调试操作以查看路由器是否正在生成和发送必要的 BGP 数据包，以及是否正在接收相关数据包。不过，对于 NX-OS 来说，不需要执行调试操作，因为跟踪 BGP 的操作过程就能提供足够的故障调试信息。NX-OS 支持多种 BGP 调试命令，可以根据 BGP 被卡的状态来选择适当的调试命令。

对于 BGP 对等关系中断故障来说，最重要的调试操作就是 BGP KEEPALIVE 调试。可以通

过命令 **debug bgp keepalives** 来启用 BGP KEEPALIVE 调试。查看调试输出结果时，需要清楚确保 BGP 对等关系成功建立的两个重要因素。

- 是否定期生成了 BGP KEEPALIVE 消息。
- 是否定期收到了 BGP KEEPALIVE 消息。

如果定期生成了 BGP KEEPALIVE 消息，但 BGP 对等关系仍然处于中断状态，那么就可能是 BGP KEEPALIVE 消息无法到达对端，或者到达了对等路由器却未被处理或者被丢弃。此时，BGP KEEPALIVE 调试就显得非常有用，可以通过调试命令 **debug bgp keepalives** 来验证是否正在发送和接收 BGP KEEPALIVE 消息。例 11-13 给出了 BGP KEEPALIVE 调试示例，第一个输出结果表明每 60s 生成了一条 BGP KEEPALIVE 消息，第二个输出结果表明从远程对等体 192.168.1.1 收到了 BGP KEEPALIVE 消息。

例 11-13 BGP KEEPALIVE 调试

```
NX-4# debug logfile bgp
NX-4# debug bgp keepalives

NX-4# show debug logfile bgp | grep "192.168.1.1 sending"
05:37:13.870261 bgp: 100 [9544] (default) ADJ: 192.168.1.1 sending KEEPALIVE
05:38:13.890290 bgp: 100 [9544] (default) ADJ: 192.168.1.1 sending KEEPALIVE
05:39:13.900376 bgp: 100 [9544] (default) ADJ: 192.168.1.1 sending KEEPALIVE
05:40:13.920290 bgp: 100 [9544] (default) ADJ: 192.168.1.1 sending KEEPALIVE
05:41:13.940395 bgp: 100 [9544] (default) ADJ: 192.168.1.1 sending KEEPALIVE
05:42:13.960350 bgp: 100 [9544] (default) ADJ: 192.168.1.1 sending KEEPALIVE
05:43:13.980363 bgp: 100 [9544] (default) ADJ: 192.168.1.1 sending KEEPALIVE

NX-4# show debug logfile bgp | grep 192.168.1.1
05:37:13.870160 bgp: 100 [9544] (default) ADJ: 192.168.1.1 keepalive timer fired
05:37:13.870236 bgp: 100 [9544] (default) ADJ: 192.168.1.1 keepalive
 timer fired for peer
05:37:13.870261 bgp: 100 [9544] (default) ADJ: 192.168.1.1 sending KEEPALIVE
05:37:13.870368 bgp: 100 [9544] (default) ADJ: 192.168.1.1 next keep
alive expiry due in 00:00:59
05:37:13.946248 bgp: 100 [9544] (default) ADJ: Peer 192.168.1.1 has
pending data on socket during recv, extending expiry timer
05:37:13.946387 bgp: 100 [9544] (default) ADJ: 192.168.1.1 KEEPALIVE rcvd
```

11.3.2 揭秘 BGP NOTIFICATION

BGP NOTIFICATION 在理解和解决 BGP 对等关系中断或对等关系翻动故障方面发挥着至关重要的作用。检测到相关差错之后，BGP 发话端就会向对等体发送 BGP NOTIFICATION 消息，根据具体差错类型，可以在 BGP 会话建立之前或建立之后发送 NOTIFICATION 消息。每条 NOTIFICATION 消息都有一个定长报头，报头之后可能有（也可能没有）数据部分（取决于消息类型），图 11-3 给出了 NOTIFICATION 消息的报头字段信息。

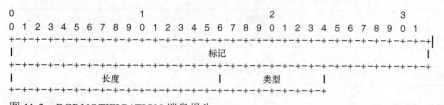

图 11-3 BGP NOTIFICATION 消息报头

除定长 BGP 消息报头之外，NOTIFICATION 消息还包含了其他信息，如图 11-4 所示。

图 11-4 BGP 报头中的 NOTIFICATION 信息部分

差错代码和子差错代码定义在 RFC 4271 中，表 11-3 列出了所有差错代码和子差错代码信息。

表 11-3　　　　　BGP NOTIFICATION 消息的差错代码和子差错代码

差错代码	子差错代码	描述
01	00	消息报头差错
01	01	消息报头差错——连接未同步
01	02	消息报头差错——消息长度错误
01	03	消息报头差错——消息类型错误
02	00	OPEN 消息差错
02	01	OPEN 消息差错——不支持的版本号
02	02	OPEN 消息差错——对等 AS 错误
02	03	OPEN 消息差错——BGP 标识符错误
02	04	OPEN 消息差错——不支持的选项参数
02	05	OPEN 消息差错——已废除
02	06	OPEN 消息差错——不可接受的保持时间
03	00	UPDATE 消息差错
03	01	UPDATE 消息差错——属性列表异常
03	02	UPDATE 消息差错——无法识别的周知属性
03	03	UPDATE 消息差错——周知属性缺失
03	04	UPDATE 消息差错——属性标志错误
03	05	UPDATE 消息差错——属性长度错误
03	06	UPDATE 消息差错——无效路由来源属性
03	07	（已废除）
03	08	UPDATE 消息差错——无效 NEXT_HOP 属性
03	09	UPDATE 消息差错——可选属性错误
03	0A	UPDATE 消息差错——无效网络字段
03	0B	UPDATE 消息差错——AS_PATH 异常
04	00	保持定时器到期
05	00	有限状态机差错
06	00	终止
06	01	终止——达到最大前缀数量
06	02	终止——管理性中断
06	03	终止——对等体配置撤销
06	04	终止——管理性重置
06	05	终止——连接拒绝
06	06	终止——其他配置变更
06	07	终止——连接冲突解决
06	08	终止——资源耗尽

由于生成 NOTIFICATION 消息时，始终包含差错代码和子差错代码，因而这些 NOTIFICATION 消息对于解决对等关系中断故障或对等关系翻动故障来说非常有用。

11.3.3　IPv6 对等关系故障排查

随着 IPv4 地址的耗尽，IPv6 地址的应用开始大幅增加。大多数服务提供商已经升级或计划将其基础设施升级为双栈模式，以支持 IPv4 和 IPv6 流量，并为企业客户提供 IPv6 就绪服务。此外，业界已经开发了大量与 IPv6 兼容或完全运行在 IPv6 上的新应用程序。在这样的推进速度下，迫切需要有适当的技术来排查 IPv6 BGP 邻居故障。

IPv6 BGP 对等关系故障排查的方法与 IPv4 BGP 对等关系相同，解决 IPv6 BGP 邻居对等关系中断故障的步骤如下。

- **第 1 步**：验证是否正确配置了对等 IPv6 地址、AS 号、更新源接口、认证密码和 EBGP 多跳配置。
- **第 2 步**：使用 **ping ipv6** *ipv6-neighbor-address* [**source** *interface-id* | *ipv6-address*]验证可达性。
- **第 3 步**：在 NX-OS 上通过命令 **show socket connection tcp** 验证 TCP 连接。对于 IPv6 来说，需要检查 TCP 连接的源和目的 IPv6 地址以及端口 179。
- **第 4 步**：验证路径中的 IPv6 ACL。与 IPv4 一样，路径中的 IPv6 ACL 应该允许端口 179 上的 TCP 连接以及 ICMPv6 数据包，从而验证可达性。
- **第 5 步**：调试。在 NX-OS 交换机上，使用调试命令 **debug bgp ipv6 unicast neighbors** *ipv6-neighbor-address* 抓取 IPv6 BGP 包。启用调试操作之前，需要为 BGP 调试启用 **debug logfile**。如果希望过滤特定 IPv6 邻居的调试结果，那么就可以通过 IPv6 ACL 来过滤该邻居的调试输出。

11.3.4　BGP 对等关系翻动问题

BGP 会话中断之后，会话状态将始终无法进入 Established 状态，将一直在 Idle 与 Active 之间翻动。但是，BGP 对等关系出现翻动故障之后，意味着该会话在建立后正在不断改变其状态，此时的 BGP 状态始终都在 Idle 状态与 Established 状态之间翻动。BGP 可能存在以下两种状态翻动情形。

- **Idle/Active**：如 11.3.3 节所述。
- **Idle/Established**：更新错误、TCP 问题（多跳部署方案中的 MSS 大小）。

BGP 对等体出现状态翻动的原因包括以下几种，接着对前 3 个原因进行介绍。

- 错误的 BGP 更新。
- 保持定时器到期。
- MTU 不匹配问题。
- 高 CPU 利用率。
- 不当的控制平面策略。

1. 错误的 BGP 更新

错误的 BGP 更新指的是从对等体收到了损坏的更新包。这种情况不是正常情况，出现这种情况的主要原因通常包括以下几种。

- 携带更新消息的链路出现故障或者硬件损坏。
- BGP 更新消息出现了打包问题。
- 攻击者（黑客）生成了恶意更新消息或者修改了更新包。

如果 BGP 更新消息出现了损坏，那么就会生成一条差错代码为 3 的 BGP NOTIFICATION 消息（见表 11-3）。如果发现 BGP 更新出现了差错，那么 BGP 就会生成该错误更新消息的 Hexdump

（十六进制转储文件），进一步解码之后就可以了解哪部分更新消息出现了损坏。除 Hexdump 之外，BGP 还会生成一条日志消息，以说明出现的更新差错类型，如例 11-14 所示。

例 11-14　损坏的 BGP 更新消息

```
22:10:13.366354 bgp: 65000 [14982] Hexdump at 0xd5893430, 19 bytes:
22:10:13.366362 bgp: 65000 [14982]     FFFFFFFF FFFFFFFF FFFFFFFF FFFFFFFF
22:10:13.366368 bgp: 65000 [14982]     001302
22:10:13.366379 bgp: 65000 [14982] (default) UPD: Badly formatted UPDATE message
  from peer 10.46.1.4, illegal length for withdrawn routes 65001 [afi/safi: 1/1]
22:10:13.366393 bgp: 65000 [14982] (default) UPD: Sending NOTIFY bad msg length
  error of length 2 to peer 10.46.1.4
22:10:13.366403 bgp: 65000 [14982] Hexdump at 0xd7eaa5fc, 23 bytes:
22:10:13.366413 bgp: 65000 [14982]     FFFFFFFF FFFFFFFF FFFFFFFF FFFFFFFF
22:10:13.366426 bgp: 65000 [14982]     00170301 02FFFF
22:10:13 %BGP-5-ADJCHANGE: bgp-65000 [14982] (default) neighbor 10.46.1.4 Down -
  bad msg length error
```

可以通过命令 **debug bgp packets** 查看 Hexdump 中的 BGP 消息，从而进一步了解相关信息。如果 NX-OS 设备交换的 BGP 更新和消息太多，那么最好通过 Ethanalyzer 或 SPAN 抓取错误的 BGP 更新包并加以分析。

注：可以通过一些在线工具对 BGP 消息中的 Hexdump 进行进一步分析。

2．保持定时器到期

保持定时器到期（Hold Timer Expired）是 BGP 对等关系翻动故障的主要原因，保持定时器到期意味着路由器没有收到或处理 KEEPALIVE 消息（或 UPDATE 消息），因而发送通知消息 4/0（保持定时器到期）并关闭会话。因保持定时器到期而导致的 BGP 对等关系翻动的主要原因如下。

- 接口/平台丢包。
- MTS 队列被阻塞。
- 控制平面策略丢包。
- BGP KEEPALIVE 消息生成问题。
- MTU 不匹配问题。

一种可能原因是接口/平台丢包。各种接口问题（如物理层问题或接口丢包）都可能导致 BGP 会话因保持定时器到期而出现翻动。如果接口承载的流量过大或者线卡本身出现过载或繁忙状态，那么数据包就可能会被接口或线卡 ASIC 丢弃。如果此时丢弃的是 BGP KEEPALIVE 或 UPDATE 包，那么 BGP 就会通知对等体保持定时器到期。

另一种可能原因是 MTS 队列被阻塞。有时 BGP KEEPALIVE 虽然到达了 TCP 接收队列，但是并没有被处理并移到 BGP InQ。如果 BGP InQ 队列为空且 BGP 邻居因保持定时器到期而中断，那么就会出现这个问题。Nexus 交换机出现这种场景的最常见原因是，MTS 队列被卡在 BGP 或 TCP 进程上，其中，MTS 是 NX-OS 将信息从一个组件传送到另一个组件的主要组件。在这种情况下，可能会有多个 BGP 对等体出现系统级影响，此时可能需要切换或重新加载管理引擎才能恢复该故障。

此外，CoPP 策略引起的丢包也可能是一个主要原因。CoPP 策略旨在防止 CPU 处理过多且非期望流量，但是设计欠佳的 CoPP 策略可能会导致控制平面协议出现翻动。如果 CoPP 策略无法处理所有 BGP 控制平面数据包以及路由器的所有 BGP 对等体，那么就可能会丢弃这些数据包。在这种情况下，用户就可能会发现因 CoPP 策略丢弃了某些数据包而出现随机的 BGP

翻动现象。

> **注：** 有关 MTS、CoPP 以及其他平台故障排查技术的详细信息请参见第 3 章。

BGP KEEPALIVE 消息生成问题

网络中有时可能会出现随机性的 BGP 对等关系翻动问题。除包丢失或控制平面策略丢包等原因之外，还可能有其他原因导致 BGP 对等关系出现翻动或保持定时器到期，其中，未及时生成 BGP KEEPALIVE 消息就是可能的原因之一。排查此类故障时，需要首先了解 BGP 翻动的特征，此时可以通过以下问题来收集相关信息。

- 每天在什么时候出现 BGP 翻动问题？
- 出现翻动的频率是什么？
- 出现翻动问题时，接口/系统的流量负载情况怎么样？
- 出现翻动问题时，CPU 利用率是否很高？如果是，那么是由流量过大引起的还是由特定进程引起的？

上述问题不但能够确定 BGP 翻动的特征，而且能执行一定的故障排查操作。为了进一步排查故障，还需要理解 BGP 翻动的两个主要原因。

- 定期生成了 KEEPALIVE 消息，但是未离开路由器，或者没有发送给对端。
- 未定期生成 KEEPALIVE 消息。

如果定期生成了 KEEPALIVE 消息，但是未离开路由器，那么就需要注意 BGP 对等体的 OutQ 是否一直在递增。如果 OutQ 因 KEEPALIVE 的生成而不断增加，但 MsgSent 却未增长，那么就表明 KEEPALIVE 消息可能被卡在了 OutQ 中。例 11-15 给出了该应用场景示例，此时的 BGP KEEPALIVE 以固定间隔生成，但是却没有离开路由器，从而因保持定时器到期而导致 BGP 翻动。请注意，本例中的 OutQ 值从 10 增加到了 12，但 MsgSent 计数器却停在了 3938。本例中的 BGP 对等关系在每个 BGP 保持定时器定期的时候都会出现翻动问题。

例 11-15 发送的 BGP 消息和 OutQ

```
NX-4# show bgp ipv4 unicast summary
BGP summary information for VRF default, address family IPv4 Unicast
BGP router identifier 192.168.4.4, local AS number 65000
BGP table version is 19, IPv4 Unicast config peers 2, capable peers 2
4 network entries and 4 paths using 576 bytes of memory
BGP attribute entries [4/576], BGP AS path entries [1/6]
BGP community entries [0/0], BGP clusterlist entries [0/0]


Neighbor        V    AS MsgRcvd MsgSent   TblVer  InQ OutQ Up/Down   State/PfxRcd
10.46.1.6       4 65001    3933    3938       19    0   10 14:30:46 1
192.168.1.1     4 65000     997    1009       19    0    0 15:02:52 1
NX-4# show bgp ipv4 unicast summary
! Output omitted for brevity

Neighbor        V    AS MsgRcvd MsgSent   TblVer  InQ OutQ Up/Down   State/PfxRcd
10.46.1.6       4 65001    3933    3938       19    0   12 14:30:46 1
192.168.1.1     4 65000     997    1009       19    0    0 15:02:52 1
```

但是，如果设备出现随机的 BGP 翻动问题且出现间隔不定，那么就表明可能定期生成了 BGP KEEPALIVE 消息，即使仍然频繁出现翻动问题，如每隔 4～10min 就出现一次 BGP 对等关系翻动。此类场景的故障排查较为复杂，可能需要使用除 **show** 命令之外的其他故障排查技术。困难之处就在于很难定位哪些设备未及时生成 KEEPALIVE，或者即便及时生成了 KEEPALIVE 消息，但是到达远程对等体的时延过大。排查此类故障时，需要在 BGP 连接的两端执行以下操作。

- **第1步**：在两台路由器上启用BGP KEEPALIVE调试以及调试日志文件。
- **第2步**：在两台路由器上启用Ethanalyzer。

启用Ethanalyzer或其他抓包工具（基于底层平台）的目的是希望确定BGP KEEPALIVE可能及时到达了对端，但是到达BGP进程的过程却出现了时延。有了BGP KEEPALIVE调试结果以及远端设备的Ethanalyzer输出结果之后，就可以确定导致BGP翻动故障的时延发生位置。总的来说，可能是BGP进程导致了KEEPALIVE消息生成出现了延时，也可能是与BGP交互的其他组件导致KEEPALIVE消息处理出现了延时。

3. MTU不匹配问题

一般来说，MTU（Maximum Transmission Unit，最大传输单元）不是阻碍BGP邻接关系建立的主要原因，但MTU不匹配可能会导致BGP会话出现翻动问题。网络设备出现MTU设置不匹配的主要原因包括以下几种。

- 规划和网络设计不当。
- 设备不支持巨型MTU或某些MTU值。
- 因应用程序需求而改变了MTU值。
- 因最终客户需求而改变了MTU值。

BGP根据TCP计算的MSS（Maximum Segment Size，最大报文长度）值发送更新消息。如果未启用PMTUD（Path-MTU-Discovery，路径MTU发现），那么BGP MSS值将默认为RFC 879规定的536字节。问题是，如果两台路由器以536字节的MSS值交换大量更新消息，那么就会出现收敛问题并导致网络出现低效使用问题。出现该问题的原因是，MTU为1500字节的接口几乎能够发送3倍于MSS的数据，如果接口支持巨型MTU，那么就可以发送更多的数据，但此时却必须将更新消息分成大小为536字节的报文段。

RFC 1191引入了PMTUD，希望减少在传输路径上对IP包进行分段的概率，从而加快收敛速度。使用了PMTUD之后，源端就能识别出到达目的端的路径上的最小MTU，从而更好地确定所要发送的数据包大小。

那么PMTUD的工作方式如何？源端生成数据包的时候，将MTU大小设置为与出站接口相同，并设置DF（Do-Not-Fragment，不分段）比特。如果收到该数据包的中间设备的出站接口MTU值小于收到的数据包，那么就会丢弃该数据包并发送一条类型为3（目的端不可达）、代码为4（需要分段且设置DF比特）的ICMP差错消息，同时还在下一跳MTU字段中包含该接口的MTU信息并发送给源端。源端收到目的端不可达的ICMP差错消息之后，会将出站数据包的MTU值修改为上述下一跳MTU字段的指定值。该过程将持续进行，直至数据包成功到达最终目的端。

BGP也支持PMTUD。PMTUD允许BGP路由器沿着去往邻居的路径发现最佳MTU的大小，从而最有效地交换数据包。启用PMTUD机制之后，两个邻居之间初始TCP协商的MSS值等于（IP MTU—20字节IP报头—20字节TCP报头）并设置DF比特。因此，如果IP MTU值为1500（等于接口MTU），那么MSS值就是1460。如果路径中的设备的MTU较小或者目的端路由器的MTU较小（如1400字节），那么MSS值就会协商为1400 − 40 = 1360字节。MSS的计算公式如下。

- 无MPLS的MSS=MTU—IP报头（20字节）—TCP报头（20字节）。
- MSS over MPLS=MTU—IP报头—TCP报头—n*4字节（其中，n是标签栈中的标签数）。
- 跨GRE隧道的MSS=MTU—IP报头（内层）—TCP报头—[IP报头（外层）+GRE报头（4字节）]。

> **注**：MPLS VPN提供商应该至少将MPLS MTU增加到1508字节（假定最少2个标签）或者将MPLS MTU增加到1516字节（以容纳最多4个标签）。

现在的问题是为什么 MTU 不匹配会导致 BGP 会话出现翻动问题？路由器在建立 BGP 连接的时候会通过 TCP 会话协商 MSS 值，生成 BGP 更新时，会将 BGP 更新打包到 BGP 更新消息中（将前缀和报头信息填充到 MSS 的最大允许容量），然后将这些 BGP 更新消息发送到设置了 DF 比特的远程对等体。如果路径中的设备或者目的端设备无法接受拥有较大 MTU 的数据包，那么就会向 BGP 发话端发送 ICMP 差错消息。目的端路由器将等待 BGP KEEPALIVE 或 BGP UPDATE 包以更新自己的保持定时器，180s 之后，目的端路由器将向源端返回一条携带 "Hold Time expired" 差错信息的 NOTIFICATION 消息。

> **注**：BGP 路由器向 BGP 邻居发送更新时，并不会单独发送 BGP KEEPALIVE，但是会更新邻居的 KEEPALIVE 定时器。在 BGP 更新过程中，BGP 发话端将 UPDATE 消息视为 KEEPALIVE。

例 11-16 说明了路径存在 MTU 不匹配问题时的 BGP 对等体翻动故障，以图 11-2 所示拓扑结构中的相同设备集 NX-1、NX-2、NX-4 和 NX-6 为例，假设该拓扑中的设备在接口上禁用了 ICMP 不可达特性，NX-6 向 NX-4 宣告了 10000 条前缀，NX-4 又进一步宣告给了 NX-1。NX-1 和 NX-4 的接口 MTU 均设置为 9100，而 NX-2 面向 NX-1 的接口的 MTU 设置为默认值（1500）。由于启用了 PMTUD 特性，因而 MSS 值将被协商为 9060。由于 NX-1 未收到 NX-2 接口设置的较低 MTU 值，因此 ICMP 不可达消息将被拒绝。

例 11-16 因 MSS 问题引发的 BGP 翻动故障

```
NX-4
NX-4# show bgp ipv4 unicast summary
! Output omitted for brevity

Neighbor          V    AS   MsgRcvd MsgSent   TblVer  InQ OutQ Up/Down  State/PfxRcd
10.46.1.6         4 65001   10475   10482       26      0    0 1d17h    10000
192.168.1.1       4 65000    2643    2659       26      0    0 00:01:59 1

NX-1
NX-1# show bgp ipv4 unicast summary
BGP summary information for VRF default, address family IPv4 Unicast
BGP router identifier 192.168.1.1, local AS number 65000
BGP table version is 37, IPv4 Unicast config peers 1, capable peers 1
4 network entries and 4 paths using 576 bytes of memory
BGP attribute entries [4/576], BGP AS path entries [1/6]
BGP community entries [0/0], BGP clusterlist entries [0/0]

Neighbor          V    AS   MsgRcvd MsgSent   TblVer  InQ OutQ Up/Down  State/PfxRcd
192.168.4.4       4 65000    2579    2566       37      0    0 00:02:49 0

NX-1# show sockets connection tcp foreign 192.168.4.4 detail
Total number of tcp sockets: 4
[host]: Local host: 192.168.1.1 (22543), Foreign host: 192.168.4.4 (179)
  Protocol: tcp, type: stream, ttl: 64, tos: 0xc0, Id: 19
  Options: none, pcb flags unknown, state: | NBIO
  MTS: sap 10486
  Receive buffer:
    cc: 0, hiwat: 17184, lowat: 1, flags: none
  Send buffer:
    cc: 0, hiwat: 17184, lowat: 2048, flags: none
  Sequence number state:
    iss: 987705410, snduna: 987705603, sndnxt: 987705603, sndwnd: 17184
    irs: 82840884, rcvnxt: 82841199, rcvwnd: 17184, sndcwnd: 4296
  Timing parameters:
```

```
       srtt: 3200 ms, rtt: 0 ms, rttv: 0 ms, krtt: 1000 ms
       rttmin: 1000 ms, mss: 9060, duration: 43800 ms
   State: ESTABLISHED
   Flags:  | SENDCCNEW
No MD5 peers  Context: default
```

```
NX-1
! Logs showing BGP flap after hold timer expiry
00:56:27.873 NX-1 %BGP-5-ADJCHANGE: bgp-65000 [6884] (default) neighbor 192.168.4.4
 Down - holdtimer expired error
00:57:26.627 NX-1 %BGP-5-ADJCHANGE: bgp-65000 [6884] (default) neighbor 192.168.4.4 Up
```

如果对等体之间交换的前缀数量较少，那么就不会出现 BGP 翻动问题，因为 BGP 包的大小少于 1460 字节。由于 MSS/MTU 问题导致的 BGP 翻动故障的一个表征就是因保持定时器到期而出现重复性的 BGP 翻动。

因 MTU 不匹配而导致 BGP 会话翻动的原因可能包括以下几种。

- 两台对等路由器的接口 MTU 不匹配。
- 两台对等路由器之间的二层路径存在不一致的 MTU 设置。
- PMTUD 没有为 TCP BGP 会话计算正确的 MSS。
- 因路径中的路由器或防火墙阻塞了 ICMP 消息而导致 BGP PMTUD 失效。

如果要验证路径中是否存在 MTU 不匹配问题，那么就可以执行扩展 ping 测试，将数据包的大小设置为出站接口 MTU 值并设置 DF 比特。同时还要确保路径中的设备不会阻塞 ICMP 消息，以确保 PMTUD 工作正常。此外，要检查设备配置，以确保全网 MTU 值的一致性。

例 11-17 给出了对远程对等体发起的 ping 测试结果（将数据包大小设置为接口 MTU，并设置 DF 比特）。

例 11-17 设置了 DF 比特的 PING 测试

```
NX-1# ping 192.168.4.4 source 192.168.1.1 packet-size 1500 df-bit
PING 192.168.4.4 (192.168.4.4) from 192.168.1.1: 1500 data bytes
Request 0 timed out
Request 1 timed out
Request 1 timed out
--- 192.168.4.4 ping statistics ---
3 packets transmitted, 0 packets received, 100.00% packet loss

NX-1# ping 192.168.4.4 source 192.168.1.1 packet-size 1472 df-bit
PING 192.168.4.4 (192.168.4.4) from 192.168.1.1: 1472 data bytes
1480 bytes from 192.168.4.4: icmp_seq=0 ttl=253 time=5.298 ms
1480 bytes from 192.168.4.4: icmp_seq=1 ttl=253 time=3.494 ms
1480 bytes from 192.168.4.4: icmp_seq=2 ttl=253 time=4.298 ms
1480 bytes from 192.168.4.4: icmp_seq=3 ttl=253 time=4.528 ms
1480 bytes from 192.168.4.4: icmp_seq=4 ttl=253 time=3.606 ms

--- 192.168.4.4 ping statistics ---
5 packets transmitted, 5 packets received, 0.00% packet loss
round-trip min/avg/max = 3.494/4.244/5.298 ms
```

注：以 MTU 大小执行 ping 测试时，Nexus 平台会添加 28 字节（20 字节 IP 报头+8 字节 ICMP 报头），因而在设置了 DF 比特之后执行 ping 测试的时候，将数据包大小设置为 1500 字节将会导致 ping 测试失败。因此，为了确保以接口 MTU 值为大小且设置了 DF 比特之后的 ping 测试成功，必须将接口 MTU 值减去 28 字节，即 1500−28=1472（字节）。

11.4 BGP 路由处理与路由传播

建立了 BGP 对等关系之后，BGP 对等体之间就需要交换网络前缀和路径属性。与 IGP 不同，BGP 允许 AS 中的每个对等体都拥有不同的路由策略。BGP 对网络前缀的入站和出站交换的路由处理方式可以采用一种非常简单的方式加以理解，如图 11-5 所示，BGP 路由器收到对等体的路由之后，BGP 会通过入站策略（如果已配置）过滤这些路由并安装到 BGP 表中。如果 BGP 表包含了同一前缀的多条路径，那么就会选择最佳路径，并将最佳路径安装到路由表中。与此相似，BGP 宣告前缀时，也仅将最佳路由宣告给对等设备。如果配置了出站策略，那么在将前缀宣告给远程对等体之前还要对前缀进行过滤。

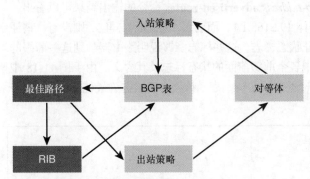

图 11-5　BGP 路由处理方式

接下来将详细讨论路由宣告的相关内容，相应的示例拓扑结构如图 11-6 所示。

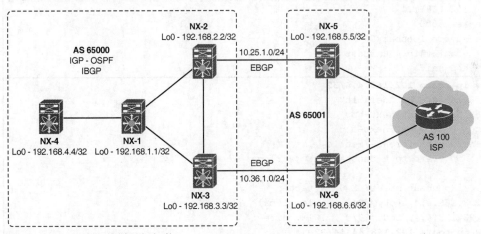

图 11-6　BGP 路由传播示例拓扑

11.4.1　BGP 路由宣告

BGP 通过显式配置将 BGP 前缀注入 BGP 表中进行宣告。将 BGP 前缀注入 BGP 表中有以下 4 种方法。

- **network 语句**：使用命令 **network** *ip-address/length*。
- **路由重分发**：可以重分发直连链路、静态路由和 IGP 路由，如 RIP（Routing Information Protocol，路由信息协议）、OSPF（Open Shortest Path First，开放最短路径优先）、EIGRP（Enhanced Interior Gateway Routing Protocol，增强内部网关路由协议）、IS-IS（Intermediate System to Intermediate System，中间系统到中间系统）和 LISP（Locator ID Separation

Protocol，定位与身份分离协议）。从其他路由协议（包括直连链路）重分发前缀时，需要用到路由映射。

- **聚合路由**：汇总路由（虽然成员路由必须位于 BGP 表中）。
- **默认路由**：使用默认信息来源命令。

1. network 语句

通过 BGP 宣告 BGP 前缀的一种方式是使用 **network** 语句。为了保证 **network** 语句的正常操作，必须确保路由位于路由表中。如果路由表中没有该路由，那么 **network** 语句既不会在 BGP 表中安装该路由，也不向 BGP 对等体宣告该路由。例 11-18 给出了通过 **network** 语句宣告两条前缀的配置示例，其中一条前缀在路由器本地配置了环回接口，而另一条前缀则在路由表中没有相应的路由。从 **show bgp ipv4 unicast neighbors** *ip-address* **advertised-routes** 命令的输出结果可以看出，前缀 192.168.4.4/32 已经宣告给了 BGP 对等体 192.168.1.1，而前缀 192.168.44.44/32 则没有。请注意，查看所有地址簇的 BGP 表时，必须验证状态标志，这些标志能够说明路由器学到这些前缀的方式。本例以高亮方式显示了列在 BGP 表中各个前缀前面的状态标志及其含义。由于例 11-18 中的前缀是本地前缀，因而状态标志为 L，同时还有标志 *>，表示该路由已被选为最佳路由。

例 11-18 使用 network 命令宣告前缀

```
NX-4
router bgp 65000
  router-id 192.168.4.4
  log-neighbor-changes
  address-family ipv4 unicast
    network 192.168.4.4/32
    network 192.168.44.44/32
  neighbor 192.168.1.1
    remote-as 65000
    update-source loopback0
    address-family ipv4 unicast
      next-hop-self
NX-4# show ip route 192.168.4.4/32
IP Route Table for VRF "default"
'*' denotes best ucast next-hop
'**' denotes best mcast next-hop
'[x/y]' denotes [preference/metric]
'%<string>' in via output denotes VRF <string>

192.168.4.4/32, ubest/mbest: 2/0, attached
    *via 192.168.4.4, Lo0, [0/0], 1w1d, local
    *via 192.168.4.4, Lo0, [0/0], 1w1d, direct
NX-4# show ip route 192.168.44.44/32
Route not found
NX-4# show bgp ipv4 unicast
BGP routing table information for VRF default, address family IPv4 Unicast
BGP table version is 27, local router ID is 192.168.4.4
Status: s-suppressed, x-deleted, S-stale, d-dampened, h-history, *-valid, >-best
Path type: i-internal, e-external, c-confed, l-local, a-aggregate, r-redist,
I-injected
Origin codes: i - IGP, e - EGP, ? - incomplete, | - multipath, & - backup
   Network            Next Hop         Metric     LocPrf     Weight Path
*>i192.168.1.1/32     192.168.1.1                 100            0 i
*>l192.168.4.4/32     0.0.0.0                     100        32768 i
  l192.168.44.44/32   0.0.0.0                     100        32768 i
```

```
NX-4# show bgp ipv4 unicast neighbors 192.168.1.1 advertised-routes
! Output omitted for brevity

   Network          Next Hop         Metric    LocPrf     Weight Path
*>1192.168.4.4/32   0.0.0.0                    100        32768 i
```

2. 路由重分发

将路由重分发到 BGP 是填充 BGP 表的常用方法。以图 11-6 所示拓扑结构为例，图中的路由器 NX-1 将 OSPF 路由重分发到 BGP 中，将路由从 OSPF 重分发到 BGP 时，虽然路由表仅从 NX-4 学到了前缀 192.168.4.4/32，但路由映射却允许前缀 192.168.4.4/32 和 192.168.44.44/32。例 11-19 解释了将路由重分发到 BGP 的过程，从输出结果可以看出，前缀 192.168.4.4/32 有一个标志 r，表示这是一条重分发前缀。此外，重分发前缀的 AS 路径列表中还有一个问号（?）。

例 11-19　BGP 和 IGP 重分发

```
NX-1
router bgp 65000
  address-family ipv4 unicast
    redistribute ospf 100 route-map OSPF-BGP
!
ip prefix-list OSPF-BGP seq 5 permit 192.168.4.4/32
ip prefix-list OSPF-BGP seq 10 permit 192.168.44.44/32
!
route-map OSPF-BGP permit 10
  match ip address prefix-list OSPF-BGP
    redistribute ospf 100 route-map OSPF-BGP

NX-1# show ip route ospf
192.168.4.4/32, ubest/mbest: 1/0
    *via 10.14.1.4, Eth2/1, [110/41], 00:30:27, ospf-100, intra

NX-1# show bgp ipv4 unicast
BGP routing table information for VRF default, address family IPv4 Unicast
BGP table version is 6, local router ID is 192.168.1.1
Status: s-suppressed, x-deleted, S-stale, d-dampened, h-history, *-valid, >-best
Path type: i-internal, e-external, c-confed, l-local, a-aggregate, r-redist,
I-injected
Origin codes: i - IGP, e - EGP, ? - incomplete, | - multipath, & - backup

   Network           Next Hop         Metric    LocPrf     Weight Path
*>i192.168.2.2/32    192.168.2.2                100             0 i
*>r192.168.4.4/32    0.0.0.0          41        100         32768 ?
```

注： 从例 11-19 可以看出，其他路由协议以及静态路由和直连链路的重分发过程完全相同。

执行 OSPF 和 IS-IS 重分发操作时，需要注意以下几点。

- **OSPF**：将 OSPF 重分发到 BGP 时，默认仅重分发 OSPF 内部路由。重分发 OSPF 外部路由时，需要在路由映射下执行路由类型（route-type）的条件式匹配。
- **IS-IS**：IS-IS 不将直连子网重分发到目的路由协议中，为了解决这个问题，可以将直连网络重分发到 BGP 中。

例 11-20 显示了路由映射可用的多种路由类型匹配选项，这些路由类型选项适用于 OSPF 和 IS-IS 路由类型。

例 11-20　match route-type 命令选项

```
NX-1(config-route-map)# match route-type ?
  external       External route (BGP, EIGRP and OSPF type 1/2)
  inter-area     OSPF inter area route
  internal       Internal route (including OSPF intra/inter area)
  intra-area     OSPF intra area route
  level-1        IS-IS level-1 route
  level-2        IS-IS level-2 route
  local          Locally generated route
  nssa-external  Nssa-external route (OSPF type 1/2)
  type-1         OSPF external type 1 route
  type-2         OSPF external type 2 route
```

3. 聚合路由

网络中的设备并非都足够强大，能够承载通过 BGP 或其他路由协议学到的所有路由。此外，如果网络存在多条路径，那么就会消耗更多的 CPU 和内存资源。为了解决这个问题，可以执行路由聚合或路由汇总操作。BGP 的路由聚合命令是 **aggregate-address** *aggregate-prefix/length* [**advertise-map** | **as-set** | **attribute-map** | **summary-only** | **suppress-map**]，表 11-4 列出了 **aggregate-address** 命令的所有可用选项信息。

表 11-4　aggregate-address 命令选项

选项	描述
advertise-map *map-name*	用于从明细路由选择属性信息
as-set	从起作用的路由生成 AS_SET 路径信息和团体信息
attribute-map *map-name*	用于为明细路由设置属性信息，从而允许更改聚合路由的属性
summary-only	过滤更新消息中的明细路由，仅宣告汇总路由
suppress-map *map-name*	有条件地过滤路由映射中指定的明细路由

例 11-21 解释了 **aggregate-address** 命令的 **summary-only** 选项的使用方式，可以看出，NX-2 有 3 个前缀，但是仅向 NX-5 宣告了一条聚合前缀。在 NX-2 上运行了 **summary-only** 命令之后，就抑制了明细路由。

例 11-21　聚合路由

```
NX-2
router bgp 65000
  address-family ipv4 unicast
    network 192.168.2.2/32
    aggregate-address 192.168.0.0/16 summary-only

NX-2# show bgp ipv4 unicast
BGP routing table information for VRF default, address family IPv4 Unicast
BGP table version is 19, local router ID is 192.168.2.2
Status: s-suppressed, x-deleted, S-stale, d-dampened, h-history, *-valid, >-best
Path type: i-internal, e-external, c-confed, l-local, a-aggregate, r-redist,
I-injected
Origin codes: i - IGP, e - EGP, ? - incomplete, | - multipath, & - backup

   Network            Next Hop         Metric     LocPrf     Weight Path
*>a192.168.0.0/16     0.0.0.0                     100        32768 i
s>i192.168.1.1/32     192.168.1.1                 100            0 i
s>l192.168.2.2/32     0.0.0.0                     100        32768 i
s>i192.168.4.4/32     192.168.4.4                 100            0 i
```

```
NX-5# show bgp ipv4 unicast

 Network              Next Hop           Metric     LocPrf     Weight Path
*>e192.168.0.0/16     10.25.1.2                                     0 65000 i
```

4. 默认路由

不是每条外部路由都能在网络中进行重分发和宣告。网关或边缘设备通常利用路由协议向网络的其他部分宣告一条默认路由。如果要通过 BGP 宣告默认路由，那么就需要在邻居配置模式下使用命令 **default-information originate**。需要注意的是，仅当路由表中存在默认路由时，该命令才能宣告默认路由，如果不存在默认路由，那么就需要创建指向 null0 接口的默认路由。

11.4.2 BGP 最佳路径计算

BGP 的路由宣告包括 NLRI（Network Layer Reachability Information，网络层可达性信息）和 PA（Path Attribute，路径属性），NLRI 包含网络前缀和前缀长度，BGP 属性（如 AS-Path、路由来源等）存储在路径属性中。

BGP 使用下面 3 个表来维护路由的网络前缀和 PA。

- **Adj-RIB-in**：包含了执行入站路由策略之前的原始 NLRI。为了节省内存资源，处理完所有路由策略之后将清除该表。
- **Loc-RIB**：包含了所有源自本地或者从其他 BGP 对等体收到的 NLRI。这些 NLRI 通过有效性和下一跳可达性检查之后，BGP 最佳路径算法就可以为特定前缀选择最佳 NLRI。Loc-RIB 表是向 IP 路由表提供路由的表。
- **Adj-RIB-out**：包含了执行出站路由策略之后的 NLRI。BGP 路由可能包含多条去往到同一目的网络的路径，路由器在选择最佳路径时，路径属性会影响每条路由的可取性。BGP 路由器仅宣告去往邻居路由器的最佳路径。

BGP 的 Loc-RIB 表负责维护所有路由及其路径属性以及计算出来的最佳路径。最佳路径安装在路由器的 RIB 中。如果没有可用最佳路径，那么路由器就会利用现有路径快速确定新的最佳路径。如果出现了下列 4 种情形，那么 BGP 就会重新计算前缀的最佳路径。

- BGP 下一跳可达性出现变化。
- 连接到 EBGP 对等体的接口出现故障。
- 重分发出现变化。
- 收到路由的新路径。

BGP 最佳路径选择算法会影响流量进入或离开 AS（Autonomous System，自治系统）的方式。BGP 不通过度量来确定网络的最佳路径，而是使用路径属性来确定最佳路径。不过，在 BGP 使用 PA 来影响最佳路径的选择之前，路由器会在 RIB 中查找路由的最长前缀匹配，并优先将该路由安装到 FIB（Forwarding Information Base，转发信息库）中。

在接收或宣告路由时修改 BGP 的路径属性，可以影响本地 AS 或邻居 AS 的路由选择。BGP 流量工程的一个基本规则是，修改出站路由策略以影响入站流量，修改入站路由策略以影响出站流量。

BGP 自动将收到的第一条路径安装为最佳路径，收到其他路径之后，会将新路径与当前最佳路径进行对比，如果两者相同，那么就执行后续对比处理，直到确定最佳路径。

表 11-5 列出了 BGP 最佳路径算法在最佳路由选择进程中用到的属性信息，按表中所列顺序依次处理这些属性。

表 11-5　BGP 属性

BGP 属性	范围
权重（Weight）	仅路由器，值越大越优
本地优先级（Local Preference）	AS 边界内部，值越大越优
源自本地（Locally Originated）	**Network** 或 **redistribute** 命令优于本地聚合（**aggregate-address** 命令）
AIGP（Accumulated Interior Gateway Protocol, 累积的内部网关协议）	AIGP 路径属性
AS_PATH	AS_PATH 越短越优 ■ 如果配置了命令 **bgp bestpath as-path ignore**，那么就忽略该属性 ■ 将 AS_SET 算作 1 ■ 不计算 CONFED 部分
路由来源类型（Origin Type）	IGP < EGP < Incomplete（不完全的），越小越优
MED（Multi-Exit Discriminator，多出口鉴别器）	仅当多条路径的 AS_SEQUENCE 中的第一个 AS 相同时才比较该属性
EBGP over IBGP	外部 BGP 路径优于内部 BGP 路径
到下一跳的度量（Metric to Next Hop）	到达 BGP 下一跳的 IGP 开销，度量越小越优
最早外部路径（Oldest External）	如果两条路径都是外部路径，那么优选第一条（最早）路径
BGP RID（Router ID，路由器 ID）	优选 BGP RID 最小的路径
CLUSTER_LIST	优选 CLUSTER_LIST 长度最短的路由
邻居地址（Neighbor Address）	优选从最小邻居地址收到的路径（邻居配置了命令 **neighbor** *ip-address*）

最佳路径算法通过修改 BGP 路由器的各种路径属性来控制特定路由的网络流量模式，改变 BGP PA 能够影响进入、离开和跨越 AS 的流量。根据对 BGP PA 的控制情况，每个组织机构的 BGP 路由策略都有所不同。由于某些 PA 是可传递性的，能够从一个 AS 传递到另一个 AS，因而更改这些 PA 能够影响其他 SP 的下游路由；而有些 PA 则是非传递性的，只能影响本组织机构内部的路由策略。根据各种因素（如 AS-Path 长度、特定 ASN、BGP 团体或其他属性）对网络前缀进行条件式匹配。

以图 11-6 所示拓扑结构为例，NX-5 和 NX-6 向 AS65000 宣告其环回地址。NX-1 通过 NX-2 和 NX-3 收到这些环回地址，不过仅将其中的一条路径选为最佳路径。从 **show bgp** *afi safi ip-address/length* 命令的输出结果可以看出，虽然收到了两条路径，但是仅将其中的一条路径选为最佳路径，如例 11-22 所示。请注意，本例中的 NX-1 最初选择经 NX-2 的路径去往 192.168.5.5/32（因为其 RID 最小），但是后来在 NX-3 上部署了入站策略，将其本地优先级设置得更高之后，就将经 NX-3 的路径选择为了最佳路径。

例 11-22　BGP 最佳路径选择

```
NX-1# show bgp ipv4 unicast 192.168.5.5/32
BGP routing table information for VRF default, address family IPv4 Unicast
BGP routing table entry for 192.168.5.5/32, version 32
Paths: (2 available, best #1)
Flags: (0x08001a) on xmit-list, is in urib, is best urib route, is in HW,

  Advertised path-id 1
  Path type: internal, path is valid, is best path
  AS-Path: 65001 , path sourced external to AS
    192.168.2.2 (metric 41) from 192.168.2.2 (192.168.2.2)
      Origin IGP, MED not set, localpref 100, weight 0

  Path type: internal, path is valid, not best reason: Router Id
```

```
    AS-Path: 65001 , path sourced external to AS
      192.168.3.3 (metric 41) from 192.168.3.3 (192.168.3.3)
        Origin IGP, MED not set, localpref 100, weight 0

  Path-id 1 advertised to peers:
    192.168.3.3      192.168.4.4

NX-3(config)# route-map LP permit 10
NX-3(config-route-map)# set local-preference 200
NX-3(config-route-map)# exit
NX-3(config)# router bgp 65000
NX-3(config-router)# neighbor 10.36.1.6
NX-3(config-router-neighbor)# address-family ipv4 unicast
NX-3(config-router-neighbor-af)# route-map LP in
NX-3(config-router-neighbor-af)# end

NX-1# show bgp ipv4 unicast 192.168.5.5/32
BGP routing table information for VRF default, address family IPv4 Unicast
BGP routing table entry for 192.168.5.5/32, version 38
Paths: (2 available, best #2)
Flags: (0x08001a) on xmit-list, is in urib, is best urib route, is in HW,

  Path type: internal, path is invalid, not best reason: Local Preference, is de
leted, no labeled nexthop
    AS-Path: 65001 , path sourced external to AS
      192.168.2.2 (metric 41) from 192.168.2.2 (192.168.2.2)
        Origin IGP, MED not set, localpref 100, weight 0

  Advertised path-id 1
  Path type: internal, path is valid, is best path
    AS-Path: 65001 , path sourced external to AS
      192.168.3.3 (metric 41) from 192.168.3.3 (192.168.3.3)
        Origin IGP, MED not set, localpref 200, weight 0

  Path-id 1 advertised to peers:
    192.168.2.2      192.168.4.4

NX-1# show bgp ipv4 unicast 192.168.5.5/32
BGP routing table information for VRF default, address family IPv4 Unicast
BGP routing table entry for 192.168.5.5/32, version 38
Paths: (1 available, best #1)
Flags: (0x08001a) on xmit-list, is in urib, is best urib route, is in HW,

  Advertised path-id 1
  Path type: internal, path is valid, is best path
    AS-Path: 65001 , path sourced external to AS
      192.168.3.3 (metric 41) from 192.168.3.3 (192.168.3.3)
        Origin IGP, MED not set, localpref 200, weight 0

  Path-id 1 advertised to peers:
    192.168.2.2      192.168.4.4
```

注：从 BRIB（BGP RIB）中删除前缀之后，该前缀会被标记为已删除，且该路径也不再用于转发流量。等到 BGP 更新完成之后，BRIB 就不再显示已删除路径/前缀。

11.4.3 BGP 多路径

BGP 的默认行为是仅向 RIB 宣告最佳路径，意味着将网络流量转发给目的端时仅使用该网络前缀的一条路径。BGP 多路径特性允许向 RIB 提供多条路径，这样就可以同时通过多条路径将流量转发给指定网络前缀。BGP 多路径（BGP multipath）是 BGP 多归属（BGP multihoming）的一种增强形式。

> **注：** BGP 多路径与 BGP 多归属的主要区别就在于负载均衡的实现方式。BGP 多路径以动态方式分发流量负载，而 BGP 多归属实现流量分发的方式在某种程度上是由 BGP 最佳路径算法决定的，但是为了在多条链路之间实现更均匀的流量分发，还需要在入站/出站方向实施路由策略。

BGP 支持 3 种类型的 ECMP（Equal Cost MultiPath，等价多路径）：EBGP 多路径、IBGP 多路径或 eiBGP 多路径。对于这 3 种 BGP 多路径机制来说，必须确保以下 BGP 路径属性（PA）匹配才能满足多路径条件。

- 权重。
- 本地优先级。
- AS-Path 长度及内容（联盟可以包含不同的 AS_CONFED_SEQ 路径）。
- 路由来源。
- MED。
- 宣告方法必须匹配（IBGP 或 EBGP）；如果前缀是通过 IBGP 宣告学到的，那么 IGP 开销必须匹配才能认为等价。

> **注：** 截至本书写作之时，NX-OS 还不支持 eiBGP 多路径功能。

EBGP 和 IBGP 多路径

在 NX-OS 上启用 EBGP 多路径特性时需要使用 BGP 配置命令 **maximum-paths** *number-paths*，其中的 *number-paths* 表示允许安装到 RIB 中的 EBGP 路径数。请注意，EBGP 多路径仅允许将外部路径选为多路径的最佳路径。如果是内部路径，那么就需要启用 IBGP 多路径功能，此时可以通过 **maximum-paths ibgp** *number-paths* 命令设置将要安装到 RIB 中的 IBGP 路由数。需要注意的是，这些命令都要运行在适当的地址簇下。

仍然以图 11-6 所示拓扑结构为例，NX-1 从 NX-2 和 NX-3 学到了相同前缀。由于 NX-1、NX-2 和 NX-3 之间存在 IBGP 对等关系，因而通过 NX-1 学到的路径都是内部路径。因此，为了在 RIB 和 BRIB 中安装多条 BGP 路径，需要在 NX-1 上配置 IBGP 多路径功能特性，如例 11-23 所示。

例 11-23 IBGP 多路径

```
NX-1# show bgp ipv4 unicast 192.168.5.5/32
BGP routing table information for VRF default, address family IPv4 Unicast
BGP routing table entry for 192.168.5.5/32, version 32
Paths: (2 available, best #1)
Flags: (0x08001a) on xmit-list, is in urib, is best urib route, is in HW,

  Advertised path-id 1
  Path type: internal, path is valid, is best path
  AS-Path: 65001 , path sourced external to AS
    192.168.2.2 (metric 41) from 192.168.2.2 (192.168.2.2)
      Origin IGP, MED not set, localpref 100, weight 0

  Path type: internal, path is valid, not best reason: Router Id
  AS-Path: 65001 , path sourced external to AS
    192.168.3.3 (metric 41) from 192.168.3.3 (192.168.3.3)
      Origin IGP, MED not set, localpref 100, weight 0
```

```
      Path-id 1 advertised to peers:
        192.168.3.3          192.168.4.4
NX-1(config)# router bgp 65000
NX-1(config-router)# address-family ipv4 unicast
NX-1(config-router-af)# maximum-paths ibgp 2
NX-1# show bgp ipv4 unicast
BGP routing table information for VRF default, address family IPv4 Unicast
BGP table version is 65, local router ID is 192.168.1.1
Status: s-suppressed, x-deleted, S-stale, d-dampened, h-history, *-valid, >-best
Path type: i-internal, e-external, c-confed, l-local, a-aggregate, r-redist,
  I-injected
Origin codes: i - IGP, e - EGP, ? - incomplete, | - multipath, & - backup

   Network            Next Hop            Metric     LocPrf     Weight Path
*>l192.168.1.1/32    0.0.0.0                         100        32768 i
*>i192.168.2.2/32    192.168.2.2                     100            0 i
*>i192.168.3.3/32    192.168.3.3                     100            0 i
*>i192.168.4.4/32    192.168.4.4                     100            0 i
*>i192.168.5.5/32    192.168.2.2                     100            0 65001 i
*|i                  192.168.3.3                     100            0 65001 i
*>i192.168.6.6/32    192.168.2.2                     100            0 65001 i
*|i                  192.168.3.3                     100            0 65001 i

NX-1# show bgp ipv4 unicast 192.168.5.5
BGP routing table information for VRF default, address family IPv4 Unicast
BGP routing table entry for 192.168.5.5/32, version 59
Paths: (2 available, best #1)
Flags: (0x08001a) on xmit-list, is in urib, is best urib route, is in HW,
Multipath: iBGP

  Advertised path-id 1
  Path type: internal, path is valid, is best path
  AS-Path: 65001 , path sourced external to AS
    192.168.2.2 (metric 41) from 192.168.2.2 (192.168.2.2)
      Origin IGP, MED not set, localpref 100, weight 0

  Path type: internal, path is valid, not best reason: Router Id, multipath
  AS-Path: 65001 , path sourced external to AS
    192.168.3.3 (metric 41) from 192.168.3.3 (192.168.3.3)
      Origin IGP, MED not set, localpref 100, weight 0

  Path-id 1 advertised to peers:
    192.168.3.3          192.168.4.4

NX-1# show ip route 192.168.5.5/32 detail

192.168.5.5/32, ubest/mbest: 2/0
    *via 192.168.2.2, [200/0], 00:45:02, bgp-65000, internal, tag 65001,
        client-specific data: a
        recursive next hop: 192.168.2.2/32
        extended route information: BGP origin AS 65001 BGP peer AS 65001
    *via 192.168.3.3, [200/0], 00:02:22, bgp-65000, internal, tag 65001,
        client-specific data: a
        recursive next hop: 192.168.3.3/32
        extended route information: BGP origin AS 65001 BGP peer AS 65001
```

可以通过 BGP 事件历史日志来验证添加到 URIB（Unicast Routing Information Base，单播路由信息库）中的次优路径，可以通过命令 **show bgp event-history detail** 查看添加到 URIB 中的前

缀的最佳路径和次优路径的详细信息，如例 11-24 所示。从例 11-24 可以看出，NX-1 首先选择了最佳路径（经 192.168.2.2 的路径），然后又向 URIB 添加了另一条路径（通过下一跳 192.168.3.3 学到的）。

例 11-24　BGP 多路径的事件历史记录日志

```
NX-1# show bgp event-history detail | in 192.168.5.5
16:48:55.864118: (default) RIB: [IPv4 Unicast] Adding path (0x18) to
 192.168.5.5/32 via 192.168.3.3 in URIB (table-id 0x1, flags 0x10, nh 192.168.3.
3) extcomm-len=0, preference=200
16:48:55.864112: (default) RIB: [IPv4 Unicast]: adding route 192.168.5.5/32 via
   192.168.3.3
16:48:55.864108: (default) RIB: [IPv4 Unicast] Sending route 192.168.5.5/32 to URIB
16:48:55.864101: (default) RIB: [IPv4 Unicast] No change (0x80038) in best path
 for 192.168.5.5/32 , resync with RIB, backup/multipath changed
16:48:55.864093: (default) RIB: [IPv4 Unicast] Begin select bestpath for
 192.168.5.5/32, adv_all=0, cal_nth=0, install_to_rib=0, flags=0x80038
16:48:55.863833: (default) RIB: [IPv4 Unicast] Triggering bestpath s
election for 192.168.5.5/32 , flags=0x8003a

! Output omitted for brevity

16:06:15.704376: (default) BRIB: [IPv4 Unicast] 192.168.5.5/32, no Label AF
16:06:15.704373: (default) RIB: [IPv4 Unicast] 192.168.5.5/32 path#1
: set to rid=192.168.2.2 nh=192.168.2.2, flags=0x12, changed=1
16:06:15.704369: (default) RIB: [IPv4 Unicast] Selected new bestpath
 192.168.5.5/32 flags=0x880018 rid=192.168.2.2 nh=192.168.2.2
```

11.4.4　BGP 更新生成过程

　　NX-OS 的更新生成过程与思科 IOS 和 IOS XR 平台不完全一样。与 IOS 和 IOS XR 不同，NX-OS 没有更新组（update-group）的概念。BGP 处理从对等体收到的路由更新消息时，通过已配置的入站策略处理前缀和属性，并将新路径安装到 BRIB（BGP RIB）中。更新了 BRIB 中的路由之后，BGP 会标记该路由以进一步生成更新。BGP 在打包这些前缀之前，需要利用已配置的出站策略进行处理，将标记后的路由放到更新消息中并发送给对等体。例 11-25 解释了 NX-OS 的 BGP 更新生成过程。如果要更好地理解更新生成过程，可以使用调试命令 **debug ip bgp update** 和 **debug ip bgp brib**。从例 11-25 的调试输出可以看出，NX-1 从 NX-4（192.168.4.4）收到的更新消息中包含了前缀 192.168.44.44/32 的宣告信息，然后又在 BRIB 中进行了更新。接下来是 NX-4，然后是为对等体 NX-2 和 NX-3 生成更新消息，请注意，是分别为 NX-2（192.168.2.2）和 NX-3（192.168.3.3）生成更新。

例 11-25　调试 BGP 更新以及在 BRIB 中安装路由

```
NX-1# debug logfile bgp
NX-1# debug ip bgp update
NX-1# debug ip bgp brib
NX-1# show debug logfile bgp

! Receiving an update from peer for 192.168.44.44/32

22:40:31.707254 bgp: 65000 [10739] (default) UPD: Received UPDATE message from
  192.168.4.4
22:40:31.707422 bgp: 65000 [10739] (default) UPD: 192.168.4.4 parsed UPDATE
 message from peer, len 55 , withdraw len 0, attr len 32, nlri len 0
```

```
22:40:31.707499 bgp: 65000 [10739] (default) UPD: Attr code 1, length 1,
 Origin: IGP
22:40:31.707544 bgp: 65000 [10739] (default) UPD: Attr code 5, length 4,
 Local-pref: 100
22:40:31.707601 bgp: 65000 [10739] (default) UPD: Peer 192.168.4.4 nexthop
 length in MP reach: 4
22:40:31.707672 bgp: 65000 [10739] (default) UPD: Recvd NEXTHOP 192.168.4.4
22:40:31.707716 bgp: 65000 [10739] (default) UPD: Attr code 14, length 14,
 Mp-reach
22:40:31.707787 bgp: 65000 [10739] (default) UPD: [IPv4 Unicast] Received prefix
 192.168.44.44/32 from peer 192.168.4.4, origin 0, next hop 192.168.4.4,
 localpref 100, med 0
22:40:31.707859 bgp: 65000 [10739] (default) BRIB: [IPv4 Unicast] Installing
 prefix 192.168.44.44/32 (192.168.4.4) via 192.168.4.4 into BRIB with extcomm
22:40:31.707915 bgp: 65000 [10739] (default) BRIB: [IPv4 Unicast] Created new
 path to 192.168.44.44/32 via 0.0.0.0 (pflags=0x0)
22:40:31.707962 bgp: 65000 [10739] (default) BRIB: [IPv4 Unicast]
 (192.168.44.44/32 (192.168.4.4)): bgp_brib_add: handling nexthop
22:40:31.708054 bgp: 65000 [10739] (default) BRIB: [IPv4 Unicast]
 (192.168.44.44/32 (192.168.4.4)): returning from bgp_brib_add, new_path: 1,
 change : 1, undelete: 0, history: 0, force: 0, (pflags=0x2010), reeval=0
22:40:31.708292 bgp: 65000 [10739] (default) BRIB: [IPv4 Unicast]
 192.168.44.44/32, no Label AF

! Generating update for peer 192.168.2.2

22:40:31.709476 bgp: 65000 [10739] (default) UPD: [IPv4 Unicast] Starting
 update run for peer 192.168.2.2 (#65)
22:40:31.709514 bgp: 65000 [10739] (default) UPD: [IPv4 Unicast] consider
 sending 192.168.44.44/32 to peer 192.168.2.2, path-id 1, best-ext is off
22:40:31.709553 bgp: 65000 [10739] (default) UPD: 192.168.2.2 Sending attr
 code 1, length 1, Origin: IGP
22:40:31.709581 bgp: 65000 [10739] (default) UPD: 192.168.2.2 Sending attr
 code 5, length 4, Local-pref: 100
22:40:31.709613 bgp: 65000 [10739] (default) UPD: 192.168.2.2 Sending attr
 code 9, length 4, Originator: 192.168.4.4
22:40:31.709654 bgp: 65000 [10739] (default) UPD: 192.168.2.2 Sending attr
 code 10, length 4, Cluster-list
22:40:31.709700 bgp: 65000 [10739] (default) UPD: 192.168.2.2 Sending attr
 code 14, length 14, Mp-reach
22:40:31.709744 bgp: 65000 [10739] (default) UPD: 192.168.2.2 Sending nexthop
 address 192.168.4.4 length 4
22:40:31.709789 bgp: 65000 [10739] (default) UPD: [IPv4 Unicast] 192.168.2.2
 Created UPD msg (len 69) with prefix 192.168.44.44/32 ( Installed in HW
 ) path-id 1 for peer
22:40:31.709820 bgp: 65000 [10739] (default) UPD: [IPv4 Unicast] 192.168.2.2:
 walked 0 nodes and packed 0/0 prefixes
22:40:31.709859 bgp: 65000 [10739] (default) UPD: [IPv4 Unicast] (#66) Finished
 update run for peer 192.168.2.2 (#66)

! Generating update for peer 192.168.3.3

22:40:31.709891 bgp: 65000 [10739] (default) UPD: [IPv4 Unicast] Starting update
 run for peer 192.168.3.3 (#65)
22:40:31.709917 bgp: 65000 [10739] (default) UPD: [IPv4 Unicast] consider
 sending 192.168.44.44/32 to peer 192.168.3.3, path-id 1, best-ext is off
22:40:31.709948 bgp: 65000 [10739] (default) UPD: 192.168.3.3 Sending attr
```

```
 code 1, length 1, Origin: IGP
22:40:31.709974 bgp: 65000 [10739] (default) UPD: 192.168.3.3 Sending attr
 code 5, length 4, Local-pref: 100
22:40:31.709998 bgp: 65000 [10739] (default) UPD: 192.168.3.3 Sending attr
 code 9, length 4, Originator: 192.168.4.4
22:40:31.710149 bgp: 65000 [10739] (default) UPD: 192.168.3.3 Sending attr
 code 10, length 4, Cluster-list
22:40:31.710180 bgp: 65000 [10739] (default) UPD: 192.168.3.3 Sending attr
 code 14, length 14, Mp-reach
22:40:31.710204 bgp: 65000 [10739] (default) UPD: 192.168.3.3 Sending nexthop
 address 192.168.4.4 length 4
22:40:31.710231 bgp: 65000 [10739] (default) UPD: [IPv4 Unicast] 192.168.3.3
 Created UPD msg (len 69) with prefix 192.168.44.44/32 ( Installed in HW)
 path-id 1 for peer
22:40:31.710261 bgp: 65000 [10739] (default) UPD: [IPv4 Unicast] 192.168.3.3:
 walked 0 nodes and packed 0/0 prefixes
22:40:31.710286 bgp: 65000 [10739] (default) UPD: [IPv4 Unicast] (#66)
 Finished update run for peer 192.168.3.3 (#66)
```

对于 NX-OS 来说，不一定非要通过调试操作来理解更新生成过程。可以使用 **show bgp event-history detail** 命令查看详细的事件日志，其中，选项 **detail** 默认不可用，需要在 **router bgp** 配置下使用命令 **event-history detail [size** *large* | *medium* | *small*]。例 11-26 给出了 BGP 事件历史日志的详细输出结果，显示了相同的更新过程。本例为 NX-3 生成了更新消息。如果事件历史记录被滚动更新，且问题仍然不断发生，那么就需要启用调试操作，如例 11-25 所示。

例 11-26　BGP 更新生成过程的事件历史日志

```
NX-1# show bgp event-history detail
BGP event-history detail
22:40:31.710283: (default) UPD: [IPv4 Unicast] (#66) Finished update
 run for peer 192.168.3.3 (#66)
22:40:31.710258: (default) UPD: [IPv4 Unicast] 192.168.3.3: walked 0 nodes and
 packed 0/0 prefixes
22:40:31.710226: (default) UPD: [IPv4 Unicast] 192.168.3.3 Created UPD msg
 (len 69) with prefix 192.168.44.44/32 ( Installed in HW) path-id 1 for peer
22:40:31.710201: (default) UPD: 192.168.3.3 Sending nexthop address
 192.168.4.4 length 4
22:40:31.710177: (default) UPD: 192.168.3.3 Sending attr code 14, length 14,
 Mp-reach
22:40:31.710145: (default) UPD: 192.168.3.3 Sending attr code 10, length 4,
 Cluster-list
22:40:31.709995: (default) UPD: 192.168.3.3 Sending attr code 9, length 4,
 Originator: 192.168.4.4
22:40:31.709971: (default) UPD: 192.168.3.3 Sending attr code 5, length 4,
 Local-pref: 100
22:40:31.709945: (default) UPD: 192.168.3.3 Sending attr code 1, length 1,
 Origin: IGP
22:40:31.709913: (default) UPD: [IPv4 Unicast] consider sending 192.
168.44.44/32 to peer 192.168.3.3, path-id 1, best-ext is off
22:40:31.709887: (default) UPD: [IPv4 Unicast] Starting update run for peer
 192.168.3.3 (#65)
! Output omitted for brevity
```

11.4.5　BGP 收敛

BGP 的收敛过程与多种因素有关，BGP 的收敛速度取决于以下几点。

- 与大量对等体建立的会话。
- 本地生成的所有 BGP 路径（通过 **network** 语句或者重分发静态路由/直连路由/IGP 路由）和/或来自其他地址簇的其他组件（如来自多播的 MVPN[Multicast VPN，多播 VPN]、来自 L2VPN 管理器的 L2VPN 等）。
- 发送和接收的多个 BGP 表，也就是说，与每个对等体之间收发的 BGP 地址簇不同。
- 从对等体收到所有路径之后，执行最佳路径计算，以找到最佳路径和/或多路径、附加路径及备份路径。
- 将最佳路径安装到多个路由表中（如默认或 VRF 路由表）。
- 导入和导出机制。
- 对于其他地址簇（如 L2VPN 或多播）来说，需要将路径计算结果传递给不同的底层组件。

BGP 需要使用大量 CPU 周期来处理 BGP 更新，而且需要耗费内存资源来维护 BGP 表中的 BGP 对等体和路由。根据 BGP 路由器在网络中的功能角色，需要选择合适的硬件设备。路由器拥有的内存越多，所能支持的路由数就越多，就像路由器的 CPU 速度越快，所能支持的对等体数量也就越多一样。

> **注：** BGP 更新依赖于 TCP、路由器资源（如内存）以及 TCP 会话参数（如 MSS[Maximum Segment Size，最大报文长度]、路径 MTU 发现、接口输入队列、TCP 窗口大小等），需要优化这些参数来改善 BGP 的路由收敛速度。

应该通过不同的步骤来验证 BGP 是否已收敛以及路由是否已经安装到 BRIB 中。

如果出现了流量丢失问题，那么就需要在 BGP 完成给定地址簇的收敛操作之前，验证 URIB 中的路由信息和 FIB 中的转发信息。例 11-27 给出了 BGP 路由刷新示例，本例通过 **show bgp event-history** [**event** | **detail**]命令验证了前缀已经安装到了 BRIB 表中，通过 **show routing internal event-history** [**add-route** | **modify-route** | **delete-route**]命令验证了路由已经安装到了 URIB 中。同时，还在 URIB 中验证了路由下载到 URIB 的时间戳。如果前缀刚刚下载到 URIB 中，那么就可能会导致路由刷新事件。此外，还可以通过在 BRIB 中安装前缀与将前缀进一步下载到 URIB 中的时间差来了解 BGP 收敛时间。

例 11-27　BRIB 和 URIB 路由安装

```
NX-1# show bgp event-history detail
BGP event-history detail
! Output omitted for brevity
22:40:31.707849: (default) BRIB: [IPv4 Unicast] Installing prefix 19
2.168.44.44/32 (192.168.4.4) via 192.168.4.4  into BRIB with extcomm
NX-1# show routing internal event-history add-route | grep 192.168.44.44
22:40:31.708531 urib: "bgp-65000": 192.168.44.44/32 xri info for rnh
 192.168.4.4/32: origin AS fde8 peer AS fde8
22:40:31.708530 urib: "bgp-65000": 192.168.44.44/32, new rnh 192.168
.4.4/32, metric [200/0] route-type internal tag 0x0000fde8 flags 0x0000080e
22:40:31.708496 urib: "bgp-65000": 192.168.44.44/32 add rnh 192.168.
4.4/32 epoch 1 recursive
22:40:31.708495 urib: "bgp-65000": 192.168.44.44/32, adding rnh 192.
168.4.4/32, metric [200/0] route-type internal tag 0x0000fde8 flags 0x00000010
```

可以通过命令 **show bgp convergence detail vrf all** 检查相关地址簇的 BGP 收敛情况。例 11-28 显示了 **show bgp convergence detail vrf all** 命令的输出结果，可以看到最佳路径选择进程的启动时间和完成时间，同时还可以看到前缀融合到 URIB 所花费的时间，有助于从 BGP 和 URIB 收敛的角度来了解设备运行情况。

例 11-28 show bgp convergence detail 命令输出结果

```
NX-1# show bgp convergence detail vrf all
Global settings:
BGP start time 1 day(s), 04:38:39 ago
Config processing completed 0.068404 after start
BGP out of wait mode 0.068493 after start
LDP convergence not required
Convergence to ULIB not required

Information for VRF default
Initial-bestpath timeout: 300 sec, configured 0 sec
BGP update-delay-always is not enabled
First peer up 00:06:14 after start
Bestpath timer not running

  IPv4 Unicast:
  First bestpath signalled 0.068443 after start
  First bestpath completed 0.069397 after start
  Convergence to URIB sent 0.082041 after start
  Peer convergence after start:
   192.168.2.2        (EOR after bestpath)
   192.168.3.3        (EOR after bestpath)
   192.168.4.4        (EOR after bestpath)

  IPv6 Unicast:
  First bestpath signalled 0.068467 after start
  First bestpath completed 0.069574 after start
```

注：如果该节点尚未运行 BGP 最佳路径选择进程，那么就表明故障问题可能与该节点的 BGP 无关。

如果在收到 EOR 之前已经运行了最佳路径选择进程，或者对等体还没有发送 EOR 标记，那么就可能产生流量丢失问题。在这种情况下，可以考虑对 BGP 更新启用携带 VRF、地址簇和对等体等调试过滤器的调试操作，如例 11-29 所示。

例 11-29 携带过滤器的调试命令

```
debug logfile bgp
debug bgp events updates rib brib import
debug-filter bgp address-family ipv4 unicast
debug-filter bgp neighbor 192.168.4.4
debug-filter bgp prefix 192.168.44.44/32
```

可以检查调试输出结果中的事件日志来了解时间戳信息，确定最近发送给对等体的 EOR 的时间。此外，调试结果还能显示发送 EOR 之前已经向对等体宣告了多少条路由。如果对等体提前刷新了过时的路由，那么过早向对等体发送 EOR 也可能会导致流量丢失。

如果发现 URIB 中的路由尚未下载，那么就需要执行进一步检查，因为故障根源可能不是 BGP。此时，可以运行下列命令来了解 URIB 的活动信息，以排查流量丢失原因。

- **show routing internal event-history ufdm**。
- **show routing internal event-history ufdm-summary**。
- **show routing internal event-history recursive** 缩放 BGP。

11.5 BGP 的扩展性

BGP 是功能特性极为丰富的路由协议之一，通过路由策略实现了路由和控制的易用性。虽然 BGP 提供了大量内置功能特性，可以很好地实现协议扩展，但这些增强功能却始终未能得到正确利用，导致在扩展性要求很高的环境中部署 BGP 时，出现了诸多挑战。

BGP 是一种重量级协议，需要占用路由器的大量 CPU 和内存资源。BGP 需要耗用越来越多资源的原因有很多，下面 3 种因素是 BGP 需要耗用大量内存资源的主要因素。

- 前缀。
- 路径。
- 属性。

BGP 可以保存大量前缀，而每个前缀都要占用一定的内存。不过，如果通过多条路径学到同一个前缀，那么这些信息也都会保存在 BGP 表中，每条路径都要占用内存资源。由于 BGP 被设计为通过不同的属性来控制每个 AS 的流量流，因而每个前缀在每条路径上都可以拥有不同的属性，可以将其视为一个数学函数，其中，N 表示前缀数，M 表示给定前缀的路径数，L 表示与给定前缀相关联的属性。

- 前缀：$(O(N))$。
- 路径：$(O(M \times N))$。
- 属性：$(O(L \times M \times N))$。

11.5.1 优化 BGP 内存

如前所述，为了减少或优化 BGP 内存消耗，需要调整导致 BGP 内存消耗较多的 3 个主要因素。接下来将详细讨论每种因素的调整措施。

1. 前缀

如果 BGP 保存了大量前缀甚至是 Internet 路由表，那么 BGP 的内存消耗将变得非常大。在大多数情况下，并不需要网络中的所有 BGP 路由器都维护全部 BGP 前缀。为了减少前缀数量，可以考虑执行以下操作。

- 聚合。
- 过滤。
- 维护部分路由表而非全路由表。

路由聚合可以将多条特定路由聚合为一条路由。但是，在完全部署的运行网络上尝试聚合操作也是一项不小的挑战。网络启动并运行之后，必须查看完整的 IP 寻址方案才能正确执行聚合操作。聚合对于新网络的部署来说是一个非常好的选择，新网络可以在部署时采取更多的 IP 地址控制机制，使聚合操作也更加容易。

路由过滤可以控制保留在 BGP 表中或宣告给 BGP 对等体的前缀数量。BGP 可以根据前缀、BGP 属性以及团体属性进行路由过滤。需要记住的是，复杂的路由过滤或者对大量前缀应用路由过滤操作，虽然有助于减少内存耗用，但是会给路由器的 CPU 利用率带来较大冲击。

很多部署方案并不要求所有的 BGP 发话端都维护完整的 BGP 路由表，特别是企业和数据中心应用场景，这些场景不需要拥有完整的 Internet 路由表，BGP 发话端只要维护包含最需要、最相关前缀的部分路由表即可，甚至只要维护一条指向 Internet 网关的默认路由即可。这种设计模式不但能够大大减少全网资源的耗用程度，而且能大大提高网络的扩展性。

2. 路径

在某些情况下，虽然 BGP 表包含的前缀数量并不多，但是由于存在多路径情形，因而仍然需要耗用很多的内存资源。虽然可以从多条路径学到同一个前缀，但是路由表中安装的只是最佳路径或多条最佳路径。为了减少 BGP 因多路径而导致的内存消耗，可以采取以下解决方案。

- 减少对等关系数量。

- 使用 RR，而不是全网状 IBGP。

多个 BGP 对等关系带来的直接影响就是多 BGP 路径，特别是在全网状 IBGP 环境中，BGP 会话的数量呈现指数级增长趋势，导致路径数量急剧增加。很多客户通过增加 IBGP 邻居数来获得更多的冗余路径，但实际上，两条路径就足以实现冗余性。从会话数和 BGP 内存占用率的角度来看，增加对等关系数量会带来扩展性问题。

众所周知，IBGP 要求全网状连接，图 11-7 给出了一个 IBGP 全网状拓扑结构示意图，可以看出，如果 IBGP 全网状网络包含 n 个节点，那么每个 BGP 发话端就需要建立 $n*(n-1)/2$ 条 IBGP 会话。

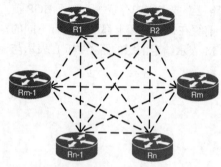

图 11-7　IBGP 全网状拓扑结构

这种全网状拓扑不但影响单个节点或路由器的可扩展性，而且影响整个网络的可扩展性。为了提高 IBGP 网络的可扩展性，可以采用以下两种设计方法。

- 联盟（Confederation）。
- 路由反射器（Route Reflector）。

注：有关 BGP 联盟和路由反射器的详细内容将在后面的章节讨论。

3．属性

BGP 路由就是一组属性，每个 BGP 前缀都包含一些自动分配的默认或强制属性（如 Next-Hop 或 AS-PATH）或者由客户手动配置的属性（如 MED 等）。附加到前缀的每个属性都要占用一部分内存资源。除属性之外，团体（标准和扩展团体属性）也会增加内存消耗。为了减少各种属性和团体造成的 BGP 内存消耗，可以采用以下解决方案。

- 减少属性数量。
- 过滤标准或扩展团体。
- 限制本地团体。

可以在 NX-OS 上使用 **show bgp private attr detail** 命令查看附加到 BGP 前缀的各种属性。例 11-30 显示了 NX-1 的各种全局 BGP 属性，这些属性是通过各种前缀学到的，包括附加在从 NX-4 学到的前缀上的团体。

例 11-30　BGP 属性详细信息

```
NX-1# show bgp private attr detail
BGP Global attributes vxlan-enable:0 nve-api-init:0 nve-up:0 mac:0000.0000.0000

BGP attributes information
Number of attribute entries     : 4
HWM of attribute entries        : 5
Bytes used by entries           : 400
Entries pending delete          : 0
```

```
HWM of entries pending delete   : 0
BGP paths per attribute HWM     : 4
BGP AS path entries             : 1
Bytes used by AS path entries   : 26

BGP as-path traversal count     : 20

Attribute 0x64c1b8bc : Hash: 191, Refcount: 1, Attr ID 10
  origin        : IGP
  as-path       : 65001
                : 0 (path hash)
                : 1 (path refcount)
                : 20 (path marker)
  localpref     : 100
  weight        : 0
      Extcommunity presence mask: (nil)

Attribute 0x64c1b7dc : Hash: 2649, Refcount: 1, Attr ID 5
  origin        : IGP
  as-path       :
  localpref     : 100
  weight        : 32768
      Extcommunity presence mask: (nil)

Attribute 0x64c1b84c : Hash: 2651, Refcount: 1, Attr ID 2
  origin        : IGP
  as-path       :
  localpref     : 100
  weight        : 0
      Extcommunity presence mask: (nil)

Attribute 0x64c1b6fc : Hash: 3027, Refcount: 1, Attr ID 12
  origin        : IGP
  as-path       :
  localpref     : 100
  weight        : 0
      Community: 65000:44
      Extcommunity presence mask: (nil)
```

虽然没有方法删除默认 BGP 属性，但是可以控制对其他属性的使用。请注意，如果使用属性之后使事情变得更加复杂，那么将毫无意义。例如，使用 MED 以及与 MED 相关的命令（如 **bgp always-compare-med** 或 **bgp deterministic-med**），可能会对网络造成不利影响，导致路由不稳定或出现路由环路。此外，用户分配的属性会消耗更多的 BGP 内存，这一点实际上很容易避免。

4．BGP 配置扩展性

可以利用 BGP 模板为多个邻居分配相同的策略和属性，如 AS 号或源接口等。如果需要为多个邻居配置相同的策略，那么就可以大大节省输入工作量。对等体模板的 NX-OS 实现包括 3 种模板类型：**peer-policy**、**peer-session** 和 **peer** 模板。

peer-policy 模板可以为对等体定义与地址簇相关的策略，包括入站和出站策略、过滤器列表（filter-list）、前缀列表（prefix-list）以及软重置（soft-reconfiguration）等。**peer-session** 模板可以定义会话属性，如详细传输信息和会话定时器等。**peer-policy** 和 **peer-session** 模板都是可继承的，也就是说，**peer-policy** 或 **peer-session** 可以从其他 **peer-policy** 或 **peer-session** 继承属性。**peer** 模板可以将 **peer-policy** 和 **peer-session** 组合在一起，从而实现批量化的邻居策略定义。例 11-31 给出

了 NX-1 的 BGP 模板配置示例。

例 11-31　BGP 模板配置

```
NX-1(config)# router bgp 65000
! Configure peer-policy template
NX-1(config-router)# template peer-policy PEERS-V4
NX-1(config-router-ptmp)# route-reflector-client
NX-1(config-router-ptmp)# exit
! Configure peer-session template
NX-1(config-router)# template peer-session PEER-DEFAULT
NX-1(config-router-stmp)# remote-as 65530
NX-1(config-router-stmp)# update-source loopback0
NX-1(config-router-stmp)# password cisco
NX-1(config-router-stmp)# exit
! Configure peer template
NX-1(config-router)# template peer IBGP-RRC
NX-1(config-router-neighbor)# inherit peer-session PEER-DEFAULT
NX-1(config-router-neighbor)# address-family ipv4 unicast
NX-1(config-router-neighbor-af)# inherit peer-policy PEERS-V4 10
! Applying Peer Template to BGP peers
NX-1(config-router)# neighbor 192.168.4.4
NX-1(config-router-neighbor)# inherit peer IBGP-RRC
NX-1(config-router-neighbor)# exit
NX-1(config-router)# neighbor 192.168.3.3
NX-1(config-router-neighbor)# inherit peer IBGP-RRC
NX-1(config-router-neighbor)# exit
NX-1(config-router)# neighbor 192.168.2.2
NX-1(config-router-neighbor)# inherit peer IBGP-RRC
NX-1(config-router-neighbor)# exit
```

11.5.2　入站软重置与路由刷新

调整入站 BGP 策略时，需要对等体重新发送 BGP 更新。BGP 更新采取的是增量更新方式，也就是说，完成初始更新之后仅接收新的变更。因此，要求重置 BGP 会话，请求对等体发送包含所有 NLRI 的 BGP 更新消息，从而可以通过新的过滤器重新运行这些更新。实现会话重置的方法有两种。

- **硬重置**：删除并重建 BGP 会话，相应的命令是 **clear bgp** *afi safi* [* | *ip-address*]。
- **软重置**：软重置使用存储在内存中的过滤后的前缀来重新配置并激活 BGP 路由表，而不破坏 BGP 会话，相应的命令是 **clear bgp** *afi safi* [* | *ip address*] **soft** [**in** | **out**]。

BGP 会话的硬重置会中断运行中的网络。如果 BGP 会话因 BGP 策略的多次变更而在短时间内重复重置，那么很可能会导致网络中的其他路由器抑制前缀，从而导致目的地不可达、流量被黑洞化。

软重置是一种传统的允许在入站 BGP 路由更新上应用路由策略的方法。可以在邻居配置模式下使用命令 **soft-reconfiguration inbound** 启用 BGP 软重置。启用了软重置之后，BGP 会存储从对等体收到的所有未经修改的路由副本，即使路由策略并未频繁变更。启用软重置意味着路由器在应用任何策略之前还要存储收到的前缀/属性，从而导致路由器出现额外的内存和 CPU 开销。

如果要手动执行软重置，可以使用命令 **clear bgp ipv4 unicas** [* | *ip-address*] **soft** [**in** | **out**]。如果操作人员希望在应用入站策略之前知道哪些前缀已经发送给了路由器，那么软重置功能特性将非常有用。

为了解决软重置入站配置带来的挑战问题，RFC 2918 定义了 BGP 路由刷新功能。BGP 路由刷新功能的功能代码为 2、功能长度为 0。使用了路由刷新功能之后，路由器可以向对等体发送路由刷新请求，从而能够再次从对等体获得全路由表。路由刷新功能的一个优点是无须预先启用该功能特性。ROUTE-REFRESH 消息是一种新的 BGP 消息类型，如图 11-8 所示。

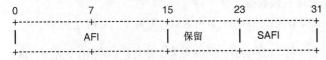

图 11-8 BGP ROUTE-REFRESH 消息

ROUTE-REFRESH 消息中的 AFI 和 SAFI 指向对等体正在协商路由刷新功能的地址簇。保留比特未使用，发送端将该比特设置为 0，接收端则忽略该比特。

BGP 发话端与对等体协商了路由刷新功能之后，就会发送路由刷新消息，意味着所有参与路由器都应该支持路由刷新功能。路由器向对等体发送路由刷新请求（*REFRESH_REQ*），BGP 发话端收到路由刷新请求之后，会将消息中携带的 AFI 和 SAFI 重新宣告给对等体。如果 BGP 发话端配置了出站路由过滤策略，那么就会对更新进行过滤。路由刷新功能要求对等体接收过滤后的路由。

命令 **clear ip bgp** *ip-address* **in** 或 **clear bgp** *afi safi ip-address* **in** 的作用是告诉对等体通过发送路由刷新请求来重新发送完整的 BGP 宣告，而命令 **clear bgp** *afi safi ip-address* **out** 的作用则是向对等体重新发送完整的 BGP 宣告，不过不会发起路由刷新请求。可以通过命令 **show bgp** *afi safi* **neighbors** *ip-address* 验证路由刷新功能，例 11-32 显示了两个 BGP 对等体协商的路由刷新功能。

例 11-32 BGP 路由刷新功能

```
NX-1
NX-1# show bgp ipv4 unicast neighbors 192.168.2.2
BGP neighbor is 192.168.2.2, remote AS 65000, ibgp link, Peer index 1
  Inherits peer configuration from peer-template IBGP-RRC
  BGP version 4, remote router ID 192.168.2.2
  BGP state = Established, up for 01:10:46
  Using loopback0 as update source for this peer
  Last read 00:00:43, hold time = 180, keepalive interval is 60 seconds
  Last written 00:00:31, keepalive timer expiry due 00:00:28
  Received 77 messages, 0 notifications, 0 bytes in queue
  Sent 80 messages, 0 notifications, 0 bytes in queue
  Connections established 1, dropped 0
  Last reset by us never, due to No error
  Last reset by peer never, due to No error

  Neighbor capabilities:
  Dynamic capability: advertised (mp, refresh, gr) received (mp, refresh, gr)
  Dynamic capability (old): advertised received
  Route refresh capability (new): advertised received
  Route refresh capability (old): advertised received
  4-Byte AS capability: advertised received
  Address family IPv4 Unicast: advertised received
  Graceful Restart capability: advertised received
! Output omitted for brevity
```

注： 配置软重置功能特性时，不会使用 BGP 路由刷新功能（即使已协商该功能），软重置功能可以控制路由刷新的处理或发起操作。

下列情况下会发送 BGP 刷新请求（REFRESH_REQ）。

- 运行命令 **clear bgp** *afi safi* [* | *ip-address*] **in**。

- 运行命令 **clear bgp** *afi safi* [* | *ip-address*] **soft in**。
- 通过路由映射在 BGP 邻居上添加或更改入站过滤策略。
- 为 BGP 邻居配置 **allowas-in**。
- 为 BGP 邻居配置 **soft-reconfiguration inbound**。
- 在 MPLS VPN 中向 VRF 增加 **route-target import**（AFI/SAFI 值为 1/128 或 2/128）。

> **注**：如果在通过入站路由映射过滤或修改前缀属性之前，需要知道这些原始前缀属性，那么就建议仅在 EBGP 对等关系上使用 **soft-reconfiguration inbound**，而不建议在接收大量被交换前缀（如 Internet 路由表）的路由器上配置该命令。

11.5.3 利用路由反射器实现 BGP 的扩展性

由于 BGP 不能将学自 IBGP 对等体的前缀宣告给其他 IBGP 对等体，因而可能会在 AS 内部出现扩展性问题。公式 $n*(n-1)/2$ 表示需要的会话数，其中的 n 表示路由器数量。也就是说，包含 5 台路由器的全网状拓扑需要 10 条会话，包含 10 台路由器的全网状拓扑需要 45 条会话。因此，IBGP 的扩展性已成为大型网络的一个重要问题。

RFC1966 引入了一种扩展机制，可以将 IBGP 对等关系配置为将路由反射给其他 IBGP 对等体，负责反射路由的路由器被称为 RR（Route Reflector，路由反射器），接收被反射路由的路由器则被称为路由反射器客户端。RR 设计模式可以将 IBGP 全网状拓扑转换为星型（Hub-and-Spoke）设计模式，其中的 RR 是 Hub 路由器。RR 客户端可以是普通的 IBGP 对等体（客户端不相互直连），也可以将这些 RR 客户端设计成全网状互联。路由反射器和路由反射的 3 个基本设计规则如下。

- **规则 1**：如果 RR 从非 RR 客户端收到了 NLRI，那么 RR 就会将该 NLRI 宣告给 RR 客户端，但是不会将 NLRI 宣告给非 RR 客户端。
- **规则 2**：如果 RR 从 RR 客户端收到了 NLRI，那么就会将该 NLRI 宣告给 RR 客户端和非 RR 客户端。即使发送该路由宣告的 RR 客户端收到了路由副本，也会丢弃该 NLRI，因为它可以看出自己是路由发起端。
- **规则 3**：如果 RR 从 EBGP 对等体收到了路由，那么就会将该路由宣告给 RR 客户端和非 RR 客户端。此时只有路由反射器能够感知这种行为变化，因为路由反射器客户端未执行任何额外的 BGP 配置。BGP 路由反射的配置与每个特定的地址簇相关，需要在 NX-OS 设备的邻居地址簇配置下应用命令 **route-reflector-client**。

图 11-9 给出了两种 RR 设计方案。(a)拓扑结构中的 R1 充当 RR，R2、R3 和 R4 是 RR 客户端。(b)拓扑结构与(a)拓扑结构相似，区别在于此时的 RR 客户端之间是全网状互联。

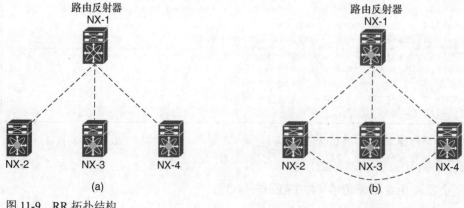

图 11-9　RR 拓扑结构

RR 和客户端对等体构成了一个簇（Cluster），不需要全网状连接。由于（b）拓扑结构配置了 RR，同时还配置了全网状连接的 IBGP 客户端对等体，实际上违背了 RR 的配置目的，因而应禁用 BGP 的 RR 反射行为。可以通过命令 **no bgp client-to-client reflection** 禁用 BGP RR 客户端到客户端的反射行为，该命令只要配置在 RR 客户端上即可，不需要配置在 RR 客户端上。例 11-33 给出了禁用 BGP 客户端到客户端反射行为的配置示例。

例 11-33　禁用 BGP 客户端到客户端的反射行为

```
NX-1(config)# router bgp 65000
NX-1(config-router)# address-family ipv4 unicast
NX-1(config-router-af)# no bgp client-to-client reflection
```

路由反射器的环路避免

解决了 IBGP 拓扑结构的全网状互联需求之后，可能会导致潜在的路由环路问题。当初起草 RFC1966 时，为了避免此类环路问题，特意定义了两种 BGP 路由反射器属性。

（1）ORIGINATOR_ID

该可选非传递性 BGP 属性由第一台路由反射器创建，且属性值被设置为将该路由注入/宣告到 AS 中的路由器的 RID。如果 NLRI 已经有了 ORIGINATOR_ID，那么就不应该覆盖该属性值。

路由器收到 NLRI 之后，如果发现自己的 RID 位于 ORIGINATOR_ID 属性中，那么就会丢弃该 NLRI。

（2）CLUSTER_LIST

该非传递性 BGP 属性由路由反射器负责更新，路由反射器将自己的 cluster-id 追加（而不是覆盖）到该属性中。默认情况下，这是 BGP 标识符，可以通过 BGP 配置命令 **cluster-id** 来设置 cluster-id。

路由反射器收到 NLRI 之后，如果发现自己的 cluster-id 位于 CLUSTER_LIST 属性中，那么就会丢弃该 NLRI。

例 11-34 给出了路由反射器 NX-1（见图 11-9）反射的路由的前缀情况，可以看出，ORIGINATOR_ID 是宣告路由器且 CLUSTER_LIST 包含了路由反射器 ID，CLUSTER_LIST 包含的路由反射器是该前缀穿越过的最后一台宣告该路由的路由反射器。

例 11-34　RR 反射的前缀

```
NX-4# show bgp ipv4 unicast 192.168.5.5/32
BGP routing table information for VRF default, address family IPv4 Unicast
BGP routing table entry for 192.168.5.5/32, version 52
Paths: (1 available, best #1)
Flags: (0x08001a) on xmit-list, is in urib, is best urib route, is in HW,

  Advertised path-id 1
  Path type: internal, path is valid, is best path
  AS-Path: 65001 , path sourced external to AS
  192.168.2.2 (metric 81) from 192.168.1.1 (192.168.1.1)
      Origin IGP, MED not set, localpref 100, weight 0
      Originator: 192.168.2.2 Cluster list: 192.168.1.1

  Path-id 1 not advertised to any peer
```

如果拓扑结构包含两台 RR，且两台 RR 均配置了不同的 cluster-id，那么第二台 RR 将保留第一台 RR 的所有路径，因而会消耗更多的内存和 CPU 资源。因此，无论是单个 cluster-id，还是多个 cluster-id，都各有缺点。

- **配置不同的 cluster-id**：RR 需要消耗更多的内存和 CPU 开销。

- 配置相同的 **cluster-id**：冗余路径较少。

如果簇内的 RR 客户端存在全网状连接，那么就需要在 RR 上启用 **no bgp client-to-client reflection** 命令。

11.5.4 最大前缀数量

在默认情况下，BGP 对等体会保存对等路由器宣告的所有路由。可以在本地路由器的入站方向或者对等路由器的出站方向对路由进行过滤，从而减少路由数量。但是在某些情况下，路由条数仍然有可能超过路由器的需求量或者能够处理的数量。

NX-OS 支持 BGP 最大前缀数量（maximum-prefix）功能特性，可以限制每个对等体的前缀数量。一般来说，该功能特性主要用于 EBGP 会话，但也可以用于 IBGP 会话。虽然该功能特性有助于提高扩展性，防止网络路由条数过多，但必须掌握该功能特性的使用方式。以下情况可以启用 BGP 最大前缀数量功能特性。

- 知道对等体可能发送的 BGP 路由数量。
- 知道路由条数超量后需要采取的措施，应该重置 BGP 连接或者记录告警消息。

如果要限制前缀数量，那么就要为每个邻居都使用命令 **maximum-prefix** *maximum* [*threshold*] [**restart** *restart-interval* | **warning-only**]，表 11-6 列出了该命令的详细选项信息。

表 11-6　　　　　　　　　　　BGP maximum-prefix 命令选项

maximum	定义最大前缀极限
threshold	定义生成告警所要达到的阈值比例
restart *restart-interval*	默认行为，超过指定的前缀极限后将重置 BGP 连接，*restart-interval* 以分钟为单位进行配置。BGP 会在指定的时间间隔之后尝试重新建立对等关系。如果设置了 **restart** 选项，那么就会向邻居发送终止通知，并关闭 BGP 连接
warning-only	超过指定的前缀极限后仅发送一条告警消息

需要记住的是，如果在 **maximum-prefix** 命令中配置了 **restart** 选项，那么除等待 *restart-interval* 定时器到期之后才能重建 BGP 连接之外，唯一的方法就是使用 **clear bgp** *afi safi ip-address* 命令手动重置对等体。

例 11-35 给出了 **maximum-prefix** 命令的使用示例。NX-2 从邻居 10.25.1.5 收到了 10 个以上的前缀，但是该设备将最大前缀数量设置为 10 个前缀。此时，设置了最大前缀数量功能特性的设备将关闭 BGP 对等关系，而远端对等体则保持在 Idle 状态。排查 BGP 对等关系故障时，需要执行 **show bgp** *afi safi* **neighbors** *ip-address* 命令以验证最后一次重置原因。

例 11-35　最大前缀数量

```
NX-2(config)# router bgp 65000
NX-2(config-router)# neighbor 10.25.1.5
NX-2(config-router-neighbor)# address-family ipv4 unicast
NX-2(config-router-neighbor-af)# maximum-prefix 10
NX-2# show bgp ipv4 unicast summary
BGP summary information for VRF default, address family IPv4 Unicast
BGP router identifier 192.168.2.2, local AS number 65000
BGP table version is 257, IPv4 Unicast config peers 2, capable peers 1
106 network entries and 108 paths using 15424 bytes of memory
BGP attribute entries [16/2304], BGP AS path entries [11/360]
BGP community entries [0/0], BGP clusterlist entries [2/8]


Neighbor        V    AS    MsgRcvd MsgSent   TblVer  InQ OutQ Up/Down   State/PfxRcd
10.25.1.5       4 65001      7349    7354        0    0    0 00:00:08  Shut (PfxCt)
```

```
192.168.1.1     4 65000      7781       7778       257     0     0 3d05h 5
NX-5# show bgp ipv4 unicast neighbors 10.25.1.2
BGP neighbor is 10.25.1.2, remote AS 65000, ebgp link, Peer index 1
  BGP version 4, remote router ID 0.0.0.0
  BGP state = Idle, down for 00:32:34, retry in 00:00:01
  Last read never, hold time = 180, keepalive interval is 60 seconds
  Last written never, keepalive timer not running
  Received 7354 messages, 1 notifications, 0 bytes in queue
  Sent 7379 messages, 0 notifications, 0 bytes in queue
  Connections established 1, dropped 1
  Connection attempts 28
  Last reset by us 00:01:27, due to session closed
  Last reset by peer 00:32:34, due to maximum prefix count error
  Message statistics:
                          Sent               Rcvd
   Opens:                   31                  1
   Notifications:            0                  1
   Updates:                 13                 20
   Keepalives:            7333               7330
   Route Refresh:            0                  0
   Capability:               2                  2
   Total:                 7379               7354
   Total bytes:         140687             140306
   Bytes in queue:           0                  0
! Output omitted for brevity
```

11.5.5 BGP Max-AS

BGP 默认为每条 BGP 前缀都分配各种属性，能够附加到单个前缀的属性长度最多可以达到 64 KB，如此一来，就可能会产生 BGP 扩展性问题和收敛问题。

很多时候为了实现路径优选，都会通过 AS-PATH 追加选项来增加 AS-PATH 列表，从而优选 AS-PATH 列表较短的路径。虽然该操作本身并没有什么负面影响，但是从 Internet 的角度来看，AS-PATH 列表过长不但会产生收敛问题，而且还可能产生安全漏洞。事实上，AS-PATH 列表指示了路由器在 Internet 上的位置。

为了限制网络支持的最大 AS-PATH 长度，引入了命令 **maxas-limit**。如果在 NX-OS 中使用了 **maxas-limit** *1-512* 命令，那么 AS-PATH 长度超过指定数量的路由都将被丢弃。

11.6 BGP 路由过滤和路由策略

除扩展性之外，BGP 还能提供路由过滤、流量工程和流量负载共享等功能。BGP 可以通过定义路由策略和路由过滤器来提供这些功能，可以采用以下 3 种方式来定义路由过滤机制。

- 前缀列表（Prefix-list）。
- 过滤器列表（Filter-list）。
- 路由映射（Route-map）。

与前缀列表和过滤器列表相比，BGP 路由映射能够提供更加动态化的处理能力，不但能够执行路由过滤，而且允许网络运营商定义策略和设置属性，从而进一步控制网络内的流量。后面将详细讨论这些路由过滤和路由策略方法。

例 11-36 给出了图 11-9 拓扑结构中的 Nexus 交换机 NX-2 的 BGP 表信息，下面将以 NX-2 交换机为例来解释上述过滤技术。

例 11-36 NX-2 的 BGP 表

```
NX-2# show bgp ipv4 unicast | b Network
   Network            Next Hop        Metric    LocPrf    Weight  Path
*>e100.1.1.0/24       10.25.1.5                                0  65001 100 {220} e
*>e100.1.2.0/24       10.25.1.5                                0  65001 100 {220} e
*>e100.1.3.0/24       10.25.1.5                                0  65001 100 {220} e
*>e100.1.4.0/24       10.25.1.5                                0  65001 100 {220} e
*>e100.1.5.0/24       10.25.1.5                                0  65001 100 {220} e
*>e100.1.6.0/24       10.25.1.5                                0  65001 100 {220} e
*>e100.1.7.0/24       10.25.1.5                                0  65001 100 {220} e
*>e100.1.8.0/24       10.25.1.5                                0  65001 100 {220} e
*>e100.1.9.0/24       10.25.1.5                                0  65001 100 {220} e
*>e100.1.10.0/24      10.25.1.5                                0  65001 100 {220} e
*>e100.1.11.0/24      10.25.1.5                                0  65001 100 292 {218 230}
*>e100.1.12.0/24      10.25.1.5                                0  65001 100 292 {218 230}
*>e100.1.13.0/24      10.25.1.5                                0  65001 100 292 {218 230}
*>e100.1.14.0/24      10.25.1.5                                0  65001 100 292 {218 230}
*>e100.1.15.0/24      10.25.1.5                                0  65001 100 292 {218 230}
*>e100.1.16.0/24      10.25.1.5                                0  65001 100 292 {218 230}
*>e100.1.17.0/24      10.25.1.5                                0  65001 100 292 {218 230}
*>e100.1.18.0/24      10.25.1.5                                0  65001 100 292 {218 230}
*>e100.1.19.0/24      10.25.1.5                                0  65001 100 292 {218 230}
*>e100.1.20.0/24      10.25.1.5                                0  65001 100 292 {218 230}
*>e100.1.21.0/24      10.25.1.5                                0  65001 100 228 274 {300 243}
*>e100.1.22.0/24      10.25.1.5                                0  65001 100 228 274 {300 243}
*>e100.1.23.0/24      10.25.1.5                                0  65001 100 228 274 {300 243}
*>e100.1.24.0/24      10.25.1.5                                0  65001 100 228 274 {300 243}
*>e100.1.25.0/24      10.25.1.5                                0  65001 100 228 274 {300 243}
*>e100.1.26.0/24      10.25.1.5                                0  65001 100 228 274 {300 243}
*>e100.1.27.0/24      10.25.1.5                                0  65001 100 228 274 {300 243}
*>e100.1.28.0/24      10.25.1.5                                0  65001 100 228 274 {300 243}
*>e100.1.29.0/24      10.25.1.5                                0  65001 100 228 274 {300 243}
*>e100.1.30.0/24      10.25.1.5                                0  65001 100 228 274 {300 243}
*>i192.168.1.1/32     192.168.1.1               100              0  i
*>l192.168.2.2/32     0.0.0.0                   100          32768  i
*>i192.168.3.3/32     192.168.3.3               100              0  i
*>i192.168.4.4/32     192.168.4.4               100              0  i
*>e192.168.5.5/32     10.25.1.5                                0  65001 i
*>e192.168.6.6/32     10.25.1.5                                0  65001 i
*>i192.168.44.0/24    192.168.4.4               100              0  i
```

11.6.1 基于前缀列表的过滤技术

如第 10 章所述，前缀列表是在路由协议中识别网络的方法之一，不但可以识别特定 IP 地址、网络或网络范围，而且可以利用前缀匹配规范选择拥有不同前缀长度（子网掩码）的多个网络。

可以将前缀列表直接用于 BGP 对等体，也可以作为路由映射的匹配语句。前缀列表的配置命令是 **ip prefix-list** *name* [**seq** *sequence-number*] [**permit** *ip-address/length* | **deny** *ip-address/ length*] [**le** *length* | **ge** *length* | **eq** *length*]。仍然以图 11-6 所示拓扑结构为例，例 11-37 给出了 NX-2 利用前缀列表配置的 BGP 入站和出站路由过滤策略，入站前缀列表允许 5 个网络，而出站前缀列表允许的主机网络表项则是与子网 192.168.0.0/16 相匹配的/32 前缀。配置了前缀列表之后，可以使用命令 **how bgp** *afi safi* **neighbor** *ip-address* 验证是否为邻居应用了前缀列表。

例 11-37 基于前缀列表的路由过滤

```
NX-2(config)# ip prefix-list Inbound permit 100.1.1.0/24
NX-2(config)# ip prefix-list Inbound permit 100.1.2.0/24
NX-2(config)# ip prefix-list Inbound permit 100.1.3.0/24
NX-2(config)# ip prefix-list Inbound permit 100.1.4.0/24
NX-2(config)# ip prefix-list Inbound permit 100.1.5.0/24
NX-2(config)#
NX-2(config)# ip prefix-list Outbound permit 192.168.0.0/16 eq 32
NX-2(config)# router bgp 65000
NX-2(config-router)# neighbor 10.25.1.5
NX-2(config-router-neighbor)# address-family ipv4 unicast
NX-2(config-router-neighbor-af)# prefix-list Inbound in
NX-2(config-router-neighbor-af)# prefix-list Outbound out
NX-2(config-router-neighbor-af)# end

NX-2# show bgp ipv4 unicast neighbors 10.25.1.5
BGP neighbor is 10.25.1.5,   remote AS 65001, ebgp link, Peer index 2
  BGP version 4, remote router ID 192.168.5.5
  BGP state = Established, up for 2d00h
! output omitted for brevity
  For address family: IPv4 Unicast
  BGP table version 1085, neighbor version 1085
  5 accepted paths consume 400 bytes of memory
  4 sent paths
  Inbound ip prefix-list configured is Inbound, handle obtained
  Outbound ip prefix-list configured is Outbound, handle obtained
  Last End-of-RIB received 1d23h after session start

  Local host: 10.25.1.2, Local port: 58236
  Foreign host: 10.25.1.5, Foreign port: 179
  fd = 74
```

配置了前缀列表并附加到 BGP 邻居 10.25.1.5 之后的 BGP 表信息如例 11-38 所示，从 NX-2 交换机的输出结果可以看出，邻居 10.25.1.5 只能看到 5 个前缀。从 NX-5 的输出结果可以看出，除 192.168.44.0/24 之外，AS 65000 中的所有节点的环回地址都被宣告了。

例 11-38 配置了前缀列表之后的 BGP 表输出结果

```
NX-2# show bgp ipv4 unicast
BGP routing table information for VRF default, address family IPv4 Unicast
BGP table version is 1085, local router ID is 192.168.2.2
Status: s-suppressed, x-deleted, S-stale, d-dampened, h-history, *-valid, >-best
Path type: i-internal, e-external, c-confed, l-local, a-aggregate, r-redist,
  I-injected
Origin codes: i - IGP, e - EGP, ? - incomplete, | - multipath, & - backup

   Network            Next Hop         Metric     LocPrf     Weight Path
*>e100.1.1.0/24       10.25.1.5                   0 65001    100 220
*>e100.1.2.0/24       10.25.1.5                   0 65001    100 220
*>e100.1.3.0/24       10.25.1.5                   0 65001    100 220
*>e100.1.4.0/24       10.25.1.5                   0 65001    100 220
*>e100.1.5.0/24       10.25.1.5                   0 65001    100 220
*>i192.168.1.1/32     192.168.1.1                 100            0 i
*>l192.168.2.2/32     0.0.0.0                     100        32768 i
*>i192.168.3.3/32     192.168.3.3                 100            0 i
*>i192.168.4.4/32     192.168.4.4                 100            0 i
NX-5# show bgp ipv4 unicast neighbors 10.25.1.2 routes
```

```
Peer 10.25.1.2 routes for address family IPv4 Unicast:
BGP table version is 1209, local router ID is 192.168.5.5
Status: s-suppressed, x-deleted, S-stale, d-dampened, h-history, *-valid, >-best
Path type: i-internal, e-external, c-confed, l-local, a-aggregate, r-redist,
  I-injected
Origin codes: i - IGP, e - EGP, ? - incomplete, | - multipath, & - backup

   Network            Next Hop         Metric      LocPrf     Weight Path
*>e192.168.1.1/32     10.25.1.2                               0 65000 i
*>e192.168.2.2/32     10.25.1.2                               0 65000 i
*>e192.168.3.3/32     10.25.1.2                               0 65000 i
*>e192.168.4.4/32     10.25.1.2                               0 65000 i
```

在 NX-2 的入站方向上，可以通过命令 **show bgp event-history detail** 查看与入站前缀列表相匹配的前缀详细信息。根据匹配情况，前缀将被允许或拒绝。如果前缀列表中没有特定的前缀表项，那么 BGP 就会删除该表项，该表项也将无法成为 BGP 表的一部分。例 11-39 显示了 BGP 事件历史记录的详细信息，可以看出 BGP 前缀列表拒绝或丢弃了前缀 100.1.30.0/24，同时允许了前缀 100.1.5.0/24。

例 11-39 入站前缀的 BGP 事件历史记录

```
! Event-History output for incoming prefixes
NX-2# show bgp event-history detail
14:54:41.278141: (default) UPD: [IPv4 Unicast] 10.25.1.5 processing EOR update from
  peer
14:54:41.278138: (default) UPD: 10.25.1.5 parsed UPDATE message from peer, len 29 ,
  withdraw len 0, attr len 6, nlri len 0
14:54:41.278135: (default) UPD: Received UPDATE message from 10.25.1.5
14:54:41.278131: (default) UPD: [IPv4 Unicast] Dropping prefix 100.1.30.0/24 from
  peer 10.25.1.5, due to prefix policy rejected
14:54:41.278129: (default) UPD: [IPv4 Unicast] Prefix 100.1.30.0/24 from peer
  10.25.1.5 rejected by inbound policy
14:54:41.278126: (default) UPD: [IPv4 Unicast] 10.25.1.5 Inbound ip prefix-list
  Inbound, action deny
14:54:41.278124: (default) UPD: [IPv4 Unicast] Received prefix 100.1.30.0/24 from
  peer 10.25.1.5, origin 1, next hop 10.25.1.5, localpref
0, med 0
14:54:41.278119: (default) UPD: [IPv4 Unicast] Dropping prefix 100.1.29.0/24 from
  peer 10.25.1.5, due to prefix policy rejected
14:54:41.278116: (default) UPD: [IPv4 Unicast] Prefix 100.1.29.0/24 from peer
  10.25.1.5 rejected by inbound policy

! output omittied for brevity

14:54:41.277740: (default) BRIB: [IPv4 Unicast] (100.1.5.0/24 (10.25.1.5)):
  returning from bgp_brib_add, new_path: 0, change: 0, undelete:
 0, history: 0, force: 0, (pflags=0x28), reeval=0
14:54:41.277737: (default) BRIB: [IPv4 Unicast] 100.1.5.0/24 from 10.25.1.5 was
  already in BRIB with same attributes
14:54:41.277734: (default) BRIB: [IPv4 Unicast] (100.1.5.0/24 (10.25.1.5)): bgp_
  brib_add: handling nexthop
14:54:41.277731: (default) BRIB: [IPv4 Unicast] Path to 100.1.5.0/24 via 192.168.5.5
  already exists, dflags=0x8001a
14:54:41.277728: (default) BRIB: [IPv4 Unicast] Installing prefix 100.1.5.0/24
  (10.25.1.5) via 10.25.1.5 into BRIB with extcomm
14:54:41.277723: (default) UPD: [IPv4 Unicast] 10.25.1.5 Inbound ip prefix-list
```

```
   Inbound, action permit
14:54:41.277720: (default) UPD: [IPv4 Unicast] Received prefix 100.1.5.0/24 from
   peer 10.25.1.5, origin 1, next hop 10.25.1.5, localpref 0 , med 0
```

在出站方向上，可以通过命令 **show bgp event-history detail** 显示 BGP 表中被允许和拒绝的前缀信息（根据名为 *Outbound* 的出站前缀列表的匹配情况）。执行了路由过滤操作之后，这些前缀将随同相关属性一起宣告给 BGP 对等体，如例 11-40 所示。

例 11-40　出站前缀的 BGP 事件历史记录

```
NX-2# show bgp event-history detail
BGP event-history detail

17:53:22.110665: (default) UPD: [IPv4 Unicast] 10.25.1.5 192.168.44.0/24 path-id 1
   not sent to peer due to: outbound policy
17:53:22.110659: (default) UPD: [IPv4 Unicast] 10.25.1.5 Outbound ip prefix-list
   Outbound, action deny
17:53:22.110649: (default) UPD: [IPv4 Unicast] 10.25.1.5 Created UPD msg (len 54)
   with prefix 192.168.4.4/32 ( Installed in HW) path-id 1 for peer
17:53:22.110643: (default) UPD: 10.25.1.5 Sending nexthop address 10.25.1.2 length 4
17:53:22.110638: (default) UPD: 10.25.1.5 Sending attr code 14, length 14, Mp-reach
17:53:22.110631: (default) UPD: 10.25.1.5 Sending attr code 2, length 6, AS-Path:
   <65000 >
17:53:22.110624: (default) UPD: 10.25.1.5 Sending attr code 1, length 1, Origin: IGP
17:53:22.110614: (default) UPD: [IPv4 Unicast] 10.25.1.5 Outbound ip prefix-list
   Outbound, action permit
17:53:22.110605: (default) UPD: [IPv4 Unicast] consider sending 192.168.4.4/32 to
   peer 10.25.1.5, path-id 1, best-ext is off
```

NX-OS 还支持通过 CLI 来验证前缀列表基于策略的统计信息，统计信息可以显示入站和出站方向隐含的策略并显示各个方向允许和拒绝的前缀数。可以通过命令 **show bgp** *afi safi* **policy statistics neighbor** *ip-address* **prefix-list [in | out]** 查看应用于 BGP 邻居的前缀列表的策略统计信息。策略统计命令的计数器会随着 BGP 邻居的翻动或软清除而不断递增。例 11-41 在 BGP 对等体 10.25.1.5 的入站和出站方向使用了策略统计命令，希望了解入站和出站方向所允许和丢弃的前缀数。本例在出站方向执行了软清除操作，可以看出，出站前缀列表策略统计信息的计数器对于允许的前缀来说增加了 4，对于丢弃的前缀来说增加了 1。

例 11-41　前缀列表的 BGP 策略统计

```
NX-2# show bgp ipv4 unicast policy statistics neighbor 10.25.1.5 prefix-list in
Total count for neighbor rpm handles: 1

C: No. of comparisons, M: No. of matches

ip prefix-list Inbound seq 5 permit 100.1.1.0/24            M: 3
ip prefix-list Inbound seq 10 permit 100.1.2.0/24           M: 3
ip prefix-list Inbound seq 15 permit 100.1.3.0/24           M: 3
ip prefix-list Inbound seq 20 permit 100.1.4.0/24           M: 3
ip prefix-list Inbound seq 25 permit 100.1.5.0/24           M: 3
Total accept count for policy: 15
Total reject count for policy: 81

NX-2# show bgp ipv4 unicast policy statistics neighbor 10.25.1.5 prefix-list out
Total count for neighbor rpm handles: 1

C: No. of comparisions, M: No. of matches
```

```
ip prefix-list Outbound seq 5 permit 192.168.0.0/16 eq 32    M: 17

Total accept count for policy: 17
Total reject count for policy: 3
! Perform soft clear out on neighbor 10.25.1.5
NX-2# clear bgp ipv4 unicast 10.25.1.5 soft out
NX-2# show bgp ipv4 unicast policy statistics neighbor 10.25.1.5 prefix-list out
Total count for neighbor rpm handles: 1

C: No. of comparisons, M: No. of matches

ip prefix-list Outbound seq 5 permit 192.168.0.0/16 eq 32    M: 21

Total accept count for policy: 21
Total reject count for policy: 4
```

如果发现BGP邻居的入站或出站前缀列表出现了问题，那么就可以验证NX-OS的RPM（Route Policy Manager，路由策略管理器）组件。验证操作的第一步是验证该前缀列表是否关联了BGP进程，验证命令是 **show system internal rpm ip-prefix-list**，该命令可以显示前缀列表的名称及其客户端信息，同时还能显示前缀列表中的表项数。

验证了前缀列表及其客户端之后，需要通过命令 **show system internal rpm event-history rsw** 验证前缀列表是否已正确绑定了BGP进程。如果绑定错误或绑定事件历史记录日志缺失，那么就表明前缀列表没有与BGP进程或BGP邻居进行正确关联。

例11-42给出了这两条命令的输出结果示例。

例11-42 前缀列表的RPM客户端信息以及事件历史记录日志

```
NX-2# show system internal rpm ip-prefix-list
Policy name: Inbound              Type: ip prefix-list
Version: 6                        State: Ready
Ref. count: 1                     PBR refcount: 0
Stmt count: 5                     Last stmt seq: 25
Set nhop cmd count: 0             Set vrf cmd count: 0
Set intf cmd count: 0             Flags: 0x00000003
PPF nodeid: 0x00000000            Config refcount: 0
PBR Stats: No
Clients:
    bgp-65000 (Route filtering/redistribution)    ACN version: 0

Policy name: Outbound             Type: ip prefix-list
Version: 2                        State: Ready
Ref. count: 1                     PBR refcount: 0
Stmt count: 1                     Last stmt seq: 5
Set nhop cmd count: 0             Set vrf cmd count: 0
Set intf cmd count: 0             Flags: 0x00000003
PPF nodeid: 0x00000000            Config refcount: 0
PBR Stats: No
Clients:
    bgp-65000 (Route filtering/redistribution) ACN version: 0
NX-2# show system internal rpm event-history rsw

Routing software interaction logs of RPM
1) Event:E_DEBUG, length:81, at 104214 usecs after Sun Sep 17 06:01:47 2017
    [120] [5736]: Bind ack sent - client bgp-65000 uuid 0x0000011b for policy
```

```
Outbound
2) Event:E_DEBUG, length:76, at 104179 usecs after Sun Sep 17 06:01:47 2017
   [120] [5736]: Bind request - client bgp-65000 uuid 0x0000011b policy Outbound
3) Event:E_DEBUG, length:80, at 169619 usecs after Sun Sep 17 06:01:42 2017
   [120] [5736]: Bind ack sent - client bgp-65000 uuid 0x0000011b for policy
Inbound
4) Event:E_DEBUG, length:75, at 169469 usecs after Sun Sep 17 06:01:42 2017
   [120] [5736]: Bind request - client bgp-65000 uuid 0x0000011b policy Inbound
```

11.6.2 过滤器列表

BGP 过滤器列表可以根据 AS-Path 列表对前缀进行过滤，可以将 BGP 过滤器列表用于入站和出站方向。可以在邻居地址簇配置模式下，使用命令 **filter-list** *as-path-list- name* [**in** | **out**]配置 BGP 过滤器列表。例 11-43 给出了图 11-6 拓扑结构中的 NX-2 交换机的过滤器列表配置示例，可以看出，入站过滤器列表被配置为允许 AS_PATH 列表中有 AS 274 的前缀。本例的第二部分输出结果表明该过滤器列表的应用方向是入站方向。

例 11-43　BGP 过滤器列表

```
NX-2(config)# ip as-path access-list ALLOW_274 permit 274
NX-2(config)# router bgp 65000
NX-2(config-router)# neighbor 10.25.1.5
NX-2(config-router-neighbor)# address-family ipv4 unicast
NX-2(config-router-neighbor-af)# filter-list ALLOW_274 in
NX-2(config-router-neighbor-af)# end
NX-2# show bgp ipv4 unicast neighbors 10.25.1.5
BGP neighbor is 10.25.1.5, remote AS 65001, ebgp link, Peer index 2
  BGP version 4, remote router ID 192.168.5.5
  BGP state = Established, up for 2d00h
! output omitted for brevity
  For address family: IPv4 Unicast
  BGP table version 1085, neighbor version 1085
  5 accepted paths consume 400 bytes of memory
  4 sent paths
  Inbound as-path-list configured is ALLOW_274, handle obtained
  Outbound ip prefix-list configured is Outbound, handle obtained
  Last End-of-RIB received 1d23h after session start

  Local host: 10.25.1.2, Local port: 58236
  Foreign host: 10.25.1.5, Foreign port: 179
  fd = 74
```

注：有关 AS-Path 访问列表的详细信息将在本章后面进行讨论。

执行了过滤器列表操作之后，例 11-44 显示了 BGP 表中从对等体 10.25.1.5 收到的前缀信息。可以看出，BGP 表中显示的所有前缀的 AS_PATH 列表都有 AS 274。

例 11-44　应用了过滤器列表之后的 BGP 表

```
! Output after configuring filter-list
NX-2# show bgp ipv4 unicast neighbor 10.25.1.5 routes
! Output omitted for brevity
   Network          Next Hop         Metric  LocPrf  Weight Path
*>e100.1.21.0/24    10.25.1.5                         0 65001 100 228 274 {300
   243}
*>e100.1.22.0/24    10.25.1.5                         0 65001 100 228 274 {300
```

```
                           243}
*>e100.1.23.0/24           10.25.1.5                    0 65001 100 228 274 {300
                           243}
*>e100.1.24.0/24           10.25.1.5                    0 65001 100 228 274 {300
                           243}
*>e100.1.25.0/24           10.25.1.5                    0 65001 100 228 274 {300
                           243}
*>e100.1.26.0/24           10.25.1.5                    0 65001 100 228 274 {300
                           243}
*>e100.1.27.0/24           10.25.1.5                    0 65001 100 228 274 {300
                           243}
*>e100.1.28.0/24           10.25.1.5                    0 65001 100 228 274 {300
                           243}
*>e100.1.29.0/24           10.25.1.5                    0 65001 100 228 274 {300
                           243}
*>e100.1.30.0/24           10.25.1.5                    0 65001 100 228 274 {300
                           243}
```

> **注：** 如果 BGP 对等体配置了 **soft-reconfiguration inbound** 命令，那么还可以通过命令 **show bgp** *afi safi* **neighbor** *ip-address* **received-routes** 查看接收到的 BGP 前缀。

验证哪些前缀被允许或被拒绝的最简单方式就是使用命令 **show bgp event-history detail**，但是如果没有在 **router bgp** 配置下启用命令 **event-history detail**，那么就需要通过调试操作来验证更新信息，可以通过命令 **debug bgp updates** 验证入站和出站更新情况。例 11-45 通过命令 **debug bgp updates** 验证了被允许或被拒绝的前缀信息，允许或拒绝操作始终基于 AS-path 列表中的表项信息。

例 11-45 debug bgp updates 命令输出结果

```
NX-2# debug logfile bgp
NX-2# debug bgp updates
NX-2# clear bgp ipv4 unicast 10.25.1.5 soft in
NX-2# show debug logfile bgp
21:39:01.721587 bgp: 65000 [10743] (default) UPD: [IPv4 Unicast] 10.25.1.5 Inbound
 as-path-list ALLOW_274, action deny
21:39:01.721622 bgp: 65000 [10743] (default) UPD: [IPv4 Unicast] Received prefix
 100.1.1.0/24 from peer 10.25.1.5, origin 1, next hop 10.25.1.5, localpref 0, med 0
21:39:01.721649 bgp: 65000 [10743] (default) UPD: [IPv4 Unicast] Dropping prefix
 100.1.1.0/24 from peer 10.25.1.5, due to attribute policy rejected
21:39:01.721678 bgp: 65000 [10743] (default) UPD: [IPv4 Unicast] Received prefix
 100.1.2.0/24 from peer 10.25.1.5, origin 1, next hop 10.25.1.5, localpref 0, med 0
21:39:01.721702 bgp: 65000 [10743] (default) UPD: [IPv4 Unicast] Dropping prefix
 100.1.2.0/24 from peer 10.25.1.5, due to attribute policy rejected
! Output omittied for brevity
21:39:01.723538 bgp: 65000 [10743] (default) UPD: [IPv4 Unicast] 10.25.1.5 Inbound
 as-path-list ALLOW_274, action permit
21:39:01.723592 bgp: 65000 [10743] (default) UPD: [IPv4 Unicast] Received prefix
 100.1.21.0/24 from peer 10.25.1.5, origin 1, next hop 10.25.1.5, localpref 0, med 0
21:39:01.723687 bgp: 65000 [10743] (default) UPD: [IPv4 Unicast] Received prefix
 100.1.22.0/24 from peer 10.25.1.5, origin 1, next hop 10.25.1.5, localpref 0, med 0
```

与前缀列表的策略统计信息类似，这些统计信息也可用于过滤器列表表项。执行了 **show bgp** *afi safi* **policy statistics neighbor** *ip-address* **filter-list [in | out]** 命令之后，可以看出 AS-path 访问列表被引用为过滤器列表的一部分，同时还能显示每个表项的匹配数。此外，该输出结果还可以显示过滤器列表所允许和拒绝的前缀数，如例 11-46 所示。

例 11-46　BGP 过滤器列表

```
NX-2# show bgp ipv4 unicast policy statistics neighbor 10.25.1.5 filter-list in
Total count for neighbor rpm handles: 1

C: No. of comparisons, M: No. of matches

ip as-path access-list ALLOW_274 permit "274"          C: 5       M: 1

Total accept count for policy: 1
Total reject count for policy: 4
```

由于过滤器列表使用了 AS-Path 访问列表，因而可以通过命令 **show system internal rpm as-path-access-list** *as-path-acl-name* 验证 RPM 信息来了解 AS-Path 访问列表，该命令可以确认 AS-Path 访问列表是否与 BGP 进程进行了关联。可以通过命令 **show system internal rpm event-history rsw** 验证 AS-Path 访问列表是否绑定了 BGP 进程。例 11-47 给出了这两条命令的输出结果。

例 11-47　BGP 过滤器列表

```
NX-2# show system internal rpm as-path-access-list ALLOW_274
Policy name: ALLOW_274              Type: as-path-list
Version: 2                          State: Ready
Ref. count: 1                       PBR refcount: 0
Stmt count: 1                       Last stmt seq: 1
Set nhop cmd count: 0               Set vrf cmd count: 0
Set intf cmd count: 0               Flags: 0x00000003
PPF nodeid: 0x00000000              Config refcount: 0
PBR Stats: No
Clients:
    bgp-65000 (Route filtering/redistribution)    ACN version: 0

! RPM Event-History
NX-2# show system internal rpm event-history rsw

Routing software interaction logs of RPM
1) Event:E_DEBUG, length:82, at 684846 usecs after Sun Sep 17 19:46:46 2017
    [120] [5736]: Bind ack sent - client bgp-65000 uuid 0x0000011b for policy
 ALLOW_274
2) Event:E_DEBUG, length:77, at 684797 usecs after Sun Sep 17 19:46:46 2017
    [120] [5736]: Bind request - client bgp-65000 uuid 0x0000011b policy ALLOW_274
```

11.7　BGP 路由映射

BGP 通过路由映射（Route-Map）来提供路由过滤和流量工程能力，方法是为前缀设置特定属性，从而有效控制入站和出站流量。路由映射通常通过某种形式的条件匹配来拒绝或允许某些前缀。最简单的情况是，路由映射可以像 AS-Path 过滤器/前缀列表那样过滤网络，不过，路由映射还能通过添加或修改网络属性来提供更多的额外功能。路由映射可以被特定的路由宣告或被 BGP 邻居引用，需要指定路由宣告的方向（入站/出站方向）。路由映射是 BGP 的一个关键组件，能够以逐个邻居为基础应用唯一的路由策略。

例 11-48 为 BGP 邻居应用了多序列路由映射。可以看出，本例中的路由映射序列 10 被配置为匹配前缀列表，以匹配指定前缀集。序列 20 被配置为匹配 AS-Path 访问列表。请注意，本例没有配置序列 30。对于路由映射来说，缺少的任何表项都将充当隐式拒绝语句并拒绝所有前缀。

例 11-48　BGP 路由映射配置

```
NX-2(config)# route-map Inbound-RM permit 10
NX-2(config-route-map)# match ip address prefix-list Inbound
NX-2(config-route-map)# set local-preference 200
NX-2(config-route-map)# exit
NX-2(config)# route-map Inbound-RM permit 20
NX-2(config-route-map)# match as-path AlLOW_274
NX-2(config-route-map)# set local-preference 300
NX-2(config-route-map)# exit
! The above referenced Prefix-list and AS-Path Access-list were shown in previous
! examples
NX-2(config)# router bgp 65000
NX-2(config-router)# neighbor 10.25.1.5
NX-2(config-router-neighbor)# address-family ipv4 unicast
NX-2(config-router-neighbor-af)# route-map Inbound-RM in
NX-2(config-router-neighbor-af)# end
```

执行了入站路由映射过滤操作之后的 BGP 表信息如例 11-49 所示，可以看出，前缀 100.1.1.0/24 ~ 100.1.5.0/24 的本地优先级被设置为 200，而与 AS-Path 列表中 AS 274 相匹配的前缀的本地优先级则被设置为 300。由于没有匹配序列 30 的路由映射表项，因而其他前缀均被入站路由映射过滤机制拒绝。

例 11-49　执行了路由映射过滤操作之后的 BGP 表

```
NX-2# show bgp ipv4 unicast neighbor 10.25.1.5 routes
BGP routing table information for VRF default, address family IPv4 Unicast
BGP table version is 1141, local router ID is 192.168.2.2
Status: s-suppressed, x-deleted, S-stale, d-dampened, h-history, *-valid, >-best
Path type: i-internal, e-external, c-confed, l-local, a-aggregate, r-redist, I-injected
Origin codes: i - IGP, e - EGP, ? - incomplete, | - multipath, & - backup

   Network            Next Hop          Metric     LocPrf    Weight Path
*>e100.1.1.0/24       10.25.1.5                    200           0 65001 100 {220} e
*>e100.1.2.0/24       10.25.1.5                    200           0 65001 100 {220} e
*>e100.1.3.0/24       10.25.1.5                    200           0 65001 100 {220} e
*>e100.1.4.0/24       10.25.1.5                    200           0 65001 100 {220} e
*>e100.1.5.0/24       10.25.1.5                    200           0 65001 100 {220} e
*>e100.1.21.0/24      10.25.1.5                    300           0 65001 100 228 274 {300 243}
*>e100.1.25.0/24      10.25.1.5                    300           0 65001 100 228 274 {300 243}
*>e100.1.26.0/24      10.25.1.5                    300           0 65001 100 228 274 {300 243}
*>e100.1.27.0/24      10.25.1.5                    300           0 65001 100 228 274 {300 243}
*>e100.1.28.0/24      10.25.1.5                    300           0 65001 100 228 274 {300 243}
*>e100.1.29.0/24      10.25.1.5                    300           0 65001 100 228 274 {300 243}
*>e100.1.30.0/24      10.25.1.5                    300           0 65001 100 228 274 {300 243}
```

此时可以再次使用命令 **show bgp event-history detail** 来验证路由映射策略所允许或拒绝的前缀信息。路由映射将根据 **match** 语句，执行相应的 **set** 操作（如果有）。例 11-50 显示了事件历史记录的详细输出结果，可以看出路由映射所允许和拒绝的前缀信息。

例 11-50　BGP 事件历史记录

```
NX-2# show bgp event-history detail
04:36:32.954809: (default) BRIB: [IPv4 Unicast] Installing prefix 100.1.21.0/24
  (10.25.1.5) via 10.25.1.5 into BRIB with extcomm
04:36:32.954796: (default) UPD: [IPv4 Unicast] 10.25.1.5 Inbound route-map Inbound-
  RM, action permit
04:36:32.954763: (default) UPD: [IPv4 Unicast] Received prefix 100.1.21.0/24 from
```

```
  peer 10.25.1.5, origin 1, next hop 10.25.1.5, localpref 0, med 0
! Output omitted for brevity
04:36:32.954690: (default) UPD: [IPv4 Unicast] Dropping prefix 100.1.20.0/24 from
  peer 10.25.1.5, due to prefix policy rejected
04:36:32.954684: (default) UPD: [IPv4 Unicast] Prefix 100.1.20.0/24 from peer
  10.25.1.5 rejected by inbound policy
04:36:32.954679: (default) UPD: [IPv4 Unicast] 10.25.1.5 Inbound route-map Inbound-
  RM, action deny
04:36:32.954647: (default) UPD: [IPv4 Unicast] Received prefix 100.1.20.0/24 from
  peer 10.25.1.5, origin 1, next hop 10.25.1.5, localpref 0, med 0
```

此外，还可以像前缀列表和过滤器列表那样，验证路由映射的策略统计信息。可以通过命令 **show bgp ipv4 unicast policy statistics neighbor** *ip-address* **route-map [in | out]** 显示每条路由映射序列所匹配的前缀列表、路径访问列表或其他属性，以及相应的匹配统计信息，如例 11-51 所示。

例 11-51 路由映射的 BGP 策略统计信息

```
NX-2# show bgp ipv4 unicast policy statistics neighbor 10.25.1.5 route-map in
Total count for neighbor rpm handles: 1

C: No. of comparisons, M: No. of matches

route-map Inbound-RM permit 10
  match ip address prefix-list Inbound                    C: 52        M: 5
route-map Inbound-RM permit 20
  match as-path ALLOW_274                                 C: 47        M: 30


Total accept count for policy: 35
Total reject count for policy: 17
```

路由映射需要用到各种条件匹配功能特性，如前缀列表、正则表达式（RegEx）、AS-Path 访问列表、BGP 团体以及团体列表。如果在同一个邻居下配置了多种过滤机制，那么入站和出站过滤的优先顺序如下。

- 入站过滤。
  - 路由映射。
  - 过滤器列表。
  - 前缀列表、分发列表。
- 出站过滤。
  - 过滤器列表。
  - 路由映射。
  - 宣告映射（条件式宣告）。
  - 前缀列表、分发列表。

虽然可以通过各种路由协议属性实现网络前缀的条件式匹配，不过接下来将主要讨论常用的条件匹配前缀技术。

11.7.1 正则表达式

很多时候，条件匹配网络前缀显得过于复杂，最好能标识来自特定组织机构的所有路由。如此一来，就可以通过 BGP AS-Path 来选择路径。

为了解析大量可用 ASN（4294967295），可以使用正则表达式（RegEx）。正则表达式可以根据查询修饰符选择相应的内容。可以在 Nexus 交换机上通过命令 **show bgp** *afi safi* **regexp** "regex-pattern"以正则表达式来解析BGP表。

注：NX-OS设备要求将regex-pattern放在双引号（""）内。

表11-7列出了常见正则表达式的查询修饰符信息。

表11-7　　　　　　　　　　　　RegEx查询修饰符

修饰符	描述
_（下画线）	匹配空格
^（脱字符）	表示字符串的开始
$（美元符号）	表示字符串的结束
[]（方括号）	匹配指定范围内的单个字符或嵌套
-（连字符）	表示圆括号中的数字范围
[^]（方括号中的脱字符）	不包含方括号中的字符
()（圆括号）	用于嵌套搜索模式
|（管道符）	为查询操作提供or功能
.（句点）	匹配单个字符（包括空格）
*（星号）	匹配零个或多个字符或模式
+（加号）	一个或多个字符或模式实例
?（问号）	匹配一个或多个字符或模式实例

注：.^$*+()[]?等字符是特殊控制字符，必须结合反斜杠转义字符(\)才能使用。例如，如果要匹配输出结果中的*，那么就要使用语法*。

为了更好地理解这些RegEx修饰符，接下来将给出不同的示例任务。例11-52给出的参考BGP表显示了每个RegEx查询修饰符的场景，目的是查询通过图11-10学到的前缀。

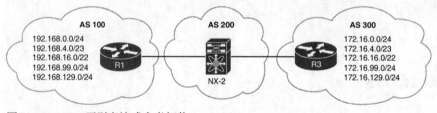

图11-10　BGP正则表达式参考拓扑

例11-52　用于RegEx查询的BGP表

```
NX-2# show bgp ipv4 unicast
! Output omitted for brevity
   Network          Next Hop      Metric LocPrf Weight Path
*>e172.16.0.0/24    172.32.23.3       0             0 300 80 90 21003 2100 i
*>e172.16.4.0/23    172.32.23.3       0             0 300 878 1190 1100 1010 i
*>e172.16.16.0/22   172.32.23.3       0             0 300 779 21234 45 i
*>e172.16.99.0/24   172.32.23.3       0             0 300 145 40 i
*>e172.16.129.0/24  172.32.23.3       0             0 300 10010 300 1010 40 50 i
*>e192.168.0.0/16   172.16.12.1       0             0 100 80 90 21003 2100 i
*>e192.168.4.0/23   172.16.12.1       0             0 100 878 1190 1100 1010 i
*>e192.168.16.0/22  172.16.12.1       0             0 100 779 21234 45 i
*>e192.168.99.0/24  172.16.12.1       0             0 100 145 40 i
*>e192.168.129.0/24 172.16.12.1       0             0 100 10010 300 1010 40 50 i
```

> 注：出于特殊目的，前缀 172.16.129.0/24 的 AS-Path 中出现了两个不连续的 AS 300。实际应用中不会出现这种情况，因为这种情况表示存在路由环路。

1. 下画线_

查询修饰符功能：匹配空格。

场景：仅显示穿越 AS 100 的 AS。第一个假设是例 11-53 显示的语法 **show bgp ipv4 unicast regex "100"** 是理想的，该 RegEx 查询包括以下非期望 ASN：1100、2100、21003 和 10010。

例 11-53 AS 100 的 BGP RegEx 查询

```
NX-2# show bgp ipv4 unicast regex "100"
! Output omitted for brevity
   Network          Next Hop        Metric LocPrf Weight Path
*>e172.16.0.0/24    172.32.23.3      0             0 300 80 90 21003 2100 i
*>e172.16.4.0/23    172.32.23.3      0             0 300 878 1190 1100 1010 i
*>e172.16.129.0/24  172.32.23.3      0             0 300 10010 300 1010 40 50 i
*>e192.168.0.0/16   172.16.12.1      0             0 100 80 90 21003 2100 i
*>e192.168.4.0/23   172.16.12.1      0             0 100 878 1190 1100 1010 i
*>e192.168.16.0/22  172.16.12.1      0             0 100 779 21234 45 i
*>e192.168.99.0/24  172.16.12.1      0             0 100 145 40 i
*>e192.168.129.0/24 172.16.12.1      0             0 100 10010 300 1010 40 50 i
```

例 11-54 使用下画线（_）来表示 100 左边的空格，作用是去掉不需要的 ASN，该 RegEx 查询包含以下非期望 ASN：10010。

例 11-54 AS_100 的 BGP RegEx 查询

```
NX-2# show bgp ipv4 unicast regexp "_100"
! Output omitted for brevity
   Network          Next Hop        Metric LocPrf Weight Path
   Network          Next Hop        Metric LocPrf Weight Path
*>e172.16.129.0/24  172.32.23.3      0             0 300 10010 300 1010 40 50 i
*>e192.168.0.0/16   172.16.12.1      0             0 100 80 90 21003 2100 i
*>e192.168.4.0/23   172.16.12.1      0             0 100 878 1190 1100 1010 i
*>e192.168.16.0/22  172.16.12.1      0             0 100 779 21234 45 i
*>e192.168.99.0/24  172.16.12.1      0             0 100 145 40 i
*>e192.168.129.0/24 172.16.12.1      0             0 100 10010 300 1010 40 i
```

例 11-55 给出了最终查询方式，在 ASN（100）的前面和后面均使用下画线（_）来完成对穿越 AS 100 的路由的查询操作。

例 11-55 AS _100_ 的 BGP RegEx 查询

```
NX-2# show bgp ipv4 unicast regexp "_100_"
! Output omitted for brevity
   Network          Next Hop        Metric LocPrf Weight Path
*>e192.168.0.0/16   172.16.12.1      0             0 100 80 90 21003 2100 i
*>e192.168.4.0/23   172.16.12.1      0             0 100 878 1190 1100 1010 i
*>e192.168.16.0/22  172.16.12.1      0             0 100 779 21234 45 i
*>e192.168.99.0/24  172.16.12.1      0             0 100 145 40 i
*>e192.168.129.0/24 172.16.12.1      0             0 100 10010 300 1010 40 50 i
```

2. 脱字符 ^

查询修饰符功能：表示字符串的开始。

场景：仅显示从 AS 300 宣告的路由。乍看起来，命令 **show bgp ipv4 unicast regex "_300_"** 似乎可用，不过例 11-56 还包含了路由 192.168.129.0/24。

例 11-56 AS 300 的 BGP RegEx 查询

```
NX-2# show bgp ipv4 unicast regexp "_300_"
! Output omitted for brevity
   Network          Next Hop        Metric  LocPrf  Weight  Path
*>e172.16.0.0/24    172.32.23.3     0               0       300 80 90 21003 2100 i
*>e172.16.4.0/23    172.32.23.3     0               0       300 878 1190 1100 1010 i
*>e172.16.16.0/22   172.32.23.3     0               0       300 779 21234 45 i
*>e172.16.99.0/24   172.32.23.3     0               0       300 145 40 i
*>e172.16.129.0/24  172.32.23.3     0               0       300 10010 300 1010 40 50 i
*>e192.168.129.0/24 172.16.12.1     0               100     10010 300 1010 40 50 i
```

由于 AS 300 是直连 AS，因而更有效的方式是确保 AS 300 是列表中的第一个 ASN。例 11-57 显示了 RegEx 模式中的脱字符（^）。

例 11-57 使用脱字符的 BGP RegEx 查询

```
NX-2# show bgp ipv4 unicast regexp "^300_"
! Output omitted for brevity
   Network          Next Hop        Metric  LocPrf  Weight  Path
*>e172.16.0.0/24    172.32.23.3     0               0       300 80 90 21003 2100 i
*>e172.16.4.0/23    172.32.23.3     0               0       300 878 1190 1100 1010 i
*>e172.16.16.0/22   172.32.23.3     0               0       300 779 21234 45 i
*>e172.16.99.0/24   172.32.23.3     0               0       300 145 40 i
*>e172.16.129.0/24  172.32.23.3     0               0       300 10010 300 1010 40 50 i
```

3. 美元符号 $

查询修饰符功能：表示字符串的结束。

场景：仅显示源自 AS 40 的路由。例 11-58 使用了 RegEx 模式"_40_"，但不幸的是，仍然包含了源自 AS 50 的路由。

例 11-58 使用 AS 40 的 BGP RegEx 查询

```
NX-2# show bgp ipv4 unicast regexp "_40_"
! Output omitted for brevity
   Network          Next Hop        Metric  LocPrf  Weight  Path
*>e172.16.99.0/24   172.32.23.3     0               0       300 145 40 i
*>e172.16.129.0/24  172.32.23.3     0               0       300 10010 300 1010 40 50 i
*>e192.168.99.0     172.16.12.1     0               100     145 40 i
*>e192.168.129.0    172.16.12.1     0               100     10010 300 1010 40 50 i
```

例 11-59 提供了使用美元符号（$）的解决方案，即 RegEx 模式"_40$"。

例 11-59 使用美元符号的 BGP RegEx 查询

```
NX-2# show bgp ipv4 unicast regexp "_40$"
! Output omitted for brevity
   Network          Next Hop        Metric  LocPrf  Weight  Path
*>e172.16.99.0/24   172.32.23.3     0               0       300 145 40 i
*>e192.168.99.0     172.16.12.1     0       100     100     145 40 i
```

4. 方括号 [ ]

查询修饰符功能：匹配指定范围内的单个字符或嵌套。

场景：仅显示 ASN 内包含 11 或 14 的路由。例 11-60 使用了 RegEx 过滤器"1[14]"。

例 11-60 使用方括号的 BGP RegEx 查询

```
NX-2# show bgp ipv4 unicast regexp "1[14]"
! Output omitted for brevity
```

```
     Network          Next Hop         Metric LocPrf Weight Path
*>e172.16.4.0/23      172.32.23.3      0             0 300 878 1190 1100 1010 i
*>e172.16.99.0/24     172.32.23.3      0             0 300 145 40 i
*>e192.168.4.0/23     172.16.12.1      0             0 100 878 1190 1100 1010 i
*>e192.168.99.0       172.16.12.1      0             0 100 145 40 i
```

5. 连字符-

查询修饰符功能：表示圆括号中的数字范围。

场景：仅显示 AS 号的最后两位数字是 40、50、60、70 或 80 的路由。例 11-61 使用了 RegEx 查询 "[4-8]0_"。

例 11-61 使用连字符的 BGP RegEx 查询

```
NX-2# show bgp ipv4 unicast regexp "[4-8]0_"
! Output omitted for brevity
     Network          Next Hop         Metric LocPrf Weight Path
*>e172.16.0.0/24      172.32.23.3      0             0 300 80 90 21003 2100 i
*>e172.16.99.0/24     172.32.23.3      0             0 300 145 40 i
*>e172.16.129.0/24    172.32.23.3      0             0 300 10010 300 1010 40 50 i
*>e192.168.0.0        172.16.12.1      0             0 100 80 90 21003 2100 i
*>e192.168.99.0       172.16.12.1      0             0 100 145 40 i
*>e192.168.129.0      172.16.12.1      0             0 100 10010 300 1010 40 50 i
```

6. 方括号中的脱字符[^]

查询修饰符功能：不包含方括号中的字符。

场景：仅显示第二个 AS 号是 AS 100 或 AS 300 且后面不以 3、4、5、6、7 或 8 开头的路由。该 RegEx 查询的第一个组件是 RegEx 查询 "^[13]00_"，负责将 AS 限制为 AS 100 或 300；第二个组件是 RegEx 过滤器 "_[^3-8]"，负责将 3~8 开头的 AS 号过滤掉，完整的 RegEx 查询是 "^[13]00_[^3-8]"，如例 11-62 所示。

例 11-62 使用方括号中的脱字符的 BGP RegEx 查询

```
NX-2# show bgp ipv4 unicast regexp "^[13]00_[^3-8]"
! Output omitted for brevity
     Network          Next Hop         Metric LocPrf Weight Path
*>e172.16.99.0/24     172.32.23.3      0             0 300 145 40 i
*>e172.16.129.0/24    172.32.23.3      0             0 300 10010 300 1010 40 50 i
*>e192.168.99.0       172.16.12.1      0             0 100 145 40 i
*>e192.168.129.0      172.16.12.1      0             0 100 10010 300 1010 40 50 i
```

7. 圆括号()和管道符|

查询修饰符功能：搜索模式嵌套并提供 or 功能。

场景：仅显示 AS_PATH 以 AS 40 或 45 结尾的路由，例 11-63 给出了 RegEx 过滤器 "_4(5|0)$" 示例。

例 11-63 使用圆括号的 BGP RegEx 查询

```
NX-2# show bgp ipv4 unicast regexp "_4(5|0)$"
! Output omitted for brevity
     Network          Next Hop         Metric LocPrf Weight Path
*>e172.16.16.0/22     172.32.23.3      0             0 300 779 21234 45 i
*>e172.16.99.0/24     172.32.23.3      0             0 300 145 40 i
*>e192.168.16.0/22    172.16.12.1      0             0 100 779 21234 45 i
*>e192.168.99.0       172.16.12.1      0             0 100 145 40 i
```

8. 句点.

查询修饰符功能：匹配单个字符（包括空格）。

场景：仅显示源自 AS 1–99 的路由。例 11-64 的 RegEx 查询"..$"需要一个空格，其后是任意字符（包括其他空格）。

例 11-64 使用句点的 BGP RegEx 查询

```
NX-2# show bgp ipv4 unicast regexp "_..$"
! Output omitted for brevity
   Network              Next Hop         Metric LocPrf Weight Path
*>e172.16.16.0/22       172.32.23.3      0                 0 300 779 21234 45 i
*>e172.16.99.0/24       172.32.23.3      0                 0 300 145 40 i
*>e172.16.129.0/24      172.32.23.3      0                 0 300 10010 300 1010 40 50 i
*>e192.168.16.0/22      172.16.12.1      0      100        0 100 779 21234 45 i
*>e192.168.99.0         172.16.12.1      0      100        0 100 145 40 i
*>e192.168.129.0        172.16.12.1      0      100        0 100 10010 300 1010 40 50 i
```

9. 加号+

查询修饰符功能：一个或多个字符或模式实例。

场景：仅显示 AS-Path 中至少包含一个或多个"11"的路由。例 11-65 给出了 RegEx 模式"(11)+"示例。

例 11-65 使用加号的 BGP RegEx 查询

```
NX-2# show bgp ipv4 unicast regexp "(10)+[^(100)]"
! Output omitted for brevity
   Network              Next Hop         Metric LocPrf Weight Path
*>e172.16.4.0/23        172.32.23.3      0                 0 300 878 1190 1100 1010 i
*>e192.168.4.0/23       172.16.12.1      0                 0 100 878 1190 1100 1010 i
```

10. 问号?

查询修饰符功能：匹配一个或多个字符或模式实例。

场景：仅显示来自相邻 AS 或直连 AS 的路由（也就是说，限制在两个 AS 之外）。例 11-66 给出的查询示例更加复杂，首先需要定义一个初始查询来标识 AS，即"[0-9]+"；第二个查询组件包括空格和可选的第二个 AS。"？"的作用是限制 AS 匹配一个或两个 AS。

注：输入"？"之前必须使用 Ctrl+V 组合键转义序列。

例 11-66 使用问号的 BGP RegEx 查询

```
NX-2# show bgp ipv4 unicast regexp "^[0-9]+ ([0-9]+)?$"
! Output omitted for brevity
   Network              Next Hop         Metric LocPrf Weight Path
*>e172.16.99.0/24       172.32.23.3      0                 0 300 40 i
*>e192.168.99.0         172.16.12.1      0      100        0 100 40 i
```

11. 星号*

查询修饰符功能：匹配零个或多个字符或模式。

场景：显示来自任意 AS 的所有路由。该操作看起来似乎没什么用，但是使用 AS-Path 访问列表时可能就是一个有效要求，具体将在本章后面进行解释，如例 11-67 所示。

例 11-67 使用星号的 BGP RegEx 查询

```
NX-2# show bgp ipv4 unicast regexp ".*"
! Output omitted for brevity
```

```
   Network          Next Hop        Metric LocPrf Weight Path
*>e172.16.0.0/24    172.32.23.3        0           0    300 80 90 21003 2100 i
*>e172.16.4.0/23    172.32.23.3        0           0    300 1080 1090 1100 1110 i
*>e172.16.16.0/22   172.32.23.3        0           0    300 11234 21234 31234 i
*>e172.16.99.0/24   172.32.23.3        0           0    300 40 i
*> 172.16.129.0/24  172.32.23.3        0           0    300 10010 300 30010 30050 i
*>e192.168.0.0      172.16.12.1        0     100   0    100 80 90 21003 2100 i
*>e192.168.4.0/23   172.16.12.1        0     100   0    100 1080 1090 1100 1110 i
*>e192.168.16.0/22  172.16.12.1        0     100   0    100 11234 21234 31234 i
*>e192.168.99.0     172.16.12.1        0     100   0    100 40 i
*>e192.168.129.0    172.16.12.1        0     100   0    100 10010 300 30010 30050 i
```

11.7.2 AS-Path 访问列表

如果要在路由映射中使用 AS-Path 来选择路由，那么就需要定义 AS-Path 访问列表（AS-Path ACL）。按照自上而下的顺序进行处理，找到第一个匹配项之后就执行相应的 **permit** 或 **deny** 操作。AS-Path ACL 的末尾都包含一条隐式拒绝语句。IOS 最多支持 500 条 AS-Path ACL，可以通过命令 **ip as-path access-list** *acl-number* {**deny** | **permit**} *regex-query* 创建 AS-Path 访问列表。

例 11-68 给出了两个 AS-Path 访问列表示例。AS-Path access-list 1 匹配所有本地 IBGP 前缀，或者穿越 AS 300 且由 AS-Path access-list 2 提供更复杂的 AS-Path 访问控制的前缀，AS-Path access-list 2 匹配 16 比特的私有 ASN 区间（64512–65536）。

例 11-68　配置 AS-Path 访问列表

```
ip as-path access-list 1 permit _300_
ip as-path access-list 1 permit ^$
ip as-path access-list 2 permit _(6451[2-9])_
ip as-path access-list 2 permit _(645[2-9][0-9])_
ip as-path access-list 2 permit _(64[6-9][0-9][0-9])_
ip as-path access-list 2 permit _(65[0-4][0-9][0-9])_
ip as-path access-list 2 permit _(655[0-2][0-9])_
ip as-path access-list 2 permit _(6553[0-6])_
```

11.7.3 BGP 团体

BGP 团体提供了额外的路由标记能力，包括周知或私有 BGP 团体。路由器的路由策略使用私有 BGP 团体实现条件匹配，可以在入站或出站路由策略处理过程当中对路由施加影响。仅影响出站路由宣告的 4 个周知 BGP 团体如下。

- **No_Advertise:** No_Advertise 团体（0xFFFFFF02 或 4294967042）指定不要将携带该团体的路由宣告给任何 BGP 对等体。可以从上游 BGP 对等体宣告 BGP 团体 No_Advertise，也可以使用入站 BGP 策略在本地进行宣告，这两种方法都要在 BGP Loc-RIB 表中设置 No_Advertise 团体，都会影响出站路由宣告。
- **No_Export:** 如果收到的路由携带了 No_Export 团体（0xFFFFFF01 或 4294967041），那么就不向任何 EBGP 对等体宣告该路由。如果收到携带 No_Export 团体的路由的路由器是联盟成员，那么就会将该路由宣告给联盟中的其他 sub-AS。
- **Local-AS:** No_Export_Subcommunity（0xFFFFFF03 或 4294967043）被称为 Local-AS 团体，携带该团体的路由不能宣告到本地 AS 之外。如果收到携带 Local-AS 团体的路由的路由器是联盟成员，那么仅在该 sub-AS（成员 AS）内部宣告该路由，而不会在成员 AS 之间进行宣告。

- **Internet**：向 Internet 团体和所有属于 Internet 团体的路由器宣告该路由。

私有团体的数值格式是（*AS 号:16 比特数字*）。条件匹配 BGP 团体允许根据路由的路径属性中的 BGP 团体选择路由，从而在路由映射中执行选择性处理。

NX-OS 设备默认不向对等体宣告 BGP 团体。可以在邻居的地址簇配置下使用 BGP 地址簇配置命令 **send-community [standard | extended | both]**，以逐个邻居的方式启用 BGP 团体，此后将默认仅发送标准团体，除非使用可选关键字 **extended** 或 **both**。

在 NX-OS 设备上执行条件匹配需要创建团体列表。团体列表与 ACL（标准或扩展 ACL）的结构相似，可以通过编号或名称加以引用。标准团体列表可以匹配周知团体或私有团体号（*AS 号:16 比特数字*），而扩展团体列表则使用 RegEx 模式。

仍然以图 11-6 所示拓扑结构为例，NX-5 为 AS_PATH 列表中存在 AS 274 的前缀分配团体值 65001:274。例 11-69 给出了 NX-5 为前缀分配团体值的配置示例。

例 11-69　宣告团体值

```
NX-5(config)# ip as-path access-list ASN_274 permit 274
NX-5(config)# route-map set-Comm
NX-5(config-route-map)# match as-path ASN_274
NX-5(config-route-map)# set community 65001:274
NX-5(config-route-map)# route-map set-Comm per 20
NX-5(config-route-map)# exit
NX-5(config)# router bgp 65001
NX-5(config-router)# neighbor 10.25.1.2
NX-5(config-router-neighbor)# address-family ipv4 unicast
NX-5(config-router-neighbor-af)# route-map set-Comm out
NX-5(config-router-neighbor-af)# send-community
NX-5(config-router-neighbor-af)# end

NX-2# show bgp ipv4 unicast 100.1.25.0/24
BGP routing table information for VRF default, address family IPv4 Unicast
BGP routing table entry for 100.1.25.0/24, version 1195
Paths: (1 available, best #1)
Flags: (0x08001a) on xmit-list, is in urib, is best urib route, is in HW,

  Advertised path-id 1
  Path type: external, path is valid, is best path
  AS-Path: 65001 100 228 274 {300 243} , path sourced external to AS
    10.25.1.5 (metric 0) from 10.25.1.5 (192.168.5.5)

      Origin EGP, MED not set, localpref 100, weight 0
      Community: 65001:274

  Path-id 1 advertised to peers:
    192.168.1.1
```

对于 NX-2 来说，如果操作人员希望根据匹配的团体值设置 BGP 属性，那么就可以在路由映射的匹配语句中使用团体列表。例 11-70 给出了通过 BGP 团体值影响路由策略的配置示例。

例 11-70　使用 BGP 团体影响路由策略

```
NX-2(config)# ip community-list standard Comm-65001:274 permit 65001:274
NX-2(config)# route-map Match-Comm per 10
NX-2(config-route-map)# match community Comm-65001:274
NX-2(config-route-map)# set local-preference 200
NX-2(config-route-map)# route-map Match-Comm per 20
NX-2(config-route-map)# exit
```

```
NX-2(config)# router bgp 65000
NX-2(config-router)# neighbor 10.25.1.5
NX-2(config-router-neighbor)# address-family ipv4 unicast
NX-2(config-router-neighbor-af)# route-map Match-Comm in
NX-2(config-router-neighbor-af)# end
NX-2# show bgp ipv4 unicast neighbor 10.25.1.5 routes
BGP routing table information for VRF default, address family IPv4 Unicast
BGP table version is 1141, local router ID is 192.168.2.2
Status: s-suppressed, x-deleted, S-stale, d-dampened, h-history, *-valid, >-best
Path type: i-internal, e-external, c-confed, l-local, a-aggregate, r-redist, I-injected
Origin codes: i - IGP, e - EGP, ? - incomplete, | - multipath, & - backup
   Network            Next Hop         Metric    LocPrf   Weight Path
! Output omittied for brevity
*>e100.1.21.0/24      10.25.1.5        200                0 65001 100 228 274 {300 243}
*>e100.1.22.0/24      10.25.1.5        200                0 65001 100 228 274 {300 243}
*>e100.1.23.0/24      10.25.1.5        200                0 65001 100 228 274 {300 243}
*>e100.1.24.0/24      10.25.1.5        200                0 65001 100 228 274 {300 243}
*>e100.1.25.0/24      10.25.1.5        200                0 65001 100 228 274 {300 243}
*>e100.1.26.0/24      10.25.1.5        200                0 65001 100 228 274 {300 243}
*>e100.1.27.0/24      10.25.1.5        200                0 65001 100 228 274 {300 243}
*>e100.1.28.0/24      10.25.1.5        200                0 65001 100 228 274 {300 243}
*>e100.1.29.0/24      10.25.1.5        200                0 65001 100 228 274 {300 243}
*>e100.1.30.0/24      10.25.1.5        200                0 65001 100 228 274 {300 243}
```

11.8 Looking Glass 和路由服务器

亲身体验对于学习 RegEx 等技术来说非常有用，用户可以登录 Looking Glass 服务器或路由服务器等公用设备来查看 BGP 表，这类设备绝大多数是思科路由器，当然也有其他设备商的设备。这些服务器不但允许网络工程师们查看他们是否正在按照设计预期向 Internet 宣告路由，而且提供了一种极其有用的在 Internet BGP 表上尝试正则表达式的方法。在网上搜索相关词汇即可发现大量 Looking Glass 服务器和路由服务器网站列表。

11.9 日志采集

出现 BGP 故障之后，可以采集以下 **show tech** 日志。

- **show tech bgp**。
- **show tech netstack**。

如果出现了 BGP 路由策略问题，那么就可以采集包括 **show tech bgp** 在内的以下日志。

- **show tech rpm**。

如果路由未安装到路由表中，但位于 BGP 表中，那么还可以采集以下 **show tech** 命令输出结果。

- **show tech routing ipv4 unicast [brief]**。

采集了这些日志信息之后，可以与思科 TAC 共享，从而快速找出故障根源。

11.10 本章小结

BGP 是一种非常强大的路径向量路由协议，提供了其他路由协议所无法比拟的扩展性和灵活性。BGP通过TCP端口179建立邻居，允许BGP与直连路由器或多跳之外的路由器建立会话。

虽然最早的BGP主要在组织机构之间路由IPv4前缀，但多年来在功能和特性增强方面获得了显著成效，目前BGP已经从一个Internet路由协议扩展到网络的方方面面，包括数据中心。

BGP为叠加拓扑（包括MPLS VPN、IPSec SA和VXLAN）提供了可扩展的控制平面信令，这些叠加拓扑可以跨越广泛使用的可扩展的控制平面，为服务提供商业务和数据中心提供三层服务（如L3VPN）或二层服务（如eVPN）。每个AFI/SAFI组合都可以维护一个独立的BGP表和路由策略，使BGP成为理想的控制平面协议。

本章重点介绍了因MTU不匹配或BGP更新错误而产生的BGP对等关系故障和对等关系翻动故障的排查技术，深入探讨了BGP路由处理和收敛问题，讨论了包括BGP更新生成、路由宣告、最佳路径计算和多路径在内的大量路由处理概念。此外，本章还介绍了包括BGP路由反射器在内的多种BGP扩展技术。

最后，本章详细介绍了前缀列表、过滤器列表和路由映射等路由过滤技术，分析了路由映射可用的多种匹配条件，如前缀列表、团体列表和正则表达式。

第四部分

高可用性故障排查

第 12 章　高可用性

第 12 章

高可用性

本章主要讨论如下主题。
- BFD（Bidirectional Forwarding Detection，双向转发检测）。
- Nexus 高可用性。
- GIR（Graceful Insertion and Removal，平滑插拔）。

NX-OS（Nexus OS）是一种极具弹性的操作系统，在系统级别、网络级别和进程级别都采取了高可用性设计范式。某些 Nexus 交换机通过冗余硬件（如冗余交换矩阵、管理引擎和电源）提供高可用性，网络级高可用性则通过 vPC（virtual Port-Channel，虚拟端口通道）和 FHRP（First Hop Redundancy Protocol，第一跳冗余协议）等功能特性来提供，这些功能特性可以在主路径发生故障后为用户提供故障切换的备用路径。NX-OS 利用多种系统组件来实现进程重启与虚拟化能力，从而提供进程级的高可用性。本章将详细讨论 NX-OS 提供的网络高可用性的主要功能特性及组件。

12.1 BFD

BFD（Bidirectional Forwarding Detection，双向转发检测）是一种简单的定长 Hello 协议，用于快速检测故障。BFD 提供了一种低开销、短持续时间的故障检测机制，可以检测相邻转发引擎之间的路径故障。BFD 定义在 RFC 5880 ~ RFC 5884 中，支持自适应检测次数和三向握手，能够确保两端系统都感知所有变化情况。BFD 控制报文包含了发送端所需的 tx 和 rx 间隔。例如，如果节点无法处理高速 BFD 包，那么就可以设定较大的期望 rx 间隔，从而确保邻居无法以较小的时间间隔发送数据包。BFD 的以下功能特性使其成为最理想的故障检测协议。
- 亚秒级故障检测。
- 介质无关性（以太网、POS、串行接口等）。
- 能够运行在 UDP（User Data Protocol，用户数据协议）之上，与数据协议无关（IPv4、IPv6、LSP[Label Switched Path，标签交换路径]）。
- 应用与 IGP（Interior Gateway Protocol，内部网关协议）无关，支持隧道及 FRR（Fast Reroute，快速重路由）触发机制。

BGP（Border Gateway Protocol，边界网关协议）、OSPF（Open Shortest Path First，开放最短路径优先）等应用在创建或修改 BFD 会话时，应提供以下信息。
- 接口句柄（单跳会话）。
- 邻居地址。
- 本地地址。
- 期望间隔。
- 倍数。

期望间隔与倍数的乘积表示期望的故障检测间隔。对于给定的协议 P 来说，BFD 的操作过程如下。

- 用户在物理接口上为协议 P 配置 BFD。
- 协议 P 开始创建 BFD 会话。
- 创建 BFD 会话之后，协商定时器。
- BFD 定期向对等体发送控制报文。
- 如果链路出现故障，那么 BFD 就可以在期望的故障检测间隔（期望间隔 × 倍数）内检测到故障，并将故障告知给对等体及本地 BFD 客户端（如 BGP）。
- 协议 P 的会话将立即关闭，而不用等待保持定时器超时。

BFD 支持两种运行模式。

- 异步模式（Asynchronous mode）。
- 查询模式（Demand mode）。

> **注：** 思科平台不支持查询模式。对于查询模式来说，会话建立之后无须交换任何 Hello 包，该模式下的 BFD 假定存在其他验证两个端点之间连接性的方法。当然，任何一台主机都能根据需要发送 Hello 包，但通常并不交换这些 Hello 包。

12.1.1 异步模式

异步模式是 BFD 的主要运行模式，也是 BFD 的强制支持选项。该模式下的每个系统都要定期发送 BFD 控制报文。例如，图 12-1 中的路由器 R1 发送的控制报文的源地址是 NX-1，目的地址是路由器 NX-2。

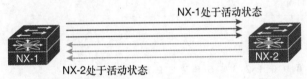

图 12-1　BFD 异步模式

每个 BFD 控制报文流都是独立的，不遵循请求—响应周期。如果对端系统未连续收到已配置数量的数据包（基于 BFD 定时器和倍数），那么就会宣告会话关闭。如果邻居发送数据包的速度慢于其通告速度，那么就会使用自适应故障检测时间以避免故障误报。

BFD 在 UDP 端口 3784 上发送异步数据包，BFD 源端口必须位于 49152～65535。表 12-1 列出了 BFD 控制报文的相关字段信息。

表 12-1　　　　　　　　　　　BFD 控制报文字段

控制报文字段	描述
版本（Version）	BFD 控制报头的版本
诊断（Diag）	诊断代码指定本地系统最近一次会话出现状态变更、检测时间超时、回显失败等问题的原因
状态（State）	传输系统感知的当前 BFD 会话状态
P（Poll Bit，轮询比特）	轮询比特。如果设置了该比特，那么就要求传输系统验证连接性或参数变更情况，并要求收到携带 F（Final，结束）比特的数据包作为响应
F（Final Bit，结束比特）	结束比特。如果设置了该比特，那么就表明传输系统正在响应收到的 BFD 控制报文（设置了 P 比特）
检测倍数（Detect Multiplier）	检测时间倍数。协商的发送间隔乘以该数值，为异步模式下的发送系统提供检测时间

控制报文字段	描述
本端标识符（My Discriminator）	由传输系统生成的唯一的非零标识符，用于在同一对系统之间复用多个 BFD 会话
远端标识符（Your Discriminator）	从对应的远端系统收到的标识符，该字段反映接收到的本端标识符值，如果该值未知，那么本字段为零
期望的最小发送间隔（Desired Min TX Interval）	本地系统发送 BFD 控制报文时希望使用的最小间隔（μs）
期望的最小接收间隔（Desired Min RX Interval）	系统能够支持的接收 BFD 控制报文的最小间隔（μs）
需要的最小回显接收间隔（Required Min Echo RX Interval）	系统能够支持的接收 BFD 回显报文的最小间隔（μs）

图 12-2 显示了 IETF 定义的 BFD 控制报文格式

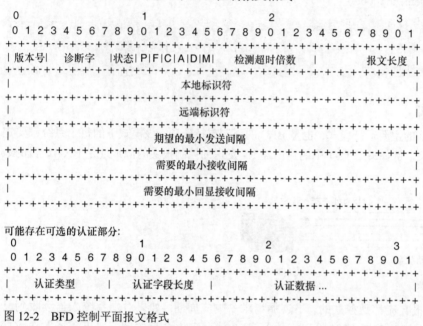

图 12-2　BFD 控制平面报文格式

注： 从 NX-OS Release 5.2 开始，BFD 支持键控 SHA-1 认证机制。

12.1.2　带回显功能的异步模式

带回显功能的异步模式仅测试转发路径，而不测试远端系统的主机栈。请注意，必须在启用了 BFD 会话之后才能启用回显功能。BFD 回显报文的发送方式是由对端直接将回显报文按转发路径返送回来。例如，路由器 NX-1 发送的数据包的源地址和目的地址均属于 NX-1，如图 12-3 所示。

图 12-3　带回显功能的 BFD 异步模式

由于远端应用或主机栈无须处理回显报文，因而该功能可用于主动检测定时器。使用回显功能的另一个好处是，发送端可以完全控制响应时间。为了保证回显功能的正常工作，远端节点也应该支持回显功能。启用了回显功能的 BFD 控制报文将以 UDP 数据包方式进行发送，源和目的端口为 3785。

12.1.3 配置和验证 BFD 会话

如果要在 Nexus 设备上启用 BFD，那么就需要配置命令 **feature bfd**。在 Nexus 交换机上启用了 **feature bfd** 命令之后，设备将在所有启用了 IPv4 和 IPv6 BFD 的接口上打印一条通知消息，以禁用 ICMP（Internet Control Message Protocol，Internet 控制消息协议）和 ICMPv6 重定向。例 12-1 显示了启用 BFD 功能特性之后打印的终端消息。

例 12-1 启用 BFD 功能特性

```
NX-1(config)# feature bfd
Please disable the ICMP / ICMPv6 redirects on all IPv4 and IPv6 interfaces
running BFD sessions using the command below

'no ip redirects '
'no ipv6 redirects '
```

必须在路由协议配置模式下以及将要参与 BFD 的接口下启用 BFD 配置。如果要在路由协议（如 OSPF）下启用 BFD 配置，那么就可以在 **router ospf** 配置下使用命令 **bfd**。在接口下配置 BFD 时需要定义两个非常重要的 BFD 参数。

- BFD 间隔。
- BFD 回显功能。

可以在接口和全局配置模式下定义 BFD 间隔，配置命令是 **bfd interval** *tx-interval* **min_rx** *rx-interval* **multiplier** *number*。系统默认启用 BFD 回显功能，如果要禁用或启用 BFD 回显功能，那么就可以使用命令[**no**] **bfd echo**。例 12-2 显示了为 OSPF 启用 BFD 的配置示例。

例 12-2 为 OSPF 配置 BFD

```
NX-1
NX-1(config)# int e4/1
NX-1(config-if)# no ip redirects
NX-1(config-if)# no ipv6 redirects
NX-2(config)# bfd interval 300 min_rx 300 multiplier 3
NX-1(config-if)# ip ospf bfd
NX-1(config-if)# no bfd echo
NX-1(config-if)# exit
NX-1(config)# router ospf 100
NX-1(config-router)# bfd
```

注：如果要为其他路由协议启用 BFD，那么可以参阅思科文档以获得不同 Nexus 设备的配置方法。

启用了 BFD 功能之后，就可以建立 BFD 会话。可以通过命令 **show bfd neighbors [detail]** 验证 BFD 的状态，**show bfd neighbors** 命令可以显示 BFD 邻居的状态以及接口、本地和远端标识符、VRF（Virtual Routing and Forwarding，虚拟路由转发）等详细信息。如果使用了关键字 **detail**，那么输出结果还将显示 BFD 控制报文的所有字段，对于调试操作来说非常有用，可以查看导致 BFD 会话震荡的不匹配原因。需要确保状态比特被设置为 *Up*，而不是 *AdminDown*。此外，该命令的输出结果还可以显示回显功能的启用或禁用情况。例 12-3 显示了命令 **show bfd neighbors [detail]** 的输出结果示例。

例 12-3 验证 BFD 邻居

```
NX-1# show bfd neighbors detail

OurAddr      NeighAddr    LD/RD                   RH/RS  Holdown(mult)  State  Int
  Vrf
10.1.12.1    10.1.12.2    1090519044/1107296259   Up     667(3)         Up     Eth4/1
```

```
    default
Session state is Up and not using echo function
Local Diag: 0, Demand mode: 0, Poll bit: 0, Authentication: None
MinTxInt: 300000 us, MinRxInt: 300000 us, Multiplier: 3
Received MinRxInt: 300000 us, Received Multiplier: 3
Holdown (hits): 900 ms (0), Hello (hits): 300 ms (47)
Rx Count: 47, Rx Interval (ms) min/max/avg: 0/1600/260 last: 134 ms ago
Tx Count: 47, Tx Interval (ms) min/max/avg: 236/236/236 last: 136 ms ago
Registered protocols: ospf
Uptime: 0 days 0 hrs 36 mins 9 secs
Last packet: Version: 1              - Diagnostic: 0
             State bit: Up            - Demand bit: 0
             Poll bit: 0              - Final bit: 0
             Multiplier: 3            - Length: 24
             My Discr.: 1107296259    - Your Discr.: 1090519044
             Min tx interval: 300000  - Min rx interval: 300000
             Min Echo interval: 50000 - Authentication bit: 0
Hosting LC: 4, Down reason: None, Reason not-hosted: None
```

排查与 BFD 相关的故障问题之前，必须首先验证 BFD 功能特性的状态。可以通过命令 **show system internal feature-mgr feature** *feature-name* **current status** 验证 BFD 的当前状态。如果进程出现了问题（如进程未运行或崩溃），那么进程的状态就不会显示为 Running。例 12-4 显示了 BFD 功能特性的状态信息，可以看出，此时的 BFD 处于 Running 状态。

例 12-4　BFD 功能特性的状态

```
NX-1# show system internal feature-mgr feature bfd current status
Feature Name      State      Feature  ID UUID   SAP    PID    Service State
--------------    -------    -------  -------   -----  -----  -------------
bfd               enabled    87       706       121    2574   Running
```

与其他功能特性一样，BFD 也要维护内部事件历史日志，可以通过这些日志来调试与状态机相关的故障问题或 BFD 翻动故障。BFD 的事件历史记录提供了多种命令行选项，如果要查看 BFD 事件历史记录，那么就可以使用命令 **show system internal bfd event-history [all | errors | logs | msgs | session** [*discriminator*]]。选项 **all** 的作用是显示所有事件历史记录（所有事件和差错事件历史记录日志），选项 **errors** 的作用是仅显示与 BFD 相关的差错。选项 **logs** 的作用是显示 BFD 的所有事件，选项 **msgs** 的作用是显示与 BFD 相关的消息，选项 **session** 的作用是查看与特定会话相关的差错、日志消息以及应用程序事件日志。

例 12-5 显示 module 4 上的接口启用了 BFD 会话之后的 BFD 事件历史记录日志（标识符为 0x41000004），本例有助于理解信息交换过程以及系统启动 BFD 会话所经历的具体步骤。

- 第 1 步：会话首先处于 *AdminDown* 状态。
- 第 2 步：BFD 客户端（BFDC）利用将要建立 BFD 会话的设备的接口和 IP 地址添加 BFD 会话。
- 第 3 步：BFD 组件向线卡上的 BFDC 组件发送 MTS 消息。
- 第 4 步：BFD 将收到的会话通告发送给客户端。

注：BFD 进程运行在管理引擎上，BFDC 运行在线卡上。

例 12-5　BFD 事件历史日志

```
NX-1# show system internal bfd event-history logs

1) Event:E_DEBUG, length:95, at 686796 usecs after Sat Oct 28 14:07:10 2017
```

```
      [102] bfd_mts_send_msg_to_bfdc(4848): opc 116747 length 36 sent to host_module 4
rrtok 0x27de77d


2) Event:E_DEBUG, length:95, at 676629 usecs after Sat Oct 28 13:14:38 2017
      [102] bfd_mts_send_msg_to_bfdc(4848): opc 116747 length 36 sent to host_module 4
rrtok 0x27ceb4e


3) Event:E_DEBUG, length:95, at 506685 usecs after Sat Oct 28 13:14:11 2017
      [102] bfd_mts_send_msg_to_bfdc(4848): opc 116747 length 36 sent to host_module 4
rrtok 0x27ce825


4) Event:E_DEBUG, length:106, at 92550 usecs after Sat Oct 28 13:14:08 2017
      [102] bfd_mts_sess_change_state_notif_cb(2379): Received sess 0x41000004 notif 3
reason_code No Diagnostic


5) Event:E_DEBUG, length:151, at 92524 usecs after Sat Oct 28 13:14:08 2017
      [102] bfd_mts_sess_change_state_notif_cb(2366): notif 0x41000004 3: [1 if Eth4/1
0x1a180000 iod 0x26 10c010a:0:0:0=10.1.12.1 -> 20c010a:0:0:0=10.1
.12.2]


6) Event:E_DEBUG, length:96, at 315822 usecs after Sat Oct 28 13:14:06 2017
      [102] bfd_mts_send_msg_to_bfdc(4848): opc 116745 length 396 sent to host_module 4
rrtok 0x27ce7ef


7) Event:E_DEBUG, length:81, at 315599 usecs after Sat Oct 28 13:14:06 2017
      [102] bfd_fu_timer_cancel_app_client_expiry(244): disc 0x41000004 app
1090519321:1


8) Event:E_DEBUG, length:167, at 315560 usecs after Sat Oct 28 13:14:06 2017
      [102] bfd_sess_create_session(1713): Client Add 1:1090519321 to session
0x41000004 [1 if Eth4/1 0x1a180000 iod 0x26 10c010a:0:0:0=10.1.12.1 -> 20
c010a:0:0:0=10.1.12.2]


9) Event:E_DEBUG, length:114, at 399344 usecs after Sat Oct 28 13:14:02 2017
      [102] bfd_mts_sess_change_state_notif_cb(2379): Received sess 0x41000004 notif 1
reason_code Administratively Down
```

例 12-6 使用 **show system internal bfd event-history session** *discriminator* 命令显示了详细的会话信息，其中的 *discriminator* 值可以由 LD 或*对端标识符*（来自 **show bfd neighbors detail** 命令输出结果）计算得到，*discriminator* 值以十六进制形式进行计算（见例 12-6），并与 **event-history** 命令输出结果一起使用。可以通过 **event-history session** 命令查看与特定 BFD 会话相关的差错、日志记录（如参数交换和状态变更信息）以及应用事件。

例 12-6　基于 BFD 会话的事件历史日志

```
NX-1# hex 1090519044
0x41000004

NX-1# show system internal bfd event-history session 0x41000004
```

```
Start of errors for session 0x41000004
1:1365   292509 usecs after Sat Oct 28 13:13:15 2017
         : Code 0x1 0x0 0x0 0x0

End of errors for session 0x41000004

Start of Logs for session 0x41000004
1:2455   612556 usecs after Sat Oct 28 13:14:08 2017
         : Session active params changed: State 3(Up), TX(300000), RX(300000),
Mult(3)
2:2455   332770 usecs after Sat Oct 28 13:14:08 2017
         : Session active params changed: State 3(Up), TX(300000), RX(300000),
Mult(3)
3:649    92566 usecs after Sat Oct 28 13:14:08 2017
         : Session Up
4:628    92526 usecs after Sat Oct 28 13:14:08 2017
         : Session state changed: 1(Down) -> 3(Up), New diag: 0(No Diagnostic),
After: 6 secs
5:2523   92282 usecs after Sat Oct 28 13:14:08 2017
         : Session remote disc changed: 0(0x0) -> 1107296261(0x42000005)
6:2523   732472 usecs after Sat Oct 28 13:14:02 2017
         : Session remote disc changed: 1107296260(0x42000004) -> 0(0x0)
7:2523   732438 usecs after Sat Oct 28 13:14:02 2017
         : Session remote disc changed: 0(0x0) -> 1107296260(0x42000004)
8:2523   452189 usecs after Sat Oct 28 13:14:02 2017
         : Session remote disc changed: 1107296260(0x42000004) -> 0(0x0)
9:2523   452163 usecs after Sat Oct 28 13:14:02 2017
         : Session remote disc changed: 0(0x0) -> 1107296260(0x42000004)
10:707   399365 usecs after Sat Oct 28 13:14:02 2017
         : Session Down Diag 7(Administratively Down)
11:628   399285 usecs after Sat Oct 28 13:14:02 2017
         : Session state changed: 3(Up) -> 1(Down), New diag: 7(Administratively
Down), After: 44 secs
12:2523  398654 usecs after Sat Oct 28 13:14:02 2017
         : Session remote disc changed: 1107296260(0x42000004) -> 0(0x0)
13:2455  49895 usecs after Sat Oct 28 13:13:19 2017
         : Session active params changed: State 3(Up), TX(300000), RX(300000),
Mult(3)
14:2455  49490 usecs after Sat Oct 28 13:13:19 2017
         : Session active params changed: State 3(Up), TX(300000), RX(300000),
Mult(3)
15:2455  770894 usecs after Sat Oct 28 13:13:18 2017
         : Session active params changed: State 3(Up), TX(300000), RX(300000),
Mult(3)
16:649   769776 usecs after Sat Oct 28 13:13:18 2017
         : Session Up
17:628   769732 usecs after Sat Oct 28 13:13:18 2017
         : Session state changed: 1(Down) -> 3(Up), New diag: 0(No Diagnostic),
After: 3 secs
18:2523  59514 usecs after Sat Oct 28 13:13:17 2017
         : Session remote disc changed: 0(0x0) -> 1107296260(0x42000004)
19:1396  347952 usecs after Sat Oct 28 13:13:15 2017
         : ACL installed
20:602   293945 usecs after Sat Oct 28 13:13:15 2017
         : Session installed on LC 4
21:112   292529 usecs after Sat Oct 28 13:13:15 2017
         : Session Created if 0x1a180000 iod 38 (Eth4/1) src 10.1.12.1, dst 10.1.12.2
```

```
22:1364   292508 usecs after Sat Oct 28 13:13:15 2017
        : Code 0x1 0x0 0x0 0x0

End of Logs for session 0x41000004

Start of app-events for session 0x41000004
1:958    315615 usecs after Sat Oct 28 13:14:06 2017
        : Client Add type 1, 1090519321 in state 14
2:1709   292536 usecs after Sat Oct 28 13:13:15 2017
        : Client Add type 1, 1090519321 in state 10
3:1363   292507 usecs after Sat Oct 28 13:13:15 2017
        : Code 0x1 0x0 0x0 0x0

End of app-events for session 0x41000004
```

命令 **show system internal bfd transition-history** 可以显示 BFD 会话经历的与内部状态机相关的各种事件，如例 12-7 所示。请注意，BFD 会话的最终状态应为 BFD_SESS_ST_SESSION_UP。如果 BFD 会话卡在其他状态，那么就可以通过该命令显示 BFD 会话被卡的状态。

例 12-7　BFD 的状态迁移历史日志

```
NX-1# show system internal bfd transition-history

>>>>FSM: <Proto  Sess 0x41000004> has 8 logged transitions<<<<<

1) FSM:<Proto  Sess 0x41000004> Transition at 292788 usecs after Sat Oct 28 13:13:15
2017
    Previous state: [BFD_SESS_ST_INIT]
    Triggered event: [BFD_SESS_EV_INTERFACE]
    Next state: [BFD_SESS_ST_INSTALLING_SESSION]

2) FSM:<Proto  Sess 0x41000004> Transition at 293898 usecs after Sat Oct 28 13:13:15
2017
    Previous state: [BFD_SESS_ST_INSTALLING_SESSION]
    Triggered event: [BFD_SESS_EV_SESSION_INSTALL_SUCCESS]
    Next state: [BFD_SESS_ST_INSTALLING_ACL]

3) FSM:<Proto  Sess 0x41000004> Transition at 347878 usecs after Sat Oct 28 13:13:15
2017
    Previous state: [BFD_SESS_ST_INSTALLING_ACL]
    Triggered event: [BFD_SESS_EV_ACL_RESPONSE]
    Next state: [FSM_ST_NO_CHANGE]

4) FSM:<Proto  Sess 0x41000004> Transition at 347948 usecs after Sat Oct 28 13:13:15
2017
    Previous state: [BFD_SESS_ST_INSTALLING_ACL]
    Triggered event: [BFD_SESS_EV_ACL_INSTALL_SUCCESS]
    Next state: [BFD_SESS_ST_SESSION_DOWN]

5) FSM:<Proto  Sess 0x41000004> Transition at 769773 usecs after Sat Oct 28 13:13:18
2017
    Previous state: [BFD_SESS_ST_SESSION_DOWN]
    Triggered event: [BFD_SESS_EV_SESSION_UP]
    Next state: [BFD_SESS_ST_SESSION_UP]

6) FSM:<Proto  Sess 0x41000004> Transition at 399361 usecs after Sat Oct 28 13:14:02
2017
    Previous state: [BFD_SESS_ST_SESSION_UP]
```

```
        Triggered event: [BFD_SESS_EV_SESSION_DOWN]
        Next state: [BFD_SESS_ST_SESSION_DOWN]

7) FSM:<Proto  Sess 0x41000004> Transition at 315593 usecs after Sat Oct 28 13:14:06
2017
        Previous state: [BFD_SESS_ST_SESSION_DOWN]
        Triggered event: [BFD_SESS_EV_CLIENT_ADD]
        Next state: [FSM_ST_NO_CHANGE]

8) FSM:<Proto  Sess 0x41000004> Transition at 92563 usecs after Sat Oct 28 13:14:08
2017
        Previous state: [BFD_SESS_ST_SESSION_DOWN]
        Triggered event: [BFD_SESS_EV_SESSION_UP]
        Next state: [BFD_SESS_ST_SESSION_UP]

Curr state: [BFD_SESS_ST_SESSION_UP]
```

配置了 BFD 会话之后，就会在硬件中安装访问列表。可以通过命令 **show system internal access-list interface** *interface-id* **module** *sl* 加以验证。如果要显示硬件 ACL（Access Control List，访问控制列表）的相关统计信息，可以使用命令 **show system internal access-list input statistics module** *slot*。需要注意的是，在接口上启用了 BFD 之后，系统就会在硬件中为 IPv4 和 IPv6 安装 ACL。例 12-8 给出了 Nexus 7000 交换机在硬件中为 BFD 编程的 ACL 情况。

例 12-8　在硬件中为 BFD 安装 ACL

```
NX-1# show system internal access-list interface ethernet 4/1 module 4
Policies in ingress direction:
        Policy type                     Policy Id           Policy name
---------------------------------------------------------------
        QoS                                 3
        BFD                                 6

No Netflow profiles in ingress direction

INSTANCE 0x0
---------------

    Tcam 1 resource usage:
    ----------------------
    Label_b = 0x2
    Bank 0
    ------
      IPv4 Class
        Policies: BFD()   [Merged]
        Netflow profile: 0
        Netflow deny profile: 0
        4 tcam entries
      IPv6 Class
        Policies: BFD()   [Merged]
        Netflow profile: 0
        Netflow deny profile: 0
        2 tcam entries

    0 l4 protocol cam entries
    0 mac etype/proto cam entries
    2 lous
    0 tcp flags table entries
```

```
      1 adjacency entries

! Output omitted for brevity
NX-1# show system internal access-list input statistics module 4
              VDC-1 Ethernet4/1 :
              =====================

INSTANCE 0x0
--------------

  Tcam 1 resource usage:
  ---------------------
  Label_b = 0x2
   Bank 0
   ------
     IPv4 Class
       Policies: BFD()  [Merged]
       Netflow profile: 0
       Netflow deny profile: 0
       Entries:
         [Index] Entry [Stats]
         ---------------------
  [0008:0408:0006] prec 1 redirect(0x40001)-routed udp 0.0.0.0/0 0.0.0.0/0 eq 3785
 ttl eq 254  flow-label 3785   [0]
  [0009:0508:0007] prec 1 redirect(0x40001)-routed udp 0.0.0.0/0 0.0.0.0/0 eq 3784
 ttl eq 255  flow-label 3784   [26874]
  [000a:0608:0008] prec 1 permit-routed ip 0.0.0.0/0 0.0.0.0/0    [1641]
  [000b:0488:0009] prec 1 permit-routed ip 0.0.0.0/0 0.0.0.0/0 fragment    [0]
     IPv6 Class
       Policies: BFD()  [Merged]
       Netflow profile: 0
       Netflow deny profile: 0
       Entries:
         [Index] Entry [Stats]
         ---------------------
  [000c:0509:000a] prec 1 redirect(0x40001)-routed udp 0x0/0 0x0/0 eq 3784 ttl eq
 255  flow-label 3784   [0]
  [000d:0409:000b] prec 1 redirect(0x40001)-routed udp 0x0/0 0x0/0 eq 3785 ttl eq
 254  flow-label 3785   [0]
```

注：硬件ACL编程取决于底层线卡硬件和Nexus硬件平台，不同的Nexus硬件平台可能有不同的处理行为。

如果要启用BFD回显功能，那么就要在接口下配置 **bfd echo** 命令。使用回显功能配置了BFD会话之后，BFD会话将按照2s的慢间隔以异步模式进行启动。会话启动之后，如果客户端指定的间隔小于2s，那么就会激活回显功能（假设远端对等体已经启用了回显功能）。

例12-9给出了NX-1与NX-2之间BFD回显功能配置示例，同时还显示了BFD会话建立之后的 **show bfd neighbors detail** 命令输出结果。

例12-9 配置和验证带回显功能的BFD

```
NX-1(config)# interface Ethernet4/1
NX-1(config-if)# bfd echo
NX-1(config-if)# bfd interval 50 min_rx 50 multiplier 3
NX-1# show bfd neighbors detail

OurAddr      NeighAddr     LD/RD            RH/RS  Holdown(mult)  State  Int
 Vrf
```

```
 10.1.12.1    10.1.12.2   1090519047/1107296265       Up       667(3)         Up    Eth4/1
 default

Session state is Up and using echo function with 50 ms interval
Local Diag: 0, Demand mode: 0, Poll bit: 0, Authentication: None
MinTxInt: 50000 us, MinRxInt: 2000000 us, Multiplier: 3
Received MinRxInt: 2000000 us, Received Multiplier: 3
Holdown (hits): 6000 ms (0), Hello (hits): 2000 ms (1690)
Rx Count: 1690, Rx Interval (ms) min/max/avg: 0/1880/1807 last: 268 ms ago
Tx Count: 1690, Tx Interval (ms) min/max/avg: 1806/1806/1806 last: 269 ms ago
Registered protocols:  ospf
Uptime: 0 days 0 hrs 50 mins 52 secs
Last packet: Version: 1              - Diagnostic: 0
             State bit: Up            - Demand bit: 0
             Poll bit: 0              - Final bit: 0
             Multiplier: 3            - Length: 24
             My Discr.: 1107296265    - Your Discr.: 1090519047
             Min tx interval: 50000   - Min rx interval: 2000000
             Min Echo interval: 50000 - Authentication bit: 0
Hosting LC: 4, Down reason: None, Reason not-hosted: None
```

如果出现了故障，那么 NX-OS 就会记录 BFD 故障的系统日志消息、故障原因代码和会话标识符值。例 12-10 显示了 NX-1 交换机出现 BFD 故障后的系统日志消息，可以看出，故障原因是 0x2，表示 "Echo Function Failed"（回显功能故障）。

例 12-10　BFD 故障日志

```
02:42:01 NX-1 BFD-5-SESSION_STATE_DOWN BFD session 1107296259 to neighbor
  10.1.12.2 on interface Eth4/1 has gone down. Reason: 0x2.
02:44:01 NX-1 BFD-5-SESSION_STATE_DOWN BFD session 1090519047 to neighbor
  10.1.12.2 on interface Eth4/1 has gone down. Reason: 0x2.
```

表 12-2 列出了 BFD 故障原因代码及其描述信息。

表 12-2　　　　　　　　　　　BFD 故障原因代码及其描述信息

故障原因代码	描述
0	无诊断
1	控制报文检测定时器超时
2	回显功能故障
3	邻居告知会话中断
4	转发平面重置
5	路径中断
6	级联路径中断
7	管理性中断
8	反向级联路径中断

> **注：** 如果出现了与 BFD 相关的故障事件，那么建议在 BFD 会话出现翻动之后立即抓取命令 **show tech bfd** 的输出结果，同时还应该抓取与 BFD 相关联的功能特性的 **show tech** *feature* 命令输出结果（例如，如果是 OSPF，那么就是 **show tech ospf**）。

此外，Nexus 还支持 L3 端口通道 BFD 或 L2 端口通道上的 SVI 接口 BFD。对这两种场景来说，都必须为端口通道接口启用 LACP（Link Aggregation Control Protocol，链路聚合控制协议）。在 L3 端口通道接口上启用 BFD 的方法如下。

- BFD 单链路模式（BFD per-link）。
- 微 BFD 会话模式（Micro BFD session）。

如果要启用 BFD 单链路模式，那么需要在端口通道接口下使用命令 **bfd per-link** 和 **no ip redirects**，这样就可以为 L3 端口通道接口上的客户端协议启用 BFD。采用 BFD 单链路模式时，BFD 会为端口通道中的每条链路都创建一个会话，并向客户端协议提供累积或聚合结果。例 12-11 给出了在端口通道接口上配置 BFD 单链路模式的示例，并通过 **show bfd neighbors [detail]**命令加以验证。命令 **show port-channel summary** 可以验证端口通道接口的成员端口。

例 12-11 在端口通道上配置 BFD 单链路模式

```
NX-1
NX-1(config)# interface port-channel1
NX-1(config-if)# no ip redirects
NX-1(config-if)# bfd per-link
NX-1(config-if)# ip router ospf 100 area 0.0.0.0
NX-1(config-if)# ip ospf network point-to-point
NX-1(config-if)# exit
NX-1(config)# router ospf 100
NX-1(config-router)# bfd

NX-1# show port-channel summary
! Output omitted for brevity
--------------------------------------------------------------------------------
Group Port-        Type     Protocol   Member Ports
      Channel
--------------------------------------------------------------------------------
1     Po1(RU)      Eth      LACP       Eth4/1(P)

NX-1# show bfd neighbors detail

OurAddr         NeighAddr      LD/RD                   RH/RS   Holdown(mult)   State   Int
  Vrf
10.1.12.1       10.1.12.2      1090519048/0            Up      N/A(3)          Up      Po1
  default

Session state is Up
Local Diag: 0
Registered protocols:  ospf
Uptime: 0 days 0 hrs 0 mins 9 secs
Hosting LC: 0, Down reason: None, Reason not-hosted: None
Parent session, please check port channel config for member info

OurAddr         NeighAddr      LD/RD                   RH/RS   Holdown(mult)   State   Int
  Vrf
10.1.12.1       10.1.12.2      1090519049/1107296267   Up      148(3)          Up      Eth4/1
  default

Session state is Up and not using echo function
Local Diag: 0, Demand mode: 0, Poll bit: 0, Authentication: None
MinTxInt: 50000 us, MinRxInt: 50000 us, Multiplier: 3
Received MinRxInt: 50000 us, Received Multiplier: 3
Holdown (hits): 150 ms (0), Hello (hits): 50 ms (176)
Rx Count: 176, Rx Interval (ms) min/max/avg: 0/2133/72 last: 1 ms ago
Tx Count: 176, Tx Interval (ms) min/max/avg: 48/48/48 last: 2 ms ago
Registered protocols:
Uptime: 0 days 0 hrs 0 mins 9 secs
Last packet: Version: 1            - Diagnostic: 0
             State bit: Up         - Demand bit: 0
             Poll bit: 0           - Final bit: 0
```

```
                    Multiplier: 3                  - Length: 24
                    My Discr.: 1107296267          - Your Discr.: 1090519049
                    Min tx interval: 50000         - Min rx interval: 50000
                    Min Echo interval: 50000       - Authentication bit: 0
Hosting LC: 4, Down reason: None, Reason not-hosted: None
Member session under parent interface Po1
```

此外，根据 RFC 7130 定义，Nexus 9000 还可以在每个 LAG（Link Aggregation Group，链路聚合组）成员接口上启用 BFD，该方法被称为 IETF 微 BFD 会话。微 BFD 会话不支持回显功能。使用微 BFD 会话的好处是，如果某个成员端口出现了故障，那么就会从转发表中删除该成员端口，避免成员链路上的流量出现中断。

如果要配置微 BFD 会话，需要在激活的 L3 端口通道接口上配置命令 **port-channel bfd track-member-link** 和 **port-channel bfd destination** *ip-address*。例 12-12 给出了 Nexus 9000 交换机 N9k-1 和 N9k-2 的微 BFD 会话配置示例。

例 12-12 端口通道 BFD（配置微 BFD 会话）

```
N9k-1
N9k-1(config)# interface port-channel2
N9k-1(config-if)# port-channel bfd track-member-link
N9k-1(config-if)# port-channel bfd destination 172.16.0.1

N9k-2
N9k-2(config)# interface port-channel2
N9k-2(config-if)# port-channel bfd track-member-link
N9k-2(config-if)# port-channel bfd destination 172.16.0.0
```

通过验证可以发现，端口通道的每个成员端口均已建立了 BFD 会话。该配置方法中的 BFD 客户端就是端口通道本身。例 12-13 以微 BFD 会话模式在端口通道接口上启用了 BFD 会话，请注意，此时的 BFD 客户端是以太网端口通道。

例 12-13 验证端口通道 BFD

```
N9k-1
N9k-1# show bfd neighbors

OurAddr        NeighAddr      LD/RD                   RH/RS   Holdown(mult)   State   Int
  Vrf
172.16.0.0     172.16.0.1     1090519044/0            Up      N/A(3)          Up      Po2
  default
172.16.0.0     172.16.0.1     1090519045/1090519045   Up      121(3)          Up      Eth1/3
  default
N9k-1# show bfd neighbors details

OurAddr        NeighAddr      LD/RD                   RH/RS   Holdown(mult)   State   Int
  Vrf
172.16.0.1     172.16.0.0     1090519044/0            Up      N/A(3)          Up      Po2
  default
Session state is Up
Local Diag: 0
Registered protocols:  eth_port_channel
Uptime: 0 days 0 hrs 9 mins 56 secs
Hosting LC: 0, Down reason: None, Reason not-hosted: None
Parent session, please check port channel config for member info

172.16.0.1     172.16.0.0     1090519045/1090519045   Up      121(3)          Up      Eth1/3
  default
```

```
Session state is Up and not using echo function
Local Diag: 0, Demand mode: 0, Poll bit: 0, Authentication: None
MinTxInt: 50000 us, MinRxInt: 50000 us, Multiplier: 3
Received MinRxInt: 50000 us, Received Multiplier: 3
Holdown (hits): 150 ms (0), Hello (hits): 50 ms (12619)
Rx Count: 12357, Rx Interval (ms) min/max/avg: 1/1987/48 last: 25 ms ago
Tx Count: 12619, Tx Interval (ms) min/max/avg: 47/47/47 last: 32 ms ago
Registered protocols:   eth_port_channel
Uptime: 0 days 0 hrs 9 mins 56 secs
Last packet: Version: 1              - Diagnostic: 0
             State bit: Up            - Demand bit: 0
             Poll bit: 0              - Final bit: 0
             Multiplier: 3            - Length: 24
             My Discr.: 1090519045    - Your Discr.: 1090519045
             Min tx interval: 50000   - Min rx interval: 50000
             Min Echo interval: 50000 - Authentication bit: 0
Hosting LC: 1, Down reason: None, Reason not-hosted: None
Member session under parent interface Po2
```

> **注：**如果出现了与 BFD 单链路或微 BFD 会话相关的故障问题，那么就需要收集 **show tech bfd** 和 **show tech lacp all** 命令的输出结果，并与思科技术支持中心（TAC）共享相关日志。

12.2 Nexus 高可用性

NX-OS 的核心基础是 HA（High Availability，高可用性）和虚拟化。由于 Nexus 设备主要面向数据中心和企业网环境，因而 NX-OS 体系架构在提供大量功能特性和丰富能力的同时，还具备强大的高可用性机制，从而大大降低了设备中断时间。

- SSO（Stateful Switchover，状态化切换）。
- ISSU（In-Service Software Upgrade，不中断软件升级）。
- GIR（Graceful Insertion and Removal，平滑插拔）。

本节将详细讨论这些功能特性以及为 Nexus 设备提供 HA 能力的方式。

12.2.1 SSO

所有的 Nexus 平台（包括 Nexus 7000、Nexus 7700 和 Nexus 9500）都支持交换矩阵及管理引擎冗余机制。硬件冗余的好处是，如果主用硬件（交换矩阵或管理引擎卡）出现了故障，那么备用硬件将接管主用硬件的角色并避免出现流量和服务中断问题。此外，某些基于软件的 HA 功能（如 NSR[Nonstop Routing，不间断路由]、NSF[Nonstop Forwarding，不间断转发]和 GR[Graceful Restart，平滑重启]）仅在具备冗余管理引擎时才能启用，此时主用管理引擎必须能够将状态同步给备用管理引擎，且备用管理引擎能够在主用管理引擎出现故障后无缝接管主用角色。

配置了冗余硬件之后，管理引擎必须运行在主用/备用模式。可以通过命令 **show module** 验证管理引擎的状态，该命令可以显示机箱中安装的所有管理引擎、线卡和交换矩阵卡。例 12-14 给出了 Nexus 7000 交换机的 **show module** 命令输出结果。可以看出，插槽 1 中的管理引擎处于 HA 备用（ha-standby）状态，插槽 2 中的管理引擎处于主用状态。

例 12-14　show module 命令输出结果

```
NX-1# show module
Mod  Ports  Module-Type                          Model              Status
---  -----  -----------------------------------  -----------------  ----------
```

```
1    0    Supervisor Module-2                   N7K-SUP2E         ha-standby
2    0    Supervisor Module-2                   N7K-SUP2E         active *
3    32   10 Gbps Ethernet XL Module            N7K-M132XP-12L    ok
4    32   1/10 Gbps Ethernet Module             N7K-F132XP-15     powered-dn
5    48   10/100/1000 Mbps Ethernet XL Module   N7K-M148GT-11L    ok
6    48   1/10 Gbps Ethernet Module             N7K-F248XP-25E    ok
7    32   10 Gbps Ethernet XL Module            N7K-M132XP-12L    ok
8    48   1/10 Gbps Ethernet Module             N7K-F248XP-25     ok
! Output omitted for brevity
```

此外，还可以使用命令 **show system redundancy status** 来验证 HA 的状态。备用管理引擎在启动的时候或者执行管理引擎切换操作（主用管理引擎转为备用角色）之后，并不会立即达到 ha-standby 状态。备用管理引擎需要与主用管理引擎进行状态同步，该操作可以通过主用管理引擎上的 sysmgr（system manager，系统管理器）组件来实现，sysmgr 组件将启动一个将主用管理引擎状态同步给备用管理引擎的 gsync（global sync，全局同步）操作，同步过程中的状态被称为 *HA synchronization in progress*（正在进行 HA 同步）。需要注意的是，备用管理引擎不应该在该状态停留太长时间，否则将被认为出现了故障和其他问题。

如果主用管理引擎与备用管理引擎的所有组件和状态都完成了同步操作，那么就会通知 Module-Manager（模块管理器）：备用管理引擎已启动。此后，Module-Manager 将备用管理引擎的可用性告知主用管理引擎上的所有软件组件并进行配置。该事件被称为 *Standby Sup Insertion Sequence*（备用管理引擎插入序列），该序列期间出现的任何差错，都会导致备用管理引擎重启。

例 12-15 显示了系统的冗余状态，理想的冗余状态是主用/备用状态。可以看出，备用管理引擎已经与插槽 2 中的主用管理引擎实现了状态同步。

例 12-15 系统冗余状态

```
NX-1# show system redundancy status
Redundancy mode
---------------
       administrative: HA
          operational: HA

This supervisor (sup-2)
-----------------------
    Redundancy state: Active
    Supervisor state: Active
      Internal state: Active with HA standby

Other supervisor (sup-1)
------------------------
    Redundancy state: Standby
    Supervisor state: HA standby
      Internal state: HA synchronization in progress
```

注：如果设备在 *Standby Sup Insertion Sequence*（备用管理引擎插入序列）期间出现了故障，那么建议收集以下命令的输出结果以帮助排查故障位置。

- **show logging [nvram]**。
- **show module internal exception-log**。
- **show system reset-reason**。
- **show module internal event-history module** *slot*。

对于支持 VDC（Virtual Device Context，虚拟设备上下文）的 Nexus 7000 或 Nexus 7700 平

台来说，还要在系统上为所有 VDC 配置 HA 状态。可以通过命令 **show system redundancy ha status** 加以验证。例 12-16 验证了所有 VDC 的系统冗余状态。

例 12-16 系统冗余 HA 状态

```
NX-1# show system redundancy ha status
VDC No     This supervisor              Other supervisor
------     ---------------              ----------------
vdc 1      Active with HA standby       HA standby
vdc 2      Active with HA standby       HA standby
```

由于状态同步是通过 sysmgr 组件实现的，因而也可以利用 **show system internal sysmgr state** 命令验证状态信息。从例 12-17 可以看出，sysmgr 状态被设置为主用/热备状态。此外，该命令还显示了主用管理引擎卡的当前状态，此时的状态为主用状态（SYSMGR_CARDSTATE_ACTIVE）。

例 12-17 sysmgr 状态信息

```
NX-1# show system internal sysmgr state
The master System Manager has PID 4967 and UUID 0x1.
Last time System Manager was gracefully shutdown.
The state is SRV_STATE_MASTER_ACTIVE_HOTSTDBY entered at time Thu Oct 26 13:20:5
4 2017.

The '-b' option (disable heartbeat) is currently disabled.

The '-n' (don't use rlimit) option is currently disabled.

Hap-reset is currently enabled.

Process restart capability is currently disabled.

Watchdog checking is currently enabled.

Watchdog kgdb setting is currently disabled.


        Debugging info:

The trace mask is 0x00000000, the syslog priority enabled is 3.
The '-d' option is currently disabled.
The statistics generation is currently enabled.



        HA info:

slotid = 2      supid = 0
cardstate = SYSMGR_CARDSTATE_ACTIVE .
cardstate = SYSMGR_CARDSTATE_ACTIVE (hot switchover is configured enabled).
Configured to use the real platform manager.
Configured to use the real redundancy driver.
Redundancy register: this_sup = RDN_ST_AC, other_sup = RDN_ST_SB.
EOBC device name: veobc.
Remote addresses:  MTS - 0x00000101/3       IP - 127.1.1.1
MSYNC done.
Remote MSYNC not done.
Module online notification received.
Local super-state is: SYSMGR_SUPERSTATE_STABLE
```

```
Standby super-state is: SYSMGR_SUPERSTATE_STABLE
Swover Reason : SYSMGR_UNKNOWN_SWOVER
Total number of Switchovers: 0
Swover threshold settings: 5 switchovers within 4800 seconds
Switchovers within threshold interval: 0
Last switchover time: 0 seconds after system start time
Cumulative time between last 0 switchovers: 0
Start done received for 1 plugins, Total number of plugins = 1


        Statistics:

Message count:           0
Total latency:           0          Max latency:           0
Total exec:              0          Max exec:              0
```

如果系统处于 HA 或冗余状态,那么从主用管理引擎切换到备用管理引擎不会对服务产生太大影响,通常都是在需要执行升级操作、MTS(Message and Transaction Service,消息和事务服务)(MTS)队列被卡住、管理引擎出现编程差错等情况时才执行切换操作。可以通过 **system switchover** 命令执行手动切换操作,执行该命令之后,备用管理引擎将接管主用角色,原来的主用管理引擎将重启。请注意,执行管理引擎切换操作之后,某些协议(无状态协议)可能会出现翻动故障,但并不影响流量转发。例 12-18 给出了管理引擎的手动切换示例,可以看出,系统因存在冗余管理引擎模块而实现了高可用性。

例 12-18 冗余模块切换

```
NX-1 SUP-1
NX-1# system switchover
NX-1#
User Access Verification
NX-1 login:
User Access Verification
NX-1 login:
>>>
>>>
>>>
NX7k SUP BIOS version ( 2.12 ) : Build - 05/29/2013 11:58:20
PM FPGA Version : 0x00000025
Power sequence microcode revision - 0x00000009 : card type - 10156EEA0
Booting Spi Flash : Primary
  CPU Signature - 0x000106e4: Version - 0x000106e0
  CPU - 2 : Cores - 4 : HTEn - 1 : HT - 2 : Features - 0xbfebfbff
  FSB Clk - 532 Mhz : Freq - 2140 Mhz - 2128 Mhz
  MicroCode Version : 0x00000002
  Memory - 32768 MB : Frequency - 1067 MHZ
  Loading Bootloader: Done
  IO FPGA Version     : 0x1000d
  PLX Version         : 861910b5
Bios digital signature verification - Passed
USB bootflash status : [1-1:1-1]

Reset Reason Registers: 0x1 0x0
 Filesystem type is ext2fs, partition type 0x83


             GNU GRUB   version 0.97
```

```
Autobooting bootflash:/n7000-s2-kickstart.7.3.2.D1.1.bin bootflash:/n7000-s2-dk
9.7.3.2.D1.1.bin...
 Filesystem type is ext2fs, partition type 0x83
! Output omitted for brevity
NX-1 SUP-2
NX-1 login: admin
Password:


Cisco Nexus Operating System (NX-OS) Software
TAC support: http://www.cisco.com/tac
! Output omitted for brevity
NX-1#
```

> **注**：手动切换过程中，原始主用管理引擎将重启以接管备用角色，如果新的主用管理引擎出现崩溃或重新加载，那么可能会导致整个系统重新加载并导致严重停机。因此，应该始终在计划维护时段执行手动切换操作。

12.2.2 ISSU

在任何网络（尤其是在大型数据中心和企业网络）中执行升级操作都不是一件轻松的事情，对于大多数情况来说，如果要升级设备，则必须将业务和流量转移到备用或冗余设备上并设置启动变量，然后通过 **reload** 命令关闭该设备以执行升级操作。对于 Nexus 7000 等设备来说，该操作将更具挑战性，因为设备运行了多个 VDC，这些 VDC 不但相当于多台设备，而且要承担不同的角色。为了克服网络升级带来的操作挑战，可以使用 ISSU 功能特性。

ISSU 并不是一个新概念，包括 Nexus 4500 和 Nexus 6500 交换机在内的多款思科 Catalyst 交换机都支持该功能特性。Nexus 7000 系列设备也支持相同的 ISSU 功能，整个 ISSU 进程包括以下步骤。

- 第 1 步：升级管理引擎和线卡模块上的 BIOS（Basic Input and Output System，基本输入输出系统）。
- 第 2 步：用新映像启动备用管理引擎。
- 第 3 步：从主用管理引擎切换到运行了新映像的备用管理引擎。
- 第 4 步：利用新映像启动原来的主用管理引擎卡。
- 第 5 步：执行无中断的线卡升级操作（每次一块线卡）。
- 第 6 步：升级 CMP（Connectivity Management Processor，连接管理处理器）。

> **注**：从 NX-OS 版本 5.2(1)开始，Nexus 交换机可以同时升级多块线卡，从而大大缩短了 ISSU 升级时间。

执行 ISSU 之前，特别是对软件进行降级的时候，需要对系统运行的现有软件版本与将要被降级到的旧映像之间的配置兼容性进行完整性检查，帮助网络管理员了解新版本支持但旧版本不支持的功能特性及配置信息，从而删除这些配置。从例 12-19 的 **how incompatibilityall system nx-os-file-name** 命令输出结果可以看出，存在配置不兼容问题。

例 12-19 验证配置不兼容

```
NX-1# show incompatibility-all system bootflash:n7000-s2-dk9.7.3.2.D1.1.bin

Checking incompatible configuration(s) for vdc 'NX-1':
------------------------------------------------------
No incompatible configurations
Checking dynamic incompatibilities for vdc 'NX-1':
------------------------------------------------------
```

```
No incompatible configurations

Checking incompatible configuration(s) for vdc 'TEST':
-------------------------------------------------------------
No incompatible configurations

Checking dynamic incompatibilities for vdc 'TEST2':
-------------------------------------------------------------
No incompatible configurations

Checking incompatible configuration(s) for vdc 'TEST3':
-------------------------------------------------------------
No incompatible configurations

Checking dynamic incompatibilities for vdc 'TEST4':
-------------------------------------------------------------
No incompatible configurations
```

可以通过命令 **install all kickstart** *kickstart-image* **system** *system-image* **[parallel]**执行 ISSU 升级操作，关键字 **parallel** 的作用是可以与 I/O 模块执行并行升级操作。ISSU 可以实现无中断软件升级，可以在不影响数据平面的情况下升级 Nexus 交换机的软件。对于无中断升级操作来说，该软件必须兼容各种版本，如果映像不兼容，那么升级操作就可能是中断性的。例 12-20 从映像 6.2(16)升级到 7.3(2)D1(1)的操作就是中断性操作，从输出结果可以看出，这两种映像不兼容，因而导致升级操作出现中断问题。

例 12-20   ISSU 升级

```
NX-1# install all kickstart bootflash:n7000-s2-kickstart.7.3.2.D1.1.bin system
  bootflash:n7000-s2-dk9.7.3.2.D1.1.bin
Installer will perform compatibility check first. Please wait.

Verifying image bootflash:/n7000-s2-kickstart.7.3.2.D1.1.bin for boot variable
  "kickstart".
[####################] 100% -- SUCCESS

Verifying image bootflash:/n7000-s2-dk9.7.3.2.D1.1.bin for boot variable "system".
[####################] 100% -- SUCCESS

Performing module support checks.
[####################] 100% -- SUCCESS

Verifying image type.
[####################] 100% -- SUCCESS

Extracting "system" version from image bootflash:/n7000-s2-dk9.7.3.2.D1.1.bin.
[####################] 100% -- SUCCESS

Extracting "kickstart" version from image bootflash:/n7000-s2-
kickstart.7.3.2.D1.1.bin.
[####################] 100% -- SUCCESS

Extracting "bios" version from image bootflash:/n7000-s2-dk9.7.3.2.D1.1.bin.
[####################] 100% -- SUCCESS

Extracting "lcflnn7k" version from image bootflash:/n7000-s2-dk9.7.3.2.D1.1.bin.
[####################] 100% -- SUCCESS
```

```
Notifying services about system upgrade.
[####################] 100% -- SUCCESS




Compatibility check is done:
Module  bootable         Impact   Install-type  Reason
------  --------   -------------  ------------  ------------------
    1      yes       disruptive      reset      Incompatible image
    2      yes       disruptive      reset      Incompatible image
    3      yes       disruptive      reset      Incompatible image
    4      yes       disruptive      reset      Incompatible image



Images will be upgraded according to following table:
Module     Image            Running-Version(pri:alt)        New-Version      Upg-Required
------  ----------    -------------------------------------  -----------     ------------
    1      system                                  6.2(16)    7.3(2)D1(1)        yes
    1    kickstart                                 6.2(16)    7.3(2)D1(1)        yes
    1        bios   v2.12.0(05/29/2013):v2.12.0(05/29/2013) v2.12.0(05/29/2013)   no
    2      system                                  6.2(16)    7.3(2)D1(1)        yes
    2    kickstart                                 6.2(16)    7.3(2)D1(1)        yes
    2        bios   v2.12.0(05/29/2013):v2.12.0(05/29/2013) v2.12.0(05/29/2013)   no
    3     lcflnn7k                                 6.2(16)    7.3(2)D1(1)        yes
    3        bios   v3.0.29(12/15/2015):v3.0.29(12/15/2015) v3.0.29(12/15/2015)   no
    4     lcflnn7k                                 6.2(16)    7.3(2)D1(1)        yes
    4        bios   v3.0.29(12/15/2015):v3.0.29(12/15/2015) v3.0.29(12/15/2015)   no


Switch will be reloaded for disruptive upgrade.
Do you want to continue with the installation (y/n)?  [n] y

Install is in progress, please wait.

Performing runtime checks.
[####################] 100% -- SUCCESS

Syncing image bootflash:/n7000-s2-kickstart.7.3.2.D1.1.bin to standby.
[####################] 100% -- SUCCESS

Syncing image bootflash:/n7000-s2-dk9.7.3.2.D1.1.bin to standby.
[####################] 100% -- SUCCESS

Setting boot variables.
[####################] 100% -- SUCCESS

Performing configuration copy.
[####################] 100% -- SUCCESS

Module 1:  Upgrading bios/loader/bootrom.
Warning: please do not remove or power off the module at this time.
[####################] 100% -- SUCCESS

Module 2:  Upgrading bios/loader/bootrom.
Warning: please do not remove or power off the module at this time.
[####################] 100% -- SUCCESS

Module 3:  Upgrading bios/loader/bootrom.
Warning: please do not remove or power off the module at this time.
```

```
[####################] 100% -- SUCCESS

Module 4:  Upgrading bios/loader/bootrom.
Warning: please do not remove or power off the module at this time.
[####################] 100% -- SUCCESS

Finishing the upgrade, switch will reboot in 10 seconds.
NX-1#
>>>
>>>
>>>
NX7k SUP BIOS version ( 2.12 ) : Build - 05/29/2013 11:58:20
PM FPGA Version : 0x00000025
Power sequence microcode revision - 0x00000009 : card type - 10156EEA0
Booting Spi Flash : Primary
  CPU Signature - 0x000106e4: Version - 0x000106e0
  CPU - 2 : Cores - 4 : HTEn - 1 : HT - 2 : Features - 0xbfebfbff
  FSB Clk - 532 Mhz : Freq - 2144 Mhz - 2128 Mhz
  MicroCode Version : 0x00000002
  Memory - 32768 MB : Frequency - 1067 MHZ
  Loading Bootloader: Done
  IO FPGA Version   : 0x1000d
  PLX Version       : 861910b5
Bios digital signature verification - Passed
USB bootflash status : [1-1:1-1]

Reset Reason Registers: 0x10 0x0
 Filesystem type is ext2fs, partition type 0x83


             GNU GRUB   version 0.97

Autobooting bootflash:/n7000-s2-kickstart.7.3.2.D1.1.bin bootflash:/n7000-s2-dk
9.7.3.2.D1.1.bin...
 Filesystem type is ext2fs, partition type 0x83
Booting kickstart image: bootflash:/n7000-s2-kickstart.7.3.2.D1.1.bin....
................................................................
................................................................
Kickstart digital signature verification Successful
Image verification OK

INIT: version 2boot device node /dev/sda
Bootflash firmware upgrade not required
boot device node /dev/sda
boot mirror device node /dev/sdb
Bootflash mirror firmware upgrade not required
boot mirror device node /dev/sdb
obfl device node /dev/sdc
OBFL firmware upgrade not required
obfl device node /dev/sdc
slot0 flash device node /dev/sdd
Checking obfl filesystem.
Checking all filesystems..r.r.r.retval=[1]
r done.
Starting mcelog daemon
Creating logflash directories
Loading system software
/bootflash//n7000-s2-dk9.7.3.2.D1.1.bin read done
```

```
System image digital signature verification successful.
Uncompressing system image: bootflash:/n7000-s2-dk9.7.3.2.D1.1.bin Sun Mar 5
  09:19:07 UTC 2017
blogger: nothing to do.
C
..done Sun Mar 5 09:19:12 UTC 2017
INIT: Entering runlevel: 3
Starting portmap daemon...
creating NFS state directory: done

System is coming up ... Please wait ...
System is coming up ... Please wait ...
System is coming up ... Please wait ...
System is coming up ... Please wait ...
```

软件升级的各个阶段都可能出现 ISSU 故障。
- 升级前和 BIOS 升级。
- 备用组件启动和切换。
- 线卡升级。

如果 ISSU 升级失败，那么最重要的操作就是要定位故障组件。此时，首先要收集以下日志信息。
- 安装程序日志。

    show system internal log install [details]
- Sysmgr 以及与 HA 相关的事件历史记录日志。

    show system internal log sysmgr state

    show system internal log sysmgr event-history errors
- Module-Manager 日志。

    show module internal event module *slot*
- 线卡升级日志。

    show system internal log sysmgr rtdbctrl

抓取相关日志信息之后，需要立即从 ISSU 故障状态恢复服务，此时可以使用命令 **install all**，该命令可以确保系统使用运行映像恢复正常，且确保所有模块均运行同一映像。

请注意，ISSU 升级机制可能无法兼容所有方案场景，如 OTV（对某些版本来说）、LACP 快速速率（Fast rate）以及网络中的连续 TCN。因此，在执行升级操作之前，必须仔细了解 CCO 提供的 ISSU 说明。

> **注**：如果出现了 ISSU 故障，那么在恢复服务之前请收集 **show tech-support issu** 和 **show tech-support ha** 命令的输出结果。

12.3 GIR

对于所有网络部署方案来说，为了排查故障根源，网络工程师都要执行硬件替换、硬件和软件替换或调试操作，此时，网络工程师都不希望影响网络上运行的任何服务。一般来说，提前规划维护窗口并将流量迁移到备用路径或冗余设备，可以最大程度地减少对现网服务的影响，但这是一项非常烦琐的任务。NX-OS 提供了 GIR（Graceful Insertion and Removal，平滑插拔）功能特性，可以将设备置于维护模式，在不影响任何服务的情况下执行上述操作。GIR 的目的是通过一组简单的命令来实现交换机与网络的隔离，而不用手动关闭接口或更改度量。也就是，可以将 GIR 理解为宏，能够自动执行一系列手动操作，实现交换机与网络的隔离。

GIR 支持两种操作模式。
- 维护模式（Maintenance mode）。
- 正常模式（Normal mode）。

在维护模式下（也称为平滑拔出阶段），所有数据流量都会绕过该节点，可以通过并行路径保证 GIR 的正常运行。如果没有可用的并行路径，那么就可能会导致网络服务出现中断。维护模式通常用于执行与维护相关的操作，如软件/硬件升级、替换故障硬件或者在节点上执行其他中断性操作。此后，节点可以返回正常模式（也称为"平滑插入"阶段）。

为了更好地理解 GIR 功能，下面以图 12-4 拓扑结构为例。该拓扑结构是典型的叶脊拓扑结构，包含两个脊节点和 6 个叶节点，叶节点与脊节点之间通过 OSPF 进行连接。

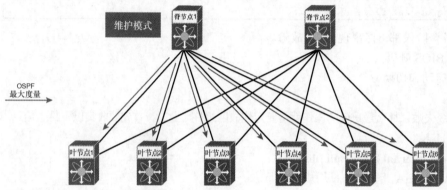

图 12-4　典型的叶脊拓扑结构

假设该拓扑中的脊节点 1 被设置为维护模式以执行软件升级操作。GIR 的第一步是在路由协议中宣告高开销度量，因而脊节点 1 向所有 OSPF 邻居宣告 OSPF 最大度量。叶节点收到最大度量之后，会改变转发路径，通过脊节点 2 来转发所有流量。此时，脊节点 1 与所有 6 个叶节点（假设处于默认的隔离模式，具体内容将在后面讨论）之间的 OSPF 邻居状态仍处于正常状态，但没有任何数据通过脊节点 1 进行转发。

Nexus 7000 和 Nexus 7700 系列平台从 Release 7.2.0 开始，Nexus 5500/5600 平台从 Release 7.1.0 开始，均支持维护模式。维护模式的配置命令是 **system mode maintenance [shutdown]**，配置了 **system mode maintenance** 命令之后，将以默认模式（也称为隔离模式[Isolate mode]）启用 GIR。该模式下的协议邻居关系仍处于维持状态，但流量均被转移到备用或并行路径上。配置了命令 **system mode maintenance shutdown** 之后，将以关闭模式（Shutdown mode）启用 GIR，该模式下的协议处于关闭状态，链路也处于关闭状态，流量也会出现丢失。通常建议 GIR 使用隔离模式，而非关闭模式。

隔离和关闭维护模式的配置方式不同，例 12-21 给出了这两种模式的功能特性配置示例。在这两种模式下，命令 **show system mode** 显示的系统模式均为 *Maintenance*。系统在进入维护模式之前，NX-OS 将设备的当前状态保存为名为 *before_maintenance* 的快照。

例 12-21　隔离和关闭维护模式

```
N7k-1(config)# system mode maintenance
Following configuration will be applied:

router bgp 100
  isolate
router eigrp 100
  isolate
router ospf 100
```

```
    isolate
router isis IS-IS
    isolate

Do you want to continue (yes/no)? [no] yes

Generating a snapshot before going into maintenance mode

Starting to apply commands...

Applying : router bgp 100
Applying :     isolate
Applying : router eigrp 100
Applying :     isolate
Applying : router ospf 100
Applying :     isolate
Applying : router isis IS-IS
Applying :     isolate

Maintenance mode operation successful.
N7k-1(config)#
2017 Mar 5 20:40:45 N7k-1 %$ VDC-2 %$ %MMODE-2-MODE_CHANGED: System changed to
  "maintenance" mode.

N7k-1# show system mode
System Mode: Maintenance
Maintenance Mode Timer: not running
N7k-1(config)# system mode maintenance shutdown

Following configuration will be applied:

router bgp 100
  shutdown
router eigrp 100
  shutdown
  address-family ipv6 unicast
    shutdown
router ospf 100
  shutdown
router isis IS-IS
  shutdown
system interface shutdown

NOTE: 'system interface shutdown' will shutdown all interfaces excluding mgmt 0
Do you want to continue (yes/no)? [no] yes

Generating a snapshot before going into maintenance mode

Starting to apply commands...

Applying : router bgp 100
Applying :     shutdown
Applying : router eigrp 100
Applying :     shutdown
Applying :     address-family ipv6 unicast
Applying :         shutdown
Applying : router ospf 100
Applying :     shutdown
```

```
Applying : router isis IS-IS
Applying :    shutdown
Applying : system interface shutdown

Maintenance mode operation successful.
```

系统进入维护模式之后，受维护模式影响的进程会将运行状态更改为 *Isolate* 或 *Shutdown*。例 12-22 显示了不同的路由协议进程及其在系统中的当前状态。

例 12-22 维护模式期间的路由协议状态

```
N7k-1# show bgp process

BGP Process Information
BGP Process ID                       : 20105
BGP Protocol Started, reason:        : configuration
BGP Protocol Tag                     : 100
BGP Protocol State                   : Running (Isolate)
BGP MMODE                            : Initialized
BGP Memory State                     : OK
BGP asformat                         : asplain
! Output omitted for brevity
N7k-1# show ip eigrp
IP-EIGRP AS 100 ID 0.0.0.0 VRF default
  Process-tag: 100
  Instance Number: 1
  Status: running (isolate)
  Authentication mode: none
  Authentication key-chain: none
! Output omitted for brevity
  Redistributed max-prefix: Disabled
  MMODE: Initialized
  Suppress-FIB-Pending Configured
N7k-1# show isis protocol

ISIS process : IS-IS
 Instance number :  1
 UUID: 1090519320
 Process ID 20143
VRF: default
  System ID : 0000.0000.0001  IS-Type : L1-L2
  SAP : 412   Queue Handle : 15
  Maximum LSP MTU: 1492
  Stateful HA enabled
  Graceful Restart enabled. State: Inactive
  Last graceful restart status : none
  Start-Mode Complete
  BFD IPv4 is globally disabled for ISIS process: IS-IS
  BFD IPv6 is globally disabled for ISIS process: IS-IS
  Topology-mode is base
  Metric-style : advertise(wide), accept(narrow, wide)
  Area address(es) :
    49.0001
  Process is up and running (isolate)
! Output omitted for brevity
N7k-1# show ip ospf internal

ospf 100 VRF default
```

```
ospf process tag 100
ospf process instance number 1
ospf process uuid 1090519321
ospf process linux pid 20064
ospf process state running(isolate)
System uptime 05:18:06
SUP uptime 2 05:18:06
Server up         : L3VM|IFMGR|RPM|AM|CLIS|URIB|U6RIB|IP|IPv6|SNMP|BGP|MMODE
Server required  : L3VM|IFMGR|RPM|AM|CLIS|URIB|IP|SNMP
Server registered: L3VM|IFMGR|RPM|AM|CLIS|URIB|IP|SNMP|BGP|MMODE
Server optional  : BGP|MMODE
Early hello : OFF
Force write PSS: FALSE
OSPF mts pkt sap 324
OSPF mts base sap 320
```

执行了维护操作之后，可以通过 **no system mode maintenance** 命令让系统退出维护模式。配置该命令之后，系统将回滚到正常模式，并且在隔离或关闭维护模式期间所做的所有配置变更也都将进行回滚。例 12-23 解释了系统从维护模式切换到正常模式的过程，此后生成了一个名为 *after_maintenance* 的快照。

例 12-23　从维护模式切换到正常模式

```
N7k-1(config)# no system mode maintenance

Following configuration will be applied:

router isis IS-IS
  no isolate
router ospf 100
  no isolate
router eigrp 100
  no isolate
router bgp 100
  no isolate

Do you want to continue (yes/no)? [no] yes

Starting to apply commands...

Applying : router isis IS-IS
Applying :   no isolate
Applying : router ospf 100
Applying :   no isolate
Applying : router eigrp 100
Applying :   no isolate
Applying : router bgp 100
Applying :   no isolate
Maintenance mode operation successful.

The after_maintenance snapshot will be generated in 120 seconds
After that time, please use 'show snapshots compare before_maintenance after_
  maintenance' to check the health of the system
```

系统返回正常模式之后，需要验证服务是否已恢复正常，包括 RIB（Routing Information Base，路由信息库）中的路由以及 VLAN 等。可以通过同一条命令来验证维护前后的快照信息，命令 **show snapshots** 可以验证当前可用快照，如果同时存在维护前后的快照，那么就可以通过命令 **show** snapshots compare *before_maintenance after_maintenance* [summary]来对比维护前后的差异情况。例12-24给出了维护前后的快照对比信息。

例12-24 维护前后的快照对比

```
N7k-1# show snapshots
Snapshot Name              Time                          Description
--------------------------------------------------------------------------------
after_maintenance          Wed Nov  1 02:42:07 2017      system-internal-snapshot
before_maintenance         Wed Nov  1 02:38:01 2017      system-internal-snapshot
N7k-1# show snapshots compare before_maintenance after_maintenance summary

====================================================================================
Feature                            before_maintenance after_maintenance changed
====================================================================================
basic summary
  # of interfaces                          63                 63
  # of vlans                                1                  1
  # of ipv4 routes vrf default             43                 43
  # of ipv4 paths  vrf default             46                 46
  # of ipv4 routes vrf management           9                  9
  # of ipv4 paths  vrf management           9                  9
  # of ipv6 routes vrf default              3                  3
  # of ipv6 paths  vrf default              3                  3
interfaces
  # of eth interfaces                      60                 60
  # of eth interfaces up                    7                  7
  # of eth interfaces down                 53                 53
  # of eth interfaces other                 0                  0
  # of vlan interfaces                      1                  1
  # of vlan interfaces up                   0                  0
  # of vlan interfaces down                 1                  1
  # of vlan interfaces other                0                  0
```

大多数生产环境对于维护时长有一定的限制要求。如果要为维护窗口设置系统时限，那么就可以通过命令 **system mode maintenance timeout** *time-in-minutes* 为维护模式设置超时时间。如果达到了超时值，那么系统就会自动从维护模式回滚到正常模式。例12-25将维护超时时间配置为30min，并通过命令 **show maintenance timeout** 加以验证。

例12-25 设置维护模式超时时间

```
N7k-1(config)# system mode maintenance timeout 30
Timer will be started for 30 minutes when the system switches to maintenance mode.
N7k-1# show maintenance timeout
Maintenance mode timeout value: 30 minutes
```

并非所有维护窗口都应该是非中断性的，某些维护窗口可能需要重新加载系统，某些维护窗口可能需要重现故障问题而必须自动重启交换机。因此，进入维护模式之前，必须要为系统的重新加载定义重置原因（reset-reason），可以通过命令 **ystem mode maintenance on-reload reset-reason** *options* 为维护窗口内计划的各种重新加载情形设置选项，可以为重置原因设置多种选项。例12-26显示了所有可用的重置原因选项，并解释了如何为维护模式设置多种重置原因选项。可以通过命令 **show maintenance on-reload reset-reasons** 验证维护窗口内重新加载事件所定义的重置原因。

例 12-26 配置和验证重新加载事件的重置原因

```
Spine2(config)# system mode maintenance on-reload reset-reason ?
  HW_ERROR         Hardware Error
  SVC_FAILURE      Critical service failure
  KERN_FAILURE     Kernel panic
  WDOG_TIMEOUT     Watchdog reset
  FATAL_ERROR      Fatal errors
  LC_FAILURE       LC failure
  MANUAL_RELOAD    Manual reload
  MAINTENANCE      Maintenance mode
  ANY_OTHER        Any other reset
  MATCH_ANY        Any of the above listed reasons
Spine2(config)# system mode maintenance on-reload reset-reason MANUAL_RELOAD
Spine2(config)# system mode maintenance on-reload reset-reason MAINTENANCE
Spine2# show maintenance on-reload reset-reasons
Reset reasons for on-reload maintenance mode:
---------------------------------------------
MANUAL_RELOAD
MAINTENANCE

bitmap = 0xc0
```

> **注：** 如果出现了与维护模式相关的故障问题，那么就可以在故障期间或故障之后立即收集命令 **show tech support mmode** 的输出结果。

自定义维护配置文件

　　隔离模式 GIR 和关闭模式 GIR 各有优势，但并非始终有效。例如，如果 Nexus 交换机充当 BGP 路由反射器且不在数据路径中，那么好的处理方式可能是不关闭该设备的 BGP 或者将该设备与 BGP 进程相隔离，此时关闭维护模式或隔离维护模式都有可能影响服务。对于这类场景来说，可以创建自定义维护配置文件。

　　目前支持两类维护配置文件。

- 维护模式（Maintenance-mode）。
- 正常模式（Normal-mode）。

　　系统进入维护模式并切换回正常模式之后，将首先生成这些配置文件的配置信息。创建自定义配置文件时，虽然配置文件的名称保持不变，但可以修改配置文件中的配置信息。创建自定义配置文件时，命令将附加到现有维护配置文件中。因此，第一步是检查是否已定义了维护配置文件，可以通过命令 **show maintenance profile** 加以验证，如例 12-27 所示。

例 12-27 验证维护模式和正常模式配置文件的配置信息

```
N7k-1# show maintenance profile
[Normal Mode]
router isis IS-IS
  no isolate
router ospf 100
  no isolate
router eigrp 100
  no isolate
router bgp 100
  no isolate

[Maintenance Mode]
router bgp 100
```

```
    isolate
router eigrp 100
    isolate
router ospf 100
    isolate
router isis IS-IS
    isolate
```

如果维护模式和正常模式配置文件不为空,那么最好删除现有的维护配置文件,然后重新创建自定义配置文件。可以通过命令 **no configure maintenance profile [maintenance- mode | normal-mode]**删除维护配置文件,需要通过 Exec 模式执行该命令。删除现有配置文件的配置内容之后,可以通过命令 **configure maintenance profile [maintenance-mode | normal-mode]**从配置模式配置自定义配置文件。如果同时配置了自定义维护模式和正常模式配置文件,那么还需要配置命令 **system mode maintenance always-usecustom-profile**,以确保不生成和使用系统生成的配置文件的配置信息。例 12-28 给出了自定义维护模式和正常模式配置文件的配置步骤。本例配置了维护模式配置文件,以隔离 BGP 和 IS-IS 协议,但关闭 OSPF、EIGRP 以及接口 Ethernet 3/1。除配置自定义维护配置文件之外,还必须保存这些配置,使即便重新加载之后仍能保留这些自定义配置文件。

例 12-28 配置自定义维护配置文件

```
N7k-1# no configure maintenance profile maintenance-mode
Maintenance mode profile maintenance-mode successfully deleted
Enter configuration commands, one per line.  End with CNTL/Z.
Exit maintenance profile mode.
N7k-1# no configure maintenance profile normal-mode
Maintenance mode profile normal-mode successfully deleted
Enter configuration commands, one per line.  End with CNTL/Z.
Exit maintenance profile mode.
N7k-1(config)# configure maintenance profile maintenance-mode
Please configure 'system mode maintenance always-use-custom-profile' if you want to
  use custom profile always for maintenance mode.
N7k-1(config-mm-profile)#
N7k-1(config-mm-profile)# router bgp 100
N7k-1(config-mm-profile-router)# isolate
N7k-1(config-mm-profile-router)# router ospf 100
N7k-1(config-mm-profile-router)# shutdown
N7k-1(config-mm-profile-router)# router eigrp 100
N7k-1(config-mm-profile-router)# shutdown
N7k-1(config-mm-profile-router)# router isis IS-IS
N7k-1(config-mm-profile-router)# isolate
N7k-1(config-mm-profile-router)# interface e3/1
N7k-1(config-mm-profile-if-verify)# shutdown
N7k-1(config-mm-profile-if-verify)# end

N7k-1(config)# configure maintenance profile normal-mode
Please configure 'system mode maintenance always-use-custom-profile' if you want to
  use custom profile always for maintenance mode.
N7k-1(config-mm-profile)# router ospf 100
N7k-1(config-mm-profile-router)# no shutdown
N7k-1(config-mm-profile-router)# router eigrp 100
N7k-1(config-mm-profile-router)# no shutdown
N7k-1(config-mm-profile-router)# router isis IS-IS
N7k-1(config-mm-profile-router)# no isolate
N7k-1(config-mm-profile-router)# router bgp 100
N7k-1(config-mm-profile-router)# no isolate
```

```
N7k-1(config-mm-profile-router)# interface ethernet 3/1
N7k-1(config-mm-profile-if-verify)# no shutdown
N7k-1(config-mm-profile-if-verify)# exit
N7k-1(config-mm-profile)# exit
N7k-1(config)# system mode maintenance always-use-custom-profile

N7k-1# copy running-config startup-config
[#########################################] 100%
N7k-1# show maintenance profile
[Normal Mode]
router ospf 100
  no shutdown
router eigrp 100
  no shutdown
router isis IS-IS
  no isolate
router bgp 100
  no isolate
interface Ethernet3/1
  no shutdown
[Maintenance Mode]
router bgp 100
  isolate
router ospf 100
  shutdown
router eigrp 100
  shutdown
router isis IS-IS
  isolate
interface Ethernet3/1
  shutdown
```

注：可以通过命令 show running-config mmode 验证与维护模式相关的所有配置信息。

如果要利用自定义配置文件激活维护方式，那么就可以配置命令 **system mode maintenance dont-generate-profile**，该命令使用 Nexus 交换机创建的自定义配置文件的配置信息进入维护模式。例 12-29 给出了以自定义配置文件激活维护模式的示例。

例 12-29 利用自定义配置文件激活维护模式

```
N7k-1(config)# system mode maintenance dont-generate-profile

Following configuration will be applied:

router bgp 100
  isolate
router ospf 100
  shutdown
router eigrp 100
  shutdown
router isis IS-IS
  isolate
interface Ethernet3/1
  shutdown

Do you want to continue (yes/no)? [no] yes

Generating a snapshot before going into maintenance mode
```

```
Starting to apply commands...
Applying : router bgp 100
Applying :   isolate
Applying : router ospf 100
Applying :   shutdown
Applying : router eigrp 100
Applying :   shutdown
Applying : router isis IS-IS
Applying :   isolate
Applying : interface Ethernet3/1
Applying :   shutdown

Maintenance mode operation successful.
```

> **注**：如果要调试维护模式，那么就可以使用命令 **debug mmode logfile**。启用该调试操作的同时，还会将调试日志记录到日志文件中（可以使用命令 **show system internal mmode logfile** 查看该日志文件）。出现了 GIR 故障之后，建议收集 **show tech-support mmode** 命令的输出结果。

12.4 本章小结

作为优秀的数据中心交换机操作系统，NX-OS 在设计之初就全面融入了 HA（High Availability，高可用性）理念。本章重点介绍了 Nexus 交换机的常见高可用性功能，包括通过 BFD 实现高可用性，该功能特性可以与各种路由协议及功能特性协同使用。本章详细讨论了通过硬件编程及事件历史日志排查 BFD 故障的方法，排查 BFD 会话故障时应验证以下信息。

- 确保在接口上启用了 **no ip redirects** 或 **no ipv6 redirects** 命令。
- 验证故障原因代码，找出 BFD 故障原因。
    - 0：无诊断。
    - 1：控制报文检测定时器超时。
    - 2：回显功能故障。
    - 3：邻居告知会话中断。
    - 4：转发平面重置。
    - 5：路径中断。
    - 6：级联路径中断。
    - 7：管理性中断。
    - 8：反向级联路径中断。

此外，本章还介绍了多种系统 HA 功能特性，如 SSO 和 ISSU，这些功能特性对于生产环境来说至关重要。与通过 **reload** 命令执行升级操作相比，无中断的增量 ISSU 升级方式要好得多。

接下来本章讨论了 GIR 功能特性，解释了通过 GIR 执行网络维护操作的方式及优点。可以通过以下两种模式下启用 GIR 的维护模式。

- 隔离模式。
- 关闭模式。

建议使用隔离模式 GIR。最后，本章还详细介绍了如何为维护窗口创建和使用自定义配置文件，而不是使用系统生成的配置文件。

第五部分

多播网络流量

第 13 章 多播故障排查

第 13 章

多播故障排查

本章主要讨论如下主题。
- 多播基础知识。
- NX-OS 多播体系架构。
- IGMP（Internet Group Management Protocol，Internet 组管理协议）协议操作。
- IGMP 配置和验证。
- PIM（Protocol Independent Multicast，协议无关多播）协议操作。
- PIM 配置和验证。
- 多播和 vPC（virtual Port-Channel，虚拟端口通道）。
- 多播 Ethanalyzer 案例。

当前几乎所有的网络都存在多播流量。多播通信的概念很容易理解，主机发送一条旨在供多个接收端使用的消息，这些接收端可以侦听感兴趣的多播流量，而忽略其他流量，从而能够更加有效地利用系统资源。不过，将这个简单概念引入现代网络时可能会产生混淆和误解。本章将讨论如何利用思科 NX-OS 实现多播通信，讨论完基本概念之后，将通过一些示例来说明如何验证控制平面和数据平面是否按预期运行。由于多播内容非常宽泛，因而不可能提供所有功能特性的示例。本章将主要讨论较常见的 IPv4 多播部署技术，不涉及 IPv6 多播通信。

13.1 多播基础知识

通常可以将网络通信描述为以下类型。
- 单播（一对一）。
- 广播（一对全部）。
- 任播（一对最近的一个）。
- 多播（一对多）。

简单来说，单播流量就是单台源主机将数据包发送给单台目的主机。任播是另一种单播流量，由多台目的设备共享相同的网络层地址，流量则来自拥有目的任播地址的单台主机，数据包按照单播路由到达最近的任播主机，由路由度量来确定最近的设备。

广播和多播都提供了一种在网络上进行一对多通信的方法。多播通信与广播通信的不同之处在于，广播通信中的每台主机都必须接收和处理广播流量，导致设备必须消耗系统资源来处理那些最终被丢弃的数据帧，而多播流量仅由对接收流量感兴趣的设备进行处理。此外，多播流量可以跨越三层（L3）子网边界进行路由，而广播流量则仅限于本地子网。图 13-1 给出了广播通信与多播通信之间的行为差异。

图 13-1 中的 NX-2 被配置为在两个 L3 子网之间进行路由，主机 H3 发送了一个广播数据包，目的 IP 地址为 255.255.255.255，目的 MAC 地址为 ff:ff:ff:ff:ff:ff。图中以黑色箭头来表示广播流

量，广播包在 L2 交换机的所有端口进行泛洪，子网 10.12.1.0/24 中的每台设备都会收到该广播包。主机 H1 是唯一运行了需要接收该广播包的应用程序的设备，其他设备接收该广播包都只会浪费带宽和数据包处理资源。NX-2 收到该广播包之后，并不会将广播包转发给子网 10.12.2.0/24，该操作行为将广播通信范围限定在同一个广播域或 L3 子网内的设备。从图 13-1 可以看出，如果子网内的部分主机不需要接收这些数据包，那么使用广播通信方式会带来潜在的效率低下问题。

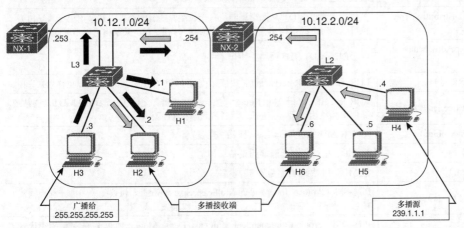

图 13-1 多播通信与广播通信

主机 H4 正在将由浅色箭头表示的多播流量发送给组地址 239.1.1.1，L2 交换机对这些多播包的处理方式有所不同，仅泛洪给主机 H6 和 NX-2，其中，NX-2 充当 L3 多播路由器（MROUTER）。NX-2 执行多播路由并将流量转发给 L2 交换机，最终由 L2 交换机将数据包转发给主机 H2。由于 NX-1 没有接收多播流量，因而 L2 交换机不将其视为 MROUTER。如果将 NX-1 重新配置多播路由器，连接了对多播流量感兴趣的接收端，那么 NX-1 就会接收这些多播包并路由给子网上的接收端。这里解释 NX-1 操作行为的目的是说明多播包的传播范围受到多播源在 IP 报头设置的 TTL（Time To Live，生存时间）值的限制，而不像广播那样受到 L3 子网边界的限制。此外，多播范围还受到管理边界、ACL（Access Control List，访问控制列表）或其他与协议相关的过滤技术的限制。

13.1.1 多播术语

在探讨具体的多播技术之前，有必要首先明确描述多播状态和多播行为的一些专用术语。表 13-1 列出了在本章将要使用的多播术语及其定义。

表 13-1　　　　　　　　　　　　　　　多播术语

术语	定义
MROUTE（多播路由）	MRIB（Multicast Routing Information Base，多播路由信息库）中的表项，不同类型的 MROUTE 表项与有源树或共享树相关联
IIF（Incoming Interface，入接口）	设备上将要接收多播流量的接口
OIF（Outgoing Interface，出接口）	设备上将要向外（发送给接收端）多播流量的接口
OIL（Outgoing Interface List，出接口列表）	设备上将多播流量发送给对特定 MROUTE 表项感兴趣的接收端的一组接口
组地址（Group address）	多播组的目的 IP 地址
源地址（Source address）	多播源的单播地址，也称为发送端地址
L2 复制（L2 replication）	沿多播分发树的分支点处复制多播包的行为。在 L2 复制多播流量，不需要重写源 MAC 地址或递减 TTL，且数据包位于同一广播域内

续表

术语	定义
L3 复制（L3 replication）	沿多播分发树的分支点复制多播包的行为。在 L3 复制多播流量，需要 PIM 状态和多播路由，多播路由器会更新源 MAC 地址并递减 TTL
RPF（Reverse Path Forwarding，反向路径转发）检查	将多组流量的 IIF 与源 IP 地址或 RP（Rendezvous Point，聚合点）地址的路由表表项进行对比，确保该多播流量来自源端
MDT（Multicast Distribution Tree，多播分发树）	多播流量通过 MDT 从源端流向所有接收端。所有源都可以共享该树（共享树），也可以为每个源都构建一棵单独的分发树（有源树）。共享树可以是单向或双向的
PIM（Protocol Independent Multicast，协议无关多播）	用于创建 MDT 的多播路由协议
RPT（RP Tree，RP 树）	LHR（Last-Hop Router，最后一跳路由器）与 PIM RP 之间的 MDT，也称为共享树
SPT（Shortest-Path Tree，最短径树）	LHR 与 FHR（First-Hop Router，第一跳路由器）之间去往源端的 MDT，通常遵循由单播路由度量确定的最短路径，也称为有源树
分叉点（Divergence point）	RPT 和 SPT 去往不同上游设备的分叉位置
上游（Upstream）	沿 MDT 相对更靠近源端的设备
下游（Downstream）	沿 MDT 相对更靠近接收端的设备
稀疏模式（Sparse mode）	PIM-SM（Protocol Independent Multicast, Sparse Mode，协议无关多播—稀疏模式）依赖于 PIM 邻居的显式加入，然后向接收端发送流量
密集模式（Dense mode）	PIM-DM（Protocol Independent Multicast, Dense Mode，协议无关多播-密集模式）依赖于泛洪/剪除转发行为，源端会向所有可能的接收端发送流量，直至收到对多播流量不感兴趣的下游 PIM 邻居的剪除消息为止。NX-OS 不支持 PIM-DM
RP（Rendezvous Point，聚合点）	多播路由器，是 PIM-SM 共享多播分发树的根
加入（Join）	一种 PIM 消息，一般来说，指的是下游设备请求特定组或源端流量的操作，该操作可以将接口添加到 OIL 中
剪除（Prune）	一种 PIM 消息，一般来说，指的是下游设备表明接收端不再请求该组或源端的流量，如果没有其他下游 PIM 邻居，那么该操作就可以将接口从 OIL 中删除
FHR（First-Hop Router，第一跳路由器）	与多播源直接相邻的 L3 路由器。FHR 通过 PIM RP 执行源注册操作
LHR（Last-Hop Router，最后一跳路由器）	与多播接收端直接相邻的 L3 路由器。LHR 向 PIM RP 发起加入操作，并从 RPT 切换到 SPT
中间路由器（Intermediate router）	启用 L3 多播的路由器，为 MDT 转发数据包

图 13-2 的多播拓扑结构很好地解释了表 13-1 列出的相关术语。

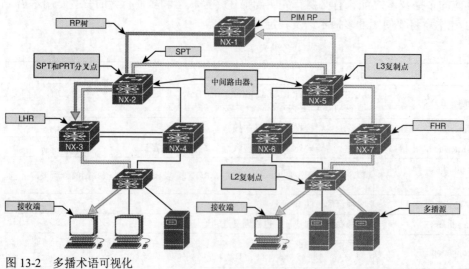

图 13-2　多播术语可视化

图 13-2 给出了 PIM-SM ASM（Any-Source Multicast，任意源多播）典型部署方案。为了实现从源端到接收端的端到端的流量流，需要通过以下中间步骤来构建 MDT。

- 第 1 步：向 PIM RP 注册源端。
- 第 2 步：建立从 RP 到接收端的 RPT。
- 第 3 步：建立从源端到接收端的 SPT。

排查多播故障时，可以根据当前网络状态来确定已经完成了上述中间步骤中的哪些步骤。本章将深入分析这些中间步骤所包含的各种检查、条件和协议状态机等内容。

> 注：出于解释目的，图 13-2 同时显示了 RP 树和有源树。但这种状态并不会持续存在，因为 NX-3 会将自己从 RP 树中剪除，并从有源树接收多播组流量。

13.1.2 二层多播地址

主机在 L2 使用 MAC（Media Access Control，介质访问控制）地址进行通信，MAC 地址的长度为 48 比特，是 LAN 网段上的 NIC（Network Interface Card，网络接口卡）的唯一标识符。MAC 地址采用 0012.3456.7890 或 00:12:34:56:78:90 等格式的 12 位十六进制数字表示。

主机使用的 MAC 地址通常由制造商分配，也称为 BIA（Burned-In-Address，烧录地址）。同一个 IP 子网中的两台主机进行通信时，L2 帧的目的地址将设置为目的设备的 MAC 地址。收到数据帧之后，如果目的 MAC 地址与主机的 BIA 相匹配，那么就接收该帧并传递给高层应用进行进一步处理。

主机之间的广播消息发送给保留地址 FF:FF:FF:FF:FF:FF，接收广播消息的主机必须处理该帧并传递给高层执行进一步处理，高层应用可能会丢弃该帧，也可能对该帧进行处理。如前所述，如果应用程序不要求网络中的所有主机都接收这些流量，那么与广播通信方式相比，多播通信方式的效率更高。

多播通信需要某种方法来识别二层帧，这些帧虽然不是广播帧，但仍然可以由 LAN 网段上的一台或多台主机进行处理，既要允许对该流量感兴趣的主机处理这些帧，又要允许对该流量不感兴趣的主机丢弃这些帧，从而节约设备的处理和缓存资源。

多播 MAC 地址可以在二层将多播与单播或广播帧区分开。RFC 1112 指定了多播 MAC 地址的保留范围是 01:00:5E:00:00:00 ~ 01:00:5E:7F:FF:FF，前 24 位始终是 01:00:5E。第一个字节包含了 I/G（Individual/Group，单个/组）比特，该比特设置为 1，则表明是多播 MAC 地址。第 25 比特始终为 0，因而还剩下 23 比特，三层组地址将映射到这 23 比特，从而构成完整的多播 MAC 地址，如图 13-3 所示。

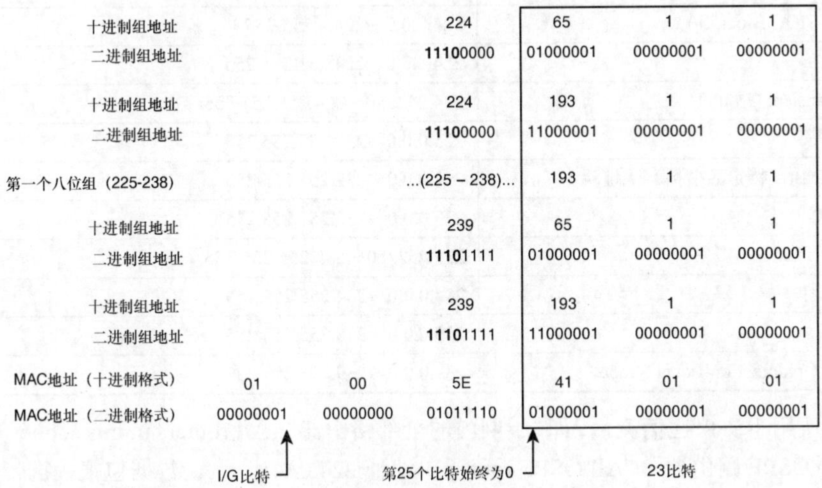

十进制组地址		224	65	1	1
二进制组地址		11100000	01000001	00000001	00000001
十进制组地址		224	193	1	1
二进制组地址		11100000	11000001	00000001	00000001
第一个八位组 (225-238)		...(225 – 238)...	193	1	1
十进制组地址		239	65	1	1
二进制组地址		11101111	01000001	00000001	00000001
十进制组地址		239	193	1	1
二进制组地址		11101111	11000001	00000001	00000001
MAC地址（十进制格式）	01	00	5E	41	01	01
MAC地址（二进制格式）	00000001	00000000	01011110	01000001	00000001	00000001

↑ I/G比特　　↑ 第25个比特始终为0　　　23比特

图 13-3 三层组地址到多播 MAC 地址的映射

将地址扩展成二进制格式，就能明显看出，多个 L3 组地址将映射为同一个多播 MAC 地址。实际上，32 个 L3 多播组地址映射为一个多播 MAC 地址，这是因为 L3 组地址中有 9 个比特未映射多播 MAC 地址，第一个八位组的 4 个高阶比特始终为 1110，其余 4 比特可变。需要记住的是，多播组 IP 地址的第一个八位组的取值范围是 224~239。L3 组地址映射到多播 MAC 地址时，第三个八位组的第一个高阶比特将被忽略。多播 MAC 地址的第 25 比特始终设置为零。综上所述，这 5 比特的潜在取值范围有 32（2^5）种，这就是 32 个多播组地址映射为同一个多播 MAC 地址的原因。

对于主机来说，如果配置 NIC 监听特定多播 MAC 地址，那么这种地址重叠问题就会导致主机收到多个多播组的帧。例如，假设某个多播源在 LAN 网段上处于活动状态，且正在生成去往 233.65.1.1、239.65.1.1 和 239.193.1.1 的多播组流量，这些多播组都映射为同一个多播 MAC 地址。如果主机仅对 239.65.1.1 的多播包感兴趣，那么将无法在 L2 区分不同的多播组，必须将所有帧都传送给高层，由高层应用丢弃不需要的帧，并将感兴趣的帧发送给应用程序进行处理。因此，确定多播组编址方案时必须考虑 32:1 的重叠问题。此外，最好避免使用组 X.0.0.Y 和 X.128.0.Y，因为这些组的多播 MAC 地址与 224.0.0.X 重叠，交换机会在同一 VLAN 的所有端口上泛洪多播帧。

13.1.3　三层多播地址

可以从第一个八位组来识别 IPv4 多播地址，多播地址的第一个八位组位于 224.0.0.0~239.255.255.255 之间，也称为 D 类地址。以二进制格式来看，就可以发现多播地址的第一个八位组的前 4 个比特始终为 1110。多播不存在子网划分的概念，因为每个多播地址都表示一个单独的多播组地址。不过，多播地址 224.0.0.0/4 范围内的各种地址块都有不同的特定用途。表 13-2 列出了 IANA（Internet Assigned Numbers Authority，Internet 号码分配机构）定义的多播地址范围。

表 13-2　IPv4 多播地址空间注册情况

名称	多播地址空间
本地网络控制地址块（Local Network Control Block）	224.0.0.0~224.0.0.255
网络互联控制地址块（Internetwork Control Block）	224.0.1.0~224.0.1.255
AD-HOC 第 II 地址块（AD-HOC Block I）	224.0.2.0~224.0.255.255
保留	224.1.0.0~224.1.255.255
SDP/SAP 地址块(SDP/SAP Block)	224.2.0.0~224.2.255.255
AD-HOC 第 II 地址块（AD-HOC Block II）	224.3.0.0~224.4.255.255
保留	224.5.0.0~224.251.255.255
DIS 临时组地址块（DIS Transient Group）	224.252.0.0~224.255.255.255
保留	225.0.0.0~231.255.255.255
SSM（Source-Specific Multicast，特定源多播）地址块	232.0.0.0~232.255.255.255
GLOP 地址块（GLOP Block）	233.0.0.0~233.251.255.255
AD-HOC 第 III 块（AD-HOC Block III）	233.252.0.0~233.255.255.255
基于单播前缀的 IPv4 多播地址块	234.0.0.0~234.255.255.255
保留	235.0.0.0~238.255.255.255
组织机构本地范围地址块（Organization-Local Scope）	239.0.0.0~239.255.255.255

本地网络控制地址块用于协议通信流量，如"子网内的全部路由器"（All routers in this subnet）地址 224.0.0.2 和"全部 OSPF 路由器"（All OSPF routers）地址 224.0.0.5。无论数据包报头携带的 TTL 值是什么，多播路由器都不应该转发该地址范围内的数据包。事实上，使用本地网络控

制地址块的协议数据包在发送时总是将 TTL 设置为 1。

网络互联控制地址块用于多播路由器在子网之间转发的协议通信流量或者转发给 Internet 的协议通信流量，如 Cisco-RP-Announce 地址 224.0.1.39、Cisco-RP-Discovery 地址 224.0.1.40 和 NTP 地址 224.0.1.1。

表 13-3 列出了本地网络控制地址块和网络互联控制地址块中由控制平面协议使用的周知多播地址。请注意，为了在排查控制平面故障时更快地识别这些地址，必须熟悉这些特殊保留地址。

表 13-3　　　　　　　　　　　　　　保留的周知多播地址

名称	多播地址
子网内的全部主机（All Host in this subnet）（all-host）	224.0.0.1
子网内的全部路由器（All Routers in this subnet）（all-router）	224.0.0.2
全部 OSPF 路由器（All OSPF router）（AllSPFRouter）	224.0.0.5
全部 OSPF DR（All OSPF DR）（AllDRouter）	224.0.0.6
全部 RIPv2 路由器（All RIPv2 router）	224.0.0.9
全部 EIGRP 路由器（All EIGRP router）	224.0.0.10
全部 PIM 路由器（All PIM router）	224.0.0.13
VRRP	224.0.0.18
IGMPv3	224.0.0.22
HSRPv2 和 GLBP	224.0.0.102
NTP	224.0.1.1

SSM（Source-Specific Multicast，特定源多播）地址块用于 SSM，SSM 是 PIM-SM 的扩展，具体内容将在本章后面进行介绍。如果主机应用能够感知多播组的特定源 IP 地址，那么就可以对一对多应用进行优化。知道了源地址之后，就无须使用 PIM RP，也不需要多播路由器维护共享树的状态。

组织机构本地范围地址块也称为管理范围地址块（Administratively Scoped Block），这些地址是与 RFC 1918 单播 IP 地址相对应的多播地址（组织机构可以根据需要自行分配这些地址），IANA 不公开路由或管理这些多播地址。

13.2　NX-OS 多播体系架构

NX-OS 的多播体系架构与操作系统的设计原理一脉相承，所有组件进程都采用了模块化设计方式，为 HA（High Availability，高可用性）、可靠性和可扩展性奠定了坚实基础。

NX-OS HA 体系架构支持状态化进程重启和 ISSU（In-Service Software Upgrade，不中断软件升级），可以将数据平面的影响最小化。从图 13-4 可以看出，NX-OS 的多播体系架构采用了分布式架构，包括运行在管理引擎模块上的 PI（Platform-Independent，平台无关）组件以及运行在 I/O 模块或系统 ASIC（Application-Specific Integrated Circuit，专用集成电路）上负责流量转发的专用硬件组件。

该通用架构适用于所有 NX-OS 平台，但每种平台实现的转发组件有所不同，具体取决于特定硬件 ASIC 的能力。

每种协议（如 IGMP[Internet Group Management Protocol，Internet 组管理协议]、PIM[Protocol

Independent Multicast，协议无关多播]和 MSDP[Multicast Source Discovery Protocol，多播源发现协议])都采用独立运行方式，拥有自己的进程状态，且使用 NX-OS PSS（Persistent Storage Service，持久存储服务）存储这些进程状态。此外，还通过 MTS（Message and Transactional Service，消息和事务服务）与其他服务（如 MRIB[Multicast Routing Information Base，多播路由信息库])进行通信并交换协议状态消息。

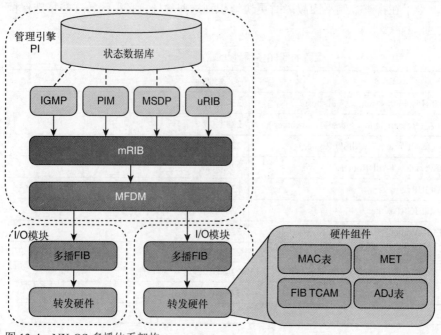

图 13-4 NX-OS 多播体系架构

MRIB 由客户端协议（如 PIM、IGMP 和 MSDP）进行填充，以创建多播路由状态表项。这些 MROUTE 状态描述了路由器与特定 MDT 的关系，并由各种 MRIB 客户端协议（如 IGMP、PIM 和 IP）进行填充。MRIB 创建了 MROUTE 状态之后，会将这些状态推送给 MFDM（Multicast Forwarding Distribution Manager，多播转发分发管理器）。

MRIB 与 URIB（Unicast Routing Information Base，单播路由信息库）进行交互以获取路由协议度量和下一跳信息，从而在 RPF（Reverse Path Forwarding，反向路径转发）查找期间使用这些信息。此外，由管理引擎在软件转发路径中路由的多播包也由 MRIB 负责处理。

MFDM 是 MRIB 与平台转发组件之间的中间组件，负责从 MRIB 获取 MROUTE 状态，并为每个表项分配平台资源。MFDM 可以将 MRIB 转换为平台组件能够理解的数据结构。如果是分布式平台（如 Nexus 7000 系列），那么 MFDM 就会将数据结构推送给每个 I/O 模块；如果是非模块化平台，那么 MFDM 就会将信息分发给平台转发组件。

MFIB（Multicast Forwarding Information Base，多播转发信息库）负责将从 MFDM 收到的 (*,G)、(S,G) 以及 RPF 表项编程到硬件转发表中，这些硬件转发表被称为 FIB TCAM（Ternary Content-Addressable Memory，三态内容寻址存储器）。TCAM 是一种高速存储空间，用于存储指向邻接关系的指针，可以通过这些邻接关系获取 MET（Multicast Expansion Table，多播扩展表）索引，MET 索引包含了 OIF 信息以及将数据包复制并转发给每个下游接口的方式等信息。很多平台和 I/O 模块都有专用的复制 ASIC。请注意，这里描述的处理步骤与具体平台所使用的硬件类型有关，而且执行这种深层故障排查操作通常需要寻求思科 TAC 的技术支持。表 13-4 列出了常见的多播组件以及验证这些组件进程状态的相关命令。

表 13-4　每种多播组件的 CLI 命令

组件	CLI 命令
IGMP	show ip igmp route show ip igmp groups show ip igmp snooping groups
PIM	show ip pim route
MSDP	show ip msdp route show ip msdp sa-cache
URIB	show ip route
MRIB	show routing ip multicast [group] [source] show ip mroute
MFDM	show forwarding distribution ip multicast route show forwarding distribution ip igmp snooping
MFIB	show forwarding ip multicast route module [module number]
转发硬件	show system internal forwarding ip multicast route show system internal ip igmp snooping
TCAM、MET、ADJ 表	与具体的平台和硬件相关

如果 Nexus 7000 系列使用了 VDC（Virtual Device Context，虚拟设备上下文），那么前面提到的所有 PI 组件对于 VDC 来说都是唯一的。每个 VDC 都有自己的 PIM、IGMP、MRIB 和 MFDM 进程。不过，对于每个 I/O 模块来说，都需要在不同的 VDC 之间共享系统资源。

13.2.1　复制

多播通信效率高的主要原因是，可以对源端发送的单个数据包在沿 MDT 去往多个接收端（位于 MDT 树的不同分支）时进行多次复制。如果多个接收端位于同一个 VLAN 的不同接口上，那么就可以在 L2 复制数据包；如果多个下游 PIM 邻居从不同的 OIF 加入 MDT，那么就可以在 L3 复制数据包。

多播流量的复制操作由专用硬件完成，不同 Nexus 平台的专用复制硬件有所不同。对于拥有不同 I/O 模块的分布式平台来说，使用的是出站复制（egress replication）方式，如图 13-5 所示。

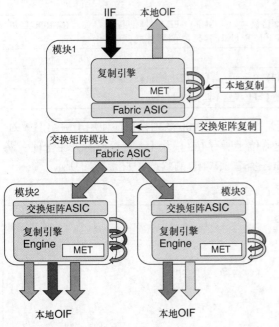

图 13-5　出站多播复制

出站复制的好处在于，允许系统的所有模块共享数据包复制产生的负荷，从而提高平台的转发能力和可扩展性。流量从 IIF 到达之后，将发生以下操作。

- 将数据包复制到本地模块上的所有接收端。
- 将数据包副本发送给交换矩阵模块。
- 交换矩阵模块复制额外的数据包副本（每个拥有 OIF 的模块一份）。
- 在每个出口模块上，根据 MET 表的内容为每个本地接收端创建数据包副本。

每个模块上的 MET 表都包含一个本地 OIF 列表。为了提高可扩展性，每个模块都要维护自己的 MET 表。此外，共享相同 OIF 的多播转发表项可以共享相同的 MET 表项，从而进一步提高系统的扩展性。

13.2.2 保护 CPU

出于多种因素，可以将某些多播流量定向到管理引擎 CPU（Central Processing Unit，中央处理器），这些可能的流量包括以下几种。

- 用于生成 PIM Assert（声明）消息的非 RPF 流量。
- TTL 在传输过程中已经到期的数据包。
- 新源端发送的用于创建 PIM Register 消息的初始数据包。
- 用于在监听表中创建表项的 IGMP Membership Report（成员关系报告）消息。
- 用于 PIM 或 IGMP 的多播控制平面数据包。

NX-OS 使用 CoPP（Control Plane Policing，控制平面策略）策略来保护管理引擎 CPU 免受过量流量的冲击。虽然用于多播流量的 CoPP 策略类别与具体的平台类型有关，但作用都非常重要：保护设备。建议用户始终启用 CoPP，当然，在某些特殊情况下可能还需要修改策略类别或警管速率。可以通过命令 **show policy-map interface control-plane** 查看当前应用的 CoPP 策略。表 13-5 列出了与多播流量有关的默认 CoPP 类别的相关信息。

表 13-5　　　　　　　　　　　　　　多播 CoPP 类别

CoPP 类别	描述
copp-system-p-class-multicast-router	将多播控制平面协议（如 MSDP）、PIM 消息匹配到 ALL-PIM-ROUTER（224.0.0.13）和 PIM Register 消息（单播）
copp-system-p-class-multicast-host	匹配 IGMP 包
copp-system-p-class-normal	匹配用于构造 PIM Register 消息的直连多播源的流量
Class-default	匹配所有与其他 CoPP 类别不匹配的数据包

除 CoPP 之外（负责警管到达管理引擎的流量），Nexus 7000 系列设备还使用了一组 HWRL（Hardware Rate Limiter，硬件限速器），硬件限速器位于所有 I/O 模块上，负责控制去往管理引擎的流量。可以通过命令 **show hardware rate-limiter** 查看 HWRL 的状态，如例 13-1 所示。

例 13-1　Nexus 7000 硬件限速器

```
NX-1# show hardware rate-limiter
! Output omitted for brevity

Units for Config: packets per second
Allowed, Dropped & Total: aggregated since last clear counters
rl-1: STP and Fabricpath-ISIS
rl-2: L3-ISIS and OTV-ISIS
rl-3: UDLD, LACP, CDP and LLDP
rl-4: Q-in-Q and ARP request
```

```
  rl-5: IGMP, NTP, DHCP-Snoop, Port-Security, Mgmt and Copy traffic


Module: 3

Rate-limiter PG Multiplier: 1.00

  R-L Class             Config      Allowed           Dropped           Total
+-------------------+---------+----------------+----------------+----------------+
  L3 mtu                500         0                 0                 0
  L3 ttl                500         12                0                 12
  L3 control            10000       0                 0                 0
  L3 glean              100         1                 0                 1
  L3 mcast dirconn      3000        13                0                 13
  L3 mcast loc-grp      3000        2                 0                 2
  L3 mcast rpf-leak     500         0                 0                 0
  L2 storm-ctrl         Disable
  access-list-log       100         0                 0                 0
  copy                  30000       7182002           0                 7182002
  receive               30000       27874374          0                 27874374
  L2 port-sec           500         0                 0                 0
  L2 mcast-snoop        10000       34318             0                 34318
  L2 vpc-low            4000        0                 0                 0
  L2 l2pt               500         0                 0                 0
  L2 vpc-peer-gw        5000        0                 0                 0
  L2 lisp-map-cache     5000        0                 0                 0
  L2 dpss               100         0                 0                 0
  L3 glean-fast         100         0                 0                 0
  L2 otv                100         0                 0                 0
  L2 netflow            48000       0                 0                 0
  L3 auto-config        200         0                 0                 0
  Vxlan-peer-learn      100         0                 0                 0
```

表 13-6 列出了多播 HWRL 信息。

表 13-6 多播 HWRL

R-L 类别	描述
L3 mcast dirconn	与源端直连的数据包，这些数据包被发送到 CPU 以创建 PIM Register 消息
L3 mcast loc-grp	在 LHR 位置发送给 CPU 以触发 SPT 切换的数据包
L3 mcast rpf-leak	发送给 CPU 以创建 PIM Assert 消息的数据包
L2 mcast-snoop	转储给 CPU 以执行 IGMP 监听操作的 IGMP Membership Report、查询及 PIM Hello 数据包

与 CoPP 策略一样，不建议用户禁用任何已默认启用的 HWRL。对于大多数部署方案来说，无须修改默认的 CoPP 或 HWRL 配置。

如果怀疑到达 CPU 的流量过多，那么就可以在特定 CoPP 类别或 HWRL 中增加匹配或丢弃量，从而对到达 CPU 的流量提供提示。如果希望了解更多详细信息，可以通过 Ethanalyzer 抓取达到 CPU 的流量信息，以进行进一步的故障排查。

13.2.3 部署 NX-OS 多播

考虑到很多网络环境都混合部署了思科 NX-OS 设备及其他平台，因而了解思科 NX-OS 设备与 IOS 设备之间的默认行为差异至关重要。NX-OS 的差异化主要表现在以下几个方面。

- 无须在全局范围内启用多播。

- 必须在配置前先启用某些功能特性（PIM、MSDP）。启用 PIM 之后，将自动启用 IGMP。
- 删除指定功能特性会删除与该功能特性相关的所有配置。
- 不支持 PIM-DM。
- 默认启用多路径支持特性，该特性允许在 ECMP（Equal-Cost Multipath，等价多路径）路由之间均衡多播流量。
- 默认不复制已转储的多播包（仅在需要时通过命令 **ip routing multicast software-replicate** 进行配置）。
- 支持 PIM IPsec AH-MD5 邻居认证。
- 不支持 PIM 监听。
- IGMP 监听默认使用基于 IP 的转发表，也可以将 IGMP 监听配置为基于 MAC 地址表进行查找。
- NX-OS 平台可能需要为多播路由分配 TCAM 空间。

1. 静态加入

一般来说，只要正确配置了多播通信，就无须使用静态加入。但是在某些情况下，静态加入对于故障排查来说还是一种非常有用的选项。例如，如果接收端不可用，那么就可以使用静态加入机制在网络中建立多播状态。

NX-OS 提供了接口命令 **ip igmp join-group** [*group*] [*source*]，可以将 NX-OS 设备配置为多播组的多播接收端。除非加入操作针对 IGMPv3，否则无须提供源地址。该命令强制 NX-OS 发出 IGMP Membership Report 消息并以主机身份加入该多播组。收到该组地址的所有数据包都将在设备的控制平面进行处理，该命令能够阻止将数据包复制到其他 OIF，因而必须谨慎使用。

第二种方式是使用接口命令 **ip igmp static-oif** [*group*] [*source*]，该命令可以将 OIF 静态添加到现有 MROUTE 表项中，并将数据包转发给硬件中的 OIF。选项 *source* 仅用于 IGMPv3。需要注意的是，如果将该命令添加到 VLAN 接口，那么还必须利用 VLAN 配置命令 **ip igmp snooping static-group** [*group*] [*source*] **interface** [*interface name*]配置一条静态 IGMP 监听表项，以实际转发数据包。

2. 清除 MROUTE 表项

清除与多播路由表项相关联的数据结构的常用方法就是使用 **clear ip mroute** 命令。对于思科 IOS 平台来说，该命令可有效清除表项。但是，对于 NX-OS 来说，与特定 MROUTE 表项相关联的数据结构可能来自任何 MRIB 客户端协议。NX-OS 提供了清除单个 MRIB 客户端表项所需的命令，NX-OS 7.3 增强了 **clear ip mroute *** 命令，可以自动清除各个客户端协议以及 MRIB 表项。早期的 NX-OS 版本还需要使用其他命令，才能从 MRIB 和所有相关联的客户端协议中彻底清除 MROUTE 表项。

- **clear ip mroute ***：清除 MRIB 中的表项。
- **clear ip pim route ***：清除由 PIM Join（加入）消息创建的 PIM 表项。
- **clear ip igmp route ***：清除由 IGMP Membership Report 消息创建的 IGMP 表项。
- **clear ip mroute data-created ***：清除通过接收多播包创建的 MRIB 表项。

3. 多播边界与过滤

NX-OS 没有与思科 IOS 多播边界相对应的概念，思科 IOS 中的多播边界命令是应用于接口的过滤器，可以创建管理范围边界，并在边界执行多播流量的过滤操作。可以通过以下控制平面和数据平面过滤技术在 NX-OS 中创建管理边界。

- 过滤 PIM Join 消息：**ip pim jp-policy** [*route-map*] [*in* | *out*]。
- 过滤 IGMP Membership Report 消息：**ip igmp report-policy** [*route-map*]。

- 过滤数据流量：**ip access-group** [*ACL*] [*in* | *out*]。

此外，还可以在接口上配置 **ip pim border** 命令，以防止转发任何 Auto-RP、引导或 C-RP（Candidate-RP，候选 RP）消息。

4．事件历史记录与 show 技术

NX-OS 提供了事件历史记录，完整记录了系统启用的功能特性的重要进程事件。在很多情况下，事件历史记录日志对于故障排查来说完全够用，不需要再执行额外的调试操作。本章从故障排查的角度出发，引用了很多与多播协议及进程有关的事件历史记录日志。由于协议消息量通常很大，因而某些故障排查场景需要增大默认事件历史记录的大小。可以增加各种事件历史记录的大小（与其类型无关）。对于 PIM 来说，可以通过配置命令 **ip pim event-history** [*event type*] **size** [*small* | *medium* | *large*] 来增加事件历史记录的大小。

NX-OS 中与转发多播流量有关的每种功能特性或服务都有自己的 **show tech-support** [*feature*] 命令，可以通过这些命令在单一输出结果中收集与故障相关的大部分信息，供离线或事后分析。技术支持文件包含了特定功能特性的配置、数据结构和事件历史记录输出结果。如果遇到了故障问题且收集信息的时间有限，那么就可以抓取以下 NX-OS 技术支持命令的输出结果，并将输出结果重定向到 Bootflash 中，以供事后分析。

- **show tech-support ip multicast**。
- **show tech-support forwarding l2 multicast vdc-all**。
- **show tech-support forwarding l3 multicast vdc-all**。
- **show tech-support pixm**。
- **show tech-support pixmc-all**。
- **show tech-support module all**。

由于了解故障的可能发生时间对于故障排查操作来说非常重要，因而需要在事件历史记录中关联各种系统消息和协议事件。如果故障发生在过去，那么某些或所有事件历史缓冲区都可能已经被覆盖，导致与故障相关的事件信息可能丢失。在这种情况下，增加某些事件历史记录的缓存大小对于再现故障来说非常有用。

收集了所有相关数据之后，可以将这些文件组合成单个归档文件并进行压缩，供思科技术支持团队排查故障原因。

虽然无法一一列出所有可能情况的命令列表，不过上面列出的 **show** 命令已足以将故障范围缩小到一定程度（即使还没有找出故障根源）。需要记住的是，通常很难将多播故障隔离到单台设备，这意味着可能还需要从对等设备或 PIM 邻居收集相关信息。

13.3 IGMP

主机使用 IGMP 协议通过 LHR 动态加入和离开多播组。有了 IGMP 之后，主机可以随时加入或离开多播组。如果没有 IGMP，那么多播路由器将无法知道何时有感兴趣的接收端驻留在接口上，也不知道这些接收端何时不再对组播流量感兴趣。因而可以很明显地看出，如果没有 IGMP，那么多播网络中的带宽和资源利用效率将大幅降低。设想一下，如果每台多播路由器都要为每个接口上的每个多播组发送流量，那将是一种什么样的场景！因此，如果配置主机和路由器支持多播通信，那么就必须要求它们支持 IGMP。对于 NX-OS 的 IGMP 实现来说，单个 IGMP 进程可以为所有 VRF（Virtual Routing and Forwarding，虚拟路由转发）实例提供服务。如果使用了 VDC（Virtual Device Context，虚拟设备上下文），那么每个 VDC 都要运行自己的 IGMP 进程。

IGMPv1 定义在 RFC 1112 中，通过向本地路由器发送 IGMP Membership Report 消息来提供

状态机以及主机加入和离开多播组所需的消息机制。虽然现代网络很少通过 IGMPv1 来查找设备，但是为了更好地理解 IGMPv2 和 IGMPv3 的差异及其发展历史，这里也对 IGMPv1 做简单描述。

配置了 IGMPv1 的多播路由器会定期向 All-Host（全部主机）地址 224.0.0.1 发送查询消息，此后主机将等待一个随机时间间隔（在报告时延定时器范围内），并使用组地址作为 Membership Report 消息的目的地址来发送 Membership Report。多播路由器收到该消息之后，就知道应该发送特定多播组的流量。路由器收到 Membership Report 之后，判断出网段上的主机是多播组的当前成员，并将组流量转发到网段上。使用组地址作为 Membership Report 消息目的地址的原因是，主机可以知道同一网络上是否还存在该多播组的其他接收端，从而允许主机抑制自己的报告消息，进而减少网段上的 IGMP 流量。多播路由器只要收到一条 Membership Report 消息，就要向该网段发送流量。

如果主机希望加入新的多播组，那么就可以立即发送该多播组的 Membership Report 消息，而不必等待多播路由器发送的查询消息。不过，如果主机希望离开多播组，那么由于 IGMPv1 并没有提供相应的方法向本地多播路由器通告该需求，因而此时的主机所能做的只是停止响应查询消息。如果多播路由器没有继续收到 Membership Report 消息，那么就会发送 3 条查询消息，此后就会从 OIL 剪除接口并确定不再有感兴趣的接收端。

13.3.1 IGMPv2

IGMPv2 定义在 RFC 2236 中，在 IGMPv1 的基础上增加了一些额外功能。IGMPv2 需要定义新消息来实现这些新功能，图 13-6 给出了 IGMPv2 消息格式。

8比特	8比特	16比特
类型	最大响应时间	校验和
组地址		

图 13-6 IGMPv2 消息格式

IGMPv2 消息的字段信息如下。

- 类型。
    - 0x11=Membership Query（成员关系查询）消息（通用查询或特定组查询）。
    - 0x12=版本 1 Membership Report 消息（保持后向兼容性）。
    - 0x16=版本 2 Membership Report 消息。
    - 0x17=Leave Group（离开组）消息。
- 最大响应时间：仅用于 Membership Query 消息，其他消息将该字段置为 0。用于调整主机的响应时间以及最后一个成员决定离开多播组时遵守的离开时延。
- 校验和：用于确保 IGMP 消息的完整性。
- 组地址：通用查询消息将该字段设置为零，发送特定组查询消息时将该字段设置为组地址。Membership Report 或 Leave Group 消息中的组地址被设置为正在报告或离开的组。

> 注：携带 IGMP 消息的 IP 数据包将 TTL 设置为 1，并在 IP 报头中设置路由器告警选项，从而强制路由器检查数据包的内容。

对于 IGMPv2 来说，如果网段上存在一台以上的多播路由器，那么就会启动选举进程来确定 IGMP 查询路由器。启动时，多播路由器会向 All-Host 组地址 224.0.0.1 发送 IGMP 通用查询消息。路由器收到其他多播路由器发来的通用查询消息之后，将执行检查操作，IP 地址最小的路由器将承担查询路由器的角色。此后，查询路由器将负责在网段上发送查询消息。

IGMPv2 加入多播组的过程与 IGMPv1 相似。主机使用 Membership Report 消息来响应通用查询以及特定组查询，主机在实现时可以选择一个随机时间来响应查询消息，这个随机时间可以介于零秒与查询消息中发送的最大响应间隔（Max-response-interval）之间。主机加入新组以初始化网段上的多播流量时，也可以发送非请求的 Membership Report 消息。

为了解决 IGMPv1 中的主机决定离开组时无法显式通知网络的问题，IGMPv2 定义了 Leave Group 消息。如果网段不再需要特定多播组且所有成员均已离开该多播组，那么就可以利用该消息通知路由器。如果某主机是在网段上发送 Membership Report 消息的最后一个成员，那么在该主机不再希望接收组流量的时候，就可以发送 Leave Group 消息，Leave Group 消息的目的地址是 All-Router 多播地址 224.0.0.2。查询路由器收到该消息之后，将发送特定组查询消息作为响应（这也是对 IGMPv1 的增强型功能）。特定组查询消息使用多播组的目的 IP 地址，以确保监听该多播组的所有主机都能收到该查询消息，这些消息的发送间隔基于最后一个成员查询间隔。如果未收到 Membership Report 消息，那么路由器就会将接口从 OIL 中剪除。

13.3.2 IGMPv3

IGMPv3 定义在 RFC 3376 中，提供了主机支持 SSM（Source Specific Multicast，特定源多播）所需的所有功能。SSM 多播不但允许接收端明确加入特定多播组地址，而且能加入特定组的源地址，这样一来，运行在多播接收端主机上的应用程序就能请求特定多播源。

对于 IGMPv3 来说，主机的接口状态包含过滤模式和源列表。其中，过滤模式可以是包含（include）模式，也可以是排除（exclude）模式。如果过滤模式是包含模式，那么仅从源列表中的源端请求流量；如果过滤模式是排除模式，那么就从源列表之外的其他源端请求流量。源列表是 IP 单播源地址的无序列表，可以与过滤模式相结合以实现特定源逻辑，从而允许 IGMPv3 在协议消息中仅向接收端告知感兴趣的源端。

图 13-7 列出了 IGMPv3 Membership Query 消息格式，与 IGMPv2 Membership Query 消息相比，IGMPv3 Membership Query 消息包含了几个新字段，但消息类型保持不变（0x11）。

8比特	8比特	16比特
类型 = 0x11	最大响应时间	校验和
组地址		
保留 | S | QRV | QQIC		源数量(N)
源地址 [1]		
源地址 [2]		
.		
.		
源地址 [N]		

图 13-7　IGMPv3 Membership Query 消息格式

IGMPv3 Membership Query 消息的字段信息如下。

- **类型 0x11**：表示 Membership Query 消息，包括通用查询、特定组查询或特定组和源查询消息，通过组地址和源地址字段的内容区分这些消息。
- **最大响应代码**：主机发送响应报告所允许的最大时间。运营商可以据此调整 IGMP 流量的突发性和离开时延。
- **校验和**：确保 IGMP 消息的完整性。该字段由整个 IGMP 消息计算得到。

- **组地址**：通用查询消息将该字段设置为零，特定组查询或特定组和源查询消息将该字段设置为组地址。
- **保留**：设置为零，收到后将被忽略。
- **S 标志**：如果设置为 1，那么路由器收到查询消息之后将禁止执行常规的定时器更新操作。
- **QRV（Querier's Robustness Variable，查询路由器的健壮性变量）**：用于解决潜在的丢包问题。该字段允许主机发送多个 Membership Report 消息以确保查询路由器能够收到这些消息。
- **QQIC（Querier's Query Interval Code，查询路由器的查询间隔代码）**：提供 QQI（Querier's Query Interval，查询路由器的查询间隔）。
- **源数量**：指定查询消息中存在多少多播源。
- **源地址**：指定源单播 IP 地址。

与 IGMPv2 相比，IGMPv3 的 Membership Query 消息有一些差异，最重要的是具备特定组和源查询能力，可以向特定的多播组源端发送查询消息。

IGMPv3 的 Membership Report 消息由消息类型 0x22 加以标识，而且与 IGMPv2 的 Membership Report 消息相比做了一些修改。接收端主机使用该消息向本地多播路由器报告其接口的当前成员关系状态以及成员关系状态的变更情况。主机使用组 IP 目的地址 224.0.0.22 将该消息发送给多播路由器。图 13-8 列出了 IGMPv3 Membership Report 消息的格式。

8比特	8比特	16比特
类型 = 0x22	保留	校验和
保留		组记录数量(M)
组记录 [1]		
组记录 [2]		
⋮		
组记录 [M]		

图 13-8 IGMPv3 Membership Report 消息格式

Membership Report 消息中的每条组记录都使用图 13-9 所示的格式。

IGMPv3 Membership Report 消息的字段信息如下。

- **类型 0x22**：表示 IGMPv3 Membership Report 消息。
- **保留**：发送时设置为零，接收端忽略该字段。
- **校验和**：验证消息的完整性。
- **组记录数量**：提供该 Membership Report 消息中的组记录数量。
- **组记录**：是一组字段，提供了发送该报告消息接口上的单个多播组发送端的成员关系。

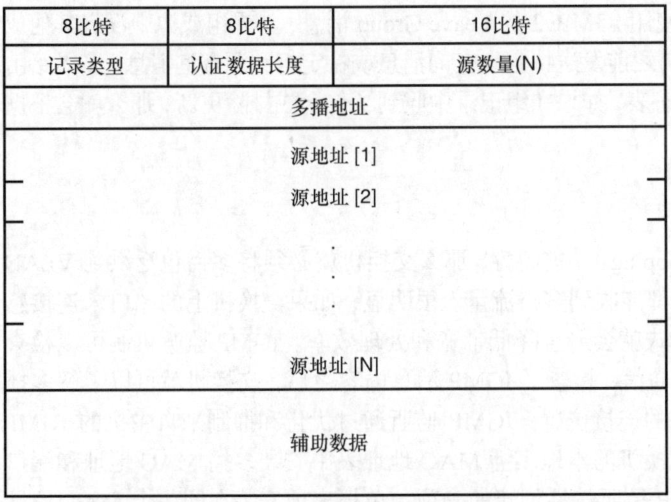

8比特	8比特	16比特
记录类型	认证数据长度	源数量(N)
多播地址		
源地址 [1]		
源地址 [2]		
. . .		
源地址 [N]		
辅助数据		

图 13-9　IGMPv3 Membership Report 组记录格式

每条组记录的字段信息如下。

- **记录类型**：组记录的类型。
  - **当前状态记录**：接口的当前接收状态。
    - Mode_is_include：过滤模式为包含（include）模式。
    - Mode_is_exclude：过滤模式为排除（exclude）模式。
  - **过滤模式变更记录**：表示过滤模式已变更。
    - Change_to_Include_Mode：过滤模式变更为包含（include）模式。
    - Change_to_Exclude_Mode：过滤模式变更为排除（exclude）模式。
  - **源列表变更记录**：表示源列表已变更，而不是过滤模式。
    - Allow_New_Sources：列出正在请求的新多播源。
    - Block_Old_Sources：列出不再请求的多播源。
- **辅助数据长度**：组记录中的辅助数据长度。
- **源数量**：表示本条组记录存在多少多播源。
- **多播地址**：表示本条组记录所属的多播组。
- **源地址**：表示该组多播源的单播 IP 地址。
- **辅助数据**：表示没有为 IGMPv3 定义辅助数据。应该将辅助数据长度字段设置为零，收到后应忽略辅助数据。
- **其他数据**：IGMP 校验和考虑该字段，但最后一条组记录以外的所有数据都将被忽略。

与 IGMPv2 Membership Report 相比，IGMPv3 Membership Report 消息最重要的区别在于包含了组记录块数据。这是 IGMPv3 实现过滤模式和源列表等特定功能的重要基础。

IGMPv3 与早期的 IGMP 版本保持向后兼容，且遵循相同的通用状态机机制。如果检测到网络中存在运行了旧版 IGMP 的主机或路由器，那么所有的查询和报告消息都将从 IGMPv2 转换为对应的 IGMPv3 消息。例如，与多播组 239.1.1.1 的 IGMPv2 Membership Report 消息相对应的 IGMPv3 兼容形式包含了 IGMPv3 中的所有多播源。

与 IGMPv2 一样，IGMPv3 的通用查询消息也从查询路由器发送到 All-Host 组 224.0.0.1。主机以 Membership Report 消息作为响应，响应消息的源列表中包含了指定多播源，而且记录类型字段还列出了包含或排除逻辑。希望加入新多播组或多播源的主机，可以使用非请求的 Membership Report 消息。如果要离开组或特定多播源，那么主机就可以发送更新后的当前状态组记录消息以指示状态的变化情况。IGMPv3 没有使用 IGMPv2 的 Leave Group 消息，如果组或源中没有其他成员，那么查询路由器就会在剪除有源树之前发送特定组查询消息或特定组和源查询消息。多播路由器会为每个组和源都维护一个接口状态表，如果在组记录中收到了包含或排除更新，那么就会根据需要更新接口状态表。

13.3.3 IGMP 监听

如果没有 IGMP 监听（IGMP Snooping）功能特性，那么交换机就必须将多播包泛洪到 VLAN 的所有端口，以确保所有潜在组成员都能收到多播流量。很明显，如果交换机上的端口未连接感兴趣的多播接收端，那么这种处理方式就会严重降低带宽和处理效率。IGMP 监听机制可以检查（或"侦听"）穿越交换机的高层协议通信，检查了 IGMP 消息内容之后，交换机就可以了解多播路由器以及对多播组感兴趣的接收端的连接位置。IGMP 监听通过优化和抑制来自主机的 IGMP 消息来实现控制平面操作，通过在交换机的本地多播 MAC 地址表中安装多播 MAC 地址和端口映射表项来实现数据平面操作。IGMP 监听特性创建的表项与单播表项安装在同一个 MAC 地址表中，虽然需要通过不同的命令来查看由常规单播学习和 IGMP 监听机制安装的表项，但这些表项都共享由 MAC 地址表提供的相同的硬件资源。

IGMP 监听交换机将侦听 IGMP 查询消息和 PIM Hello 消息，以确定哪些端口连接了 MROUTER。端口被确定为 MROUTER 端口之后，就会接收 VLAN 中的所有多播流量，从而在 MROUTER 上创建相应的控制平面状态，并向 PIM RP 注册多播源（如果适用）。此外，IGMP 监听交换机还会将 IGMP Membership Report 转发给 MROUTER，以发起到组成员的多播流量。

可以通过侦听 IGMP Membership Report 消息来发现主机端口。评估 IGMP Membership Report 消息可以确定所要请求的组和源，并将适当的转发表项添加到多播 MAC 地址表或 IP 转发表中。请注意，IGMP 监听交换机不应该将 IGMP Membership Report 转发给主机，因为这样会导致主机抑制自己的 IGMPv1 和 IGMPv2 Membership Report 消息。

收到网络控制块 224.0.0.0/24 的多播包之后，可能需要在所有端口上泛洪这些数据包，这是因为设备无须发送该组的 Membership Report 即可侦听该范围内的多播组，而且抑制这些数据包可能会导致控制平面协议出现中断。

IGMP 监听与 IGMP 控制平面进程是两个相互独立的不同进程，NX-OS 默认启用 IGMP 监听进程，用户无须配置即可在设备上运行基本的 IGMP 监听功能。例 13-2 给出了验证特定 VLAN 的 IGMP 监听状态和查找模式的示例。

例 13-2 验证 IGMP 监听

```
NX-2# show ip igmp snooping vlan 115
Global IGMP Snooping Information:
  IGMP Snooping enabled
  Optimised Multicast Flood (OMF) enabled
  IGMPv1/v2 Report Suppression enabled
  IGMPv3 Report Suppression disabled
  Link Local Groups Suppression enabled

IGMP Snooping information for vlan 115
  IGMP snooping enabled
  Lookup mode: IP
  Optimised Multicast Flood (OMF) enabled
  IGMP querier present, address: 10.115.1.254, version: 2, i/f Po1
  Switch-querier disabled
  IGMPv3 Explicit tracking enabled
```

```
IGMPv2 Fast leave disabled
IGMPv1/v2 Report suppression enabled
IGMPv3 Report suppression disabled
Link Local Groups suppression enabled
Router port detection using PIM Hellos, IGMP Queries
Number of router-ports: 1
Number of groups: 1
VLAN vPC function disabled
Active ports:
  Po1   Po2       Eth3/19
```

能够以逐个 VLAN 方式将设备配置为使用基于 MAC 地址的转发机制，虽然这样做可能会因为地址重叠而导致次优转发。例 13-3 在 VLAN 配置子模式下配置了该选项。

例 13-3　启用 MAC 地址查找模式

```
NX-2(config)# vlan configuration 115
NX-2(config-vlan-config)# layer-2 multicast lookup mac
```

如果到达的多播组流量并不是主机通过 Membership Report 消息请求的多播组，那么默认仅将这些数据包转发给 MROUTER 端口，NX-OS 将该功能特性称为 OMF（Optimised Multicast Flood，优化的多播泛洪）。从例 13-2 可以看出，NX-OS 默认启用该功能特性。如果禁用了该功能特性，那么就会将未知多播组流量泛洪到 VLAN 中的所有端口。

> **注：** 应该在 IPv6 网络中禁用 OMF 功能特性，以避免出现与 ND（Neighbor Discovery，邻居发现）相关的故障问题，后者特别依赖多播通信。可以在 VLAN 配置模式下使用命令 **no ip igmp snooping optimised-multicast-flood** 禁用该功能特性。

为了减少 MROUTER 收到的消息数量，NX-OS 默认抑制 IGMP Membership Report 消息。前面曾经说过，MROUTER 只要收到一台主机发送的 Membership Report 消息，即可将接口添加到多播组的 OIL 中。

配置 IGMP 监听功能特性时，NX-OS 提供了多种可用选项。虽然大多数配置是按照逐个 VLAN 方式应用的，但某些参数只能在全局范围内进行配置，全局配置值适用于所有 VLAN。表 13-7 列出了交换机全局应用的 IGMP 监听功能特性的默认配置参数。

表 13-7　　　　　　　　　　　IGMP 监听特性的全局配置参数

参数	CLI 命令	描述
ICMP 监听	**ip igmp snooping**	在活动 VDC 上启用 IGMP 监听 注：如果全局设置是禁用该特性，那么所有 VLAN 都将禁用该特性，与是否在 VLAN 上启用该特性无关
事件历史记录	**ip igmp snooping eventhistory { vpc | igmp-snoop-internal | mfdm | mfdm-sum | vlan | vlan-events } size** *buffer-size*	配置 IGMP 监听历史缓存的大小，默认为 small（小）
组超时	**ip igmp snooping group-timeout {** *minutes* **| never }**	为设备上的所有 VLAN 配置组成员关系超时参数
链路本地组抑制	**ip igmp snooping link-local-groups-suppression**	在设备上配置链路本地组抑制机制，默认启用该机制
OMF（Optimised Multicast Flood，优化的多播泛洪）	**ip igmp optimise-multicast-flood**	在所有 VLAN 上配置 OMF，默认启用该特性
代理	**ip igmp snooping proxy general-inquiries [mrt** *seconds* **]**	启用监听功能代理答复多播路由器常规查询消息，同时还使用指定的 MRT 值在每个交换机端口上发送轮询通用查询消息。默认值为 5s

参数	CLI 命令	描述
报告抑制	**ip igmp snooping report-suppression**	在设备上限制发送给支持多播机制的路由器的 Membership Report 流量,如果禁用了报告抑制特性,那么所有的 IGMP 报告消息都会按原样发送给支持多播机制的路由器。默认启用该特性
IGMPv3 报告抑制	**ip igmp snooping v3-report-suppression**	配置 IGMPv3 报告抑制机制并在设备上代理发送报告消息,默认禁用该特性

表 13-8 列出了以逐个 VLAN 方式配置的 IGMP 监听参数,可以在 **vlan configuration** [*vlan-id*] 子模式下应用这些 VLAN 配置。

表 13-8　　　　　　　　　IGMP 监听特性的 VLAN 配置参数

参数	CLI 命令	描述
ICMP 监听	**ip igmp snooping**	以逐个 VLAN 方式启用 IGMP 监听。默认启用该特性
显式跟踪	**ip igmp snooping explicit-tracking**	以逐个 VLAN 方式跟踪每台主机为每个端口发送的 IGMPv3 Membership Report 消息。默认启用该特性
快速离开	**ip igmp snooping fast-leave**	允许软件在收到 IGMP 离开报告后立即删除组状态,而不需要发送 IGMP 查询消息。该参数用于 IGMPv2 主机,此时每个 VLAN 端口上只有一台主机。默认设置禁用该参数
组超时	**ip igmp snooping group-timeout** { *minutes* | **never** }	如果 3 条通用查询消息之后都未收到新的组成员关系消息,那么 IGMP 监听组成员关系将超时,该参数的目的是修改超时后的默认行为
最后一个成员查询间隔	**ip igmp snooping last-member-query-interval** *seconds*	设置软件在发送 IGMP 查询后等待的时间间隔,以验证网段不再具有希望接收特定多播组的主机。如果在最后一个成员查询间隔到期之前没有主机响应,则软件将从关联的 VLAN 端口中删除该组。数值范围为 1~25s。默认值为 1s
OMF(Optimised Multicast Flood,优化的多播泛洪)	**ip igmp optimise-multicast-flood**	在指定 VLAN 上配置 OMF,默认启用该特性
代理	**ip igmp snooping proxy general-inquiries** [**mrt** *seconds*]	启用监听功能代理答复多播路由器常规查询消息,同时还使用指定的 MRT 值在每个交换机端口上发送轮询通用查询消息。默认值为 5s
监听查询路由器	**ip igmp snooping querier** *ip-address*	如果不希望路由多播流量而没有启用 PIM,那么就可以在接口上配置监听查询路由器
查询超时	**ip igmp snooping querier-timeout** *seconds*	IGMPv2 的查询超时值,默认值为 255s
查询间隔	**ip igmp snooping query-interval** *seconds*	发送查询消息的间隔,默认值为 125s
查询消息的最大响应时间	**ip igmp snooping query-max-response-time** *seconds*	查询消息的最大响应时间,默认值为 10s
启动次数	**ip igmp snooping startup-query-count** *value*	启动时发送的查询次数,默认值为 2
启动间隔	**ip igmp snooping startup-query-interval** *seconds*	启动时发送查询消息的间隔,默认值为 31s
鲁棒性变量	**ip igmp snooping robustness-variable** *value*	为指定 VLAN 配置鲁棒性数值,默认值为 2
报告抑制	**ip igmp snooping report-suppression**	以逐个 VLAN 方式限制发送给支持多播机制的路由器的 Membership Report 流量,如果禁用了报告抑制特性,那么所有的 IGMP 报告消息都会按原样发送给支持多播机制的路由器。默认启用该特性

参数	CLI 命令	描述
静态 MROUTER 端口	ip igmp snooping mrouter interface	配置去往多播路由器的静态连接，连接路由器的接口必须位于选定的 VLAN 中
二层静态组	ip igmp snooping static-group group-ip-addr [source source-ip-addr] interface interface	将 VLAN 的二层端口配置为多播的静态成员
链路本地组抑制	ip igmp snooping link-local-groups-suppression	以逐个 VLAN 方式配置链路本地组抑制机制，默认启用该机制
IGMPv3 报告抑制	ip igmp snooping v3-report-suppression	配置 IGMPv3 报告抑制机制并以逐个 VLAN 方式代理发送报告消息，默认以逐个 VLAN 方式启用该特性
版本	ip igmp snooping version value	为指定 VLAN 配置 IGMP 版本号

对于纯 L2 多播部署环境来说，必须配置一台监听查询路由器。这一点适用于以下场景：没有在任何接口上启用 PIM、没有 MROUTER、没有在 VLAN 之间路由多播流量。

注：如果 vPC 配置了 IGMP 监听特性，那么建议在两个 vPC 对等体上配置相同的 IGMP 参数。可以通过 CFS（Cisco Fabric Service，思科交换矩阵服务）在 vPC 对等体之间实现 IGMP 状态的同步。

13.3.4 验证 IGMP

如果接口启用了 PIM，那么将默认启用 IGMP。排查 IGMP 故障时通常包括以下场景：LHR 缺少 IGMP 安装的 MROUTE 表项，需要将该故障问题隔离到 LHR、L2 基础设施或主机本身。此时，通常需要验证 IGMP 监听特性是否正常，这是因为系统默认启用 IGMP 监听特性，该特性在将查询消息传递给主机以及将 Membership Report 消息传递给 MROUTER 的过程中起着非常重要的作用。

图 13-10 拓扑结构中的 NX-1 充当 VLAN 115 和 VLAN 116 的接收端的 LHR，同时也是两个 VLAN 的 IGMP 查询路由器。NX-2 是一台不执行任何多播路由的 IGMP 监听交换机。所有 L3 设备都配置了 PIM ASM，且 NX-3 与 NX-4 共享同一个 Anycast-RP（任播 RP）地址。

如果接收端未收到多播组的流量，那么就需要验证 IGMP 的当前状态和运行情况。请注意，在执行验证操作之前，必须了解以下信息：

- 多播组地址：239.215.215.1。
- 源端 IP 地址：10.215.1.1。
- 接收端 IP 地址：10.115.1.4。
- LHR：NX-1。
- 故障范围：运行不正常的组、源和接收端。

IGMP 的作用是告知 LHR 对组业务感兴趣的接收端。在最基本的层面来说，这一点是通过接收端发送的 Membership Report 消息进行通信的，而且应该在 LHR 上创建(*,G)状态。对于大多数情况来说，检查 LHR 的 MROUTE，如果发现存在(*,G)，那么就足以证实至少收到了一条 Membership Report 消息。MROUTE 的 OIL 应该包含收到该 Membership Report 消息的接口。如果通过了该检查，那么通常就可以沿 MDT 继续排查 PIM RP 或多播源，以确定流量为何未到达接收端。

接下来的示例并没有真的出现 IGMP 故障，因为 NX-1 拥有(*,G)状态。这里的目的不是排查特定故障，而是希望查看 IGMP 的协议状态并解释验证 IGMP 所要用到的命令输出结果、进程事件以及验证方法。

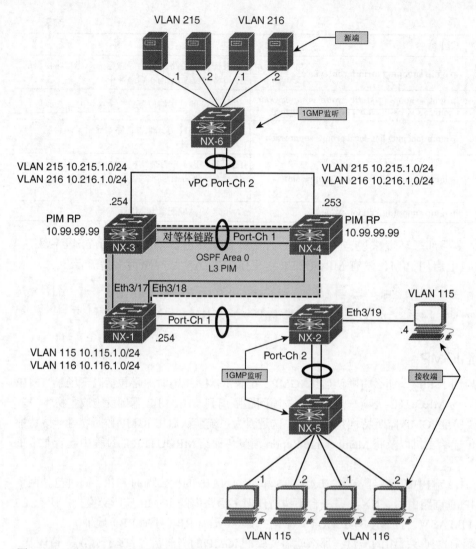

图 13-10 验证 IGMP 的示例拓扑结构

验证操作首先从 NX-2（是连接接收端 10.115.1.4 的 IGMP 监听交换机）开始，然后沿 L2 网络到达 MROUTER NX-1。例 13-4 给出了 **show ip igmp snooping vlan 115** 命令的输出结果（是接收端连接 NX-2 的位置），从输出结果可以证实已经启用了 IGMP 监听且检测到了 MROUTER 端口。

例 13-4 VLAN 115 的 IGMP 监听状态

```
NX-2# show ip igmp snooping vlan 115
Global IGMP Snooping Information:
  IGMP Snooping enabled
  Optimised Multicast Flood (OMF) enabled
  IGMPv1/v2 Report Suppression enabled
  IGMPv3 Report Suppression disabled
  Link Local Groups Suppression enabled

IGMP Snooping information for vlan 115
  IGMP snooping enabled
  Lookup mode: IP
  Optimised Multicast Flood (OMF) enabled
  IGMP querier present, address: 10.115.1.254, version: 2, i/f Po1
```

```
  Switch-querier disabled
  IGMPv3 Explicit tracking enabled
  IGMPv2 Fast leave disabled
  IGMPv1/v2 Report suppression enabled
  IGMPv3 Report suppression disabled
  Link Local Groups suppression enabled
  Router port detection using PIM Hellos, IGMP Queries
  Number of router-ports: 1
  Number of groups: 1
  VLAN vPC function disabled
  Active ports:
    Po1 Po2      Eth3/19
```

Number of groups（组数）字段表明存在一个多播组，可以通过命令 **show ip igmp snooping groups vlan 115** 获取该多播组的详细信息，如例 13-5 所示。

例 13-5　VLAN 115 的 IGMP 监听组成员关系

```
NX-2# show ip igmp snooping groups vlan 115
Type: S - Static, D - Dynamic, R - Router port, F - Fabricpath core port

Vlan  Group Address      Ver  Type  Port list
115   */*                -    R     Po1
115   239.215.215.1      v2   D     Eth3/19
```

使用关键字 **detail** 可以看到最后一台报告设备，如例 13-6 所示。

例 13-6　VLAN 115 的 IGMP 监听组成员关系详细信息

```
NX-2# show ip igmp snooping groups vlan 115 detail
IGMP Snooping group membership for vlan 115
  Group addr: 239.215.215.1
    Group ver: v2 [old-host-timer: not running]
    Last reporter: 10.115.1.4
    Group Report Timer: 0.000000
    IGMPv2 member ports:
    IGMPv1/v2 memb ports:
      Eth3/19 [0 GQ missed], cfs:false, native:true
    vPC grp peer-link flag: include
    M2RIB vPC grp peer-link flag: include
```

注： 如果为 VLAN 115 配置了基于 MAC 地址的多播转发机制，那么就可以通过 **show hardware mac address-table** [*module*] [*VLAN identifier*]命令确认多播 MAC 表的表项，而命令 **show mac address-table multicast** [*VLAN identifier*]的输出结果却没有预期的软件 MAC 表的表项。

NX-2 的 IGMP 监听被配置为使用 IP 查找机制。例 13-7 中的 **show forwarding distribution ip igmp snooping vlan** [*VLAN identifier*]命令用于查找平台索引（platform index），可以将帧定向到正确的输出接口。这里所说的平台索引也被称为 LTL（Local Target Logic，本地目标逻辑）索引。该命令可以提供 MFDM（Multicast Forwarding Distribution Manager，多播转发分发管理器）表项，有关 MFDM 表项的详细内容请见本章的 13.2 节。

例 13-7　IGMP 监听 MFDM 表项

```
NX-2# show forwarding distribution ip igmp snooping vlan 115 group 239.215.215.1
  detail
Vlan: 115, Group: 239.215.215.1, Source: 0.0.0.0
  Route Flags: 0
```

```
    Outgoing Interface List Index: 13
    Reference Count: 2
    Platform Index: 0x7fe8
    Vpc peer link exclude flag clear
    Number of Outgoing Interfaces: 2
      port-channel1
      Ethernet3/19
```

Ethernet3/19 接口是由接收端的 Membership Report 消息填充进去的。Port-channel 1 接口包含在出站接口列表中的原因是该接口是 MROUTER 端口。例 13-8 验证了平台索引，目的是确保存在正确的接口且与先前的 MFDM 输出相匹配。命令 **show system internal pixm info ltl** [*index*] 可以从 PIXM（Port Index Manager，端口索引管理器）获得输出结果，例中的 IFIDX/RID 为 0xd，与出接口列表索引（Outgoing Interface List Index）13 相匹配。

例 13-8　验证平台 LTL 索引

```
NX-2# show system internal pixm info ltl 0x7fe8
MCAST LTLs allocated for VDC:1
==============================================
LTL     IFIDX/RID   LTL_FLAG  CB_FLAG
0x7fe8  0x0000000d  0x00      0x0002

mi | v5_f3_fpoe | v4_fpoe | v5_fpoe | clp_v4_l2 | clp_v5_l2 | clp20_v4_l3
 | clp_cr_v4_l3 | flag | proxy_if_index
0x3 | 0x3 | 0x0 | 0x3 | 0x0 | 0x3 | 0x3 | 0x3 | 0x0 | none

Member info
------------------
IFIDX              LTL
-------------------------------
Eth3/19            0x0012
Po1                0x0404
```

注： 如果感兴趣的 IFIDX 是端口通道，那么检查端口通道的 LTL 索引即可找到物理接口。第 5 章详细讨论了端口通道负载均衡散列以及查找传输数据包的端口通道成员链路的方法。

至此，除多播组的转发平面状态之外，已经通过 **show** 命令全面验证了 IGMP 监听的控制平面。此外，NX-OS 还为 IGMP 和其他多播协议提供了一些有用的事件历史记录，事件历史记录采集了进程的重要事件并存储在循环缓冲区中。大多数情况下，多播协议的事件历史记录可以提供与进程调试相当的细节信息。

可以通过命令 **show ip igmp snooping internal event-history vlan** 显示 VLAN 115 以及感兴趣的多播组 239.215.215.1 的 IGMP 监听事件序列。例 13-9 显示了从 Port-channel 1 收到的通用查询消息以及 Eth3/19 从 10.115.1.4 收到的 Membership Report 消息。

例 13-9　IGMP 监听 VLAN 事件历史记录

```
NX-2# show ip igmp snooping internal event-history vlan | inc
239.215.215.1|General
! Output omitted for brevity
02:19:33.729983 igmp [7177]: [7314]: SN: <115> Forwarding report for
(*, 239.215.215.1) came on Eth3/19
02:19:33.729973 igmp [7177]: [7314]: SN: <115> Updated oif Eth3/19 for
(*, 239.215.215.1) entry
02:19:33.729962 igmp [7177]: [7314]: SN: <115> Received v2 report:
```

```
group 239.215.215.1 from 10.115.1.4 on Eth3/19
02:19:33.721639 igmp [7177]: [7314]: SN: <115> Report timer not running.
..starting with MRT expiry 10 for group: 239.215.215.1
02:19:33.721623 igmp [7177]: [7314]: SN: <115> Received v2 General query
from 10.115.1.254 on Po1
```

Ethanalyzer 工具提供了一种在 NX-OS 的 netstack 组件级别抓取数据包的方法，这是一种适用于所有控制平面协议交换故障排查的有用工具。例 13-10 中的 Ethanalyzer 以过滤方式抓取了 IGMP 包，可以看出 NX-2 收到了通用查询消息以及 10.115.1.4 发送的 Membership Report 消息。可以通过 write 选项将 Ethanalyzer 的输出结果定向到本地存储，此后可以根据需要复制该存储文件，进而执行详细的协议检查。

例 13-10　在 NX-2 上通过 Ethanalyzer 抓取 IGMP 消息

```
NX-2# ethanalyzer local interface inband-in capture-filter "igmp"
! Output omitted for brevity
Capturing on inband
1 02:29:24.420135 10.115.1.254 -> 224.0.0.1    IGMPv2 Membership Query, general
2 02:29:24.421061 10.115.1.254 -> 224.0.0.1    IGMPv2 Membership Query, general
3 02:29:24.430482 10.115.1.4   -> 239.215.215.1 IGMPv2 Membership Report group
  239.215.215.1
```

NX-OS 在全局和接口级别维护了 IGMP 监听统计信息，可以通过命令 **show ip igmp snooping statistics global** 或 **show ip igmp snooping statistics vlan** [*VLAN identifier*]查看这些统计信息。例 13-11 显示了 NX-2 的 VLAN 115 统计信息，可以看出，VLAN 统计信息还包含了全局统计信息，对于确认 VLAN 收到了多少以及何种类型的 IGMP 和 PIM 消息来说非常有用。如果需要更多有关数据包级别的详细信息，那么通常建议在 Ethanalyzer 中使用适当的过滤器。

例 13-11　NX-2 的 VLAN 115 IGMP 监听统计信息

```
NX-2# show ip igmp snooping statistics vlan 115
Global IGMP snooping statistics: (only non-zero values displayed)
  Packets received: 3783
  Packets flooded: 1882
  vPC PIM DR queries fail: 2
  vPC PIM DR updates sent: 6
  vPC CFS message response sent: 19
  vPC CFS message response rcvd: 16
  vPC CFS unreliable message sent: 403
  vPC CFS unreliable message rcvd: 1632
  vPC CFS reliable message sent: 16
  vPC CFS reliable message rcvd: 19
  STP TCN messages rcvd: 391
  IM api failed: 1
VLAN 115 IGMP snooping statistics, last reset: never (only non-zero values
  displayed)
  Packets received: 666
  IGMPv2 reports received: 242
  IGMPv2 queries received: 267
  IGMPv2 leaves received: 4
  PIM Hellos received: 1065
  IGMPv2 reports suppressed: 1
  IGMPv2 leaves suppressed: 2
  Queries originated: 2
  IGMPv2 proxy-leaves originated: 1
  Packets sent to routers: 242
```

```
    STP TCN received: 18
  vPC Peer Link CFS packet statistics:
      IGMP packets (sent/recv/fail): 300/150/0
IGMP Filtering Statistics:
Router Guard Filtering Statistics:
```

验证了 NX-2 之后，接下来可以继续检查 LHR（NX-1），NX-1 是 VLAN 115 的 MROUTER 和 IGMP 查询路由器。可以通过 **show ip igmp interface vlan 115** 命令验证 NX-1 的 IGMP 状态，如例 13-12 所示。

例 13-12　NX-1 的 IGMP 接口 VLAN 115 的状态

```
NX-1# show ip igmp interface vlan 115
IGMP Interfaces for VRF "default"
Vlan115, Interface status: protocol-up/link-up/admin-up
  IP address: 10.115.1.254, IP subnet: 10.115.1.0/24
  Active querier: 10.115.1.254, version: 2, next query sent in: 00:00:06
  Membership count: 1
  Old Membership count 0
  IGMP version: 2, host version: 2
  IGMP query interval: 125 secs, configured value: 125 secs
  IGMP max response time: 10 secs, configured value: 10 secs
  IGMP startup query interval: 31 secs, configured value: 31 secs
  IGMP startup query count: 2
  IGMP last member mrt: 1 secs
  IGMP last member query count: 2
  IGMP group timeout: 260 secs, configured value: 260 secs
  IGMP querier timeout: 255 secs, configured value: 255 secs
  IGMP unsolicited report interval: 10 secs
  IGMP robustness variable: 2, configured value: 2
  IGMP reporting for link-local groups: disabled
  IGMP interface enable refcount: 1
  IGMP interface immediate leave: disabled
  IGMP VRF name default (id 1)
  IGMP Report Policy: None
  IGMP State Limit: None
  IGMP interface statistics: (only non-zero values displayed)
    General (sent/received):
      v2-queries: 999/1082, v2-reports: 0/1266, v2-leaves: 0/15
    Errors:
  Interface PIM DR: Yes
  Interface vPC SVI: No
  Interface vPC CFS statistics:
    DR queries sent: 1
    DR queries rcvd: 1
    DR updates sent: 1
    DR updates rcvd: 3
```

Port-channel 1 收到了 NX-2 转发的由主机发送的 Membership Report 消息，可以通过 **show ip igmp internal event-history debugs** 命令查看查询消息和 Membership Report 消息，如例 13-13 所示。收到 Membership Report 消息之后，NX-1 确定需要创建(*,G)状态。

例 13-13　NX-1 的 IGMP 调试事件历史记录

```
NX-1# show ip igmp internal event-history debugs
! Output omitted for brevity
```

```
debugs events for IGMP process
04:39:34.349013 igmp [7011]: : Processing report for (*, 239.215.215.1)
[i/f Vlan115], entry not found, creating
 04:39:34.348973 igmp [7011]: : Received v2 Report for 239.215.215.1 from
10.115.1.4 (Vlan115)
 04:39:34.336092 igmp [7011]: : Received General v2 Query from 10.115.1.254
(Vlan115), mrt: 10 sec
 04:39:34.335543 igmp [7011]: : Sending SVI query packet to IGMP-snooping module
 04:39:34.335541 igmp [7011]: : Send General v2 Query on Vlan115 (mrt:10 sec)
```

IGMP 可以根据 VLAN 115 中收到的 Membership Report 创建路由表项，例 13-14 显示了 IGMP 路由表项信息。

例 13-14　NX-1 的 IGMP 路由表项

```
NX-1# show ip igmp route
IGMP Connected Group Membership for VRF "default" - 1 total entries
Type: S - Static, D - Dynamic, L - Local, T - SSM Translated
Group Address      Type Interface          Uptime    Expires   Last Reporter
239.215.215.1       D   Vlan115            01:59:49  00:03:49  10.115.1.4
```

此外，IGMP 还必须通知 MRIB 以创建适当的 MROUTE 表项，可以通过 **show ip igmp internal event-history igmp-internal** 命令加以查看，如例 13-15 所示。IGMP 通过 MTS（Message and Transactional Service，消息和事务服务）将更新消息发送给 MRIB 进程缓存。可以看出，IGMP 收到的 MRIB 通告表明已经处理了该消息且回收了消息缓存。

例 13-15　IGMP 内部事件的事件历史记录

```
NX-1# show ip igmp internal event-history igmp-internal
! Output omitted for brevity

 igmp-internal events for IGMP process
 04:39:34.354419 igmp [7011]: [7564]: MRIB: Processing ack: reclaiming buffer
0x0x967cbe4, xid 0xffff000c, count 1
 04:39:34.354416 igmp [7011]: [7564]: Received Message from MRIB minor 16
 04:39:34.353742 igmp [7011]: [7566]: default: Sending IGMP update-route buffer
0x0x967cbe4, xid 0xffff000c, count 1 to MRIB
 04:39:34.353738 igmp [7011]: [7566]: default: Moving MRIB txlist member marker
to version 12
 04:39:34.353706 igmp [7011]: [7566]: Inserting IGMP update-update for
(*, 239.215.215.1) (context 1) into MRIB buffer
```

可以通过消息标识符 0xffff000c 跟踪 MRIB 进程事件中的消息，例 13-16 通过命令 **show routing ip multicast event-history rib** 显示了该消息的 MRIB 处理过程。

例 13-16　MRIB 创建(*,G)状态

```
NX-1# show routing ip multicast event-history rib
! Output omitted for brevity

04:39:34.355736 mrib [7170]::RPF change for (*, 239.215.215.1/32) (10.99.99.99)
, iif: Ethernet3/18 (iod 64), RPF nbr: 10.1.13.3
04:39:34.355730 mrib [7170]::RPF lookup for route (*, 239.215.215.1/32)
RPF Source 10.99.99.99 is iif: Ethernet3/18 (iod 64), RPF nbr: 10.1.13.3,  pa
04:39:34.354481 mrib [7170]::Inserting add-op-update for (*, 239.215.215.1/32)
(context 1) from txlist into MFDM route buffer
04:39:34.354251 mrib [7170]::Copy oifs to all (Si,G)s for "igmp"
04:39:34.354246 mrib [7170]::Doing multi-route add for "igmp"
```

```
04:39:34.354126 mrib [7170]::       OIF : Vlan115
04:39:34.354099 mrib [7170]::"igmp" add route (*, 239.215.215.1/32)
(list-00000000)[1],rpf Null 0.0.0.0(0.0.0.0), iod 0, mdt_encap_index 0, bidir: 0
, multi-route
04:39:34.353994 mrib [7170]::update IPC message (type:mts) from "igmp", 1 routes
present: [xid: 0xffff000c]
```

MRIB 进程收到 ICMP 的 MTS 消息之后，就可以为(*,239.215.215.1/32)创建 MROUTE 表项并通知 MFDM，接下来就可以确认面向 PIM RP（10.99.99.99）的 RPF 并将其添加到表项中。

从例 13-17 的 **show ip mroute** 命令输出结果可以看出，IGMP 已经创建了(*,G)表项且填充了 OIF。

例 13-17　ICMP 在 NX-1 上创建了 MROUTE 表项

```
NX-1# show ip mroute
IP Multicast Routing Table for VRF "default"

(*, 232.0.0.0/8), uptime: 10:08:39, pim ip
  Incoming interface: Null, RPF nbr: 0.0.0.0
  Outgoing interface list: (count: 0)

(*, 239.215.215.1/32), uptime: 01:59:08, igmp ip pim
  Incoming interface: Ethernet3/18, RPF nbr: 10.1.13.3
  Outgoing interface list: (count: 1)
    Vlan115, uptime: 01:59:08, igmp

(10.215.1.1/32, 239.215.215.1/32), uptime: 02:14:30, pim mrib ip
  Incoming interface: Ethernet3/17, RPF nbr: 10.2.13.3
  Outgoing interface list: (count: 1)
    Vlan115, uptime: 01:59:08, mrib
```

注：多播源 10.215.1.1 发出的流量到达之后，还会触发其他操作。RP 发出的数据流量到达之后，会向多播源发起 PIM 加入操作并创建(S,G) MROUTE。具体内容将在本章后面的 13.4.4 节进行讨论。

13.4　PIM 多播

PIM 是构建共享树和最短路径树的多播路由协议，可以在 L3 网络中分发多播流量。顾名思义，PIM 与具体协议无关。从本质上来说，PIM 是在底层单播路由拓扑提供的可用信息的基础上创建了一个多播叠加网络。术语"与协议无关"（protocol independent）指的是 PIM 可以使用任何源协议（如 EIGRP、OSPF 或 BGP）提供的 RIB（Routing Information Base，路由信息库）中的单播路由信息。单播路由表可以为 PIM 提供源端、RP（Rendezvous Point，聚合点）以及接收端的相对位置，对于构建无环 MDT 来说至关重要。

PIM 支持两种操作模式：DM（Dense Mode，密集模式）或 SM（Sparse Mode，稀疏模式）。密集模式的操作基础是假设接收端在整个网络中呈现密集分布模式。密集模式假设所有 PIM 邻居都应该接收流量，该模式下的多播流量将泛洪给所有下游邻居。如果不需要组流量，那么邻居就可以将自己从树中剪除。将密集模式称为推送模型的原因是所有多播流量都是从树根推送给树叶的，此处假设存在很多树叶且这些树叶都希望接收多播流量。NX-OS 不支持 PIM-DM，因为 PIM-SM 更具优势，也是现代数据中心最受欢迎的操作模式。

PIM-SM 基于拉取模型。拉取模型假设接收端在整个网络中呈现稀疏分布模式，因而仅将流量转发给显式请求流量的 PIM 邻居更加有效。无论接收端在拓扑结构中呈现稀疏分布方式还是密集分布方式，PIM-SM 都能很好地完成多播分发操作。由于 PIM-SM 拥有显式加入特性，因而已成为多播通信的首选部署方式。

将多播流量从源端分发到接收端的过程中，PIM 的作用如下。

- 向 PIM RP（ASM）注册多播源。
- 将感兴趣的接收端加入 MDT。
- 代表接收端确定应该加入哪棵树。
- 如果同一 L3 网络存在多台 PIM 路由器，那么就确定由哪台 PIM 路由器转发流量。

本节将介绍 PIM 协议以及 PIM 用来构建 MDT 并创建转发状态的消息。此外，还将讨论 PIM-SM 的多种操作模型，包括 ASM、SSM 以及 BiDir（Bi-Directional，双向）PIM。

> 注：RFC 2362 最初将 PIM 定义为实验性协议，后来被 RFC 4601 废除。当前 RFC 4601 已被 RFC 7761 更新。NX-OS 的 PIM 实现基于 RFC 4601。

13.4.1 PIM 协议状态和多播树

在深入探讨 PIM 协议机制和消息类型之前，有必要理解各种不同的多播树。PIM 使用 RPT 和 SPT 来构建无环转发路径，从而将多播流量传递到接收端。RPT 以 PIM RP 为树根，SPT 以多播源为树根。PIM-SM 中的这两种树都是单向树，流量从树根流向连接接收端的树叶，如果需要将流量分流到其他分支以到达树叶，那么就需要执行复制操作。

讨论多播转发过程时，常常需要用到 MROUTE 状态。对于 PIM 多播来说，(*,G)状态由位于 LHR 的接收端创建，表示 RPT 与接收端的关系。收到多播流量之后就会创建(S,G)状态，(S,G)状态表示 SPT 与多播源之间的关系。

数据包到达多播路由器之后，将根据到达树根的单播路由检查这些数据包，称为 RPF（Reverse Path Forwarding，反向路径转发）检查。RPF 检查可以确保 MDT 保持无环状态。路由器发送 PIM Join-Prune（加入/剪除）消息以创建 MROUTE 状态时，就会从 RPF 接口（由去往树根的最佳单播路由确定）发送给树根。图 13-11 解释了 MROUTE 状态以及 PIM MDT 等概念。

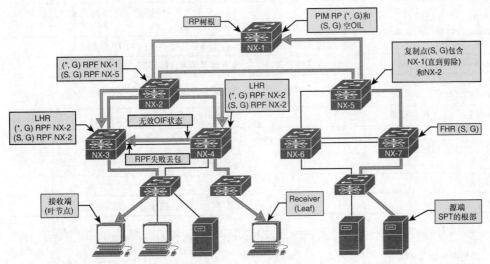

图 13-11　PIM MDT 和 MROUTE 状态

13.4.2 PIM 消息类型

PIM 定义了多种消息类型，可以通过这些消息发现邻居并构建 MDT。所有 PIM 消息都承载在 IP 包中并使用 IP 协议 103。有些消息（如 Register[注册]、Register-Stop[注册终止]）使用单播目的地址，能够跨越多跳 L3 节点从源端到达目的端。不过，其他消息（如 Hello 和 Join-Prune 消息）都是通过多播通信传递的，依赖于周知多播地址 ALL-PIM-ROUTERS 224.0.0.13，且 TTL 值为 1。无论是通过多播包还是单播包进行传递，所有 PIM 消息的报文格式都相同。图 13-12 列出了 PIM 控制消息的报头格式。

4比特	4比特	8比特	16比特
PIM Ver	Type	Reserved	Checksum

图 13-12 PIM 控制消息报头格式

PIM 控制消息的报头字段定义如下。

- **PIM 版本**：PIM 版本号为 2。
- **类型**：指的是 PIM 消息类型（见表 13-9）。
- **保留**：该字段在发送时设置为零，接收端忽略该字段。
- **校验和**：该字段是对整个 PIM 消息计算得到的（Register 消息的多播包部分除外）。

控制消息报头的类型字段负责标识所要发送的 PIM 消息的类型。表 13-9 列出了 RFC 6166 定义的各种 PIM 消息类型。

表 13-9　　　　　　　　　　　PIM 控制消息类型

类型	消息类型	目的地址	描述
0	Hello	224.0.0.13	用于邻居发现
1	Register（注册）	RP 地址（单播）	由 FHR 发送给 RP 以注册多播源，仅用于 PIM SM
2	Register-Stop（注册终止）	FHR（单播）	由 FHR 发送给 RP 以响应 Register 消息，仅用于 PIM SM
3	Join-Prune（加入/剪除）	224.0.0.13	加入 MDT 或者从 MDT 中剪除，不适用于 PIM-DM
4	Bootstrap（引导）	224.0.0.13	引导路由器逐跳发送以传播域中的 RP 映射信息，用于 PIM SM 和 BiDIR
5	Assert（声明）	224.0.0.13	在 LAN 网段上检测到多个转发路由器之后选举单个转发路由器
6	Graft（嫁接）	单播去往 RPF 邻居	重新加入先前被剪除的 MDT 分支
7	Graft-ACK（嫁接确认）	单播去往嫁接发起端	向下游邻居确认嫁接消息
8	Candidate-RP-Advertisement（C-RP 宣告）	BSR 地址（单播）	发送给 BSR 以通告 RP 的候选状态
9	State Refresh（状态刷新）	224.0.0.13	由 FHR 逐跳发送以刷新剪除状态，仅用于 PIM DM
10	DF-Election（DF 选举）	224.0.0.13	用于 PIM BiDIR，负责选举转发路由器。子类型包括 Offer（提议）、Winner（胜出方）、Backoff（退避）和 Pass（转交）
11~14	未分配	-	-
15	保留	-	RFC 6166，用于将来进一步扩展类型字段

注：由于 NX-OS 不支持 PIM-DM，因而本章将不再详细讨论专用于 PIM-DM 的 PIM 消息。如果希望了解与 PIM-DM 相关的详细信息，可以参阅 RFC 3973。

1. PIM Hello 消息

所有启用了 PIM 功能的接口都要定期发送 PIM Hello 消息，目的是发现邻居并建立 PIM 邻居邻接关系。PIM Hello 消息由 PIM 消息类型 0 标识。

DR 优先级选项数值用于 DR（Designated Router，指派路由器）选举进程，默认值为 1，优先级数值较高的邻居将被选为 PIM DR。如果 DR 优先级相同，那么 IP 地址较大的将赢得选举进程。PIM DR 负责向 PIM RP 注册多播源，并代表接口上的多播接收端加入 MDT。

Hello 消息以 TLV（Type, Length, Value，类型—长度—值）格式来提供不同的选项类型。常见的 Hello 消息选项类型包括以下几种。

- **选项类型 1**：保持时间是邻居保持可达的时间。值 0xffff 表示该邻居始终不会超时，值 0 则表示该邻居即将宕机或者已经更改其 IP 地址。
- **选项类型 2**：使用 LAN 剪除时延在多路接入 LAN 网络上调整剪除传播时延。仅当 LAN 上的所有路由器都支持该选项时，才能使用该选项，上游路由器可以利用该选项来确定剪除接口之前应该等待加入覆盖消息多长时间。
- **选项类型 3 ~ 16**：保留给将来使用。
- **选项类型 18**：已废除，不再使用。
- **选项类型 19**：在 DR 选举期间使用 DR 优先级。
- **选项类型 20**：GENID（Generation ID，生成 ID）是发送 Hello 消息的接口的 32 比特随机值，该值在接口上重新启动 PIM 之前保持不变。
- **选项类型 24**：通过地址列表告知邻居有关接口上的辅助 IP 地址的信息。

2. PIM Register 消息

PIM DR 以单播包方式向 PIM RP 发送 PIM Register（注册）消息，Register 消息的作用是告诉 PIM RP 多播源正在将多播流量主动发送给组地址，实现方式是将封装后的多播包从 Register 消息中的多播源发送给 RP。收到源端发送的数据流量之后，PIM DR 将执行以下操作。

- 将源端到达的多播包发送给管理引擎。
- 管理引擎为多播组创建硬件转发状态、构建 Register 消息并将 Register 消息发送给 PIM RP。
- 建立了硬件转发状态之后，路由器从源端收到的后续数据包将不再发送给管理引擎以创建 Register 消息。这样做的目的是限制发送给管理引擎控制平面的流量。

与此相反，思科 IOS PIM DR 会持续发送 Register 消息，直至收到 PIM RP 发送的 Register-Stop 消息。NX-OS 提供了全局配置命令 **ip pim register-until-stop**，可以将 NX-OS 的默认行为修改为思科 IOS 方式。不过，大多数情况不需要修改 NX-OS 的默认行为。

PIM Register 消息包含了以下字段。

- **类型（Type）**：Register 消息的类型值为 1。
- **B 比特（The Border Bit，边界比特）**：该比特在发送时被设置为零，接收后被忽略（RFC 7761）。RFC 4601 描述了 PMBR（PIM Multicast Border Router，PIM 多播边界路由器）功能，PMBR 功能利用该比特指定本地源（该比特设置为 0 时）或者在 PMBR 上指定直连云中的源端（该比特设置为 1 时）。
- **空注册比特（The Null-Register Bit）**：如果数据包是空 Register 消息，那么就将该比特设置为 1。空 Register 消息封装的是来自源端的 *虚拟*（*dummy*）IP 报头，而不是 Register 消息中完整封装的数据包。
- **多播数据包（Multicast Data Packet）**：Register 消息中的该字段是源端发送的原始数据包。封装到 Register 消息中之前，会递减原始数据包的 TTL。如果数据包是一个空 Register 消息，那么 Register 消息的该字段就是一个包含源地址和组地址的虚拟 IP 报头。

3. PIM Register-Stop 消息

PIM Register-Stop（注册终止）消息是 PIM RP 在收到 Register 消息之后以单播包方式发送给 PIM DR 的响应消息。Register-Stop 消息的目的地址是发送 Register 消息的 PIM DR 使用的源地址。

Register-Stop 消息的作用是通知 DR 停止将封装后的多播包发送给 PIM RP 并确认已收到 Register 消息。Register-Stop 消息包括以下编码字段。

- 类型（Type）：Register-Stop 消息的类型值为 2。
- 组地址（Group Address）：是封装在 Register 消息中的多播包的组地址。
- 源地址（Source Address）：是 Register 消息已封装的多播数据包中的源 IP 地址。

4. PIM Join-Prune 消息

PIM 路由器使用 ALL-PIM-ROUTERS 多播地址 224.0.0.13 将 PIM Join-Prune（加入—剪除）消息发送给去往源端或 PIM RP 的上游邻居。此时会发送 Join 消息以构建去往 PIM RP 的 RPT（RP Tree，RP 树）（共享树），或构建去往多播源的 SPT（Shortest-Path Tree，最短路径树）（有源树）。Join-Prune 消息包含了希望加入的组和源的编码列表以及希望剪除的源列表，通常被称为组集（Group Set）和源列表（Source List）。

目前存在两种类型的组集，而且这两种组集都有一个加入源列表和一个剪除源列表。通配符组集表示整个多播组区间（224.0.0.0/4），特定组集表示一个有效的多播组地址。单条 Join-Prune 消息可以包含多个特定组集，但是只能包含单个通配符组集实例。请注意，同一 Join-Prune 消息允许存在单个通配符组集和一个或多个特定组集。Join-Prune 消息包含以下字段。

- 类型（Type）：Join-Prune 消息的类型值为 3。
- 单播邻居上游地址（Unicast Neighbor Upstream Address）：作为消息目的端上游邻居的地址。
- 保持时间（Holdtime）：使 Join-Prune 状态保持激活状态的时间。
- 组数（Number of Groups）：消息中包含的多播组集的数量。
- 多播组地址（Multicast Group Address）：多播组地址负责标识组集，可以是通配符或特定组。
- 加入的源数量（Number of Joined Sources）：该组已加入的源数量。
- 加入的源地址 1 .. n（Joined Source Address 1 .. n）：该组正在加入的多播源的源列表，该字段包含 3 个标志。
    - S（Sparse bit，稀疏比特）：对于 PIM-SM 来说，此标志被设置为 1。
    - W（Wildcard bit，通配符比特）：设置为 1 时，表示已编码的源地址代表(*,G)表项中的通配符；设置为 0 时，表示已编码的源地址代表(S,G)表项的源地址。
    - R（RP Bit，RP 比特）：设置为 1 时，表示将 Join 消息发送给 PIM RP；设置为 0 时，表示将 Join 消息发送给多播源。
- 剪除的源数量（Number of Pruned Source）：该组已剪除的源数量。
- 剪除的源地址 1 .. n（Pruned Source Address 1 .. n）：该组已剪除的多播源的源列表。该字段也包含了与加入的源地址字段相同的 3 个标志(S,W,R)。

> **注**：从理论上说，组集的数量可能超过最大 IP 包大小 65535，此时需要使用多条 Join-Prune 消息。必须确保 PIM 邻居拥有匹配的 L3 MTU 大小，因为邻居发送的 Join-Prune 消息可能过大以至于接口无法容纳，进而导致接收端 PIM 邻居出现多播状态丢失以及 MDT 损坏问题。

5. PIM Bootstrap 消息

PIM Bootstrap（引导）消息由 BSR（Bootstrap Router，引导路由器）发出，提供了一个包含 group-to-RP（组到 RP）映射信息的 RP 集。Bootstrap 消息的目的地址是 ALL-PIM- ROUTERS 地址 224.0.0.13，在整个多播域中逐跳转发。PIM 路由器收到 Bootstrap 消息之后，将处理消息之后的内容并构建新的数据包，从而将 Bootstrap 消息转发给每个接口的所有 PIM 邻居。Bootstrap 消息可能会被分段成多个 BSMF（Bootstrap Message Fragment，引导消息分段），每个分段使用的消息格式都与 Bootstrap 消息相同。PIM Bootstrap 消息包含了以下字段。

- 类型（Type）：Bootstrap 消息的类型值为 4。
- 不转发比特（No-Forward Bit）：指示不应该转发该 Bootstrap 消息。
- 分段标签（Fragment Tag）：随机生成的数字，用于区分属于同一 Bootstrap 消息的 BSMF。每个分段携带的该字段值相同。
- 散列掩码长度（Hash Mask Length）：散列函数使用的掩码长度（以比特为单位）。
- BSR 优先级（BSR Priority）：发端 BSR 的优先级值，取值访问是 0~255（值越大越优）。
- BSR 地址（BSR Address）：该域的引导路由器地址。
- 组地址 1..n（Group Address 1..n）：与 C-RP（Candidate-RP，候选 RP）相关联的组区间。
- RP Count 1..n（RP Count 1..n）：整个 Bootstrap 消息包含的对应组区间的 C-RP 地址数。
- 分段 RP 数 1..m（Frag RP Count 1..m）：Bootstrap 消息分段中包含的对应组区间的 C-RP 地址数。
- RP 地址 1..m（RP Address 1..m）：对应组区间的 C-RP 地址。
- RP1..m 保持时间（RP1..m Holdtime）：对应 RP 的保持时间（以秒为单位）。
- RP1..m 优先级（RP1..m Priority）：对应 RP 和组地址的优先级。从 C-RP 宣告消息复制该字段，最高优先级为零（对于每个 RP 和每个组地址来说）。

6．PIM Assert 消息

PIM Assert（声明）消息用于解决同一网段上存在多台路由器时的转发路由器冲突问题，该消息的目的地址是 ALL-PIM-ROUTERS 地址 224.0.0.13。如果路由器在其正常情况下应该向外发送数据包的接口上收到了多播包，那么就会发送 Assert 消息。如果两台或多台路由器都将流量发送给同一网段，那么就会发生这种情况。此外，如果从另一台路由器收到了 Assert 消息，那么也要发送 Assert 消息作为响应。Assert 消息允许两台发送路由器根据度量值以及去往源或 RP 地址的管理距离，来确定由哪台路由器继续转发以及哪台路由器应该停止转发。Assert 消息以特定组(*,G)或特定源(S,G)的形式发送，表示流量从所有源端到特定组，或者从特定源端到达特定组。Assert 消息包含以下字段。

- 类型（Type）：PIM Assert 消息的类型值为 5。
- 组地址（Group Address）：需要解决转发路由器冲突问题的组地址。
- 源地址（Source Address）：需要解决转发路由器冲突问题的源地址。值为零表示(*,G)声明。
- RPT 比特（RPT-Bit）：对于(*,G)声明消息来说，该值被设置为 1；对于(S,G)声明消息来说，该值被设置为零。
- 度量优先级（Metric Preference）：分配给单播路由协议（为源或 PIM RP 提供路由）的优先级值，该值指的是单播路由协议的管理距离。
- 度量（Metric）：去往源或 PIM RP 的路由的单播路由表度量。

7．PIM Candidate-RP-Advertisement 消息

如果 PIM 域被配置为使用 BSR 的 RP 宣告方法，那么每个 C-RP 都会以单播方式定期向 BSR 发送 PIM Candidate-RP-Advertisement（候选 RP 宣告）消息。该消息的目的是通知 BSR，C-RP 愿意充当宣告消息中包含的多播组的 RP。PIM Candidate-RP-Advertisement 消息包含以下字段。

- 类型（Type）：Candidate-RP-Advertisement 消息的类型值为 8。
- 前缀数（Prefix Count）：消息中包含的组地址数，不能为零。
- 优先级（Priority）：对应组地址包含的 RP 的优先级，最高优先级为零。

- **保持时间（Holdtime）**：该宣告消息的有效时间（以秒为单位）。
- **RP 地址（RP Address）**：宣告为 C-RP 的接口的地址。
- **组地址 1 .. n（Group Address 1 .. n）**：与 C-RP 相关联的组区间。

  8. PIM DF-Election 消息

对于 PIM BiDIR 来说，DF（Designated Forwarder，指派转发路由器）选举进程负责在网段上选举最佳路由器，从而沿多播树将流量从 RPL（Rendezvous Point Link，聚合点链路）向下转发到网段上。此外，DF 还负责沿上游方向从本地网段向 RPL 发送数据包。DF 的选举依据是到达 RPA（Rendezvous Point Address，聚合点地址）的单播路由度量。同一网段上的路由器利用 PIM DF-Election（DF 选举）消息来确定哪台路由器是 DF（逐个 RPA）。路由器在 Offer（提议）、Winner（胜出方）、Backoff（退避）和 Pass（转交）消息中宣告其度量，这些都是 DF-Election 消息的子消息类型。PIM DF-Election 消息包含以下字段。

- **类型（Type）**：PIM DF-Election 消息的类型值为 10，包含以下 4 种子类型。
  - **Offer（提议）**：子类型 1。由路由器发送，这些路由器认为它们到 RPA 的度量比目前在 Offer 消息中看到的度量更优。
  - **Winner（胜出方）**：子类型 2。路由器承担 DF 角色之后或者对较差的 Offer 重新声明时发送该消息。
  - **Backoff（退避）**：子类型 3。DF 利用该消息来确认更优的 Offer，指示其他拥有相等或较差 Offer 的路由器等待，直至 DF 将职责转交给 Offer 发送端。
  - **Pass（转交）**：子类型 4。旧 DF 利用该消息将职责转交给先前发送 Offer 的路由器。Old-DF-Metric 是 DF 发送 Pass 消息时的当前度量。
- **RP 地址（RP Address）**：正在执行选举操作的 RPA。
- **发送端度量优先级（Sender Metric Preference）**：分配给单播路由协议（负责提供到 RPA 路由）的优先级值，该值指的是单播路由协议的管理距离。
- **发送端度量（Sender Metric）**：消息发送端用于到达 RPA 的单播路由表的度量。

Backoff 消息在通用选举消息格式的基础上增加了以下字段。

- **提议地址（Offering Address）**：提供最新（最佳）提议的路由器的地址。
- **提议度量优先级（Offering Metric Preference）**：分配给单播路由协议（提议路由器利用该路由协议去往 RPA）的优先级值。
- **提议度量（Offering Metric）**：提议路由器用于到达 RPA 的单播路由表度量。
- **间隔（Interval）**：度量较差的路由器（与提议路由器相比）使用的退避间隔（以毫秒为单位）。

Pass 消息在通用选举消息格式的基础上增加了以下字段。

- **新胜出方地址（New Winner Address）**：提供最新（最佳）提议的路由器的地址。
- **新胜出方度量优先级（New Winner Metric Preference）**：分配给单播路由协议（提议路由器利用该路由协议去往 RPA）的优先级值。
- **新获胜者度量（New Winner Metric）**：提议路由器用于到达 RPA 的单播路由表度量。

13.4.3 PIM 接口和邻居验证

NX-OS 需要安装 LAN_ENTERPRISE_SERVICES_PKG 许可才能启用 **feature pim**。在安装许可并启用该 PIM 功能特性之前，用户无法使用各种 PIM 配置命令。

例 13-18 使用 **ip pim sparse-mode** 命令在接口上启用了 PIM。

例 13-18　在接口上配置 PIM-SM

```
NX-1# show run pim
! Output omitted for brevity
!Command: show running-config pim

version 7.2(2)D1(2)
feature pim

interface Vlan115
  ip pim sparse-mode

interface Vlan116
  ip pim sparse-mode

interface Ethernet3/17
  ip pim sparse-mode

interface Ethernet3/18
  ip pim sparse-mode
```

在接口上启用了 PIM 之后，如果链路上还存在其他也启用了 PIM 功能特性的路由器，那么就会发送 Hello 包以建立 PIM 邻居关系。

注：PIM 的 Hello 间隔以毫秒为单位进行配置，最小可接受间隔为 1000ms（等于 1s）。如果希望 Hello 间隔小于默认值以便检测故障 PIM 邻居，那么就可以在 PIM 中使用 BFD，而不是降低 Hello 间隔。

从例 13-19 的输出结果可以看出，NX-1 与 NX-3 和 NX-4 建立了 PIM 邻居关系。输出结果显示了邻居是否支持 BiDIR 功能，同时还提供了用于 DR 选举进程的每个邻居的优先级值。

例 13-19　NX-1 的 PIM 邻居

```
NX-1# show ip pim neighbor

PIM Neighbor Status for VRF "default"
Neighbor          Interface          Uptime    Expires    DR         Bidir-    BFD
                                                          Priority   Capable   State
10.2.13.3         Ethernet3/17       4d21h     00:01:34   1          yes       n/a
10.1.13.3         Ethernet3/18       4d21h     00:01:19   1          yes       n/a
```

PIM 提供了多个与接口相关的参数，可以通过这些参数来确定协议的运行方式。可以通过命令 **show ip pim interface** [*interface identifier*]查看所有启用了 PIM 功能特性的接口详细信息，如例 13-20 所示。从故障排查角度来看，该命令输出结果最有用的地方就在于能够以逐个接口的方式提供统计信息，这些统计信息可以为不同的 PIM 消息类型以及与 Hello 包相关的字段提供非常有用的计数器。DR 选举状态对于确定哪台设备在支持 PIM-SM 的网段上注册了多播源以及哪台设备将流量转发给接收端（通过 IGMP Membership Report 消息获知）来说也非常有用。

例 13-20　NX-1 的 PIM 接口参数

```
NX-1# show ip pim interface e3/18

PIM Interface Status for VRF "default"
Ethernet3/18, Interface status: protocol-up/link-up/admin-up
  IP address: 10.1.13.1, IP subnet: 10.1.13.0/24
  PIM DR: 10.1.13.3, DR's priority: 1
  PIM neighbor count: 1
```

```
      PIM hello interval: 30 secs, next hello sent in: 00:00:10
      PIM neighbor holdtime: 105 secs
      PIM configured DR priority: 1
      PIM configured DR delay: 3 secs
      PIM border interface: no
      PIM GenID sent in Hellos: 0x2cc432ed
      PIM Hello MD5-AH Authentication: disabled
      PIM Neighbor policy: none configured
      PIM Join-Prune inbound policy: none configured
      PIM Join-Prune outbound policy: none configured
      PIM Join-Prune interval: 1 minutes
      PIM Join-Prune next sending: 1 minutes
      PIM BFD enabled: no
      PIM passive interface: no
      PIM VPC SVI: no
      PIM Auto Enabled: no
      PIM Interface Statistics, last reset: never
        General (sent/received):
          Hellos: 19246/19245 (early: 0), JPs: 8246/8, Asserts: 0/0
          Grafts: 0/0, Graft-Acks: 0/0
          DF-Offers: 0/0, DF-Winners: 0/0, DF-Backoffs: 0/0, DF-Passes: 0/0
        Errors:
          Checksum errors: 0, Invalid packet types/DF subtypes: 0/0
          Authentication failed: 0
          Packet length errors: 0, Bad version packets: 0, Packets from self: 0
          Packets from non-neighbors: 0
              Packets received on passiveinterface: 0
          JPs received on RPF-interface: 0
          (*,G) Joins received with no/wrong RP: 0/0
          (*,G)/(S,G) JPs received for SSM/Bidir groups: 0/0
          JPs filtered by inbound policy: 0
          JPs filtered by outbound policy: 0
```

除逐个接口的统计信息之外，NX-OS 还提供了整个 PIM 路由器进程的汇总统计信息（全局统计信息），可以通过命令 **show ip pim statistics** 查看该统计信息，如例 13-21 所示。这些统计信息对于排查与 PIM RP 相关的消息处理故障来说非常有用。

例 13-21　PIM 全局统计信息

```
NX-1# show ip pim statistics

PIM Global Counter Statistics for VRF:default, last reset: never
  Register processing (sent/received):
    Registers: 1/3, Null registers: 1/293, Register-Stops: 4/2
    Registers received and not RP: 1
    Registers received for SSM/Bidir groups: 0/0
  BSR processing (sent/received):
    Bootstraps: 0/0, Candidate-RPs: 0/0
    BSs from non-neighbors: 0, BSs from border interfaces: 0
    BS length errors: 0, BSs which RPF failed: 0
    BSs received but not listen configured: 0
    Cand-RPs from border interfaces: 0
    Cand-RPs received but not listen configured: 0
  Auto-RP processing (sent/received):
    Auto-RP Announces: 0/0, Auto-RP Discoveries: 0/0
    Auto-RP RPF failed: 0, Auto-RP from border interfaces: 0
    Auto-RP invalid type: 0, Auto-RP TTL expired: 0
```

```
    Auto-RP received but not listen configured: 0
  General errors:
    Control-plane RPF failure due to no route found: 2
    Data-plane RPF failure due to no route found: 0
    Data-plane no multicast state found: 0
    Data-plane create route state count: 5
```

如果接口未建立特定的 PIM 邻居关系，那么就可以通过 NX-OS 提供的事件历史记录或 Ethanalyzer 工具排查故障原因。从例 13-22 的 **show ip pim internal event-history hello** 命令输出结果可以确认，NX-1 发送了 PIM Hello 消息，而且在 Ethernet 3/18 上收到了 NX-3 的 Hello 消息。

例 13-22　Hello 消息的 PIM 事件历史记录

```
NX-1# show ip pim internal event-history hello
! Output omitted for brevity
02:19:48.277885 pim [31641]: :   GenID Option: 0x2da27857
02:19:48.277882 pim [31641]: :   Bidir Option present
02:19:48.277881 pim [31641]: :   DR Priority Option: 1
02:19:48.277878 pim [31641]: :   Holdtime Option: 105 secs
02:19:48.277875 pim [31641]: : Received Hello from 10.1.13.3 on Ethernet3/18,
length: 30
02:19:42.688032 pim [31641]: : iod = 64 - Send Hello on Ethernet3/18 from
10.1.13.1, holdtime: 105 secs, genID: 0x2cc432ed, dr-priority: 1, vpc: 0
02:19:41.714660 pim [31641]: : iod = 259 - Send Hello on Vlan116 from
10.116.1.254, holdtime: 105 secs, genID: 0xfb8dc7c, dr-priority: 1, vpc: 0
02:19:38.268071 pim [31641]: : iod = 258 - Send Hello on Vlan115 from
10.115.1.254, holdtime: 105 secs, genID: 0x2fd1ac5d, dr-priority: 1, vpc: 0
```

如果希望获取与 PIM 消息内容有关的更多详细信息，那么就可以使用 Ethanalyzer 工具抓取数据包，如例 13-23 所示。此时，可以通过选项 *detail* 在本地检查数据包的详细信息，也可以通过选项 *write* 保存抓取到的数据包以进行离线分析。

例 13-23　通过 PIM Ethanalyzer 抓取 PIM Hello 消息

```
NX-1# ethanalyzer local interface inband-in capture-filter "pim" detail
! Output omitted for brevity

Capturing on inband
Frame 1: 64 bytes on wire (512 bits), 64 bytes captured (512 bits)
    Encapsulation type: Ethernet (1)
    Arrival Time: Oct 29, 2017 00:48:35.186687000 UTC
    [Time shift for this packet: 0.000000000 seconds]
    Epoch Time: 1509238115.186687000 seconds
    [Time delta from previous captured frame: 0.029364000 seconds]
    [Time delta from previous displayed frame: 0.029364000 seconds]
    [Time since reference or first frame: 3.751505000 seconds]
    Frame Number: 5
    Frame Length: 64 bytes (512 bits)
    Capture Length: 64 bytes (512 bits)
    [Frame is marked: False]
    [Frame is ignored: False]
    [Protocols in frame: eth:ip:pim]
<>
Internet Protocol Version 4, Src: 10.1.13.3 (10.1.13.3), Dst: 224.0.0.13
  (224.0.0.13)
<>
```

```
Protocol Independent Multicast
    0010 .... = Version: 2
    .... 0000 = Type: Hello (0)
    Reserved byte(s): 00
    Checksum: 0x3954 [correct]
    PIM options: 4
        Option 1: Hold Time: 105s
            Type: 1
            Length: 2
            Holdtime: 105s
        Option 19: DR Priority: 1
            Type: 19
            Length: 4
            DR Priority: 1
        Option 22: Bidir Capable
            Type: 22
            Length: 0
        Option 20: Generation ID: 765622359
            Type: 20
            Length: 4
            Generation ID: 765622359
```

注：NX-OS 支持 PIM 邻居认证以及 BFD 故障检测机制，有关这些功能特性的详细信息，可参阅 NX-OS 配置指南。

13.4.4 PIM ASM

目前最常见的 PIM-SM 部署形式就是 ASM（Any-Source Multicast，任意源多播），ASM 使用源于 PIM RP 的 RPT（RP Tree，RP 树）和源于多播源的 SPT（Shortest Path Trees，最短路径树）将多播流量分发给接收端，从而实现多播流量的分发。任意源的名称表示接收端加入多播组之后，就加入了可能向该多播组发送流量的任意多播源。虽然听起来似乎很直观，但大家必须清楚地了解 ASM 与 SSM（Source Specific Multicast，特定源多播）之间的区别。

PIM ASM 中的所有源端均由本地 FHR 注册到 PIM RP，这使 PIM RP 成为拓扑结构中的设备，知道所有多播源的信息。接收端加入多播组之后，本地路由器（LHR）也就加入了 RPT。多播流量从 RPT 到达 LHR 的时候，就知道了该多播组的源地址，从而向源端发送 PIM Join 消息以加入该 SPT，通常将该过程称为 SPT 切换。收到 SPT 的流量之后，就将 LHR 从 RPT 中剪除，从而仅接收 SPT 的流量。这些事件在 MROUTE 表中都有对应的状态，可以利用这些状态为接收端确定 MDT 的当前状态。为便于理解，图 13-13 给出了支持 PIM ASM 的示例拓扑结构。

图 13-13 的处理过程如下。

- **第 1 步**：源 10.115.1.4 开始向组 239.115.115.1 发送流量，NX-2 收到流量并为（10.115.1.4，239.115.115.1）创建一条(S,G) MROUTE 表项。
- **第 2 步**：NX-2 向 PIM RP NX-1（10.99.99.1）注册源，PIM RP 创建一条(S,G) MROUTE 并发送 Register-Stop 消息作为响应。只要数据流量从源端到达，NX-2 就会一直定期向 PIM RP 发送空 Register 消息。
- **第 3 步**：接收端 10.215.1.1 发送 IGMP Membership Report 消息以加入 239.115.115.1。NX-4 收到报告之后，就会为（*，239.115.115.1）创建一条(*,G) MROUTE 表项。
- **第 4 步**：NX-4 向 PIM RP NX-1 发送 PIM Join 消息，流量到达 RPT。
- **第 5 步**：NX-4 从 RPT 接收流量，然后向 NX-2 发送 PIM Join 消息以切换到 SPT。NX-2 收到

该 PIM Join 消息之后，将包含 Eth3/17 的 OIF 添加到(S,G) MROUTE 表项中。
- **第 6 步**：虽然图 13-13 并未明确显示，但 NX-4 将会从 RPT 中剪除，流量将从 NX-2 沿 SPT 进行分发。

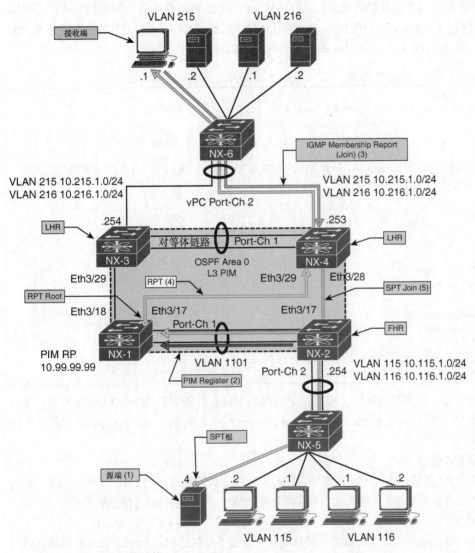

图 13-13　PIM ASM 拓扑结构

如果接收端在多播源激活之前就加入了 RPT，那么上述步骤的顺序可能会有所不同，但这些步骤都是必需的且必然会发生。如果接收端收不到流量，那么了解这些强制性事件以及 FHR、LHR、PIM RP 和中间路由器的 MROUTE 状态，就能准确确定 MDT 的故障位置。需要记住的是，MROUTE 状态的创建是由 IGMP 和 PIM 的控制平面事件以及数据平面收到多播流量触发的。

注：SPT 切换对于 PIM ASM 来说是可选的，可以通过命令 **ip pim spt-threshold infinity** 将设备强制留在 RPT 上。

1. PIM ASM 配置

PIM ASM 的配置非常简单，每个属于多播域的接口都要配置命令 **ip pim sparse-mode**，包括路由器之间的 L3 接口以及连接接收端的所有接口。为了简化起见和保持一致性，最佳实践是使

用 **ip pim sparse-mode** 启用 PIM RP 环回接口（虽然某些平台可能不需要这么做）。请注意，必须在所有 PIM 路由器上都配置 PIM RP 地址，并建立多播组到特定 RP 地址映射关系。NX-OS 支持 BSR 和 Auto-RP，可以自动配置 PIM 域中的 RP 地址，具体内容请参见 13.4.6 节。例 13-24 显示了 NX-1 的 PIM 配置信息，此时的 NX-1 充当 PIM RP。其他 PIM 路由器的配置相似，但是没有 Loopback99 接口。Loopback99 是在 NX-1 上配置 PIM RP 地址的接口。可以在网络中配置多个 PIM RP，并通过选项 *group-list* 或 *prefix-list* 控制将哪些多播组映射到特定 RP。

例 13-24　NX-1 的 PIM ASM 配置

```
NX-1# show run pim
!Command: show running-config pim

feature pim

ip pim rp-address 10.99.99.99 group-list 224.0.0.0/4
ip pim ssm range 232.0.0.0/8



interface Vlan1101
  ip pim sparse-mode

interface loopback99
  ip pim sparse-mode

interface Ethernet3/17
  ip pim sparse-mode

interface Ethernet3/18
  ip pim sparse-mode
```

排查与 PIM 相关的网络故障时，根据网络规模的不同，可能需要增加 PIM 事件历史记录日志的大小，此时可以通过命令 **ip pim event-history** [*event type*] **size** [*event-history size*]增加各类事件历史记录的大小。

2. PIM ASM 验证

排查 PIM ASM 多播路由故障时，最好首先验证连接了故障接收端的 LHR 的多播状态。这是因为验证 LHR 是否通过 IGMP 了解到接收端非常重要，该步骤不但能够确定故障源于 L2（IGMP）还是 L3 多播路由（PIM），而且能将接下来的故障排查步骤引导到 RPT 或 SPT 上。

如果 LHR 存在(*,G)状态，那么就表示接收端已经发送了有效的 Membership Report 消息，LHR 也向 PIM RP 发送了 RPT Join 消息（PIM RP 通过单播路由选择接口）。请注意，(*,G)的存在表示只有一台接收端发送了 Membership Report，这就意味着故障接收端可能并未发送 Membership Report。接下来验证所有承载了 VLAN 的交换机的 IGMP 监听转发表，以确定接收端的端口是否被配置为接收多播流量。如果同一 VLAN 中的其他接收端能够收到组流量，那么就可以确认接收端主机或 L2 转发出现了问题。

如果 LHR 只有一条(*,G)表项，那么通常表示流量不是来自 RPT。在这种情况下，就需要验证 LHR 与 PIM RP 之间以及多播树上的所有中间 PIM 路由器的 MROUTE 状态。如果 PIM RP 拥有去往 LHR 的有效 OIF 且数据包计数器一直都处于递增状态，那么就表明数据平面可能出现了问题，阻止流量到达 RPT 上的 LHR，或者数据包的 TTL 在传输过程中到期。此时，可以利用 SPAN（Switch Port Analyzer，交换端口分析器）、ACL 命中计数器或 ELAM（Embedded Logic Analyzer Module，嵌入式逻辑分析器模块）等工具将故障隔离到 RPT 上的特定设备。

流量到达 RPT 上的 LHR 之后，将尝试切换到 SPT。这一步需要在路由表中查找源地址，以确定在哪个 PIM 接口上发送 SPT Join 消息。此时的 LHR 拥有 SPT 的(S,G)状态，且 OIL 包含了指向接收端的接口。SPT 的 IIF 与 RPT 的 IIF 可以不同，但并非必须如此。

LHR 向源端发送 PIM SPT Join 消息，路径上的每台中间路由器都有(S,G)状态，且 OIF 指向 LHR、IIF 指向 SPT 源端。FHR 上的 IIF 是连接源端的接口，OIF 包含了收到 PIM SPT Join 消息的接口（指向 LHR）。

可以采用相同的方法沿 SPT 排查多播转发故障。此时需要确定是否所有接收端（可能位于 SPT 的其他分支）都能收到多播流量，确定 SPT 中的哪台设备是故障分支与正常分支的交汇点，该设备的 MROUTE 状态应该指示这两个分支的接口都在 OIL 中。如果不是，那么就要验证 PIM 以确定为何未收到 SPT Join 消息。如果 OIL 确实包含了两个 OIF，那么就表明故障问题可能与数据平面丢包有关，此时可以考虑采用 SPAN、ACL 或 ELAM 等工具隔离故障设备。将故障隔离到多播树上的特定设备之后，就可以验证控制平面及特定平台的硬件转发表项，以确定故障根源。

3. PIM ASM 事件历史记录和 MROUTE 状态验证

验证已经发送和接收了哪些 PIM 消息的主要方法就是查看 NX-OS 的 PIM 事件历史记录，该记录可以显示调试级别的 PIM 进程和消息处理过程的详细信息，而且不会给系统资源带来任何影响。接下来将以图 13-13 所示拓扑结构为例来验证每台设备的 PIM 消息和 MROUTE 状态，此时新的源端变为活动状态且接收端加入了该多播组。

源 10.115.1.4 开始向 239.115.115.1 发送流量，流量到达 VLAN 115 上的 NX-2。收到该多播流量之后，NX-2 会创建一条(S,G) MROUTE（见例 13-25），MROUTE 中的标志 *ip* 表示该状态是由接收流量创建的。

例 13-25 存在活动多播源时的 NX-2 MROUTE 状态

```
NX-2# show ip mroute 239.115.115.1
! Output omitted for brevity
IP Multicast Routing Table for VRF "default"

(10.115.1.4/32, 239.115.115.1/32), uptime: 00:00:04, ip pim
  Incoming interface: Vlan115, RPF nbr: 10.115.1.4
  Outgoing interface list: (count: 0)
```

接下来 NX-2 向 PIM RP NX-1（10.99.99.99）注册该多播源，发送的 PIM Register 消息携带了封装后的源数据包。从例 13-26 的 **show ip pim internal event-history null-register** 命令输出结果可以看出，NX-1 收到了该 Register 消息，第一条 Register 消息的 pktlen（包长）为 84，且在 PIM RP 上创建了 MROUTE 状态，后续不携带封装源数据包的空 Register 消息的 pktlen 只有 20 字节。NX-1 通过 Register-Stop 消息来响应 Register 消息。

例 13-26 NX-1 收到 Register 消息

```
NX-1# show ip pim internal event-history null-register
! Output omitted for brevity
null-register events for PIM process
16:36:33.724154 pim [31641]::Send Register-Stop to 10.115.1.254 for
(10.115.1.4/32, 239.115.115.1/32)
16:36:33.724133 pim [31641]::Received NULL Register from 10.115.1.254
for (10.115.1.4/32, 239.115.115.1/32) (pktlen 20)
16:34:35.177572 pim [31641]::Send Register-Stop to 10.115.1.254
for (10.115.1.4/32, 239.115.115.1/32)
16:34:35.177543 pim [31641]::Add new route (10.115.1.4/32, 239.115.115.1/32)
to MRIB, multi-route TRUE
```

```
16:34:35.177508 pim [31641]::Create route for (10.115.1.4/32, 239.115.115.1/32)
16:34:35.177398 pim [31641]::Received Register from 10.115.1.254 for
(10.115.1.4/32, 239.115.115.1/32) (pktlen 84)
```

注：NX-OS 可以为收到的携带封装数据包的 Register 消息创建单独的事件历史记录（取决于具体版本），显示命令是 **show ip pim internal event-history data-register-receive**。早期的 NX-OS 版本需要使用 **debug ip pim data-register send** 和 **debug ip pim data-register receive** 命令来调试 PIM 注册进程。

由于当前的 PIM 域没有接收端，因而 NX-1 添加了一条带有空 OIL 的(S,G) MROUTE（见例 13-27），IIF 是 NX-1 与 NX-2 之间的 L3 接口（VLAN 1101 承载在 Port-channel 1 上）。MROUTE 中的标志 *pim* 表示该 MROUTE 状态是由 PIM 创建的。

例 13-27　无接收端时的 NX-1 MROUTE 状态

```
NX-1# show ip mroute 239.115.115.1
! Output omitted for brevity
IP Multicast Routing Table for VRF "default"

(10.115.1.4/32, 239.115.115.1/32), uptime: 00:00:09, pim ip
  Incoming interface: Vlan1101, RPF nbr: 10.1.11.2, internal
  Outgoing interface list: (count: 0)
```

添加了 MROUTE 表项之后，NX-1 向 NX-2 发送 Register-Stop 消息，如例 13-28 所示。此后 NX-2 将抑制其第一条空 Register 消息，因为 NX-2 刚刚收到了响应 Register 消息的 Register-Stop 消息。NX-2 收到 Register-Stop 消息之后会启动一个 Register-Suppression（注册抑制）定时器，定时器即将到期之前，NX-2 会发送一条空 Register 消息。如果定时器到期但未收到 RP 发送的 Register-Stop 消息，那么 DR 就会重新发送完整的封装数据包。

例 13-28　收到 NX-1 的 Register-Stop 消息

```
NX-2# show ip pim internal event-history null-register
! Output omitted for brevity

null-register events for PIM process
16:36:29.667674 pim [10076]::Received Register-Stop from 10.99.99.99 for
(10.115.1.4/32, 239.115.115.1/32)
16:36:29.666010 pim [10076]::Send Null Register to RP 10.99.99.99 for
(10.115.1.4/32, 239.115.115.1/32)
16:35:29.466161 pim [10076]::Suppress Null Register for
(10.115.1.4/32, 239.115.115.1/32) due to recent data Register sent
16:34:31.121180 pim [10076]::Received Register-Stop from 10.99.99.99 for
(10.115.1.4/32, 239.115.115.1/32)
```

此时，已经将源端成功注册到了 PIM RP。该状态将一直持续直至接收端加入多播组，NX-2 则通过空 Register 消息定期通知 NX-1：源端一直都在向组地址发送流量。

连接在 NX-4 上的 VLAN 215 中的接收端发送 Membership Report 消息，希望接收多播组 239.115.115.1 的流量。消息到达 NX-4 之后，将触发 IGMP 创建(*,G) MROUTE 表项（见例 13-29），其中，OIL 包含了 VLAN 215，IIF Ethernet 3/29 则是到达 NX-1 上的 PIM RP 地址的接口。

例 13-29　存在接收端时的 NX-4 MROUTE 状态

```
NX-4# show ip mroute 239.115.115.1
! Output omitted for brevity
IP Multicast Routing Table for VRF "default"
```

```
 (*, 239.115.115.1/32), uptime: 00:01:12, igmp ip pim
   Incoming interface: Ethernet3/29, RPF nbr: 10.2.13.1
   Outgoing interface list: (count: 1)
     Vlan215, uptime: 00:01:12, igmp
```

该 MROUTE 表项与 NX-4 发送给 NX-1 的 PIM RPT Join 消息相对应，如例 13-30 所示。

例 13-30　NX-4 发送给 NX-1 的 PIM RPT Join 消息

```
NX-4# show ip pim internal event-history join-prune
! Output omitted for brevity
16:36:32.630520 pim [13449]::Send Join-Prune on Ethernet3/29, length: 34
16:36:32.630489 pim [13449]::Put (*, 239.115.115.1/32), WRS in join-list for
 nbr 10.2.13.1
16:36:32.630483 pim [13449]::wc_bit = TRUE, rp_bit = TRUE
```

NX-1 收到 NX-4 发送的 RPT Join 消息之后，就会将 OIF Ethernet 3/17 添加到该 MROUTE 的 OIL 中，如例 13-31 所示。

例 13-31　NX-1 收到的 PIM RPT Join 消息

```
NX-1# show ip pim internal event-history join-prune
! Output omitted for brevity
16:36:36.688773 pim [31641]::Add Ethernet3/17 to all (S,G)s for group
239.115.115.1
16:36:36.688652 pim [31641]::No (*, 239.115.115.1/32) route exists, to us
16:36:36.688643 pim [31641]::pim_receive_join: We are target comparing with iod
16:36:36.688604 pim [31641]::pim_receive_join: route: (*, 239.115.115.1/32),
wc_bit: TRUE, rp_bit: TRUE
16:36:36.688593 pim [31641]::Received Join-Prune from 10.2.13.3 on Ethernet3/17
length: 34, MTU: 9216, ht: 210
```

收到 Join 消息之后，就会触发 NX-1 创建 (*,G) MROUTE 状态，同时还会触发 NX-1 通过 VLAN 1101 向 NX-2 发送该多播源的 Join 消息，如例 13-32 所示。

例 13-32　NX-1 向 NX-2 发送 PIM Join 消息

```
NX-1# show ip pim internal event-history join-prune
! Output omitted for brevity
16:36:36.690787 pim [31641]::Send Join-Prune on loopback99, length: 34
16:36:36.690481 pim [31641]::Send Join-Prune on Vlan1101, length: 34
16:36:36.690227 pim [31641]::Put (10.115.1.4/32, 239.115.115.1/32),
S in join-list for nbr 10.1.11.2
16:36:36.690220 pim [31641]::wc_bit = FALSE, rp_bit = FALSE
16:36:36.690158 pim [31641]::Put (10.115.1.4/32, 239.115.115.1/32),
RS in prune-list for nbr 10.99.99.99
16:36:36.690150 pim [31641]::wc_bit = FALSE, rp_bit = TRUE
16:36:36.690078 pim [31641]::(*, 239.115.115.1/32) we are RPF nbr
```

NX-1 向 NX-2 发送 PIM Join 消息之后，NX-2 就会添加 OIF VLAN 1101，如例 13-33 所示。

例 13-33　NX-2 收到 NX-1 发送的 PIM Join 消息

```
NX-2# show ip pim internal event-history join-prune
! Output omitted for brevity
16:36:32.634207 pim [10076]::(10.115.1.4/32, 239.115.115.1/32) route exists,
RPF if Vlan115, to us
16:36:32.634186 pim [10076]::pim_receive_join: We are target comparing with iod
16:36:32.634142 pim [10076]::pim_receive_join: route:
```

```
(10.115.1.4/32, 239.115.115.1/32), wc_bit: FALSE, rp_bit: FALSE
16:36:32.634125 pim [10076]::Received Join-Prune from 10.1.11.1 on Vlan1101,
 length: 34, MTU: 9216, ht: 210
```

至此，多播流量就可以从源端经 NX-2 流向 NX-1，NX-1 将接收流量并通过 RPT 转发给 NX-4。从例 13-34 显示的 PIM 事件历史记录可以看出，NX-4 在 RPT 上收到了流量，且发生了 SPT 切换操作，NX-4 首先向 NX-2（10.2.23.2）发送 SPT Join 消息，然后将自己从去往 NX-1（10.2.13.1）的 RPT 上剪除。

例 13-34 NX-4 的 SPT 切换

```
NX-4# show ip pim internal event-history join-prune
! Output omitted for brevity
16:36:33.256859 pim [13449]:: Send Join-Prune on Ethernet3/29, length: 34 in context 1
16:36:33.256735 pim [13449]::Put (10.115.1.4/32, 239.115.115.1/32), RS in prune-list
   for nbr 10.2.13.1
16:36:33.256729 pim [13449]::wc_bit = FALSE, rp_bit = TRUE
16:36:33.255153 pim [13449]::Send Join-Prune on Ethernet3/28, length: 34 in context 1
16:36:33.253999 pim [13449]::Put (10.115.1.4/32, 239.115.115.1/32), S in join-list
   for nbr 10.2.23.2
16:36:33.253991 pim [13449]::wc_bit = FALSE, rp_bit = FALSE
```

此时，NX-4 的 MROUTE 状态已经创建了(S,G)，且 OIL 包含了 VLAN215。(S,G)的 IIF 指向 NX-2，而(*,G)的 IIF 指向 PIM RP（NX-1）。例 13-35 显示了 NX-4 的 **show ip mroute** 命令的输出结果。

例 13-35 SPT 切换之后的 NX-4 MROUTE 状态

```
NX-4# show ip mroute 239.115.115.1
! Output omitted for brevity
IP Multicast Routing Table for VRF "default"

(*, 239.115.115.1/32), uptime: 00:01:12, igmp ip pim
  Incoming interface: Ethernet3/29, RPF nbr: 10.2.13.1
  Outgoing interface list: (count: 1)
    Vlan215, uptime: 00:01:12, igmp

(10.115.1.4/32, 239.115.115.1/32), uptime: 00:01:11, ip mrib pim
  Incoming interface: Ethernet3/28, RPF nbr: 10.2.23.2
  Outgoing interface list: (count: 1)
    Vlan215, uptime: 00:01:11, mrib
```

NX-2 有一条(S,G) MROUTE，IIF 为 VLAN 115，OIF 为连接 NX-4 的 Ethernet3/17。例 13-36 显示了 NX-2 的 MROUTE 状态。

例 13-36 SPT 切换之后的 NX-2 MROUTE 状态

```
NX-2# show ip mroute 239.115.115.1
! Output omitted for brevity
IP Multicast Routing Table for VRF "default"

(10.115.1.4/32, 239.115.115.1/32), uptime: 00:03:09, ip pim
  Incoming interface: Vlan115, RPF nbr: 10.115.1.4
  Outgoing interface list: (count: 1)
    Ethernet3/17, uptime: 00:01:07, pim
```

NX-1 有一条来自 NX-4 的(*,G)状态，但是没有(S,G)状态的 OIF。例 13-37 显示了 SPT 切换之后的 NX-1 MROUTE 表。(*,G)的 IIF 是 Loopback99 的 RP 接口（是 RPT 的根）。

例 13-37　SPT 切换之后的 NX-1 MROUTE 状态

```
NX-1# show ip mroute 239.115.115.1
! Output omitted for brevity
IP Multicast Routing Table for VRF "default"
 (*, 239.115.115.1/32), uptime: 03:34:42, pim ip
  Incoming interface: loopback99, RPF nbr: 10.99.99.99
  Outgoing interface list: (count: 1)
    Ethernet3/17, uptime: 03:34:42, pim

(10.115.1.4/32, 239.115.115.1/32), uptime: 03:36:44, pim ip
  Incoming interface: Vlan1101, RPF nbr: 10.1.11.2, internal
  Outgoing interface list: (count: 0)
```

如上节所述，NX-OS 提供的 MROUTE 状态和事件历史记录不但能够帮助确定故障与 RPT 还是 SPT 有关，而且能帮助确定树上的故障设备。

4．PIM ASM 平台验证

排查 PIM 故障时可能需要验证多播路由表项的硬件编程情况。如果控制平面的 PIM 消息和 MROUTE 表指示数据包应该离开接口，但下游 PIM 邻居未收到流量，那么就必须执行该操作。

接下来以 NX-2（配置了 F3 模块的 Nexus 7700）为例来解释该验证操作。在到达 I/O（Input/Output，输入/输出）模块之前，这里提供的验证步骤与其他 NX-OS 平台相似。如果故障排查操作到达了 I/O 级别，那么验证命令将会因不同的平台而异。

不同的 NX-OS 平台的 PI（Platform-Independent，平台无关）组件都相似，如 MROUTE 表、MROUTE 表客户端（PIM、IGMP 和 MSDP）以及 MFDM（Multicast Forwarding Distribution Manager，多播转发分发管理器）等。不过，将这些表项编程到转发和复制 ASIC 的方式有所不同，如果需要排查 ASIC 编程级别的故障，那么最好交给思科 TAC，这是因为如果用户对 NX-OS 平台的 PD（Platform-Dependent，平台相关）架构理解不到位的话，就很容易误解输出结果中显示的信息。例 13-38 验证了 NX-2 的当前 MROUTE 状态。

例 13-38　验证 NX-2 的 MROUTE

```
NX-2# show ip mroute 239.115.115.1
! Output omitted for brevity
IP Multicast Routing Table for VRF "default"

(10.115.1.4/32, 239.115.115.1/32), uptime: 00:00:31, ip pim
  Incoming interface: Vlan115, RPF nbr: 10.115.1.4
  Outgoing interface list: (count: 1)
    Ethernet3/17, uptime: 00:00:31, pim
```

MROUTE 提供了 IIF 和 OIF 信息，可以帮助确定接下来需要验证哪些模块，知道这些信息非常重要，因为 Nexus 7000 系列设备执行多播流量的出站复制。对于出站复制操作来说，数据包到达入站模块之后，会将一份数据包副本发送给同一 I/O 模块上的所有本地接收端，同时将另一份数据包副本发送给交换矩阵，去往该 MROUTE OIL 中的接口 I/O 模块。数据包到达出站模块后，将再次执行查找操作并将数据包复制到出站接口。

OIL 包含了 L3 接口 Ethernet 3/17，IIF 为 VLAN115。为了确认流量到达 VLAN 115 中的哪个物理接口，需要检查多播源的 ARP 缓存和 MAC 地址表表项。可以通过 **show ip arp** 命令显示多播源的 MAC 地址，如例 13-39 所示。

例 13-39　多播源的 ARP 表项

```
NX-2# show ip arp 10.115.1.4
! Output omitted for brevity
```

```
Flags: * - Adjacencies learnt on non-active FHRP router
       + - Adjacencies synced via CFSoE
       # - Adjacencies Throttled for Glean
       D - Static Adjacencies attached to down interface

IP ARP Table
Total number of entries: 1
Address         Age       MAC Address      Interface
10.115.1.4      00:10:53  64a0.e73e.12c2   Vlan115
```

接下来检查 MAC 地址表，以确认多播源 10.115.1.4 发送的数据包到达了哪个接口。例 13-40 显示了 MAC 地址表的输出结果。

例 13-40 多播源的 MAC 地址表表项

```
NX-2# show mac address-table dynamic vlan 115
! Output omitted for brevity

Note: MAC table entries displayed are getting read from software.
 Use the 'hardware-age' keyword to get information related to 'Age'

 Legend:
        * - primary entry, G - Gateway MAC, (R) - Routed MAC, O - Overlay MAC
        age - seconds since last seen,+ - primary entry using vPC Peer-Link, E -
EVPN entry
        (T) - True, (F) - False ,   ~~~ - use 'hardware-age' keyword to retrieve
age info
   VLAN/BD    MAC Address     Type      age      Secure NTFY Ports/SWID.SSID.LID
---------+-----------------+--------+---------+------+----+-------------------
* 115       64a0.e73e.12c2   dynamic   ~~~        F     F   Eth3/19
```

至此已经确认数据包到达了 NX-2 的 Ethernet 3/19，并通过 Ethernet 3/17 去往 NX-4。接下来的验证操作是检查多播组的 MFDM 表项，以确定该多播组是否具有正确的 IIF 和 OIL，如例 13-41 所示。

例 13-41 验证 NX-2 的 MFDM

```
NX-2# show forwarding distribution ip multicast route group 239.115.115.1
! Output omitted for brevity
show forwarding distribution ip multicast route group 239.115.115.1

  (10.115.1.4/32, 239.115.115.1/32), RPF Interface: Vlan115, flags:
    Received Packets: 18 Bytes: 1862
    Number of Outgoing Interfaces: 1
    Outgoing Interface List Index: 30
      Ethernet3/17
```

MFDM 表项看起来没有问题。接下来的验证操作需要通过 LC 控制台执行，可以通过 **attach mod** [*module number*] 命令访问 LC 控制台。如果在非默认 VDC 中执行验证操作，那么就需要在登录模块之后使用 **vdc** [*vdc number*] 命令进入正确的上下文。登录正确的入站模块之后，就可以确认 L3LKP ASIC。

> **注：** 可以通过 **slot** [*module number*] **quoted** [*LC CLI command*] 命令直接获取模块的输出结果，而不需要登录 I/O 模块。

F3 模块采用了 SOC（Switch-On-Chip，片上交换）架构，前面板端口组均由单个 SOC 提供服务。例 13-42 显示了 **show hardware internal dev-port-map** 命令的输出结果。

例 13-42　确定 NX-2 Module 3 的 SoC 实例

```
NX-2# attach mod 3
! Output omitted for brevity
Attaching to module 3 ...
To exit type 'exit', to abort type '$.'
module-3# show hardware internal dev-port-map
--------------------------------------------------------------
CARD_TYPE:        48 port 10G
>Front Panel ports:48
--------------------------------------------------------------
 Device name             Dev role                Abbr num_inst:
--------------------------------------------------------------
> Flanker Eth Mac Driver DEV_ETHERNET_MAC        MAC_0  6
> Flanker Fwd Driver     DEV_LAYER_2_LOOKUP      L2LKP  6
> Flanker Xbar Driver    DEV_XBAR_INTF           XBAR_INTF 6
> Flanker Queue Driver   DEV_QUEUEING            QUEUE  6
> Sacramento Xbar ASIC   DEV_SWITCH_FABRIC       SWICHF 1
> Flanker L3 Driver      DEV_LAYER_3_LOOKUP      L3LKP  6
> EDC                    DEV_PHY                 PHYS   7
+------------------------------------------------------------+
+-----------------+++FRONT PANEL PORT TO ASIC INSTANCE MAP+++--------------+
+------------------------------------------------------------+
FP port | PHYS | MAC_0 | L2LKP | L3LKP | QUEUE |SWICHF
   17      2       2       2       2       2      0
   18      2       2       2       2       2      0
   19      2       2       2       2       2      0
   20      2       2       2       2       2      0
   21      2       2       2       2       2      0
```

本例的入站端口和出站端口使用相同的 SOC 实例（2），且位于同一模块上。如果模块或 SOC 实例不同，那么就需要验证每个模块的每个 SOC，以确保所有信息的正确性。

确认了入站和出站接口的 SOC 号之后，接下来需要检查 I/O 模块的转发表项。从例 13-43 可以看出，该转发表项拥有正确的 IIF（Vlan115）和 OIL（包含接口 Ethernet 3/17）。验证输出数据包计数器，以确保计数器一直都在周期性递增。

例 13-43　在 Module 3 上验证 I/O 模块 MFIB

```
Module-3# show forwarding ip multicast route group 239.115.115.1
! Output omitted for brevity

  (10.115.1.4/32, 239.115.115.1/32), RPF Interface: Vlan115, flags:
    Received Packets: 1149 Bytes: 117224
    Number of Outgoing Interfaces: 2
    Outgoing Interface List Index: 31
      Vlan1101 Outgoing Packets:0 Bytes:0
      Ethernet3/17 Outgoing Packets:1148 Bytes:117096
```

从目前的所有信息可以看出，IIF 和 OIF 均正确，因而验证操作的最后一步是检查 SOC 的编程情况，如例 13-44 所示。

例 13-44 在 Module 3 上验证硬件转发

```
Module-3# show system internal forwarding multicast route source 10.115.1.4
group 239.115.115.1 detail
! Output omitted for brevity
Hardware Multicast FIB Entries:
 Flags Legend:
  * - s_star_priority
  S - sg_entry
  D - Non-RPF Drop
  B - Bi-dir route  W - Wildcard route

(10.115.1.4/32, 239.115.115.1/32), Flags: *S
  Dev: 2, HWIndex: 0x6222 Priority: 0x4788, VPN/Mask: 0x1/0x1fff
  RPF Interface: Vlan115, LIF: 0x15
  MD Adj Idx: 0x5c, MDT Idx: 0x1, MTU Idx: 0x0, Dest Idx: 0x2865
  PD oiflist Idx: 0x1, EB MET Ptr: 0x1
  Dev: 2 Index: 0x70      Type: OIF      elif: 0x5       Ethernet3/17
                          Dest Idx: 0x10          SMAC: 64a0.e73e.12c1
module-3#
```

思科 TAC 会解析这些字段信息，这些字段表示指针，指向本地复制多播包所需的各种表查找操作，或者指向交换矩阵（如果出站接口位于不同模块或 SOC 上）。验证这些索引需要在不同的转发查找和复制阶段执行多次 ELAM 抓包操作。

13.4.5 PIM BiDIR

PIM BiDIR（PIM Bidirectional，双向 PIM）是另一种 PIM-SM 版本，对传统 ASM 操作行为做了一定程度的修改。PIM ASM 和 PIM BiDIR 之间的主要区别如下。

- BiDIR 使用双向共享树，而 ASM 依赖单向共享树和有源树。
- BiDIR 不使用(S,G)状态，而 ASM 则必须为所有将流量发送给组地址的多播源维持(S,G)状态。
- BiDIR 不需要源注册进程，因而可以降低处理开销。
- ASM 和 BiDIR 都必须将每个组映射到 RP，但 BiDIR 中的 RP 实际上并不执行任何数据包处理操作，BiDIR 中的 RPA（Rendezvous Point Address，聚合点地址）只是一个路由向量，用作沿共享树向上或向下转发流量的参考点。
- BiDIR 使用 DF（Designated Forwarder，指派转发路由器）概念，需要在 PIM 域中的每条链路上选举 DF。

由于 BiDIR 不需要任何(S,G)状态，因而只需要一条(*,G) MROUTE 表项即可代表一个多播组。与 ASM 相比，可以大大减少具有大量多播源的网络中的 MROUTE 表项数量。随着 MROUTE 表项的减少，网络的扩展性更强，因为任何路由器平台在耗尽资源之前都只能存储有限数量的表项。扩展能力的增强带来的影响就是失去了对单个多播源流量的可见性，因为没有(S,G)状态可以跟踪这些流量。不过，对于超大规模的多对多网络环境来说，减少状态数量和无须注册进程完全能够弥补这一弊端。

在进一步研究 BiDIR 的工作方式之前，有必要了解 BiDIR 的术语定义（详见表 13-10）。

表 13-10 　　　　　　　　　　　　　PIM BiDIR 术语

术语	定义
RPA（Rendezvous Point Address，聚合点地址）	该地址用作映射到它的所有多播组的 MDT 的根。RPA 对于 PIM 域中的所有路由器来说都必须可达，用于 RPA 的地址不需要在 PIM 域中的任何路由器接口上进行配置
RPL（Rendezvous Point Link，聚合点链路）	用于到达 RPA 的物理链路。所有映射到 RPA 的多播组数据包都从 RPL 向外转发。RPL 是唯一不会发生 DF 选举的接口

术语	定义
DF（Designated Forwarder，指派转发路由器）	需要在每条链路上为每个 RPA 选举一个 DF，DF 的选举依据是到达 RPA 的单播路由度量。DF 负责沿多播树向下将多播流量发给链路，同时还负责向上将流量从链路发送给 RPA。此外，DF 还负责根据本地接收端或 PIM 邻居的状态向上将 PIM Join-Prune 消息发送给 RPA
RPF 接口	用于到达某地址的接口（基于单播路由协议度量）
RPF 邻居	用于到达某地址的 PIM 邻居（基于单播路由协议度量）。对于 BiDIR 来说，RPF 邻居可能并不是应该接收 Join-Prune 消息的路由器，所有 Join-Prune 消息都要定向给选中的 DF

支持 BiDIR 的 PIM 邻居会在 PIM Hello 消息中设置 *BiDIR 使能*（*BiDIR capable*）比特，这是启用 BiDIR 的基本要求。等到所有路由器都开始运行 PIM 进程之后，就可以通过静态配置或 Auto-RP 或 BSR 建立 group-to-RP（组到 RP）映射表。路由器知道了 RPA 之后，就可以确定去往 RPA 的单播路由度量，从而进入下一阶段，即在所有接口上选举 DF。

刚开始的时候，所有路由器都要发送携带子类型 Offer 的 PIM DF-Election 消息，Offer 消息包含了发送端路由器到达 RPA 的单播路由度量。交换了这些消息之后，链路上的所有路由器都能感知到对方，且知道每台路由器到达 RPA 的路由度量。如果路由器收到度量更优的 Offer 消息，那么就不再发送 Offer 消息，从而允许度量更优的路由器被选举为 DF。但是，如果没有进行 DF 选举，那么就会重启 DF 选举进程。最初的 DF 选举结果应该是，除拥有最佳路由度量的路由器之外，所有路由器都将停止发送 Offer 消息。这样一来，拥有最佳路由度量的路由器就可以在发送 3 条 Offer 消息且没有收到其他邻居发送的 Offer 消息之后，承担 DF 角色。成为 DF 之后，路由器将发送携带子类型 Winner 的 DF-Election 消息，该消息将告诉链路上的所有路由器哪台设备是 DF 以及获胜度量是多少。

在正常操作期间，可能会有新的路由器加入，也可能到达 RPA 的度量发生了变化，从而触发向当前 DF 发送 Offer 消息。如果当前 DF 仍然拥有到达 RPA 的最佳度量，那么就会以 Winner 消息进行响应。如果收到的度量优于当前 DF，那么当前 DF 就会发送 Backoff 消息，通过 Backoff 消息告诉挑战路由器在承担 DF 角色之前需要等待一段时间，使链路上的所有路由器都有机会发送 Offer 消息。在此期间，原来的 DF 仍然充当 DF。选举出新 DF 之后，旧 DF 就会发送携带子类型 Pass 的 DF-Election 消息，将 DF 责任转交给新获胜者。选出 DF 之后，PIM BiDIR 网络就已经准备好使用以 RPA 为根的共享树双向转发多播包。

来自下游链路的数据包将一直向上游转发，直至到达拥有 RPL 的路由器（包含 RPA）。由于 PIM BiDIR 没有注册进程，也不会切换到 SPT，因而 RPA 无须位于路由器上。虽然听起来可能会让人感到困惑，但确实可行，因为数据包是通过 RPL 转发给 RPA 的，而且(*,G)状态是通过连接源端的所有 FHR 以及连接了感兴趣接收端的 LHR 向 RPA 建立的。也就是说，BiDIR 中的数据包不需要像 ASM 中那样实际穿越 RP，双向(*,G)树的交叉分支可以直接在源和接收端之间分发多播包。

对于 NX-OS 来说，每个 VRF 最多支持 8 个 BiDIR RPA，通过虚拟 RP（phantom RP）的概念实现 RPA 的冗余性。使用该术语的原因是并没有将 RPA 分配给 PIM 域中的任何路由器。例如，假设 RPA 地址为 10.1.1.1，NX-1 可以在 Loopback10 接口上配置 10.1.1.0/30，NX-3 可以在 Loopback10 接口配置 10.1.1.0/29。由于 PIM 域中的路由器遵循路由表最长前缀匹配规则，因而优选 NX-1。如果 NX-1 发生故障，那么在单播路由协议收敛之后，NX-3 将成为去往 RPL 和 RP 的优选路径。

接下来将以图 13-14 为例解释 PIM BiDIR 的配置和故障排查操作。

连接在 NX-4 上的 VLAN 215 的接收端加入 239.115.115.1 之后，就会在 NX-4 上创建一条(*,G) MROUTE 表项。在 NX-4 与 NX-1 之间的链路上，NX-1 被选举为 DF，因为 NX-1 到达 RPA 的单播路由度量更优。因此，来自 NX-4 的(*,G) Join 消息将被发送给上游 NX-1 并去往主用 RPA。

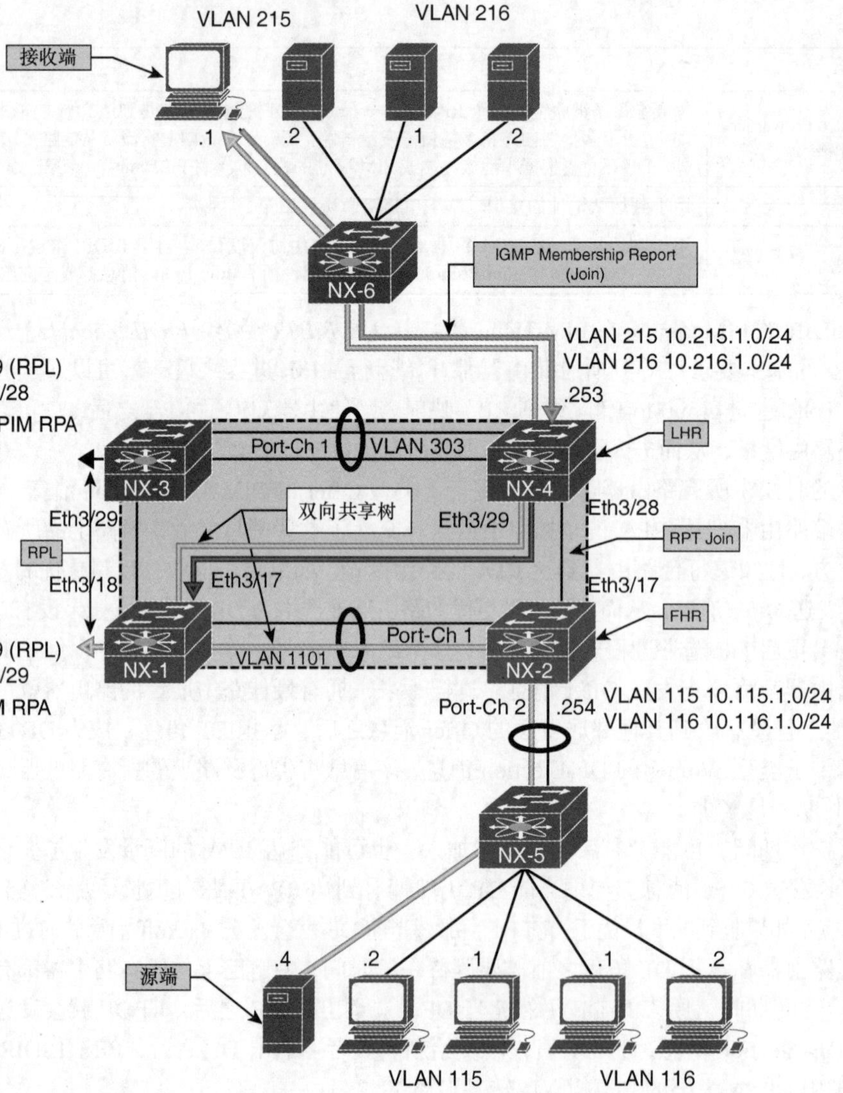

图 13-14　PIM BiDIR 拓扑结构

虽然 NX-1 和 NX-3 都配置了一条去往虚拟 RP 10.99.99.99 的链路（Loopback99），但 NX-1 拥有去往 RPA 的更明细路由（通过其 RPL），因而拓扑结构中的所有路由器都通过 NX-1 去往 RPA。

10.115.1.4 开始向 239.115.115.1 发送多播流量之后，流量将到达 NX-2 的 VLAN 115。由于 NX-2 被选举为 VLAN 115 的 DF，因而 NX-2 通过 RPF 接口（VLAN 1101）沿上游方向将流量发送给 RPA。NX-1 是 NX-2 与 NX-1 之间的 VLAN 1101 的 DF（因为 NX-1 到达 RPA 的路由度量更优），NX-1 收到 NX-2 的流量之后，将根据(*,G) MROUTE 表项的 OIL 转发该流量，该 OIL 不但包含了去往 NX-4 的 Ethernet 3/17 链路，而且包含了 Loopback99 接口（RPL）。多播流量从源端流向接收端的时候，将端到端地使用共享树，NX-4 将永远也不会使用其直连链路去往 NX-2，因为 BiDIR 不会发生 SPT 切换操作。由于多播组的所有流量都沿着共享树流动，因而不需要向 PIM RP 注册多播源，也不需要创建(S,G)状态。

1. BiDIR 配置

PIM BiDIR 的配置与 PIM ASM 相似，必须在所有接口上启用 PIM-SM。由于 PIM Hello 消息默认将 *BiDIR 使能*（*BiDIR capable*）比特设置为 1，因而不需要使用接口级命令来专门启用 PIM

BiDIR。在 **ip pim rp-address** [*RP address*] **group-range** [*groups*] **bidir** 命令中使用关键字 **bidir** 配置了 RP 之后，就可以将其指定为 BiDIR RPA。

例 13-45 给出了前面所说的虚拟 RPA 配置示例。Loopback99 是 RPL，配置了一个包含 RPA 的子网。请注意，实际上并没有在拓扑结构中的任何路由器上配置 RPA，这也是 PIM BiDIR 与 PIM ASM 之间的主要区别。本例通过 OSPF 将 RPA 通告给 PIM 域，由于本例希望 OSPF 将链路宣告为 10.99.99.96/29，因而使用了命令 **ip ospf network point-to-point**，该命令将强制 NX-1 的 OSPF 在 Type 1 LSA（Link-State Advertisement，链路状态通告）中将该链路宣告为末梢链路（stub-link）。

例 13-45 在 NX-1 上配置 PIM BiDIR

```
NX-1# show run pim
! Output omitted for brevity
!Command: show running-config pim

feature pim

ip pim rp-address 10.99.99.99 group-list 224.0.0.0/4 bidir
ip pim ssm range 232.0.0.0/8

interface Vlan1101
  ip pim sparse-mode

interface loopback0
  ip pim sparse-mode

interface loopback99
  ip pim sparse-mode

interface Ethernet3/17
  ip pim sparse-mode

interface Ethernet3/18
  ip pim sparse-mode
NX-1# show run interface loopback99
! Output omitted for brevity

!Command: show running-config interface loopback99

interface loopback99
  ip address 10.99.99.98/29
  ip ospf network point-to-point
  ip router ospf 1 area 0.0.0.0
  ip pim sparse-mode
NX-1# show ip pim group-range 239.115.115.1
PIM Group-Range Configuration for VRF "default"
Group-range         Action Mode RP-address      Shrd-tree-range     Origin

224.0.0.0/4         -    Bidir 10.99.99.99      -                   Static
NX-1# show ip pim rp
PIM RP Status Information for VRF "default"
BSR disabled
Auto-RP disabled
BSR RP Candidate policy: None
BSR RP policy: None
Auto-RP Announce policy: None
```

```
Auto-RP Discovery policy: None

RP: 10.99.99.99, (1),
 uptime: 22:29:39 priority: 0,
 RP-source: (local),
 group ranges:
 224.0.0.0/4 (bidir)
```

注：拓扑结构中的其他路由器都拥有相同的 BiDIR 配置，通过关键字 **bidir** 配置静态 RPA。NX-1 和 NX-3 配置了去往 RPA 的 RPL 的路由器。

2. BiDIR 验证

为了更好地理解 MROUTE 状态和 BiDIR 事件，接下来将从 NX-4 开始进行验证操作（接收端连接在 NX-4 上的 VLAN 215 中）。例 13-46 给出了 NX-4（LHR）的 **show ip mroute** 命令输出结果，创建(*,G) MROUTE 的原因是收到了接收端的 IGMP Membership Report 消息。由于这是一棵双向共享树，因而用于到达 RPA 的 RPF 接口 Ethernet 3/29 也包含在该 MROUTE 的 OIL 中。

例 13-46　NX-4 的 PIM BiDIR MROUTE 表项

```
NX-4# show ip mroute
! Output omitted for brevity
IP Multicast Routing Table for VRF "default"

(*, 224.0.0.0/4), bidir, uptime: 00:06:39, pim ip
  Incoming interface: Ethernet3/29, RPF nbr: 10.2.13.1
  Outgoing interface list: (count: 1)
    Ethernet3/29, uptime: 00:06:39, pim, (RPF)

(*, 239.115.115.1/32), bidir, uptime: 00:04:08, igmp ip pim
  Incoming interface: Ethernet3/29, RPF nbr: 10.2.13.1
  Outgoing interface list: (count: 2)
    Ethernet3/29, uptime: 00:04:08, pim, (RPF)
    Vlan215, uptime: 00:04:08, igmp
```

BiDIR 中的 DF 选举进程负责确定每个接口上的哪台 PIM 路由器将负责发送 Join-Prune 消息，并负责在双向共享树上将数据包从上游路由到下游，反之亦然。**show ip pim df** 命令可以显示每个启用了 PIM 功能的接口上的当前 DF 状态的摘要视图，如例 13-47 所示。NX-4 在 VLAN 215 上是 DF，但是在指向 RPA 的 RPF 接口上并不是 DF，因为对等体拥有更优的到达 RPA 的路由度量。

例 13-47　NX-4 的 PIM BiDIR DF 状态

```
NX-4# show ip pim df
! Output omitted for brevity
Bidir-PIM Designated Forwarder Information for VRF "default"

RP Address (ordinal)    RP Metric         Group Range
10.99.99.99 (1)         [110/5]           224.0.0.0/4

  Interface       DF Address      DF State    DF Metric DF   Uptime
  Vlan303         10.2.33.2       Winner      [110/5]        00:22:28
  Vlan216         10.216.1.254    Loser       [110/5]        00:22:29
  Vlan215         10.215.1.253    Winner      [110/5]        00:19:58
  Lo0             10.2.2.3        Winner      [110/5]        00:22:29
  Eth3/28         10.2.23.2       Loser       [110/2]        00:22:29
  Eth3/29         10.2.13.1       Loser       [0/0]          00:22:29 (RPF)
```

如果希望了解 BiDIR DF 选举进程的更多详细信息，可以使用 **show ip pim internal event-history bidir** 命令，该命令可以显示接口状态机以及收到 PIM DF-Election 消息之后的响应情况。例 13-48 显示了 NX-4 的事件历史记录，从 VLAN 215 的 DF 选举过程可以看出，由于没有收到其他 Offer 消息，因而 NX-4 胜出。但是对于 Ethernet 3/29 来说，由于 NX-4（10.2.13.3）的度量（-1/-1）劣于当前 DF（10.2.13.1），且没有以 Offer 消息进行应答，因而 NX-1 成为 Ethernet 3/29 上的 DF。

例 13-48　NX-4 的 PIM BiDIR 事件历史记录

```
NX-4# show ip pim internal event-history bidir
! Output omitted for brevity

bidir events for PIM process
20:32:46.269627 pim [10572]:: pim_update_df_state: vrf: default: rp:
10.99.99.99 iod Ethernet3/29 prev_state 2 Notify IGMP
20:32:46.269623 pim [10572]:: Entering Lose state on Ethernet3/29
20:32:46.269439 pim [10572]:: pim_update_df_state: vrf: default: rp:
10.99.99.99 iod Ethernet3/29 prev_state 2 Notify IGMP
20:32:46.269433 pim [10572]:: Our metric: -1/-1 is worse than received
metric: 0/0 RPF Ethernet3/29 old_winner 10.2.13.1
20:32:46.269419 pim [10572]:: Received DF-Winner from 10.2.13.1 on Ethernet3/29
RP 10.99.99.99, metric 0/0
20:32:40.205960 pim [10572]:: Add RP-route for RP 10.99.99.99,
Bidir-RP Ordinal:1, DF-interfaces: 00000000
20:32:40.205947 pim [10572]:: pim_df_expire_timer: Entering Winner state
on Vlan215
20:32:40.205910 pim [10572]:: Expiration timer fired in Offer state for RP
10.99.99.99 on Vlan215
```

由于 NX-4 是 VLAN 215 上的 DF 选举进程的胜出方，因而向 RPF 接口 Ethernet 3/29 上的 DF 发送共享树的 PIM Join 消息，可以通过命令 **show ip pim internal event-history join-prune** 查看这些事件，如例 13-49 所示。

例 13-49　NX-4 的 PIM BiDIR 加入剪除事件历史记录

```
NX-4# show ip pim internal event-history join-prune
! Output omitted for brevity

join-prune events for PIM process
20:34:34.286181 pim [10572]:: Keep bidir (*, 239.115.115.1/32) entry alive due
to joined oifs exist
20:33:40.056128 pim [10572]: [10739]: skip sending periodic join not having
any oif
20:33:40.056116 pim [10572]:: Keep bidir (*, 224.0.0.0/4) prefix-entry alive
20:33:34.016224 pim [10572]:: Send Join-Prune on Ethernet3/29, length:
34 in context 1
20:33:34.016186 pim [10572]:: Put (*, 239.115.115.1/32), WRS in join-list for
nbr 10.2.13.1
20:33:34.016179 pim [10572]:: wc_bit = TRUE, rp_bit = TRUE
```

除这些详细的事件历史记录之外，还可以检查接口统计信息以查看已交换的 BiDIR 消息总数，如例 13-50 所示。

例 13-50　NX-4 的 PIM BiDIR 接口计数器

```
NX-4# show ip mroute
! Output omitted for brevity
show ip pim interface ethernet 3/29
```

```
PIM Interface Status for VRF "default"
Ethernet3/29, Interface status: protocol-up/link-up/admin-up
  IP address: 10.2.13.3, IP subnet: 10.2.13.0/24
  PIM DR: 10.2.13.3, DR's priority: 1
  PIM neighbor count: 1
  PIM hello interval: 30 secs, next hello sent in: 00:00:22
  PIM neighbor holdtime: 105 secs
  PIM configured DR priority: 1
  PIM configured DR delay: 3 secs
  PIM border interface: no
  PIM GenID sent in Hellos: 0x140c2403
  PIM Hello MD5-AH Authentication: disabled
  PIM Neighbor policy: none configured
  PIM Join-Prune inbound policy: none configured
  PIM Join-Prune outbound policy: none configured
  PIM Join-Prune interval: 1 minutes
  PIM Join-Prune next sending: 0 minutes
  PIM BFD enabled: no
  PIM passive interface: no
  PIM VPC SVI: no
  PIM Auto Enabled: no
  PIM Interface Statistics, last reset: never
    General (sent/received):
      Hellos: 4880/2121 (early: 0), JPs: 378/0, Asserts: 0/0
      Grafts: 0/0, Graft-Acks: 0/0
      DF-Offers: 1/3, DF-Winners: 0/381, DF-Backoffs: 0/0, DF-Passes: 0/0
    Errors:
      Checksum errors: 0, Invalid packet types/DF subtypes: 0/0
      Authentication failed: 0
      Packet length errors: 0, Bad version packets: 0, Packets from self: 0
      Packets from non-neighbors: 0
         Packets received on passiveinterface: 0
      JPs received on RPF-interface: 0
      (*,G) Joins received with no/wrong RP: 0/0
      (*,G)/(S,G) JPs received for SSM/Bidir groups: 0/0
      JPs filtered by inbound policy: 0
      JPs filtered by outbound policy: 0
```

双向共享树的下一跳是 NX-1,是 NX-4 去往 RPA 的 RPF 邻居。从加入—剪除事件历史记录可以确认已经从 NX-4 收到了(*,G) Join 消息,如例 13-51 所示。

例 13-51　NX-1 的 PIM BiDIR 加入剪除事件历史记录

```
NX-1# show ip pim internal event-history join-prune
! Output omitted for brevity

bidir events for PIM process
20:33:34.020037 pim [7851]:: -----
20:33:34.020020 pim [7851]:: (*, 239.115.115.1/32) route exists, RPF if
loopback99, to us
20:33:34.020008 pim [7851]:: pim_receive_join: We are target comparing with iod
20:33:34.019968 pim [7851]:: pim_receive_join: route: (*, 239.115.115.1/32),
 wc_bit: TRUE, rp_bit: TRUE
20:33:34.019341 pim [7851]:: Received Join-Prune from 10.2.13.3 on
Ethernet3/17, length: 34, MTU: 9216, ht: 210
```

NX-1 的 MROUTE 状态包含了 Ethernet3/17 和 Loopback99(例 13-52 中的 RPL),所有映射

到该 RPA 的多播组都通过 RPL 转发给 RPA。

例 13-52　NX-1 的 PIM BiDIR MROUTE 表项

```
NX-1# show ip mroute
! Output omitted for brevity

IP Multicast Routing Table for VRF "default"

(*, 224.0.0.0/4), bidir, uptime: 00:13:22, pim ip
  Incoming interface: loopback99, RPF nbr: 10.99.99.99
  Outgoing interface list: (count: 1)
    loopback99, uptime: 00:13:22, pim, (RPF)

(*, 239.115.115.1/32), bidir, uptime: 00:14:13, pim ip
  Incoming interface: loopback99, RPF nbr: 10.99.99.99
  Outgoing interface list: (count: 2)
    Ethernet3/17, uptime: 00:08:47, pim
    loopback99, uptime: 00:13:22, pim, (RPF)
```

例 13-53 显示了 **show ip pim df** 命令的输出结果。由于 RPL 对于 NX-1 来说是本地链路，因而 NX-1 是除 RPL 之外的所有接口的 DF 胜出方。PIM BiDIR 中的 RPL 未选举任何 DF。

例 13-53　NX-1 的 PIM DF 状态

```
NX-1# show ip pim df
! Output omitted for brevity
Bidir-PIM Designated Forwarder Information for VRF "default"

RP Address (ordinal)    RP Metric        Group Range
10.99.99.99 (1)         [0/0]            224.0.0.0/4

  Interface             DF Address       DF State    DF Metric    DF Uptime
  Vlan1101              10.1.11.1        Winner      [0/0]        00:14:43
  Po3                   10.1.12.2        Winner      [0/0]        00:14:43
  Lo0                   10.1.1.1         Winner      [0/0]        00:14:43
  Lo99                  0.0.0.0          Loser       [0/0]        00:14:43    (RPF)
  Eth3/17               10.2.13.1        Winner      [0/0]        00:14:43
  Eth3/18               10.1.13.1        Winner      [0/0]        00:14:43
```

与 PIM ASM 一样，从 RPA 到多播源没有(S,G)加入。BiDIR 中所有来自源端的流量都从 NX-2（FHR）转发到 RPA。因此，NX-1 不需要向 NX-2 发起加入操作即可通过 VLAN 1101 将流量拉到 NX-1。这一事实突显了排查 BiDIR 故障时的不利因素。由于没有(S,G)状态，因而从 RPA 到 FHR 是无法了解该特定多播源的。

可以在 NX-1 上通过 ELAM 抓包操作来验证流量是否来自 NX-2。另一种有用技术是在 ACL 中配置 **permit** 语句来匹配流量，利用 **statistics per-entry** 配置 ACL，该命令可以提供一个计数器来验证流量是否已到达。从例 13-54 的输出结果可以看出，本例配置了一个名为 *verify* 的 ACL 以匹配连接在 NX-2 上的多播源，该 ACL 应用在 VLAN 1101（是多播流量到达的接口）的入站方向。

例 13-54　在 NX-1 配置匹配多播流量的 ACL

```
NX-1# show run | sec verify
! Output omitted for brevity
ip access-list verify
  statistics per-entry
  10 permit ip 10.115.1.4/32 239.115.115.1/32
```

```
           20 permit ip any any
NX-1# show running-config interface Vlan1101
! Output omitted for brevity

interface Vlan1101
  description L3 to 7009-B-NX-2
  no shutdown
  mtu 9216
  ip access-group verify in
  no ip redirects
  ip address 10.1.11.1/30
  no ipv6 redirects
  ip ospf cost 1
  ip router ospf 1 area 0.0.0.0
  ip pim sparse-mode
NX-1# show access-list verify

IP access list verify
        statistics per-entry
        10 permit ip 10.115.1.4/32 239.115.115.1/32 [match=448]
        20 permit ip any any [match=108]
```

本例中的源端连接在 NX-2 上，因而可以验证 MROUTE 表项以确定 OIL 包含了连接 NX-1 的 VLAN 1101。例 13-55 显示了 NX-2 的 MROUTE 表项信息，可以看出，这些 MROUTE 表项涵盖了所有映射到 RPA 的多播组。

例 13-55　NX-2 的 PIM BiDIR MROUTE 表项

```
NX-2# show ip pim df
! Output omitted for brevity
Bidir-PIM Designated Forwarder Information for VRF "default"

RP Address (ordinal)    RP Metric          Group Range
10.99.99.99 (1)         [110/2]            224.0.0.0/4

  Interface             DF Address         DF State      DF Metric     DF Uptime
  Vlan1101              10.1.11.1          Loser         [0/0]         00:08:49   (RPF)
  Vlan116               10.116.1.254       Winner        [110/2]       00:08:49
  Vlan115               10.115.1.254       Winner        [110/2]       00:08:49
  Eth3/17               10.2.23.2          Winner        [110/2]       00:08:49
  Eth3/18               10.1.23.2          Winner        [110/2]       00:08:48


NX-2# show ip mroute
IP Multicast Routing Table for VRF "default"

(*, 224.0.0.0/4), bidir, uptime: 2d12h, pim ip
  Incoming interface: Vlan1101, RPF nbr: 10.1.11.1
  Outgoing interface list: (count: 1)
    Vlan1101, uptime: 2d12h, pim, (RPF)
```

由于 NX-2 是 VLAN 115 的 DF 胜出方，因而负责将多播流量从 VLAN 115 转发给位于 VLAN 1101 上的 RPA 的 RPF 接口。对于 BiDIR 来说，NX-2 不需要将源端注册到 RPA，只要将 VLAN 115 的流量沿双向共享树向上转发即可。

本节讨论了 PIM BiDIR 的基本内容以及所有参与双向共享树的多播路由器验证 DF 和 MROUTE 表项的方式。BiDIR 和 ASM 在多播状态和转发行为方面有一定的区别，排查 BiDIR 故

13.4.6 PIM RP 配置

为 ASM 或 BiDIR 配置 PIM-SM 时，每个多播组都必须映射到一个 PIM RP 地址。必须确保该映射关系在网络中的一致性，而且 PIM 域中的每台路由器都必须知道 RP 地址到多播组的映射关系。可以通过以下 3 种方式在多播网络中配置 PIM RP 地址。

- **静态 PIM RP**：在每台路由器上静态配置 RP-to-group（RP 到多播组）映射关系。
- **Auto-RP**：PIM RP 将自己宣告给映射代理，由映射代理将 RP-to-group 映射关系宣告给 PIM 域中的所有路由器。思科在 PIM BSR 机制实现标准化之前开发了 Auto-RP 机制。
- **BSR**：C-RP 将自己宣告给引导路由器，由引导路由器在 Bootstrap 消息中将 RP-to-group 映射关系宣告给 PIM 域中的所有路由器。

1. 静态 RP 配置

静态 RP 是最简单的部署方式，只要在域中的每台路由器上都配置 PIM RP 地址即可，如例 13-56 所示。

例 13-56　NX-3 的 PIM 静态 RP 配置示例

```
NX-3# show run pim
! Output omitted for brevity

!Command: show running-config pim

feature pim

ip pim rp-address 10.99.99.99 group-list 224.0.0.0/4
ip pim ssm range 232.0.0.0/8

interface Vlan215
  ip pim sparse-mode

interface Vlan216
  ip pim sparse-mode

interface Vlan303
  ip pim sparse-mode

interface Ethernet3/28
  ip pim sparse-mode

interface Ethernet3/29
  ip pim sparse-mode
```

不过，简单也有简单的缺点。一旦 RP-to-group 映射出现了任何变化，网络运营商就必须更新每台路由器的配置。除此以外，如果注册了成百上千个多播源，那么单台静态 PIM RP 就会成为扩展性瓶颈。当然，如果网络规模较小，或者所有多播组都使用单个 PIM RP 地址，那么静态 RP 也是一种不错的配置选择。

注：如果配置了静态 RP 且收到了动态 RP-to-group（RP 到多播组）映射，则路由器将使用动态获知的地址（如果更具体）。如果组掩码长度相等，那么就使用较高的 IP 地址。关键字 *override* 的作用是强制静态 RP 胜过 Auto-RP 或 BSR。

2. Auto-RP 配置与验证

Auto-RP 使用 C-RP（Candidate RP，候选 RP）和候选映射代理的概念。C-RP 在 RP-Announce（RP 通告）消息中发送它们已配置的多播组区间，RP-Announce 消息以多播方式发送给 224.0.1.39。映射代理侦听 RP-Announce 消息，并将 RP-to-group 映射数据收集到本地表中。解析出映射数据中的冲突之后，将通过 RP-Discovery（RP 发现）消息将列表传递给网络，RP-Discovery 消息的目的地址是多播地址 224.0.1.40。网络中的路由器则被配置为侦听由选举出的映射代理发送的 RP-Discovery 消息，收到 RP-Discovery 消息之后，PIM 域中的每台路由器都会更新其本地 RP-to-group 映射表。

网络中可能存在多个映射代理，因而需要通过某种确定性的方法来确定路由器应该侦听哪个映射代理。网络中的路由器使用 IP 地址最大的映射代理来填充它们的 group-to-RP（组到 RP）映射表。接下来以图 13-15 为例来说明 Auto-RP 的操作与验证方式。

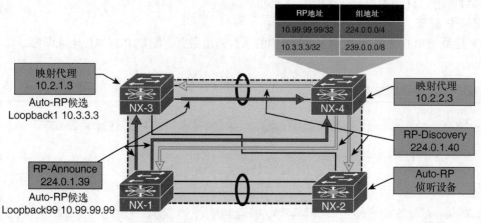

图 13-15 PIM Auto-RP 拓扑结构

图 13-15 拓扑结构中的 NX-1 被配置为以 RP 地址 10.99.99.99 发送 224.0.0.0/4 的 RP-Announce 消息，NX-3 被配置为以 RP 地址 10.3.3.3 发送 239.0.0.0/8 的 RP-Announce 消息，此外，NX-3 还被配置为 Auto-RP 映射代理（地址为 10.2.1.3）。NX-4 被配置为 Auto-RP 映射代理（地址为 10.2.2.3），而 NX-2 只是侦听 Auto-RP 发现消息以填充本地 RP-to-group 映射信息。本例的目的是说明多个 C-RP（和多个映射代理）可以共存的事实。

如果 PIM 域存在重叠或冲突信息（如两个 C-RP 通告同一个多播组），那么映射代理就必须确定应该在 RP-Discovery 消息中通告哪个 RP。判断规则如下。

- 选择通告了更明细组地址的 RP。
- 如果以相同数量的掩码比特通告多播组，那么就选择 IP 地址较大的 RP。

对于本例来说，NX-1 通告的多播组是 224.0.0.0/4，而 NX-3 通告的则是更明细的 239.0.0.0/8，因而 NX-3 被选为多播组 239.0.0.0/8 的 RP，而 NX-1 则被选为其他所有多播组的 RP。如果配置了多个 Auto-RP 映射代理，那么 NX-OS 将选择侦听 IP 地址较大的映射代理发送的 RP-Discovery 消息。

例 13-57 显示了 NX-1 的 PIM 配置示例。例中配置了命令 **ip pim auto-rp rp-candidate**，作用是让 NX-1 以 TTL 值 16 为所有多播组发送 Auto-RP RP-Announce 消息。NX-OS 默认不侦听或转发 Auto-RP 消息。**ip pim auto-rp forward listen** 命令指示设备侦听并转发 Auto-RP 组 224.0.1.39 和 224.0.1.40。可以通过命令 **show ip pim rp** 显示本地 PIM RP-to-group 映射信息，该命令显示了每个 RP 的当前组映射信息以及 RP-Source，即映射代理 NX-4（10.2.2.3）。

例 13-57　NX-1 的 PIM Auto-RP C-RP 配置

```
NX-1# show run pim
! Output omitted for brevity

!Command: show running-config pim

feature pim

ip pim rp-address 10.99.99.99 group-list 224.0.0.0/4
ip pim auto-rp rp-candidate 10.99.99.99 group-list 224.0.0.0/4 scope 16
ip pim ssm range 232.0.0.0/8
ip pim auto-rp forward listen

interface Vlan1101
  ip pim sparse-mode

interface loopback99
  ip pim sparse-mode

interface Ethernet3/17
  ip pim sparse-mode

interface Ethernet3/18
  ip pim sparse-mode


NX-1# show ip pim rp
PIM RP Status Information for VRF "default"
BSR disabled
Auto-RP RPA: 10.2.2.3, uptime: 00:55:41, expires: 00:02:28
BSR RP Candidate policy: None
BSR RP policy: None
Auto-RP Announce policy: None
Auto-RP Discovery policy: None

RP: 10.3.3.3, (0), uptime: 00:48:46, expires: 00:02:28,
  priority: 0, RP-source: 10.2.2.3 (A), group ranges:
      239.0.0.0/8
RP: 10.99.99.99*, (0), uptime: 1w5d, expires: 00:02:28 (A),
  priority: 0, RP-source: 10.2.2.3 (A), (local), group ranges:
      224.0.0.0/4
```

如果希望更加精细化地配置组区间，那么就可以使用 *group-list*、*prefix-list* 或 *route-map* 等选项。

注： 如果接口被用作 Auto-RP C-RP 或映射代理，那么就必须配置 **ip pim sparse-mode**。

例 13-58 显示了 NX-4 的 Auto-RP 映射代理配置示例，该配置让 NX-4 以 TTL 值 16 发送 RP-Discovery 消息。从 **show ip pim rp** 命令的输出结果可以看出，由于 NX-4 是当前映射代理，因而显示了一个定时器以指示何时将发送下一条 RP-Discovery 消息。

例 13-58　NX-4 的 Auto-RP 映射代理配置

```
NX-4# show run pim
! Output omitted for brevity

!Command: show running-config pim
```

```
feature pim

ip pim auto-rp mapping-agent loopback0 scope 16
ip pim ssm range 232.0.0.0/8
ip pim auto-rp listen forward
interface Vlan215
  ip pim sparse-mode

interface Vlan216
  ip pim sparse-mode

interface Vlan303
  ip pim sparse-mode

interface loopback0
  ip pim sparse-mode

interface Ethernet3/28
  ip pim sparse-mode

interface Ethernet3/29
  ip pim sparse-mode
NX-4# show ip pim rp
PIM RP Status Information for VRF "default"
BSR disabled
Auto-RP RPA: 10.2.2.3*, next Discovery message in: 00:00:29
BSR RP Candidate policy: None
BSR RP policy: None
Auto-RP Announce policy: None
Auto-RP Discovery policy: None

RP: 10.3.3.3, (0),
 uptime: 01:18:01   priority: 0,
 RP-source: 10.3.3.3 (A),
 group ranges:
 239.0.0.0/8     , expires: 00:02:37 (A)
RP: 10.99.99.99, (0),
 uptime: 01:20:27 priority: 0,
 RP-source: 10.99.99.99 (A),
 group ranges:
 224.0.0.0/4     , expires: 00:02:36 (A)
```

注：不要将任播 IP 地址用作映射代理地址，否则可能会导致网络中的 RP 映射出现频繁刷新问题。

NX-3 被同时配置为 Auto-RP C-RP 和映射代理，例 13-59 显示了 NX-3 的配置信息。请注意，接口 Loopback0 被用作映射代理地址，而 Loopback1 被用作 C-RP 地址，且两者均配置了 **ip pim sparse-mode** 命令。

例 13-59　NX-3 的 Auto-RP 配置

```
NX-3# show run pim
! Output omitted for brevity

!Command: show running-config pim

feature pim
```

```
ip pim rp-address 10.3.3.3 group-list 239.0.0.0/8
ip pim auto-rp rp-candidate 10.3.3.3 group-list 239.0.0.0/8 scope 16
ip pim auto-rp mapping-agent loopback0 scope 16
ip pim ssm range 232.0.0.0/8
ip pim auto-rp listen forward

interface Vlan215
  ip pim sparse-mode

interface Vlan216
  ip pim sparse-mode

interface Vlan303
  ip pim sparse-mode

interface loopback0
  ip pim sparse-mode

interface loopback1
  ip pim sparse-mode

interface Ethernet3/28
  ip pim sparse-mode

interface Ethernet3/29
  ip pim sparse-mode
NX-3# show ip pim rp
PIM RP Status Information for VRF "default"
BSR disabled
Auto-RP RPA: 10.2.2.3, uptime: 01:21:50, expires: 00:02:49
BSR RP Candidate policy: None
BSR RP policy: None
Auto-RP Announce policy: None
Auto-RP Discovery policy: None

RP: 10.3.3.3*, (0), uptime: 01:16:28, expires: 00:02:49 (A),
  priority: 0, RP-source: 10.2.2.3 (A), (local), group ranges:
      239.0.0.0/8
RP: 10.99.99.99, (0), uptime: 01:18:18, expires: 00:02:49,
  priority: 0, RP-source: 10.2.2.3 (A), group ranges:
      224.0.0.0/4
```

最后，NX-2 被配置为 Auto-RP 侦听路由器和转发路由器。例 13-60 显示了该配置，该配置允许 NX-2 从 NX-4 和 NX-3 接收 Auto-RP RP 发现消息。

例 13-60 NX-2 的 Auto-RP 侦听路由器配置

```
NX-2# show run pim
! Output omitted for brevity

!Command: show running-config pim

feature pim

ip pim ssm range 232.0.0.0/8
ip pim auto-rp listen forward
```

```
interface Vlan115
  ip pim sparse-mode

interface Vlan116
  ip pim sparse-mode

interface Vlan1101
  ip pim sparse-mode

interface Ethernet3/17
  ip pim sparse-mode

interface Ethernet3/18
  ip pim sparse-mode
NX-2# show run pim
PIM RP Status Information for VRF "default"
BSR disabled
Auto-RP RPA: 10.2.2.3, uptime: 00:07:29, expires: 00:02:25
BSR RP Candidate policy: None
BSR RP policy: None
Auto-RP Announce policy: None
Auto-RP Discovery policy: None

RP: 10.3.3.3, (0),
 uptime: 00:00:34   priority: 0,
 RP-source: 10.2.2.3 (A),
 group ranges:
 239.0.0.0/8    , expires: 00:02:25 (A)
RP: 10.99.99.99, (0),
 uptime: 00:02:59   priority: 0,
 RP-source: 10.2.2.3 (A),
 group ranges:
 224.0.0.0/4    , expires: 00:02:25 (A)
```

由于 Auto-RP 消息受到其配置的 TTL 范围的约束，因而必须确保所有的 RP-Announce 消息都能到达网络中的所有映射代理。此外，还要确保 RP-Discovery 消息的范围足够大，以保证 PIM 域中的所有路由器都能收到该消息。如果存在多个映射代理且 TTL 配置错误，那么就可能会导致 PIM 域中的 RP-to-group 映射不一致（取决于与映射代理的距离）。

NX-OS 为排查 Auto-RP 消息故障提供了非常有用的事件历史记录。例 13-61 显示了 NX-4 的 **show ip pim internal event-history rp** 命令输出结果，虽然输出结果有点冗长，不过可以看出 NX-4 将自己选举为映射代理，之后所有启用了 PIM 功能特性的接口都向外发送了 Auto-RP 发现消息。此外，该输出结果还表明 Auto-RP 消息需要通过 RPF 检查，如果检查失败，那么就会丢弃该消息。最后，NX-4 收到了 NX-3 的 RP-Announce 消息，从而安装了新的 PIM RP-to-group 映射。

例 13-61　NX-4 的 Auto-RP 事件历史记录

```
NX-4# show ip pim internal event-history rp
! Output omitted for brevity
02:34:30.112521 pim [13449]::Scan MRIB to process RP change event
02:34:30.112255 pim [13449]::RP 10.1.1.1, group range 239.0.0.0/8 cached
02:34:30.112248 pim [13449]::(default) pim_add_rp: RP:10.1.1.1 rp_change:yes
change_flag: yes bidir:no, group:239.0.0.0/8 rp_priority: -1,
static: no, action: Permit, prot_souce: 4 hash_len: 181
02:34:30.112138 pim [13449]::(default) pim_add_rp: Added the following in pt_rp_
```

```
    cache_by_group: group: 239.0.0.0/8, pcib->pim_rp_change: yes
02:34:30.112133 pim [13449]::Added group range: 239.0.0.0/8 from
pim_rp_cache_by_group
02:34:30.112127 pim [13449]::(default) pim_add_rp: Added the following in pt_rp_
    cache_by_rp: RP: 10.1.1.1, rp_priority: 0, prot_souce: 4, pcib->pim_
rp_change: yes
02:34:30.112070 pim [13449]::(default) pim_add_rp: Received rp_entry from
caller: RP: 10.1.1.1 bidir:no, group:239.0.0.0/8 rp_priority: -1, static:
no, prot_souce: 4 override: no hash_len: 181 holdtime: 180
02:34:30.112030 pim [13449]::RPF interface is Ethernet3/29, RPF check passed
02:34:30.111913 pim [13449]::Received Auto-RP v1 Announce from
10.1.1.1 on Ethernet3/29, length: 20, ttl: 15, ht: 180
02:34:30.110112 pim [13449]::10.2.2.3 elected new RP-mapping Agent,
    old RPA: 10.2.2.3
02:34:30.110087 pim [13449]::RPF interface is loopback0, RPF check passed
02:34:30.110064 pim [13449]::Received Auto-RP v1 Discovery from
10.2.2.3 on loopback0, length: 8, ttl: 16, ht: 180
02:34:30.109856 pim [13449]::Send Auto-RP Discovery message on Vlan216,
02:34:30.109696 pim [13449]::Send Auto-RP Discovery message on Vlan215,
02:34:30.109496 pim [13449]::Send Auto-RP Discovery message on Vlan303,
02:34:30.109342 pim [13449]::Send Auto-RP Discovery message on Ethernet3/29,
02:34:30.107940 pim [13449]::Send Auto-RP Discovery message on Ethernet3/28,
02:34:30.107933 pim [13449]::Build Auto-RP Discovery message, holdtime: 180
02:34:30.107900 pim [13449]::Elect ourself as new RP-mapping Agent
```

请注意，Auto-RP 状态是动态变化的，必须通过在网络中发送和接收 RP-Announce 和 RP-Discovery 消息来定期刷新 Auto-RP 的状态。如果设备的 RP 状态丢失或不正确，那么就要跟踪适当的 Auto-RP 消息直至回溯到源端，以找出错误配置。NX-OS 提供的事件历史记录和 Ethanalyzer 实用工具是查找这类故障根源的主要工具。

3. BSR 配置与验证

动态 RP 配置方法 BSR 是在思科创建 Auto-RP 之后出现的，目前定义在 RFC 4601 和 RFC 5059 中。BSR 和 Auto-RP 都可以在整个 PIM 域中自动分发 PIM RP 信息，不过，BSR 是 IETF 标准，而 Auto-RP 是思科专有协议。

BSR 依赖于 C-RP（Candidate-RP，候选 RP）和 BSR（Bootstrap Router，引导路由器），其中，BSR 是根据最高优先级选举出来的。如果优先级相等，那么就将 IP 地址最大的路由器选举为单一 BSR。如果路由器被配置为 C-BSR（Candidate-BSR，候选 BSR），那么就会开始发送 Bootstrap（引导）消息，以允许所有 C-BSR 都能互相侦听到对方并确定谁应该成为当选 BSR。选出 BSR 之后，该路由器将成为 PIM 域中唯一发送 Bootstrap 消息的路由器。

C-RP 负责侦听 BSR 发出的 Bootstrap 消息，以发现 BSR 正在使用的单播地址，从而允许 C-RP 通过发送单播 C-RP 消息将自己通告给 BSR。C-RP 发出的消息包含 RP 地址以及其希望成为 RP 的多播组及其他详细信息（如 RP 优先级），BSR 收到所有 C-RP 发来的 RP 信息之后，就会构建 PIM Bootstrap 消息，将 RP 信息宣告给网络的其余部分。此外，C-BSR 还利用相同的 Bootstrap 消息宣告网络中的 group-to-RP 映射列表，从而确定当选 BSR。该模式允许活动 BSR 在因某种原因而停止发送 Bootstrap 消息的情况下，由其他 C-BSR 承担当选 BSR 的角色。

到目前为止，上述进程与 Auto-RP 看起来非常类似。不过，与 Auto-RP 的映射代理不同的是，BSR 不会尝试选择 RP-to-group 映射以包含到 Bootstrap 消息中，而是在 Bootstrap 消息中包含从所有 C-RP 收到的数据。

所有启用了 PIM 功能特性的接口都将 Bootstrap 消息发送给 ALL-PIM-ROUTERS 多播地址 224.0.0.13。如果路由器被配置为侦听和转发 BSR，那么就会检查其收到的 Bootstrap 消息内容，然后构建一个新数据包，通过所有已启用 PIM 功能特性的接口向外发送相同的 BSR 消息。BSR 消息就以该方式在 PIM 域中进行逐跳传播，使每台路由器都拥有完全一致的 C-RP 到多播组映射数据列表。网络中的所有路由器都对 BSR 消息中的数据应用相同的算法以确定 group-to-RP 映射，从而实现全网范围内的一致性。

路由器收到 BSR 的 Bootstrap 消息之后，必须确定每个多播组区间所要使用的 RP 地址。具体步骤如下：

- 对组区间和掩码长度执行最长匹配以获取 RP 列表。
- 从列表中找出优先级最高的 RP。
- 如果只剩下一个 RP，那么该组区间的 RP 选举进程就结束。
- 如果列表中存在多个 RP，那么就使用 PIM 散列函数选择 RP。

如果特定组区间的多个 RP 都拥有相同的最长匹配掩码长度和优先级，那么就要使用散列函数。由于域中每台路由器的散列函数都会返回相同的计算结果，因而可以实现全网范围内的 group-to-RP 映射数据的一致性。

RFC 4601 的 4.7.2 节给出的散列函数如下：

Value(G,M,C(i))=

(1103515245 * ((1103515245 * (G&M) + 12345) XOR C(i)) + 12345) mod 2^31

计算公式中的变量输入如下。

- G = 多播组地址。
- M = BSR 的 Bootstrap 消息提供的散列长度。
- C(i)=C-RP 地址。

需要对所有与组区间相匹配的 C-RP 都进行上述计算，并返回所要使用的 RP 地址。拥有最大散列计算值的 RP 将被选为该多播组的 RP，如果两个 C-RP 都拥有相同的散列计算结果，那么就使用 IP 地址较大的 RP。由于默认散列长度为 30，因而会出现 4 个连续多播组地址映射为同一个 RP 地址。

接下来以图 13-16 所示拓扑结构为例来说明 BSR 的配置和验证步骤。

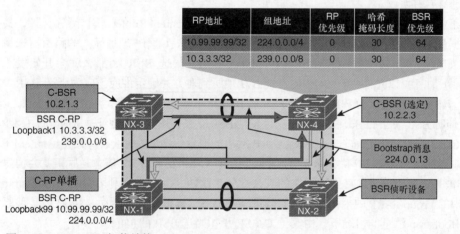

RP 地址	组地址	RP 优先级	哈希掩码长度	BSR 优先级
10.99.99.99/32	224.0.0.0/4	0	30	64
10.3.3.3/32	239.0.0.0/8	0	30	64

图 13-16 PIM BSR 拓扑结构

NX-1 被配置为多播组区间 224.0.0.0/4 的 C-RP，如例 13-62 所示。由于路由器默认不侦听或转发 BSR 消息，因而需要使用 **ip pim bsr listen forward** 命令进行配置。NX-1 通过收到的 Bootstrap 消息得知 BSR 地址之后，就开始发送单播 C-RP 消息，宣告自己愿意成为 224.0.0.0/4 的 RP。

例 13-62 NX-1 的 BSR 配置

```
NX-1# show run pim
! Output omitted for brevity

!Command: show running-config pim

feature pim

ip pim bsr rp-candidate loopback99 group-list 224.0.0.0/4 priority 0
ip pim ssm range 232.0.0.0/8
ip pim bsr listen forward
interface Vlan1101
  ip pim sparse-mode

interface loopback0
  ip pim sparse-mode

interface loopback99
  ip pim sparse-mode

interface Ethernet3/17
  ip pim sparse-mode

interface Ethernet3/18
  ip pim sparse-mode

NX-1# show ip pim rp

PIM RP Status Information for VRF "default"
BSR: 10.2.2.3, uptime: 06:36:03, expires: 00:02:00,
     priority: 64, hash-length: 30
Auto-RP disabled
BSR RP Candidate policy: None
BSR RP policy: None
Auto-RP Announce policy: None
Auto-RP Discovery policy: None

RP: 10.3.3.3, (0), uptime: 06:30:44, expires: 00:02:20,
  priority: 0, RP-source: 10.2.2.3 (B), group ranges:
    239.0.0.0/8
RP: 10.99.99.99*, (0), uptime: 06:30:15, expires: 00:02:20,
  priority: 0, RP-source: 10.2.2.3 (B), group ranges:
    224.0.0.0/4
```

show ip pim rp 的输出结果显示了正在使用的 RP-to-group 映射选择结果（根据从当选 BSR 发送的 Bootstrap 消息中收到的信息）。

胜选的 BSR 是 NX-4，因为其 BSR IP 地址大于 NX-3 的 BSR IP 地址（10.2.2.3 与 10.2.1.3）；两台 C-BSR 的默认优先级都是 64。命令 **ip pim bsr-candidate loopback0** 将 NX-4 配置为 C-BSR，且允许其定期发送 Bootstrap 消息。**show ip pim rp** 命令的输出结果证实本地设备是当前 BSR，同时还提供了一个指示何时发送下一条 Bootstrap 消息的定时器值。虽然散列长度的默认值为 30，不过用户可以在 0～32 范围内进行灵活配置。例 13-63 显示了 NX-4 的配置及 RP 映射信息。

例 13-63　NX-4 的 BSR 配置

```
NX-4# show run pim
! Output omitted for brevity

!Command: show running-config pim

feature pim

ip pim bsr-candidate loopback0
ip pim ssm range 232.0.0.0/8
ip pim bsr listen forward

interface Vlan215
  ip pim sparse-mode

interface Vlan216
  ip pim sparse-mode

interface Vlan303
  ip pim sparse-mode

interface loopback0
  ip pim sparse-mode

interface Ethernet3/28
  ip pim sparse-mode

interface Ethernet3/29
  ip pim sparse-mode
NX-4# show ip pim rp
PIM RP Status Information for VRF "default"
BSR: 10.2.2.3*, next Bootstrap message in: 00:00:53,
     priority: 64, hash-length: 30
Auto-RP disabled
BSR RP Candidate policy: None
BSR RP policy: None
Auto-RP Announce policy: None
Auto-RP Discovery policy: None

RP: 10.3.3.3, (0),
 uptime: 06:30:36   priority: 0,
 RP-source: 10.3.3.3 (B),
 group ranges:
 239.0.0.0/8    , expires: 00:02:11 (B)
RP: 10.99.99.99, (0),
 uptime: 06:30:07   priority: 0,
 RP-source: 10.99.99.99 (B),
 group ranges:
 224.0.0.0/4    , expires: 00:02:28 (B)
```

例 13-64 列出了 NX-3 的配置信息，可以看出，NX-3 被同时配置为 239.0.0.0/8 的 C-RP 和 C-BSR。由于 NX-3 的 C-BSR IP 地址小于 NX-4，因而在 BSR 选举过程失败后不再发送任何 Bootstrap 消息。

例 13-64 NX-3 的 BSR 配置

```
NX-3# show run pim
! Output omitted for brevity
feature pim

ip pim bsr-candidate loopback0
ip pim bsr rp-candidate loopback1 group-list 239.0.0.0/8 priority 0
ip pim ssm range 232.0.0.0/8
ip pim bsr listen forward

interface Vlan215
  ip pim sparse-mode

interface Vlan216
  ip pim sparse-mode

interface Vlan303
  ip pim sparse-mode

interface loopback0
  ip pim sparse-mode

interface loopback1
  ip pim sparse-mode

interface Ethernet3/28
  ip pim sparse-mode

interface Ethernet3/29
  ip pim sparse-mode
NX-3# show ip pim rp
PIM RP Status Information for VRF "default"
BSR: 10.2.2.3, uptime: 07:05:30, expires: 00:02:05,
     priority: 64, hash-length: 30
Auto-RP disabled
BSR RP Candidate policy: None
BSR RP policy: None
Auto-RP Announce policy: None
Auto-RP Discovery policy: None

RP: 10.3.3.3*, (0), uptime: 00:00:04, expires: 00:02:25,
  priority: 0, RP-source: 10.2.2.3 (B), group ranges:
     239.0.0.0/8
RP: 10.99.99.99, (0), uptime: 06:59:41, expires: 00:02:25,
  priority: 0, RP-source: 10.2.2.3 (B), group ranges:
     224.0.0.0/4
```

最后需要验证的路由器是 NX-2，NX-2 仅充当 BSR 侦听路由器和转发路由器。NX-2 收到并检查了 NX-4 的 Bootstrap 消息内容之后，将为每个组区间选择 RP-to-group 映射并将表项安装到本地 RP 缓存中。请注意，虽然 NX-4、NX-3 和 NX-1 也是 BSR 客户端，但它们还同时充当了 C-RP 或 C-BSR。例 13-65 显示了 NX-2 的配置和 RP 映射信息。

例 13-65 NX-2 的 BSR 配置

```
NX-2# show run pim
! Output omitted for brevity
```

```
!Command: show running-config pim

feature pim

ip pim ssm range 232.0.0.0/8
ip pim bsr listen forward

interface Vlan115
  ip pim sparse-mode

interface Vlan116
  ip pim sparse-mode

interface Vlan1101
  ip pim sparse-mode

interface Ethernet3/17
  ip pim sparse-mode

interface Ethernet3/18
  ip pim sparse-mode
NX-2# show ip pim rp
PIM RP Status Information for VRF "default"
BSR: 10.2.2.3, uptime: 07:11:35, expires: 00:01:39,
     priority: 64, hash-length: 30
Auto-RP disabled
BSR RP Candidate policy: None
BSR RP policy: None
Auto-RP Announce policy: None
Auto-RP Discovery policy: None

RP: 10.3.3.3, (0),
 uptime: 07:06:15   priority: 0,
 RP-source: 10.2.2.3 (B),
 group ranges:
 239.0.0.0/8    , expires: 00:01:59 (B)
RP: 10.99.99.99, (0),
 uptime: 07:05:47   priority: 0,
 RP-source: 10.2.2.3 (B),
 group ranges:
 224.0.0.0/4    , expires: 00:01:59 (B)
```

与 Auto-RP 不同，BSR 消息不受已配置的 TTL 范围的约束。对于复杂的 BSR 设计方案来说，可能需要定义允许哪些 C-RP 与特定 BSR 进行通信，此时需要使用命令 **ip pim bsr** [*bsr-policy* | *rp-candidate-policy*]来过滤 Bootstrap 消息和 RP-Candidate（RP 候选）消息，同时还要使用路由映射进行过滤。

与 Auto-RP 相似，这里也使用 **show ip pim internal event-history rp** 命令监控路由器上的 C-BSR、C-RP 和 Bootstrap 消息活动情况。例 13-66 给出了该事件历史记录示例。

例 13-66 带有 BSR 的 NX-4 RP 的 PIM 事件历史记录

```
NX-4# show ip pim internal event-history rp
! Output omitted for brevity

rp events for PIM process
02:50:51.766388 pim [13449]::Group range 239.0.0.0/8 cached
```

```
02:50:51.766385 pim [13449]::(default) pim_add_rp: RP:10.3.3.3
rp_change:no change_flag: yes
 bidir:no, group:239.0.0.0/8 rp_priority: 0, static: no, action: Permit,
prot_souce: 2 hash_len: 30
02:50:51.766325 pim [13449]::(default) pim_add_rp: Received rp_entry from
caller: RP: 10.3.3.3 bidir:no, group:239.0.0.0/8 rp_priority: 0, static:
no, prot_souce: 2 override: no hash_len: 30 holdtime: 150
02:50:51.766304 pim [13449]::RP 10.3.3.3, prefix count: 1, priority: 0,
holdtime: 150
02:50:51.766297 pim [13449]::Received Candidate-RP from 10.3.3.3, length: 76
02:50:09.705668 pim [13449]::Group range 224.0.0.0/4 cached
02:50:09.705664 pim [13449]::(default) pim_add_rp: RP:10.99.99.99 rp_change:no
  change_flag:
yes bidir:no, group:224.0.0.0/4 rp_priority: 0, static: no, action: Permit,
prot_souce: 2 hash_len: 30
02:50:09.705603 pim [13449]::(default) pim_add_rp: Received rp_entry from
caller: RP: 10.99.99.99 bidir:no, group:224.0.0.0/4 rp_priority: 0, static:
no, prot_souce: 2 override: no hash_len: 30 hold time: 150
02:50:09.705581 pim [13449]::RP 10.99.99.99, prefix count: 1, priority: 0,
holdtime: 150
02:50:09.705574 pim [13449]::Received Candidate-RP from 10.99.99.99, length: 76
02:50:03.996080 pim [13449]::Send Bootstrap message on Vlan216
02:50:03.996039 pim [13449]::Send Bootstrap message on Vlan215
02:50:03.995995 pim [13449]::Send Bootstrap message on Vlan303
02:50:03.995940 pim [13449]::Send Bootstrap message on Ethernet3/29
02:50:03.995894 pim [13449]::Send Bootstrap message on Ethernet3/28
02:50:03.995863 pim [13449]:: RP 10.3.3.3, priority: 0, holdtime 150
02:50:03.995860 pim [13449]::Group range 239.0.0.0/8, RPs:
02:50:03.995857 pim [13449]:: RP 10.99.99.99, priority: 0, holdtime 150
02:50:03.995853 pim [13449]::Group range 224.0.0.0/4, RPs:
02:50:03.995847 pim [13449]::Build Bootstrap message, priority: 64, hash-len: 30
```

除事件历史记录之外，**show ip pim statistics** 命令对于查看与 BSR 相关的各类消息的设备级聚合计数器以及故障排查操作来说也非常有用。例 13-67 给出了 NX-4 的输出结果示例。

例 13-67　带有 BSR 的 NX-4 PIM 统计信息

```
NX-4# show ip pim statistics
! Output omitted for brevity
PIM Global Counter Statistics for VRF:default, last reset: never
  Register processing (sent/received):
    Registers: 0/0, Null registers: 0/0, Register-Stops: 0/0
    Registers received and not RP: 0
    Registers received for SSM/Bidir groups: 0/0
  BSR processing (sent/received):
    Bootstraps: 2025/1215, Candidate-RPs: 0/796
    BSs from non-neighbors: 0, BSs from border interfaces: 0
    BS length errors: 0, BSs which RPF failed: 0
    BSs received but not listen configured: 0
    Cand-RPs from border interfaces: 0
    Cand-RPs received but not listen configured: 0
  Auto-RP processing (sent/received):
    Auto-RP Announces: 0/0, Auto-RP Discoveries: 0/0
    Auto-RP RPF failed: 0, Auto-RP from border interfaces: 0
    Auto-RP invalid type: 0, Auto-RP TTL expired: 0
    Auto-RP received but not listen configured: 0
  General errors:
```

```
    Control-plane RPF failure due to no route found: 9
    Data-plane RPF failure due to no route found: 0
    Data-plane no multicast state found: 0
    Data-plane create route state count: 10
  vPC packet stats:
    rpf-source metric requests sent: 11
    rpf-source metric requests received: 483
    rpf-source metric request send error: 0
    rpf-source metric response sent: 483
    rpf-source metric response received: 11
    rpf-source metric response send error: 0
    rpf-source metric rpf change trigger sent: 2
    rpf-source metric rpf change trigger received: 13
    rpf-source metric rpf change trigger send error: 0
```

如果特定组区间存在多个 C-RP，那么确定特定组区间与 RP 的映射关系可能也是一大挑战。NX-OS 提供了两条命令来帮助用户解决这个问题，如例 13-68 所示。

例 13-68　NX-2 的 PIM group-to-RP 映射信息

```
NX-2# show ip pim group-range 239.1.1.1

PIM Group-Range Configuration for VRF "default"
Group-range        Action Mode  RP-address      Shrd-tree-range  Origin
239.0.0.0/8        -      ASM   10.3.3.3        -                BSR
NX-2# show ip pim rp-hash 239.1.1.1

PIM Hash Information for VRF "default"
PIM RPs for group 239.1.1.1, using hash-length: 30 from BSR: 10.2.2.3
  RP 10.99.99.99, hash: 645916811
  RP 10.3.3.3, hash: 1118649067 (selected)
```

第一条命令是 **show ip pim group-range** [*group address*]，该命令可以显示多播组的当前 PIM 模式、RP 地址以及获取 RP 地址的方法。第二条命令是 **show ip pim rp-hash** [*group-address*]，该命令可以按需运行 PIM 散列函数，并提供散列计算结果以及指定组区间内从所有 C-RP 选出的 RP。

NX-OS 不支持在同一 PIM 域中同时运行 Auto-RP 和 BSR。Auto-RP 和 BSR 都能为网络提供动态和冗余 RP 映射功能。如果 PIM 域中还有第三方设备，那么就可以选择 IETF 标准 BSR，可以实现多供应商之间的互操作性。

4. Anycast-RP 配置与验证

冗余性是现代网络设计的一个重要特征。PIM RP 对于整个多播网络的重要性要远远超过其他任何设备。前面讨论了 Auto-RP 和 BSR，它们在提供冗余能力的同时都带来了额外的复杂性，包括选举进程以及在网络中分发多播 group-to-RP 信息。

幸运的是，如果管理员既希望拥有静态 PIM RP 的简单性，又同时拥有 RP 的冗余性，那么还可以使用另一种方法。Anycast-RP（任播 RP）允许多台 PIM 路由器共享单个公有 IP 地址，需要使用/32 掩码在 Loopback 接口配置 IP 地址。所有配置了任播地址的路由器都会将直连主机的地址宣告到网络选定的路由协议中，PIM 域中的所有路由器都使用任播地址作为 RP。如果 FHR 需要注册多播源，那么网络的单播路由协议就会自动将 PIM 消息路由到最近的配置了任播地址的设备，这样就能实现多台设备对 PIM Register 消息的负荷分担，并在 RP 出现故障后提供冗余能力。

很明显，在多台设备上配置相同 IP 地址时一定要足够慎重。例如，对于所有可能错误地将任播环回地址用作路由器 ID 或源地址的路由协议或管理功能来说，都应该将其配置为始终使用不同的接口。解决了这个问题之后，就可以安全地使用任播地址了，对于大型和多区域多播网络来说也是一种常见选项。

配置 Anycast-RP 功能的方法包括以下几种。

- 基于 MSDP（Multicast Source Discovery Protocol，多播源发现协议）的 Anycast-RP。
- 基于 RFC 4610 的 PIM Anycast-RP。

接下来将分别介绍这两种配置方式。

5. 基于 MSDP 的 Anycast-RP

MSDP 协议定义了一种 PIM RP 相互宣告处于激活状态且已注册的多播源的方法。最初的 MSDP 旨在连接多个独立的 PIM 域，这些 PIM 域均使用自己的 PIM RP。不过，后来该协议也被选为 RFC 3446 中的 Anycast-RP 规范的一部分。

MSDP 允许所有配置了 Anycast-RP 地址的 PIM RP 都能独立运行，同时还能与域中的其他 Anycast-RP 共享活动多播源信息。以图 13-17 中的拓扑结构为例，图中的 FHR 可以向 Anycast-RP NX-3 注册多播源，此后接收端就可以通过 Anycast-RP NX-4 加入该多播组。通过 RPT 收到多播流量之后，就会在 LHR 上发生常规的 PIM SPT 切换操作。

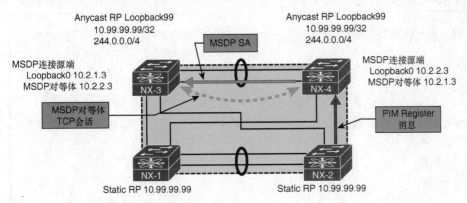

图 13-17 基于 MSDP 的 Anycast-RP

基于 MSDP 的 Anycast-RP 要求每个 Anycast-RP 都要与其他所有 Anycast-RP 建立 MSDP 对等关系，MSDP 对等会话是通过 TCP 端口 639 建立的。建立了 TCP 会话之后，MSDP 就可以在对等体之间发送 TLV 格式的保持激活消息和 SA（Source-Active，源激活）消息。

Anycast-RP 知道新多播源之后，就使用 SA 消息来通知所有 MSDP 对等体有关该多播源的信息。SA 消息包含以下信息。

- 多播源的单播地址。
- 多播组地址。
- PIM RP 的 IP 地址（originator-id）。

对等体收到 MSDP SA 之后，将对 SA 消息执行 RPF 检查，将 SA 消息中的 PIM RP 的 IP 地址与 MSDP 对等体地址进行对比，该地址必须是 MSDP 对等体上的唯一 IP 地址，且不能是任播地址。可以通过 NX-OS 提供的 **ip msdp originator-id** [*address*]命令来配置在 SA 消息中使用的始发端 RP 地址。

> 注：MSDP SA 消息 RPF 检查的详细内容与本章的 MSDP 案例无关，有关 MSDP SA 消息 RPF 检查的详细内容请参阅 RFC 3618 第 10 节。

如果接收该 SA 消息，那么就会将该消息发送给所有其他 MSDP 对等体（收到该 SA 消息的对等体除外)。如果需要为大量 Anycast-RP 配置 MSDP 对等关系,那么就可以通过网状组(mesh group)概念来减少 SA 消息泛洪。网状组是一组 MSDP 对等体,且与其他网状组对等体都建立了 MSDP 邻居关系。因此,从网状组对等体收到的任何 SA 消息都不需要转发给网状组中的任何对等体,因为所有对等体都已经从始发端收到了相同的消息。

MSDP 支持 SA 过滤器,可以通过消息过滤机制来强制执行指定的设计参数。SA 过滤器的配置命令是 **ip msdp sa- policy** [*peer address*] [*route-map | prefix-list*],还可以通过 **ip msdp sa-limit** [*peer address*] [*number of SAs*]命令来限制对等体发送的 SA 消息总数。

图 13-17 的示例网络在 NX-3 与 NX-4 之间配置了 Anycast-RP 和 MSDP。NX-3 和 NX-4 都在 Loopback99 接口上配置了 Anycast-RP 地址 10.99.99.99。NX-3 和 NX-4 的 Loopback0 接口用于建立 MSDP 对等关系。此外,NX-1 和 NX-2 还被静态配置为使用 Anycast-RP 地址 10.99.99.99。

例 13-69 显示了 NX-3 基于 MSDP 的 Anycast-RP 配置示例。与 PIM 一样,在配置 MSDP 之前,必须先使用 **feature msdp** 命令启用该功能特性。请注意,originator-id 和 MSDP connect-source 使用的都是接口 Loopback0 上配置的唯一 IP 地址,而 PIM RP 则被配置为使用 Loopback99 的任播 IP 地址。MSDP 对等体地址是 NX-4 的 Loopback0 接口地址。

例 13-69　NX-3 基于 MSDP 的 Anycast-RP

```
NX-3# show run pim
! Output omitted for brevity

!Command: show running-config pim

feature pim

ip pim rp-address 10.99.99.99 group-list 224.0.0.0/4
ip pim ssm range 232.0.0.0/8

interface Vlan215
  ip pim sparse-mode

interface Vlan216
  ip pim sparse-mode

interface Vlan303
  ip pim sparse-mode

interface loopback0
  ip pim sparse-mode

interface loopback99
  ip pim sparse-mode

interface Ethernet3/28
  ip pim sparse-mode

interface Ethernet3/29
  ip pim sparse-mode
NX-3# show run msdp
! Output omitted for brevity
!Command: show running-config msdp

feature msdp
```

```
ip msdp originator-id loopback0
ip msdp peer 10.2.2.3 connect-source loopback0
NX-3# show run interface lo0 ; show run interface lo99
! Output omitted for brevity

show running-config interface lo0

interface loopback0
  ip address 10.2.1.3/32
  ip router ospf 1 area 0.0.0.0
  ip pim sparse-mode

!Command: show running-config interface loopback99

interface loopback99
  ip address 10.99.99.99/32
  ip router ospf 1 area 0.0.0.0
  ip pim sparse-mode
```

NX-4 的配置与 NX-3 相似，唯一的区别在于 Loopback0 IP 地址和 MSDP 对等体 IP 地址（是 NX-3 的 Loopback0 地址）。例 13-70 给出了 NX-4 基于 MSDP 的 Anycast-RP 配置示例。

例 13-70　NX-4 基于 MSDP 的 Anycast-RP

```
NX-4# show run pim
! Output omitted for brevity

!Command: show running-config pim

feature pim

ip pim rp-address 10.99.99.99 group-list 224.0.0.0/4
ip pim ssm range 232.0.0.0/8

interface Vlan215
  ip pim sparse-mode

interface Vlan216
  ip pim sparse-mode

interface Vlan303
  ip pim sparse-mode

interface loopback0
  ip pim sparse-mode

interface loopback99
  ip pim sparse-mode

interface Ethernet3/28
  ip pim sparse-mode

interface Ethernet3/29
  ip pim sparse-mode
NX-3# show run msdp
! Output omitted for brevity
```

```
!Command: show running-config msdp

feature msdp

ip msdp originator-id loopback0
ip msdp peer 10.2.1.3 connect-source loopback0
NX-3# show run interface lo0 ; show run interface lo99
! Output omitted for brevity

show running-config interface lo0

interface loopback0
  ip address 10.2.2.3/32
  ip router ospf 1 area 0.0.0.0
  ip pim sparse-mode
!Command: show running-config interface loopback99

interface loopback99
  ip address 10.99.99.99/32
  ip router ospf 1 area 0.0.0.0
  ip pim sparse-mode
```

应用了上述配置之后，NX-3 和 NX-4 就可以通过 TCP 端口 639 在 Loopback0 接口之间建立 MSDP 对等会话，可以利用 **show ip msdp peer** 命令确认 MSDP 对等状态，如例 13-71 所示。输出概述了 MSDP 对等体状态以及建立对等体的时间。它还列出所有已配置的 SA 策略过滤器或限制，并提供与对等体交换的 MSDP 消息数量的计数器。

例 13-71　NX-4 的 MSDP 对等状态

```
NX-4# show ip msdp peer

MSDP peer 10.2.1.3 for VRF "default"
AS 0, local address: 10.2.2.3 (loopback0)
  Description: none
  Connection status: Established
    Uptime(Downtime): 00:13:34
    Password: not set
  Keepalive Interval: 60 sec
  Keepalive Timeout: 90 sec
  Reconnection Interval: 10 sec
  Policies:
    SA in: none, SA out: none
    SA limit: unlimited
  Member of mesh-group: no
  Statistics (in/out):
    Last messaged received: 00:00:55
    SAs: 0/13, SA-Requests: 0/0, SA-Responses: 0/0
    In/Out Ctrl Msgs: 0/12, In/Out Data Msgs: 0/1
    Remote/Local Port 14/13
    Keepalives: 0/0, Notifications: 0/0
  Remote/Local Port 65205/639
  RPF check failures: 0
  Cache Lifetime: 00:03:30
  Established Transitions: 1
  Connection Attempts: 0
  Discontinuity Time: 00:13:34
```

如本章前面案例所述，多播源 10.115.1.4 通过 VLAN 115 接口连接了 NX-2，10.115.1.4 为多播组 239.115.115.1 发送流量的时候，NX-2 会向其 RP 地址 10.99.99.99 发送 PIM Register 消息，由于该地址是任播地址，因而 NX-3 和 NX-4 也拥有该地址。本例中的 NX-2 将 Register 消息发送给 NX-4，收到 Register 消息之后，NX-4 会应答一条 Register-Stop 消息，并创建一条(S,G) MROUTE 表项。此外，NX-4 还会创建一条发送给 NX-3 的 MSDP SA 消息，在消息的 RP 字段中设置相应的源 IP 地址、组地址以及 originator-id。NX-3 收到 SA 消息之后，会进行 RPF 检查并应用各种过滤器，如果检查通过，那么就会将该表项添加到 SA 缓存中，同时还会将 MSDP 创建的(S,G) MROUTE 状态添加到 SA 缓存中，并将 MSDP 创建的(S,G) MROUTE 状态添加到 MROUTE 表中，如例 13-72 所示。

例 13-72　NX-3 的 MSDP SA 状态和 MROUTE 状态

```
NX-3# show ip msdp count

SA State per ASN, VRF "default" - 1 total entries
 note: only asn below  65536
  <asn>: <(S,G) count>/<group count>
       0:     1/1
NX-3# show ip msdp sa-cache

MSDP SA Route Cache for VRF "default" - 1 entries
Source          Group              RP              ASN         Uptime
10.115.1.4      239.115.115.1      10.2.2.3        0           01:21:30
NX-3# show ip mroute
! Output omitted for brevity

IP Multicast Routing Table for VRF "default"

(*, 239.115.115.1/32), uptime: 16:41:50, igmp ip pim
  Incoming interface: loopback99, RPF nbr: 10.99.99.99
  Outgoing interface list: (count: 1)
    Vlan215, uptime: 16:41:50, igmp

(10.115.1.4/32, 239.115.115.1/32), uptime: 01:23:25, ip mrib msdp pim
  Incoming interface: Ethernet3/28, RPF nbr: 10.1.23.2
  Outgoing interface list: (count: 1)
```

常见的 Anycast-RP 故障通常与状态缺失或者为活动多播源配置的 RP 之间未实现同步有关。解决这类问题的第一步就是要确定多个 Anycast-RP 中的哪个 Anycast-RP 从 FHR 向故障多播源和组发送了 Register 消息。其次，需要确定所有 Anycast-RP 之间是否都建立了 MSDP 对等会话。如果始发 RP 存在(S, G) MROUTE 表项，那么就表明故障原因可能是 MSDP 未通过 SA 消息通告源和组。可以通过 NX-OS 事件历史记录日志或 Ethanalyzer 来确定当前 MSDP 对等体发送给下一个 MSDP 对等体的消息信息。

多播源 10.115.1.4 开始将流量发送给 239.115.115.1 的时候，NX-2 会向 NX-4 发送 PIM Register 消息。注册多播源之后，例 13-73 显示了 **show ip msdp internal event-history routes** 和 **show ip msdp internal event-history tcp** 命令的输出结果，事件历史记录显示了以下关键信息。

- SA 消息在 04:06:14 和 04:13:27 被添加到 SA 缓存中。
- MSDP TCP 事件历史记录可以与下列时间戳相关联。
    - 104 字节消息是一条被封装的数据包 SA 消息。
    - 20 字节消息是一条空 Register 数据包 SA 消息。
    - 3 字节消息是来去对等体的保持激活消息。

例 13-73 NX-4 的 MSDP 事件历史记录

```
NX-4# show ip msdp internal event-history routes
! Output omitted for brevity

 routes events for MSDP process
2017 Nov  1 04:13:27.815880 msdp [1621]: : Add (10.115.1.4, 239.115.115.1, RP:
   10.99.99.99) to SA buffer
2017 Nov  1 04:12:47.969879 msdp [1621]: : Processing for (*, 239.115.115.1/32)
2017 Nov  1 04:12:47.967291 msdp [1621]: : Processing for (10.115.1.4/32,
   239.115.115.1/32)
2017 Nov  1 04:12:47.967286 msdp [1621]: : Processing for (*, 239.115.115.1/32)
2017 Nov  1 04:06:14.875895 msdp [1621]: : Add (10.115.1.4, 239.115.115.1, RP:
   10.99.99.99) to SA buffer
2017 Nov  1 04:06:04.758524 msdp [1621]: : Processing for (10.115.1.4/32,
   239.115.115.1/32)
NX-4# show ip msdp internal event-history tcp
! Output omitted for brevity

 tcp events for MSDP process
04:13:27.816367 msdp [1621]: : TCP at peer 10.2.1.3 accepted 20 bytes,
0 bytes left to send from buffer, total send bytes: 0
04:13:27.815998 msdp [1621]: : 20 bytes enqueued for send (20 bytes in buffer)
to peer 10.2.1.3
04:06:04.659887 msdp [1621]: : TCP at peer 10.2.1.3 accepted 104 bytes, 0 bytes
 left to send from buffer, total send bytes: 0
04:06:04.659484 msdp [1621]: : 104 bytes enqueued for send (104 bytes in buffer)
to peer 10.2.1.3
04:05:17.778269 msdp [1621]: : Read 3 bytes from TCP with peer 10.2.1.3 ,
buffer offset 0
04:05:17.736188 msdp [1621]: : TCP at peer 10.2.1.3 accepted 3 bytes, 0 bytes
left to send from buffer, total send bytes: 0
04:04:20.111337 msdp [1621]: : Connection established on passive side
04:04:13.085442 msdp [1621]: : We are listen (passive) side of connection, using
   local address 10.2.2.3
```

即使正确生成了 MSDP SA 消息并宣告给了对等体，但是由于 RPF 失败、SA 失败或 SA 限制，仍然有可能会丢弃该 SA 消息。此时，可以利用对等体上的相同事件历史记录来确定 MSDP 收到 SA 消息之后立即丢弃的原因。需要记住的是，PIM RP 是 RPT 的根。如果 LHR 拥有故障多播源和组的(S,G)状态，那么就表明故障原因可能在于以多播源为根的 SPT。

虽然前面的案例采用的都是静态 PIM RP 配置模式，但实际上，基于 MSDP 的 Anycast-RP 也完全可以与 Auto-RP 或 BSR 协同使用，以实现动态 group-to-RP 映射并充分利用 Anycast-RP 的各种优势。

6. PIM Anycast-RP

RFC 4610 规定了 PIM Anycast-RP。PIM Anycast-RP 的设计目标是消除对 MSDP 的依赖性，希望仅使用 PIM 协议来实现 Anycast-RP 功能。这样做的好处是，可以让端到端的进程减少一个控制平面协议，也就减少了一个故障点或错误配置点。

PIM Anycast-RP 利用 Anycast-RP 之间的 PIM Register 消息和 Register-Stop 消息来实现前面 MSDP 提供的相同功能。PIM 任播的设计需求包括以下几种。

- 每个 Anycast-RP 都要配置相同的 Anycast-RP 地址。
- 每个 Anycast-RP 都要为 Anycast-RP 之间的 PIM 消息配置唯一的地址。

- 每个 Anycast-RP 都要配置其他所有 Anycast-RP 的地址。

接下来以图 13-18 为例来解释 PIM Anycast-RP 的配置和故障排查操作。

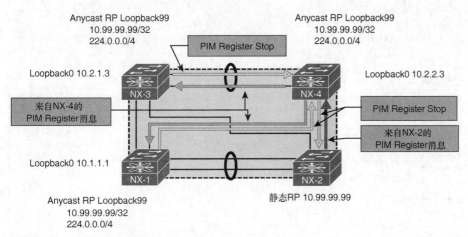

图 13-18　PIM Anycast-RP

与本章前面的案例一样，多播源 10.115.1.4 通过 VLAN 115 连接 NX-2 并向多播组 239.115.115.4 发送流量。为清楚起见，图 13-18 未对此加以说明。NX-2 是 FHR，负责向 RP 注册多播源。NX-2 构建了 Register 消息之后，会在单播路由表中执行查找操作以找到 Anycast-RP 地址 10.99.99.99。NX-1、NX-3 和 NX-4 均配置了 Anycast-RP 地址 10.99.99.99，它们都是同一个 Anycast-RP 集的成员。此后，按照路由表中的最佳路由将 Register 消息发送给 NX-4。

NX-4 收到 Register 消息之后，PIM Anycast-RP 功能需要对收到的消息进行额外检查和处理。与其他 PIM RP 一样，NX-4 也会建立其(S,G)状态。不过，NX-4 还会查看 Register 消息的源端，并确定由于该地址不是 Anycast-RP 集的一部分，因而必须是 FHR。此后，NX-4 还必须构建一个源自其 Loopback0 地址的 Register 消息，并发送给已配置的 Anycast-RP 集中的所有 Anycast-RP。接着，NX-4 向 FHR NX-2 发送 Register-Stop 消息。NX-1 和 NX-3 收到 NX-4 的 Register 消息之后，会在 MROUTE 表中建立(S,G)状态，并通过 Register-Stop 消息应答 NX-4。由于 NX-4 是 NX-1 和 NX-3 的 Anycast-RP 集的一部分，因而 NX-1 和 NX-3 将 NX-4 识别为 Anycast-RP 集的成员，无须构建额外的 Register 消息。

PIM Anycast-RP 的配置采用 FHR 与 RP 之间的标准 Register 和 Register-Stop PIM 消息，并将其应用于 Anycast-RP 集的所有成员。构建 Register 消息以通知其他 Anycast-RP 的操作基于 Register 消息的源地址，如果不是 Anycast-RP 集的成员，那么消息的发送端就肯定是 FHR，因而将 Register 消息发送给 Anycast-RP 集的其他成员。该方法非常简单直接。

例 13-74 显示了 NX-4 的配置信息。可以看出，域中的每台 PIM 路由器都配置了多播组 224.0.0.0/4 的静态 RP 地址 10.99.99.99；NX-1、NX-3 和 NX-4 的 Anycast-RP 集完全相同，且包含所有 Anycast-RP Loopback0 接口地址（包括本地设备自己的 IP 地址）。

例 13-74　NX-4 的 PIM Anycast-RP 配置

```
NX-4# show run pim
! Output omitted for brevity

!Command: show running-config pim

feature pim
```

```
ip pim rp-address 10.99.99.99 group-list 224.0.0.0/4
ip pim ssm range 232.0.0.0/8
ip pim anycast-rp 10.99.99.99 10.1.1.1
ip pim anycast-rp 10.99.99.99 10.2.1.3
ip pim anycast-rp 10.99.99.99 10.2.2.3

interface Vlan215
  ip pim sparse-mode

interface Vlan216
  ip pim sparse-mode

interface Vlan303
  ip pim sparse-mode

interface loopback0
  ip pim sparse-mode

interface loopback99
  ip pim sparse-mode

interface Ethernet3/28
  ip pim sparse-mode

interface Ethernet3/29
  ip pim sparse-mode
```

PIM Anycast-RP 集也可以采用与 PIM 源注册进程相同的调试方法。可以通过 **show ip pim internal event-history null-register** 和 **show ip pim internal event-history data- header- register** 命令显示 Anycast-RP 集与向设备发送 Register 消息的 FHR 之间交换的消息记录。

例 13-75 显示了 NX-4 的事件历史记录，可以看出，10.115.1.254 的空 Register 消息来自 NX-2，即 FHR。添加了 MROUTE 表项之后，NX-4 将 Register 消息转发给 Anycast-RP 集的其他成员，然后收到作为响应消息的 Register-Stop 消息。

例 13-75　NX-4 的 PIM 空 Register 事件历史记录

```
NX-4# show ip pim internal event-history null-register
! Output omitted for brevity

04:26:04.289082 pim [31641]:: Received Register-Stop from 10.2.1.3 for
 (10.115.1.4/32, 239.115.115.1/32)
04:26:02.289082 pim [31641]:: Received Register-Stop from 10.1.1.1 for
 (10.115.1.4/32, 239.115.115.1/32)
04:25:02.126926 pim [31641]:: Send Register-Stop to 10.115.1.254 for
 (10.115.1.4/32, 239.115.115.1/32)
04:25:02.126909 pim [31641]:: Forward Register to Anycast-RP member 10.2.1.3
04:25:02.126885 pim [31641]:: Forward Register to Anycast-RP member 10.1.1.1
04:25:02.126874 pim [31641]:: RP 10.99.99.99 is an Anycast-RP
04:25:02.126866 pim [31641]:: Add new route (10.115.1.4/32, 239.115.115.1/32)
 to MRIB, multi-route TRUE
04:25:02.126715 pim [31641]:: Create route for (10.115.1.4/32, 239.115.115.1/32)
04:25:02.126600 pim [31641]:: Received NULL Register from 10.115.1.254 for
  (10.115.1.4/32, 239.115.115.1/32) (pktlen 20)
```

虽然前面的案例采用的都是静态 PIM RP 配置模式，但实际上，PIM Anycast-RP 也完全可以与 Auto-RP 或 BSR 协同使用，以实现动态 group-to-RP 映射并充分利用 Anycast-RP 的各种优势。

13.4.7　PIM SSM

PIM SSM（Source Specific Multicast，特定源多播）服务模型定义在 RFC 4607 中，允许接收端在没有 PIM RP 的情况下直接加入有源树。这种多播分发方式对于一对多通信进行了优化，广泛应用于流式视频应用（如 IPTV）。此外，SSM 还广泛用于通过 L3 VPN（MVPN）分发 IP 多播流量的提供商多播组。

SSM 可以在无 PIM RP 的情况下正常工作，因为接收端知道所有需要加入的源和组地址。这些信息可以预先配置在应用程序中，如通过 DNS（Domain Name System，域名系统）查询方式，也可以在 LHR 上进行映射。由于 SSM 中没有 PIM RP，因而也就没有了 RPT 或共享树的概念，也没有了 SPT 切换操作。与 PIM ASM 相比，SSM 取消了向 RP 注册源的进程，因而效率更高、开销更低。

PIM SSM 将(S,G)组合称为一个唯一可识别的频道。在 PIM ASM 模式下，所有多播源都可以将流量发送给多播组，接收端也可以隐式加入将流量发送给组地址的任意多播源。在 SSM 模式下，接收端需要通过 IGMPv3 Membership Report 显式请求每个多播源，从而允许不同的应用通过使用唯一的源地址来共享相同的多播组地址。由于 NX-OS 默认实现基于 IP 的 IGMP 监听表，因而主机可以仅接收被请求的多播源的流量。基于 MAC 的 IGMP 监听表无法区分向同一个多播组发送流量的不同源地址。

> **注：** 由于源地址对于接收端来说是已知的，因而 SSM 可以在本地加入其他 PIM 域中的多播源。PIM ASM 和 BiDIR 还需要使用其他协议和配置才能实现域间多播。

接下来以图 13-19 为例来解释 PIM SSM 的配置和验证操作。

VLAN 215 中的接收端加入(10.115.1.4,232.115.115.1)之后，将生成 IGMPv3 Membership Report，该加入消息包含了接收端感兴趣的频道的组和源地址。LHR（NX-4）收到该加入消息并向源端查找了 RPF 接口之后，将构建(S,G) MROUTE 表项。此后，将向 NX-2 发送 SPT PIM Join 消息，同时也将创建(S,G)状态。

NX-2 收到 NX-4 的 PIM Join 消息或者收到多播源发送的数据流量之后（取决于首先发生的事件），就可以创建(S,G)状态。如果特定 SSM 组没有接收端，那么 FHR 就会静默丢弃流量，且 MROUTE 的 OIL 为空。建立了(S,G)SPT 状态之后，流量流就可以从多播源 10.115.1.4 沿下游方向直接到达 SSM 组 232.115.115.1 的接收端。

1. SSM 配置

配置 PIM SSM 时，必须在参与多播转发的所有接口上配置命令 **ip pim sparse-mode**。虽然不需要定义 PIM RP，但是所有连接接收端的接口都必须配置 **ip igmp version 3**。命令 **ip pim ssm-range** 命令默认将地址范围配置为 IANA 保留范围 232.0.0.0/8，也可以配置其他地址范围，但必须在整个 PIM 域中保持一致，否则将导致转发中断，因为配置错误的路由器会认为这是一个 ASM 组，没有有效的 PIM RP-to-group 映射。

可以通过命令 **ip igmp ssm-translate** [*group*] [*source*]将不包含源地址的 IGMPv1 或 IGMPv2 Membership Report 转换为与 IGMPv3 相兼容的状态表项。如果连接在接口上的所有主机都支持 IGMPv3，那么就不需要这样做。

例 13-76 显示了 NX-2 的 SSM 完整配置信息。

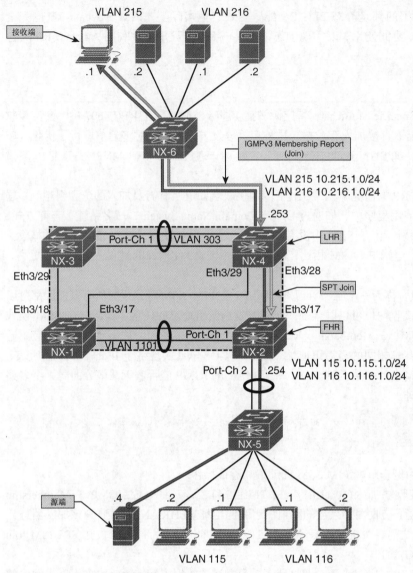

图 13-19 PIM SSM 拓扑结构

例 13-76 NX-2 的 PIM SSM 配置

```
NX-2# show run pim ; show run | inc translate
! Output omitted for brevity

!Command: show running-config pim

feature pim

ip pim ssm range 232.0.0.0/8

interface Vlan115
  ip pim sparse-mode

interface Vlan116
  ip pim sparse-mode
```

```
interface Vlan1101
  ip pim sparse-mode

interface Ethernet3/17
  ip pim sparse-mode

interface Ethernet3/18
  ip pim sparse-mode

ip igmp ssm-translate 232.1.1.1/32 10.215.1.1
NX-2# show run interface vlan115

!Command: show running-config interface Vlan115

interface Vlan115
  no shutdown
  no ip redirects
  ip address 10.115.1.254/24
  ip ospf passive-interface
  ip router ospf 1 area 0.0.0.0
  ip pim sparse-mode
  ip igmp version 3
```

NX-4 的配置与 NX-2 相似，如例 13-77 所示。

例 13-77 NX-4 的 PIM SSM 配置

```
NX-4# show run pim
! Output omitted for brevity

!Command: show running-config pim

feature pim

ip pim ssm range 232.0.0.0/8

interface Vlan215
  ip pim sparse-mode

interface Vlan216
  ip pim sparse-mode
interface Vlan303
  ip pim sparse-mode

interface loopback0
  ip pim sparse-mode

interface Ethernet3/28
  ip pim sparse-mode

interface Ethernet3/29
  ip pim sparse-mode

NX-4# show run interface vlan215

!Command: show running-config interface Vlan215

interface Vlan215
```

```
no shutdown
no ip redirects
ip address 10.215.1.253/24
ip ospf passive-interface
ip router ospf 1 area 0.0.0.0
ip pim sparse-mode
ip igmp version 3
```

NX-1 和 NX-3 的配置方式相似。由于 NX-1 和 NX-3 在本例中不负责转发流量，因而未显示它们的配置。

2. SSM 验证

如果要验证 SSM 中使用的 SPT，那么最好从连接接收端的 LHR 开始。如果接收端发送了 IGMPv3 Membership Report，那么 LHR 就存在(S,G)状态。如果没有该表项，那么就需要检查主机的配置是否正确。SSM 要求主机必须知道源地址，为了保证 SSM 正常运行，必须确保主机知道所要加入的多播源，或者在接收端未使用 IGMPv3 时配置了正确的消息转换机制。

如果对主机是否正在发送正确的 Membership Report 有疑惑，那么就可以在 LHR 上执行 Ethanalyzer 抓包操作。此外，还可以通过 **show ip igmp groups** 和 **show ip igmp snooping groups** 命令确认接口已收到有效的 Membership Report 消息。例 13-78 显示了 NX-4 的输出结果，由于使用的是 IGMPv3，且 NX-OS 使用了基于 IP 的 IGMP 监听表，因而同时提供了源和组信息。

例 13-78 验证 NX-4 的 IGMPv3

```
NX-4# show ip igmp groups
IGMP Connected Group Membership for VRF "default" - 1 total entries
Type: S - Static, D - Dynamic, L - Local, T - SSM Translated
Group Address     Type Interface           Uptime    Expires   Last Reporter
232.115.115.1
  10.115.1.4      D    Vlan215             01:26:41  00:02:06  10.215.1.1
NX-4# show ip igmp snooping groups
Type: S - Static, D - Dynamic, R - Router port, F - Fabricpath core port

Vlan  Group Address       Ver   Type  Port list
215   */*                 -     R     Vlan215 Po2
215   232.115.115.1       v3
        10.115.1.4              D     Po2
216   */*                 -     R     Vlan216
303   */*                 -     R     Vlan303 Po1
```

NX-4 收到 Membership Report 之后，将创建一条(S,G) MROUTE 表项。创建(S,G) MROUTE 状态的原因是，接收端已经知道将要加入的多播组的精确源地址。与此相反，PIM ASM 建立的是(*,G)状态，因为 LHR 并不知道多播源。例 13-79 显示了 NX-4 的 MROUTE 表信息。

例 13-79 NX-4 的 PIM SSM MROUTE 表项

```
NX-4# show ip mroute
IP Multicast Routing Table for VRF "default"

(*, 232.0.0.0/8), uptime: 00:02:07, pim ip
  Incoming interface: Null, RPF nbr: 0.0.0.0
  Outgoing interface list: (count: 0)

(10.115.1.4/32, 232.115.115.1/32), uptime: 00:00:33, igmp ip pim
  Incoming interface: Ethernet3/28, RPF nbr: 10.2.23.2
  Outgoing interface list: (count: 1)
    Vlan215, uptime: 00:00:33, igmp
```

去往 10.115.1.4 的 RPF 接口是 Ethernet 3/28，该接口直连 NX-2。可以通过 **show ip pim internal event-history join-pru** 命令确认 NX-4 已经发送了 SPT Join 消息。例 13-80 显示了该命令的输出结果。

例 13-80　NX-4 的 PIM SSM Join-Prune 消息事件历史记录

```
NX-4# show ip pim internal event-history join-prune
! Output omitted for brevity

03:44:55.372584 pim [10572]:: Send Join-Prune on Ethernet3/28, length: 34
03:44:55.372553 pim [10572]:: Put (10.115.1.4/32, 239.115.115.1/32),
S in join-list for nbr 10.2.23.2
03:44:55.372548 pim [10572]:: wc_bit = FALSE, rp_bit = FALSE
```

NX-2 收到 PIM Join 消息之后，将更新 MROUTE 表项的 OIL 以包含与 NX-4 直连的接口 Ethernet 3/17。例 13-81 显示了 PIM Join-Prune 消息的事件历史记录以及 NX-2 的 MROUTE 表项。

例 13-81　NX-2 的 PIM SSM Join-Prune 消息事件历史记录

```
NX-2# show ip pim internal event-history join-prune
! Output omitted for brevity

join-prune events for PIM process

03:44:55.429867 pim [7192]: : (10.115.1.4/32, 232.115.115.1/32) route exists
, RPF if Vlan115, to us
03:44:13.429837 pim [7192]: : pim_receive_join: We are target comparing with iod
03:44:13.429794 pim [7192]: : pim_receive_join: route:
(10.115.1.4/32, 232.115.115.1/32), wc_bit: FALSE, rp_bit: FALSE
03:44:13.429780 pim [7192]: : Received Join-Prune from 10.2.23.3 on Ethernet3/17
, length: 34, MTU: 9216, ht: 210
NX-2# show ip mroute

IP Multicast Routing Table for VRF "default"

(*, 232.0.0.0/8), uptime: 00:00:47, pim ip
  Incoming interface: Null, RPF nbr: 0.0.0.0
  Outgoing interface list: (count: 0)

(10.115.1.4/32, 232.115.115.1/32), uptime: 00:00:46, ip pim
  Incoming interface: Vlan115, RPF nbr: 10.115.1.4
  Outgoing interface list: (count: 1)
    Ethernet3/17, uptime: 00:00:15, pim
```

与 PIM ASM 或 BiDIR 的故障排查操作相比，SSM 的故障排查更为直观。SSM 不需要 PIM RP，因而不存在与动态 RP、Anycast-RP 以及 group-to-RP 映射相关的配置差错和协议复杂性。此外，由于 SSM 没有源注册进程以及 RPT 或 SPT 切换操作，因而能够一步降低故障排查难度。

大多数 SSM 故障的根源是部分设备配置了错误的 SSM 多播组区间，或者接收端主机配置出错，又或者接收端主机试图加入错误的源地址。SSM 的故障排查方法与 PIM ASM 的 SPT 故障排查过程相似：从接收端开始，逐跳遍历网络，直至到达与多播源相连的 FHR。可以通过抓包工具（如 ELAM、ACL 或 SPAN）来隔离多播树上的路由器的包转发故障。

13.5 多播和 vPC

端口通道是多个物理成员链路接口的逻辑捆绑，可以让上层协议将多个物理接口视为单个接口。vPC（virtual Port-Channel，虚拟端口通道）是一种特殊的端口通道，可以让一对对等交换机在连接其他设备时看起来就像单台交换机一样。

该体系架构可以在 L2 提供无环冗余机制，实现方式是在 vPC 对等体之间同步转发状态和 L2 控制平面信息。为了避免环路和重复数据包问题，需要对 vPC 接口发送的流量部署严格的转发规则。

虽然可以通过 CFS（Cisco Fabric Service，思科交换矩阵服务）实现 vPC 对等体之间的 L2 状态同步，但对等体都拥有独立的 L3 控制平面。与标准端口通道一样，vPC 也利用散列表来确定通过哪条成员链路来转发特定流量流的数据包。对于从 vPC 接口连接的主机发出的流量来说，任一 vPC 对等体都有可能收到，具体取决于散列结果。正因为如此，两个对等体都必须能够将流量转发给通过 vPC 接口连接的主机，也必须能够接收主机发送的流量。NX-OS 支持多播源和接收端都连接在 vPC 端口上。vPC 支持多播流量时的要求如下。

- 利用 CFS 协议实现对等体之间的 IGMP 同步，这样就可以用相同的信息填充两个 vPC 对等体的 IGMP 监听转发表。PIM 和 MROUTE 则不需要通过 CFS 进行同步。
- vPC 对等链路是 IGMP 监听表中的一个 MROUTER 端口，这就意味着 vPC VLAN 收到的所有多播包都将通过对等链路转发给 vPC 对等体。
- 从 vPC 成员端口收到并通过对等链路发送的数据包，不会从收端 vPC 对等体的任何 vPC 成员端口向外发送。
- 如果通过 vPC 连接源端，那么两个 vPC 对等体都能将多播流量转发给 L3 OIF。
- 如果通过 vPC 连接接收端，那么就由到达源端的单播度量最优的 vPC 对等体负责转发数据包。如果度量相同，那么就由运行中的 vPC 主设备转发数据包。该 vPC 声明机制是通过 CFS 协议实现的。
- vPC 不支持 PIM SSM 和 PIM BiDIR，因为可能存在错误的转发行为。

注：虽然 vPC 支持多播源和接收端流量，但是不支持从 vPC 对等体到 vPC 连接的多播路由器的 L3 PIM 邻居。

13.5.1 通过 vPC 连接源端

接下来将以图 13-20 为例来解释通过 vPC 连接源源时的配置与验证操作。

图 13-20 中的多播组 239.215.215.1 的多播源包括 VLAN 215 中的 10.215.1.1 和 VLAN 216 中的 10.216.1.1，这两个多播源都连接在 L2 交换机 NX-6 上，NX-6 使用本地散列算法选择成员链路并将流量转发给该链路。NX-3 和 NX-4 是 vPC 对等体，充当 VLAN 215 和 VLAN 216 的 FHR，并通过 vPC 与 NX-6 进行中继连接。

接收端连接在 NX-2（充当 LHR）的 VLAN 115 上。此外，网络还在 NX-1 和 NX-2 的 Loopback99 上配置了静态 PIM Anycast-RP 地址 10.99.99.99。

配置完 vPC 之后，无须其他特殊配置，vPC 就能支持多播机制。vPC 默认支持多播转发机制且自动启用该功能特性。CFS 负责处理 IGMP 的状态同步，除要在 vPC VLAN 接口上启用 **ip pim sparse-mode** 命令之外，用户不需要为 PIM 启用任何特定的 vPC 配置。

例 13-82 显示了 NX-4 的 PIM 和 vPC 配置示例。

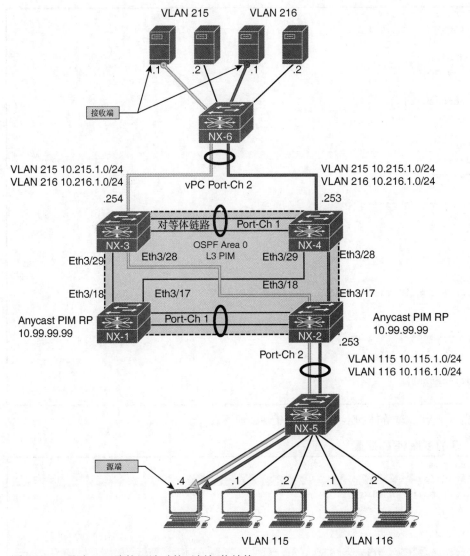

图 13-20 通过 vPC 连接源端时的示例拓扑结构

例 13-82 NX-4 的多播 vPC 配置

```
NX-4# show run pim
! Output omitted for brevity

!Command: show running-config pim

feature pim

ip pim rp-address 10.99.99.99 group-list 224.0.0.0/4
ip pim ssm range 232.0.0.0/8

interface Vlan215
  ip pim sparse-mode

interface Vlan216
  ip pim sparse-mode
```

```
interface Vlan303
  ip pim sparse-mode

interface loopback0
  ip pim sparse-mode

interface Ethernet3/28
  ip pim sparse-mode

interface Ethernet3/29
  ip pim sparse-mode
NX-4# show run vpc

!Command: show running-config vpc

feature vpc

vpc domain 2
  peer-switch
  peer-keepalive destination 10.33.33.1 source 10.33.33.2 vrf peerKA
  peer-gateway

interface port-channel1
  vpc peer-link

interface port-channel2
  vpc 2
```

例 13-83 显示了 vPC 对等体 NX-3 的 PIM 和 vPC 配置。

例 13-83　NX-3 的多播 vPC 配置

```
NX-3# show run pim
! Output omitted for brevity
!Command: show running-config pim

feature pim

ip pim rp-address 10.99.99.99 group-list 224.0.0.0/4
ip pim ssm range 232.0.0.0/8

interface Vlan215
  ip pim sparse-mode

interface Vlan216
  ip pim sparse-mode

interface Vlan303
  ip pim sparse-mode

interface loopback0
  ip pim sparse-mode

interface Ethernet3/28
  ip pim sparse-mode
interface Ethernet3/29
  ip pim sparse-mode
```

```
NX-3# show run vpc

!Command: show running-config vpc

feature vpc

vpc domain 2
  peer-switch
  peer-keepalive destination 10.33.33.2 source 10.33.33.1 vrf peerKA
  peer-gateway

interface port-channel1
  vpc peer-link

interface port-channel2
  vpc 2
```

完成上述配置之后，接下来需要验证 PIM 和 IGMP 在 vPC 对等体上的运行是否正常。从 NX-4 的 **show ip pim interface** 命令输出结果可以看出，VLAN 215 是 vPC VLAN，如例 13-84 所示。请注意，NX-3（10.215.1.254）是 PIM DR，负责处理与 PIM RP 的源注册操作。至于 NX-3 和 NX-4 非 vPC 接口的 PIM 邻居验证过程以及 NX-1 和 NX-2 的相关验证操作，与 13.4.4 节所示案例相同。

例 13-84　NX-4 的多播 vPC PIM 接口

```
NX-4# show ip pim interface vlan215
! Output omitted for brevity
PIM Interface Status for VRF "default"
Vlan215, Interface status: protocol-up/link-up/admin-up
  IP address: 10.215.1.253, IP subnet: 10.215.1.0/24
  PIM DR: 10.215.1.254, DR's priority: 1
  PIM neighbor count: 2
  PIM hello interval: 30 secs, next hello sent in: 00:00:12
  PIM neighbor holdtime: 105 secs
  PIM configured DR priority: 1
  PIM configured DR delay: 3 secs
  PIM border interface: no
  PIM GenID sent in Hellos: 0x29002074
  PIM Hello MD5-AH Authentication: disabled
  PIM Neighbor policy: none configured
  PIM Join-Prune inbound policy: none configured
  PIM Join-Prune outbound policy: none configured
  PIM Join-Prune interval: 1 minutes
  PIM Join-Prune next sending: 0 minutes
  PIM BFD enabled: no
  PIM passive interface: no
  PIM VPC SVI: yes
  PIM Auto Enabled: no
  PIM vPC-peer neighbor: 10.215.1.254
  PIM Interface Statistics, last reset: never
    General (sent/received):
      Hellos: 14849/4299 (early: 0), JPs: 0/13, Asserts: 0/0
      Grafts: 0/0, Graft-Acks: 0/0
      DF-Offers: 1/3, DF-Winners: 2/13, DF-Backoffs: 0/0, DF-Passes: 0/0
    Errors:
      Checksum errors: 0, Invalid packet types/DF subtypes: 0/0
```

```
    Authentication failed: 0
    Packet length errors: 0, Bad version packets: 0, Packets from self: 0
    Packets from non-neighbors: 0
        Packets received on passiveinterface: 0
    JPs received on RPF-interface: 13
    (*,G) Joins received with no/wrong RP: 0/0
    (*,G)/(S,G) JPs received for SSM/Bidir groups: 0/0
    JPs filtered by inbound policy: 0
    JPs filtered by outbound policy: 0
```

例 13-85 中的 **show ip igmp interface** 命令输出结果表明 VLAN 215 是 vPC VLAN，同时还表明 PIM DR 是 vPC 对等体，而不是本地接口。

例 13-85　NX-4 的多播 vPC IGMP 接口

```
NX-4# show ip igmp interface vlan215
! Output omitted for brevity
IGMP Interfaces for VRF "default"
Vlan215, Interface status: protocol-up/link-up/admin-up
  IP address: 10.215.1.253, IP subnet: 10.215.1.0/24
  Active querier: 10.215.1.1, expires: 00:04:10, querier version: 2
  Membership count: 0
  Old Membership count 0
  IGMP version: 2, host version: 2
  IGMP query interval: 125 secs, configured value: 125 secs
  IGMP max response time: 10 secs, configured value: 10 secs
  IGMP startup query interval: 31 secs, configured value: 31 secs
  IGMP startup query count: 2
  IGMP last member mrt: 1 secs
  IGMP last member query count: 2
  IGMP group timeout: 260 secs, configured value: 260 secs
  IGMP querier timeout: 255 secs, configured value: 255 secs
  IGMP unsolicited report interval: 10 secs
  IGMP robustness variable: 2, configured value: 2
  IGMP reporting for link-local groups: disabled
  IGMP interface enable refcount: 1
  IGMP interface immediate leave: disabled
  IGMP VRF name default (id 1)
  IGMP Report Policy: None
  IGMP State Limit: None
  IGMP interface statistics: (only non-zero values displayed)
    General (sent/received):
      v2-queries: 2867/2908, v2-reports: 0/2898, v2-leaves: 0/31
      v3-queries: 15/1397, v3-reports: 0/1393
    Errors:
      Packets with Local IP as source: 0, Source subnet check failures: 0
      Query from non-querier:1
      Report version mismatch: 4, Query version mismatch: 0
      Unknown IGMP message type: 0
  Interface PIM DR: vPC Peer
  Interface vPC SVI: Yes
  Interface vPC CFS statistics:
    DR queries rcvd: 1
    DR updates rcvd: 4
```

与传统 PIM ASM 一样，找出哪台设备充当指定 VLAN 的 PIM DR 也非常重要，因为该设备将负责向 RP 注册多播源。与传统源注册操作相比，vPC 的源注册操作的差异在于 DR 接收源端数据包的接口，数据包可以直接到达 vPC 成员链路，也可以来自对等链路，由于对等链路在 IGMP 监听中被编程为 MROUTER 端口，因而可以在对等链路上转发数据包，如例 13-86 所示。

例 13-86　NX-4 的 vPC IGMP 监听状态

```
NX-4# show ip igmp snooping mrouter
! Output omitted for brevity
Type: S - Static, D - Dynamic, V - vPC Peer Link
      I - Internal, F - Fabricpath core port
      C - Co-learned, U - User Configured
      P - learnt by Peer
Vlan  Router-port  Type   Uptime     Expires
215   Vlan215      I      21:52:05   never
215   Po1          SV     00:43:00   never
215   Po2          D      00:36:25   00:04:59
216   Vlan216      ID     3d06h      00:04:33
216   Po1          SV     00:43:00   never
303   Vlan303      I      4d21h      never
303   Po1          SVD    3d13h      00:04:28
```

　　VLAN 216 中的多播源开始向 239.215.215.1 发送流量，流量将到达 NX-4。NX4 将创建一条 (S,G) MROUTE 表项，并将数据包通过对等链路转发给 NX-3。NX-3 收到数据包之后，将创建一条(S,G) MROUTE 表项，并将源端注册到 RP 上。VLAN 215 中的 10.215.1.1 流量到达 NX-3 的 vPC 成员链路之后，NX-3 将创建一条(S,G) MROUTE，然后通过对等链路将数据包副本转发给 NX-4，NX-4 从对等链路收到流量之后，也会创建一条(S,G) MROUTE 表项。

　　例 13-87 显示了 NX-3 和 NX-4 的 MROUTE 表项，可以看出，虽然来自 10.216.1.1 的多播组 239.215.215.1 流量仅被散列到了 NX-4，但两个 vPC 对等体都创建了(S,G)状态，其原因就在于通过对等链路收到了数据包。

例 13-87　NX-3 和 NX-4 的多播 vPC 源 MROUTE 表项

```
NX-4# show ip mroute
! Output omitted for brevity

IP Multicast Routing Table for VRF "default"

 (10.215.1.1/32, 239.215.215.1/32), uptime: 00:00:14, ip pim
  Incoming interface: Vlan215, RPF nbr: 10.215.1.1
  Outgoing interface list: (count: 0)

 (10.216.1.1/32, 239.215.215.1/32), uptime: 00:00:14, ip pim
  Incoming interface: Vlan216, RPF nbr: 10.216.1.1
  Outgoing interface list: (count: 0)
NX-3# show ip mroute
! Output omitted for brevity
IP Multicast Routing Table for VRF "default"

 (10.215.1.1/32, 239.215.215.1/32), uptime: 00:00:51, ip pim
  Incoming interface: Vlan215, RPF nbr: 10.215.1.1
  Outgoing interface list: (count: 0)

 (10.216.1.1/32, 239.215.215.1/32), uptime: 00:00:51, ip pim
  Incoming interface: Vlan216, RPF nbr: 10.216.1.1
  Outgoing interface list: (count: 0)
```

　　NX-3 和 NX-4 创建了(S,G) MROUTE 表项之后，均意识到多播源与其直连。接下来两台设备需要为这些多播源确定转发路由器。由于本例中的多播源通过 vPC 进行连接，因而两个多播源的转发状态均为 Win-force (forwarding)。可以通过命令 **show ip pim internal vpc rpf-source** 显示转发路由器的选举结果（见例 13-88），输出结果可以显示哪个 vPC 对等体负责转发特定源地址的数据包。对于本例来说，两者等价；由于多播源是通过 vPC 直接相连的，因而收到 PIM Join 消息或 IGMP Membership Report 消息之后，NX-3 和 NX-4 都可以转发数据包。

例 13-88 NX-3 和 NX-4 的 PIM vPC RPF-Source 缓存表

```
NX-4# show ip pim internal vpc rpf-source
! Output omitted for brevity

PIM vPC RPF-Source Cache for Context "default" - Chassis Role Primary

Source: 10.215.1.1
  Pref/Metric: 0/0
  Ref count: 1
  In MRIB: yes
  Is (*,G) rpf: no
  Source role: primary
  Forwarding state: Win-force (forwarding)
  MRIB Forwarding state: forwarding

Source: 10.216.1.1
  Pref/Metric: 0/0
  Ref count: 1
  In MRIB: yes
  Is (*,G) rpf: no
  Source role: primary
  Forwarding state: Win-force (forwarding)
  MRIB Forwarding state: forwarding
NX-3# show ip pim internal vpc rpf-source
! Output omitted for brevity
PIM vPC RPF-Source Cache for Context "default" - Chassis Role Secondary

Source: 10.215.1.1
  Pref/Metric: 0/0
  Ref count: 1
  In MRIB: yes
  Is (*,G) rpf: no
  Source role: secondary
  Forwarding state: Win-force (forwarding)
  MRIB Forwarding state: forwarding

Source: 10.216.1.1
  Pref/Metric: 0/0
  Ref count: 1
  In MRIB: yes
  Is (*,G) rpf: no
  Source role: secondary
  Forwarding state: Win-force (forwarding)
  MRIB Forwarding state: forwarding
```

注： 可以通过命令 **show ip pim internal event-history vpc** 查看历史 vPC RPF-Source 缓存创建事件。

NX-3 是 VLAN 215 和 VLAN 216 的 PIM DR，负责向 PIM RP（NX-1 和 NX-2）注册多播源。NX-3 将 PIM Register 消息发送给 NX-1（见例 13-89 的 **show ip pim internal event-history**

null-register 命令输出结果）。由于 NX-1 是 Anycast-RP 集的一部分，因而将 Register 消息转发给 NX-2 并向 NX-3 发送 Register-Stop 消息。此时，两个 vPC 对等体都有了两个多播源的(S,G)，两个 Anycast-RP 也都有了(S,G)状态。

例 13-89　NX-3 的多播 vPC 源注册

```
NX-3# show ip pim internal event-history null-register
! Output omitted for brevity
04:18:55.957833 pim [10975]:: Received Register-Stop from
10.99.99.99 for (10.216.1.1/32, 239.215.215.1/32)
04:18:55.956223 pim [10975]:: Send Null Register to RP 10.99.99.99
for (10.216.1.1/32, 239.215.215.1/32)
04:17:55.687544 pim [10975]:: Received Register-Stop from
10.99.99.99 for (10.215.1.1/32, 239.215.215.1/32)
04:17:55.686261 pim [10975]:: Send Null Register to RP 10.99.99.99
for (10.216.1.1/32, 239.215.215.1/32)
```

将多播源注册到 RP 之后，VLAN 115 中的接收端将发送 IGMP Membership Report 消息以请求组 239.215.215.1 的所有多播源（到达 NX-2）。NX-2 加入了 RPT，在收到第一个数据包之后将会发起切换到 SPT 的操作。NX-2 到达目的端有两条等价路由（见例 13-90），选择经 NX-3 加入 10.215.1.1，经 NX-4 加入 10.216.1.1。由于 NX-OS 默认启用多路径多播机制，因而 NX-2 加入 SPT 的时候，可以在任一有效 RPF 接口上向多播源发送 PIM Join 消息。

例 13-90　NX-2 到达 VLAN 215 和 VLAN 216 的单播路由

```
NX-2# show ip route 10.215.1.0
IP Route Table for VRF "default"
'*' denotes best ucast next-hop
'**' denotes best mcast next-hop
'[x/y]' denotes [preference/metric]
'%<string>' in via output denotes VRF <string>

10.215.1.0/24, ubest/mbest: 2/0
    *via 10.1.23.3, Eth3/18, [110/44], 02:49:13, ospf-1, intra
    *via 10.2.23.3, Eth3/17, [110/44], 02:49:13, ospf-1, intra
NX-2# show ip route 10.216.1.0
IP Route Table for VRF "default"
'*' denotes best ucast next-hop
'**' denotes best mcast next-hop
'[x/y]' denotes [preference/metric]
'%<string>' in via output denotes VRF <string>

10.216.1.0/24, ubest/mbest: 2/0
    *via 10.1.23.3, Eth3/18, [110/44], 02:49:18, ospf-1, intra
    *via 10.2.23.3, Eth3/17, [110/44], 02:49:18, ospf-1, intra
```

从 **show ip pim internal event-history join-prune** 命令的输出结果可以看出，NX-2 经 NX-3 加入了 VLAN 215 多播源，经 NX-4 加入了 VLAN 216 多播源，如例 13-91 所示。

例 13-91　NX-2 通过 PIM SPT 加入 vPC 连接的源端

```
NX-2# show ip pim internal event-history join-prune
! Output omitted for brevity

03:29:44.703690 pim [7192]:: Send Join-Prune on Ethernet3/18, length: 34
03:29:44.703666 pim [7192]:: Put (10.215.1.1/32, 239.215.215.1/32), S in
join-list for nbr 10.1.23.3
```

```
03:29:44.703661 pim [7192]:: wc_bit = FALSE, rp_bit = FALSE
03:29:44.702673 pim [7192]:: Send Join-Prune on Ethernet3/17, length: 34
03:29:44.702648 pim [7192]:: Put (10.216.1.1/32, 239.215.215.1/32), S in
join-list for nbr 10.2.23.3
03:29:44.702641 pim [7192]:: wc_bit = FALSE, rp_bit = FALSE
```

这些 PIM Join 消息到达 NX-3 和 NX-4 之后，NX-3 和 NX-4 都能将数据包从 VLAN 215 和 VLAN 216 转发到 SPT 上的接收端。由于 NX-2 选择经 NX-4 加入(10.216.1.1, 239.215.215.1)，因而其 OIL 填充了 Ethernet 3/28，且 NX-3 转发了(10.215.1.1, 239.215.215.1)以响应 NX-2 的 PIM Join 消息。例 13-92 显示了 NX-3 和 NX-4 收到 NX-2 的 SPT Join 消息之后的 MROUTE 表项信息。

例 13-92　NX-3 和 NX-4 收到 SPT Join 消息之后的 MROUTE 表项

```
NX-3# show ip mroute
! Output omitted for brevity
IP Multicast Routing Table for VRF "default"

(10.215.1.1/32, 239.215.215.1/32), uptime: 00:01:14, ip pim
  Incoming interface: Vlan215, RPF nbr: 10.215.1.1
  Outgoing interface list: (count: 1)
    Ethernet3/28, uptime: 00:01:14, pim

(10.216.1.1/32, 239.215.215.1/32), uptime: 00:01:14, ip pim
  Incoming interface: Vlan216, RPF nbr: 10.216.1.1
  Outgoing interface list: (count: 0)

NX-4# show ip mroute
! Output omitted for brevity
IP Multicast Routing Table for VRF "default"

(10.215.1.1/32, 239.215.215.1/32), uptime: 00:01:21, ip pim
  Incoming interface: Vlan215, RPF nbr: 10.215.1.1
  Outgoing interface list: (count: 0)

(10.216.1.1/32, 239.215.215.1/32), uptime: 00:01:21, ip pim
  Incoming interface: Vlan216, RPF nbr: 10.216.1.1
  Outgoing interface list: (count: 1)
    Ethernet3/28, uptime: 00:01:21, pim
```

对于通过 vPC 连接源端的场景来说，最后一个案例是解释通过 vPC 连接的接收端加入多播组之后的操作过程。为了在两个 vPC 对等体上创建(S,G)状态，10.216.1.1 会发起 IGMP Membership Report 消息以加入多播组 239.215.215.1，Membership Report 消息通过 L2 交换机 NX-6 发送给 NX-3 或 NX-4。IGMP Membership Report 消息到达 NX-3 或 NX-4 的 vPC Port-channel 2 之后，将发生以下两个事件。

- 由于 vPC 对等体是 MROUTER，因而 IGMP Membership Report 消息通过 vPC 对等链路进行转发。
- 将 CFS 消息发送给对等体。CFS 消息通知 vPC 对等体将 vPC Port-channel 2 编程到 IGMP OIF 中，其中，vPC Port-channel 2 是收到原始 IGMP Membership Report 的接口。

上述事件将在 NX-3 和 NX-4 上利用 IGMP OIF 创建一条同步的(*,G) MROUTE（见例 13-93），OIF 也将被添加到前面已经存在的(S,G) MROUTE 中。

例 13-93　通过 IGMP 加入多播组之后的 NX-3 和 NX-4 的 MROUTE 表项

```
NX-3# show ip mroute
! Output omitted for brevity
```

```
(*, 239.215.215.1/32), uptime: 00:00:05, igmp pim ip
  Incoming interface: Ethernet3/29, RPF nbr: 10.1.13.1
  Outgoing interface list: (count: 1)
    Vlan216, uptime: 00:00:05, igmp

(10.215.1.1/32, 239.215.215.1/32), uptime: 00:57:01, ip pim mrib
  Incoming interface: Vlan215, RPF nbr: 10.215.1.1
  Outgoing interface list: (count: 2)
    Ethernet3/28, uptime: 00:00:05, pim
    Vlan216, uptime: 00:00:05, mrib

(10.216.1.1/32, 239.215.215.1/32), uptime: 00:57:01, ip pim mrib
  Incoming interface: Vlan216, RPF nbr: 10.216.1.1
  Outgoing interface list: (count: 2)
    Ethernet3/29, uptime: 00:00:05, pim
    Vlan216, uptime: 00:00:05, mrib, (RPF)
NX-4# show ip mroute
! Output omitted for brevity
(*, 239.215.215.1/32), uptime: 00:00:11, igmp ip pim
  Incoming interface: Ethernet3/28, RPF nbr: 10.2.23.2
  Outgoing interface list: (count: 1)
    Vlan216, uptime: 00:00:11, igmp

(10.215.1.1/32, 239.215.215.1/32), uptime: 00:57:11, ip pim mrib
  Incoming interface: Vlan215, RPF nbr: 10.215.1.1
  Outgoing interface list: (count: 2)
    Vlan216, uptime: 00:00:11, mrib
    Ethernet3/29, uptime: 00:00:12, pim

(10.216.1.1/32, 239.215.215.1/32), uptime: 00:57:11, ip pim mrib
  Incoming interface: Vlan216, RPF nbr: 10.216.1.1
  Outgoing interface list: (count: 1)
    Vlan216, uptime: 00:00:11, mrib, (RPF)
```

目前已经有了(*,G)表项，因为已经收到了 IGMP Membership Report 消息，而且两条(S,G) MROUTE 的 OIL 中都包含了 VLAN 216。在这种情况下，NX-6 会将数据包从源 10.215.1.1 散列到 NX-3。NX-3 收到该流量之后，将发生以下事件。

- NX-3 在 VLAN 215 中通过对等链路转发数据包。
- NX-3 复制多播流量，并按照 MROUTE 表项将数据包从 VLAN 215 多播路由到 VLAN 216。
- NX-3 将数据包发送给 Port-channel 2（vPC）上的 VLAN 216 中的接收端。
- NX-4 通过对等链路收到 VLAN 215 中的 NX-3 的数据包，然后将数据包转发给所有非 vPC 接收端，但并不将数据包转发出 vPC VLAN。

(10.216.1.1, 239.215.215.1) MROUTE 表项上的标志（RPF）表表示源和接收端位于同一个 VLAN 中。

13.5.2 通过 vPC 连接接收端

本节仍然使用通过 vPC 连接源端的场景示例拓扑，来解释通过 vPC 连接接收端场景的操作方式。虽然此时的源端和接收端的位置发生了变化，但拓扑结构的其余部分保持不变，如图 13-21 所示。

除一条命令之外，不需要对 vPC 连接源端的案例做任何修改。NX-4 和 NX-3 均配置了这条 **ip pim pre-build-spt** 命令，配置了该命令之后，两个 vPC 对等体都会为所有源端发起 SPT 加入操作，但是只有当选的转发路由器才将流量转发给 vPC 连接的接收端。该命令的作用是在因故障而导致当前 vPC 转发路由器突然停止发送流量的情况下，实现更快的故障切换操作。

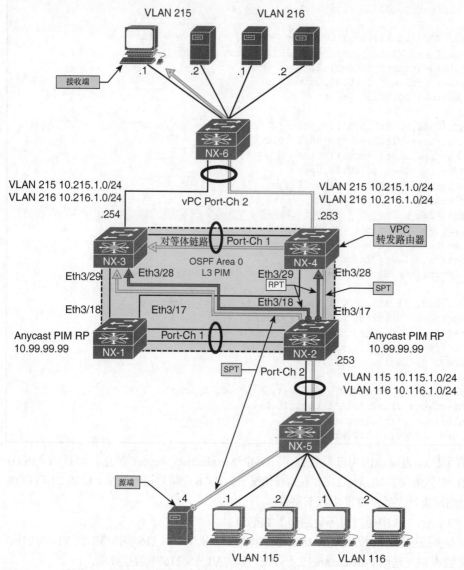

图 13-21 通过 vPC 连接接收端的示例拓扑结构

该配置需要消耗额外的网络带宽并进行额外的流量复制操作，因为非转发路由器并不将自己从 SPT 中剪除，而是持续接收和丢弃流量，直至检测出当前转发路由器发生了故障。如果发生了这种情况，那么也不会因为必须加入 SPT 而产生任何时延，流量已经为故障事件做好了准备。对于大多数部署环境来说，这样做的好处始终大于开销，因而通常建议在 vPC 环境中配置 **ip pim pre-build-spt** 命令。

多播源 10.115.1.4 开始向组 239.115.115.1 发送流量之后，由 L2 交换机 NX-5 将流量转发给 NX-2。收到流量之后，NX-2 会为该多播流量创建一条(S,G)表项。由于目前还没有接收端，因而此时的 OIL 为空。但是，由于 NX-1 和 NX-2 被配置为同一 RP 集的 PIM Anycast-RP，因而 NX-2 通过 PIM Register 消息将源端信息通知给 NX-1。

本例的接收端是 10.215.1.1，连接在 vPC VLAN 215 的网络上。NX-6 将 IGMP Membership Report

消息转发给 Port-channel 2 上的 MROUTER 端口，可以将该消息散列给 NX-3 或 NX-4。NX-4 收到消息之后，IGMP 会创建一条(*,G) MROUTE 表项，然后将来自接收端的 Membership Report 消息以及相应的 CFS 消息通过对等链路发送给 NX-3。收到消息之后，NX-3 也会创建一条(*,G) MROUTE 表项。例 13-94 显示了 NX-4 的 IGMP 监听状态、IGMP 组状态以及 MROUTE 信息。

例 13-94　NX-4 的 IGMP 状态

```
NX-4# show ip igmp snooping groups
! Output omitted for brevity

Type: S - Static, D - Dynamic, R - Router port, F - Fabricpath core port

Vlan   Group Address      Ver   Type   Port list
215    */*                 -     R     Vlan215 Po1 Po2
215    239.115.115.1       v2    D     Po2
NX-4# show ip igmp groups

IGMP Connected Group Membership for VRF "default" - 1 total entries
Type: S - Static, D - Dynamic, L - Local, T - SSM Translated
Group Address       Type  Interface        Uptime    Expires    Last Reporter
239.115.115.1        D    Vlan215          2d18h     00:04:19   10.215.1.1
NX-4# show ip igmp internal vpc
IGMP vPC operational state UP
IGMP ES operational state DOWN
IGMP is registered with vPC library
IGMP is registered with MCEC_TL/CFS
VPC peer link is configured on port-channel1 (Up)
IGMP vPC Operating Version: 3 (mcecm ver:100)
IGMP chassis role is known
IGMP chassis role: Primary (cached Primary)
IGMP vPC Domain ID: 2
IGMP vPC Domain ID Configured: TRUE
IGMP vPC Peer-link Exclude feature enabled
IGMP emulated-switch id not configured
VPC Incremental type: no vpc incr upd, no proxy reporting, just sync (2)
    Configured type: none (0)
VPC Incremental Once download: False
IGMP Vinci Fabric Forwarding DOWN
Implicit adding router for Vinci: Enabled
IGMP single DR: FALSE
NX-4# show ip mroute

IP Multicast Routing Table for VRF "default"

(*, 239.115.115.1/32), uptime: 00:00:04, igmp ip pim
  Incoming interface: Ethernet3/28, RPF nbr: 10.2.23.2
  Outgoing interface list: (count: 1)
    Vlan215, uptime: 00:00:04, igmp
```

例 13-95 显示了 NX-3 收到 NX-4 的 CFS 消息之后的 IGMP 状态，可以看出，两个 vPC 对等体的 IGMP 状态已经实现了同步，而且 IGMP 已经在 vPC 管理器进程中进行了正确注册。

例 13-95　NX-3 的 IGMP 状态

```
NX-3# show ip igmp snooping groups
! Output omitted for brevity
```

```
Type: S - Static, D - Dynamic, R - Router port, F - Fabricpath core port

Vlan  Group Address       Ver   Type  Port list
215   */*                 -     R     Vlan215 Po1 Po2
215   224.0.1.40          v2    D     Po2
215   239.115.115.1       v2    D     Po2
NX-3# show ip igmp groups
IGMP Connected Group Membership for VRF "default" - 1 total entries
Type: S - Static, D - Dynamic, L - Local, T - SSM Translated
Group Address     Type  Interface            Uptime    Expires    Last Reporter
239.115.115.1     D     Vlan215              2d18h     00:04:13   10.215.1.1
NX-3# show ip igmp internal vpc
IGMP vPC operational state UP
IGMP ES operational state DOWN
IGMP is registered with vPC library
IGMP is registered with MCEC_TL/CFS
VPC peer link is configured on port-channel1 (Up)
IGMP vPC Operating Version: 3 (mcecm ver:100)
IGMP chassis role is known
IGMP chassis role: Secondary (cached Secondary)
IGMP vPC Domain ID: 2
IGMP vPC Domain ID Configured: TRUE
IGMP vPC Peer-link Exclude feature enabled
IGMP emulated-switch id not configured
VPC Incremental type: no vpc incr upd, no proxy reporting, just sync (2)
    Configured type: none (0)
VPC Incremental Once download: False
IGMP Vinci Fabric Forwarding DOWN
Implicit adding router for Vinci: Enabled
IGMP single DR: FALSE
NX-3# show ip mroute

IP Multicast Routing Table for VRF "default"

(*, 239.115.115.1/32), uptime: 00:00:09, igmp ip pim
  Incoming interface: Ethernet3/28, RPF nbr: 10.1.23.2
  Outgoing interface list: (count: 1)
    Vlan215, uptime: 00:00:09, igmp
```

可以通过 show ip igmp snooping statistics 命令查看 NX-3 与 NX-4 之间发送的 CFS 消息数，如例 13-96 所示。CFS 负责同步 IGMP 状态，允许 vPC 对等体进行通信并为所有源端选举转发路由器。

例 13-96　NX-4 的 IGMP 监听统计信息

```
NX-4# show ip igmp snooping statistics
! Output omitted for brevity

Global IGMP snooping statistics: (only non-zero values displayed)
   Packets received: 43815
   Packets flooded: 21828
   vPC PIM DR queries fail: 3
   vPC PIM DR updates sent: 6
   vPC CFS message response sent: 15
   vPC CFS message response rcvd: 11
   vPC CFS unreliable message sent: 3688
   vPC CFS unreliable message rcvd: 28114
```

```
vPC CFS reliable message sent: 11
vPC CFS reliable message rcvd: 15
STP TCN messages rcvd: 588
```

> **注：** 可以通过 **show ip igmp snooping internal event-history vpc** 命令查看 IGMP 控制平面的数据包活动信息。

NX-3 和 NX-4 都向 RP 发送了 PIM Join 消息，可以通过 **show ip pim internal event-history join-prune** 命令查看该信息，如例 13-97 所示。

例 13-97　NX-4 和 NX-3 发送的(*,G)加入消息

```
NX-4# show ip pim internal event-history join-prune
! Output omitted for brevity

21:31:32.075044 pim [10572]:: Send Join-Prune on Ethernet3/28, length: 34
21:31:32.075016 pim [10572]:: Put (*, 239.115.115.1/32), WRS in join-list
for nbr 10.2.23.2
21:31:32.075010 pim [10572]:: wc_bit = TRUE, rp_bit = TRUE
NX-3# show ip pim internal event-history join-prune
! Output omitted for brevity
21:31:32.193623 pim [10975]:: Send Join-Prune on Ethernet3/28, length: 34
21:31:32.193593 pim [10975]:: Put (*, 239.115.115.1/32), WRS in join-list
for nbr 10.1.23.2
21:31:32.193586 pim [10975]:: wc_bit = TRUE, rp_bit = TRUE
```

收到 NX-3 和 NX-4 的(*,G)加入消息之后，NX-2 将更新 MROUTE 表项，在 OIL 中加入去往 NX-3 和 NX-4 的 Ethernet 3/17 和 Ethernet 3/18 接口，然后通过 RPT 向外发送流量。

流量到达 RPT 上的 NX-3 和 NX-4 之后，多播组流量的源地址将变为已知，从而触发创建(S,G) MROUTE 表项。接下来 NX-3 和 NX-4 将利用 CFS 来确定由哪台设备充当多播源的转发路由器。可以通过 **show ip pim internal event-history vpc** 命令查看转发路由器的选举过程。由于 NX-3 和 NX-4 到达源端的路由度量和路由优先级都相同，因而等价。但是，由于 NX-4 是运行中的 vPC 主设备，因而胜过 NX-3，从而充当 10.115.1.4 的转发路由器。

产生选举结果之后，就会在 vPC RPF-Source 缓存中创建表项，可以通过 **show ip pim internal vpc rpf-source** 命令查看该表项信息。例 13-98 显示了 NX-4 和 NX-3 的 PIM vPC 转发路由器选举结果。

例 13-98　NX-3 和 NX-4 的 PIM vPC 转发路由器选举

```
NX-4# show ip pim internal event-history vpc
! Output omitted for brevity
21:31:33.795807 pim [10572]: Sending RPF source updates for 1 entries to MRIB
21:31:33.795803 pim [10572]: RPF-source 10.115.1.4 state changed to
forwarding, our pref/metric: 110/44, peer's pref/metric: 110/44, updating MRIB
21:31:33.744941 pim [10572]: Updated RPF-source for local pref/metric: 110/44
for source 10.115.1.4, rpf-interface Ethernet3/28
21:31:33.743829 pim [10572]: Trigger handshake for rpf-source metrices for VRF
default upon MRIB notification
21:31:33.743646 pim [10572]: Ref count increased to 1 for vPC rpf-source
10.115.1.4
21:31:33.743639 pim [10572]: Created vPC RPF-source entry for 10.115.1.4 upon
creation of new (S,G) or (*,G) route in PIM

NX-3# show ip pim internal event-history vpc
! Output omitted for brevity
```

```
21:31:33.913558 pim [10975]: RPF-source 10.115.1.4 state changed to
not forwarding, our pref/metric: 110/44, updating MRIB
21:31:33.913554 pim [10975]: Updated RPF-source for local pref/metric: 110/44
for source 10.115.1.4, rpf-interface Ethernet3/28
21:31:33.912607 pim [10975]: Trigger handshake for rpf-source metrices for VRF
default upon MRIB notification
21:31:33.912508 pim [10975]: Ref count increased to 1 for vPC rpf-source
10.115.1.4
21:31:33.912501 pim [10975]: Created vPC RPF-source entry for 10.115.1.4 upon
creation of new (S,G) or (*,G) route in PIM
NX-4# show ip pim internal vpc rpf-source
! Output omitted for brevity
PIM vPC RPF-Source Cache for Context "default" - Chassis Role Primary

Source: 10.115.1.4
  Pref/Metric: 110/44
  Ref count: 1
  In MRIB: yes
  Is (*,G) rpf: no
  Source role: primary
  Forwarding state: Tie (forwarding)
  MRIB Forwarding state: forwarding
NX-3# show ip pim internal vpc rpf-source
PIM vPC RPF-Source Cache for Context "default" - Chassis Role Secondary

Source: 10.115.1.4
  Pref/Metric: 110/44
  Ref count: 1
  In MRIB: yes
  Is (*,G) rpf: no
  Source role: secondary
  Forwarding state: Tie (not forwarding)
  MRIB Forwarding state: not forwarding
```

为了保证选举进程的正常运行，必须在 vPC 管理器进程中注册 PIM，如例 13-99 所示。

例 13-99 NX-4 的 PIM vPC 状态

```
NX-4# show ip pim internal vpc
! Output omitted for brevity

PIM vPC operational state UP
PIM emulated-switch operational state DOWN
PIM's view of VPC manager state: up
PIM is registered with VPC manager
PIM is registered with MCEC_TL/CFS
PIM VPC peer CFS state: up
PIM VPC CFS reliable send: no
PIM CFS sync start: yes
VPC peer link is up on port-channel1
PIM vPC Operating Version: 2
PIM chassis role is known
PIM chassis role: Primary (cached Primary)
PIM vPC Domain ID: 2
PIM emulated-switch id not configured
PIM vPC Domain Id Configured: yes
```

配置了 **ip pim pre-build-spt** 之后，NX-3 和 NX-4 就会沿着去往源端的 RPF 路径向 NX-2 发起

(S,G)加入操作。不过,由于 NX-3 不是转发路由器,因而将简单地丢弃在 SPT 上收到的数据包。NX-4 则将数据包转发给 vPC 接收端,并通过对等链路转发给 NX-3。

例 13-100 显示了(S,G) MROUTE 状态以及 NX-3 和 NX-4 发起的 PIM SPT 加入操作,可以看出,只有 NX-4 的 OIL 包含了(S,G) MROUTE 表项的 VLAN 215。

例 13-100　PIM (S,G)加入事件及 MROUTE 状态

```
NX-4# show ip pim internal event-history join-prune
! Output omitted for brevity
21:31:33.745236 pim [10572]:: Send Join-Prune on Ethernet3/28, length: 34
21:31:33.743825 pim [10572]:: Put (10.115.1.4/32, 239.115.115.1/32), S in
join-list for nbr 10.2.23.2
21:31:33.743818 pim [10572]:: wc_bit = FALSE, rp_bit = FALSE
NX-3# show ip pim internal event-history join-prune
! Output omitted for brevity
21:31:33.913795 pim [10975]:: Send Join-Prune on Ethernet3/28, length: 34
21:31:33.912603 pim [10975]:: Put (10.115.1.4/32, 239.115.115.1/32), S in
join-list for nbr 10.1.23.2
21:31:33.912597 pim [10975]:: wc_bit = FALSE, rp_bit = FALSE
NX-4# show ip mroute
! Output omitted for brevity
IP Multicast Routing Table for VRF "default"

(*, 239.115.115.1/32), uptime: 00:07:08, igmp ip pim
  Incoming interface: Ethernet3/28, RPF nbr: 10.2.23.2
  Outgoing interface list: (count: 1)
    Vlan215, uptime: 00:07:08, igmp

(10.115.1.4/32, 239.115.115.1/32), uptime: 00:07:06, ip mrib pim
  Incoming interface: Ethernet3/28, RPF nbr: 10.2.23.2
  Outgoing interface list: (count: 1)
    Vlan215, uptime: 00:07:06, mrib
NX-3# show ip mroute
! Output omitted for brevity
IP Multicast Routing Table for VRF "default"

(*, 239.115.115.1/32), uptime: 00:06:05, igmp ip pim
  Incoming interface: Ethernet3/28, RPF nbr: 10.1.23.2
  Outgoing interface list: (count: 1)
    Vlan215, uptime: 00:06:05, igmp

(10.115.1.4/32, 239.115.115.1/32), uptime: 00:06:03, ip mrib pim
  Incoming interface: Ethernet3/28, RPF nbr: 10.1.23.2
  Outgoing interface list: (count: 1)
```

有关 MROUTE 状态的更多详细信息,可以查看 **show routing ip multicast source-tree detail** 命令的输出结果,该命令提供了更多配置验证信息。从例 13-101 可以确认,NX-4 是该(S,G)表项的 RPF-Source 转发路由器。虽然 NX-3 也有相同的 OIL,但是其状态被设置为"*inactive*",表示处于非转发状态。

例 13-101　NX-4 和 NX-3 的多播有源树信息

```
NX-4# show routing ip multicast source-tree detail
! Output omitted for brevity
IP Multicast Routing Table for VRF "default"
```

```
Total number of routes: 3
Total number of (*,G) routes: 1
Total number of (S,G) routes: 1
Total number of (*,G-prefix) routes: 1

(10.115.1.4/32, 239.115.115.1/32) Route ptr: 0x5ced35b4 , uptime: 00:14:50,
ip(0) mrib(1) pim(0)
  RPF-Source: 10.115.1.4 [44/110]
  Data Created: Yes
  VPC Flags
    RPF-Source Forwarder
  Stats: 422/37162 [Packets/Bytes], 352.000 bps
  Stats: 422/37162 [Packets/Bytes], 352.000 bps
  Incoming interface: Ethernet3/28, RPF nbr: 10.2.23.2
  Outgoing interface list: (count: 1)
    Vlan215, uptime: 00:14:50, mrib (vpc-svi)
NX-3# show routing ip multicast source-tree detail
IP Multicast Routing Table for VRF "default"

Total number of routes: 3
Total number of (*,G) routes: 1
Total number of (S,G) routes: 1
Total number of (*,G-prefix) routes: 1

(10.115.1.4/32, 239.115.115.1/32) Route ptr: 0x5cfd46b0 , uptime: 00:15:14,
ip(0) mrib(1) pim(0)
  RPF-Source: 10.115.1.4 [44/110]
  Data Created: Yes
  Stats: 440/38746 [Packets/Bytes], 352.000 bps
  Stats: 440/38746 [Packets/Bytes], 352.000 bps
  Incoming interface: Ethernet3/28, RPF nbr: 10.1.23.2
  Outgoing interface list: (count: 1) (inactive: 1)
    Vlan215, uptime: 00:15:14, mrib (vpc-svi)
```

为了有效解决 vPC 环境下的多播问题，就必须理解 vPC 环境与传统多播环境的行为差异。排查 vPC 环境下的多播故障时，为了确定应该关注哪个 vPC 对等体，必须借助(*,G)和(S,G) MROUTE 状态并了解 IIF 和 OIL 的填充方式。

13.5.3 多播流量的 vPC 注意事项

本节将讨论 vPC 环境下的多播流量注意事项，这里提到的注意事项可能并不适用于所有网络，但确实非常常见，因而在联合部署 vPC 和多播通信时应特别关注。

1. 多播包重复

在某些网络环境下，将多播流量与 vPC 结合使用，有可能会瞬时观察到重复帧，一般都是在状态切换发起期间（如切换到 SPT 树）出现重复帧。如果网络应用对此非常敏感，且无法处理重复帧，那么建议执行以下操作。

- 通过 **ip pim sg-expiry-timer** 命令加大 PIM SG-Expiry 定时器，该定时器数值应该足够大，以确保(S,G)状态在工作时间内不会超时。
- 配置 **ip pim pre-build-spt** 命令。
- 在每个工作日之前，使用多播源生成的探测数据包在网络中填充(S,G)状态。

上述操作的目的是在每天发送关键业务数据之前构建 SPT 树，加大(S,G)超时定时器的目的是允许(S,G)状态在关键时间内保持稳定，避免状态超时以及为间歇性的多播发送端重建(S,G)状态。这样就能避免状态切换操作，从而避免潜在的重复流量问题。

2．保留 VLAN

配置 vPC 时，Nexus 5500 和 Nexus 6000 系列平台会利用保留 VLAN 进行多播路由。通过 vPC 连接的源端发送的流量到达之后，将发生以下事件。

- 将多播流量复制到同一个 VLAN 中的所有接收端，包括对等链路。
- 将多播流量路由到不同 vPC VLAN 中的所有接收端。
- 使用保留 VLAN 通过对等链路发送多播流量副本。

数据包从 vPC 对等体的对等链路到达之后，如果流量是从保留 VLAN 以外的其他 VLAN 收到的，那么就不会进行多播路由。如果两个 vPC 对等体均未配置 **vpc bind-vrf** [*vrf name*] **vlan** [*VLAN ID*]，那么孤立端口（Orphan Port）或者通过 L3 连接的接收端将不会接收流量，因而必须为所有参与多播路由的 VRF 配置该命令。

13.6 多播 Ethanalyzer 案例

本章讨论的各种故障排查步骤都依赖 NX-OS 的 Ethanalyzer 工具来抓取控制平面协议消息。表 13-11 列出了故障排查过程中利用 Ethanalyzer 抓取协议消息的案例。通常情况下，执行 Ethanalyzer 抓包操作时，需要决定是否在会话中显示数据包、在会话中解码数据包或者写入本地文件以进行离线分析，相应的命令语法格式是 **ethanalyzer local interface** [*inband*] **capture-filter** [*filter-string in quotes*] **write** [*location:filename*]，该命令存在很多变体，具体取决于使用的选项。

表 13-11 Ethanalyzer 抓包案例

抓取的信息	Ethanalyzer 抓包过滤器
PIM 及来去主机 10.2.23.3 的数据包	"pim && host 10.2.23.3"
单播 PIM 包，如注册消息或候选 RP 宣告消息	"pim && not host 224.0.0.13"
来自 10.1.1.1 的 MSDP 消息	"src host 10.1.1.1 && tcp port 639"
IGMP 通用查询消息	"igmp && host 224.0.0.1"
IGMP 特定组查询消息或报告消息	"igmp && host 239.115.115.1"
IGMP 离开消息	"igmp && host 224.0.0.2"
从 10.115.1.4 发送给管理引擎的多播数据包	"src host 10.115.1.4 && dst host 239.115.115.1"

Ethanalyzer 命令语法可能会因不同的平台而略有不同。例如，某些 NX-OS 平台（如 Nexus 3000）拥有 inband-hi 和 inband-lo 接口，对于大多数控制平面协议来说，要在 inband-hi 接口上进行抓包。但是，如果无法抓取任何数据包，那么就可能需要尝试其他接口选项。

13.7 本章小结

本章详细讨论了 NX-OS 的多播通信机制，在深入分析 NX-OS 多播体系架构之前，首先介绍了多播转发的基本概念，接着详细讨论了 IGMP 和 PIM 协议，为后面的配置验证案例打下了很好的基础。本章详细讨论了多种应用广泛的 PIM 操作模式（ASM、BiDIR 和 SSM），包括每种操作模式使用的消息类型以及各种多播分发树的验证方式。最后，本章回顾了多播和 vPC 的基本知识，解释了 vPC 环境下支持多播通信所需的差异化协议处理方式。本章的目的不是涵盖所有可能的多播转发方案，而是希望提供一些基本的故障排查工具箱，让用户能够更好地适应复杂多变的多播环境故障排查需求。

第六部分

Nexus 隧道故障排查

第 14 章　OTV 故障排查

第 14 章

OTV 故障排查

本章主要讨论如下主题。
- OTV 基础知识。
- OTV 控制平面故障排查。
- OTV 数据平面故障排查。
- OTV 高级功能特性。

OTV（Overlay Transport Virtualization，叠加传输虚拟化）是一种 MAC-in-IP 叠加封装方式，允许在被三层（L3）路由网络分隔的站点之间进行二层（L2）通信。OTV 可以在多个数据中心之间扩展 L2 应用，在不更改现有网络设计的情况下极大地扩展了网络连接能力。本章将重点讨论 OTV 的基础知识、OTV 控制平面和数据平面进程以及 OTV 故障排查方法。

14.1 OTV 基础知识

L2 层面连接数据中心站点的需求主要来自于 VM（Virtual Machine，虚拟机）、工作流的移动性以及地理环境的多样性冗余需求。关键网络甚至可能需要建立一个完全镜像的灾难恢复站点，而且需要在站点之间实现数据和服务的同步。如果能够将多个位置的服务放到同一个 VLAN 中，那么在主机或服务器出现位置变更之后，就可以在不重配网络层编址方案的情况下实现数据中心之间的移动性。在 L2 层面连接两个或多个数据中心时，需要考虑如下挑战和注意事项。
- 可用传输网络类型。
- 实现冗余机制的多归属站点。
- 允许各个站点彼此独立。
- 创建故障隔离边界。
- 确保可以将网络扩展到将来位置而不破坏现有站点。

在 OTV 出现之前，L2 DCI（Data Center Interconnect，数据中心互联）的主要实现方式是使用配置为 L2 中继的直连光纤链路、IEEE 802.1Q 隧道（Q-in-Q）、EoMPLS（Ethernet over MPLS，基于 MPLS 的以太网）或 VPLS（Virtual Private LAN Service，虚拟专用 LAN 服务），这些实现方式严重依赖于传输服务提供商的复杂配置，增加新站点时，必须由传输服务提供商完成必需的网络配置操作。

与传统方式不同，OTV 可以利用底层 L3 路由网络在站点之间提供 L2 叠加网络。由于 OTV 封装在 IP 数据包中进行传输，因而能够完全利用 L3 路由优势，如实现负载分担和冗余机制的 IP ECMP（Equal Cost Multipath，等价多路径）路由、基于路由协议度量实现 OTV ED（Edge Device，边缘设备）之间的最佳数据包路径等。此外，OTV 的故障排查过程也相对较为简单，因为传输网络中的流量是大家非常熟悉的传统 IP 流量，相应的故障排查技术也是大家熟悉的 IP 故障排查技术。

Q-in-Q、EoMPLS 和 VPLS 等 L2 DCI 解决方案要求服务提供商必须对站点流量执行某种形式的封装和解封装操作。对于 OTV 来说，叠加封装的边界从服务提供商转移到了 OTV 站点，从而给网络运营商带来了更大的可视性和控制能力。用户可以根据需要灵活调整叠加配置，无须与底层服务提供商进行任何交互，也不依赖于底层服务提供商。常见的叠加配置调整操作包括增加新的 OTV 站点或者更改通过 OTV 叠加网络扩展的 VLAN。

前面提到的传输协议依赖于静态或状态化隧道。OTV 在封装叠加流量时，需要根据 OTV 的 IS-IS（Intermediate System to Intermediate System，中间系统到中间系统）控制平面提供的 MAC 地址到 IP 下一跳信息进行动态封装，通常将这种方式称为 MAC 地址路由，具体内容将在本章后面进行讨论。需要注意的是，OTV 使用控制平面协议将 MAC 地址动态映射为远程 IP 下一跳。

虽然多归属对于冗余性来说非常必要，但是如果从不使用这些冗余链路和设备，那么效率就比较低。对于传统 L2 交换场景来说，必须仔细规划和配置多归属机制，以避免产生 L2 环路和 STP（Spanning-Tree Protocol，生成树协议）阻塞端口。OTV 在协议设计之初就考虑到了内置多归属特性。例如，可以在单个站点中部署多台 OTV 边缘设备，而且每台设备都能主动转发不同 VLAN 的流量。数据中心之间存在多条 L3 路由链路，可以在数据中心站点内的 OTV 边缘设备之间提供 L3 ECMP 冗余性和负载均衡。

仅当冗余数据中心位于不同的故障域中且一个数据中心的问题不会影响另一个数据中心的时候，冗余数据中心才有价值。这就意味着每个数据中心都必须在 STP 方面进行隔离，而且必须避免站点之间的流量转发环路。对于通过 OTV 扩展的 VLAN 来说，OTV 允许每个数据中心站点都包含一个独立的 STP 根网桥，其原因是 OTV 不会通过叠加网络转发 STP BPDU（Bridge Protocol Data Unit，桥接协议数据单元），从而允许每个站点都能独立运行。

14.1.1 泛洪控制与广播优化

传统的 L2 交换机会在帧到达端口之后学习 MAC 地址，并保留源 MAC 地址以及相关的接口映射，直至 MAC 地址老化或者在新接口上学到该 MAC 地址。如果 L2 交换机不知道目的 MAC 地址，那么就会执行单播洪泛操作，出现这种情况时，为了保证流量能够到达正确的目的端，需要在 VLAN 的所有端口上泛洪该未知单播流量。与此相反，OTV 通过 IS-IS 控制平面协议从远程数据中心学习 MAC 地址，而且不会通过叠加网络泛洪任何未知单播流量。ARP（Address Resolution Protocol，地址解析协议）流量是传统交换网络中的另一种泛洪流量。如果启用了 OTV，那么就能以受控方式泛洪 ARP 流量，而且 OTV 边缘设备还能侦听 ARP 响应并存储在本地 ARP ND（Neighbor Discovery，邻居发现）缓存中，此后向主机发起的 ARP 请求，就由 OTV 边缘设备代表主机进行应答，从而大幅减少跨叠加网络的广播流量。

VLAN 中的广播和多播流量必须到达所有远程数据中心。OTV 依靠底层传输网络中的 IP 多播通信以有效、可扩展的方式分发这类流量。使用了 IP 多播传输机制之后，OTV 无须边缘设备为每个远程边缘设备都执行头端复制（head-end replication）操作。头端复制意味着始发 OTV 边缘设备需要为每个远程边缘设备都创建一份帧副本，如果有大量 OTV 站点且数据包速率很高，那么就会产生巨大负担。采用了 IP 多播传输机制之后，OTV 边缘设备只需要创建单个数据包，数据包遍历多播树到达接收端（远程 OTV 边缘设备）之后，底层传输网络中的多播路由器就可以自动执行复制操作。

14.1.2 支持 OTV 的平台

Nexus 7000 系列设备均支持 OTV，但需要安装 TRS（Transport Service，传输服务）许可。大多数部署方案利用 VDC（Virtual Device Context，虚拟设备上下文）在单个机箱中实现路由与 OTV 功能的逻辑隔离。

> 注：思科 ASR1000 系列路由器也支持 OTV，虽然协议功能相似，但实现方式有所差异。本章仅讨论 Nexus 7000 系列交换机的 OTV 功能特性。

分布层交换机聚合了 VLAN 流量之后，通过 L2 中继连接到专用 OTV VDC 中。VLAN 中需要到达远程数据中心的流量都将切换到 OTV VDC，并由边缘设备进行封装。此后数据包将作为 L3 IP 包穿越路由 VDC，并路由到远程 OTV 边缘设备进行解封装。需要 L3 路由的流量则从 L2 分布层交换机传送给路由 VDC。路由 VDC 通常会配置 HSRP（Hot Standby Routing Protocol，热备路由协议）或 VRRP（Virtual Router Redundancy Protocol，虚拟路由器冗余协议）等 FHRP（First Hop Redundancy Protocol，第一跳冗余协议），为所连接的 VLAN 中的主机提供默认网关地址并执行 VLAN 间路由。

> 注：配置多个 VDC 时可能需要安装其他许可，具体取决于部署需求和 VDC 的数量。

14.1.3　OTV 术语

图 14-1 给出了一个 OTV 网络拓扑示例，图中的两个数据中心站点通过一个启用了 IP 多播机制的 L3 路由网络进行连接。为了保证 OTV 的正常运行，该 L3 路由网络必须在 OTV 边缘设备之间提供 IP 连接。OTV ED 的布放位置非常灵活，只要 OTV ED 能够收到需要通过 OTV 进行扩展的 VLAN 的 L2 帧即可，通常将 OTV ED 连接在 L2 与 L3 的边界位置。

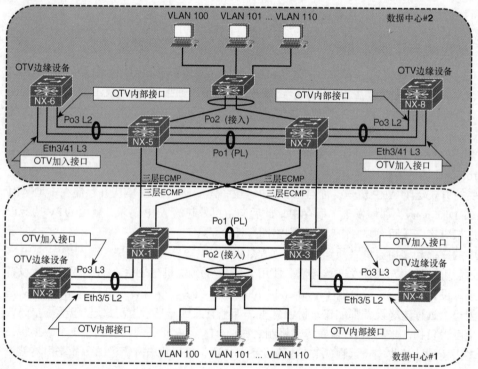

图 14-1　OTV 拓扑结构示例

数据中心#1 包含了冗余 OTV VDC（NX-2 和 NX-4），它们都是边缘设备。NX-1 和 NX-3 执行路由和 L2 VLAN 聚合功能，并将接入交换机连接到 OTV VDC 内部接口。OTV 加入接口（Join interface）是连接路由 VDC 的三层接口。数据中心#2 被配置为数据中心#1 的镜像，但 Port-channel 3 接口被用作 OTV 内部接口，而不像数据中心#1 那样被用作 OTV 加入接口。本例通过 OTV 叠加网络将 VLAN 100～VLAN 110 扩展到了两个数据中心。

表 14-1 列出了图 14-1 中的 OTV 术语。

表 14-1　　　　　　　　　　　　　　　OTV 术语

术语	定义
ED（Edge Device，边缘设备）	负责为需要通过 OTV 扩展的 VLAN 将以太网帧动态封装到 L3 IP 包中
AED（Authoritative Edge Device，权威边缘设备）	为通过 OTV 进行扩展的 VLAN 转发流量。通过 OTV IS-IS 控制平面向远程站点宣告活动 VLAN 的 MAC 地址可达性。根据次序值（ordinal value）0 或 1 来确定 AED，次序值为 0 的 AED 将成为所有编号为偶数的 VLAN 的 AED，而次序值为 1 的 AED 则成为所有编号为奇数的 VLAN 的 AED，该次序值无法手工配置，是两台边缘设备建立邻接关系时确定的
内部接口（Internal Interface）	OTV 边缘设备连接本地站点的接口。该接口提供了从 ED 到内部网络的传统 L2 接口，并在收到流量时学习 MAC 地址。内部接口是 L2 中继接口，负责承载由 OTV 扩展的 VLAN 流量
加入接口（Join Interface）	OTV 边缘设备连接 L3 路由网络的接口，是 OTV 封装流量的源端。可以是环回接口、L3 点对点接口或 L3 端口通道接口，也可以使用子接口。多个叠加接口可以使用同一个加入接口
叠加接口（Overlay Interface）	OTV ED 上的接口。叠加接口负责将扩展 VLAN 的 L2 流量动态封装到 IP 包中，以传输到远程 OTV 站点。边缘设备支持多个叠加接口
站点 VLAN（Site VLAN）	本地站点中的 VLAN，负责在 L2 连接 OTV 边缘设备。站点 VLAN 用于发现本地站点中的其他边缘设备，并允许这些边缘设备建立邻接关系。建立邻接关系之后，就要选举每个 VLAN 的 AED。站点 VLAN 应该是 OTV 专用 VLAN，不能跨叠加网络进行扩展。对于所有 OTV 站点来说，站点 VLAN 的 VLAN 号都应该相同
站点标识符（site-id，Site Identifier）	对于属于同一站点的所有边缘设备来说，站点标识符必须相同。site-id 的取值范围是 0x1～0xffffffff。site-id 由 IS-IS 进行宣告，允许边缘设备标识属于同一站点的边缘设备。边缘设备可以在叠加接口和站点 VLAN 上建立邻接关系（双重邻接[Dual Adjacency]），这样一来，即使站点 VLAN 邻接关系因连接问题而中断，也能维持站点中的边缘设备之间的邻接关系
站点邻接关系（Site Adjacency）	同一站点内的 OTV 边缘设备之间通过站点 VLAN 建立的邻接关系。如果站点 VLAN 的 OTV ED 收到的 IS-IS Hello 包的站点标识符与本地路由器不同，那么就会禁用叠加功能。这样做的目的是防止 OTV 内部接口和叠加接口之间出现环路，这也是建议每个站点都设置相同的 OTV 内部 VLAN 的原因
叠加邻接关系（Overlay Adjacency）	在 OTV 加入接口建立的 OTV 邻接关系。可以在站点之间以及同一站点的边缘设备之间的叠加接口上建立邻接关系。边缘设备建立双重邻接关系（站点邻接关系和叠加邻接关系）的目的是实现弹性能力。为了确保同一站点中的设备能够建立叠加邻接关系，site-id 必须匹配

14.1.4　部署 OTV

OTV 边缘设备的配置包括 OTV 内部接口、加入接口和叠加虚拟接口。配置 OTV 之前，必须全面了解传输网络的能力并保证传输网络的配置能够支持 OTV 部署模型。

1. OTV 部署模型

根据传输网络的能力情况，目前主要存在以下两种 OTV 部署模型。

- **启用多播的传输网络（Multicast Enabled Transport）**：将控制平面封装在 IP 多播包中。该模式让每个 OTV ED 都通过传输网络加入多播控制组，从而实现动态邻居发现。OTV ED 只要发送单个多播包，就可以沿传输网络中的多播树将数据包复制给每个远程 OTV ED。
- **邻接服务器模式（Adjacency Server Mode）**：必须手工为叠加接口配置邻居，需要为每个邻居都创建单播控制平面包并通过传输网络进行路由。

验证了传输网络能力之后，需要在规划阶段确定 OTV 的部署模型。如果传输网络支持多播通信，那么就建议采用多播部署模型。如果传输网络不支持多播，那么就可以采用邻接服务器部署模型。

传输网络必须为 OTV ED 之间的单播和多播通信提供 IP 路由连接，只要是 L3 路由协议就能满足单播连接的需求。如果 OTV ED 没有与数据中心建立动态路由邻接关系，那么就需要配置静态路由以到达对端 OTV ED 的加入接口。

必须将传输网络中的多播路由配置为支持 PIM（Protocol Independent Multicast，协议无关多播）协议，ASM（Any Source Multicast，任意源多播）组用于 OTV 控制组（control-group），PIM SSM（Source Specific Multicast，特定源多播）组用于 OTV 数据组（data-group）。此外，还需要在 OTV ED 的加入接口上启用 IGMPv3。

> **注**：为了实现弹性能力，建议在传输网络中部署冗余 PIM RP（Rendezvous Point，聚合点）。

2. OTV 站点 VLAN

每个 OTV 站点都应该配置一个 OTV 站点 VLAN。站点 VLAN 应该从数据中心 L2 交换网络中继到每个 OTV ED 的 OTV 内部接口。虽然并不必需，但强烈建议所有 OTV 站点都使用相同的站点 VLAN，以避免 OTV 站点之间出现站点 VLAN 的意外泄露。

确定了部署模型并通过已安装的 TRANSPORT_SERVICES_PKG 许可创建了 OTV VDC 之后，就可以通过以下步骤启用 OTV 功能特性（以下示例基于启用了多播机制的传输网络）。

3. OTV 配置

在输入 OTV 配置之前，必须首先通过命令 **feature otv** 启用 OTV 功能特性。例 14-1 显示了与 OTV 内部接口相关联的配置示例，该内部接口是 L2 中继端口，负责处理数据中心网络的传统交换功能。通过 OTV 进行扩展的 VLAN 是 VLAN 100 ~ VLAN 110。两个数据中心的站点 VLAN 都是 VLAN 10，与 VLAN 100 ~ VLAN 110 一起通过 OTV 内部接口进行中继。

例 14-1　OTV 内部接口配置

```
NX-2# show run | no-more
! Output omitted for brevity
feature otv

vlan 1,10,100-110

interface Ethernet3/5
 description To NX-1 3/19, OTV internal interface
 switchport
 switchport mode trunk
 mtu 9216
 no shutdown
```

设计数据中心的 STP 域时，应该将 OTV 内部接口视为接入交换机。

配置了 OTV 内部接口之后，接下来就可以配置 OTV 加入接口。可以在 M1、M2、M3 或 F3 模块上配置 OTV 加入接口，加入接口可以是环回接口或 L3 点对点链路，也可以使用 L3 端口通道或子接口（取决于具体部署需求）。例 14-2 显示了 OTV 加入接口的相关配置。

例 14-2　OTV 加入接口配置

```
NX-2# show run | no-more
! Output omitted for brevity
feature otv

interface port-channel3
 description To NX-1 Po3, OTV Join interface
 mtu 9216
 ip address 10.1.12.1/24
```

```
ip router ospf 1 area 0.0.0.0
ip igmp version 3

interface Ethernet3/7
 description To NX-1 Eth3/22, OTV Join interface
 mtu 9216
 channel-group 3 mode active
 no shutdown

interface Ethernet3/8
 description To NX-1 Eth3/23, OTV Join interface
 mtu 9216
 channel-group 3 mode active
 no shutdown
```

本例的 OTV 加入接口是一个配置了 IGMPv3 的三层点对点接口。配置 IGMPv3 的原因是确保 OTV ED 能够加入 OTV 功能所需的控制组和数据组。

本例使用的路由协议是 OSPF（Open Shortest Path First，开放最短路径优先），两个数据中心使用的都是 OSPF。OTV ED 通过 OSPF 学习单播路由以到达所有其他 OTV ED。例中的数据中心在所有基础设施链路上都配置了 MTU 9216，以保证应用程序能够在不分段的情况下传送完整的 1500 字节帧。

从 NX-OS Release 8.0(1)开始，就可以将环回接口用作 OTV 加入接口。如果使用了该配置选项，那么配置方式将与本例有所不同（本例使用的是 L3 点对点接口）。OTV ED 至少要通过一个 L3 路由接口连接数据中心网络，并通过该 L3 接口建立 PIM 邻居。此外，还要为 OTV ED 配置正确的 PIM RP 和 SSM 地址区间，以匹配路由式数据中心设备及传输网络。最后，将环回接口用作加入接口，还必须配置 **ip pim sparse-mode** 命令，以确保该接口能够同时充当 OTV 控制组和数据组的源端和接收端。除此此外，还必须将环回接口包含在数据中心三层连接所用的动态路由协议当中，以确保其他 OTV ED 都能到达该环回接口。

注： 通过 IP 传输网络进行传输时，OTV 封装会增加 L2 帧的大小。本章稍后将讨论 OTV MTU 的相关注意事项。

配置了 OTV 内部接口和加入接口之后，接下来可以配置被称为叠加接口的逻辑接口并与加入接口进行绑定。叠加接口负责动态封装 OTV 站点之间的 VLAN 流量。请注意，对于叠加网络中的所有 OTV ED 来说，应该为它们的叠加接口分配相同的编号。虽然同一个 OTV ED 可以存在多个叠加接口，但是通过叠加网络扩展的 VLAN 一定不能重叠。

OTV 站点 VLAN 负责与位于同一站点中的其他 OTV ED 建立站点邻接关系。即使是单个 OTV ED 站点，也必须配置站点 VLAN 才能启动叠加接口。虽然并不必需，但仍然建议所有 OTV 站点都配置相同的站点 VLAN，从而允许 OTV 检测出 OTV 站点是故意合并的还是错误合并的。请注意，站点 VLAN 不应该包含在 OTV 扩展 VLAN 列表中。应该将属于同一站点的所有 OTV ED 的站点标识符都配置为相同值。可以通过命令 **otv join-interface** [*interface*]将叠加接口绑定到加入接口，加入接口负责发送和接收用于建立邻接关系的 OTV 多播控制平面消息，并从其他 OTV ED 学习 MAC 地址。

由于本例使用的是支持多播通信的传输网络，因而通过 **otv control-group** [*group number*]命令声明了哪个 IP PIM ASM 组将用于 OTV 控制平面组。控制平面组将在整个传输过程中承载 OTV 控制平面流量（如 IS-IS Hello），并允许 OTV ED 进行通信。所有 OTV ED 的组号都必须匹配，且在传输网络中进行多播路由时，每个 OTV ED 都充当多播组的源端和接收端。

本例通过 **otv data-group** [*group number*]命令配置将使用哪些 SSM 组来承载叠加网络上的多

播数据流量。数据组负责通过站点之间的 OTV 叠加网络传输 VLAN 中的多播流量。对于数据组包含的多播组数量来说，需要在优化与扩展性之间做出平衡。如果使用单个多播组，那么即便站点上没有任何接收端，所有 OTV ED 也都要接收叠加网络上的所有多播流量。如果定义了大量多播组，那么就能实现多播流量的最佳转发，但是在传输网络中配置多个多播组可能会存在扩展性问题。目前，OTV 最多支持 256 个多播数据组。

完成上述配置之后，必须确保不关闭 Overlay0 接口。如果底层网络配置了正确的单播和多播路由，那么 OTV ED 之间就可以建立正常的 OTV 邻接关系。例 14-3 的配置示例包括了 NX-2 的 Overlay0 接口、站点 VLAN 以及 site-id 的配置。

例 14-3　OTV 叠加接口配置

```
NX-2# show running-config | no-more
! Output omitted for brevity
feature otv

otv site-vlan 10

interface Overlay0
 description Site A
 otv join-interface port-channel3
 otv control-group 239.12.12.12
 otv data-group 232.1.1.0/24
 otv extend-vlan 100-110
 no shutdown

otv site-identifier 0x1
```

注：如果要部署多归属环境，那么建议首先在每个站点都启用一个 OTV ED，验证了 OTV 功能特性之后，再启用第二个 OTV ED。采用这种分阶段配置方法有助于简化故障排查过程。

14.2　理解和验证 OTV 控制平面

OTV 不依赖传统 L2 交换机的数据包泛洪和数据平面 MAC 学习机制，而是通过 IS-IS 控制平面在站点之间交换 MAC 地址可达性信息，这样做的好处是可以消除未知单播地址的数据包泛洪问题（假设没有静默主机）。

OTV 尽可能地利用了 IS-IS 的现有功能，包括邻居关系的建立以及通过 LSP 和 PDU 交换可达性信息等。OTV ED 通过 IS-IS Hello 包发现对方，并在站点 VLAN 以及叠加网络上建立邻接关系，如图 14-2 所示。

IS-IS 使用 TLV（Type-Length-Value，类型—长度—值）对邻居之间的消息进行编码，因而非常灵活且拥有非常好的扩展性。通过长时间的大量功能增强，IS-IS 可以通过定义新的 TLV 来携带多种协议的可达性信息。OTV 使用被称为 *MAC-Reachability TLV*（*MAC 可达性TLV*）的 IS-IS TLV Type 147 来承载 MAC 地址的可达性，该 TLV 包含了 Topology-ID、VLAN-ID 和 MAC 地址，使 OTV ED 可以从其他 OTV ED 学习 MAC 地址并建立 MAC 路由表。

OTV 是一种叠加协议，意味着其操作过程依赖于底层传输协议及其提供的可达性。本章在讨论控制平面的时候，大家可以很明显地看出，在排查 OTV 故障时，网络运营商必须能够对不同的协议层进行分段并理解这些协议层之间的交互关系。OTV 控制平面包括 L2 交换、L3 路由、IP 多播和 IS-IS。在传输网络中执行故障排查操作时，必须将 OTV 控制平面包视为数据平面包，其中的源主机和目的主机实际上就是 OTV ED。此外，为了解决 OTV 的故障问题，还可能需要排查传输网络的控制平面协议。

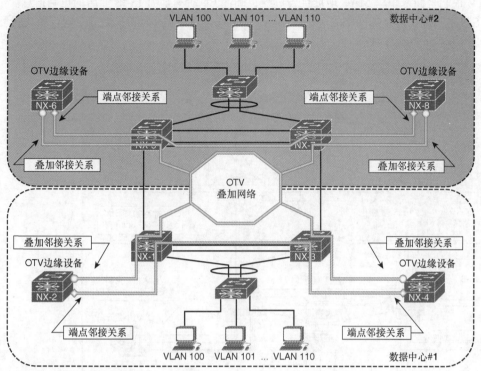

图 14-2　OTV IS-IS 邻接关系

14.2.1　OTV 多播模式

叠加接口上的 IS-IS 数据包通过 OTV IP 多播报头进行封装之后，由 OTV ED 发送给传输网络。为了清楚起见，下面将以单个 OTV ED NX-2 来解释上述过程，如图 14-3 所示。事实上，每个 OTV ED 都同时是 OTV 加入接口上的 OTV 控制组的源端和接收端。传输网络对这些数据包执行多播路由（这些数据包使用 OTV ED 加入接口的源地址和 OTV 控制组的组地址），通过传输网络沿多播树进行传送时按需进行流量复制，以确保加入 OTV 控制组的所有 OTV ED 都能收到该数据包的副本。数据包到达远程 OTV ED 之后，将删除外层 IP 多播报头封装，并将 IS-IS 数据包传递给 OTV 进行处理。

传输网络的多播能力使 OTV 能够建立 IS-IS 邻接关系，看起来就像每个 OTV ED 都连接在一个公共 LAN 网段上。也就是说，可以将控制组视为从一个 OTV ED 到所有其他 OTV ED 的逻辑多点连接。站点邻接关系是在站点 VLAN 上建立的，站点 VLAN 通过内部接口连接站点中的两个 OTV ED（采用 L2 通信机制）。

> **注**：NX-OS 从 Release 5.2(1)开始，在站点 VLAN 和叠加网络上建立双重邻接关系，在此之前，站点中的 OTV ED 仅建立站点邻接关系。

对于 OTV 使用的 IS-IS 协议来说，用户不需要做任何配置即可使用其基本功能。配置了 OTV 之后，将自动启用和配置 IS-IS。只要底层传输网络功能正常，且 OTV ED 之间配置的叠加参数兼容，那么就能建立邻接关系。

IS-IS 控制平面是 OTV 正常运行的基础，它提供了一种协议机制，可以发现本地和远程 OTV ED、建立邻接关系并在站点之间交换 MAC 地址可达性信息。IS-IS 控制平面负责学习 MAC 地址宣告，执行 SPF 计算，并根据计算结果安装 OTV MAC 路由表。排查 MAC 地址可达性故障时，需要通过 OTV 控制平面来跟踪 MAC 地址宣告过程，确保 ED 拥有所有 IS-IS 邻居的正确信息。如果主机到主机之间存在跨叠加网络的可达性问题，那么在排查数据平面故障之前，建议首先排查控制平面配置并验证其操作状态。

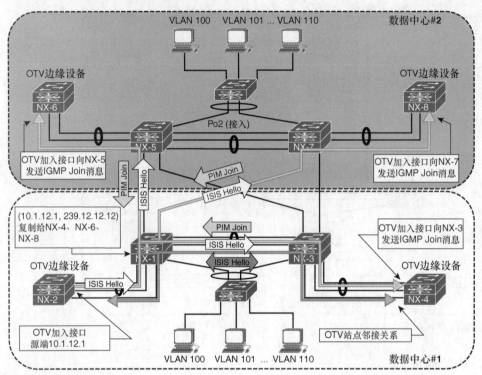

图 14-3　多播传输网络的 OTV 控制平面

14.2.2　OTV IS-IS 邻接关系验证

验证叠加接口状态是排查 OTV 邻接关系故障的第一步。例 14-4 显示了 **show otv overlay** [*overlay-identifier*]命令的输出结果，提供了排查 OTV 故障所需的大量关键信息。

例 14-4　验证叠加接口状态

```
NX-2# show otv overlay 0

show otv overlay 0

OTV Overlay Information
Site Identifier 0000.0000.0001
Encapsulation-Format ip - gre

Overlay interface Overlay0

 VPN name               : Overlay0
 VPN state              : UP
 Extended vlans         : 100-110 (Total:11)
 Control group          : 239.12.12.12
 Data group range(s)    : 232.1.1.0/24
 Broadcast group        : 239.12.12.12
 Join interface(s)      : Po3 (10.1.12.1)
```

```
  Site vlan          : 10 (up)
  AED-Capable        : Yes
  Capability         : Multicast-Reachable
```

从例 14-4 可以看出，Overlay0 接口处于运行状态，同时还显示了扩展的 VLAN 信息、OTV 控制组和数据组的传输网络多播组、加入接口、站点 VLAN 以及 AED 能力等信息。这些信息应该与本地和远程站点 OTV ED 上配置的叠加接口信息相匹配。

例 14-5 解释了在叠加接口上验证是否建立了正确 OTV IS-IS 邻接关系的过程。

例 14-5 叠加接口上的 OTV IS-IS 邻接关系

```
NX-2# show otv adjacency
Overlay Adjacency database

Overlay-Interface Overlay0 :

Hostname          System-ID        Dest Addr      Up Time     State
NX-4              64a0.e73e.12c2   10.1.22.1      03:51:57    UP
NX-8              64a0.e73e.12c4   10.2.43.1      03:05:24    UP
NX-6              6c9c.ed4d.d944   10.2.34.1      03:05:29    UP
```

例 14-6 显示的 **show otv site** 命令输出结果可以验证站点邻接关系。可以看出，NX-2 与 NX-4 的邻接关系处于 *Full*（完全邻接）状态，表示叠加邻接关系和站点邻接关系均正常（双重邻接）。

例 14-6 OTV IS-IS 站点邻接关系

```
NX-2# show otv site

Dual Adjacency State Description
  Full    - Both site and overlay adjacency up
  Partial - Either site/overlay adjacency down
  Down    - Both adjacencies are down (Neighbor is down/unreachable)
  (!)     - Site-ID mismatch detected

Local Edge Device Information:
  Hostname NX-2
  System-ID 6c9c.ed4d.d942
  Site-Identifier 0000.0000.0001
  Site-VLAN 10 State is Up

Site Information for Overlay0:
Local device is AED-Capable
Neighbor Edge Devices in Site: 1
Hostname           System-ID         Adjacency-      Adjacency-    AED-

                                     State           Uptime        Capable
--------------------------------------------------------------------------------
NX-4               64a0.e73e.12c2    Full            13:50:52      Yes
```

从例 14-5 和例 14-6 可以看出，站点邻接关系和叠加邻接关系的正常运行时间并不相同，因为它们是独立的 IS-IS 接口，而且邻接关系建立过程也完全独立。可以通过 **show otv internal adjacency** 命令找到 IS-IS 邻居的 site-id（见例 14-7），可以据此发现哪些 OTV ED 属于同一站点。

例 14-7 验证 OTV IS-IS 邻居的 site-id

```
NX-2# show otv internal adjacency
Overlay Adjacency database
```

```
Overlay-Interface Overlay0 :
System-ID      Dest Addr   Adj-State TM_State Adj-State inAS Site-ID
Version
64a0.e73e.12c2 10.1.22.1   default   default  UP        UP   0000.0000.0001*
HW-St: Default N backup (null)

64a0.e73e.12c4 10.2.43.1   default   default  UP        UP   0000.0000.0002*
HW-St: Default N backup (null)

6c9c.ed4d.d944 10.2.34.1   default   default  UP        UP   0000.0000.0002*
HW-St: Default N backup (null)
```

> **注**：OTV 提供了多种事件历史记录日志，对于故障排查来说非常有用。可以通过 **show otv isis internal event-history adjacency** 命令查看最近的邻接关系变更信息。

OTV 会为每个建立了邻接关系的 OTV ED 创建一条点对点隧道，通过这些隧道在 OTV ED 之间传输 OTV 单播包。可以通过 **show tunnel internal implicit otv brief** 命令显示 OTV ED 的可达隧道信息。例 14-8 显示了 NX-2 的相关输出结果。

例 14-8　OTV 动态单播隧道

```
NX-2# show tunnel internal implicit otv brief
--------------------------------------------------------------------------------
Interface        Status     IP Address    Encap type    MTU
--------------------------------------------------------------------------------
Tunnel16384      up         --            GRE/IP        9178
Tunnel16385      up         --            GRE/IP        9178
Tunnel16386      up         --            GRE/IP        9178
```

可以通过 **show tunnel internal implicit otv tunnel_num** [*number*] 命令显示指定隧道的详细信息。例 14-9 显示了隧道 16384 的详细输出结果，显示了 MTU 以及传输协议的源和目的地址等信息（允许将隧道映射到指定邻居）。如果特定 OTV ED 出现了问题，那么就可以验证该输出结果。

例 14-9　验证动态隧道的详细参数信息

```
NX-2# show tunnel internal implicit otv tunnel_num 16384
Tunnel16384 is up
  Admin State: up
  MTU 9178 bytes, BW 9 Kbit
  Tunnel protocol/transport GRE/IP
  Tunnel source 10.1.12.1, destination 10.2.43.1
  Transport protocol is in VRF "default"
  Rx
  0 packets input, 1 minute input rate 0 packets/sec
  Tx
  0 packets output, 1 minute output rate 0 packets/sec
  Last clearing of "show interface" counters never
```

建立了 OTV 邻接关系之后，可以使用散列函数为通过叠加网络进行扩展的每个 VLAN 确定 AED 角色。确定 AED 角色时，需要依据 OTV IS-IS System-ID、VLAN ID 以及次序值（ordinal value），拥有较低 System-ID 的设备将成为偶数 VLAN 的 AED，拥有较高 System-ID 的设备将成为奇数 VLAN 的 AED。

例 14-10 显示了 NX-2 的 **show otv vlan** 命令输出结果。*VLAN State*（VLAN 状态）列将当前

状态分为 *Active*（*活动*）或 *Inactive*（*非活动*）。状态 *Active* 表示该 OTV ED 是 VLAN 的 AED，负责通过叠加网络转发数据包并宣告 VLAN 的 MAC 地址可达性。排查故障时，可以利用该信息确定正在排查的是与特定 VLAN 相关的正确设备。

例 14-10 验证充当 AED 的 OTV ED

```
NX-2# show otv vlan

OTV Extended VLANs and Edge Device State Information (* - AED)

Legend:
(NA) - Non AED, (VD) - Vlan Disabled, (OD) - Overlay Down
(DH) - Delete Holddown, (HW) - HW: State Down
 (NFC) - Not Forward Capable

VLAN  Auth. Edge Device         Vlan State          Overlay
----  -----------------         ----------          -------
 100  NX-4                      inactive(NA)        Overlay0
 101* NX-2                      active              Overlay0
 102  NX-4                      inactive(NA)        Overlay0
 103* NX-2                      active              Overlay0
 104  NX-4                      inactive(NA)        Overlay0
 105* NX-2                      active              Overlay0
 106  NX-4                      inactive(NA)        Overlay0
 107* NX-2                      active              Overlay0
 108  NX-4                      inactive(NA)        Overlay0
 109* NX-2                      active              Overlay0
 110  NX-4                      inactive(NA)        Overlay0
```

邻接关系故障的常见原因有配置错误、传输网络中 OTV 控制组的数据包传递故障或者建立站点邻接关系的站点 VLAN 出现故障。

如果存在叠加邻接关系故障，那么就需要检查连接在 OTV ED 加入接口上的多播路由器的 IP 多播状态。每个 OTV ED 都应该有一条与控制组相对应的(S,G)多播路由。由于 OTV ED 发送了 IGMP Join（加入）消息，因而应该将多播路由器连接 OTV ED 的 L3 接口填充到(*, G)和 OTV 控制组的所有活动多播源的 OIL 中。

例 14-11 显示了 NX-1 的 **show ip mroute** [*group*]命令输出结果。可以看出，(*,239.12.12.12)表项通过 IGMP 将 Port-channel 3 填充到了 OIL 中。对于所有向 239.12.12.12 发送流量的活动多播源来说，也会用 Port-channel 3 来填充 OIL，从而允许 NX-2 接收 NX-4、NX-6 和 NX-8 的 IS-IS Hello 包和 LSP 数据包。每个(S,G)对的源地址都是其他 OTV ED 向多播组发送多播包的加入接口。

例 14-11 验证 OTV 控制组的多播路由

```
NX-1# show ip mroute 239.12.12.12
IP Multicast Routing Table for VRF "default"

(*, 239.12.12.12/32), uptime: 1w1d, pim ip igmp
 Incoming interface: loopback99, RPF nbr: 10.99.99.99
 Outgoing interface list: (count: 1)
  port-channel3, uptime: 16:17:45, igmp

(10.1.12.1/32, 239.12.12.12/32), uptime: 1w1d, ip mrib pim
 Incoming interface: port-channel3, RPF nbr: 10.1.12.1, internal
 Outgoing interface list: (count: 4)
  port-channel3, uptime: 16:17:45, mrib, (RPF)
```

```
    Vlan1101, uptime: 16:48:24, pim
    Ethernet3/17, uptime: 6d05h, pim
    Ethernet3/18, uptime: 1w1d, pim

(10.1.22.1/32, 239.12.12.12/32), uptime: 1w1d, pim mrib ip
  Incoming interface: Vlan1101, RPF nbr: 10.1.11.2, internal
  Outgoing interface list: (count: 1)
    port-channel3, uptime: 16:17:45, mrib

(10.2.34.1/32, 239.12.12.12/32), uptime: 1w1d, pim mrib ip
  Incoming interface: Ethernet3/18, RPF nbr: 10.1.13.3, internal
  Outgoing interface list: (count: 1)
    port-channel3, uptime: 16:17:45, mrib

(10.2.43.1/32, 239.12.12.12/32), uptime: 1w1d, pim mrib ip
  Incoming interface: Ethernet3/17, RPF nbr: 10.2.13.3, internal
  Outgoing interface list: (count: 1)
    port-channel3, uptime: 16:17:45, mrib
```

如果存在 IGMP 为多播组创建的(*,G)，那么就表示路由器至少收到了一条 IGMP Join 消息，而且接口上至少有一个感兴趣的接收端。PIM Join 消息由最后一跳路由器发送给 PIM RP，去往 PIM RP 的多播树应该存在(*,G)加入状态。如果最后一跳路由器（本例为 NX-1）在共享树上收到了多播组的数据包，那么就会向多播源发送 PIM (S,G) Join 消息，该消息的发送会触发建立有源树（去往连接多播源的第一跳路由器），只要有接收端对多播组感兴趣，那么有源树就会始终存在。

例 14-12 通过 **show ip mroute summary** 命令验证了多播流量的接收情况，显示了每个多播源的数据包计数器和比特率数值。

例 14-12 验证 OTV 控制组的当前比特率

```
NX-1# show ip mroute 239.12.12.12 summary
IP Multicast Routing Table for VRF "default"

Total number of routes: 6
Total number of (*,G) routes: 1
Total number of (S,G) routes: 4
Total number of (*,G-prefix) routes: 1
Group count: 1, rough average sources per group: 4.0

Group: 239.12.12.12/32, Source count: 4
Source        packets     bytes         aps pps    bit-rate    oifs
(*,G)         3           4326          1442 0     0.000 bps   1
10.1.12.1     927464      193003108     208 2      3.154 kbps  4
10.1.22.1     872869      173599251     198 3      3.844 kbps  1
10.2.34.1     1060046     203853603     192 3      3.261 kbps  1
10.2.43.1     1000183     203775760     203 3      3.466 kbps  1
```

由于叠加网络的 IS-IS 邻接关系故障通常都是由传输网络中的多播包传递问题引起的，因而理解每台路由器的多播状态的含义非常重要。此外，还必须理解每台传输路由器的多播角色（如 FHR[First-Hop Router，第一跳路由器]、PIM RP、转接路由器或 LHR[Last-Hop Router，最后一跳路由器]），以便为多播路由表状态提供上下文信息。本例中的 NX-1 是控制组的 PIM LHR、FHR 和 RP。

如果 NX-1 没有 OTV 控制组的多播状态，那么就表明 NX-1 尚未收到 NX-2 的 IGMP Join 消息。由于 NX-1 也是该多播组的 PIM RP，因而表明还未注册任何多播源。如果存在(*,G)但不存在(S,G)，那么就表明 NX-1 收到了 NX-2 的 IGMP Join 消息，但是尚未收到 NX-4、NX-6 或 NX-8 的多播数据流量，因而不会切换到有源树。此时，就可以将故障排查对象转移到多播源和第一跳路由器，直至找出多播故障根源。

> 注：有关多播故障排查的详细内容，请参见本书第 13 章。

站点邻接关系是通过站点 VLAN 建立的。为了建立 IS-IS 邻接关系，必须确保 OTV ED 的内部接口之间具有跨数据中心网络的连接。例 14-13 给出了 **show otv site** 命令的输出结果，表明站点邻接关系处于 DOWN 状态，这一点可以从状态 *Partial*（因为 NX-4 的叠加邻接关系处于 UP 状态）看出来。

例 14-13　OTV 邻接关系（Partial）

```
NX-2# show otv site

Dual Adjacency State Description
  Full    - Both site and overlay adjacency up
  Partial - Either site/overlay adjacency down
  Down    - Both adjacencies are down (Neighbor is down/unreachable)
  (!)     - Site-ID mismatch detected

Local Edge Device Information:
  Hostname NX-2
  System-ID 6c9c.ed4d.d942
  Site-Identifier 0000.0000.0001
  Site-VLAN 10 State is Up

Site Information for Overlay0:

Local device is AED-Capable
Neighbor Edge Devices in Site: 1

Hostname          System-ID       Adjacency-     Adjacency-    AED-
                                  State          Uptime        Capable
--------------------------------------------------------------------------------
NX-4              64a0.e73e.12c2  Partial (!)    00:12:32      Yes

NX-2# show otv adjacency
Overlay Adjacency database

Overlay-Interface Overlay0 :
Hostname              System-ID       Dest Addr    Up Time   State
NX-4                  64a0.e73e.12c2  10.1.22.1    00:01:57  UP
NX-8                  64a0.e73e.12c4  10.2.43.1    00:01:57  UP
NX-6                  6c9c.ed4d.d944  10.2.34.1    00:02:09  UP
```

从 **show otv isis site** 的输出结果可以看出，站点 VLAN 的邻接关系不存在了，如例 14-14 所示。

例 14-14　验证 OTV 站点邻接关系

```
NX-2# show otv isis site

OTV-ISIS site-information for: default

  BFD: Disabled

OTV-IS-IS site adjacency local database:
```

```
 SNPA            State Last Chg Hold      Fwd-state Site-ID       Version BFD
 64a0.e73e.12c2 LOST 00:01:52  00:03:34 DOWN      0000.0000.0001 3       Disabled

OTV-IS-IS Site Group Information (as in OTV SDB):

SystemID: 6c9c.ed4d.d942, Interface: site-vlan, VLAN Id: 10, Cib: Up VLAN: Up

Overlay   State Next IIH Int Multi
Overlay0  Up    00:00:01  3   20

Overlay   Active SG       Last CSNP            CSNP Int Next CSNP
Overlay0  239.12.12.12    ffff.ffff.ffff.ff-ff 2w1d     Inactive

Neighbor SystemID: 64a0.e73e.12c2
```

IS-IS 邻接关系处于 DOWN 状态表明未在站点 VLAN 上正确交换 IS-IS Hello 包（IIH 数据包）。可以通过命令 **show otv isis internal event-history iih** 显示 IIH 数据包的发送和接收记录。例 14-15 显示 NX-2 正在发送 IIH 数据包，但是没有通过站点 VLAN 收到任何数据包。

例 14-15　NX-2 OTV IS-IS IIH 数据包事件历史记录

```
NX-2# show otv isis internal event-history iih | inc site
03:51:17.663263 isis_otv default [13901]: [13906]: Send L1 LAN IIH over site-vlan
  len 1497 prio 6,dmac 0100.0cdf.dfdf
03:51:14.910759 isis_otv default [13901]: [13906]: Send L1 LAN IIH over site-vlan
  len 1497 prio 6,dmac 0100.0cdf.dfdf
03:51:11.940991 isis_otv default [13901]: [13906]: Send L1 LAN IIH over site-vlan
  len 1497 prio 6,dmac 0100.0cdf.dfdf
03:51:08.939666 isis_otv default [13901]: [13906]: Send L1 LAN IIH over site-vlan
  len 1497 prio 6,dmac 0100.0cdf.dfdf
03:51:06.353274 isis_otv default [13901]: [13906]: Send L1 LAN IIH over site-vlan
  len 1497 prio 6,dmac 0100.0cdf.dfdf
03:51:03.584122 isis_otv default [13901]: [13906]: Send L1 LAN IIH over site-vlan
  len 1497 prio 6,dmac 0100.0cdf.dfdf
```

从上述事件历史记录日志可以看出，NX-2 已经创建了 IIH 数据包且进程正在将这些 IIH 包向外发送给站点 VLAN。可以在 NX-4 上检查相同的事件历史记录，以验证 NX-4 是否收到了 IIH 数据包。从例 14-16 的 NX-4 输出结果可以看出，NX-4 正在发送 IIH 数据包，但是没有收到 NX-2 的数据包。

例 14-16　NX-4 OTV IS-IS IIH 数据包事件历史记录

```
NX-4# show otv isis internal event-history iih | inc site
03:51:19.013078 isis_otv default [24209]: [24210]: Send L1 LAN IIH over site-vlan
  len 1497 prio 6,dmac 0100.0cdf.dfdf
03:51:16.293081 isis_otv default [24209]: [24210]: Send L1 LAN IIH over site-vlan
  len 1497 prio 6,dmac 0100.0cdf.dfdf
03:51:13.723065 isis_otv default [24209]: [24210]: Send L1 LAN IIH over site-vlan
  len 1497 prio 6,dmac 0100.0cdf.dfdf
03:51:10.813105 isis_otv default [24209]: [24210]: Send L1 LAN IIH over site-vlan
  len 1497 prio 6,dmac 0100.0cdf.dfdf
03:51:07.843102 isis_otv default [24209]: [24210]: Send L1 LAN IIH over site-vlan
  len 1497 prio 6,dmac 0100.0cdf.dfdf
```

例 14-15 和例 14-16 的输出结果证实 NX-2 和 NX-4 都在向站点 VLAN 发送 IS-IS IIH 数据包，但是双方都没有从对端 OTV ED 收到数据包。此时，应该沿着 L2 数据中心基础设施的 VLAN 开展故障排查操作，以确定 VLAN 的配置是否正确以及是否在 NX-2 与 NX-4 之间建立了中继。本例在 NX-3 上发现了一个问题，其站点 VLAN（VLAN 10）没有通过 vPC 对等链路进行中继，从而导致对等链路出现 BA（Bridge Assurance，网桥保障）不一致问题，如例 14-17 所示。

例 14-17　验证站点 VLAN 生成树

```
NX-1# show spanning-tree vlan 10 detail

 VLAN0010 is executing the rstp compatible Spanning Tree protocol
 Bridge Identifier has priority 24576, sysid 10, address 0023.04ee.be01
 Configured hello time 2, max age 20, forward delay 15
 We are the root of the spanning tree
 Topology change flag not set, detected flag not set
 Number of topology changes 2 last change occurred 0:05:26 ago
     from port-channel2
 Times: hold 1, topology change 35, notification 2
     hello 2, max age 20, forward delay 15
 Timers: hello 0, topology change 0, notification 0
 Port 4096 (port-channel1, vPC Peer-link) of VLAN0010 is broken (Bridge Assurance
  Inconsistent, VPC Peer-link Inconsistent)
  Port path cost 1, Port priority 128, Port Identifier 128.4096
  Designated root has priority 32778, address 0023.04ee.be01
  Designated bridge has priority 0, address 6c9c.ed4d.d941
  Designated port id is 128.4096, designated path cost 0
  Timers: message age 0, forward delay 0, hold 0
  Number of transitions to forwarding state: 0
  The port type is network
  Link type is point-to-point by default
  BPDU: sent 1534, received 0
```

纠正了 vPC 对等链路的中继 VLAN 配置之后，站点 VLAN 上的 OTV 站点邻接关系将恢复正常，双重邻接关系状态也将回到 *FULL* 状态。可以通过 **show otv isis internal event-history adjacency** 命令查看邻接关系的切换过程，如例 14-18 所示。

例 14-18　OTV IS-IS 邻接关系事件历史记录

```
NX-2# show otv isis internal event-history adjacency
03:52:58.909967 isis_otv default [13901]:: LAN adj L1 64a0.e73e.12c2
over site-vlan - UP T 0
03:52:58.909785 isis_otv default [13901]:: LAN adj L1 64a0.e73e.12c2
over site-vlan - INIT (New) T -1
03:52:58.909776 isis_otv default [13901]:: isis_init_topo_adj LAN
adj 1 64a0.e73e.12c2 over site-vlan - LAN MT-0
```

排查邻接关系故障的第一步是确保两个邻居都正确生成和发送了 IS-IS Hello 包。如果确实如此，那么就可以逐步跟踪传输或底层网络，直至找到连接故障。

如果已经证实站点 VLAN 在整个数据中心都能正常工作，那么排查邻接关系故障的下一步就是执行抓包操作，以确定未正确转发数据帧的设备。第 2 章介绍了 NX-OS 平台提供的多种抓包工具的使用方式，可以利用这些工具隔离故障问题。需要注意的是，虽然 NX-2 和 NX-4 上的这些数据包就是 OTV IS-IS 控制平面数据包（穿越 L3 传输网络的时候），但需要将它们作为普通的数据平面数据包进行处理。

14.2.3　OTV IS-IS 拓扑表

在叠加网络和站点 VLAN 上建立了 IS-IS 邻接关系之后，IS-IS 会发送和接收包括 LSP 在内的 PDU（Protocol Data Unit，协议数据单元），以创建 OTV MAC 路由表。每个 OTV ED 都会泛洪自己的 LSP 数据库，以确保所有邻居都拥有一致的拓扑结构视图。交换了 LSP 之后，需要运行 SPF（Shortest Path First，最短路径优先）算法，构建以 MAC 地址为叶的拓扑结构，然后将路由表项安装到 OTV MAC 路由表中，以进行流量转发。

例 14-19 给出了 OTV IS-IS 数据库示例，显示了 NX-2 IS-IS 数据库中的 NX-4 的 LSP 信息。

例 14-19　OTV IS-IS 数据库

```
NX-2# show otv isis database
OTV-IS-IS Process: default LSP database VPN: Overlay0

OTV-IS-IS Level-1 Link State Database
 LSPID              Seq Number  Checksum  Lifetime  A/P/O/T
 64a0.e73e.12c2.00-00  0x0000069F  0x643C    1198      0/0/0/1
 64a0.e73e.12c4.00-00  0x00027EBC  0x13EA    1198      0/0/0/1
 6c9c.ed4d.d942.00-00* 0x00000619  0x463D    1196      0/0/0/1
 6c9c.ed4d.d942.01-00* 0x00000003  0x2278    0 (1198)  0/0/0/1
 6c9c.ed4d.d944.00-00  0x0002AA3A  0x209E    1197      0/0/0/1
 6c9c.ed4d.d944.01-00  0x0002790A  0xD43A    1199      0/0/0/1
```

由于 Lifetime（生存期）列的取值范围是 1200~0，因而例中的 LSP 生存期只有几秒。多次重复执行该命令之后可以看出 Seq Number（序列号）字段一直都在递增，表明始发 IS-IS 邻居正在使用变更信息更新 LSP。由于 SPF 算法一直都在不断执行，因而可能会导致 OTV MAC 路由表刷新并重新安装。虽然 LSP 的刷新和更新是 IS-IS 正常操作的一部分，但如果这种更新操作持续发生，那么对于稳定状态来说就是异常行为。

排查这类故障问题时，可以检查 LSP 的内容是否随时间变化。如果希望知道哪个 OTV ED 正在宣告哪条 LSP，那么就可以检查主机名到 System-ID 的映射关系。可以通过主机名 TLV 来动态学习邻居 System-ID 到主机名的映射关系。如果希望识别哪些 IS-IS 数据库表项属于哪些邻居，那么就可以使用 **show otv isis hostname** 命令，例 14-20 中的星号(*)表示本地 System-ID。

例 14-20　OTV IS-IS 动态主机名

```
NX-2# show otv isis hostname
OTV-IS-IS Process: default dynamic hostname table VPN: Overlay0
 Level System ID      Dynamic hostname
 1     64a0.e73e.12c2 NX-4
 1     64a0.e73e.12c4 NX-8
 1     6c9c.ed4d.d942* NX-2
 1     6c9c.ed4d.d944 NX-6
```

可以通过命令 **show otv isis database detail** [*lsp-id*]验证单条 LSP 的内容。例 14-21 显示了 NX-2 从 NX-4 收到的 LSP 以及其他一些重要信息，如邻居和 MAC 地址的可达性、site-id 以及特定 VLAN 的 AED 等。

例 14-21　OTV IS-IS 数据库详细信息

```
NX-2# show otv isis database detail 64a0.e73e.12c2.00-00
OTV-IS-IS Process: default LSP database VPN: Overlay0

OTV-IS-IS Level-1 Link State Database
 LSPID              Seq Number  Checksum  Lifetime  A/P/O/T
 64a0.e73e.12c2.00-00  0x000006BB  0xAFD6    1194      0/0/0/1
   Instance     : 0x000005D0
   Area Address : 00
   NLPID        : 0xCC 0x8E
   Hostname     : NX-4         Length : 4
```

```
  Extended IS   : 6c9c.ed4d.d944.01 Metric : 40
  Vlan          : 100 : Metric   : 0
   MAC Address  : 0000.0c07.ac64
  Vlan          : 102 : Metric   : 0
   MAC Address  : 0000.0c07.ac66
  Vlan          : 104 : Metric : 0
   MAC Address  : 0000.0c07.ac68
  Vlan          : 108 : Metric : 0
   MAC Address  : 0000.0c07.ac6c
  Vlan          : 110 : Metric : 1
   MAC Address  : 0000.0c07.ac6e
  Vlan          : 106 : Metric : 1
   MAC Address  : 0000.0c07.ac6a
  Vlan          : 110 : Metric : 1
   MAC Address  : 64a0.e73e.12c1
  Vlan          : 108 : Metric : 1
   MAC Address  : 64a0.e73e.12c1
  Vlan          : 100 : Metric : 1
   MAC Address  : 64a0.e73e.12c1
  Vlan          : 104 : Metric : 1
   MAC Address  : c464.135c.6600
   MAC Address  : 64a0.e73e.12c1
  Vlan          : 106 : Metric : 1
   MAC Address  : 64a0.e73e.12c1
  Vlan          : 102 : Metric : 1
   MAC Address  : 6c9c.ed4d.d941
   MAC Address  : 64a0.e73e.12c1
  Site ID       : 0000.0000.0001
  AED-Server-ID : 64a0.e73e.12c2
Version 57
  ED Summary    :    Device ID : 6c9c.ed4d.d942 : fwd_ready : 1
  ED Summary    :    Device ID : 64a0.e73e.12c2 : fwd_ready : 1
  Site ID       : 0000.0000.0001 : Partition ID : ffff.ffff.ffff
  Device ID : 64a0.e73e.12c2 Cluster-ID   : 0
  Vlan Status   :   AED : 0 Back-up AED : 1 Fwd ready : 1 Priority : 0 Delete   : 0
  Local         : 1 Remote   : 1 Range    : 1 Version   : 9
Start-vlan : 101 End-vlan   : 109 Step      : 2
  AED : 1 Back-up AED : 0 Fwd ready : 1 Priority : 0 Delete   : 0 Local    : 1
  Remote    : 1 Range   : 1 Version   : 9
Start-vlan : 100 End-vlan   : 110 Step      : 2
  Site ID    : 0000.0000.0001 : Partition ID   : ffff.ffff.ffff
  Device ID : 64a0.e73e.12c2 Cluster-ID   : 0
  AED SVR status :    Old-AED : 64a0.e73e.12c2 New-AED : 6c9c.ed4d.d942
  old-backup-aed : 0000.0000.0000 new-backup-aed   : 64a0.e73e.12c2
  Delete-flag    : 0 No-of-range    : 1 Version      : 9
Start-vlan : 101 End-vlan   : 109 Step      : 2
  Old-AED : 64a0.e73e.12c2 New-AED : 64a0.e73e.12c2
  old-backup-aed : 0000.0000.0000 new-backup-aed    : 6c9c.ed4d.d942
  Delete-flag    : 0 No-of-range    : 1 Version      : 9
Start-vlan : 100 End-vlan   : 110 Step      : 2
  Digest Offset : 0
```

如果希望确定LSP中正在变化的信息,那么就可以使用NX-OS diff实用程序,如例14-22所示。diff实用程序的输出结果表明序列号已经更新,且LSP生存时间被刷新为1198。LSP的内容变化与通过OTV扩展的多个VLAN中的HSRP MAC地址有关。

例 14-22 OTV IS-IS LSP 频繁更新

```
NX-2# show otv isis database detail 64a0.e73e.12c2.00-00 | diff
5,6c5,6
<  64a0.e73e.12c2.00-00 0x0001CD0E   0x0FF1  1196    0/0/0/1
<    Instance    : 0x0001CC23
---
>  64a0.e73e.12c2.00-00 0x0001CD11   0x193C  1198    0/0/0/1
>    Instance    : 0x0001CC26
10a11,12
>    Vlan        : 110 : Metric    : 0
>    MAC Address : 0000.0c07.ac6e
13,16d14
<    Vlan        : 108 : Metric    : 0
<    MAC Address : 0000.0c07.ac6c
<    Vlan        : 106 : Metric    : 0
<    MAC Address : 0000.0c07.ac6a
19,22c17,18
<    Vlan        : 110 : Metric    : 1
<    MAC Address : 0000.0c07.ac6e
<    Vlan        : 102 : Metric    : 1
<    MAC Address : 0000.0c07.ac66
---
>    Vlan        : 106 : Metric    : 1
>    MAC Address : 0000.0c07.ac6a
```

来自 LSP 的 MAC 地址可达性信息将被安装到 OTV MAC 路由表中，每个 MAC 地址都会安装一个已知下一跳（通过站点 VLAN 或通过叠加接口可达的 OTV ED）。从例 14-23 的 OTV MAC 路由表可以证实 MAC 地址表项不稳定且正在刷新，多个表项的 *Uptime*（*正常运行时间*）少于 1min，而且某些表项已被抑制（携带标志 D）。

例 14-23 OTV MAC 路由表不稳定

```
NX-2# show otv route | inc 00:00
! Output omitted for brevity
OTV Unicast MAC Routing Table For Overlay0

VLAN MAC-Address      Metric Uptime   Owner    Next-hop(s)
---- --------------   ------ -------- -------- -----------
 100 0000.0c07.ac64   41     00:00:18 overlay  NX-8 (D)
 101 0000.0c07.ac65   1      00:00:07 site     Ethernet3/5
 102 0000.0c07.ac66   41     00:00:12 overlay  NX-8 (D)
 103 0000.0c07.ac67   1      00:00:07 site     Ethernet3/5
 104 0000.0c07.ac68   41     00:00:12 overlay  NX-8
 105 0000.0c07.ac69   1      00:00:07 site     Ethernet3/5
 106 0000.0c07.ac6a   41     00:00:30 overlay  NX-8
 107 0000.0c07.ac6b   41     00:00:03 overlay  NX-6
 108 0000.0c07.ac6c   41     00:00:18 overlay  NX-8 (D)
 109 0000.0c07.ac6d   1      00:00:07 site     Ethernet3/5
 110 0000.0c07.ac6e   41     00:00:12 overlay  NX-8 (D)
```

此外，还可以通过跟踪 OTV 事件来获取其他有用信息。由于对从远程 OTV ED 收到的 IS-IS LSP 的变化情况感兴趣，因而可以使用命令 **show otv isis internal event-history spf-leaf** 查看 OTV 路由表的路由变化情况以及出现路由刷新的原因，如例 14-24 所示。

例 14-24 OTV IS-IS SPF 事件历史记录

```
NX-2# show otv isis internal event-history spf-leaf | egrep "Process 0103-0000.0c07.
  ac67"
20:12:48.699301 isis_otv default [13901]: [13911]: Process 0103-0000.0c07.ac67
contained in 6c9c.ed4d.d944.00-00 with metric 0
20:12:45.060622 isis_otv default [13901]: [13911]: Process 0103-0000.0c07.ac67
contained in 6c9c.ed4d.d944.00-00 with metric 0
20:12:32.909267 isis_otv default [13901]: [13911]: Process 0103-0000.0c07.ac67
contained in 6c9c.ed4d.d944.00-00 with metric 1
20:12:30.743478 isis_otv default [13901]: [13911]: Process 0103-0000.0c07.ac67
contained in 6c9c.ed4d.d944.00-00 with metric 1
20:12:28.652719 isis_otv default [13901]: [13911]: Process 0103-0000.0c07.ac67
contained in 6c9c.ed4d.d944.00-00 with metric 0
20:12:26.470400 isis_otv default [13901]: [13911]: Process 0103-0000.0c07.ac67
contained in 6c9c.ed4d.d944.00-00 with metric 0
20:12:25.978913 isis_otv default [13901]: [13911]: Process 0103-0000.0c07.ac67
contained in 6c9c.ed4d.d944.00-00 with metric 0
20:12:13.239379 isis_otv default [13901]: [13911]: Process 0103-0000.0c07.ac67
  contained in 6c9c.ed4d.d944.00-00 with metric 0
```

至此已经清楚地看出 LSP 所发生的变化以及生存期被不断重置为 1200 的原因，度量也从 0 更改为 1。

接下来需要对通过叠加网络发起 MAC 宣告的远程 AED 做进一步排查。本例的故障原因是配置错误，OTV 错误地在叠加网络上宣告了 HSRP MAC 地址，实际上应该通过 FHRP（First Hop Routing Protocol，第一跳路由协议）本地过滤器阻塞 HSRP MAC（如本章稍后所述），但却错误地通过叠加网络进行了宣告，从而导致上述不稳定问题。

前面的示例解释了从远程 OTV ED 收到 MAC 地址宣告出现的故障问题。如果故障根源在于 MAC 地址没有从本地 AED 宣告给其他 OTV ED，那么就需要首先验证 OTV 是否将 MAC 地址传递给了 IS-IS 进程以进行宣告。此时，可以通过 **show otv isis mac redistribute route** 命令验证 OTV 是否将 MAC 地址传递给了 IS-IS，进而宣告给其他 OTV ED，如例 14-25 所示。

例 14-25 将 MAC 地址重分发给 OTV IS-IS

```
NX-2# show otv isis mac redistribute route
OTV-IS-IS process: default VPN: Overlay0
OTV-IS-IS MAC redistribute route

0101-64a0.e73e.12c1, all
 Advertised into L1, metric 1 LSP-ID 6c9c.ed4d.d942.00-00
0101-6c9c.ed4d.d941, all
 Advertised into L1, metric 1 LSP-ID 6c9c.ed4d.d942.00-00
0101-c464.135c.6600, all
 Advertised into L1, metric 1 LSP-ID 6c9c.ed4d.d942.00-00
0103-64a0.e73e.12c1, all
 Advertised into L1, metric 1 LSP-ID 6c9c.ed4d.d942.00-00
0103-6c9c.ed4d.d941, all
 Advertised into L1, metric 1 LSP-ID 6c9c.ed4d.d942.00-00
0105-64a0.e73e.12c1, all
 Advertised into L1, metric 1 LSP-ID 6c9c.ed4d.d942.00-00
0105-6c9c.ed4d.d941, all
 Advertised into L1, metric 1 LSP-ID 6c9c.ed4d.d942.00-00
0107-64a0.e73e.12c1, all
```

```
  Advertised into L1, metric 1 LSP-ID 6c9c.ed4d.d942.00-00
0109-64a0.e73e.12c1, all
  Advertised into L1, metric 1 LSP-ID 6c9c.ed4d.d942.00-00
0109-6c9c.ed4d.d941, all
  Advertised into L1, metric 1 LSP-ID 6c9c.ed4d.d942.00-00
```

IS-IS LSP 的完整性对于 OTV 控制平面的可靠性和稳定性来说极为关键。数据包在传输过程中出现损坏或丢失会直接影响 OTV IS-IS 邻接关系的建立以及 LSP 的宣告。例 14-26 和例 14-27 分别显示了叠加接口和站点 VLAN 的 IS-IS 统计信息，这些信息对于排查邻接关系或 LSP 故障来说非常有用。

例 14-26　OTV IS-IS 叠加流量统计

```
NX-2# show otv isis traffic overlay0
OTV-IS-IS process: default
VPN: Overlay0
OTV-IS-IS Traffic for Overlay0:
PDU       Received    Sent    RcvAuthErr  OtherRcvErr  ReTransmit
LAN-IIH   112327      37520   525         11           n/a
CSNP      100939      16964   0           0            n/a
PSNP      71186       19862   0           0            n/a
LSP       817782      280896  0           0            0
```

例 14-27　OTV IS-IS 站点 VLAN 统计

```
NX-2# show otv isis site statistics

OTV-ISIS site-information for: default

OTV-IS-IS Broadcast Traffic statistics for site-vlan:

OTV-IS-IS PDU statistics for site-vlan:

PDU       Received    Sent    RcvAuthErr  OtherRcvErr  ReTransmit
LAN-IIH   290557      432344  0           1            n/a
CSNP      68605       34324   0           0            n/a
PSNP      1           1       0           0            n/a
LSP       7           122     0           0            0

OTV-IS-IS Global statistics for site-vlan:

  SPF calculations:  0
  LSPs sourced:      2
  LSPs refreshed:    13
  LSPs purged:       0
```

如果接收错误或重传次数始终都在增加，那么就表明 IS-IS PDU 出现了问题，这一点可能会导致 MAC 地址可达性问题。*RcvAuthErr* 字段递增表明 OTV ED 之间的认证机制不匹配。

14.2.4　OTV IS-IS 认证

某些网络场景可能希望使用 IS-IS 认证机制。通过叠加网络建立的 OTV 邻接关系支持 IS-IS 认证功能，只要在叠加接口上配置 IS-IS 认证机制即可。例 14-28 给出了在叠加接口上进行 IS-IS 认证的配置示例。

例 14-28 配置 OTV IS-IS 认证

```
NX-2# show running-config
! Output omitted for brevity
feature otv

otv site-vlan 10
key chain OTV-CHAIN
 key 0
   key-string 7 073c046f7c2c2d
interface Overlay0
 description Site A
 otv isis authentication-type md5
 otv isis authentication key-chain OTV-CHAIN
 otv join-interface port-channel3
 otv control-group 239.12.12.12
 otv data-group 232.1.1.0/24
 otv extend-vlan 100-110
 no shutdown
otv-isis default
otv site-identifier 0x1
```

可以通过 show otv isis interface overlay [*overlay-number*]命令验证 OTV IS-IS 认证启用情况，如例 14-29 所示。

例 14-29 OTV IS-IS 认证参数

```
NX-2# show otv isis interface overlay 0
OTV-IS-IS process: default VPN: Overlay0
Overlay0, Interface status: protocol-up/link-up/admin-up
 IP address: none
 IPv6 address: none
 IPv6 link-local address: none
 Index: 0x0001, Local Circuit ID: 0x01, Circuit Type: L1
Level1
 Adjacency server (local/remote) : disabled / none
 Adjacency server capability : multicast
 Authentication type is MD5
 Authentication keychain is OTV-CHAIN
 Authentication check specified
 LSP interval: 33 ms, MTU: 1400
 Level  Metric  CSNP  Next CSNP  Hello  Multi  Next IIH
 1      40      10    Inactive   20     3      00:00:15

 Level Adjs  AdjsUp Pri Circuit ID         Since
 1     0     0      64  6c9c.ed4d.d942.01  23:40:21
```

为了建立叠加邻接关系，所有 OTV 站点都必须配置相同的认证命令。从例 14-30 可以看出，LAN-IIH 帧的 *RcvAuthErr* 不断递增，表明认证机制不匹配。

例 14-30 OTV IS-IS 认证差错统计信息

```
NX-2# show otv isis traffic overlay 0
OTV-IS-IS process: default
VPN: Overlay0
OTV-IS-IS Traffic for Overlay0:
```

```
PDU        Received    Sent    RcvAuthErr  OtherRcvErr  ReTransmit
LAN-IIH    111899      37370   260         11           n/a
CSNP       100792      16937   0           0            n/a
PSNP       71058       19832   0           0            n/a
LSP        816541      280383  0           0            0
```

命令 **show otv adjacency** 和 **show otv site** 的输出结果因不同邻接关系处于关闭状态而有所不同。由于认证配置仅应用于叠加接口，因而即使站点中的某个 OTV ED 为叠加接口配置了错误的认证机制，站点邻接关系也仍有可能处于正常状态。

例 14-31 显示叠加邻接关系处于 DOWN 状态，而站点邻接关系却处于有效状态，本例中的状态显示为 *Partial*。

例 14-31　OTV 叠加 IS-IS 邻接关系中断

```
NX-2# show otv adjacency
Overlay Adjacency database

NX-2# show otv site

Dual Adjacency State Description
  Full    - Both site and overlay adjacency up
  Partial - Either site/overlay adjacency down
  Down    - Both adjacencies are down (Neighbor is down/unreachable)
  (!)     - Site-ID mismatch detected

Local Edge Device Information:
  Hostname NX-2
  System-ID 6c9c.ed4d.d942
  Site-Identifier 0000.0000.0001
  Site-VLAN 10 State is Up

Site Information for Overlay0:

Local device is not AED-Capable (No Overlay Remote Adjacency up)
Neighbor Edge Devices in Site: 1

Hostname         System-ID         Adjacency-    Adjacency-   AED-
                                   State         Uptime       Capable
-----------------------------------------------------------------------
(null)           64a0.e73e.12c2    Partial       1w0d         Yes
```

14.2.5　邻接服务器模式

从 NX-OS Release 5.2(1) 开始，邻接服务器模式允许 OTV 通过单播传输方式运行。由于不使用支持多播的传输网络，因而邻接服务器模式下的 OTV ED 必须将 IS-IS 消息复制给每个邻居。这样做的效率较低，因为每个 OTV ED 都必须执行额外的数据包复制操作，而且要为每个远程 OTV ED 传输更新消息。

多播传输允许 ED 仅生成单个多播包，然后由传输网络进行复制，效率较高，因而应尽可能地选择多播部署模式。但是，如果只有两个站点或者无法在传输网络中启用多播机制，那么也可以使用邻接服务器模式，通过 IP 单播方式实现全功能的 OTV 部署。

采用邻接服务器部署模式时，需要为每个 ED 都部署 OTV 叠加配置以使用邻接服务器的单播 IP 地址，例 14-32 显示了部署模式。邻接服务器的角色由用户指定的 OTV ED 承担。所有的 OTV ED 都要发送 OTV IS-IS Hello 包（通过 OTV 加入接口以 OTV 封装的 IP 单播包方式进行发送），将自己注册到邻接服务器上。邻接服务器与远程 OTV ED 建立邻接关系之后，将创建动态 OTV ED 列表。邻接服务器获得已知 ED 列表之后，将宣告给所有邻居。此后，所有 ED 都可以通过某种机制来动态学习其他 OTV ED，从而创建更新消息并复制给每个远程 ED。

例 14-32　NX-4 的 OTV ED 邻接服务器部署模式

```
NX-4# show run otv
! Output omitted for brevity
otv site-vlan 10

interface Overlay0
 otv join-interface port-channel3
 otv extend-vlan 100-110
 otv use-adjacency-server 10.1.12.1 unicast-only
 no shutdown
otv site-identifier 0x1
```

例 14-33 显示了 NX-2 的配置信息，此时的 NX-2 充当邻接服务器。在邻接服务器模式下配置 OTV ED 时，必须删除前面示例中为每个 OTV ED 配置的 **otv control-group** [*multicast group*] 和 **otv data-group** [*multicast-group*]命令，然后配置 **otv use-adjacency-server** [*IP address*]命令以启用 OTV 邻接服务器模式，并通过 **otv adjacency-server unicast-only** 命令指定 NX-2 成为邻接服务器。加入接口和内部接口的配置与本章前面的示例一样。

例 14-33　NX-2 的 OTV 邻接服务器配置

```
NX-2# show run otv
! Output omitted for brevity
otv site-vlan 10

interface port-channel3
  description 7009A-Main-OTV Join
  mtu 9216
  ip address 10.1.12.1/24
  ip router ospf 1 area 0.0.0.0
  ip igmp version 3
interface Overlay0
  description Site A
 otv join-interface port-channel3
 otv extend-vlan 100-110
 otv use-adjacency-server 10.1.12.1 unicast-only
 otv adjacency-server unicast-only
 no shutdown
otv site-identifier 0x1
```

动态宣告已知 OTV ED 列表，可以让用户不必为每个 OTV ED 配置所有其他 OTV ED 地址以建立邻接关系。图 14-4 显示了邻接服务器注册进程以及 OTV 邻居列表宣告进程。请注意，此时仍然存在站点邻接关系，只是为了清楚起见而未在图中标示。

oNL（OTV Neighbor List，OTV 邻居列表）生成之后，邻接服务器就会宣告给所有 OTV ED，如图 14-5 所示。

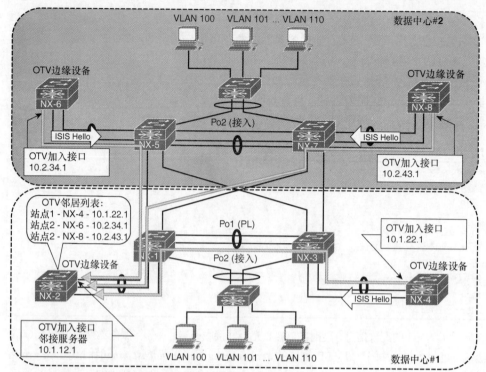

图 14-4 OTV ED 向邻接服务器注册

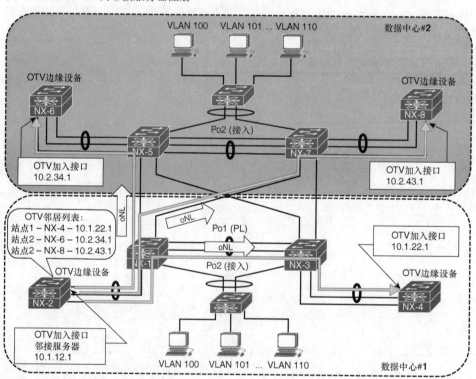

图 14-5 OTV 邻接服务器宣告邻居列表

此后每个 OTV ED 都要与其他所有 OTV ED 建立 IS-IS 邻接关系。OTV ED 通过 IP 单播包以 OTV 封装方式发送更新消息，每个 OTV ED 都必须将消息复制给其他所有邻居，详细步骤如图 14-6 所示。

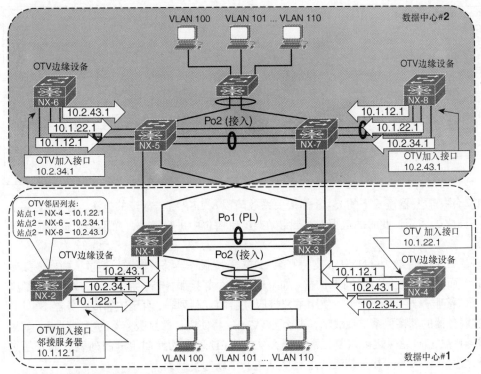

图 14-6 邻接服务器模式下的 OTV IS-IS Hello 包

例 14-34 显示了 NX-4 的 **show otv adjacency** 命令输出结果，可以看出，收到邻接服务器的 OTV 邻居列表之后，NX-4 就可以与其他所有 OTV ED 建立 IS-IS 邻接关系。

例 14-34 OTV 邻接服务器模式下的 IS-IS 邻居

```
NX-4# show otv adjacency
Overlay Adjacency database

Overlay-Interface Overlay0 :
Hostname            System-ID       Dest Addr    Up Time    State
NX-8                64a0.e73e.12c4  10.2.43.1    00:20:35   UP
NX-2                6c9c.ed4d.d942  10.1.12.1    00:20:35   UP
NX-6                6c9c.ed4d.d944  10.2.34.1    00:20:35   UP
```

从例 14-35 的 **show otv site** 命令输出结果可以看出，NX-4 也通过站点 VLAN 建立了 OTV IS-IS 站点邻接关系。

例 14-35 OTV 邻接服务器模式下的双重邻接关系

```
NX-4# show otv site

Dual Adjacency State Description
  Full    - Both site and overlay adjacency up
  Partial - Either site/overlay adjacency down
  Down    - Both adjacencies are down (Neighbor is down/unreachable)
  (!)     - Site-ID mismatch detected

Local Edge Device Information:
  Hostname NX-4
  System-ID 64a0.e73e.12c2
  Site-Identifier 0000.0000.0001
```

```
Site-VLAN 10 State is Up
Site Information for Overlay0:

Local device is AED-Capable
Neighbor Edge Devices in Site: 1

Hostname           System-ID      Adjacency-        Adjacency-      AED-
                                  State             Uptime          Capable
-----------------------------------------------------------------------------
NX-2               6c9c.ed4d.d942  Full             00:42:04        Yes
```

　　排查 OTV 邻接服务器模式下的 IS-IS 邻接关系故障以及 LSP 宣告故障的方法，与 OTV 多播模式相似。不同之处在于，此时的数据包被封装在 IP 单播包中进行发送，而不是通过传输网络进行多播传输。

　　为了实现更好的弹性能力，NX-OS 支持冗余 OTV 邻接服务器部署模式。不过，两台邻接服务器独立运行，而且彼此之间并不同步状态。如果存在多个邻接服务器，那么所有 OTV ED 都必须注册到每个邻接服务器上。OTV ED 使用主邻接服务器的复制列表，直至主邻接服务器不可用。如果与主邻接服务器的邻接关系出现故障，那么 OTV ED 将使用从邻接服务器的复制列表。如果主邻接服务器在 10min 超时之前恢复正常，那么 OTV ED 还会切换回主复制列表。如果超过了 10min，那么主邻接服务器在重新激活之后将推送一个新的复制列表。

14.2.6　OTV CoPP

　　与发送给管理引擎的其他数据包一样，为了保护交换机的有限资源，需要对 OTV 控制平面数据包实施速率限制。过多的 ARP 流量或 OTV 控制平面流量可能会影响交换机的稳定性，从而导致 CPU 或协议邻接关系故障频发，因此建议用户部署 CoPP 保护策略。

　　启用了 OTV ARP-ND-Cache 之后，就可以发现 CoPP 的重要性。OTV AED 需要侦听 ARP 应答消息并添加到本地缓存中，从而能够代表目的主机应答 ARP 请求。这些数据包必须由控制平面进行处理，如果 ARP 流量过多，那么就可能会导致策略丢包或高 CPU 利用率。有关 OTV ARP-ND-Cache 的详细内容将在本章后面进行讨论。

　　默认 VDC 的 **show policy-map interface control-plane** 命令可以提供每个控制平面流量类别的统计信息。如果存在 CoPP 丢包以及 ARP 解析失败，那么常见解决方案不是调整控制平面策略以允许更多流量，而是应该跟踪过量 ARP 流量的来源。此时，可以利用 Ethanalyzer 以及 OTV 事件历史记录等工具。

14.3　理解和验证 OTV 数据平面

　　OTV 旨在以高效、可靠的方式在站点之间传输 L2 帧。到达 OTV ED 的帧可以是单播、多播或广播帧，将这些帧封装之后，通过 OTV 控制平面提供的信息传输到目的 OTV ED。

　　OTV 的默认叠加封装方式是 GRE（见图 14-7），也称为 OTV 1.0 封装。

　　以太网帧到达内部接口之后，将执行一系列查找操作，来确定如何重写数据包以通过叠加接口进行传送。首先将原始净荷、以太类型、源 MAC 地址和目的 MAC 地址等复制到新的 OTV 封装帧中。然后删除 802.1Q 报头、插入 OTV SHIM 报头（填充报头），SHIM 报头包含了 VLAN ID 及所属叠加 ID 等信息。实际上，OTV 1.0 封装格式中的该字段就是 MPLS-in-GRE 封装，可以通过 MPLS 标签衍生出 VLAN ID，MPLS 标签等于 32+VLAN ID。对于本例来说，VLAN 101 被封装为 MPLS 标签 133。最后，还要增加外层 IP 报头，其中包含了本地 OTV ED 的源 IP 地址和远程 OTV ED 的目的 IP 地址。

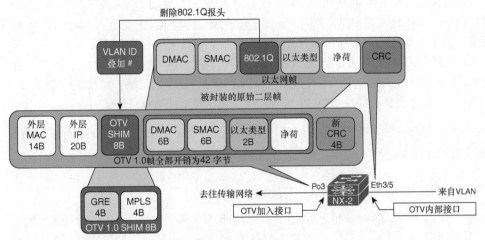

图 14-7　OTV 1.0 封装

控制平面 IS-IS 帧在跨叠加网络的 OTV ED 之间采用类似的封装方式，也要增加 42 字节 OTV 开销。用于 IS-IS 控制平面帧的 MPLS 标签是保留的标签 1，即 *Router Alert*（*路由器告警*）标签。

注：如果在传输网络中执行抓包操作，那么分析工具（如 Wireshark）就会将 OTV 1.0 封装解码为不带控制字的 MPLS 伪线。但不幸的是，截至本书写作之时，Wireshark 还无法解码 OTV 使用的所有 IS-IS PDU。

如果 Nexus 7000 系列交换机使用了 F3 或 M3 系列模块，那么 NX-OS Release 7.2(0)D1(1) 将提供不同的 OTV UDP 封装选项。图 14-8 列出了 OTV 2.5 UDP 封装格式。

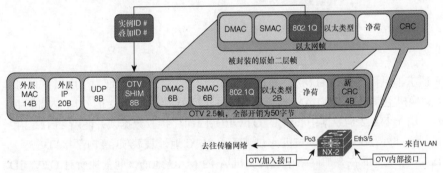

图 14-8　OTV 2.5 封装

以太网帧到达 OTV 内部接口之后，首先将原始净荷、以太类型、802.1Q 报头、源 MAC 地址和目的 MAC 地址复制到新的 OTV 2.5 封装帧中。OTV 2.5 封装使用的数据包格式与 VxLAN（Virtual Extensible LAN，虚拟可扩展 LAN）相同（详见 RFC 7348）。

OTV SHIM 报头包含了实例 ID（Instance #）和叠加 ID（Overlay #），其中，实例 ID 是一种表 ID，目的 OTV ED 可以利用实例 ID 来查找目的端；叠加 ID 则被控制平面数据包用来标识属于特定叠加网络的数据包。控制平面数据包将 VNI（VxLAN Network ID，VxLAN 网络 ID）比特设置为 False（0），而封装后的数据帧则将该比特设置为 True（1）。UDP 报头包含一个可变的源端口和目的端口 8472。

如果传输 MTU 小于 1550 字节（对于 OTV 2.5 封装）或 1542 字节（对于 OTV 1.0 封装），那么包含数据包的 OTV 帧的分段就可能有问题。出现问题的假设条件是，数据中心主机的接口 MTU 为 1500 字节，且试图发送完整 MTU 大小的帧。如此一来，增加了 OTV 封装报头之后，数据包将无法适应传输网络的可用 MTU 大小。

如果采用的是多播传输方式，那么控制平面数据包的最小传输 MTU 大小为 1442 字节；如果采用的是邻接服务器模式下的单播传输方式，那么控制平面数据包的最小传输 MTU 大小为 1450 字节。为了避免 OTV 控制平面或数据平面数据包在传输过程中被分段，OTV 会在外层 IP 报头中设置 DF（Don't Fragment，不分段）比特。如果存在 MTU 限制问题，那么在封装帧大小超过传输 MTU 之后，就可能会导致 OTV IS-IS 邻接关系无法建立或数据流量帧丢失。

注：所有站点都必须采用相同的 OTV 封装格式（GRE 或 UDP），可以通过全局配置命令 **otv encapsulation-format ip** [*gre* | *udp*] 进行配置。

14.3.1 OTV ARP 解析与 ARP-ND-Cache

主机与同一 IP 子网中的其他主机进行通信时，源主机首先需要利用 ARP 解析目的主机的 MAC 地址。图 14-9 显示了主机 A 与主机 C 之间交换的 ARP 消息，其中，主机 A 和主机 C 都是子网 10.101.0.0/16 的一部分。

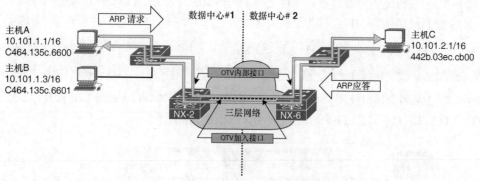

图 14-9　ARP 请求和应答

主机 A 将 ARP 请求消息广播给目的 MAC 地址 ff:ff:ff:ff:ff:ff，目地 IP 地址为 10.101.2.1。此后，该帧将通过 L2 交换机上属于同一 VLAN 的所有端口向外发送，包括 NX-2 的 OTV 内部接口和连接主机 B 的端口。由于 NX-2 是数据中心#1 的 OTV ED，因而将接收数据帧并使用 23.12.12.12 的 OTV 控制组进行封装。此外，NX-2 还会为主机 A 创建一条 MAC 地址表表项（经内部接口可达）。此后，NX-2 通过 IS-IS 控制平面在叠加网络中宣告主机 A 的 MAC 地址，向其他所有 OTV ED 提供可达性信息。

NX-2 发送的控制组多播帧将穿越底层传输网络，直至到达 NX-6。NX-6 将删除多播 OTV 封装，并将解封装后的数据帧从 OTV 内部接口发送给主机 C。主机 C 处理该广播帧，并发现目的 IP 地址是自己的 IP 地址。因此，主机 C 向主机 A 发送 ARP 应答消息，首先发送给 NX-6。由于 NX-6 收到了 IS-IS 更新消息，因而此时 NX-6 的 OTV MAC 路由表中已经拥有主机 A 的表项（IP 下一跳为 NX-2）。此外，VLAN 101 中还有一条指向叠加接口的主机 A 的 MAC 地址表表项。

NX-6 收到主机 C 的 ARP 应答消息之后，将创建一条指向 OTV 内部接口的本地 MAC 地址表表项，并通过 IS-IS 将该 MAC 地址表项宣告给所有远程 OTV ED（与 NX-2 对主机 A 的处理操作相同）。

此后，NX-6 封装该 ARP 应答消息，并通过叠加接口发送给数据中心#1 的 NX-2。NX-2 收到数据帧之后，将删除 OTV 封装，并根据 VLAN 的 MAC 地址表通过内部接口发送给主机 A。

OTV ARP-ND-Cache 是通过监听 ARP 应答消息来填充的。初始 ARP 请求通过 OTV 控制组发送给所有 OTV ED，等到 ARP 应答消息使用 OTV 控制组返回时，每个 OTV ED 都会侦听到该应答消息并在缓存中建立一个表项。如果主机 B 向主机 C 发送 ARP 请求，那么 NX-2 就会使用此前创建的缓存表项代表主机 C 答复该 ARP 请求，从而减少叠加网络上的不必要流量。

> 注：如果站点存在多个 OTV ED，那么只有 AED 会将数据包转发到叠加网络上，包括 ARP 请求和应答消息。此外，AED 还负责通过 IS-IS 控制平面向其他 OTV ED 宣告 MAC 地址可达性。

对于多播模式或邻接服务器模式来说，ARP-ND-Cache 的填充方式都相同。在邻接服务器模式下，ARP 请求和应答均被封装为 OTV 单播数据包并复制给远程 OTV ED。

如果主机无法通过叠加接口与其他主机通信，那么就需要验证 ARP-ND-Cache，以确保缓存没有包含任何过时信息。例 14-36 解释了检查 NX-2 本地 ARP-ND-Cache 的方式。

例 14-36　验证 ARP-ND-Cache

```
NX-2# show otv arp-nd-cache
OTV ARP/ND L3->L2 Address Mapping Cache

Overlay Interface Overlay0
VLAN MAC Address       Layer-3 Address   Age         Expires In
101  442b.03ec.cb00    10.101.2.1        00:02:29    00:06:07
```

此外，OTV 还会为 ARP-ND-Cache 行为保存事件历史记录，可以通过命令 **show otv internal event-history arp-nd** 查看事件历史记录。例 14-37 显示了 AED 的 VLAN 100 输出信息。

例 14-37　ARP-ND-Cache 事件历史记录

```
NX-4# show otv internal event-history arp-nd
ARP-ND events for OTV Process
02:33:17.816397 otv [9790]: [9810]: Updating arp nd cache entry in PSS TLVU.
  Overlay:249 Mac Info: 0100-442b.03ec.cb00 L3 addr: 10.100.2.1
02:33:17.816388 otv [9790]: [9810]: Caching 10.100.2.1 -> 0100-442b.03ec.cb00 ARP
  mapping
02:33:17.816345 otv [9790]: [9810]: Caching ARP Response from overlay : Overlay0
02:33:17.816337 otv [9790]: [9810]: IPv4 ARP Response packet received from source
  10.100.2.1 on interface Overlay0
02:33:17.806853 otv [9790]: [9810]: IPv4 ARP Request packet received from source
  10.100.1.1 on interface Ethernet3/5
```

OTV 的 ARP-ND-Cache 定时器可以配置为 60s～86400s，默认值为 480s 或 8min，再加上 2min 的宽限期。宽限期内，AED 会通过叠加接口转发 ARP 请求，以便应答消息刷新缓存中的表项。建议将 ARP-ND-Cache 定时器的时间值配置为小于 MAC 老化定时器，MAC 老化定时器默认值为 30min。

可以在叠加接口下通过命令 **no otv suppress-arp-nd** 禁用 OTV ARP-ND-Cache，其结果就是所有 ARP 请求都将通过叠加网络进行转发，且不会缓存任何 ARP 应答消息。

> 注：系统默认启用 ARP-ND-Cache。如果网络环境存在大量 ARP 交互行为，那么很可能会导致 OTV ED 的 CPU 利用率增高或 CoPP 丢包增多，因为管理引擎的 CPU 必须处理大量 ARP 流量以创建缓存表项。

14.3.2　广播

OTV ED 在内部接口上收到广播帧之后，由 AED 通过叠加网络转发给扩展 VLAN。首先需要将广播帧（如 ARP 请求）封装到 L3 多播包中，源地址是本地 OTV ED 的加入接口，组地址是 OTV 控制组地址。将多播包发送到传输网络上之后，由传输网络复制给所有已加入控制组的远程 OTV ED。

如果传输网络已启用多播机制，那么 OTV 就可以配置专用的 OTV 广播组（broadcast-group），如例 14-38 所示，这使操作人员可以将 OTV 控制组与广播组分离，进而简化故障排查操作，并根据组地址对数据包实施不同的处理操作。例如，为每个组定义不同的 PIM RP 点，或者为传输网络中的控制组和广播组应用不同的 QoS（Quality of Service，服务质量）策略。

例 14-38　专用 OTV 广播组

```
NX-2# show run otv
! Output omitted for brevity
interface Overlay0
 description Site A
 otv join-interface port-channel3
 otv broadcast-group 239.1.1.1
 otv control-group 239.12.12.12
 otv data-group 232.1.1.0/24
 otv extend-vlan 100-110
 no shutdown
```

如果传输网络未启用多播机制，OTV ED 运行在邻接服务器模式下，那么 OTV ED 将利用 OTV 单播包封装广播包，并通过头端复制方式将单播包副本复制给每个远程 OTV ED。

无论是多播还是单播传输，远程 OTV ED 收到数据包之后，都要删除外层 L3 数据包封装，然后由 AED 将广播帧转发给 VLAN 中所有面向内部的 L2 端口。

14.3.3　未知单播帧

OTV 的默认行为是仅将帧泛洪到内部接口上的未知单播 MAC 地址，不会跨叠加网络转发这些数据包。能够实现这种默认优化处理机制的原因是，OTV 假设不存在静默主机，且 OTV ED 最终能在内部接口看到所有主机流量。收到这些流量之后，会在 VLAN 中填充 MAC 地址表，并通过 IS-IS 将 MAC 地址宣告给所有 OTV ED。

但是，静默主机在某些情况下是无法避免的。为了保证这些主机的正常运行，OTV 提供了一个配置选项，允许从 NX-OS 6.2(2) 开始进行选择性单播洪泛。例 14-39 给出了一个配置示例，允许通过叠加网络将数据包泛洪到 VLAN 101 中的特定目的 MAC 地址。

例 14-39　选择性单播泛洪

```
NX-2# show run otv
! Output omitted for brevity

feature otv
otv site-identifier 0x1
otv flood mac C464.135C.6600 vlan 101
```

增加了该配置命令之后，就可以为 VLAN 创建一条静态 OTV 路由表项，使流量能够通过叠加网络进行传输，如例 14-40 所示。

例 14-40　配置了选择性单播泛洪功能的 OTV 路由表

```
NX-2# show otv route vlan 101

OTV Unicast MAC Routing Table For Overlay0
VLAN MAC-Address    Metric Uptime  Owner    Next-hop(s)
```

```
101  c464.135c.6600 0      00:02:38 static      Overlay0
```

14.3.4 基于多播传输网络的 OTV 单播流量

从图 14-9 可以看出，主机到主机的通信过程首先需要发送针对目的地址的 ARP 请求。完成了 ARP 请求和应答交互过程之后，所有站点的 OTV ED 都将拥有这两台主机的正确 OTV MAC 路由表和 MAC 地址表。

图 14-10 描述了 VLAN 103 中的数据中心#1 的主机 A 与数据中心#2 的主机 C 之间的通信过程。

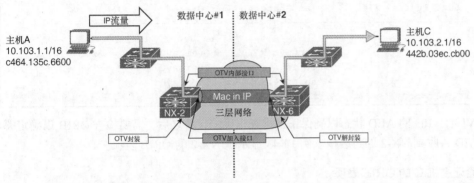

图 14-10 跨 OTV 的主机到主机的单播流量

主机 A 首先将流量发送给 L2 交换机，L2 交换机为数据包增加 VLAN 103 的 802.1Q VLAN 标签之后，按照 L2 交换机上的 MAC 地址表表项，通过中继端口将这些数据帧发送给 NX-2，到达 OTV 内部接口 Ethernet3/5。NX-2 收到数据包之后，在 VLAN 中执行 MAC 地址表查找操作，以确定如何到达主机 C 的 MAC 地址 442b.03ec.cb00。例 14-41 显示了 NX-2 的 MAC 地址表信息。

例 14-41 主机 C 的 MAC 地址表

```
NX-2# show mac address-table dynamic vlan 103
Note: MAC table entries displayed are getting read from software.
Use the 'hardware-age' keyword to get information related to 'Age'

Legend:
    * - primary entry, G - Gateway MAC, (R) - Routed MAC, O - Overlay MAC
    age - seconds since last seen,+ - primary entry using vPC Peer-Link, E - EVPN
       entry
    (T) - True, (F) - False , ~~~ - use 'hardware-age' keyword to retrieve age info
   VLAN/BD   MAC Address      Type       age     Secure NTFY Ports/SWID.SSID.LID
---------+-----------------+--------+---------+------+----+------------------
* 103      0000.0c07.ac67   dynamic    ~~~         F    F  Eth3/5
O 103      442b.03ec.cb00   dynamic    -           F    F  Overlay0
* 103      64a0.e73e.12c1   dynamic    ~~~         F    F  Eth3/5
O 103      64a0.e73e.12c3   dynamic    -           F    F  Overlay0
O 103      6c9c.ed4d.d943   dynamic    -           F    F  Overlay0
* 103      c464.135c.6600   dynamic    ~~~         F    F  Eth3/5
```

从 MAC 地址表可以看出，可以通过叠加接口到达主机 C 的 MAC 地址，意味着应该使用 OTV MAC 路由表（ORIB）来获取 IP 下一跳和详细封装信息。从 ORIB 表项可以看出如何到达通过 IS-IS 宣告去往 NX-2 的 MAC 地址的远程 OTV ED（本例中的 NX-6）。

注： 如果站点存在多个 OTV ED，那么必须按照数据路径去往 VLAN 的 AED，可以通过命令 **show otv vlan** 加以验证。正常情况下，跨 L2 网络的 MAC 转发表项应该指向 AED 的内部接口。

从例 14-42 可以看出，NX-2 是 VLAN103 的 AED。

例 14-42　验证 VLAN 103 的 AED

```
NX-2# show otv vlan
OTV Extended VLANs and Edge Device State Information (* - AED)

Legend:
(NA) - Non AED, (VD) - Vlan Disabled, (OD) - Overlay Down
(DH) - Delete Holddown, (HW) - HW: State Down
 (NFC) - Not Forward Capable
VLAN   Auth. Edge Device                 Vlan State              Overlay
----   -------------------------------   ---------------------   -------

 100   NX-4                              inactive(NA)            Overlay0
 101*  NX-2                              active                  Overlay0
 102   NX-4                              inactive(NA)            Overlay0
 103*  NX-2                              active                  Overlay0
```

验证了 VLAN 103 的 AED 状态以确保正在排查正确的设备之后，需要检查 ORIB 以确定哪个远程 OTV ED 将收到 NX-2 的封装帧。例 14-43 显示了 NX-2 的 ORIB 表项。

例 14-43　验证主机 C 的 ORIB 表项

```
NX-2# show otv route vlan 103

OTV Unicast MAC Routing Table For Overlay0

VLAN MAC-Address     Metric Uptime   Owner   Next-hop(s)
---- --------------  ------ -------- -------- -----------
 103 0000.0c07.ac67  1      00:13:43 site     Ethernet3/5
 103 442b.03ec.cb00  42     00:02:44 overlay  NX-6
 103 64a0.e73e.12c1  1      00:13:43 site     Ethernet3/5
 103 64a0.e73e.12c3  42     00:13:28 overlay  NX-6
 103 6c9c.ed4d.d943  42     00:02:56 overlay  NX-6
 103 c464.135c.6600  1      00:02:56 site     Ethernet3/5
```

如前所述，ORIB 数据是通过从 NX-6 收到的 IS-IS LSP 填充的，ORIB 表项表明 MAC 地址 442b.03ec.cb00 是其连接的一台主机。为了验证这一点，可以通过命令 **show otv adjacency** 得到 NX-6 的 system-id，然后通过命令 **show otv isis database detail** 找到正确的 LSP。

在 NX-6 上运行 **show otv isis redistribute route** 命令之后，即可验证发起地址宣告的 AED 将本地 MAC 表重分发给了 OTV IS-IS，如例 14-44 所示。

例 14-44　将 MAC 表重分发给 OTV IS-IS

```
NX-6# show otv isis redistribute route
! Output omitted for brevity
OTV-IS-IS process: default VPN: Overlay0
OTV-IS-IS MAC redistribute route
0103-442b.03ec.cb00, all
 Advertised into L1, metric 1 LSP-ID 6c9c.ed4d.d944.00-00
0103-64a0.e73e.12c3, all
 Advertised into L1, metric 1 LSP-ID 6c9c.ed4d.d944.00-00
0103-6c9c.ed4d.d943, all
 Advertised into L1, metric 1 LSP-ID 6c9c.ed4d.d944.00-00
```

至此，已经确认 NX-6 是 VLAN 103 中接收目的 MAC 地址为 442b.03ec.cb00 的数据帧的正确远程OTV ED。将数据包发送给主机C的下一步是由NX-2重写数据包以添加OTV报头，并通过加入接口将封装后的数据帧发送给传输网络。

OTV支持UDP或GRE封装，本例使用的是默认的GRE封装，为所有与本地OTV ED建立邻接关系的远程OTV ED都动态创建一条点对点隧道。可以通过命令**show tunnel internal implicit otv detail**查看这些隧道，如例14-45所示。

例14-45　NX-6的动态隧道封装

```
NX-2# show tunnel internal implicit otv detail
! Output omitted for brevity
Tunnel16389 is up
  Admin State: up
  MTU 9178 bytes, BW 9 Kbit
  Tunnel protocol/transport GRE/IP
  Tunnel source 10.1.12.1, destination 10.2.34.1
  Transport protocol is in VRF "default"
  Rx
  720357 packets input, 1 minute input rate 1024 packets/sec
  Tx
  715177 packets output, 1 minute output rate 1027 packets/sec
  Last clearing of "show interface" counters never
```

动态隧道表示OTV封装的软件转发组件，而OTV封装的硬件转发组件则需要多次通过线卡转发引擎的处理以正确重写数据包（包含OTV封装报头）。

> **注**：在硬件中重写数据包的具体过程与线卡中的转发引擎类型有关，相应的验证方式也不尽相同。如果怀疑硬件编程有问题，那么就可以先验证邻接关系、MAC地址表、ORIB和隧道状态。如果控制平面编程没问题，而且也学到了MAC地址，但仍然存在连接性故障，那么就需要联系思科TAC寻求帮助。

NX-2执行完OTV MAC-in-IP封装之后，数据包将利用附加的单播OTV报头穿越三层传输网络，源IP地址是NX-2的加入接口，目的IP地址是NX-6的加入接口。三层数据包到达NX-6的OTV加入接口之后，需要删除OTV封装并查找目的端。

外部数据包报头的目的IP地址是NX-6的OTV加入接口地址（10.2.34.1）。与OTV封装相似，删除OTV封装也要多次用到接收线卡上的转发引擎。由于外层目的IP地址属于NX-6，因而NX-6将剥离外层IP报头并查看OTV SHIM报头，以找到VLAN ID，这些查找信息均来自ORIB，ORIB包含了VLAN、MAC地址和目的接口等信息，如例14-46所示。

例14-46　NX-6关于主机C的ORIB表项

```
NX-6# show otv route
! Output omitted for brevity

OTV Unicast MAC Routing Table For Overlay0

VLAN MAC-Address      Metric Uptime   Owner    Next-hop(s)
---- --------------   ------ -------- -------- -----------
 103 0000.0c07.ac67   1      4d00h    site     port-channel3
 103 442b.03ec.cb00   1      00:44:32 site     port-channel3
 103 64a0.e73e.12c1   42     4d00h    overlay  NX-2
 103 64a0.e73e.12c3   1      4d00h    site     port-channel3
 103 6c9c.ed4d.d943   1      4d00h    site     port-channel3
 103 c464.135c.6600   42     4d00h    overlay  NX-2
```

第二次转发引擎处理需要对 VLAN MAC 地址表执行查找操作，以找到正确的出站接口和物理端口。例 14-47 显示了 NX-6 的 MAC 地址表。

例 14-47　NX6 关于主机 C 的 MAC 地址表表项

```
NX-6# show mac address-table dynamic vlan 103
Note: MAC table entries displayed are getting read from software.
Use the 'hardware-age' keyword to get information related to 'Age'

Legend:
    * - primary entry, G - Gateway MAC, (R) - Routed MAC, O - Overlay MAC
    age - seconds since last seen,+ - primary entry using vPC Peer-Link, E - EVPN
       entry
    (T) - True, (F) - False , ~~~ - use 'hardware-age' keyword to retrieve age info
   VLAN/BD   MAC Address      Type      age     Secure NTFY Ports/SWID.SSID.LID
---------+-----------------+--------+---------+------+----+------------------
* 103      0000.0c07.ac67   dynamic   ~~~       F    F   Po3
* 103      442b.03ec.cb00   dynamic   ~~~       F    F   Po3
O 103      64a0.e73e.12c1   dynamic   -         F    F   Overlay0
* 103      64a0.e73e.12c3   dynamic   ~~~       F    F   Po3
* 103      6c9c.ed4d.d943   dynamic   ~~~       F    F   Po3
O 103      c464.135c.6600   dynamic   -         F    F   Overlay0
```

数据帧离开 L2 中继上的 Port-channel 3（携带 VLAN 标记 103）。数据中心#2 的 L2 交换机收到该帧并执行 MAC 地址表查找操作，以找到连接主机 C 的端口，并将该帧转发到目的端。

> **注：** 使用邻接服务器模式时，单播数据流量的故障排查方法与启用多播的传输模式相同，区别在于 OTV ED 之间采用单播封装方式交换控制平面消息，并通过将 OTV ED 宣告给所有邻接 OTV ED 的方式复制这些消息。主机到主机的数据流量仍然采用 MAC-in-IP 单播封装方式从源端 OTV ED 传输到目的端 OTV ED。

14.3.5　基于多播传输网络的 OTV 多播流量

OTV 能够以无缝方式跨叠加网络转发多播流量，源端和接收端主机无须修改其处理行为即可在站点之间通过 OTV 网络交换 L2 多播流量。

对于传统 L2 交换网络来说，接收端主机通过发送 IGMP（Internet Group Management Protocol，Internet 组管理协议）Membership Report（成员关系报告）消息来表示自己对多播流量感兴趣。L2 交换机通常都会启用 IGMP 监听功能，负责侦听 IGMP Membership Report 消息，从而仅将多播流量泛洪给存在感兴趣接收端的端口。

此外，IGMP 监听特性还知道 MROUTER（Multicast router，多播路由器）的连接位置。必须将所有多播流量都转发给 MROUTER，以确保其他 L3 网络上感兴趣的接收端能够收到这些多播流量。如果使用的是 PIM ASM，那么 MROUTER 还需要将多播源注册到 RP 上。IGMP 监听功能通过侦听 PIM Hello 消息来发现 MROUTER，因为 Hello 消息可以指示该端口存在支持 L3 功能的 MROUTER。此后，需要更新 L2 转发表，从而将所有多播组流量都发送给 MROUTER 以及所有感兴趣的接收端。OTV ED 使用虚拟（dummy）PIM Hello 消息将多播流量和 IGMP Membership Report 消息引导到 OTV ED 的内部接口。

与维护用于单播转发的 OTV 路由表一样，OTV 也要维护用于多播转发的 MROUTE（Multicast route，多播路由）表。OTV MROUTE 表项包括 VLAN、Source 和 Group 共 3 种类型，表 14-2 列出了这些 OTV MROUTE 表项的详细信息。

表 14-2 　　　　　　　　　　　　　　OTV MROUTE 类型

类型	定义
(V,*,*)	VLAN 中存在本地 MROUTER 时（通过 IGMP 监听功能发现）创建该表项，负责将所有源和所有组的流量转发给 MROUTER
(V,*,G)	收到组 G 的 IGMP Membership Report 消息后创建该表项，并将收到 IGMP Membership Report 消息的接口添加到 MROUTE 的 OIF（Outgoing Interface，出站接口）中
(V,S,G)	源 S 向组 G 发送多播流量之后创建该表项，或者收到(S,G)的 IS-IS GMAS-TLV（Group Membership Active Source，组成员关系活动源）消息后创建该表项

有了 OTV IS-IS 控制平面协议之后，主机无须跨叠加网络发送 IGMP 消息，即可通过站点之间的扩展 VLAN 发送和接收多播流量。图 14-11 显示了一个简单的 OTV 拓扑结构，其中的主机 A 是组 239.100.100.100 的多播源，主机 C 是多播接收端。主机 A 和主机 C 都属于 VLAN 103。

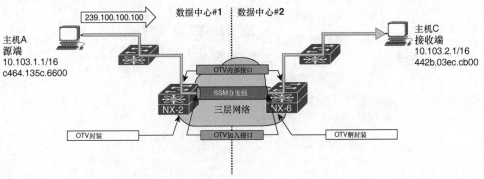

图 14-11　基于多播传输网络的跨 OTV 的多播流量

本例中的 L3 传输网络启用了 IP 多播机制。每个 OTV ED 都配置了 SSM（Source Specific Multicast，特定源多播）组区间，称为*分发组*（*Delivery Group*）或*数据组*（*data-group*），这两种说法可以互换。例 14-48 的配置示例突出显示了 NX-6 的分发组配置信息。

例 14-48　OTV SSM 数据组

```
NX-6# show running-config interface overlay 0
interface Overlay0
 description Site B
 otv join-interface Ethernet3/41
 otv control-group 239.12.12.12
 otv data-group 232.1.1.0/24
 otv extend-vlan 100-110
 no shutdown
```

分发组必须与 L3 传输网络进行协同，以确保支持 PIM SSM，并定义正确的组区间以用作 SSM 组。每个 OTV ED 都配置了相同的 OTV 数据组区间，而且每个 OTV ED 都可以是 SSM 组的源端。远程 OTV ED 加入传输网络中的 SSM 组之后，就可以从充当多播源的特定 OTV ED 接收多播帧。可以通过 OTV ED 之间的 IS-IS 宣告来发现每个站点上的活动源和接收端，从而明确所要使用的 SSM 组。

站点组（*site group*）是使用分发组跨叠加网络传输的多播组。图 14-11 中的站点组是 239.100.100.100，源端是主机 A，接收端是主机 C。从本质上来说，OTV 使用的是 multicast-in-multicast（将多播封装到多播中）OTV 封装方式，利用分发组在传输网络中跨叠加网络发送站点组流量。

将端到端的数据包分发机制分成两层（站点组和分发组）之后，可以大大简化故障排查难度。在源端站点，站点组的故障排查重点是验证源端的多播数据帧是否到达了该 VLAN 的 AED 的内部接口。在接收端站点，站点组的故障排查重点是验证接收端是否发送了 IGMP Membership Report 消息来表达自己对该多播组感兴趣。IGMP 监听必须拥有正确的端口，才能通过路径中的 L2 交换机从 OTV AED 的内部接口到达接收端。排查传输网络时，需要验证分发组是否正常，以确保充当源主机的 OTV ED 能够将 multicast-in-multicast OTV 流量发送到传输网络中，进行复制并分发给正确的 OTV ED 接收端。

为了确保主机 C 能够成功收到主机 A 发送的多播流量，必须执行一些必要步骤。L2 交换机必须将 OTV AED 的内部接口视为 MROUTER 端口，这样才能将来自接收端的 IGMP Membership Report 消息发送给 AED，同时将多播流量泛洪给 AED 的 OTV 内部接口。为了实现这一点，OTV 需要在所有通过 OTV 扩展的 VLAN 的内部接口上发送源 IP 地址为 0.0.0.0 的虚拟 PIM Hello 消息，这样做的目的不是在 VLAN 上建立 PIM 邻居，而是通过所连接的 L2 交换机强制检测 MROUTER 端口，如图 14-12 所示。

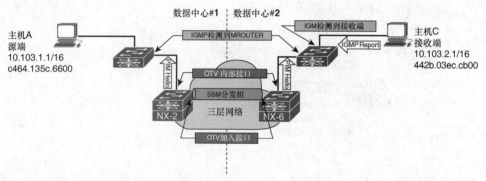

图 14-12 OTV 虚拟 PIM Hello 消息

例 14-49 显示了 Ethanalyzer 在 NX-6 的 VLAN 103 上抓取到的虚拟 PIM Hello 信息。

例 14-49 Ethanalyzer 抓取的虚拟 PIM Hello 消息

```
! Output omitted for brevity
Type: IP (0x0800)
Internet Protocol Version 4, Src: 0.0.0.0 (0.0.0.0),Dst: 224.0.0.13 (224.0.0.13)
  Version: 4
  Header length: 20 bytes
  Differentiated Services Field: 0xc0 (DSCP 0x30: Class Selector 6; ECN: 0x00:
 Not-ECT (Not ECN-Capable Transport))
    1100 00.. = Differentiated Services Codepoint: Class Selector 6 (0x30)
    .... ..00 = Explicit Congestion Notification: Not-ECT (Not ECN-Capable
Transport) (0x00)
  Total Length: 50
  Identification: 0xa51f (42271)
  Flags: 0x00
    0... .... = Reserved bit: Not set
    .0.. .... = Don't fragment: Not set
    ..0. .... = More fragments: Not set
  Fragment offset: 0
  Time to live: 1
  Protocol: PIM (103)
  Header checksum: 0x3379 [correct]
    [Good: True]
    [Bad: False]
  Source: 0.0.0.0 (0.0.0.0)
  Destination: 224.0.0.13 (224.0.0.13)
```

```
Protocol Independent Multicast
  0010 .... = Version: 2
  .... 0000 = Type: Hello (0)
  Reserved byte(s): 00
  Checksum: 0x572f [correct]
  PIM options: 4
    Option 1: Hold Time: 0s (goodbye)
      Type: 1
    Length: 2
      Holdtime: 0s (goodbye)
    Option 19: DR Priority: 0
      Type: 19
    Length: 4
      DR Priority: 0
    Option 22: Bidir Capable
      Type: 22
    Length: 0
    Option 20: Generation ID: 2882395322
      Type: 20
    Length: 4
      Generation ID: 2882395322
```

例 14-50 显示了数据中心#2 中的 L2 交换机的 IGMP 监听状态（收到 NX-6 的 VLAN 103 的虚拟 PIM Hello 消息之后）。

例 14-50　IGMP 监听将 NX-6 检测为 MROUTER 端口

```
DC2-Layer2-sw# show ip igmp snooping mrouter vlan 103
Type: S - Static, D - Dynamic, V - vPC Peer Link
   I - Internal, F - Fabricpath core port
   C - Co-learned, U - User Configured
   P - learnt by Peer
Vlan Router-port  Type   Uptime   Expires
103  Po3          D      3d09h    00:04:58
103  Po1          SVP    3d09h    never
```

主机 C 的 IGMP Membership Report 消息到达 NX-6 之后，就会在内部接口上受到监听并添加到 OTV MROUTE 表中（作为 IGMP 创建的表项）。需要记住的是，所有执行 IGMP 监听的交换机都必须将所有 IGMP Membership Report 消息转发给 MROUTER 端口。

例 14-51 显示了 NX-6 的 OTV MROUTE 表，包含了 IGMP 创建的(V,*,G)表项以及 OIF（包含收到 IGMP Membership Report 消息的 Port-channel 3）。

例 14-51　NX-6 的 OTV MROUTE 状态

```
NX-6# show otv mroute

OTV Multicast Routing Table For Overlay0

(103, *, 239.100.100.100), metric: 0, uptime: 00:00:38, igmp
 Outgoing interface list: (count: 1)
   Po3, uptime: 00:00:38, igmp
```

接下来，NX-6 将构建 IS-IS 消息，向所有 OTV ED 宣告组成员关系（GM-Update），数据中心#1 的 NX-2 收到 IS-IS GM-Update 消息，如例 14-52 所示。NX-6 由 IS-IS System-ID 6c9c.ed4d.d944 加以标识。可以通过 **show otv adjacency** 命令确认正在检查的是正确的 LSP，输出结果列出了每个 OTV ED IS-IS 邻居的 System-ID。

例 14-52　NX-2 的 OTV IS-IS MGROUP 数据库

```
NX-2# show otv isis database mgroup detail 6c9c.ed4d.d944.00-00
OTV-IS-IS Process: default LSP database VPN: Overlay0

OTV-IS-IS Level-1 Link State Database
 LSPID              Seq Number   Checksum Lifetime  A/P/O/T
 6c9c.ed4d.d944.00-00 0x00000002  0xFA73   1119      0/0/0/1
   Instance      : 0x00000000
   Group-Address :   IP Multicast : Vlan : 103     Groups : 1
             Group : 239.100.100.100 Sources : 0
   Digest Offset : 0
```

注：此时只有主机 C 加入了多播组，且没有任何活动多播源给多播组发送流量。

NX-2 收到 NX-6 的 IS-IS GM-Update 消息之后，将安装 OTV MROUTE 表项，如例 14-53 所示。其中，NX-2 的 OIF 是叠加接口，r 表示接收端可以通过叠加接口到达。

例 14-53　NX-2 的 OTV MROUTE 表项

```
NX-2# show otv mroute

OTV Multicast Routing Table For Overlay0

(103, *, 239.100.100.100), metric: 0, uptime: 00:00:47, overlay(r)
 Outgoing interface list: (count: 1)
   Overlay0, uptime: 00:00:47, isis_otv-default
```

主机 A 开始向数据中心#1 的站点组 239.100.100.100 发送流量。由于 NX-2 发送了虚拟 PIM 包，因而 L2 交换机将为该端口创建 IGMP 监听 MROUTER 表项。L2 交换机将所有多播流量都转发给 NX-2（由 OTV 内部接口接收流量）。NX-2 收到流量之后，将创建一条 OTV MROUTE 表项（见例 14-54），如果在 **show otv mroute** 命令中使用了关键字 **detail**，那么就可以看到分发组(S,G)，分发组的源端是 AED 的 OTV 加入接口，组地址是已配置的 OTV 数据组之一。

例 14-54　NX-2 的 OTV (V,S,G)MROUTE 详细信息

```
NX-2# show otv mroute detail

OTV Multicast Routing Table For Overlay0

(103, *, *), metric: 0, uptime: 00:01:02, overlay(r)
 Outgoing interface list: (count: 1)
   Overlay0, uptime: 00:01:02, isis_otv-default

(103, *, 224.0.1.40), metric: 0, uptime: 00:01:02, igmp, overlay(r)
 Outgoing interface list: (count: 2)
   Eth3/5, uptime: 00:01:02, igmp
   Overlay0, uptime: 00:01:02, isis_otv-default

(103, *, 239.100.100.100), metric: 0, uptime: 00:01:01, igmp, overlay(r)
 Outgoing interface list: (count: 2)
   Eth3/5, uptime: 00:01:01, igmp
   Overlay0, uptime: 00:01:00, isis_otv-default

(103, 10.103.1.1, 239.100.100.100), metric: 0, uptime: 00:09:20, site
 Outgoing interface list: (count: 1)
   Overlay0, uptime: 00:01:00, otv
     Local Delivery: s = 10.1.12.1, g = 232.1.1.0
```

从例 14-55 可以看出，OTV MROUTE 被自动重分发给 IS-IS，输出结果显示了 VLAN、站点组(S,G)、分发组(S,G)和 LSP-ID 等信息。

例 14-55 将 OTV MROUTE 重分发到 OTV IS-IS

```
NX-2# show otv isis ip redistribute mroute
OTV-IS-IS process: default OTV-IS-IS IPv4 Local Multicast Group database
VLAN 103: (10.103.1.1, 239.100.100.100)
AS in LSP_ID: 6c9c.ed4d.d942.00-00
[DS-10.1.12.1, DG-232.1.1.0]
```

接下来通过 IS-IS 将重分发路由宣告给所有 OTV ED。从例 14-56 可以看出，NX-6 收到了 NX-2 发起的 LSP。

例 14-56 NX-6 的 OTV MGROUP 数据库详细信息

```
NX-6# show otv isis database mgroup detail 6c9c.ed4d.d942.00-00
OTV-IS-IS Process: default LSP database VPN: Overlay0

OTV-IS-IS Level-1 Link State Database
LSPID              Seq Number   Checksum  Lifetime  A/P/O/T
 6c9c.ed4d.d942.00-00* 0x00000002   0x0110    1056      0/0/0/1
  Instance       : 0x00000004
  Active-Source  : IP Multicast : (103 - 10.1.12.1, 232.1.1.0) Groups : 1
              Group : 239.100.100.100 Sources : 1
              Source : 10.103.1.1
  Digest Offset  : 0
```

注： 排查 IS-IS 控制平面关于 OTV 多播和特定 VLAN 的组和源宣告故障时，可以使用 **show otv isis internal event-history mcast** 命令。

NX-6 根据该信息更新自己的 OTV MROUTE 表（见例 14-57），其中，s 表示源端位于叠加网络上。

例 14-57 NX-6 的 OTV (V,S,G) MROUTE 详细信息

```
NX-6# show otv mroute detail

OTV Multicast Routing Table For Overlay0

(103, *, *), metric: 0, uptime: 00:00:42, igmp, overlay(r)
 Outgoing interface list: (count: 2)
  Po3, uptime: 00:00:42, igmp
  Overlay0, uptime: 00:00:41, isis_otv-default

(103, *, 224.0.1.40), metric: 0, uptime: 00:00:42, igmp, overlay(r)
 Outgoing interface list: (count: 2)
  Po3, uptime: 00:00:42, igmp
  Overlay0, uptime: 00:00:40, isis_otv-default

(103, *, 239.100.100.100), metric: 0, uptime: 00:00:40, igmp, overlay(r)
 Outgoing interface list: (count: 2)
  Po3, uptime: 00:00:40, igmp
  Overlay0, uptime: 00:00:38, isis_otv-default

(103, 10.103.1.1, 239.100.100.100), metric: 0, uptime: 00:08:58, overlay(s)
 Outgoing interface list: (count: 0)
  Remote Delivery: s = 10.1.12.1, g = 232.1.1.0
```

可以通过 **show otv data-group** 命令验证 NX-2 和 NX-6 的站点组和分发组信息（见例 14-58），可以看出，与 **show otv mroute** 命令输出结果的内容匹配。

例 14-58　验证站点组到分发组的映射

```
NX-6# show otv data-group

Remote Active Sources for Overlay0

VLAN Active-Source   Active-Group    Delivery-Source Delivery-Group Joined-I/F
---- ---------------  ---------------  ---------------  --------------- ----------
103  10.103.1.1      239.100.100.100 10.1.12.1        232.1.1.0       Eth3/41
NX-2# show otv data-group
Local Active Sources for Overlay0
VLAN Active-Source   Active-Group   Delivery-Source  Delivery-Group  Join-IF State
---- ---------------  -------------  ---------------  ---------------  ------- -----
103  10.103.1.1      239.100.100.100 10.1.12.1       232.1.1.0        Po3     Local
```

OTV ED 充当传输网络上的分发组的源端主机和接收端主机。OTV ED 通过加入接口向传输网络发送 IGMPv3 Membership Report 消息之后，就可以接收分发组(10.1.12.1, 232.1.1.0)的数据包。

可以通过从 OTV ED 获得的 PIM SSM 分发组信息来验证传输网络的多播路由情况。AED 的加入接口是分发组的源端，AED 可以根据 OTV MROUTE 表和通过 IS-IS 控制平面收到的信息，仅加入所需的分发组多播源。该机制允许 OTV 优化传输网络中的多播流量，使每个 OTV ED 都仅接收需要的数据。使用 PIM SSM 之后，可以让每个分发组都选择性地加入特定源地址。

例 14-59 显示了传输路由器的 MROUTE 表，其中，10.1.12.1 是 NX-2 的 OTV 加入接口，该接口是分发组 232.1.1.0/32 的源端。为了通过 RPF（Reverse Path Forwarding，反向路径转发）检查，入站接口必须与去往源端的路由表路径相匹配。接口 Ethernet3/30 属于 OIF，连接 NX-6 的 OTV 加入接口。

例 14-59　验证传输网络中的 MROUTE

```
NX-5# show ip mroute 232.1.1.0
IP Multicast Routing Table for VRF "default"

(10.1.12.1/32, 232.1.1.0/32), uptime: 00:02:29, igmp ip pim
 Incoming interface: Ethernet3/29, RPF nbr: 10.1.13.1
 Outgoing interface list: (count: 1)
   Ethernet3/30, uptime: 00:02:29, igmp
```

> **注：** 排查传输网络中的 OTV ED 源端与接收端之间的多播故障时，遵循标准的分发组多播故障排查方法。OTV 将站点组封装到多播分发组中的事实并没有改变传输网络的故障排查方法，从传输网络角度来看，OTV ED 是分发组的源端主机和接收端主机。

14.3.6　基于单播传输网络的 OTV 多播流量（邻接服务器模式）

基于单播传输网络部署的 OTV 也能跨叠加网络转发扩展 VLAN 的多播流量，其做法是将站点组的多播包封装到 IP 单播 OTV 包中，再通过传输网络进行传输，如图 14-13 所示。如果多个远程站点都有感兴趣的接收端，那么源站点的 OTV ED 就必须执行多播流量的前端复制操作，将流量副本发送给每个站点，导致扩展效率低下。

例中的主机 A 和主机 C 都属于 VLAN 103，主机 A 将多播流量发送给站点组 239.100.100.100，主机 C 将 IGMP Membership Report 消息发送给数据中心#2 的 L2 交换机，L2 交换机将 IGMP Membership Report 消息转发给 NX-6（因为 NX-6 是 IGMP 监听的 MROUTER 端口）。与基于多播传输网络的部署模式相同，OTV 内部接口也采用了相同的虚拟 PIM Hello 包机制。NX-6 收到 IGMP Membership Report 消息之后，将创建一条 OTV MROUTE（见例 14-60），并将内部接口 Port-channel 3 作为 OIF。

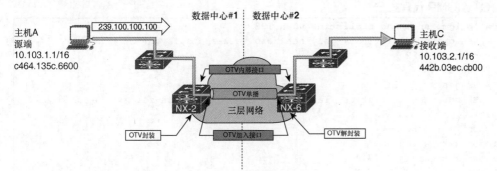

图 14-13　基于邻接服务器模式的 OTV 的多播流量

例 14-60　NX-6 的 OTV (V,*,G) MROUTE 详细信息

```
NX-6# show otv mroute detail

OTV Multicast Routing Table For Overlay0

(103, *, *), metric: 0, uptime: 00:03:25, igmp, overlay(r)
 Outgoing interface list: (count: 2)
  Po3, uptime: 00:03:25, igmp
  NX-2 uptime: 00:03:24, isis_otv-default

(103, *, 224.0.1.40), metric: 0, uptime: 00:03:25, igmp
 Outgoing interface list: (count: 1)
  Po3, uptime: 00:03:25, igmp

(103, *, 239.100.100.100), metric: 0, uptime: 00:03:23, igmp
 Outgoing interface list: (count: 1)
  Po3, uptime: 00:03:23, igmp
```

此后将 OTV MROUTE 自动重分发给 IS-IS，从而宣告给所有其他 OTV ED（见例 14-61）。请注意输出结果中的 LSP_ID，可以在 NX-2 上检查该信息（是数据中心#1 的多播源主机 A 的 OTV ED）。

例 14-61　NX-6 将 OTV MROUTE 重分发到 OTV IS-IS

```
NX-6# show otv isis ip redistribute mroute
OTV-IS-IS process: default OTV-IS-IS IPv4 Local Multicast Group database
VLAN 103: (*, *)
Receiver in LSP_ID: 6c9c.ed4d.d944.00-00
VLAN 103: IPv4 router attached
VLAN 103: (*, 224.0.1.40)
Receiver in LSP_ID: 6c9c.ed4d.d944.00-00
VLAN 103: IPv4 router attached
VLAN 103: (*, 239.100.100.100)
Receiver in LSP_ID: 6c9c.ed4d.d944.00-00
VLAN 103: IPv4 router attached
```

注： 从例 14-61 中的(*,*)表项可以看出，VLAN 103 存在启用了 PIM 功能特性的路由器。

由于叠加网络不转发 IGMP 数据包，因而将指示感兴趣接收端的 IS-IS 消息作为 IGMP 代理报告（proxy-report）。例 14-62 显示了 NX-6 的 IGMP 监听统计信息，表明代理报告是通过 IS-IS 发起的。请注意，IGMP 代理报告机制并不专用于 OTV 邻接服务器模式。

例 14-62 OTV IGMP 代理报告

```
NX-6# show ip igmp snooping statistics vlan 103
Global IGMP snooping statistics: (only non-zero values displayed)
 Packets received: 1422
 Packets flooded: 437
 STP TCN messages rcvd: 21
VLAN 103 IGMP snooping statistics, last reset: never (only non-zero values
  displayed)
 Packets received: 1350
 IGMPv2 reports received: 897
 IGMPv2 queries received: 443
 IGMPv2 leaves received: 10
 PIM Hellos received: 2598
 IGMPv2 leaves suppressed: 4
 Queries originated: 4
 IGMPv2 proxy-reports originated: 14
 IGMPv2 proxy-leaves originated: 4
 Packets sent to routers: 902
 vPC Peer Link CFS packet statistics:
IGMP Filtering Statistics:
Router Guard Filtering Statistics:
F340-35-02-N7K-7009-A-vdc_4#
```

接下来在数据中心#1 按照从接收端到源端的路径，验证 NX-2 的 IS-IS 数据库。目的是确认已经将叠加接口添加为 OTV MROUTE 的 OIF。例 14-63 包含了 NX-2 从 NX-6 收到的 GM-LSP 消息。

例 14-63 NX-2 的 OTV IS-IS MGROUP 数据库详细信息

```
NX-2# show otv isis database mgroup detail 6c9c.ed4d.d944.00-00
OTV-IS-IS Process: default LSP database VPN: Overlay0

OTV-IS-IS Level-1 Link State Database
 LSPID           Seq Number   Checksum Lifetime  A/P/O/T
 6c9c.ed4d.d944.00-00 0x00000005   0x7579   820      0/0/0/1
  Instance    : 0x00000003
  Group-Address :    IP Multicast : Vlan : 103    Groups : 2
           Group  : 239.100.100.100 Sources : 0
           Group  : 224.0.1.40    Sources : 0
  Router-capability :    Interested Vlans : Vlan Start 103 Vlan end 103
 IPv4 Router attached
  Digest Offset : 0
```

从 NX-2 的 IGMP 监听表可以看出，叠加接口已经包含在端口列表中，如例 14-64 所示。

例 14-64 NX-2 的 IGMP 监听 OTV 组

```
NX-2# show ip igmp snooping otv groups
Type: S - Static, D - Dynamic, R - Router port, F - Fabricpath core port

Vlan Group Address    Ver Type Port list
```

```
103    224.0.1.40        v3    D    Overlay0
103    239.100.100.100   v3    D    Overlay0
```

NX-2 的 OTV MROUTE 包含了(V,*,G)表项，该表项是收到 NX-6 的 IS-IS GM-LSP 消息之后触发创建的。该消息表明主机 C 是数据中心#2 中的感兴趣接收端，NX-2 应该将叠加接口添加为该多播组的 OIF。例 14-65 显示了 NX-2 的 OTV MROUTE 表信息，其中，r 表示接收端可以通过叠加接口到达。以外，表中还存在(V,S,G)表项，表明主机 A 正在向站点组 239.100.100.100 发送流量。

例 14-65　NX-2 的 OTV MROUTE 详细信息

```
NX-2# show otv mroute detail

OTV Multicast Routing Table For Overlay0

(103, *, *), metric: 0, uptime: 00:12:22, overlay(r)
 Outgoing interface list: (count: 1)
  NX-6 uptime: 00:12:21, isis_otv-default

(103, *, 224.0.1.40), metric: 0, uptime: 00:12:21, overlay(r)
 Outgoing interface list: (count: 1)
  NX-6 uptime: 00:12:21, isis_otv-default

(103, *, 239.100.100.100), metric: 0, uptime: 00:12:21, overlay(r)
 Outgoing interface list: (count: 1)
  NX-6 uptime: 00:12:21, isis_otv-default

(103, 10.103.1.1, 239.100.100.100), metric: 0, uptime: 00:12:21, site
 Outgoing interface list: (count: 1)
  NX-6 uptime: 00:10:51, otv
   Local Delivery: s = 0.0.0.0, g = 0.0.0.0
```

注：OTV MROUTE 表列出了 OTV 安装的 NX-6 的 OIF，这是邻接服务器模式使用 OTV 单播封装的结果。分发组的组地址值为全 0，如果使用的是多播传输网络，那么就会用有效分发组地址填充该信息。

NX-2 将站点组数据包封装到 OTV 单播包中（目的地址是 NX-6 的加入接口），OTV 单播包将通过传输网络到达 NX-6。数据包到达 NX-6 的 OTV 加入接口之后，NX-6 将删除外层 OTV 单播封装，然后对内层多播包执行查找操作，触发 IGMP 在 OTV 内部接口上安装 MROUTE 的 OIF。例 14-66 显示了 NX-6 的 OTV MROUTE 表，可以看出，站点组多播包离开端口 Po3 去往数据中心#2 的 L2 交换机，并最终到达主机 C。

例 14-66　NX-6 的 OTV MROUTE 详细信息

```
NX-6# show otv mroute detail
show otv mroute detail

OTV Multicast Routing Table For Overlay0

(103, *, *), metric: 0, uptime: 00:03:25, igmp, overlay(r)
 Outgoing interface list: (count: 2)
  Po3, uptime: 00:03:25, igmp
  F340-35-02-N7K-7009-A-VDC2 uptime: 00:03:24, isis_otv-default

(103, *, 224.0.1.40), metric: 0, uptime: 00:03:25, igmp
 Outgoing interface list: (count: 1)
```

```
    Po3, uptime: 00:03:25, igmp
(103, *, 239.100.100.100), metric: 0, uptime: 00:03:23, igmp
  Outgoing interface list: (count: 1)
    Po3, uptime: 00:03:23, igmp
```

在邻接服务器模式下，NX-2 不会将源宣告给其他 OTV ED，因为传输网络没有远程 OTV ED 所要加入的分发组。NX-2 只要知道叠加网络上存在感兴趣的接收端以及哪个 OTV ED 存在接收端即可，将该 OTV ED 的加入接口作为跨传输网络传送的 multicast-in-unicast（将多播封装到单播中）OTV 数据包的目的地址。从例 14-67 可以看出，站点组的多播帧是通过 OTV 单播点对点动态隧道进行封装的。

例 14-67　多播流量的动态隧道封装

```
NX-2# show tunnel internal implicit otv detail
Tunnel16390 is up
  Admin State: up
  MTU 9178 bytes, BW 9 Kbit
  Tunnel protocol/transport GRE/IP
  Tunnel source 10.1.12.1, destination 10.2.34.1
  Transport protocol is in VRF "default"
  Rx
  663 packets input, 1 minute input rate 0 packets/sec
  Tx
  156405 packets output, 1 minute output rate 0 packets/sec
  Last clearing of "show interface" counters never
```

14.4　OTV 高级功能特性

OTV 自成为 NX-OS 的功能特性以来，一直都在持续发展当中。本节将讨论 OTV 的一些高级功能特性，可以通过这些功能特性对 OTV 进行定制，以满足不同网络部署需求。

14.4.1　FHRP 本地化

常见的 FHRP（First Hop Routing Protocol，第一跳路由协议）有 HSRP（Hot Standby Routing Protocol，热备路由协议）和 VRRP（Virtual Router Redundancy Protocol，虚拟路由器冗余协议），通常为 VLAN 上的主机提供冗余默认网关。部署了 OTV 之后，VLAN 将通过叠加网络扩展到多个站点，意味着数据中心#1 中的路由器可以与数据中心#2 中的路由器建立 HSRP 邻居关系，而且数据中心#2 中的主机可能使用数据中心#1 中的默认路由器，在能够轻松实现本地路由的情况下，这样做会导致叠加网络出现大量不必要的流量。

在 OTV ED 上配置 FHRP 隔离功能之后，可以让每个站点的 FHRP 都独立运行。这样就可以过滤所有的 FHRP 协议流量以及主机为解析叠加网络上的虚拟 IP 而发送的 ARP 流量。例 14-68 显示了 NX-2 的配置示例。

例 14-68　NX-2 的 FHRP 本地化配置

```
NX-2# show running-config
! Output omitted for brevity
feature otv

ip access-list ALL_IPs
 10 permit ip any any
```

```
ipv6 access-list ALL_IPv6s
 10 permit ipv6 any any
mac access-list ALL_MACs
 10 permit any any
ip access-list HSRP_IP
 10 permit udp any 224.0.0.2/32 eq 1985
 20 permit udp any 224.0.0.102/32 eq 1985
ipv6 access-list HSRP_IPV6
 10 permit udp any ff02::66/128
mac access-list HSRP_VMAC
 10 permit 0000.0c07.ac00 0000.0000.00ff any
 20 permit 0000.0c9f.f000 0000.0000.0fff any
 30 permit 0005.73a0.0000 0000.0000.0fff any
arp access-list HSRP_VMAC_ARP
 10 deny ip any mac 0000.0c07.ac00 ffff.ffff.ff00
 20 deny ip any mac 0000.0c9f.f000 ffff.ffff.f000
 30 deny ip any mac 0005.73a0.0000 ffff.ffff.f000
 40 permit ip any mac any
vlan access-map HSRP_Localization 10
    match mac address HSRP_VMAC
    match ip address HSRP_IP
    match ipv6 address HSRP_IPV6
    action drop
vlan access-map HSRP_Localization 20
    match mac address ALL_MACs
    match ip address ALL_IPs
    match ipv6 address ALL_IPv6s
    action forward
vlan filter HSRP_Localization vlan-list 100-110

mac-list OTV_HSRP_VMAC_deny seq 10 deny 0000.0c07.ac00 ffff.ffff.ff00
mac-list OTV_HSRP_VMAC_deny seq 11 deny 0000.0c9f.f000 ffff.ffff.f000
mac-list OTV_HSRP_VMAC_deny seq 12 deny 0005.73a0.0000 ffff.ffff.f000
mac-list OTV_HSRP_VMAC_deny seq 20 permit 0000.0000.0000 0000.0000.0000

route-map OTV_HSRP_filter permit 10
 match mac-list OTV_HSRP_VMAC_deny

service dhcp

otv-isis default
 vpn Overlay0
   redistribute filter route-map OTV_HSRP_filter
otv site-identifier 0x1
ip arp inspection filter HSRP_VMAC_ARP vlan 100-110
```

以图14-1的拓扑结构为例，数据中心#1为NX-1和NX-3之间的所有VLAN都配置了HSRP，数据中心#2也为NX-5和NX-7之间的所有VLAN都配置了HSRP。例14-68的配置示例包含了3个过滤组件。

- VACL（VLAN Access Control List，VLAN访问控制列表）：用于过滤和删除HSRP Hello包。
- ARP检查过滤器（ARP Inspection Filter）：丢弃源自HSRP虚拟MAC地址的ARP流量。
- 叠加接口上的重分发过滤器路由映射（Redistribution Filter Route-Map）：过滤HSRP VMAC（Virtual MAC，虚拟MAC），以防止通过OTV IS-IS进行宣告。

FHRP 隔离特性的配置错误是常见故障原因，需要仔细配置过滤机制，以避免出现 OTV IS-IS LSP 刷新问题、IP 地址重复问题以及 HSRP VMAC 震荡问题。

14.4.2 多归属

OTV 的多归属站点指的是将站点中的两个或多个 OTV ED 配置为扩展相同的 VLAN 区间。由于 OTV 不会跨叠加网络转发 STP BPDU，因而在不选举 AED 的情况下会形成 L2 环路。

如果站点存在多个 OTV ED，那么就需要通过 OTV IS-IS system-id 和 VLAN ID 选举 AED。选举过程是通过散列函数完成的，散列结果就是次序值 0 或 1，通过次序值将每个扩展 VLAN 的 AED 角色分配给站点上具有转发功能的某个 OTV ED。

如果存在两个 OTV ED，那么 System-ID 较小的设备将成为偶数 VLAN 的 AED，System-ID 较大的设备将成为奇数 VLAN 的 AED。AED 负责通过叠加网络宣告 MAC 地址并转发扩展 VLAN 的流量。

NX-OS 从 Release 5.2(1)开始，使用双重邻接关系的概念，允许拥有相同站点标识符的 OTV ED 通过叠加网络和站点 VLAN 进行通信，从而大大降低其中某个 OTV ED 被隔离的概率，创建双活状态。此外，在配置站点标识符之前，将禁用 OTV ED 的叠加接口，从而确保 OTV 能够检测出站点标识符的任何不匹配项。如果设备不再具备 AED 能力，那么将主动通知站点上的另一个 OTV ED，以便其接管所有 VLAN 的 AED 角色。

14.4.3 入站路由优化

出站路由优化可以通过 FHRP 隔离功能来实现，入站路由优化对于某些 OTV 部署场景来说是必须考虑的另一个挑战。OTV 通过提供透明的 L2 叠加网络，将 VLAN 扩展到多个站点，但可能会导致以下情况，即多个站点可能向其他站点宣告相同的 L3 前缀，从而出现次优转发问题。

图 14-14 显示 NX-11 拥有 ECMP（Equal Cost Multipath，等价多路径）路由，可以通过 NX-9 或 NX-10 到达子网 10.103.0.0/16。根据负载共享散列算法，源自 NX-11 后面的数据包将到达数据中心#1 或数据中心#2。例如，如果流量的目的端是主机 C，而 NX-11 选择 NX-9 作为下一跳来转发流量，那么就使用了次优转发路径。此后，NX-9 将试图解析主机 C 的位置以转发流量。数据包到达 NX-2 的内部接口之后，将执行 OTV 封装，并通过叠加网络将数据包路由回主机 C。

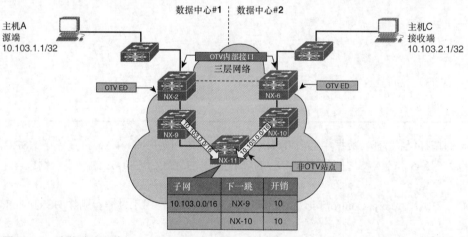

图 14-14 次优路由行为

解决该问题的常见办法是同时部署 OTV 和 LISP（Locator-ID Separation Protocol，位置与身份分离协议）。LISP 可以发现主机的位置并利用 LISP 控制平面在特定的 RLOC（Routing Locator，路由定位符）后面宣告主机位置，从而实现入站路由优化。此外，LISP 还支持站点之间的主机移动性。如果不需要部署完整的 LISP，那么 LISP 就可以在 IGP 的协助下，将路由从 LISP 重分发到 IGP 协议中。

另一种解决方案是将每个站点的明细子网路由以及/16 汇总路由宣告到路由域中，这样就能沿着明细子网路由到达数据中心#1 或数据中心#2，即使明细路由出现了问题，也仍然可以通过/16 汇总路由来引导流量。如果 OTV 在部分故障状态仍能通过后门链路进行工作，那么流量将经由叠加网络从数据中心#1 转发到数据中心#2。解决该问题的最佳方案与具体部署场景有关，与从冗余角度看两个 OTV 站点采取 Active/Standby（主备）工作方式还是 Active/Active（双活）工作方式有关。

14.4.4 VLAN 转换

对于某些网络场景来说，在某个 OTV 站点上配置的 VLAN 可能需要与其他站点上采用不同 VLAN 编号方案的 VLAN 进行通信。此时可以采用以下两种方法解决该需求。

- 在叠加接口上进行 VLAN 映射。
- 在 L2 中继端口上进行 VLAN 映射。

Nexus 7000 F3 或 M3 系列模块不支持叠加接口上的 VLAN 映射。如果 F3 或 M3 模块需要进行 VLAN 映射，那么必须在 OTV 内部接口（L2 中继端口）上进行 VLAN 映射。

例 14-69 给出了叠加接口上的 VLAN 映射配置示例。例中的 VLAN 200 是通过叠加网络扩展的 VLAN，将本地 VLAN 200 映射到另一个 OTV 站点上的 VLAN 300。

例 14-69 叠加接口上的 VLAN 映射

```
NX-2# show running-config interface overlay 0
interface Overlay0
 description Site A
 otv join-interface port-channel3
 otv control-group 239.12.12.12
 otv data-group 232.1.1.0/24
 otv extend-vlan 100-110, 200
 otv vlan mapping 200 to 300
 no shutdown

NX-2# show otv vlan-mapping
Original VLAN -> Translated VLAN
-------------------------------
    200 -> 300
```

如果使用的是 F3 或 M3 模块，那么必须在 OTV 内部接口上执行 VLAN 映射（见例 14-70）。本例将 VLAN 200 转换为 VLAN 300，通过 OTV 扩展后与远程站点 VLAN 进行互操作。

例 14-70 L2 中继端口上的 VLAN 映射

```
NX-2# show running-config interface Ethernet3/5
interface Ethernet3/5
 description 7009A-Main-VDC OTV inside
 switchport
 switchport mode trunk
 switchport vlan mapping 200 300
 mtu 9216
 no shutdown
```

14.4.5 OTV 隧道去极化

如果 L3 路由器拥有多条去往目的端的 ECMP 路由，那么就可以采用负载分担散列函数为特定流选择出接口。流通常是一个五元组，包括以下几点。

- 三层源地址。
- 三层目的地址。
- 四层协议。
- 四层协议源端口。
- 四层协议目的端口。

隧道通信存在一个典型问题，那就是在穿越多跳 L3 ECMP 网络时可能出现极化问题。通常将这些流称为"大象流"，因为这些流通常会移动大量流量，使绑定接口或 ECMP 路径中的单条链路出现饱和。由于隧道使用统一的隧道报头以及一致的源地址和目的地址，因而隧道流量的五元组是固定的，使散列算法的输入始终保持不变，这样一来，即使可以将多个不同的流封装到隧道中，但是对于散列算法来说，其输入始终相同。

如果传输网络的每一层都使用相同的散列函数，那么就会出现这种极化问题。使用相同的输入会导致每一跳产生的输出接口决策也相同。例如，如果某台路由器选择了一个偶数接口，那么下一台路由器也将选择偶数接口，再下一台路由器也同样会选择偶数接口，依此类推。

OTV 提供了一种解决方案。如果叠加网络采用了默认的 GRE/IP 封装，那么就可以在 OTV 加入接口的同一子网中配置辅助 IP 地址（见例 14-71）。这样一来，OTV 就可以在不同的地址对之间建立辅助动态隧道。辅助地址允许传输网络提供不同的散列结果，从而能够更有效地均衡叠加流量。

例 14-71 配置辅助 IP 地址以避免极化问题

```
NX-2# show running-config interface port-channel3
interface port-channel3
 description 7009A-Main-OTV Join
 mtu 9216
 no ip redirects
 ip address 10.1.12.1/24
 ip address 10.1.12.4/24 secondary
 ip router ospf 1 area 0.0.0.0
 ip igmp version 3
```

可以通过 show otv adjacency detail 命令查看辅助 OTV 邻接关系的状态，如例 14-72 所示。

例 14-72 基于辅助 IP 地址的 OTV 邻接关系

```
NX-2# show otv adjacency detail
Overlay Adjacency database

Overlay-Interface Overlay0 :
Hostname            System-ID       Dest Addr     Up Time State
NX-4                64a0.e73e.12c2 10.1.22.1      00:03:07 UP
    Secondary src/dest:    10.1.12.4    10.1.22.1            UP
HW-St: Default
NX-8                64a0.e73e.12c4 10.2.43.1      00:03:07 UP
    Secondary src/dest:    10.1.12.4    10.2.43.1            UP
HW-St: Default
NX-6                6c9c.ed4d.d944 10.2.34.1      00:03:06 UP
    Secondary src/dest:    10.1.12.4    10.2.34.1            UP
HW-St: Default
```

注： 系统默认启用 OTV 隧道去极化功能特性，可以通过全局配置命令 **otv depolarization disable** 禁用该功能特性。

如果采用了 OTV UDP 封装方式，那么将自动执行去极化操作，无须任何其他配置。此时，以太网帧封装在 UDP 中，使用可变 UDP 源端口和 UDP 目的端口 8472。由于使用了可变源端口，因而 OTV ED 能够影响传输网络的负载分担散列算法。

注： 从 NX-OS release 7.2(0)D1(1) 开始，F3 和 M3 模块支持 OTV UDP 封装方式。

14.4.6 OTV 快速故障检测

OTV 的双重邻接关系功能允许 OTV ED（携带相同的站点标识符）在站点 VLAN 和叠加网络上建立邻接关系。如果其中一台 OTV ED 不可达或出现故障，那么站点上的另一台 OTV ED 就必须接管所有 VLAN 的 AED 角色。如果能够快速检测此类故障情况，那么就能最大程度地减少过渡期间的流量丢失问题。

可以在站点 VLAN 上配置站点 VLAN IS-IS 邻接关系使用 BFD（Bidirectional Forwarding Detection，双向转发检测）来检测 IS-IS 邻居关系中断问题，BFD 对于检测站点 VLAN 出现的各种连接故障来说都非常有用。例 14-73 给出了在站点 VLAN 上启用 BFD 机制的配置示例。

例 14-73 在站点 VLAN 为 OTV IS-IS 配置 BFD

```
NX-2# show otv adjacency detail
! Output omitted for brevity
feature otv
feature bfd

otv site-vlan 10
 otv isis bfd

interface Vlan10
 no shutdown
 bfd interval 250 min_rx 250 multiplier 3
 no ip redirects
 ip address 10.111.111.1/30
```

可以通过命令 **show otv isis site** 验证站点 VLAN 的 BFD 状态，如例 14-74 所示。此外，**show bfd neighbors** 命令也能显示所有 BFD 邻居信息。

例 14-74 验证站点 VLAN 的 BFD 邻居

```
NX-2# show otv isis site

OTV-ISIS site-information for: default

 BFD: Enabled [IP: 10.111.111.1]

OTV-IS-IS site adjacency local database:

 SNPA         State Last Chg Hold     Fwd-state Site-ID         Version BFD
 64a0.e73e.12c2 UP   00:00:40 00:01:00 DOWN      0000.0000.0100  3
 Enabled [Nbr IP: 10.111.111.2]

OTV-IS-IS Site Group Information (as in OTV SDB):
```

```
 SystemID: 6c9c.ed4d.d942, Interface: site-vlan, VLAN Id: 10, Cib: Up VLAN: Up

 Overlay   State  Next IIH   Int   Multi
 Overlay1  Up     0.933427   3     20


 Overlay   Active SG      Last CSNP            CSNP Int Next CSNP
 Overlay1  0.0.2.0        ffff.ffff.ffff.ff-ff 1d14h    00:00:02

 Neighbor SystemID: 64a0.e73e.12c2
 IPv4 site groups:
   0.0.2.0
```

对于叠加邻接关系来说，可以通过跟踪到达对等 OTV ED 的加入接口的路由，来检测导致 IS-IS 邻居关系中断的可达性问题。例 14-75 给出了对 OTV ED（使用相同的站点标识符）叠加邻接关系启用下一跳邻接关系跟踪特性的配置示例。

例 14-75 配置 OTV 下一跳邻接关系跟踪特性

```
NX-2# show run otv
! Output omitted for brevity
feature otv

otv-isis default
 track-adjacency-nexthop
 vpn Overlay0
  redistribute filter route-map OTV_HSRP_filter
```

例 14-76 显示了 **show otv isis track-adjacency-nexthop** 的输出结果，验证了邻接关系跟踪特性的启用情况以及对 NX-4 下一跳可达性的跟踪情况。

例 14-76 验证 OTV 下一跳邻接关系跟踪状态

```
NX-2# show otv isis track-adjacency-nexthop
OTV-IS-IS process: default
  OTV-ISIS adjs for nexthop: 10.1.12.2, VRF: default
    Hostname: 64a0.e73e.12c2, Overlay: Overlay1
```

邻接关系跟踪特性依赖于非默认路由（通过动态路由协议学到的对等 OTV ED 的加入接口的路由）。如果路由丢失，那么 OTV IS-IS 邻接关系将直接中断，而不用等待保持定时器到期，从而允许另一台 OTV ED 承担所有 VLAN 的 AED 角色。

14.5 本章小结

本章介绍的 OTV 提供了一种有效且灵活的方法，可以将 L2 VLAN 扩展到路由式传输网络的多个站点。讨论的 MAC 路由选择和 AED 选举机制，能够在不依赖 STP 的情况下，很好地解决其他 DCI 解决方案存在的多种挑战问题。此外，本章还提供了大量有关控制平面、单播流量以及多播流量的应用示例，对于解决生产网络环境可能出现的各种连接问题都有很好的借鉴作用。

第七部分

网络可编程性

第 15 章　可编程性与自动化

第 15 章

可编程性与自动化

本章主要讨论如下主题。
- Open NX-OS 简介。
- Shell（外壳）和脚本。
- 应用程序。
- NX-API。

15.1 可编程性与自动化概述

什么是自动化？由于该问题适用于各行各业，如机器人、过程管理和 IT（Information Technology，信息技术），因而答案也千差万别。本章主要讨论计算机网络环境中的自动化机制。在讨论自动化之前，首先考虑一个案例。假定某组织机构的网络由 100 个节点组成，所有节点都运行相同的软件和硬件。经过几个月的测试和软件验证之后，网络运营团队同意将软件升级到新版本，此后团队花了将近 2 个月的时间完成了升级操作。但是升级操作之后不久，网络出现了大规模宕机和网络崩溃问题，发现根源是软件缺陷。不过，该软件提供了一种帮助恢复业务的应急措施配置命令。因此，网络运营团队面临以下两种选择。
- 升级到已经修复了该软件缺陷的版本。
- 采用应急配置解决措施。

以上两种选项都比较耗时，但第二种采用应急配置解决措施的时间相对较少。不过，在 100 个节点上应用应急解决方案也绝非易事。这就是自动化机制能够有效发挥作用的场景，如果能够自动执行保障应用程序成功运行的部署和验证操作，那么就能在所有网络节点上快速部署应急解决措施或补救程序。

可以将*网络自动化*（Network automation）简单地定义为，使用脚本、编程语言或应用程序记录某些步骤，然后在指定时间或特定时间间隔，无须人工干预即可在网络上重放记录下来的步骤。网络自动化不但能够节省时间，而且能预防脚本验证后的人为错误。自动化和编排机制充分利用网络基础设施中的设备接口，来帮助减少服务开通时间、加快服务交付速度。

可编程性（Programmability）指的是实体被编程的能力。可编程性增强了可编程设备的扩展或调整能力。与实体进行交互与对实体进行编程是两种不同的操作，这两者之间的关系很容易被人们混淆。应用程序或脚本可以在不扩展或调整网络设备现有能力的情况下与网络设备进行交互。可管理性与可编程性不同，两者面向不同的操作目的。

- **可管理性（Manageability）**：允许根据操作、配置、管理和维护方式对实体进行管理。
- **可编程性（Programmability）**：可以对实体或设备进行扩展或调整，以增加新的功能特性和属性，从而增强实体或设备的能力。

接下来将详细讨论 NX-OS 提供的多种自动化和可编程性工具，可以为网络工程师执行各种操作和运行第三方应用程序提供更多的控制机制和灵活性。

15.2 Open NX-OS 简介

虽然 NX-OS 一直都建立在 Linux 内核之上，但直到最近，很多底层 Linux 操作系统都还没有完全向用户公开。从 NX-OS Release 7.0(3)开始，最新的 Nexus 9000 和 Nexus 3000 系列交换机都可以运行思科的 *Open NX-OS*。

Linux 在计算和网络领域取得广泛成功的主要原因之一是灵活性和庞大的用户支持社区。使用了 Open NX-OS 之后，用户就可以在交换机上运行 Linux 应用程序，从而为功能丰富的 NX-OS 操作系统提供更多的功能特性，而无须封装库（wrapper library）或定制化。Open NX-OS 的主要组件包括以下几个。

- **内核版本 3.4**：这是一个 64 位内核，在功能与稳定性之间实现了很好的平衡。
- **内核栈（Kernel stack）**：早期 NX-OS 版本使用的用户空间 Netstack 进程已被替换为内核栈，允许将交换机上的接口作为标准 Linux netdev 和命名空间映射到内核，使用 Bash Shell 中的 **ifconfig** 和 **tcpdump** 等标准 Linux 命令来管理接口。
- **开放式软件包管理组件**：为交换机安装软件或打补丁以提供帮助及扩展能力的工具，如 RPM（RPM Package Manager，RPM 软件包管理器）和 YUM（Yellowdog Updater,Modified）。
- **容器支持组件**：LXC（Linux Container，Linux 容器）直接运行在平台上，可以访问基于 CentOS 7 的 Guest Shell，使用户能够在安全、隔离的环境中自定义交换机。

Open NX-OS 支持大量 Linux 功能特性（如模块化、故障隔离和弹性机制等），为实现真正的 DevOps 数据中心交换机奠定了坚实基础。

> **注**：有关 Open NX-OS 体系架构的更多详细信息，请参阅思科官网的 *Programmability and Automation with Cisco Open NX-OS* 一书。

Shell 和脚本

Shell 是一种文本用户界面，允许用户执行管理任务并与操作系统进行直接交互。Shell 通常用于执行各种管理任务，如文件系统操作、启动和终止进程、创建用户账户、执行脚本以及配置系统等。Shell 的概念最初可以追溯到 UNIX 操作系统早期，经过多年的发展和完善，Shell 已成为现代操作系统不可或缺的关键功能组件。

NX-OS 操作系统提供的 Shell 通常被称为 CLI（Command Line Interface，命令行接口），由于人们越来越多地通过脚本和网管技术来实现网络自动化，因而也越来越希望直接通过 Shell 访问 NX-OS 底层 Linux 操作系统。接下来将详细讨论 NX-OS 的 Bash Shell、Guest Shell 和 Python 功能，这些功能强大的工具可以实现大量操作任务的自动化，从而减轻管理负担。

1. Bash Shell

Bash（Bourne-Again Shell）是一款 UNIX Shell，是 Bourne Shell 的后继产品，提供了丰富的功能特性集和大量内置功能，能够与基础操作系统的低层组件进行交互。目前可以在 Nexus 9000、Nexus 3000 和 Nexus 3500 系列平台上使用 Bash Shell。Bash Shell 为用户提供了访问底层 Linux 操作系统的外壳，而 Linux 操作系统则提供了标准 NX-OS CLI 所不具备的一些重要功能。如果要在 Nexus 9000 交换机上启用 Bash Shell，那么就需要首先启用命令 **feature bash-shell**，然后通过命令 **run bash** *cli* 执行 Bash CLI 命令。此外，用户也可以通过 NX-OS CLI 命令 **run bash** 进入 Shell 模式，然后从 Bash Shell 执行各种 Bash CLI 命令。例 15-1 解释了启用 bash-shell 功能特性的方式以及通过 Bash Shell 命令 **pwd** 显示当前工作目录的示例。如果要检查是否启用了 bash-shell 功能特性，可以使用命令 **show bash-shell**。此外，例 15-1 还显示了 Bash Shell 的多种基本命令，其中，Bash 命令 **id –a** 负责验证当前用户以及 Group（组）和 Group ID（组 ID）信息，**echo** 命令可以根据脚本要求打印各种消息。

例 15-1 启用 bash-shell 功能特性并使用 Bash 命令

```
N9k-1(config)# feature bash-shell

N9k-1# show bash-shell
Bash shell is enable
N9k-1# run bash pwd
/bootflash/home/admin
N9k-1#

N9k-1# run bash
bash-4.2$ pwd
/bootflash/home/admin

bash-4.2$ id -a
uid=2002(admin) gid=503(network-admin) groups=503(network-admin)
bash-4.2$
bash-4.2$ echo "First Example on " 'uname -n' " using bash-shell " $BASH_VERSION
First Example on N9k-1 using bash-shell 4.2.10(1)-release
```

注： 建议读者掌握本节提到的 UNIX/Linux Bash Shell 命令。

对于 NX-OS 来说，只有拥有 network-admin、vdc-admin 和 dev-ops 角色的用户才能使用 Bash Shell。其他角色除非得到特别允许，否则无法使用 Bash。如果要验证被检查角色是否允许使用 Bash Shell，可以使用命令 **show role [name** *role-name*]。例 15-2 显示了 network-admin 和 dev-ops 用户角色的权限信息。

例 15-2 network-admin 和 dev-ops 用户角色权限

```
N9k-1# show role name network-admin

Role: network-admin
  Description: Predefined network admin role has access to all commands
  on the switch
  -------------------------------------------------------------------
  Rule    Perm    Type    Scope              Entity
  -------------------------------------------------------------------
  1       permit  read-write

N9k-1# show role name dev-ops

Role: dev-ops
  Description: Predefined system role for devops access. This role
  cannot be modified.
  -------------------------------------------------------------------
  Rule    Perm    Type    Scope              Entity
  -------------------------------------------------------------------
  6       permit  command                    conf t ; username *
  5       permit  command                    attach module *
  4       permit  command                    slot *
  3       permit  command                    bcm module *
```

2	permit	command	run bash *
1	permit	command	python *

启用了 NX-OS Bash Shell 功能特性之后，可以创建由多条 Bash 命令组成的 Bash Shell 脚本，这些脚本可以在底层 Linux 操作系统上依次执行。创建的 Bash 脚本以扩展名.sh 保存。此外，Bash Shell 还为用户提供了跟踪选项，可以在执行 Shell 脚本时进行调试操作，启用该功能的方式是使用选项 **-x** 和**#!/bin/bash** 语句。例 15-3 给出了创建 Shell 脚本的方式，以及通过选项 **-x** 启用脚本调试功能来验证脚本执行情况。

例 15-3 创建和调试 Bash Shell 脚本

```
bash-4.2$ pwd
/bootflash/home/admin

bash-4.2$ cat test.sh
#!/bin/bash
echo "Troubleshooting Route Flapping Issue Using Bash Shell"
counter="$(vsh -c "show ip route ospf | grep 00:00:0 | count")"
echo "Printing Counter - " $counter
if [ $counter -gt 0 ]
then
        echo "Following Routes Flapped @ " 'date'
        vsh -c "show tech ospf >> bootflash:shtechospf"
        vsh -c "show tech routing ip unicast >> bootflash:shtechrouting_unicast"
else
        echo "No Flapping Routes at this point"
bash-4.2$ /bin/bash -x test.sh
+ echo 'Troubleshooting Route Flapping Issue Using Bash Shell'
Troubleshooting Route Flapping Issue Using Bash Shell
++ vsh -c 'show ip route ospf | grep 00:00:0 | count'
+ counter=0
+ echo 'Printing Counter - ' 0
Printing Counter - 0
+ '[' 0 -gt 0 ']'
+ echo 'No Flapping Routes at this point'
No Flapping Routes at this point

bash-4.2$ vsh -c "clear ip route *"
Clearing ALL routes

bash-4.2$ /bin/bash -x test.sh
+ echo 'Troubleshooting Route Flapping Issue Using Bash Shell'
Troubleshooting Route Flapping Issue Using Bash Shell
++ vsh -c 'show ip route ospf | grep 00:00:0 | count'
+ counter=16
+ echo 'Printing Counter - ' 16
Printing Counter - 16
+ '[' 16 -gt 0 ']'
++ date
+ echo 'Following Routes Flapped @ ' Tue Nov 21 20:21:15 UTC 2017
Following Routes Flapped @ Tue Nov 21 20:21:15 UTC 2017
+ vsh -c 'show tech ospf >> bootflash:shtechospf'
+ vsh -c 'show tech routing ip unicast >> bootflash:shtechrouting_unicast'
bash-4.2$ exit
N9k-1# dir bootflash:
! Output omitted for brevity
```

```
    1175682         Nov 21 20:21:17 2017    shtechospf
    2677690         Nov 21 20:21:19 2017    shtechrouting_unicast
```

此外，还可以通过 Bash Shell 在 NX-OS 上安装 RPM 软件包。可以通过 Bash Shell 的 **yum** 命令来执行各种与 RPM 相关的操作，如安装、卸载和删除。例 15-4 解释了查看 Nexus 交换机上已安装的软件包列表以及安装和删除软件包的方式。可以看出，本例已经安装和卸载了 BFD 软件包。请注意，软件包被卸载之后，就无法再通过 NX-OS CLI 使用该功能特性，这些软件包决定了 NX-OS 的可用功能特性。

例 15-4 从 Bash Shell 安装和卸载 RPM 软件包

```
bash-4.2$ sudo yum list installed | grep n9000
base-files.n9000                        3.0.14-r74.2                installed
bfd.lib32_n9000                         2.0.0-7.0.3.I6.1            installed
bgp.lib32_n9000                         2.0.0-7.0.3.I6.1            installed
container-tracker.lib32_n9000           2.0.0-7.0.3.I6.1            installed
core.lib32_n9000                        2.0.0-7.0.3.I6.1            installed
eigrp.lib32_n9000                       2.0.0-7.0.3.I6.1            installed
eth.lib32_n9000                         2.0.0-7.0.3.I6.1            installed
fcoe.lib32_n9000                        2.0.0-7.0.3.IFD6.1          installed
isis.lib32_n9000                        2.0.0-7.0.3.I6.1            installed
lacp.lib32_n9000                        2.0.0-7.0.3.I6.1            installed
linecard2.lib32_n9000                   2.0.0-7.0.3.I6.1            installed
lldp.lib32_n9000                        2.0.0-7.0.3.I6.1            installed
ntp.lib32_n9000                         2.0.0-7.0.3.I6.1            installed
nxos-ssh.lib32_n9000                    2.0.0-7.0.3.I6.1            installed
ospf.lib32_n9000                        2.0.0-7.0.3.I6.1            installed
perf-cisco.n9000_gdb                    3.12-r0                     installed
platform.lib32_n9000                    2.0.0-7.0.3.I6.1            installed
rip.lib32_n9000                         2.0.0-7.0.3.I6.1            installed
shadow-securetty.n9000_gdb              4.1.4.3-r1                  installed
snmp.lib32_n9000                        2.0.0-7.0.3.I6.1            installed
svi.lib32_n9000                         2.0.0-7.0.3.I6.1            installed
sysvinit-inittab.n9000_gdb              2.88dsf-r14                 installed
tacacs.lib32_n9000                      2.0.0-7.0.3.I6.1            installed
task-nxos-base.n9000_gdb                1.0-r0                      installed
telemetry.lib32_n9000                   2.2.1-7.0.3.I6.1            installed
tor.lib32_n9000                         2.0.0-7.0.3.I6.1            installed
vtp.lib32_n9000                         2.0.0-7.0.3.I6.1            installed
bash-4.2$ sudo yum -y install bfd
Loaded plugins: downloadonly, importpubkey, localrpmDB, patchaction, patching,
protect-packages
groups-repo                                           | 1.1 kB     00:00 ...
localdb                                               |  951 B     00:00 ...
patching                                              |  951 B     00:00 ...
thirdparty                                            |  951 B     00:00 ...
Setting up Install Process
Resolving Dependencies
--> Running transaction check
---> Package bfd.lib32_n9000 0:2.0.0-7.0.3.I6.1 will be installed
--> Finished Dependency Resolution
Dependencies Resolved

================================================================================
 Package         Arch            Version                  Repository       Size
Installing:
 bfd             lib32_n9000     2.0.0-7.0.3.I6.1         groups-repo     483 k
```

```
Transaction Summary
================================================================================
Install       1 Package

Total download size: 483 k
Installed size: 1.8 M
Downloading Packages:
Running Transaction Check
Running Transaction Test
Transaction Test Succeeded
Running Transaction
  Installing : bfd-2.0.0-7.0.3.I6.1.lib32_n9000
1/1
starting pre-install package version mgmt for bfd
pre-install for bfd complete
starting post-install package version mgmt for bfd
post-install for bfd complete

Installed:
  bfd.lib32_n9000 0:2.0.0-7.0.3.I6.1

Complete!

N9k-1(config)# feature bfd
Please disable the ICMP / ICMPv6 redirects on all IPv4 and IPv6 interfaces
running BFD sessions using the command below

'no ip redirects '
'no ipv6 redirects '
bash-4.2$ sudo yum -y erase bfd
Loaded plugins: downloadonly, importpubkey, localrpmDB, patchaction, patching,
protect-packages
Setting up Remove Process
Resolving Dependencies
--> Running transaction check
---> Package bfd.lib32_n9000 0:2.0.0-7.0.3.I6.1 will be erased
--> Finished Dependency Resolution
Dependencies Resolved
================================================================================
Package        Arch          Version              Repository             Size
================================================================================
Removing:
bfd            lib32_n9000   2.0.0-7.0.3.I6.1     @groups-repo           1.8 M

Transaction Summary
================================================================================
Remove        1 Package

Installed size: 1.8 M
Downloading Packages:
Running Transaction Check
Running Transaction Test
Transaction Test Succeeded
Running Transaction
  Erasing    : bfd-2.0.0-7.0.3.I6.1.lib32_n9000
1/1
```

```
starting pre-remove package version mgmt for bfd
pre-remove for bfd complete

Removed:
  bfd.lib32_n9000 0:2.0.0-7.0.3.I6.1

Complete!

N9k-1(config)# feature bfd
                ^
% Invalid command at '^' marker.
```

2. Guest Shell

当前的网络范式已经从传统意义上的硬件、软件和管理网元逐步转向可扩展网元，允许网络运营商利用思科提供的 API 和类，在 NX-OS 环境中通过内置的 Python 和 Bash 执行环境运行自定义脚本，与 NX-OS 组件进行交互。但是，在某些情况下，网络运营商希望集成第三方应用并托管到 NX-OS 上。为了满足这类需求，NX-OS 提供了第三方应用托管框架，允许用户在专用的 Linux 用户空间环境托管自己的应用。网络运营商必须使用思科 ADT（Application Development Toolkit，应用开发工具包）交叉编译其软件，并与 Linux 根文件系统一起打包到思科 OVA（Open Virtual Appliance，开发虚拟应用）软件包中，然后通过应用托管功能将这些 OVA 部署到 NX-OS 网元上。

NX-OS 软件在 Nexus 9000 和 Nexus 3000 系列交换机上引入了 NX-OS Guest Shell 功能特性。Guest Shell 是一个开源的安全 Linux 环境，可以快速开发和部署第三方软件。guestshell 功能充分利用了 Python 和 Bash 执行环境以及 NX-OS 应用托管框架的优势。

Nexus 9000 和 Nexus 3000 默认启用 Guest Shell。也可以在 NX-OS 上显式启用或撤销 guestshell 功能。表 15-1 列出了常见 Guest Shell 命令。

表 15-1　　　　　　　　　　　　　常见 Guest Shell 命令

命令	描述
guestshell enable	该 CLI 命令负责安装和启用 Guest Shell 服务，启用了该命令之后，那么就可以通过命令 **guestshell** 进入 Guest Shell
guestshell disable	该 CLI 命令负责禁用 Guest Shell 服务，启用了该命令之后将无法访问 Guest Shell
guestshell destroy	该 CLI 命令负责去活并卸载当前的 Guest Shell，与该 Guest Shell 相关的所有系统资源均归还系统
guestshell reboot	该 CLI 命令负责去活并再次激活当前的 Guest Shell 服务
guestshell run *command-line*	该 CLI 命令负责执行 Guest Shell 内的应用程序、返回输出结果并退出 Guest Shell
guestshell sync	该 CLI 命令负责去活当前激活的 Guest Shell、将根文件系统同步给备用 RP，然后在活动 RP 上重新激活 Guest Shell
guestshell upgrade	该 CLI 命令负责去活并利用内嵌在引导系统映像中的 OVA 升级当前 Guest Shell，升级成功之后将重新激活该 Guest Shell
guestshell resize	该 CLI 命令负责修改 Guest Shell 的默认参数或现有参数，如 CPU、内存和根文件系统参数
guestshell	该 CLI 命令负责进入 Guest Shell

启动并运行 Guest Shell 之后，可以通过命令 **show guestshell detail** 来验证 Guest Shell 的详细信息。该命令可以显示 OVA 文件的路径、Guest Shell 服务的状态、资源保留以及 Guest Shell 的文件系统信息。例 15-5 显示了 Nexus 9000 交换机的 Guest Shell 详细信息。

例 15-5　Guest Shell 详细信息

```
N9k-1# show guestshell detail
Virtual service guestshell+ detail
  State                  : Activated
  Package information
    Name                 : guestshell.ova
    Path                 : /isanboot/bin/guestshell.ova
    Application
      Name               : GuestShell
      Installed version  : 2.1(0.0)
      Description        : Cisco Systems Guest Shell
    Signing
      Key type           : Cisco release key
      Method             : SHA-1
    Licensing
      Name               : None
      Version            : None
  Resource reservation
    Disk                 : 250 MB
    Memory               : 256 MB
    CPU                  : 1% system CPU

  Attached devices
    Type              Name            Alias
    ---------------------------------------------
    Disk              _rootfs
    Disk              /cisco/core
    Serial/shell
    Serial/aux
    Serial/Syslog                     serial2
    Serial/Trace                      serial3
```

如果没有显示 Guest Shell，那么就可以通过 **show logging logfile** 命令检查日志是否存在错误消息。排查 Guest Shell 故障问题时，可以通过命令 **show virtual-service [list]** 查看 Guest Shell 的状态以及 Guest Shell 使用的资源。例 15-6 显示了 Nexus 9000 交换机的虚拟服务列表以及当前 Guest Shell 使用的资源信息。

例 15-6　虚拟服务列表和使用的资源

```
N9k-1# show virtual-service list
Virtual Service List:

Name                    Status                  Package Name
-----------------------------------------------------------------------
guestshell+             Activated               guestshell.ova
N9k-1# show virtual-service

Virtual Service Global State and Virtualization Limits:

Infrastructure version : 1.9
Total virtual services installed : 1
Total virtual services activated : 1

Machine types supported  : LXC
Machine types disabled   : KVM
```

```
Maximum VCPUs per virtual service : 1
Resource virtualization limits:
Name                     Quota    Committed    Available
---------------------------------------------------------------
system CPU   (%)           20           1           19
memory       (MB)        3840         256         3584
bootflash    (MB)        8192         250         4031
```

注：如果无法解决 Guest Shell 故障，那么就可以收集 **show virtual-service tech-support** 命令的输出结果，并联系思科 TAC（Technical Assistance Center，技术支持中心）进行进一步排查。

3. Python

网络行业对 SDN（Software-Defined Networking，软件定义网络）的持续推动，为脚本和编程语言与网络设备的集成打开了诸多大门。Python 已成为业界公认的首选编程语言，Python 是一种功能强大且简单易学的编程语言，可以提供有效的高级数据结构和面向对象的功能特性，这些功能特性使其成为大多数平台进行快速应用开发的理想语言。

大多数 Nexus 平台集成了 Python 能力，无须安装任何特殊许可。只要键入命令 **python**，就可以通过 Nexus 平台的 CLI 调用交互式 Python 解释器。对于 Nexus 9000 和 Nexus 3000 平台来说，还可以通过 Guest Shell 使用 Python。执行了 **python** 命令之后，用户可以直接进入 Python 解释器。例 15-7 给出了基于 CLI 和 Guest Shell 方式使用 Python 解释器的示例。

例 15-7 基于 CLI 和 Guest Shell 使用 Python 解释器

```
N9k-1# python
Python 2.7.5 (default, Nov  5 2016, 04:39:52)
[GCC 4.6.3] on linux2
Type "help", "copyright", "credits" or "license" for more information.
>>> print "Hello World...!!!"
Hello World...!!!
N9k-1# guestshell
[admin@guestshell ~]$ python
Python 2.7.5 (default, Jun 17 2014, 18:11:42)
[GCC 4.8.2 20140120 (Red Hat 4.8.2-16)] on linux2
Type "help", "copyright", "credits" or "license" for more information.
>>> print "Hello Again..!!!"
Hello Again..!!!
```

注：建议读者熟悉并掌握 Python 编程语言。不过，本章的讨论重点并不是如何编写特定的 Python 程序，而是如何在 Nexus 平台上使用 Python。

除标准的 Python 库之外，NX-OS 还提供了 Cisco 库和 CLI 库，将这些库导入 Python 脚本之后，就能在 Nexus 交换机上执行思科提供的特定功能。Cisco 库提供了对思科 Nexus 组件的访问能力，CLI 库则提供了通过 Nexus CLI 执行命令并返回执行结果的能力。例 15-8 显示了 NX-OS 的 Cisco 库和 CLI Python 库的软件包信息。

例 15-8 NX-OS 的 Cisco 库和 CLI Python 库

```
>>> import cisco
>>> help(cisco)
Help on package cisco:

NAME
```

```
        cisco
FILE
    /usr/lib64/python2.7/site-packages/cisco/__init__.py

PACKAGE CONTENTS
    acl
    bgp
    buffer_depth_monitor
    check_port_discards
    cisco_secret
    dohost
    feature
    history
    interface
    ipaddress
    key
    line_parser
    mac_address_table
    nxapi
    nxcli
    ospf
    routemap
    section_parser
    ssh
    system
    tacacs
    transfer
    vlan
    vrf

! Output omitted for brevity
>>> import cli
>>> help(cli)
Help on module cli:

NAME
    cli

FILE
    /usr/lib64/python2.7/site-packages/cli.py
FUNCTIONS
    cli(cmd)

    clid(cmd)

    clip(cmd)
```

非交互式 Python 脚本将创建并保存在 bootflash:scripts/目录中，可以通过命令 **source** [*script name*]加以调用。另一种方式是通过 Guest Shell 创建和调用 Python 脚本。Python 脚本的第一行必须包含 Python 解释器的路径，即/usr/bin/env。例 15-9 给出了一个 Python 脚本示例，用于配置环回接口并列出 Nexus 交换机上处于 UP 状态的所有接口，该脚本是在 Guest Shell 环境中创建并调用的。

例 15-9 打印所有处于 UP 状态的接口的 Python 脚本

```
#!/usr/bin/env python

import sys
from cli import *
import json

cli ("conf t ; interface lo5 ; ip add 5.5.5.5/32")
print "\n***Configured interface loopback5***"
print "\n***Listing All interfaces on the device in UP state***\n"
intf_list = json.loads(clid ("show interface brief"))
i = 0
while i < len (intf_list['TABLE_interface']['ROW_interface']):
        intf = intf_list['TABLE_interface']['ROW_interface'][i]
        i += 1
        if intf['state'] == 'up':
                print intf['interface']

sys.exit(0)
[admin@guestshell ~]$ python test.py

***Configured interface loopback5***

***Listing All interfaces on the device in UP state***

mgmt0
Ethernet1/4
Ethernet1/5
Ethernet1/13
Ethernet1/14
Ethernet1/15
Ethernet1/16
Ethernet1/19
Ethernet1/32
Ethernet1/37
Ethernet2/1
port-channel10
port-channel101
port-channel600
loopback0
loopback5
loopback100
Vlan100
Vlan200
Vlan300
```

作为操作语句的一部分，还可以通过 EEM（Embedded Event Manager，嵌入式事件管理器）小程序调用 Python 脚本。由于每个事件都能执行多个操作，因而可以在不同的操作步骤中调用多个 Python 脚本，从而为构建 EEM 小程序提供了极大的灵活性。例 15-10 给出了一个 EEM 小程序配置示例，该小程序通过操作语句触发了先前配置的 Python 脚本。对于本例来说，由于 Python 脚本是在 Guest Shell 中配置的，因而 EEM 小程序中的 Python 脚本是从 Guest Shell 调用的。如果 Python 脚本位于 bootflash:source/ 目录中，那么就必须使用命令 **action** *number* **cli source python** *file-name*。

例 15-10　从 EEM 小程序调用 Python 脚本

```
N9k-1(config)# event manager applet link_monitor
N9k-1(config-applet)# event syslog pattern "IF_.*DOWN:"
N9k-1(config-applet)# action 1 cli guestshell python test.py
N9k-1(config-applet)# exit
```

15.3　NX-SDK

NX-SDK（NX-OS Software Development Kit，NX-OS 软件开发工具包）是一个 C++插件库，允许自定义的本机应用程序访问 NX-OS 功能和基础设施。可以通过 NX-SDK 创建自定义 CLI 命令、系统日志消息、事件处理程序和差错处理程序。该功能的一个常见案例就是创建自定义应用，向路由管理器进行注册，以便从 RIB（Routing Information Base，路由信息库）接收路由更新，然后根据路由的存在性采取相应的措施。使用 NX-SDK 必须满足以下 3 个基本要求。

- Docker。
- Linux 环境（Ubuntu 14.04 或更高版本，CentOS 6.7 或更高版本）。
- 思科 SDK（可选）。

注：NX-SDK 也可以与 Python 集成在一起，因而 Python 应用不需要思科 SDK。

在开发环境中使用 NX-SDK 之前，必须首先安装 NX-SDK，安装步骤如下。

- **第 1 步**：下载 Docker 映像。
- **第 2 步**：对于 32 位环境来说，设置以下环境变量。
  - **export ENXOS_SDK_ROOT=/enxos-sdk**
  - **cd $ENXOS_SDK_Root**
  - **source environment-setup-x86-linux**
- **第 3 步**：从 GitHub 克隆 NX-SDK 工具包。
  - **git clone https://github.com/CiscoDevNet/NX-SDK.git**

从 GitHub 分叉出 NX-SDK 之后，可以进一步利用 API，创建安装在 Nexus 交换机上的自定义应用程序包。

注：创建自定义应用时，请参阅 NX-SDK 的随附文档以及自定义示例应用代码。

构建了应用程序之后，可以使用 *rpm_gen.py* Python 脚本自动生成 RPM 软件包，该脚本位于 /NX-SDK/scripts 目录下。构建了 RPM 软件包之后，可以将 RPM 软件包复制到 bootflash:目录中的 Nexus Switch，然后安装到 Nexus Switch 上供进一步使用。例 15-11 解释了在 Nexus 9000 交换机上安装 RPM 软件包的步骤。本例安装了名为 *customCliApp* 的示例 RPM 软件包（该软件包是 NX-SDK 的一部分）。如果要启动自定义应用，那么就需要首先启用命令 **feature nxsdk**，然后通过命令 **nxsdk service-name** *app-name* 将自定义应用添加为服务。可以使用命令 **show nxsdk internal service** 检查应用程序的状态。

例 15-11　安装自定义 RPM 软件包

```
N9k-1# conf t
! Output omitted for brevity
Enter configuration commands, one per line. End with CNTL/Z.
N9k-1(config)# install add bootflash:customCliApp-1.0-1.0.0.x86_64.rpm
[####################] 100%
```

```
Install operation 1 completed successfully at Sun Nov 26 06:12:49 2017

N9k-1(config)# show install inactive
Boot Image:
        NXOS Image: bootflash:/nxos.7.0.3.I6.1.bin
Inactive Packages:
        customCliApp-1.0-1.0.0.x86_64

Inactive Base Packages:

N9k-1(config)# install activate customCliApp-1.0-1.0.0.x86_64
[####################] 100%
Install operation 2 completed successfully at Sun Nov 26 06:13:40 2017

N9k-1(config)# show install active
Boot Image:
        NXOS Image: bootflash:/nxos.7.0.3.I6.1.bin

Active Packages:
        customCliApp-1.0-1.0.0.x86_64

N9k-1(config)# feature nxsdk
N9k-1(config)# nxsdk service-name customCliApp
% This could take some time. "show nxsdk internal service" to check if your App
is Started & Runnning
N9k-1(config)# end
N9k-1# show nxsdk internal service

NXSDK Started/Temp unavailabe/Max services : 1/0/32
NXSDK Default App Path           : /isan/bin/nxsdk
NXSDK Supported Versions         : 1.0

Service-name    Base App      Started(PID)      Version    RPM Package
--------------------------------------------------------------------------------
customCliApp    nxsdk_app1    VSH(not running)  1.0        customCliApp-1.0-1.0.0.x86_64
! Starting the application in the background
N9k-1# run bash sudo su
bash-4.2# /isan/bin/nxsdk/customCliApp &
[1] 8887
N9k-1# show nxsdk internal service

NXSDK Started/Temp unavailabe/Max services : 2/0/32
NXSDK Default App Path           : /isan/bin/nxsdk
NXSDK Supported Versions         : 1.0

Service-name    Base App      Started(PID)      Version    RPM Package
--------------------------------------------------------------------------------
customCliApp    nxsdk_app1    VSH (8887)        1.0        customCliApp-1.0-1.0.0.x86_64
! Output omitted for brevity
```

注：例 15-11 使用 VSH（Virtual SHell，虚拟外壳）安装 RPM 软件包，也可以从 Bash Shell 安装 RPM 软件包。

安装自定义应用时，如果出现了故障或错误事件，那么就可以使用命令 **show nxsdk internal event-history [events | error]** 查看 NX-SDK 的事件历史记录日志。例 15-12 显示了 NX-SDK 的事件历史日志，并以高亮方式显示了表明已成功激活和启动应用程序 customCliApp 的日志信息。

例 15-12　NX-SDK 事件历史记录

```
N9k-1# show nxsdk internal event-history events
! Output omitted for brevity
Process Event logs of NXSDK_MGR
06:16:16 nxsdk_mgr : Added confcheck capability for 1.0, en 1, counter: 1
06:16:16 nxsdk_mgr : Adding confcheck capability for 1.0, en 0, counter: 0
06:16:16 nxsdk_mgr : Done: start service customCliApp nxsdk_app1
06:16:15 nxsdk_mgr : Heartbeat sent while start service customCliApp, nxsdk_app1
06:16:14 nxsdk_mgr : Feature nxsdk_app1 not enabled, State: 2, Error: SUCCESS
 (0x0), Reason: SUCCESS (0x40aa000a)
06:16:14 nxsdk_mgr : Config service name customCliApp
06:16:14 nxsdk_mgr : App: /isan/bin/nxsdk/customCliApp is linked to libnxsdk
06:16:14 nxsdk_mgr : App: /isan/bin/nxsdk/customCliApp, MaJor Version: 1,
Minor Version: 0
06:14:30 nxsdk_mgr : Received CLIS Done Callback
06:14:30 nxsdk_mgr : Initialized with all core components
06:14:30 nxsdk_mgr : Initialized with sdwrap
06:14:30 nxsdk_mgr : Request all commands from CLIS
06:14:30 nxsdk_mgr : Initialized with CLIS
06:14:30 nxsdk_mgr : All core components for NXSDK_MGR are UP
06:14:30 nxsdk_mgr : Core component "clis(261)" is Up
06:14:30 nxsdk_mgr : Received feature enable message from FM
06:14:30 nxsdk_mgr : Started NXSDK_MGR mts thread, pid 6864
06:14:30 nxsdk_mgr : Started NXSDK_MGR mq mts thread, pid 6865
06:14:30 nxsdk_mgr : NXSDK_MGR Main Done: await join before exiting
06:14:30 nxsdk_mgr : nxsdk_mgr_create_threads: ok
06:14:30 nxsdk_mgr : Query sysmgr for core comp status: clis(261)
06:14:30 nxsdk_mgr : Done with NXSDK_MGR stateless recovery
```

注：如果无法安装或运行自定义应用，那么除事件历史记录日志之外，还可以收集 **show tech nxsdk** 命令的输出结果。

15.4　NX-API

NX-OS 提供了一个被称为 NX-API 的 API，允许通过标准的请求/响应语言与交换机进行交互。传统的 CLI 是为人与交换机之间的交互设计的，通过输入 CLI 命令发出请求，然后通过将输出结果显示到客户端上来接收交换机的响应，这种响应数据是非结构化的，需要操作人员逐行评估输出结果以找到有用信息。对于使用传统 CLI 接口以脚本方式自动执行任务的操作人员来说，必须采用相同的数据解析方式，通过抓屏方式寻找感兴趣的数据。这样做不但效率低下，而且非常麻烦，需要执行大量的输出迭代并通过正则表达式匹配特定文本。

NX-API 的好处是面向机器间通信，优化了相应的发送请求与接收响应的过程。也就是说，使用 NX-API 进行通信之后，所有的请求和响应数据都被格式化为结构化数据。从 NX-API 收到的响应均以 XML（Extensible Markup Language，可扩展标记语言）或 JSON（JavaScript Object Notation，JavaScript 对象表示法）格式提供。与从大量人类可读的 CLI 输出结果中解析少量感兴趣数据的方式相比，这种方法的效率更高且错误率更低。可以通过 NX-API 获得 **show** 命令的输出结果、增加或删除配置，从而大大简化并实现大型网络管理与操作的自动化水平。

客户端与运行在交换机上的 NX-API 通过 TCP（Transport Control Protocol，传输控制协议）进行通信，支持 HTTP（Hypertext Transfer Protocol，超文本传输协议）或 HTTPS（Hypertext Transfer Protocol Secure，安全超文本传输协议），具体取决于应用需求。NX-API 采用 HTTP 基本认证机制，必须在请

求消息的 HTTP 报头中包含用户名和密码。认证成功后，NX-API 将使用名称 nxapi_auth 来提供基于会话的 Cookie，该会话 Cookie 应该包含在后续的 NX-API 请求中。需要检查用户的特权信息，以确认请求消息是由交换机上拥有正确用户名和密码的用户发出的，且该用户拥有通过 NX-API 执行相关命令的正确授权。

认证成功之后，可以发送请求。NX-API 请求对象支持 JSON-RPC 或思科专有格式。表 15-2 列出了 JSON-RPC 请求对象的字段信息。

表 15-2　　　　　　　　　　　JSON-RPC 请求对象字段

字段	描述
jsonrpc	该字符串指定 JSON-RPC 的协议版本，必须是 2.0
method	该字符串包含了所要调用的方法名称 NX-API 支持 "cli"：用于 show 或配置命令 "cli_ascii"：用于 show 或配置命令，输出结果不进行格式化
params	一个结构化数值，包含了方法调用期间所要使用的参数值。必须包含以下字段 "cmd"：CLI 命令 "version"：NX-API 请求版本标识符
id	由客户端建立的可选标识符，必须包含字符串、数字或 NULL 值。如果用户未指定 id 参数，那么服务器将认为该请求只是一条通知而不提供任何响应

图 15-1 给出了一个 JSON-RPC 请求对象示例，用于向交换机查询其配置的交换机名称。

JSON-RPC格式

```
[
 {
  "jsonrpc": "2.0",
  "method": "cli",
  "params": {
    "cmd": "show switchname",
    "version": 1
  },
  "id": 1
 }
]
```

图 15-1　JSON-RPC 请求对象

另一种请求对象格式是思科专有格式，可以是 XML 或 JSON。表 15-3 列出了思科专有请求对象的字段信息。

表 15-3　　　　　　　　　　　思科专有请求对象字段

字段	描述
version	指示当前 NX-API 版本
type	该字符串包含了所要执行的命令类型 cli_show：用于 show 命令 cli_show_ascii：用于 show 命令，输出结果不进行格式化 cli_conf：用于非交互式配置命令 bash：用于在启用了 Bash Shell 的设备上执行非交互式 Bash 命令
chunk	用于将大型 show 命令的输出结果分块，值为 0 表示输出结果不分块，值为 1 表示输出结果可以分成多个块
sid	对响应消息进行分块时有效。检索消息的下一个块时，用户需要发送请求，且 sid 设置为上一个响应消息中的 sid
input	输入可以是一条或多条命令，多条命令应以"；"（一个空格后面跟分号）进行分隔
ouput_format	请求消息的期望输出格式（XML 或 JSON）

图 15-2 给出了 JSON 和 XML 格式的思科专有请求对象示例，用于向交换机查询其配置的交换机名称。

JSON 格式

```
{
 "ins_api":{
   "version":"1.0",
   "type":"cli_show",
   "chunk":"0",
   "sid":"1",
   "input":"show switchname",
   "output_format":"json"
 }
}
```

XML 格式

```
<?xml version="1.0"?>
<ins_api>
  <version>1.0</version>
  <type>cli_show</type>
  <chunk>0</chunk>
  <sid>sid</sid>
  <input>show switchname</input>
  <output_format>xml</output_format>
</ins_api>
```

图 15-2 思科专有请求对象

请求对象通过已配置的 HTTP 端口（TCP 端口 80）或 HTTPS 端口（TCP 端口 443）发送给交换机，通过 Web 服务器认证接收到的请求对象，同时随请求消息一起提供相应的软件对象。此后，交换机将采用 JSON-RPC 或思科专有格式将响应对象发送给客户端。表 15-4 列出了 JSON-RPC 响应对象的字段信息。

表 15-4　　　　　　　　　　　JSON-RPC 响应对象字段

字段	描述
jsonrpc	该字符串指定 JSON-RPC 的协议版本，必须是 2.0
result	只有成功的请求才包含该字段，该字段的值包含了所请求的 CLI 输出
error	只有错误的请求才包含该字段。错误对象包含以下字段 "code"：JSON-RPC 规范指定的整数错误代码 "message"：与错误代码对应的人类可读字符串 "data"：包含对用户有用的其他信息的可选结构
id	该字段值与相应请求对象中的 id 字段相同。如果解析请求中的 id 字段时出现问题，那么该字段值为 null

图 15-3 给出了一个 JSON-RPC 响应对象示例。

JSON-RPC 响应对象

```
{
  "jsonrpc": "2.0",
  "result": {
    "body": {
      "hostname": "NX02"
    }
  },
  "id": 1
}
```

图 15-3 JSON-RPC 响应对象

表 15-5 列出了思科专有响应对象的字段信息。

表 15-5　　　　　　　　　　　思科专有响应对象字段

字段	描述
version	指示当前 NX-API 版本
type	该字符串包含了所要执行的命令类型
sid	当前相应的会话 ID（仅对响应消息进行分块时有效）

字段	描述
outputs	标记包含所有命令的输出结果
output	标记包含单条命令的输出结果,如果类型是 cli_conf 或 bash,那么就包含所有命令的输出结果
body	所请求命令的响应正文
code	命令执行的错误代码,使用标准的 HTTP 错误代码
msg	与错误代码相关的错误消息

图 15-4 给出了 JSON 和 XML 格式的思科专有响应对象示例。

JSON格式

```
{
  "ins_api":{
    "type":"cli_show",
    "version":"1.0",
    "sid":"eoc",
    "outputs":{
      "output":{
        "input":"show switchname",
        "msg":"Success",
        "code":"200",
        "body":{
          "hostname":"NX02"
        }
      }
    }
  }
}
```

XML格式

```
<?xml version="1.0"?>
<ins_api>
  <type>cli_show</type>
  <version>1.0</version>
  <sid>eoc</sid>
  <outputs>
    <output>
      <body>
        <hostname>NX02</hostname>
      </body>
      <input>show switchname</input>
      <msg>Success</msg>
      <code>200</code>
    </output>
  </outputs>
</ins_api>
```

图 15-4 思科专有响应对象

可以在一个请求中发送多条命令。对于 JSON-RPC 请求对象来说,实现方式是将无数条单个 JSON-RPC 请求链接到单个 JSON-RPC 数组中。对于思科专有请求对象来说,最多可以将 10 个用分号分隔的命令链接到输入对象中。无论使用哪种类型的请求对象,如果请求失败,那么后续请求都将不再执行。

必须在交换机的全局配置中通过 **feature nxapi** 命令启用 NX-API 功能特性,如例 15-13 所示。

例 15-13 NX-API 功能特性配置

```
NX-2# conf t
Enter configuration commands, one per line. End with CNTL/Z.
NX-2(config)# feature nxapi

NX-2# show nxapi
nxapi enabled
HTTP Listen on port 80
HTTPS Listen on port 443
```

注: 可以通过命令 **nxapi http port** 和 **nxapi https port** 更改默认的 HTTP 和 HTTPS 端口。

启用了 NX-API 功能特性之后,可以进行身份认证并将请求发送给适当的 HTTP 或 HTTPS 端口。此外,NX-OS 还提供了一种沙箱环境用于测试 API 的功能,可以通过标准的 Web 浏览器,以 HTTP 方式连接交换机管理地址来访问该沙箱环境。

排查与 NX-API 相关的故障问题时,通常需要验证负责在交换机与客户端之间传递请求和响应消息的 TCP 连接,可以通过 NX-OS 提供的 Ethanalyzer 抓包工具来解决客户端的连接问题,并确认 TCP 三次握手是否已完成，如例 15-14 所示。

例 15-14　通过 Ethanalyzer 验证客户端连接

```
NX-2# ethanalyzer local interface mgmt capture-filter "tcp port 443" limit-capturedframes
 0
Capturing on mgmt0
192.168.1.50 -> 192.168.1.201 TCP 52018 > https [SYN] Seq=0 Win=65535 Len=0
MSS=1460 WS=5 TSV=568065210 TSER=0
192.168.1.201 -> 192.168.1.50 TCP https > 52018 [SYN, ACK] Seq=0 Ack=1
Win=16768 Len=0 MSS=1460 TSV=264852 TSER=568065210
192.168.1.50 -> 192.168.1.201 TCP 52018 > https [ACK] Seq=1 Ack=1 Win=65535
Len=0 TSV=568065211 TSER=264852
192.168.1.50 -> 192.168.1.201 SSL Client Hello
192.168.1.201 -> 192.168.1.50 TLSv1.2 Server Hello, Certificate, Server Key
Exchange, Server Hello Done
192.168.1.50 -> 192.168.1.201 TCP 52018 > https [ACK] Seq=518 Ack=1294
Win=65535 Len=0 TSV=568065232 TSER=264852
192.168.1.50 -> 192.168.1.201 TLSv1.2 Client Key Exchange, Change Cipher Spec,
Hello Request, Hello Request
192.168.1.201 -> 192.168.1.50 TLSv1.2 Encrypted Handshake Message, Change
Cipher Spec, Encrypted Handshake Message
```

确认客户端已经建立了 TCP 会话之后，就可以通过 **show nxapi-server logs** 命令显示客户端的 NX-API 通信信息。例 15-15 的服务器日志显示了包括连接尝试以及接收到的请求等在内的详细信息。此外，日志文件还显示了 CLI 命令的执行情况，有助于确定特定批处理命令执行失败的原因。最后，日志文件还显示了发送给客户端的响应对象。

例 15-15　NX-API 服务器日志

```
NX-2# show nxapi-server logs
ngx_http_cookie_set:627 2017 November 17 07:18:25.292 : creating cookie
ngx_http_ins_api_post_body_handler:549 2017 November 17 02:18:25.292 : Input
Message {
  "ins_api": {
    "version": "1.0",
    "type": "cli_show",
    "chunk": "0",
    "sid": "1",
    "input": "show switchname",
    "output_format": "json"
  }
}
parse_user_from_request:41 2017 November 17 02:18:25.292 : cookie had user
'admin'
parse_user_from_request:55 2017 November 17 02:18:25.292 : auth header had user
'admin'
pterm_idle_vsh_sweep:667 2017 November 17 02:18:25.292 : pterm_idle_vsh_sweep
pterm_get_vsh:710 2017 November 17 02:18:25.292 : vsh found: child_pid = 10558,
fprd = 0x98d0800, fpwr = 0x98d0968, fd = 14, user = admin, vdc id = 1
pterm_write_to_vsh:446 2017 November 17 02:18:25.292 : In vsh [14] Writing cmd
"show switchname | xml "
pterm_write_to_vsh:522 2017 November 17 02:18:25.302 : Cmd 'show switchname | xml'
 returned with '0'
```

```
 pterm_write_to_vsh:627 2017 November 17 02:18:25.302 : Done processing vsh output
  (ret=0)
 _ins_api_cli_cmd:288 2017 November 17 02:18:25.302 : Incorrect XML data,
 replacing special characters
 _ins_api_cli_cmd:304 2017 November 17 02:18:25.302 : found ns vdc_mgr and copied
 it to blob vdc_mgr len 7
 pterm_write_to_vsh:446 2017 November 17 02:18:25.302 : In vsh [14] Writing cmd
 "end"
 pterm_write_to_vsh:522 2017 November 17 02:18:25.304 : Cmd 'end' returned with
 '0'
 pterm_write_to_vsh:627 2017 November 17 02:18:25.304 : Done processing vsh output
  (ret=0)
 ngx_http_ins_api_post_body_handler:675 2017 November 17 02:18:25.304 : Sending
 response {
     "ins_api":    {
         "type":       "cli_show",
         "version":    "1.0",
         "sid":        "eoc",
         "outputs":    {
             "output":    {
                 "input":     "show switchname",
                 "msg":       "Success",
                 "code":      "200",
                 "body":    {
                     "hostname":     "NX02"
                 }
             }
         }
     }
 }
```

> **注**：与传统 CLI 一样，NX-API 的所有活动信息都记录在交换机的记账日志中，与 NX-API 相关的用户名在记账日志中被列为 *nginx*。

除 NX-API 服务器日志之外，NX-OS 还提供了 **show tech nxapi** 命令，该命令除显示 Linux 进程的 nginx Web 服务器日志之外，还可以提供详细的服务器日志信息。

15.5 本章小结

自动化和可编程性为未来网络定义了基础构建模块。Open NX-OS 旨在满足未来的 SDN 发展需求以及用户在 Nexus 交换机上直接执行第三方应用的需求。Open NX-OS 提供了允许网络运营商和开发人员在网络设备上创建和部署自定义应用的灵活架构，集成了强大的 Bash Shell 和 Guest Shell，使在 Nexus 交换机上创建自动化任务脚本的工作变得更加容易。本章详细讨论了利用 Bash Shell 和 Guest Shell 部署第三方应用的方式。将 Python 与 NX-OS 集成之后，可以更加灵活地创建动态应用，大大增强了 Nexus 交换机的能力与可管理性。除 Python 支持能力之外，思科还提供了 NX-SDK，支持以 C++和 Python 语言构建应用程序并编译为 RPM 软件包。最后，本章还介绍了 NX-API，用户可以通过 NX-API，以标准的请求/响应语言与 Nexus 交换机进行交互。